The Changing World Religion Map

Stanley D. Brunn
Editor

The Changing World Religion Map

Sacred Places, Identities, Practices and Politics

Volume III

Donna A. Gilbreath
Assistant Editor

Editor
Stanley D. Brunn
Department of Geography
University of Kentucky
Lexington, KY, USA

Assistant Editor
Donna A. Gilbreath
UK Markey Cancer Center
Research Communications Office
Lexington, KY, USA

ISBN 978-94-017-9375-9 ISBN 978-94-017-9376-6 (eBook)
DOI 10.1007/978-94-017-9376-6

Library of Congress Control Number: 2014960060

Springer Dordrecht Heidelberg New York London

Printed on acid-free paper

Springer Science+Business Media B.V. Dordrecht is part of Springer Science+Business Media (www.springer.com)

Preface: A Continuing Journey

Religion has always been a part of my life. I am a Presbyterian PK (preacher's kid). From my father I inherited not only an interest in the histories and geographies of religions, not just Christianity, but also a strong sense of social justice, a thread that has been part of my personal and professional (teaching, research, service) life. My mother was raised as a Quaker and from her I also learned much about social justice, peace and reconciliation and being a part of an effective voice calling for ends to war, social discrimination of various types, and other injustices that seem to be a continual part of daily life on the planet. My father had churches mostly in the rural Upper Middle West. These were open country and small town congregations in Illinois, Wisconsin, Minnesota, South Dakota, Nebraska, and Missouri. The members of these congregations were Germans, Czech, Scandinavians (Norwegians, and Swedes), and English. Perhaps or probably because of these experiences, I had friendships with many young people who comprised the mosaic of the rural Middle West. Our family moved frequently when I was living at home, primarily because my father's views on social issues were often not popular with the rural farming communities. (He lost his church in northwest Missouri in 1953 because he supported the Supreme Court's decision on desegregation of schools. By the time I graduated from high school in a small town in southeastern Illinois, I had attended schools in a half-dozen states; these include one-room school house experiences as well as those in small towns.

During my childhood days my interests in religion were, of course, important in the views I had about many subjects about those of different faiths and many places on the planet. I was born in a Catholic hospital, which I always attribute to the beginning of my ecumenical experiences. The schools I attended mixes of Catholics and Protestants; I had few experiences with Native Americans, Jews and African, and Asian Americans before entering college. But that background changed, as I will explain below. My father was always interested in missionaries and foreign missions and once I considered training for a missionary work. What fascinated me most about missionaries were that they were living in distant lands, places that I just longed to know about; an atlas was always my favorite childhood book, next to a dictionary. I was always glad when missionaries visited our churches and stayed in our homes. The fascination extended to my corresponding with missionaries in Africa, Asia, and Latin America. I was curious what kind of work they did. I also found them a source for stamps, a hobby that I have pursued since primary school. Also I collected the call letters of radio stations, some which were missionary stations, especially in Latin America. (Some of these radio stations are still broadcasting.)

When I enrolled as an undergraduate student at Eastern Illinois University, a small regional university in east central Illinois, I immediately requested roommates from different countries. I very much wanted to make friends with students from outside the United States and learn about their culture. During my 3 years at EIU, I had roommates from Jordan, Samoa, Costa Rica, Ethiopia, and South Korea; these were very formative years in helping me understand cross-cultural, and especially, religious diversity. On reflection, I think that most of the Sunday services I attended were mostly Presbyterian and Methodist, not Catholic, Lutheran, or Baptist. When I entered the University of Wisconsin, Madison, for the M.A. degree, I was again exposed to some different views about religion. The Madison church that fascinated me the most was the Unitarian church, a building designed by Frank Lloyd Wright. I remember how different the services and sermons were from Protestant churches, but intellectually I felt at home. My father was not exactly pleased I found the Unitarian church a good worship experience. The UW-Madison experience also introduced me to the study of geography and religion. This was brought home especially in conversations with my longtime and good friend, Dan Gade, but also a cultural geography course I audited with Fred Simoons, whose new book on religion and food prejudices just appeared and I found fascinating. Also I had conversations with John Alexander, who eventually left the department to continue in his own ministry with the Inter Varsity Christian Fellowship. A seminar on Cultural Plant Geography co-taught with Fred Simoons, Jonathan Sauer, and Clarence Olmstead provided some opportunities to explore cultural and historical dimensions of religion, which were the major fields where geographers could study religion. The geographers I knew who were writing about religion were Pierre Deffontaines, Eric Issac, and Xavier de Phanol. That narrow focus, has, of course, changed in the past several decades, as I will discuss below.

The move to Ohio State University for my doctoral work did not have the strong religious threads that had emerged before. I attended a variety of Protestant churches, especially Presbyterian, Congregational, and Methodist. I took no formal courses in geography that dealt with religion, although I was very interested when Wilbur Zelinsky's lengthy article on church membership patterns appeared in the *Annals of the Association of American Geographers* in 1961. I felt then that this was, and would be, a landmark study in American human geography, as the many maps of denominational membership patterns plus extensive references would form the basis for future scholars interested in religion questions, apart from historical and cultural foci which were the norm at that time. My first article on religion was on religious town names; I wrote it when I was at Ohio State with another longtime friend, Jim Wheeler, who had little interest in religion. I can still remember using my knowledge of biblical place names and going through a Rand McNally atlas with Jim identifying these town names. This study appeared in *Names*, which cultural geographers acknowledge is one of the premier journals concerned with names and naming processes. Even though my dissertation on changes in the central place functions in small towns in northwest Ohio and southeast Ohio (Appalachia) did not look specifically at churches, I did tabulate the number and variety during extensive fieldwork in both areas.

My first teaching job was at the University of Florida in fall 1966. I decided once I graduated from OSU that I wanted to live in a different part of the United States where I could learn about different regional cultures and politics. I was discouraged by some former teachers about teaching in Florida, especially about the region's segregation history, recent civil rights struggles in the South and also the John Birch Society (which was also active in Columbus when I lived there). The 3 years (1966–1969) in Gainesville were also very rewarding years. These were also very formative years in developing my interests in the social geography, a new field that was just beginning to be studied in the mid-1960s. Included in the forefront of this emerging field of social geographers were Anne Buttimer, Paul Claval, Yi-Fu Tuan, Dick Morrill, Richard Peet, Bill Bunge, Wilbur Zelinsky, David Harvey, and David Smith, all who were challenging geographers to study the social geographies of race, employment, school and housing discrimination, but also poverty, environmental injustice, inequities in federal and state programs promoting human welfare, the privileges of whiteness and the minorities' participation in the voting/political process. Living in northern Florida in the late 1960s or "Wallace years" could not help but alert one to the role that religion was playing in rural and urban areas in the South. Gainesville had distinct racial landscapes. I was definitely a "northerner" and carpetbagger who was an outcast in many ways in southern culture. One vivid memory is attending a University of Florida football game (a good example of regional pride and nationalism) and being about the only person seated while the band played "Dixie." I joined a Congregational/United Church of Christ church which was attended by a small number of "northern faculty" who were supportive of initiatives to end discriminatory practices at local, university, and broader levels. At this time I also was learning about the role of the Southern Baptist Church, a bastion of segregation that was very slow to accommodate to the wishes of those seeking ends to all kinds of overt and subtle discrimination (gender, race, class) practices. The term Bible Belt was also a label that rang true; it represented, as it still does, those who adhere to a literal interpretation of the Bible, a theological position I have never felt comfortable. I soon realized that if one really wanted to make a difference in the lives of those living with discrimination, poverty, and ending racial disenfranchisement in voting, religion was a good arena to express one's feelings and work with others on coordinated efforts. Published research that emerged from my Florida experiences included studies on poverty in the United States (with Jim Wheeler), the geographies of federal outlays to states, an open housing referendum in Flint, Michigan (with Wayne Hoffman), and school levies in cities that illustrated social inequities (with Wayne Hoffman and Gerald Romsa). My Florida years also provided me the first opportunity to travel in the developing world; that was made possible with a summer grant where I visited nearly 15 different Caribbean capitals where I witnessed housing, social, and infrastructure gaps. This experience provided my first experiences with the developing world and led to a Cities of the World class I taught at Michigan State University and also co-edited several editions with the same title of a book with Jack Williams.

The 11 years at Michigan State University did not result in any major research initiatives related to religion, although it did broaden my horizons about faiths other than Christianity. I began to learn about Islam, especially from graduate students in the department from Saudi Arabia, Libya, Kuwait, and Iran. Many of these I advised on religion topics about their own cultures, especially those dealing with pilgrimages and sacred sites. Probably the main gain from living in Michigan was support for and interest in an emerging secular society. The religious "flavors" of Michigan's religious landscape ran the full gamut from those who were very traditional and conservative to those who were globally ecumenical, interfaith, and even agnostic. I continued to be active in Presbyterian and United Churches of Christ, both which were intellectually and spiritually challenging places for adult classes and singing in a choir.

When I moved to the University of Kentucky in 1980, I knew that living in the Bluegrass State would be different from Michigan in at least two respects. One is that Kentucky was considered a moderate to progressive state with many strong traditional and conservative churches, especially the Southern Baptist denomination. Zelinsky's map accurately portrayed this region as having a dominance of conservative and evangelical Protestantism. Second, I realized that for anyone interested in advancing social issues related to race and gender equality or environmental quality (especially strip mining in eastern Kentucky), there would likely be some conflicts. I also understood before coming to Kentucky that alcoholic beverage consumption was a big issue in some countries; that fact was evident in an innovative regional map Fraser Hart prepared in a small book about the South. And then there was the issue about science and religion in school curricula. With this foreknowledge, I was looking forward to living in a region where the cross-currents of religion and politics meshed, not only experiencing some of these social issues or schisms firsthand, but also having an opportunity to study them, as I did.

I realized when I moved to Lexington, it was in many ways and still is a slowly progressing socially conscious city. Southern Baptist churches, Christian churches, and Churches of Christ were dominant in the landscape and in their influences on social issues. One could not purchase alcoholic beverages on Sunday in restaurants until a couple referenda were passed in the mid-1980s that permitted sales. I think 90 of the state's 120 counties were officially dry, although everyone living in a dry county knew where to purchase liquor. One could not see the then-controversial "The Last Temptation of Christ" movie when it appeared unless one would drive three hours to Dayton. "Get Right with God" signs were prominent along rural highways. The University of Kentucky chimes in Memorial Hall on campus played religious hymns until this practice stopped sometime in the middle of the decade; I am not exactly sure why. Public schools had prayers before athletic events; some still do. Teachers in some public schools could lose their jobs if they taught evolution. Creationism was (and still is) alive and well. I was informed by university advisors that the five most "dangerous" subjects to new UK students were biology, anthropology, astronomy, geology, and physical geography. Students not used to other than literal biblical interpretations were confused and confounded by evolutional science. Betting on horses was legal, even though gambling was frowned on by some religious leaders. Cock fighting and snake handling still existed (and still do) in pockets in rural eastern Kentucky. In many ways living in Kentucky was like living "on the dark side of gray." Lexington in many ways was and still is an island or outlier. Desegregation was a slow moving process in a city with a strong southern white traditional heritage. Athletic programs were also rather slow to integrate, especially UK basketball. In short, how could one not study religion in such an atmosphere. Living in Kentucky is sort of the antipode to living in agnostic-thriving New England and Pacific Northwest. I would expect that within 100 miles of Lexington one would discover one of the most diverse religious denominational and faith belief landscapes in the United States. There are the old regular mainline denominations, new faiths that have come into the Bluegrass and also many one-of-a-kind churches, especially in rural eastern, southern, and southeastern Kentucky.

I have undertaken a number of studies related to religion in Kentucky and the South in the past three decades. Some of these have been single-authored projects, others with students and faculty at UK and elsewhere. Some were presentations at professional meetings; some resulted in publications. The topics that fascinated me were ones that I learned from my geography colleagues and those in other disciplines that were understudied. These include the history and current patterns of wet/dry counties in Kentucky, a topic that appears in local and statewide media with communities deciding whether to approve the sale of alcoholic beverages. This study I conducted with historian Tom Appleton. With regularity there were clergy of some fundamentalist denominations who decried the sale of such drinks; opposing these clergy and their supporters were often those interested in promoting tourism and attracting out-of-state traffic on interstates. Also I looked into legislation that focused on science/education interfaces in the public schools and on the types of religious books (or avoidance of such, such as dealing with Marx, Darwin, and interfaith relations) in county libraries. Craig Campbell and I published an article in *Political Geography* on Cristo Redentor (Christ of the Andes statue) as an example of differential locational harmony. At the regional level I investigated with Esther Long the mission statements of seminaries in the South, a study that led to some interesting variations not only in their statements, but course offerings and visual materials on websites. I published with Holly Barcus two articles in *Great Plains Research* about denominational changes in the Great Plains. Missionaries have also been relatively neglected in geography, so I embarked on a study with Elizabeth Leppman that looked at the contents of a leading Quaker journal in the early part of the past century. Religions magazines, as we acknowledged in our study, were (and probably still are) a very important medium for educating the public about places and cultures, especially those where most Americans would have limited first-hand knowledge. The music/religion interface has long fascinated me, not only as a regular choir member, but for the words used to convey messages about spirituality, human welfare and justice, religious traditions and promises of peace and hope.

After 11 September 2001, I collected information from a number of churches in eastern Kentucky about how that somber event was celebrated and also what hymns they sung on the tenth anniversary. As expected, some were very somber and dignified, others had words about hope, healing, and reaching across traditional religious boundaries that separate us. I also co-authored an article (mostly photos) in *Focus* on the Shankill-Falls divided between Catholic and Protestant areas of Belfast with three students in my geography of religion class at the National University of Ireland in Maynooth. The visualization theme was integral to a paper published in *Geographica Slovenica* on ecumenical spaces and the web pages of the World Council of Churches and papers I delivered how cartoonists depicted the controversial construction of a mosque at Ground Zero. How cartoonists depicted God-Nature themes (the 2011 Haitian earthquake and Icelandic volcanic eruption) were the focus of an article in *Mitteilungen der Őserreichsten Geographischen Gesellschaft*. I published in *Geographical Review* an article how the renaissance of religion in Russia is depicted on stamp issues since 1991. A major change in my thinking about the subject of religion in the South was the study that I worked on with Jerry Webster and Clark Archer, a study that appeared in the *Southeastern Geographer* in late 2011. We looked at the definition and concept of the Bible Belt as first discussed by Charles Heatwole (who was in my classes when I taught at Michigan State University) in 1978 in the *Journal of Geography*. We wanted to update his study and learn what has happened to the Bible Belt (or Belts) since this pioneering effort. What we learned using the Glenmary Research Center's county data on adherents for the past several decades was that the "buckle" has relocated. As our maps illustrated, the decline in those counties with denominations adhering to a literal interpretation of the Bible in western North Carolina and eastern Tennessee and a shift to the high concentration of Bible Belt counties in western Oklahoma and panhandle Texas. In this study using Glenmary data for 2000, we also looked at the demographic and political/voting characteristics of these counties. (In this volume we look at the same phenomenon using 2010 data and also discuss some of the visual features of the Bible Belt landscapes.)

What also was instrumental in my thinking about religion and geography interfaces were activities outside my own research agenda. As someone who has long standing interests in working with others at community levels on peace and justice issues, I worked with three other similarly committed adults in Lexington to organize the Central Kentucky Council for Peace and Justice. CKCPJ emerged in 1983 as an interfaith and interdenominational group committed to working on peace and justice issues within Lexington, in Central Kentucky especially, but also with national and global interests. The other three who were active in this initiative were Betsy Neale (from the Friends), Marylynne Flowers (active in a local Presbyterian church) and Ernie Yanarella (political scientist, Episcopalian, longtime friend, and also contributor to a very thoughtful essay on Weber in this volume). This organization is a key agent in peace/justice issues in the Bluegrass; it hosts meetings, fairs, conferences, and other events for people of all ages, plans annual marches on Martin Luther King Jr. holiday, and is an active voice on issues related to capital punishment, gun control, gay/lesbian issues, fair trade and employment, environmental responsibility and stewardship, and the rights of women, children, and minorities.

I also led adult classes at Maxwell Street Presbyterian Church where we discussed major theologians and religious writers, including William Spong, Marcus Borg, Joseph Campbell, Philip Jenkins, Diane Eck, Kathleen Norris, Diana Butler Bass, Francis Collins, Sam Harris, Paul Alan Laughlin, James Kugel, Dorothy Bass, and Garry Wills. We discuss issues about science, secularism, death and dying, interfaith dialogue, Christianity in the twenty-first century, images of God, missions and missionaries, and more. I also benefitted from attending church services in the many countries I have traveled, lived, and taught classes in the past three decades. These include services in elaborate, formal, and distinguished cathedrals in Europe, Russian Orthodox services in Central Asia, and services in a black township and white and interracial mainline churches in Cape Town. Often I would attend services where I understood nothing or little, but that did not diminish the opportunity to worship with youth and elders (many more) on Sunday mornings and listen to choirs sing in multiple languages. These personal experiences also became part of my religion pilgrimage.

While religion has been an important part of my personal life, it was less important as part of my teaching program. Teaching classes on the geography of religion are few and far between in the United States; I think the subject was accepted much more in the instructional and research arenas among geographers in Europe. I think that part of my reluctance to pursue a major book project on religion was that for a long time I considered the subject too narrowly focused, especially on cultural and historical geography. From my reading of the geography and religion literature, there were actually few studies done before 1970s. (See the bibliography at the end of Chap. 1). I took some renewed interest in the subject in the mid-1980s when a number of geographers began to examine religion/nature/environment issues. The pioneering works of Yi-Fu Tuan and Anne Buttimer were instrumental in steering the study of values, ethics, spirituality, and religion into some new and productive directions. These studies paved the way for a number of other studies by social geographers (a field that was not among the major fields until the 1970s and early 1980s). The steady stream of studies on geography and religion continued with the emergence of GORABS (Geography of Religion and Belief Systems) as a Specialty Group of the Association of American Geographers. The publication of more articles and special journal issues devoted to the geography of religion continued into the last decade of the twentieth century and first decade of this century. The synthetic works of Lily Kong that have regularly appeared in *Progress in Human Geography* further supported those who wanted to look at religion from human/environmental perspectives. These reviews not only introduced the study of religion within geography, but also to those in related scholarly disciplines.

As more and more research appeared in professional journals and more conferences included presentations on religion from different fields and subfields, it became increasing apparent that the time was propitious for a volume that looked at religion/geography interfaces from a number of different perspectives. From my own vantage point, the study of religion was one that could, should, might, and would benefit from those who have theoretical and conceptual training in many of the discipline's major subfields. The same applied to those who were regional specialists; there were topics meriting study from those who looked a political/religion issues in Southeast Asia or Central America as well as cultural/historical themes in southern Africa and continental Europe and symbolic/architectural features and built environments of religions landscapes in California, southeast Australia, and southwest Asia. Studying religions topics would not have to be limited to those in human geography, but could be seen as opportunities for those studying religion/natural disaster issues in East Asia and southeast United States as well as the spiritual roots of early and contemporary religious thinking in Central Asia, East Asia, Russia, and indigenous groups in South America. For those engaged in the study of gender, law, multicultural education, and media disciplines, there were also opportunities to contribute to the study of this emerging field. In short, there were literally "gold mines" of potential research topics in rural and urban areas everywhere on the continent.

About 7 years ago I decided to offer a class on the geography of religion in the Department of Geography at the University of Kentucky. The numbers were never larger (less than 15), but these were always enlightening and interesting, because of the views expressed by students. Their views about religion ran the gamut from very conservative to very liberal and also agnostic and atheist, which made, as one would expect, some very interesting exchanges. Students were strongly encouraged (not required, as I could not do this in a public university) to attend a half dozen different worship services during the semester. This did not mean attending First Baptist, Second Baptist, Third Baptist, etc., but different kinds of experiences. For some this course component was the first some had ever attended a Jewish synagogue, Catholic mass, Baptist service, an African American church, Unitarian church, or visited a mosque. Some students used this opportunity to attend Wiccan services, or visit a Buddhist and Hindu temple. Their write-ups about these experiences and the ensuing discussion were one of the high spots of the weekly class. In addition, the classes discussed chapters in various books and articles from the geography literature about the state of studying religion. And we always discussed current news items, using materials from the RNS (Religion News Service) website.

Another ingredient that stimulated my decision to edit a book on the geography of religion emerged from geography of religion conferences held in Europe in recent years. These were organized by my good friends Ceri Peach (Oxford), Reinhard Henkel (Heidelberg), and also Martin Baumann and Andreas Tunger-Zanetti (University of Lucerne). These miniconferences, held in Oxford, Lucerne, and Gottingen, usually attracted 20–40 junior and senior geographers and other religious scholars, and were a rich source of ideas for topics that might be studied. The opportunities for small group discussions, the field trips, and special events were conducive to learning about historical and contemporary changes in the religious landscapes of the European continent and beyond. A number of authors contributing to this volume presented papers at one or more of these conferences. Additional names came from those attending sessions at annual meetings of the Association of American Geographers.

Some of my initial thoughts and inspiration about a book came from the course I taught, conversations with friends who studied and did not study religion, and also the book I edited on megaengineering projects. This three-volume, 126-chapter book, *Engineering Earth: The Impacts of Megaengineering Projects*, was published in 2011 by Springer. There were only a few chapters in this book that had a religious content, one on megachurches, another on liberation theologians fighting megadevelopment projects in the Philippines and Guatemala. When I approached Evelien Bakker and Bernadette Deelen-Mans, my first geography editors at Springer, about a religion book, they were excited and supportive, as they have been since day one. They gave me the encouragement, certainly the latitude (and probably the longitude) to pursue the idea, knowing that I would identify significant cutting-edge topics about religion and culture and society in all major world regions. The prospectus I developed was for an innovative book that would include the contributions of scholars from the social sciences and humanities, those from different counties and those from different faiths. For their confidence and support, I am very grateful. The reviews they obtained of the prospectus were encouraging and acknowledging that there was a definite need for a major international, interdisciplinary, and interfaith volume. Springer also saw this book as an opportunity to emphasize its new directions in the social sciences and humanities. I also want to thank Stefan Einarson who came on board late in the project and shepherded the project to its completion with the usual Springer traits of professionalism, kindness, and commitment to the project's publication. And I wish to thank Chitra Sundarajan and her staff for helpful professionalism in preparing the final manuscript for publication.

The organization of the book, which is discussed in Chapter One, basically reflects the way I look at religion from a geographical perspective. I look at the subject as more than simply investigations into human geography's fields and subfields, including cultural and historical, but also economic, social, and political geography, but also human/environmental geography (dealing with human values, ethics, behavior, disasters, etc.). I also look at the study of religious topics and phenomena with respects to major concepts we use in geography; these include landscapes, networks, hierarchies, scales, regions, organization of space, the delivery of services, and virtual religion. I started contacting potential authors in September 2010. Since then I have sent or received over 15,000 emails related the volume.

I am deeply indebted to many friends for providing names of potential authors. I relied on my global network of geography colleagues in colleges and universities around the world, who not only recommended specific individuals, but also topics they deemed worthy of inclusion. Some were geographers, but many were not; some taught in universities, others in divinity schools and departments of religion around the world. Those I specifically want to acknowledge include: Barbara Ambrose, Martin Checa Artasu, Martin Baumann, John Benson, Gary Bouma, John Benson, Dwight Billings, Marion Bowman, John D. Brewer, David Brunn, David Butler, Ron Byars, Heidi Campbell, Caroline Creamer, Janel Curry, David Eicher, Elizabeth Ferris, Richard Gale, Don Gross, Wayne Gnatuk, Martin Haigh, Dan Hofrenning, Wil Holden, Hannah Holtschneider, Monica Ingalls, Nicole Karapanagiotis, Aharon Kellerman, Judith Kenny, Jean Kilde, Ted Levin, James

Munder, Alec Murphy, Tad Mutersbuagh, Garth Myers, Lionel Obidah, Sam Otterstrom, Francis Owusu, Maria Paradiso, Ron Pen, Ivan Petrella, Adam Possamai, Leonard Primiano, Craig Revels, Heinz Scheifinger, Anna Secor, Ira Sheskin, Doug Slaymaker, Patricia Solis, Anita Stasulne, Jill Stevenson, Robert Strauss, Tristan Sturm, Greg Stump, Karen Till, Andreas Tunger-Zenetti, Gary Vachicouras, Viera Vlčkova, Herman van der Wusten, Stanley Waterman, Mike Whine, Don Zeigler, Shangyi Zhou, and Matt Zook.

And I want to thank John Kostelnick who provided the GORABS Working Bibliography; most of the entries, except dissertations and theses, are included in Chap. 1 bibliography. Others who helped him prepare this valuable bibliography also need to be acknowledged: John Bauer, Ed Davis, Michael Ferber, Julian Holloway, Lily Kong, Elizabeth Leppman, Carolyn Prorock, Simon Potter, Thomas Rumney, Rana P.B. Singh, and Robert Stoddard. These are scholars who devoted their lifetimes to advancing research on geography and religion.

Finally I want to thank Donna Gilbreath for another splendid effort preparing all the chapters for Springer. She formatted the chapters and prepared all the tables and illustrations per the publisher's guidelines. Donna is an invaluable and skilled professional who deserves much credit for working with multiple authors and the publisher to ensure that all text materials were correct and in order. Also I am indebted to her husband, Richard Gilbreath, for helping prepare some of the maps and graphics for authors without cartographic services and making changes on others. As Director of the Gyula Pauer Center for Cartography and GIS, Dick's work is always first class. And, finally, thanks are much in order to Natalya Tyutenkova for her interest, support, patience, and endurance in the past several years working on this megaproject, thinking and believing it would never end.

The journey continues.

February 2014 Stanley D. Brunn

Contents of Volume III

Contributors

Jamaine Abidogun Department of History, Missouri State University, Springfield, MO, USA

Afe Adogame School of Divinity, University of Edinburgh, Edinburgh, Scotland, UK

Christopher A. Airriess Department of Geography, Ball State University, Muncie, IN, USA

Kaarina Aitamurto Aleksanteri Institute, University of Helsinki, Helsinki, Finland

Mikael Aktor Institute of History, Study of Religions, University of Southern Denmark, Odense, Denmark

Elizabeth Allison Department of Philosophy and Religion, California Institute of Integral Studies, San Francisco, CA, USA

Johan Andersson Department of Geography, King's College London, London, UK

Stephen W. Angell Earlham College, School of Religion, Richmond, IN, USA

J. Clark Archer Department of Geography, School of Natural Resources, University of Nebraska-Lincoln, Lincoln, NE, USA

Ian Astley Asian Studies, University of Edinburgh, Edinburgh, UK

Steven M. Avella Professor of History, Marquette University, Milwaukee, WI, USA

Yulier Avello COPEXTEL S.A., Ministry of Informatics and Communications, Havana, Cuba

Erica Baffelli School of Arts, Languages and Cultures, University of Manchester, Manchester, UK

Bakama BakamaNume Division of Social Work, Behavioral and Political Science, Prairie View A&M University, Prairie View, TX, USA

Josiah R. Baker Methodist University, Fayetteville, NC, USA

Economics and Geography, Methodist University, Fayetteville, USA

Holly R. Barcus Department of Geography, Macalester College, St. Paul, MN, USA

David Bassens Department of Geography, Free University Brussels, Brussels, Belgium

Ramon Bauer Wittgenstein Centre for Demography and Global Human Capital (IIASA, VID/ŐAW, WU), Vienna Institute of Demography/Austrian Academy of Sciences, Vienna, Austria

Whitney A. Bauman Department of Religious Studies, Florida International University, Miami, FL, USA

Gwilym Beckerlegge Department of Religious Studies, The Open University, Milton Keynes, UK

Michael Bégin Department of Global Studies, Pusan National University, Pusan, Republic of Korea

Demyan Belyaev Collegium de Lyon/Institute of Advanced Studies, Lyon, France

Alexandre Benod Research Division, Department of Japanese Studies, Université de Lyon, Lyon, France

John Benson School of Teaching and Learning, Minnesota State University, Moorhead, MN, USA

Sigurd Bergmann Department of Philosophy and Religious Studies, Norwegian University of Science and Technology, Trondheim, Norway

Rachel Berndtson Department of Geographical Sciences, University of Maryland, College Park, MD, USA

Martha Bettis Gee Compassion, Peace and Justice, Peace and Justice Ministries, Presbyterian Mission Agency, Presbyterian Church (USA), Louisville, KY, USA

Warren Bird Research Division, Leadership Network, Dallas, TX, USA

Andrew Boulton Department of Geography, University of Kentucky, Lexington, KY, USA

Humana, Inc., Louisville, KY, USA

Kathleen Braden Department of Political Science and Geography, Seattle Pacific University, Seattle, WA, USA

Namara Brede Department of Geography, Macalester College, St. Paul, MN, USA

John D. Brewer Institute for the Study of Conflict Transformation and Social Justice, Queen's University Belfast, Belfast, UK

Laurie Brinklow School of Geography and Environmental Studies, University of Tasmania, Hobart, Australia

Interim Co-ordinator, Master of Arts in Island Studies Program, University of Prince Edward Island, Charlottetown, PE Canada

Dave Brunn Language and Linguistics Department, New Tribes Missionary Training Center, Camdenton, MO, USA

Stanley D. Brunn Department of Geography, University of Kentucky, Lexington, KY, USA

David J. Butler Department of Geography, University of Ireland, Cork, Ireland

Anne Buttimer Department of Geography, University College Dublin, Dublin, Ireland

Éric Caron Malenfant Demography Division, Statistics Canada, Ottawa, Canada

Lori Carter-Edwards Gillings School of Global Public Health, Public Health Leadership Program, University of North Carolina, Chapel Hill, NC, USA

Clemens Cavallin Department of Literature, History of Ideas and Religion, University of Gothenburg, Göteborg, Sweden

Martin M. Checa-Artasu Department of Sociology, Universidad Autónoma Metropolitana, Unidad Iztapalapa, Mexico, DF, Mexico

Richard Cimino Department of Anthropology and Sociology, University of Richmond, Richmond, VA, USA

Paul Claval Department of Geography, University of Paris-Sorbonne, Paris, France

Paul Cloke Department of Geography, Exeter University, Exeter, UK

Kevin Coe Department of Communication, University of Utah, Salt Lake City, UT, USA

Noga Collins-Kreiner Department of Geography and Environmental Studies, Centre for Tourism, Pilgrimage and Recreation, University of Haifa, Haifa, Israel

Louise Connelly Institute for Academic Development, University of Edinburgh, Edinburgh, Scotland, UK

Thia Cooper Department of Religion, Gustavus Adolphus College, St. Peter, MN, USA

Catherine Cottrell Department of Geography and Earth Sciences, Aberystwyth University, Aberystwyth, UK

Thomas W. Crawford Department of Geography, East Carolina University, Greenville, NC, USA

Janel Curry Provost, Gordon College, Wenham, MA, USA

Seif Da'Na Sociology and Anthropology Department, University of Wisconsin-Parkside, Kenosha, WI, USA

Erik Davis Department of Religious Studies, Rice University, Houston, TX, USA

Jenny L. Davis Department of American Indian Studies, University of Illinois, Urbana-Champaign, Urbana, USA

Kiku Day Department of Ethnomusicology, Aarhus University, Aarhus, Denmark

Renée de la Torre Castellanos Centro de Investigaciones y Estudios Superiores en Antropologia Social-Occidente, Guadalajara, Jalisco, Mexico

Frédéric Dejean Institut de recherche sur l'intégration professionnelle des immigrants, Collège de Maisonneuve, Montréal (Québec), Canada

Veronica della Dora Department of Geography, Royal Holloway University of London, UK

Sergio DellaPergola The Avraham Harman Institute of Contemporary Jewry, The Hebrew University of Jerusalem, Mt. Scopus, Jerusalem, Israel

Antoinette E. DeNapoli Religious Studies Department, University of Wyoming, Laramie, WY, USA

Matthew A. Derrick Department of Geography, Humboldt State University, Arcata, CA, USA

C. Nathan DeWall Department of Psychology, University of Kentucky, Lexington, KY, USA

Jualynne Dodson Department of Sociology, American and African Studies Program, Michigan State University, East Lansing, MI, USA

David Domke Department of Communication, University of Washington, Seattle, WA, USA

Katherine Donohue M.A. Diplomacy and International Commerce, Patterson School of Diplomacy and International Commerce, University of Kentucky, Lexington, KY, USA

Lizanne Dowds Northern Ireland Life and Times Survey, University of Ulster, Belfast, UK

Kevin M. Dunn School of Social Sciences and Psychology, University of Western Sydney, Penrith, NSW, Australia

Claire Dwyer Department of Geography, University College London, London, UK

Patricia Ehrkamp Department of Geography, University of Kentucky, Lexington, KY, USA

Paul Emerson Teusner School of Media and Communication, RMIT University, Melbourne, VIC, Australia

Chad F. Emmett Department of Geography, Brigham Young University, Provo, UT, USA

Ghazi-Walid Falah Department of Public Administration and Urban Studies, University of Akron, Akron, OH, USA

Yasser Farrés Department of Philosophy, University of Zaragoza, Pedro Cerbuna, Zaragoza, Spain

Timothy Joseph Fargo Department of City Planning, City of Los Angeles, Los Angeles, CA, USA

Michael P. Ferber Department of Geography, The King's University College, Edmonton, AB, Canada

Tatiana V. Filosofova Department of World Languages, Literatures, and Cultures, University of North Texas, Denton, TX, USA

John T. Fitzgerald Department of Theology, University of Notre Dame, Notre Dame, IN, USA

Colin Flint Department of Political Science, Utah State University, Logan, UT, USA

Daniel W. Gade Department of Geography, University of Vermont, Burlington, VT, USA

Armando Garcia Chiang Department of Sociology, Universidad Autónoma Metropolitana Iztapalapa, Iztapalapa, Mexico

Jeff Garmany King's Brazil Institute, King's College London, London, UK

Martha Geores Department of Geographical Sciences, University of Maryland, College Park, MD, USA

Hannes Gerhardt Department of Geosciences, University of West Georgia, Carrolton, GA, USA

Christina Ghanbarpour History Department, Saddleback College, Mission Viejo, CA, USA

Danilo Giambra Department of Theology and Religion, University of Otago-Te Whare Wänanga o Otägo, Dunedin, New Zealand/Aotearoa

Banu Gökarıksel Department of Geography, University of North Carolina, Chapel Hill, NC, USA

Margaret M. Gold London Guildhall Faculty of Business and Law, London Metropolitan University, London, UK

Anton Gosar Faculty of Tourism Studies, University of Primorska, Portorož, Slovenia

Anne Goujon Wittgenstein Centre for Demography and Global Human Capital (IIASA, VID/ŐAW, WU), International Institute for Applied Systems Analysis (IIASA), Laxenburg, Austria

Vienna Institute of Demography/Austrian Academy of Sciences, Vienna, Austria

Alyson L. Greiner Department of Geography, Oklahoma State University, Stillwater, OK, USA

Daniel Jay Grimminger Faith Lutheran Church, Kent State University, Millersburg, OH, USA

School of Music, Kent State University, Kent, OH, USA

Zeynep B. Gürtin Department of Sociology, University of Cambridge, Cambridge, UK

Cristina Gutiérrez Zúñiga Centro Universitario de Ciencias Sociales y Humanidades, El Colegio de Jalisco, Zapopan, Jalisco, Mexico

Martin J. Haigh Department of Social Sciences, Oxford Brookes University, Oxford, UK

Anna Halafoff Centre for Citizenship and Globalisation, Deakin University, Burwood, VIC, Australia

Airen Hall Department of Theology, Georgetown University, Washington, DC, USA

Randolph Haluza-DeLay Department of Sociology, The Kings University, Edmonton, AB, Canada

Tomáš Havlíček Faculty of Science, Department of Social Geography and Regional Development, Charles University, Prague 2, Czechia

C. Michael Hawn Sacred Music Program, Perkins School of Theology, Southern Methodist University, Dallas, TX, USA

Bernadette C. Hayes Department of Sociology, University of Aberdeen, Aberdeen, Scotland, UK

Peter J. Hemming School of Social Sciences, Cardiff University, Cardiff, Wales, UK

William Holden Department of Geography, University of Calgary, Calgary, AB, Canada

Edward C. Holland Havighurst Center for Russian and Post-Soviet Studies, Miami University, Oxford, OH, USA

Beverly A. Howard School of Music, California Baptist University, Riverside, CA, USA

Martina Hupková Faculty of Science, Department of Social Geography and Regional Development, Charles University, Prague 2, Czechia

Tim Hutchings Post Doc, St. John's College, Durham University, Durham, UK

Ronald Inglehart Institute of Social Research, University of Michigan, Ann Arbor, MI, USA

World Values Survey Association, Madrid, Spain

Marcia C. Inhorn Anthropology and International Affairs, Yale University, New Haven, CT, USA

Adrian Ivakhiv Environmental Program, University of Vermont, Burlington, VT, USA

Maria Cristina Ivaldi Dipartimento di Scienze Politiche "Jean Monnet", Seconda Università degli Studi di Napoli, Caserta, Italy

Thomas Jablonsky Professor of History, Marquette University, Milwaukee, WI, USA

Maria Jaschok International Gender Studies Centre, Lady Margaret Hall, Oxford University, Norham Gardens, UK

Philip Jenkins Institute for the Study of Religion, Baylor University, Waco, TX, USA

Wesley Jetton Student, University of Kentucky, Lexington, KY, USA

Shui Jingjun Henan Academy of Social Sciences, Zhengzhou, Henan Province, China

Mark D. Johns Department of Communication, Luther College, Decorah, IA, USA

James H. Johnson Jr. Kenan-Flagler Business School and Urban Investment Strategies Center, University of North Carolina, Chapel Hill, NC, USA

Lucas F. Johnston Department of Religion and Environmental Studies, Wake Forest University, Winston-Salem, NC, USA

Peter Jordan Austrian Academy of Sciences, Institute of Urban and Regional Research, Wien, Austria

Yakubu Joseph Geographisches Institut, University of Tübingen, Tübingen, Germany

Deborah Justice Yale Institute of Sacred Music, Yale University, New Haven, CT, USA

Akel Ismail Kahera College of Architecture, Art and Humanities, Clemson University, Clemson, SC, USA

P.P. Karan Department of Geography, University of Kentucky, Lexington, KY, USA

Sya Buryn Kedzior Department of Geography and Environmental Planning, Towson State University, Towson, MD, USA

Kevin D. Kehrberg Department of Music, Warren Wilson College, Asheville, NC, USA

Laura J. Khoury Department of Sociology, Birzeit University, West Bank, Palestine

Hans Knippenberg Department of Geography, Planning and International Development Studies, University of Amsterdam, Velserbroek, The Netherlands

Katherine Knutson Department of Political Science, Gustavus Adolphus College, St. Peter, MN, USA

Miha Koderman Science and Research Centre of Koper, University of Primorska, Koper-Capodistria, Slovenia

Lily Kong Department of Geography, National University of Singapore, Singapore, Singapore

Igor Kotin Museum of Anthropology and Ethnography, Russian Academy of Sciences, St. Petersburg, Russia

Katharina Kunter Faculty of Theology, University of Bochum, Bochum, Germany

Lisa La George International Studies, The Master's College, Santa Clarita, CA, USA

Shirley Lal Wijesinghe Faculty of Humanities, University of Kelaniya, Kelaniya, Sri Lanka

Ibrahim Badamasi Lambu Department of Geography, Faculty of Earth and Environmental Sciences, Bayero University Kano, Kano, Nigeria

Michelle Gezentsvey Lamy Comparative Education Research Unit, Ministry of Education, Wellington, New Zealand

Justin Lawson Health, Nature and Sustainability Research Group, School of Health and Social Development, Deakin University, Burwood, VIC, Australia

Deborah Lee Department of Geography, National University of Singapore, Singapore, Singapore

Karsten Lehmann Senior Lecturer, Science des Religions, Bayreuth University, Fribourg, Switzerland

Reina Lewis London College of Fashion, University of the Arts, London, UK

Micah Liben Judaic Studies, Kellman Brown Academy, Voorhees, NJ, USA

Edmund B. Lingan Department of Theater, University of Toledo, Toledo, OH, USA

Rubén C. Lois-González Departamento de Xeografía, Universidade de Santiago de Compostela, Galiza, Spain

Naomi Ludeman Smith Learning and Women's Initiatives, St. Paul, MN, USA

Katrín Anna Lund Department of Geography and Tourism, Faculty of Life and Environmental Sciences, University of Iceland, Reykjavik, Iceland

Avril Maddrell Department of Geography and Environmental Sciences, University of West England, Bristol, UK

Juraj Majo Department of Human Geography and Demography, Faculty of Sciences, Comenius University in Bratislava, Bratislava, Slovak Republic

Virginie Mamadouh Department of Geography, Planning and International Development Studies, University of Amsterdam, Amsterdam, The Netherlands

Mariana Mastagar Department of Theology, Trinity College, University of Toronto, Toronto, Canada

Alberto Matarán Department of Urban and Spatial Planning, University of Granada, Granada, Spain

René Matlovič Department of Geography and Applied Geoinformatics, Faculty of Humanities and Natural Sciences, University of Prešov, Prešov, Slovakia

Kvetoslava Matlovičová Department of Geography and Applied Geoinformatics, Faculty of Humanities and Natural Sciences, University of Prešov, Prešov, Slovakia

Hannah Mayne Department of Anthropology, University of Florida, Gainesville, FL, USA

Shampa Mazumdar Department of Sociology, University of California, Irvine, CA, USA

Sanjoy Mazumdar Department of Planning, Policy and Design, University of California, Irvine, CA, USA

Andrew M. McCoy Center for Ministry Studies, Hope College, Holland, MI, USA

Daniel McGowin Department of Geology and Geography, Auburn University, Auburn, AL, USA

James F. McGrath Department of Philosophy and Religion, Butler University, Indianapolis, IN, USA

Nick Megoran Department of Geography, University of Newcastle-upon-Tyne, Newcastle, UK

Amy Messer Department of Sociology, University of Kentucky, Lexington, KY, USA

Sarah Ann Deardorff Miller Researcher, Refugee Studies Centre, Oxford, UK

Kelly Miller Centre for Integrative Ecology, School of Life and Environmental Sciences, Deakin University, Burwood, VIC, Australia

Nathan A. Mosurinjohn Center for Social Research, Calvin College, Grand Rapids, MI, USA

Sven Müller Institute for Transport Economics, University of Hamburg, Hamburg, Germany

Erik Munder Institut für Vergleichende Kulturforschung - Kultur- u. Sozialanthropologie und Religionswissenschaft, Universität Marburg, Marburg, Germany

David W. Music School of Music, Baylor University, Waco, TX, USA

Kathleen Nadeau Department of Anthropology, California State University, San Bernadino, CA, USA

Caroline Nagel Department of Geography, University of South Carolina, Columbia, SC, USA

Pippa Norris John F. Kennedy School of Government, Harvard University, Cambridge, MA, USA

Government and International Relations, University of Sydney, Sydney, Australia

Orville Nyblade Makumira University College, Usa River, Tanzania

Lionel Obadia Department of Anthropology, Université de Lyon, Lyon, France

Daniel H. Olsen Department of Geography, Brigham Young University, Provo, UT, USA

Samuel M. Otterstrom Department of Geography, Brigham Young University, Provo, UT, USA

Barbara Palmquist Department of Geography, University of Kentucky, Lexington, KY, USA

Grigorios D. Papathomas Faculty of Theology, University of Athens, Athens, Greece

Nikos Pappas Musicology, University of Alabama, Tuscaloosa, AL, USA

Mohammad Aslam Parvaiz Islamic Foundation for Science and Environment (IFSE), New Delhi, India

Valerià Paül Departamento de Xeografía, Universidade de Santiago de Compostela, Galiza, Spain

Miguel Pazos-Otón Departamento de Xeografía, Universidade de Santiago de Compostela, Galiza, Spain

David Pereyra Toronto School of Theology, University of Toronto, Toronto, Canada

Bruce Phillips Loucheim School of Judaic Studies at the University of Southern California, Hebrew Union College-Jewish Institute of Religion, Los Angeles, CA, USA

Awais Piracha School of Social Sciences and Psychology, University of Western Sydney, Penrith, NSW, Australia

Linda Pittman Department of Geography, Richard Bland College of the College of William and Mary, Petersburg, VA, USA

Richard S. Pond Department of Psychology, University of North Carolina, Wilmington, NC, USA

Carolyn V. Prorok Independent Scholar, Slippery Rock, PA, USA

Steven M. Radil Department of Geography, University of Idaho, Moscow, ID, USA

Esther Long Ratajeski Independent Scholar, Lexington, KY, USA

Daniel Reeves Faculty of Science, Department of Social Geography and Regional Development, Charles University, Prague 2, Czechia

Arthur Remillard Department of Religious Studies, St. Francis University, Loretto, PA, USA

Claire M. Renzetti Department of Sociology, University of Kentucky, Lexington, KY, USA

Friedlind Riedel Department of Musicology, Georg-August-University of Göttingen, Göttingen, Germany

Sandra Milena Rios Oyola Department of Sociology and the Compromise after Conflict Research Programme, University of Aberdeen, Aberdeen, Scotland, UK

C.K. Robertson Presiding Bishop, The Episcopal Church, New York, NY, USA

Arsenio Rodrigues School of Architecture, Prairie View A&M University, Prairie View, TX, USA

Andrea Rota Institute for the Study of Religion, University of Bern, Bern, Switzerland

Rainer Rothfuss Geographisches Institut, University of Tübingen, Tübingen, Germany

Jeanmarie Rouhier-Willoughby Department of Modern and Classical Languages, Literatures and Cultures, University of Kentucky, Lexington, KY, USA

Rex J. Rowley Department of Geography-Geology, Illinois State University, Normal, IL, USA

Bradley C. Rundquist Department of Geography, University of North Dakota, Grand Forks, ND, USA

Simon Runkel Department of Geography, University of Bonn, Bonn, Germany

Joanna Sadgrove Research Staff, United Society, London, UK

Michael Samers Department of Geography, University of Kentucky, Lexington, KY, USA

Åke Sander Department of Literature, History of Ideas and Religion, University of Gothenburg, Göteborg, Sweden

Xosé M. Santos Departamento de Xeografía, Universidade de Santiago de Compostela, Galiza, Spain

Alessandro Scafi Medieval and Renaissance Cultural History, The Warburg Institute, University of London, London, UK

Anthony Schmidt Department of Communication Studies, Edmonds Community College, Edmonds, WA, USA

Mallory Schneuwly Purdie Institut de sciences sociales des religions contemporaines, Observatoire des religions en Suisse, Université de Lausanne – Anthropole, Lausanne, Switzerland

Anna J. Secor Department of Geography, University of Kentucky, Lexington, KY, USA

Hafid Setiadi Department of Geography, University of Indonesia, Depok, West Java, Indonesia

Fred M. Shelley Department of Geography and Environmental Sustainability, University of Oklahoma, Norman, OK, USA

Ira M. Sheskin Department of Geography and Regional Studies, University of Miami, Coral Gables, FL, USA

Lia Dong Shimada Conflict Mediator, Methodist Church in Britain, London, UK

Caleb Kwang-Eun Shin ABD, Korea Baptist Theological Seminary, Daejeon, Republic of Korea

J. Matthew Shumway Department of Geography, Brigham Young University, Provo, UT, USA

Dmitrii Sidorov Department of Geography, California State University, Long Beach, Long Beach, CA, USA

Caleb Simmons Religious Studies Program, University of Arizona, Tucson, AZ, USA

Devinder Singh Department of Geography, University of Jammu, Jammu, Jammu and Kashmir, India

Rana P.B. Singh Department of Geography, Banaras Hindu University, Varanasi, UP, India

Nkosinathi Sithole Department of English, University of Zululand, KwaZulu-Natal, South Africa

Vegard Skirbekk Wittgenstein Centre for Demography and Global Human Capital (IIASA, VID/ŐAW, WU), International Institute for Applied Systems Analysis, Laxenburg, Austria

Alexander Thomas T. Smith Department of Sociology, University of Warwick, Coventry, UK

Christopher Smith Independent Scholar, Tecumseh, OK, USA

Ryan D. Smith Compassion, Peace and Justice Ministries, Presbyterian Ministry at the U.N., Presbyterian Mission Agency, Presbyterian Church (USA), New York, NY, USA

Sara Smith Department of Geography, University of North Carolina, Chapel Hill, NC, USA

Leslie E. Sponsel Department of Anthropology, University of Hawaii, Honolulu, HI, USA

Chloë Starr Asian Christianity and Theology, Yale Divinity School, New Haven, CT, USA

Jeffrey Steller Public History, Northern Kentucky University, Highland Heights, KY, USA

Christopher Stephens Southlands College, University of Roehampton, London, UK

Jill Stevenson Department of Theater Arts, Marymount Manhattan College, New York, NY, USA

Anna Rose Stewart Department of Religious Studies, University of Kent, Canterbury, UK

Nancy Palmer Stockwell Senior Contract Administrator, Enerfin Resources, Houston, TX, USA

Robert Strauss President and CEO, Worldview Resource Group, Colorado Springs, CO, USA

Tristan Sturm School of Geography, Archaeology and Palaeoecology, Queen's University Belfast, Belfast, UK

Edward Swenson Department of Anthropology, University of Toronto, Toronto, ON, Canada

Anna Swynford Duke Divinity School, Duke University, Durham, NC, USA

Jonathan Taylor Department of Geography, California State University, Fullerton, CA, USA

Francis Teeney Institute for the Study of Conflict Transformation and Social Justice, Queen's University Belfast, Belfast, UK

Mary C. Tehan Stirling College, University of Divinity, Melbourne, Australia

Andrew R.H. Thompson The School of Theology, The University of the South, Sewanee, TN, USA

Scott L. Thumma Professor, Department of Sociology, Hartford Seminary, Hartford, CT, USA

Meagan Todd Department of Geography, University of Colorado, Boulder, CO, USA

Soraya Tremayne Fertility and Reproduction Studies Group, Institute of Social and Cultural Anthropology, University of Oxford, Oxford, UK

Gill Valentine Faculty of Social Sciences, University of Sheffield, Sheffield, UK

Inge van der Welle Department of Geography, Planning and International Development Studies, University of Amsterdam, Amsterdam, The Netherlands

Herman van der Wusten Department of Geography, Planning and International Development Studies, University of Amsterdam, Amsterdam, The Netherlands

Robert M. Vanderbeck Department of Geography, University of Leeds, West Yorkshire, UK

Jason E. VanHorn Department of Geography, Calvin College, Grand Rapids, MI, USA

Viera Vlčková Department of Public Administration and Regional Development, Faculty of National Economy, University of Economics in Bratislava, Bratislava, Slovakia

Geoffrey Wall Department of Geography and Environmental Management, University of Waterloo, Waterloo, ON, Canada

Robert H. Wall Counsel, Spilman Thomas & Battle, Winston-Salem, NC, USA

Kevin Ward School of Theology and Religious Studies, University of Leeds, West Yorkshire, UK

Barney Warf Department of Geography, University of Kansas, Lawrence, KS, USA

Stanley Waterman Department of Geography and Environmental Studies, University of Haifa, Haifa, Israel

Robert H. Watrel Department of Geography, South Dakota State University, Brookings, SD, USA

Gerald R. Webster Department of Geography, University of Wyoming, Laramie, WY, USA

Paul G. Weller Research, Innovation and Academic Enterprise, University of Derby, Derby, UK

Oxford Centre for Christianity and Culture, University of Oxford, Oxford, UK

Cynthia Werner Department of Anthropology, Texas A&M University, College Station, TX, USA

Geoff Wescott Centre for Integrative Ecology, School of Life and Environmental Sciences, Deakin University, Burwood, VIC, Australia

Carroll West Center for Historic Preservation, Middle Tennessee State University, Murfreesboro, TN, USA

Gerald West School of Religion, Philosophy and Classics, University of KwaZulu-Natal, Scottsville, South Africa

Mark Whitaker Department of Anthropology, University of Kentucky, Lexington, KY, USA

Thomas A. Wikle Department of Geography, Oklahoma State University, Stillwater, OK, USA

Justin Wilford Department of Geography, University of California, Los Angeles, CA, USA

Joseph Witt Department of Philosophy and Religion, Mississippi State University, Mississippi State, MS, USA

John D. Witvliet Calvin Institute of Christian Worship, Calvin College and Calvin Theological Seminary, Grand Rapids, MI, USA

Teresa Wright Department of Political Science, California State University, Long Beach, CA, USA

Ernest J. Yanarella Department of Political Science, University of Kentucky, Lexington, KY, USA

Yukio Yotsumoto College of Asia Pacific Studies, Ritsumeikan Asia Pacific University, Beppu, Oita, Japan

Samuel Zalanga Department of Anthropology, Sociology and Reconciliation Studies, Bethel University, St. Paul, MN, USA

Donald J. Zeigler Department of Geography, Old Dominion University, Virginia Beach, VA, USA

Shangyi Zhou School of Geography, Beijing Normal University, Beijing, China

Teresa Zimmerman-Liu Departments of Asian/Asian-American Studies and Sociology, California State University, Long Beach, CA, USA

Part VII
Globalization, Diasporas and New Faces in the Global North

Chapter 69
Four Corners of the Diaspora: A Psychological Comparison of Jewish Continuity in Major Cities in New Zealand, Australia, Canada and the United States

Michelle Gezentsvey Lamy

69.1 Introduction: A Cross-Cultural Psychology Approach to the Phenomenon of Jewish Continuity

"Sound the great shofar for our freedom, and raise the ensign to gather our exiles and gather us from the four corners of the earth." This phrase, based on Isaiah 27:13, is featured in Jewish liturgy and makes reference to the ingathering of the exiles – then a prophecy, and now a reality (Rabinowitz 2008). Of world Jewry today, 42 % have returned to the land of Israel, and since 1948 have been able to live a Jewish life in a national majority context – something that was not possible since the destruction of the Second Temple by the Romans in 70 CE. This chapter briefly reviews the mechanisms that ensured the "exiles" remained Jewish over the centuries, and applies a cross-cultural psychology perspective to contribute to our understanding of how four of the remaining Diaspora communities today are ensuring Jewish continuity.

According to Asa-El (2004: 28), the Jewish nation is "so geographically stretched and culturally fluid that its dispersion set it apart from other nations even more than its distinctive laws, rules, tradition, languages, and dress." Over 2,000 years, the acculturation of Jews to different host societies at different rates and to different degrees has led to cultural heterogeneity in terms of language, food, dress, norms, and religious observance. Ashkenazi Jews lived near the Rhône basin (Franco-Germany and the Polish-Russian territories); Sephardi Jews were situated in the Iberian peninsula (Spain and Portugal); Mizrahi Jews remained in the Middle East, in North Africa (Morocco, Tunisia, Algeria), Egypt, and Asia (Persia, Syria, Lebanon, Yemen, Afghanistan, Kurdistan); Ancient Jewish communities include

---

This research was conducted as part of a doctoral thesis at Victoria University of Wellington.

M.G. Lamy (✉)
Comparative Education Research Unit, Ministry of Education,
Wellington, New Zealand
e-mail: michamie@yahoo.co.nz

S.D. Brunn (ed.), *The Changing World Religion Map*,
DOI 10.1007/978-94-017-9376-6_69

the Bene Roma in Italy, Bene Israel in India, B'nei Menashe in China, Beta Israel in Ethiopia; groups with Jewish heritage include the Lemba in South Africa, Igbo in Nigeria, and the House of Israel in Ghana; in addition, communities of Karaite Jews existed separately (Asa-El 2004; Tobin et al. 2005). While Diaspora communities were successful under regimes of tolerance, they also lived a precarious existence under regimes of oppression in both Christian and Muslim lands. Anti-Semitism in Europe, civil and international war, and massive migration patterns over the last two centuries brought thousands of Jews to new lands. This chapter focuses on the continuity of Jews (mostly Ashkenazi), who migrated to the former British colonies of the United States, Canada, Australia and New Zealand.

> The Egyptians, the Babylonians and the Persians rose, filled the planet with sound and splendour; then faded to dream-stuff and passed away; the Greeks and the Romans followed and made a vast noise, and they are gone; other peoples have sprung up and held their torch high for a time but it burned out, and they sit in twilight now, or have vanished (…) All things are mortal but the Jew; all other forces pass, but he remains. What is the secret of his immortality? Mark Twain (1897)

The phenomenon of Jewish continuity has constituted a puzzle for many. One can attempt to solve this question by examining the factors from a social science perspective that prevented the Jewish people from suffering the same fate as these grand civilizations. Jewish continuity is defined as the process by which Jews – an ethno-cultural group with shared heritage, connections to the land of Israel and distinctive religious beliefs and practices – as heterogeneous living entities, retained their uniqueness while undergoing change as they travel through time (different socio-historical contexts) and space (different societies of residence) (Gezentsvey 2008). Jews themselves have been most acutely aware of the continual threats to their survival, an awareness shared by other individuals who belong to "small peoples" that have experienced existential uncertainty and cannot take continuity for granted (Kundera 1993). Abulof (2007) examined this tension, noting how for generations Jewish individuals have seen themselves as the last link in the chain (Rawidowicz 1986), and have been characterized as existential hypochondriacs (Avidan 1986). Literature has pointed to both internal and external mechanisms as contributing factors to the longevity of the Jewish people, ensuring that there is always another link in the chain.

69.1.1 Intra and Inter-group Factors Contributing to Jewish Continuity

The long-term acculturation of Jews in the Diaspora for 2,000 years has been achieved due to internal factors such as: the canonization of core texts, the centralizing powers of the synagogue, ethnic solidarity in the Kehillah/community; a religious commitment to the covenant, the centripetal forces of collective memory, and boundary maintenance through endogamy (Yerushalmi 1982; Johnson 1987; Elazar and Cohen 1985; Neusner 1999; Salkin 1999; Sacks 2000). Endogamy, marriage among Jews, is a particularly important method of cultural transmission that fosters Jewish continuity by facilitating enculturation in a Jewish home environment (Alba 1990; Schönpflug 2001; Phillips and Chertok 2004). Biblical instructions for Jewish men not to marry the daughters of the land are found in Genesis (24 and 27), Exodus (34:16), Deuteronomy (7:3–4), and Ezra (9). While orthodox Jews maintain this as a religious obligation, many Jews outside of highly religious circles view marrying another Jew as a personal choice, an outlook shared by acculturation research in social psychology, which endeavors to examine the social factors that influence this choice.

From a cross-cultural psychology perspective, the construct of ethnolinguistic vitality can provide an understanding of community endurance, with language as a key defining feature (Giles et al. 1977). Continuity can be measured "objectively," according to factors such as relative social status of the minority group in the larger society, institutional support, demographic factors and cultural capital (Harwood et al. 1994; Berry 2003). Furthermore, perceptions of group vitality – and in the case of Jews, perceptions of collective decline – directly influence the behavior of group members, motivating individuals to boost their group's representation (Bourhis et al. 1981; Sachdev 1995).

According to a Jewish strategic perspective, forces that impact continuity can be captured by the concept of internal momentum that includes quantitative and qualitative domains such as demography (stable or increasing population), social aspects (identification, knowledge and social capital), power structures (political and communal organizations, networks) economics (financial sustainability, philanthropy) and leadership (Maimon 2007). Under positive external conditions, when internal momentum is high, the Jewish community is "thriving" and continuity ensues. However, when internal momentum is low, the community is "drifting" and continuity is lost.

Continuity has also been due to external factors such as the status of Jews as sojourners, and anti-Semitism – where discrimination enhances group solidarity (Zenner 1991; Langman 1999; MacDonald 2002). Spicer (1971) emphasized the role of inter-group opposition and a perpetual state of conflict and resistance as a driver of Jewish continuity. In the face of continuing anti-Semitism and following the destruction of one-third of world Jewry in the Shoah, the establishment of the State of Israel was seen as the political solution to ensure the survival of the Jewish people. While ontological or identity-based survival is ensured, epistemic or physical insecurity remains a concern (Abulof 2007). However, the future of Diaspora Jewish life in tolerant host societies, with less external pressure from discrimination to enhance community cohesion, rests on forces that come from within the Jewish collective (Maimon 2007).

69.1.2 Individual Choice

From a sociological perspective, individual social mobility and assimilation, an option available for non-visible minorities, pose a real threat to continuity (Dershowitz 1997; Frye Jacobson 1998). As the communal networks that bind Jews are left behind such as Jewish neighborhoods and concentrated occupations, a symbolic, individualized ethnicity remains (Gans 1979). Collective continuity is eventually undermined in the absence of such social structures (Alba 1990). Furthermore, post-modernism has had a large impact on the psychology of individual identification; whereby Jewish identity is increasingly becoming a tool that serves individuals in their personal journeys in search of meaning, rather than any collective purpose (Cohen and Eisen 2000).

Applying acculturation research from cross-cultural psychology to understand individual journeys, minority individuals in larger societies can *choose* to which degree they maintain their heritage and adapt to the culture of the larger society. As such, they can adopt an integrated (identifying as both Jewish and an American), separated (just Jewish), or assimilated (just American) acculturation strategy or become marginalized (neither Jewish nor American), see Berry (2003). The integration strategy is associated with better psychological and sociocultural adaptation (Phinney et al. 2001) and it is also the preferred strategy among an international sample of ethno-cultural youth (Berry et al. 2006).

If individuals value culture maintenance, then at the collective level, it follows that continuity of their heritage culture should be desired. For the Jewish people, ensuring collective continuity through active transmission is a biblical injunction: "And thou shalt teach them diligently to thy children, and shalt talk of them when thou sittest in thy house, and when thou walkest by the way, and when thou liest down, and when thou risest up" (Deuteronomy, 6:7). This chapter investigates how Jews today think about ensuring there is always another link in the chain and what they do about it.

69.2 Motivation for Jewish Continuity: Discourse and Measurement

A culture-general construct, Motivation for Ethno-cultural Continuity (MEC) was developed based on exploratory focus groups with small samples of three ethno-cultural groups in New Zealand[1] (Gezentsvey 2008; Gezentsvey and Ward 2008; Gezentsvey Lamy et al. 2013). This construct goes beyond traditional acculturation preferences; it measures the will of individual members to preserve and perpetuate their cultural heritage over generations, encompassing the themes of maintenance, transmission and endurance. When applied to the Jewish people, Motivation for Jewish Continuity captures an individual's motivation to maintain their Jewish heritage in their own life, raise Jewish children and ensure the survival of the Jewish people.

The transmission of life cycle rituals such as brit milah (circumcision), bar mizvah and chuppah (marriage) were very important in the Jewish focus group discussion,

[1] Jews (n=8), Chinese (n=5) and Māori (n=5).

as well as celebrating commemorative festivals like Passover. Participants emphasized that transmission is critical for Jewish continuity: "*How else does it stay alive if you can't give it to the next generation and that's where it ends*?" New Zealand Jewish participants used the favorite national sport as a metaphor for continuity: "*Everyone wants to cheer on their own team, you want to see your group carry on – let's use a rugby game for instance, you're still going to cheer on your country and don't want them to just drop out of the game half way through.*" Overall, this novel construct demonstrates that in a post-modern world, Jews continue to keep collective interests in mind during their individual journeys.

Applying methods from cross-cultural psychology, a derived-etic empirical measure was created by extracting statements from the three focus group discussions [see Gezentsvey Lamy et al. (2013) for a detailed description of the development, validation and equivalence testing of the final MEC scale]. References to specific cultures were removed so that the scale can be used across different groups. Statements were reworded with culture-general vocabulary, double-barreled items were separated, and items that lacked in face validity were eliminated. Ten items representing each theme (maintenance, transmission and endurance) were selected, resulting in a 30-statement item pool consisting of 16 positive and 14 negatively-worded items. Likert scales were used to quantify an individual's position with regards to each item, with a 7-point scale ranging from 1 (strongly disagree) to 7 (strongly agree). Item responses were averaged to give a mean score, with negatively-worded items reverse-scored prior to calculating means. High scores represent a strong motivation. The measure can be used among different samples by inserting the appropriate ethnocultural noun and adjective. The scale was tested for convergent and discriminant validity with existing psychological measures[2] in a study with 152 affiliated Jews in Sydney, Australia. Exploratory Factor Analysis revealed that the three qualitative domains are highly interrelated to produce a unifactorial measure (48.66 % of the variance was explained). The pilot scale was reduced to 18 items, with an excellent measurement reliability of $\alpha = 0.95$ (Table 69.1).

This quantitative measure was examined in a cross-national study to examine how Jewish community vitality impacts individual motivation and continuity-enhancing behavior, specifically endogamy intentions and selective dating. The predictive power of motivation was also tested. Given the tolerant environment for mixed marriages in New Zealand, Australia, Canada and the United States, it was hypothesized that Jews who do choose to date and intend to marry fellow ethnic group members do so because of their commitment to Jewish continuity.

[2] Convergent validity was found with measures of Collective Self-Esteem (Luhtanen and Crocker 1992), $r = 0.67^{**}$; Perceived Collective Continuity (Sani and Bowe 2004), $r = 0.54^{**}$ and Perceived Group Entitativity (Castano et al. 2003), $r = 0.32^{**}$. No significant correlation was found with Assimilation (Berry et al. 2006), providing evidence for discriminant validity. Thus, MJC is not simply measuring the rejection of assimilation.

Table 69.1 Pilot MEC scale (MLE, 6 iterations; N = 152; unifactorial)

No.	Item	Loading
1	I would like to keep on living according to the traditions of my Jewish heritage	0.64
2	Maintaining my Jewish traditions and language is NOT important to me[a]	0.82
3	I want to keep Jewish culture alive	0.83
4	Ultimately I would like my children to identify as Jews	0.73
5	Maintaining my Jewish heritage is NOT something I care about[a]	0.80
6	I would like to encourage my children to learn Hebrew	0.59
7	The future continuity of our Jewish community is NOT a concern of mine[a]	0.63
8	Continuing to practice my Jewish traditions and celebrations is important to me	0.79
9	Long-term, I would like my grandchildren and great grandchildren to continue our Jewish heritage	0.71
10	My Jewish heritage and traditions are something I can easily disregard[a]	0.70
11	I want to ensure the future of our Jewish heritage	0.86
12	It does NOT matter if my children don't identify with their Jewish heritage[a]	0.77
13	The endurance of Jewish people does NOT really matter[a]	0.65
14	I want to transmit to my children a love for and interest in their Jewish heritage	0.60
15	I do NOT mind setting aside the traditions of my Jewish heritage[a]	0.78
16	I do NOT care if my children are unaware of Jewish traditions and values[a]	0.75
17	I think it's good to create an environment at home where my Jewish traditions can be a normal part of life for my children	0.74
18	I do NOT really care about ensuring the future of Jewish people[a]	0.78

[a] Reverse-scored items. ***SHADED*** items were selected in the final scale. See Gezentsvey Lamy, Ward & Liu (2013) for a detailed description of the development, validation and equivalence testing of the final MEC scale.
Source: Michelle Gezentsvey Lamy

69.3 The Impact of Community Vitality on the Psychology and Behavior of Its Individual Members

The vitality of Jewish communities in Auckland and Wellington differs greatly from communities found in Melbourne, Toronto, and New York City. It is probable that Motivation for Jewish Continuity and engagement in continuity-enhancing behavior differ among Jews living in these corners of the world. Keep in mind that these communities differ not only in size, but in immigration history and country-level traits, for example, religious identification, and acculturation experiences. A brief summary of ethnographic characteristics of Jews in New Zealand (Levine 1995, 1997, 1999), Australia (Szwarc 2004; Rutland 2005), Canada (Tulchinsky 1998; Institute for International Affairs/B'nei Brith Canada 2000; Weinfeld 2001; Shahar and Rosenbaum 2006), and the United States (Mayer et al. 2001; National Jewish Population Survey 2000–2001; Sarna 2004) is presented before discussing hypothesized differences in the psychology and behavior of Jews sampled from these four communities.

69.3.1 Similarities and Differences Among the Four Jewish Communities

Jews are a minority in all four nations, representing less than 2 % of the total national population. The Australian Jewish community (101,000 Jews) is almost 15 times the size of the New Zealand community (6,800 Jews); the Canadian Jewish community (371,000 Jews) is almost four times the size of the Australian community; and the American Jewish community (5,290,000 Jews) is over 14 times the size of the Canadian community (American Jewish Committee 2005). In terms of community origins, Jews were pushed by anti-Semitism and pulled by socioeconomic prospects to these four communities. The older American Jewish community (late 1400s–1600s) and Canadian Jewish community (late 1700s–1800s) comprised of Sephardi migrants. In contrast, the initial Australian Jewish community arose from the settlement of Anglo-Jewish convicts (late 1700s–1800s) whereas New Zealand's first Jews were free British men (1800s). All four nations experienced the influx of German and Eastern European Ashkenazi Jews with the Australian and Canadian communities welcoming large numbers of Holocaust refugees.

A strength of these four Diaspora communities is that on average, Jews are highly educated and economically successful. However, in all four nations, Jews are an aging population and have low birthrates. While the major source for Jewish population growth in Canada and Australia has been immigration, as immigration declines over the following generations, communities will need to combat assimilation trends.

In terms of religious identification, the community in the United States is most heterogeneous, followed by Canada. Australia and New Zealand have a more restricted range of options. Of affiliated American Jews, a greater proportion identify as Reform (38 %), followed by Conservative (32 %) and Orthodox (10 %). Of affiliated Canadian Jews, most are Conservative (43 %), followed by Orthodox (25 %) and Reform (14 %). Of Jews in Melbourne, Australia a greater proportion identify as traditional religious/Orthodox (33 %), followed by Reform (15 %) and strictly Orthodox (6 %). Finally, of affiliated New Zealand Jews, the majority are nominally[3] Orthodox (70 %), followed by Reform (30 %).

As members of "small peoples" (Kundera 1993), Jews from all four communities have reason to be concerned about Jewish continuity at a local and global level. However, intermarriage, an indicator of assimilation, is on the rise in all four countries with rates in the United States and New Zealand near 50 %, followed by Canada with 27 % and Australia with 22 %. At the local community level, vitality limits exogamy rates such that intermarriage of Jews in greater New York is half the national rate at 23 %, and intermarriage in Toronto and Melbourne is also lower than the national rates at 15.6 and 18.4 % respectively. Vitality in terms of absolute and relative community size can impact intermarriage rates, where greater absolute

[3] Individuals are members of an Orthodox synagogue, although they may not keep traditions to orthodox standards.

numbers of Jews indicate greater resources and a larger marriage market, and a greater proportion of Jews in relation to the larger society indicates an increased likelihood of contact with fellow Jews (while small relative group size increases the likelihood of out-group contact). Research has shown that relative community size is correlated negatively to Jewish intermarriage (Rabinowitz 1989).

It is hypothesized that the above differences in community vitality impact individual acculturation journeys and the psychology and behavior of Jews such that (a) members of smaller communities who have limited resources may be more concerned about Jewish continuity than members of larger communities, but (b) because of their limited resources may not be able to engage in as much continuity-enhancing behavior as do members of larger communities. As such, the first hypothesis is that Motivation for Jewish Continuity increases as community vitality decreases, with incrementally greater concerns among the major Jewish communities sampled in New York, Toronto, Melbourne and Auckland/Wellington. The second hypothesis is that intentions for endogamy and selective dating behavior increase as community vitality increases due to the greater pool of potential Jewish dating and marriage partners, with incrementally higher endogamy intentions and selective dating behavior among the communities sampled in Auckland/Wellington, Melbourne, Toronto and New York.

69.3.2 Methodology, Sample Characteristics and Measures

Participation in the study was through a voluntary and anonymous online survey. Participants were unmarried and recruited from Jewish young adult organizations and tertiary institutions. Since the total Jewish population in New Zealand is very small, data were collected from two major cities. Participants were obtained through the Australasian Union of Jewish Students (AUJS) in Auckland and Wellington, B'nei Brith Young Adults (BBYA) in Auckland and JewNet in Wellington. Melbourne Jewish participants were recruited through social Jewish student groups AUJS and Hagshama. Jewish participants from Toronto and New York City were obtained through Hillel, the social Jewish student association. Data collection occurred December 2005 – April 2006. Table 69.2 presents sample characteristics.

A one-way ANOVA revealed a significant difference in mean age between the four Jewish communities $\underline{F}(3,461) = 13.00$, $\underline{p} < 0.001$. Chi-squared tests failed to reveal a significant relationship between community and gender, although a significant relationship was found between community and generation $X^2(9, n = 465) = 86.31$, $\underline{p} < 0.001$. The differences are consistent with the ethnographic characteristics of each of the four communities, as presented earlier. Of the aggregate sample of 465 Jews, 457 identified with a specific denomination. A chi-squared test revealed a significant relationship between country and religious identification $X^2(6, n = 465) = 87.57$, $\underline{p} < 0.001$. Because the distribution of religious denominations across these four samples was *not* representative of the

Table 69.2 Population and sample characteristics

	New Zealand Jews (Auckland and Wellington)	Australian Jews (Melbourne)	Canadian Jews (Toronto)	American Jews (New York)
Absolute population of Jews in the country	4,300	37,800	180,000	2,051,000
Relative population of Jews in the country	0.3 %	1.1 %	3.9 %	9.7 %
N participants in comparative analyses	101	101	156	107
N participants in predictive analyses	106	108	160	107
Age M(S.D)	23.7 (4.5)	21.4 (2.6)	21.4 (2.8)	22.6 (3.3)
Female/Male	59.4 %/40.6 %	63.4 %/36.6 %	66.7 %/33.3 %	64.5 %/35.5 %
First generation	45.5 %	16.8 %	25 %	10.3 %
Second generation	33.7 %	42.6 %	30.1 %	27.1 %
Third generation	16.8 %	39.6 %	28.8 %	31.8 %
Fourth generation	4.0 %	1.0 %	16.0 %	30.8 %
Orthodox	30.7 %	66.3 %	25.0 %	36.4 %
Conservative	9.9 %	12.9 %	37.8 %	39.3 %
Reform	35.6 %	5.0 %	25.6 %	15.9 %
Other Jewish	23.8 %	15.8 %	11.6 %	8.4 %

Source: Michelle Gezentsvey Lamy

actual distribution of denominations across the four Jewish community populations, they will need to be taken into account when conducting analyses. Jewish identity of participants was measured using Cameron's (2004) scale of social identity, measuring centrality to one's self-concept, positive affect towards group membership and strength of in-group ties. Overall, the strength of Jewish identity was very high (above the scale mid-point of 4, averaging 6.12 when the samples are aggregated). This is due to the sampling methodology whereby individuals were recruited from Jewish organizations. Membership is voluntary, so it is likely that affiliated persons identify more strongly with their Jewish heritage than non-affiliated people.

The dependent variables were intentions for endogamy and selective dating behavior, using statements drawn from the exploratory focus groups. Each scale consisted of four items, for example, endogamy scale item: "I intend to marry someone who is Jewish"; and selective dating scale item: "I only date people who are Jewish." Responses were averaged to give an overall scale score; high scores indicate stronger behavioral intentions to marry within their ethnic group and greater incidence of intra-ethnic dating.

Prior to examining mean group differences in Motivation for Jewish Continuity across the four Jewish community samples, the measure was subjected to a Confirmatory Factor Analysis in an aggregated dataset of all four samples, removing redundant items to shape a concise and user-friendly scale. Ten items were selected (see bold items in Table 69.1), all loading significantly on one factor. Loadings ranged from .61 to .84, and excellent reliability was obtained, $\alpha = .92$.

Goodness of fit indices of the single-factor model were acceptable: $X^2(35) = 145.75$***, $X^2/df = 4.16$, RMSEA = .08, sRMR = .03, GFI = .94, NFI = .95, and CFI = .96.

69.3.3 *Differences Across Diaspora Groups*

To determine the influence of community vitality on levels of Motivation for Jewish Continuity (MJC), endogamy intentions and selective dating behavior, mean group differences were examined once participant differences in age and religious identification[4] were controlled. One-way Analyses of Covariance were conducted where country served as the independent variable and participant age and religious identification constituted the covariates. After adjusting for participant age and religious denomination, contrary to the first hypothesis that MJC levels would increase as community vitality decreases, there was no significant difference in MJC levels across the four Jewish communities [$F(3,392) = 1.69$, $p = .17$, partial eta squared = .01]. High levels of MJC were observed (adjusted mean and std. error) among the samples of New Zealand Jews 6.36(.09), Australian Jews 6.40(.08), Canadian Jews 6.50(.06), and American Jews 6.58(.07). Thus, Jews in all four community samples are highly motivated to ensure Jewish continuity.

A strong significant effect was found for country on endogamy intentions after accounting for the effect of the covariates [$F(3,392) = 20.68$, $p < 0.001$, partial eta squared = .14]. Post-hoc tests demonstrated that New Zealand Jews had significantly lower endogamy intentions than Australian, Canadian, and American Jews. Similarly, a strong significant effect was found for country on selective dating [$F(3,392) = 31.18$, $p < 0.001$, partial eta squared = .19]. Post-hoc tests demonstrated that New Zealand Jews had lower selective dating behavior than Australian, Canadian, and American Jews, illustrated below in Figs. 69.1 and 69.2. The mean selective dating behavior of the New Zealand sample is below the scale mid-point of four, indicating that on average they do not engage in selective dating, in contrast to Australian, Canadian and American Jews who do date fellow Jews. Thus, the second hypothesis was partially supported as the New Zealand Jewish sample, representing the community with lowest vitality, exhibited the least continuity-enhancing behavior.

---

[4] Religious identification was included as a covariate since the proportion of identification with the three major denominations in each sample was not representative of the affiliation of their respective populations. In contrast, sample differences in generation do reflect actual population characteristics and as such were not included as a covariate. Gender was not included as a covariate since its distribution did not differ across the four samples.

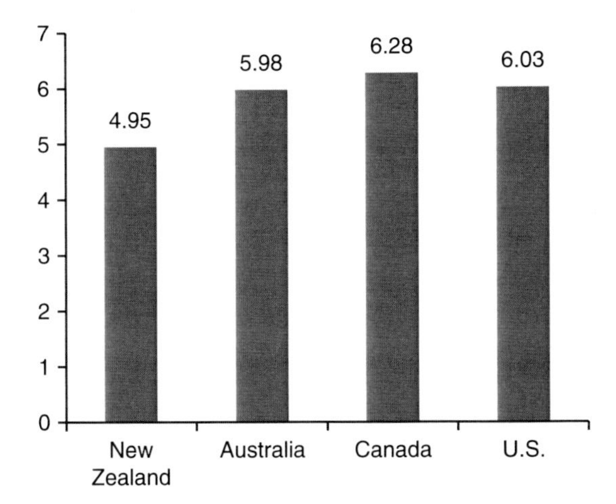

Fig. 69.1 Intentions for endogamy across four diaspora samples (Source: Michelle Gezentsvey Lamy)

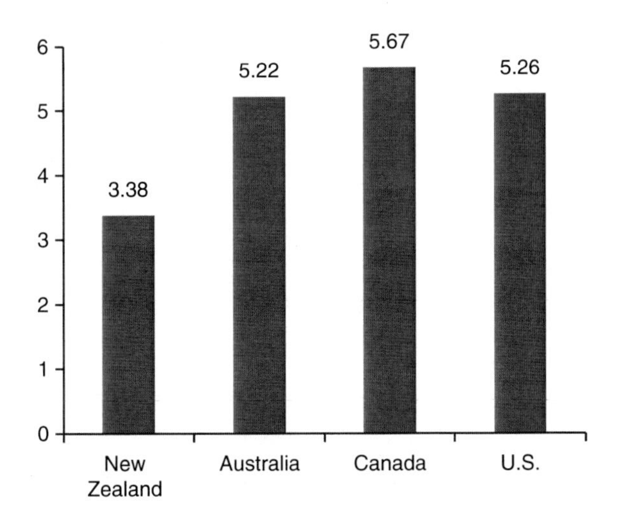

Fig. 69.2 Selective dating behavior across four diaspora samples (Source: Michelle Gezentsvey Lamy)

Despite large differences in community vitality, there were no significant differences among Australian, Canadian and American Jews all who demonstrated strong intentions for endogamy and engaged in moderate selective dating behavior. In contrast to findings by Rabinowitz (1989), where relative community size was correlated negatively to intermarriage, in this study neither absolute nor relative vitality were correlated with endogamy intentions or selective dating. Vitality did not have an incremental effect; it seems that a threshold exists between 4,300 individuals

(0.27 % of the relative population) and 37,800 individuals (1.12 %) that thwarts behavior. The small absolute population, indicative of a weak marriage market, and relative vitality, indicative of increased out-group contact, of New Zealand Jews are compounded with geographical isolation from other Jewish communities to produce low endogamy intentions and a reluctance to engage in selective dating.

The Theory of Planned Behavior provides some insight in understanding how vitality may impact behavioral intentions and behavior. Azjen (1991: 196) states that "the more resources and opportunities individuals believe they possess, and the fewer obstacles or impediments they anticipate, the greater should be their perceived behavioral control over the behavior." What could be a greater obstacle to endogamy for New Zealand Jews than the lack of available resources, that is, potential marriage partners? It is inferred that the small marriage market may induce low endogamy efficacy beliefs and perceived behavioral control among New Zealand Jews, which are in turn related to weak intentions for endogamy, and reluctance to engage in selective dating.

69.4 Investigating the Impact of Motivation for Jewish Continuity on Continuity-Enhancing Behavior

Does Motivation for Jewish Continuity as a new psychological construct, predict continuity-enhancing behavior, particularly intentions for endogamy and selective dating? Ritual observance, tertiary education, Jewish formal and informal education, synagogue affiliation, personal friendships, parental intermarriage, gender, generation and population density are all factors that have been found to predict Jewish endogamy (Sklare and Greenblum 1967; Medding 1968; Keysar et al. 1991; Waite and Friedman 1997; NJPS 2000–2001; Horowitz 2000; Cohen 2006). Research in social psychology has examined ethnocentric predictors of endogamy and selective dating: similarity, attraction, and social network approval (Liu et al. 1995). Little attention has been given to deeper collective-oriented psychological drives that can explain in addition to established socio-demographic factors why in the twenty-first century, when individual freedom reigns, more than half of American and New Zealand Jews, and two-thirds of Canadian and Australian Jews choose to marry a fellow Jew (Figs. 69.3, 69.4, 69.5, and 69.6).

The third hypothesis is that MJC predicts intentions for endogamy in addition to traditional ethnocentric predictors such as perceived similarity and attraction to Jews, and social network approval to marry a Jewish person. Assuming that individuals date fellow Jews because they intend to marry endogamously, the fourth hypothesis states that intentions for endogamy mediate the relationship between Motivation for Jewish Continuity and selective dating.

Fig. 69.3 New Zealand Jewish Wedding (Photo by Sutherland Kovach Studio, www.sutherland-kovack.com, used with permission)

69.4.1 Ethnocentric Predictors

In addition to the MJC, intentions for endogamy and selective dating behavior scales described above, Similarity, Attraction and Social Network Approval were measured using items by Liu et al. (1995). Three items assessed overall similarity to fellow ethnic group members, similarity in values and communications styles. Items such as "How similar overall do you feel to other Jewish people?" were measured on a 7-point Likert scale with scores ranging from 1 (not at all similar) to 7 (very similar). Responses were averaged to give an overall scale score where high scores indicate greater similarity among ethnic group members. The scale demonstrated good reliability ($\alpha = 0.78$–0.85).

Three items assessed subjective attraction to fellow ethnic group members in terms of physical attraction, desirability as romantic partners and sex appeal. Items such as "How physically attractive do you think Jewish people are?" were measured on a 7-point Likert scale with scores ranging from 1 (not at all) to 7 (extremely). Responses were averaged to give an overall scale score where high scores indicate greater attraction to fellow ethnic group members. Scale reliability was good ($\alpha = 0.77$–0.90).

Finally, three items assessed social network approval of endogamy regarding participants' parents, friends, and friends and family of the prospective partner. Items including "How do you think your parents would feel about your marrying someone who is Jewish?" were measured on a 7-point Likert scale with scores ranging from 1 (not at all happy) to 7 (extremely happy). Responses were averaged to give an overall scale score where high scores indicate greater approval for endogamy. Aside from the Canadian sample, the scale demonstrated good reliability ($\alpha = 0.66$–0.76).

Fig. 69.4 Australian Jewish Wedding (Photo by Rashelle Cohen, used with permission)

69.4.2 Predicting Jewish Endogamy in Diaspora Communities

To test the predictive ability of MJC, four parallel sets of regressions were performed for the New Zealand, Australian, Canadian and American samples, in which intentions for endogamy served as the criterion (dependent variable) and similarity, attraction, social network approval and MJC served as the predictor variables. As there were no gender differences between samples, there was no need to control for this variable. Although there were significant differences in age between the four samples, no significant correlation was found between age and intentions for endogamy or selective dating behavior. As such, there was no need to control for the

Fig. 69.5 Canadian Jewish Wedding (Photo by Mariusz Lasocha, www.hephoto.ca)

effects of age. Standardized regression coefficients along with significance levels are presented in Table 69.3. Across all four samples, motivation was not only a consistent predictor of intentions for endogamy, but also the strongest predictor. Social network approval was a significant predictor among the New Zealand, Australian and American samples, and Attraction was also a significant predictor among the Canadian and American samples, but the coefficients were much smaller than those for motivation. Perceived similarity to fellow Jews did not predict endogamy intentions for any sample. The results support the third hypothesis, where MJC predicted intentions for endogamy in addition to the ethnocentric variables of similarity, attraction and social network approval.

69.4.3 Predicting Selective Dating Among Diaspora Jews

To examine predictors of selective dating behavior, four parallel sets of regressions were performed with selective dating as the criterion. Similarity, attraction, social network approval and Motivation for Jewish Continuity served as the predictor variables in step one, and intentions for endogamy were added as an additional predictor variable in step two. According to Baron and Kenny (1986), three conditions must be satisfied for mediation to occur. First, the predictor variable, Motivation for Jewish Continuity, must have a significant relationship with both the mediator (intentions for endogamy), and the criterion variable (selective dating). Second, the

Fig. 69.6 American Jewish Wedding (Photo by Daniel Krieger Photography, www.danielkrieger.com, used with permission)

Table 69.3 Standardized regression coefficients with intentions for endogamy as criterion

	New Zealand	Australia	Canada	United States
Similarity	0.03	0.08	0.06	0.04
Attraction	0.03	0.02	0.18**	0.18*
Approval	0.20*	0.18*	0.04	0.18*
MJC	0.55***	0.55***	0.49***	0.51***

Source: Michelle Gezentsvey Lamy
*p<.05; **p<.01; ***p<.001

Table 69.4 Standardized regression coefficients with selective dating as criterion

	New Zealand	Australia	Canada	United States
1. Similarity	–0.07	0.39***	0.08	0.16
Attraction	0.17	–0.13	0.25**	0.13
Approval	–0.03	0.14	0.10	0.19*
MJC	0.41***	0.36***	0.34***	0.39***
2. Similarity	–0.08	0.36***	0.04	0.13
Attraction	0.16	–0.14	0.15*	0.02
Approval	–0.12	0.05	0.08	0.07
MJC	0.15	0.09	0.06	0.06
Endogamy intentions	0.47***	0.48***	0.58***	0.66***

Source: Michelle Gezentsvey Lamy
*p<.05; **p<.01; ***p<.001

mediator must affect the criterion variable. Third, when the mediator is introduced (step two), the relation between the predictor and the criterion should be reduced to non-significance, indicating full mediation. Table 69.4 presents standardized regression coefficients along with their significance levels. In step one, for all four samples, Motivation for Jewish Continuity was a significant predictor of selective dating, satisfying the first condition. Intentions for endogamy were a strong and significant predictor of selective dating in step two, satisfying the second condition. Finally, the third condition was satisfied as the relationship between MJC and selective dating was reduced to non-significance across all four samples when intentions for endogamy were added as a predictor variable. Thus, the results confirm that the relation between MJC and selective dating behavior is fully mediated by intentions for endogamy. This means that in these samples, Jews who date fellow Jews do so because they intend to marry a Jewish person, which in turn is shaped by their Motivation for Jewish Continuity. For the American sample, endogamy intentions also mediated the relation between social network approval and selective dating. In addition, among the Australian and Canadian samples, perceived similarity and attraction were direct predictors of selective dating.

69.5 Individual Agency in Collective Continuity

The Jewish people have traditionally emphasized endogamy as an internal strategy for collective continuity. Although endogamy is an established injunction in Judaism, outside of highly religious circles, such behavior is an individual choice – a choice that when multiplied by many in-group members, contributes to the endurance of the Jewish people, creating Jewish homes and building a generation that is more likely to continue to identify as Jews and participate in the Jewish community. Thus far, Jewish social science research had examined demographic, cognitive, behavioral and affective factors that lead Jewish individuals to engage in endogamy (Sklare and Greenblum 1967; Medding 1968; Keysar et al. 1991; Waite and Friedman 1997; NJPS 2000–2001; Horowitz 2000; Cohen 2006). On the other hand, social science research had interpreted endogamy and selective dating as simple manifestations of ethnocentrism. Now, perspectives from cross-cultural psychology have been applied to assert that a desire to preserve and transmit one's heritage is also basis for Jewish endogamy. This research has convincingly demonstrated that young Diaspora Jews from Oceania, Australasia and North America bear collective continuity in mind when making the most personal decisions such as whom to marry. This stands in contrast to Cohen and Eisen (2000) who conclude that Jewish identity today is a tool that serves individuals on their personal journeys in search of meaning and fulfillment, rather than any collective purpose. Unfortunately, the realities of Diaspora communal life of small communities of "small peoples" is such that despite high levels of motivation for continuity, limitations are placed on individual behavior by structural factors such as community vitality.

69.6 Limitations and Future Research

Methodological limitations arise due to the sampling from Jewish organizations where participants identified strongly with their Jewish heritage. Hence, these samples are not representative of the entire spectrum of Jewish adults, in particular the unaffiliated. In addition, caution must be exercised regarding the generalizability of the influence of community vitality. The strongest, most vibrant metropolitan communities were sampled in four countries; these are not representative of other communities within each country. Future research could sample several communities within the same country. Furthermore, only Jewish continuity in Anglo-Saxon societies was examined; the expression of Judaism within the larger society may not be equivalent in other regions, for example, in France, where anti-Jewish sentiment can impact the acculturation strategies chosen by individuals (integration, separation or assimilation), and the frequency of endogamy.

Motivation for Jewish Continuity is also being examined among a sample of Israeli Jews (Gezentsvey Lamy et al. in progress). New research questions include the influence of religiosity on levels of motivation, the relationship between motivation and perceived Jewish entitativity, and whether distinctions are made between the endurance of Jewishness in the Diaspora vs. Israel. Future research can investigate methods of enhancing Motivation for Jewish Continuity, targeting different life stages, for example, primary, secondary and tertiary school age-groups. The efficacy of programs such as Birthright and March of the Living in terms of increasing endogamy intentions and communal involvement have been examined (Helmreich 2005), however, their impact on Motivation for Jewish Continuity are yet unknown.

69.7 Conclusion

This research has demonstrated that the prevalence and predictive ability of Motivation for Jewish Continuity are not affected by community vitality. However, if behavior is what matters for continuity, small Jewish communities may be in big trouble. Despite high levels of motivation among affiliated individuals, limited communal 'resources' impede selective dating and endogamy intentions. As such, small communities may need to rely on other methods of transmission, such as Jewish education. Nonetheless, the findings of this study provide a fresh answer to Marc Twain's question: the secret of Diaspora Jewish continuity in tolerant heterogeneous societies may lie in the representation of the collective in the minds of individuals, and the exertion of individual agency in the process of continuity by engaging in continuity-enhancing behavior.

Acknowledgements This research was funded by: TEC Bright Future Top Achiever Doctoral Scholarship (2004–2006), Todd Foundation Award for Excellence (2004), Jacob Joseph Scholarship, Victoria University of Wellington PhD Completion Scholarship, and BRCSS Doctoral Scholarship (2007). The publication of this article was partially funded by the Lady Davis Fellowship and the Melton Centre for Jewish Education, at The Hebrew University of Jerusalem (2008–2009). The author would like to acknowledge Professors Colleen Ward, James Liu and Gabriel Horenczyk for their assistance.

References

Abulof, U. (2007). *"Still on the roof" Israeli-Jewish existential insecurity: The survival-identity complex*. Unpublished Ph.D. thesis, Hebrew University of Jerusalem, Department of International Relations, Jerusalem.

Alba, R. D. (1990). *Ethnic identity: The transformation of White America*. New Haven: Yale University Press.

American Jewish Committee. (2005). *Jewish geography*. New York: American Jewish Committee.

Asa-El, A. (2004). *Diaspora: The last tribes of Israel*. Westport: Hugh Lauter Levin Associates.

Avidan, D. (1986, September 5). Yediot Aharonot.

Azjen, I. (1991). The theory of planned behavior. *Organizational Behavior and Human Decision Processes, 50*, 179–211.

Baron, R. M., & Kenny, D. A. (1986). The moderator-mediator variable distinction in social psychological research: Conceptual, strategic and statistical considerations. *Journal of Personality and Social Psychology, 51*, 1173–1182.

Berry, J. W. (2003). Conceptual approaches to acculturation. In K. M. Chun, P. B. Organista, & G. Marin (Eds.), *Acculturation: Advances in theory, measurement, and applied research* (pp. 17–34). Washington, DC: American Psychological Association.

Berry, J. W., Phinney, J. S., Sam, D. L., & Vedder, P. (2006). *Immigrant youth in cultural transition*. Mahwah: Lawrence Erlbaum Associates.

Bourhis, R. Y., Giles, H., & Rosenthal, D. (1981). Notes on the construction of a 'subjective vitality questionnaire' for ethnolinguistic groups. *Journal of Multilingual and Multicultural Development, 2*, 145–155.

Cameron, J. E. (2004). A three-factor model of social identity. *Self and Identity, 3*(3), 239–262.

Castano, E., Sacchi, S., & Gries, P. H. (2003). The perception of the other in international relations: Evidence for the polarizing effect of entitativity. *Political Psychology, 24*(3), 449–468.

Cohen, S. M. (2006). *A tale of two Jewries: The "inconvenient truth" for American Jews*. New York: Steinhardt Foundation.

Cohen, S. M., & Eisen, A. M. (2000). *The Jew within: Self, family and community in America*. Bloomington: Indiana University Press.

Dershowitz, A. M. (1997). *The vanishing American Jew*. New York: Little Brown & Company.

Elazar, D. J., & Cohen, S. A. (1985). *The Jewish polity*. Bloomington: Indiana University Press.

Frye Jacobson, M. F. (1998). *Whiteness of a different color: European immigrants and the alchemy of race*. Cambridge, MA: Harvard University Press.

Gans, H. J. (1979). Symbolic ethnicity. *Ethnic and Racial Studies, 2*, 1–20.

Gezentsvey, M. A. (2008). *The long-term acculturation of ethnocultural groups: Comparisons of Maori, Chinese and Jewish Continuity*. Unpublished PhD thesis, Victoria University of Wellington, School of Psychology, New Zealand.

Gezentsvey, M. A., & Ward, C. (2008). Unveiling agency: A motivational perspective on acculturation and adaptation. In R. Sorrentino & S. Yamaguchi (Eds.), *The handbook of cognition and motivation within and across cultures* (pp. 213–235). New York: Elsevier.

Gezentsvey Lamy, M., Horenczyk, G. & HaCohen-Wolf, H. (in progress). *Motivation for Jewish continuity amongst Israeli Jews*. The Hebrew University of Jerusalem, Jerusalem.

Gezentsvey Lamy, M., Ward, C. & Liu, J. H. (2013). Motivation for ethno-cultural continuity. *Journal of Cross-cultural Psychology*.

Giles, H., Bourhis, R. Y., & Taylor, D. (1977). Towards a theory of language in ethnic group relations. In H. Giles (Ed.), *Language, ethnicity and intergroup relations* (pp. 307–348). London: Academic.

Harwood, J., Giles, H., & Bourhis, R. Y. (1994). The genesis of vitality theory: Historical patterns and discoursal dimensions. *International Journal of the Sociology of Language, 108*, 167–206.

Helmreich, W. B. (2005). *Long-range effects of the March of the living on participants*. New York: City University of New York.

Horowitz, B. (2000). *Connections and journeys: Assessing critical opportunities for enhancing Jewish Identity*. New York: United Jewish Appeal – Federation of New York.

Institute for International Affairs/B'nei Brith Canada. (2000). *From immigration to integration: The Canadian Jewish experience*.

Johnson, P. (1987). *A history of the Jews*. London: George Weidenfeld & Nicolson.

Keysar, A., Kosmin, B., Lerer, N., & Mayer, A. J. (1991). Exogamy in first marriages and remarriages. *Contemporary Jewry, 12*, 45–66.

Kundera, M. (1993). *Les testaments trahis*. Saint-Amand: Editions Gallimard.

Langman, P. F. (1999). *Jewish issues in multiculturalism: A handbook for educators and clinicians*. Northvale: Jason Aronson.

Levine, H. (1995). Migration or assimilation? The predicament of observant Jews. In S. W. Grief (Ed.), *Immigration and national identity in New Zealand: One people, two peoples, many peoples* (pp. 203–216). Palmerston North: Dunmore Press.

Levine, H. (1997). *Constructing collective identity: A comparative analysis of New Zealand Jews, Māori and urban Papua New Guineans*. Frankfurt: Peter Lang.

Levine, S. (1999). *The New Zealand Jewish community*. Lanham: Lexington Books.

Liu, J. H., Campbell, S. M., & Condie, H. (1995). Ethnocentrism in dating preferences for an American sample: The ingroup bias in social context. *European Journal of Social Psychology, 25*, 95–115.

Luhtanen, R., & Crocker, J. (1992). A collective self-esteem scale: Self-evaluation of one's social identity. *Personality and Social Psychology Bulletin, 18*(3), 302–318.

MacDonald, K. (2002). *A people that shall dwell alone: Judaism as a group evolutionary strategy, with diaspora peoples*. San Jose: Writers Club Press.

Maimon, D. (2007). *Alternative futures project summary document*. JP2030 Version 6(2). Jerusalem: JPPPI Jewish People Policy Planning Institute.

Mayer, E., Kosmin, B., & Keysar, A. (2001). *American Jewish identity survey*. New York: Centre for Jewish Studies, City University of New York.

Medding, P. Y. (1968). *From assimilation to group survival*. Melbourne: Cheshire Publishing.
National Jewish Population Survey (2000–2001). New York: United Jewish Communities. North American Jewish Data Bank [Distributor].
Neusner, J. (1999). Israel the people in Medieval and modern times. In J. Neusner, A. J. Avery-Peck, & W. S. Green (Eds.), *Encyclopedia of Judaism* (pp. 499–515). New York: Continuum Publishing Co.
Phillips, B., & Chertok, F. (2004). *Jewish identity among the adult children of intermarriages: Event horizon or navigable horizon?* Paper presented at the Association for Jewish Studies 36th annual conference, Chicago.
Phinney, J. S., Horenczyk, G., Liebkind, K., & Vedder, P. (2001). Ethnic identity, immigration and well-being: An interactional perspective. *Journal of Social Issues, 57*, 493–510.
Rabinowitz, J. (1989). The paradoxical effects of Jewish community size on Jewish communal behavior: Intermarriage, synagogue membership and giving to local Jewish federations. *Contemporary Jewry, 10*(1), 9–15.
Rabinowitz, L. I. (2008). Encyclopaedia Judaica, sourced April 2008.
Rawidowicz, S. (1986). *Israel, the ever-dying people, and other essays*. Cranbury: Associated University Presses.
Rutland, S. D. (2005). *The Jews in Australia*. Melbourne: Cambridge University Press.
Sachdev, I. (1995). Language and identity: Ethnolinguistic vitality of Aboriginal peoples in Canada. *London Journal of Canadian Studies, 11*, 41–59.
Sacks, J. (2000). *Radical then, radical now*. London: Continuum.
Salkin, J. (1999). Judaism, definition of. In J. Neusner, A. J. Avery-Peck, & W. S. Green (Eds.), *Encyclopedia of Judaism* (pp. 579–588). New York: Continuum Publishing Co.
Sani, F., & Bowe, M. (2004). *Perceived group historical continuity: Its functions and its relationship with group identification*. Paper presented at the collective remembering, collective emotions and shared representations of history: Functions and dynamics; Small meeting of EAESP, Aix-en-Provence.
Sarna, J. D. (2004). *American Judaism: A history*. New Haven: Yale University Press.
Schönpflug, U. (2001). Intergenerational transmission of values: The role of transmission belts. *Journal of Cross-Cultural Psychology, 32*(2), 174–185.
Shahar, C., & Rosenbaum, T. (2006). The Jewish Community of Toronto/Jewish life in Greater Toronto 2001 Census Analysis Series, UJA Federation of Greater Toronto; UIA Canada.
Sklare, M., & Greenblum, J. (1967). *Jewish identity on the suburban frontier: A study of group survival in the open society*. Chicago: University of Chicago Press.
Spicer, E. H. (1971, November). Persistent cultural systems. *Science, 174*, 795–800.
Szwarc, B. (2004). *A demographic profile of the Jewish community in Victoria based on the 2001 Australian Bureau of Statistics Census*. Melbourne: Jewish Community Council of Victoria.
Tobin, D., Tobin, G. A., & Rush, S. (2005). *In every tongue: The racial and ethnic diversity of the Jewish people*. San Francisco: Institute for Jewish and Community Research.
Tulchinsky, G. (1998). *Branching out: The transformation of the Canadian Jewish community*. Toronto: Stoddart Publishing Company.
Twain, M. (1897). Quoted in The National Jewish Post & Observer, June 6, 1984.
Waite, L., & Friedman, J. S. (1997). The impact of religious upbringing and marriage markets on Jewish intermarriage. *Contemporary Jewry, 18*, 1–23.
Weinfeld, M. (2001). *Like everyone else, but different: The paradoxical success of Canadian Jews*. Toronto: McClelland & Stuart.
Yerushalmi, Y. H. (1982). *Zakhor: Jewish history and Jewish memory*. Seattle: University of Washington Press.
Zenner, W. (1991). *Minorities in the middle: A cross cultural analysis*. Albany: State University of New York Press.

Chapter 70
Global Dispersion of Jews: Determinants and Consequences

Sergio DellaPergola and Ira M. Sheskin

Take ye the sum of all the congregation of the children of Israel, by their families, by their fathers' houses, according to the number of names. (Numbers 1:2)

70.1 Introduction

The art of counting Jews, describing their geographical distribution and other characteristics – and projecting their future – is as old as the Bible. Any serious discussion of Jewish demographic trends should proceed from an understanding of the broader processes that generally impact population development.

Population is a collective, macro-social concept, but population changes reflect events that mostly occur at the individual, micro-social level. All changes in *world* population size result from the balance between births and deaths (reflecting fertility rates, life expectancy, and a population's age composition). When examining a *given geographic area* where in- and out-migration is possible, population change also reflects geographical mobility. And when a subpopulation is further defined by cultural characteristics (such as religion, ethnicity, or language), a somewhat more complex *balancing equation* becomes necessary to incorporate identification changes (for example, religious conversions) over time. The important underlying principle is the continuity of a human aggregate that is not created from a vacuum (besides quite rare cases of ethnogenesis – the initial act of a new group coming into existence), but constantly evolves following a circumscribed set of drivers. Demography and geography of the Jews may thus serve as a paradigm for the more general case of

S. DellaPergola (✉)
The Avraham Harman Institute of Contemporary Jewry, The Hebrew University of Jerusalem, Mt. Scopus, Jerusalem, Israel
e-mail: sergioa@huji.ac.il

I.M. Sheskin
Department of Geography and Regional Studies, University of Miami, Coral Gables, FL 33124, USA
e-mail: isheskin@miami.edu

S.D. Brunn (ed.), *The Changing World Religion Map*,
DOI 10.1007/978-94-017-9376-6_70

subpopulations whose unfolding over time is determined not only by demographic and biological factors, but also by social, cultural, and ideational factors.

Jews and geography have been inextricably related for millennia. The story of the Jewish people can hardly be told without repeated reference to geographic location, from the story of Abraham until modern times. This chapter examines changes in the geographic distribution of the Jewish population at a number of different geographic scales (worldwide, regionally within the U.S., metropolitan, and intra-urban). This chapter will be "data based" in the sense that it will rely on data from surveys of the American Jewish population and the Israeli census as well as numerous other sources from other countries, and will examine issues of migration, demography, and religiosity.

70.2 Definition of Jewish Identity

The problem of defining who is, and who is not, a Jew is discussed in thousands of books and articles (DellaPergola 2010). Unlike for most other religious groups discussed in this volume, being Jewish is both a religious and an ethnic identity. One does not cease to be a Jew even if one becomes an atheist or agnostic and/or ceases to participate in religious services or rituals, unless, by most opinions, the same person has espoused another monotheistic religion. The 2000–01 National Jewish Population Survey (NJPS 2000–01) (Kotler-Berkowitz et al. 2003) suggests that about one-fifth of American Jews do not identify as Jewish in terms of religion. Recognizing the difficulties of defining Jewish identity, historically the Canadian Census asks a religion question in which "Jewish" is one response printed on the form and a question about ethnicity, in which "Jewish" is also one response printed on the form (Norland and Freedman 1977).

During biblical times, Jewish identity was determined by patrilineal descent. During the rabbinic period, this was changed to matrilineal descent. In the contemporary period, Orthodox and Conservative rabbis officially recognize only matrilineal descent, while Reform (as of 1983) and Reconstructionist rabbis recognize, under certain circumstances, both matrilineal and patrilineal descent. Furthermore, Orthodox rabbis only recognize as Jewish those Jews-by-Choice who have been converted by Orthodox rabbis. In Israel, the Orthodox establishment follows the matrilineal descent criterion, but the government, for the purpose of the Law of Return (which only determines immigration and citizenship rights, not religious identity), defines as eligible any person with at least one Jewish grandparent, regardless of current religion, and their spouses. In general, social scientists conducting survey research with American Jews, do not wish to choose from the competing definitions of who is a Jew, and have adopted the convention that all survey respondents who "consider themselves to be Jewish" are counted as such. Operationally, the *core Jewish population* concept, originally introduced by the NJPS 1990 analysts (Kosmin et al. 1991), addresses the self-declared or otherwise identified aggregate of persons of Jewish origin who do not hold an alternative religious identification. The underlying hypothesis is that, with all due caution and caveats, Jews *can* be counted at any given moment in time, through *mutually exclusive* definitional criteria that avoid the double count of persons with multiple identities. The *enlarged Jewish population* is the total population in households with at least one, currently or formerly, *core* Jewish individual. But, clearly the estimate of the size of the Jewish population of an area can differ depending who one counts as Jewish – and also to some extent also upon who is doing the counting. These definitional issues are one reason why differing counts of Jews, particularly in the U.S., are presented below. In this chapter, for international comparative purposes, in the sections examining world Jewish population, we have used an estimate of the U.S. Jewish population that excludes persons who describe themselves as "part Jewish." For the purposes of examining the U.S. Jewish population, "part Jews," in keeping with definitions of "being a Jew" in the United States, are included.

70.2.1 Data Sources

The U.S. and Israel combined account for more than 80 % of world Jewry (DellaPergola 2010). While Israel has accurate data on its Jewish population (Israel Central Bureau of Statistics 2011), the situation in the U.S. and most other countries is considerably more problematic, with recent U.S. estimates ranging from 5.2 million to more than 6.7 million (DellaPergola 2005, 2010, 2012, 2013; Pew Research Center 2013; Sheskin and Dashefsky 2006, 2012, 2013; Tighe et al. 2013). The main difficulty stems from the scarcity of national sources that classify population by religion (or by ethnic groups), where "Jewish" is one of the possible options. In some cases, like in the U.S., this is not feasible because of the separation of church and state. On the other hand, a growing acknowledgment exists in the social sciences that religio-ethnic identities constitute a powerful variable for analytic purposes, as such and as a correlate or determinant of other demographic, social, economic and cultural features. Hence there is a growing interest in collecting data on religious and ethnic groups in contemporary societies.

70.2.2 Estimates of the U.S. Jewish Population

In contemporary social surveys, the most common and most reliable method of estimating the U.S. Jewish population is to use random digit dialing (RDD) (Waksberg 1978). This technique basically involves generating four digit random numbers which are placed after all area code/telephone exchange code combinations in the country. These numbers are then dialed and the percentage of households reached that are Jewish is calculated. This percentage can then be applied to the number of households from the U.S. Census to derive an estimate of the number of Jewish households in the country (Sheskin 2001: 6). NJPS 2000–01 (Kotler-Berkowitz et al 2003) and the American Religious Identification Surveys (ARIS 2001 and ARIS 2008) (Mayer et al. 2001; Kosmin and Keysar 2009) have used this

RDD methodology. Caution needs to be applied in these extrapolations since the percentage of households with Jewish respondents is not the same as the percentage of Jews among the total population. Differences in family size, namely a lower proportion below age 18 and intermarried households, usually require downward weighting of the original findings.

Recently Tighe et al. (2013) performed a meta-analysis of numerous surveys completed by governmental and private groups using RDD in which a question on religion was asked. Their study suggests a total of 6.8 million Jews in 2012. Two major national studies, NJPS 2000–01 and ARIS 2001, suggested a total of 5.2 million Jews. A new reading of NJPS 2000–01, correcting for evident undercoverage of certain adult age cohorts, lead DellaPergola (2012) to an updated estimate of 5,367.000 in 2000, and 5,425,000 in 2010, with the upper boundary of the confidence interval at about 5.6 million. An alternative methodology, used by Sheskin and Dashefsky (2012, 2013), sums RDD estimates from local Jewish community studies and informant estimates from hundreds of small Jewish communities who cannot afford the RDD methodology to derive an estimate of about 6.7 million.

Because the national studies were not designed to produce estimates for the Jewish population of states and metropolitan areas, we rely here on the Sheskin & Dashefsky data for U.S. regional estimates, noting that Sheskin and Dashefsky (2006) present reasons to believe that their data somewhat overestimate the total U.S. Jewish population. On the other hand, because local surveys were not designed to provide national estimates, data consistently derived and compared from national surveys, including the U.S. Census 1957 Current Population Survey (CPS), NJPS 1970–71, and NJPS 1990, and systematic review of international migration to the U.S., Jewish birth rates and death rates result in a more reliable profile of total Jewish population change over the 65 years between 1945 and 2010 (Rosenwaike 1980; DellaPergola 2012). These alternative methodologies are a second reason why differing estimates of the U.S. Jewish population are shown in Fig. 70.1. Note that the latest estimates of U.S. Jewish population can be found at www.jewishdatabank.org.

70.2.3 Estimates of Jewish Population in Israel and Other Countries

In Israel, the Jewish population is much better documented, as Israel has a reliable Central Bureau of Statistics (CBS) which performs periodic population censuses and social surveys, including questions both on religion and ethnicity, and systematically collects data on international migration, births, deaths, and marriages, as well as on changes of religion. Several other countries have population censuses addressing religion and/or ethnicity, among these Canada, Australia, South Africa, the United Kingdom (UK), and the Republics of the former Soviet Union (FSU). On the other hand, countries with important Jewish communities like France and Argentina have long discontinued such questions from their national censuses.

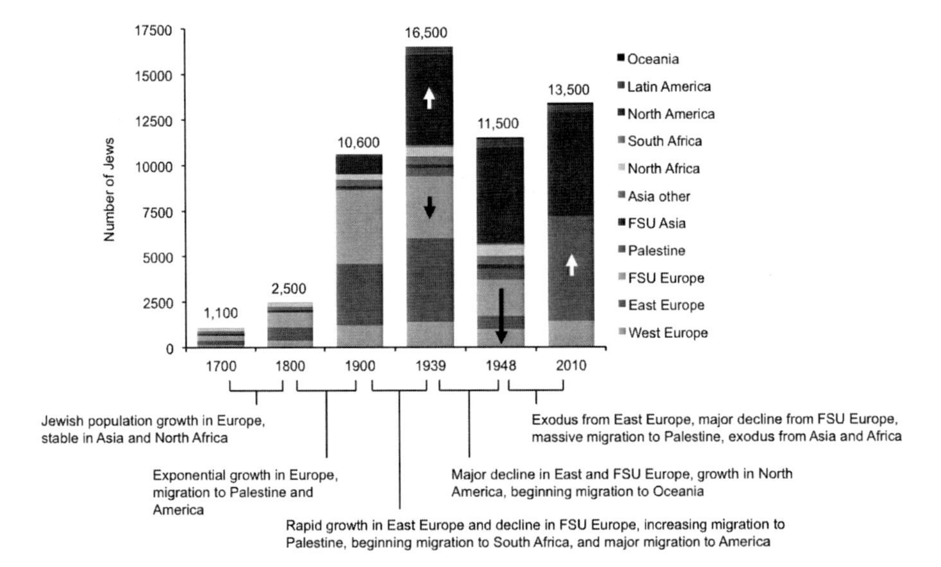

Fig. 70.1 World Jewish population by major region, 1700–2010 (Source: DellaPergola 2006, and authors' estimates)

Jewish population estimates thus have to rely on a wide array of different sources, like membership records and vital statistics in some communities, as well as independent sample surveys similar to those routinely undertaken in the U.S. The availability and comparability of Jewish population data for global synthesis are thus far from satisfactory. Nevertheless, through a keen effort of analysis and standardization, a general and quite reliable picture can be displayed of the changing Jewish population distribution worldwide.

70.3 The Global Scale

Across Jewish history, an intriguing overlap exists between Jewish *peoplehood* and *population*. History cannot be reduced to a sequence of demographic events, but the political, social, and cultural impact of mass migration and other major moments of Jewish population reduction or increase cannot be undervalued. Since the beginnings of transmitted Jewish collective memory, textual, archeological and other evidence exemplify three leading principles affecting Jewish demography in the long run:

1. significant increases and decreases in Jewish population size and the unequal pace of growth as a whole over time;
2. a differential growth of Jewish subpopulations, occasionally affecting sociodemographic composition of the whole; and
3. large scale international migration repeatedly shaping the geography and characteristics of the Jews and location of the main Jewish civilization centers.

70.3.1 Antiquity

World Jewish population in antiquity was predominantly Middle Eastern. Since the early Middle Ages it significantly expanded to Western Europe and North Africa, and after the twelfth century to Eastern Europe. The size and structure of world Jewry during the Middle Age and early Modern Period cannot be accurately assessed, but available evidence points to a range between less than one million to two million persons. While population size tended to be stable in the long term, major fluctuations reflected occasional catastrophic events such as epidemics, famine and wars – usually shared by Jews and non-Jews. Jewish population also periodically declined following massacres, mass expulsions, and forced conversions in different times and places.

70.3.2 Modern Period

Since the second half of the seventeenth century, a weakening of these negative factors and modest improvements in living standards allowed for Jewish population growth. World Jewry increased from an estimated one million around 1700 to 2.5 million around 1800 and 10.6 million around 1900 (Fig. 70.2). Jewish population during this period grew faster than most other national populations in Western and Eastern Europe, Asia, and Africa. Most of the increase after 1850 occurred in Eastern Europe. This demographic transition determined a rapid shift from an early balanced split between *Ashkenazi* (Eastern European), and *Sephardi*/Oriental (or *Mizrahi*) (Spanish, Middle Eastern, African, and Asian) communities, to an

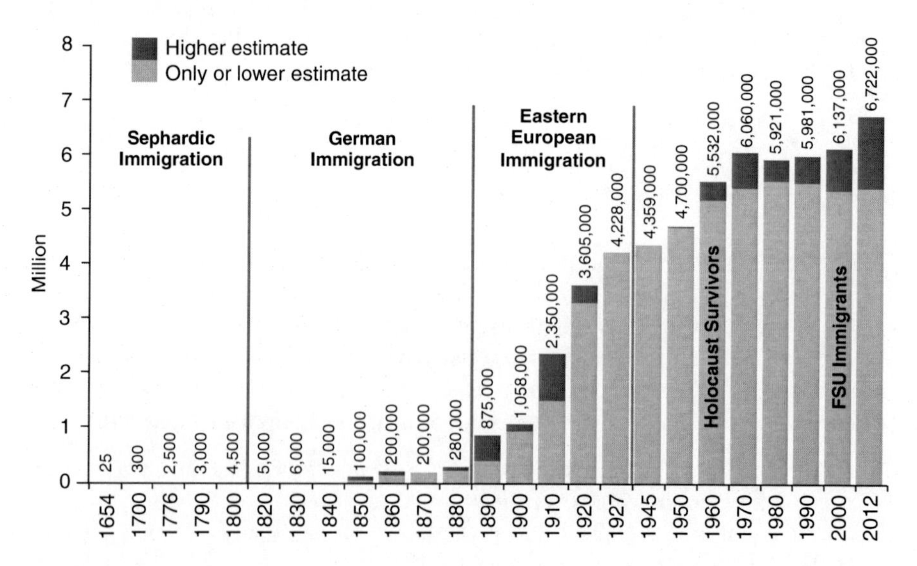

Fig. 70.2 Number of American Jews over time (Source: Sheskin and Dashefsky 2013: 144, used with permission)

overwhelming numerical predominance of Ashkenazi Jewry. Central for the onset of Jewish population growth was a comparatively early decline in mortality and low infant mortality rates in particular, in the context of nearly universal, relatively young, and homogamous Jewish marriage and high fertility levels. These demographic features reflected the influence of traditional Jewish norms, institutions and behaviors in the daily life of individuals and communities before the start of modernization. Population growth increased socioeconomic pressure among impoverished Eastern European Jewry and decisively stimulated mass westward migration from the 1880s. Eventually, cultural and social transformations of Jewish society – especially geographical and occupational mobility set into motion since the Emancipation Period (following Napoleon) – led to a diminished impact of religious norms in the life of the Jewish *Diaspora* (all Jews who live outside Israel are said to be living in the Diaspora). As in the case of mortality, Jews followed the surrounding population in the transition from higher to lower fertility levels. Jewish migrants to western countries imported the demographic models of their communities of origin, but rapidly adjusted to their new modern environments. Rates of natural increase declined, though in absolute terms, Jewish population growth was still substantial. By about 1940, the world Jewish population was estimated at 16.5 million.

Six million victims of the *Shoah* (Holocaust) during World War II meant the destruction of 36 % of prewar world Jewry, more than 60 % of European Jewry, and the virtual annihilation of large Jewish communities in Central/Eastern Europe and the Balkans. World Jewry still has not recovered its pre-World War II size. Long-lasting population imbalances reflected high Jewish child mortality and low birth rates during the *Shoah*. After 11 million Jews had survived World War II, the total number increased to 12.1 million in 1960, 12.6 million in 1970, 12.8 million in 1980, 12.9 million in 1990, 13.1 million in 2000, and 13.7 million in 2011. The more recent increase reflects the augmented role of Israel in the world total and Israel's comparatively high Jewish fertility rate and rate of natural increase.

Waves of international migration after World War II intensely affected the geography of world Jewry in subsequent decades. During the 1950s and 1960s more than 500,000 Jews left North Africa and Southwest Asia; between the late 1960s and late 1980s more than 250,000 left the FSU. Since 1990, a major exodus from the FSU involved more than a million Jews and members of their households, while the totality of Ethiopian Jewry left Ethiopia. An absolute majority of these migrants went to Israel. Of the 11 million Jews remaining in 1945, 500,000 lived in Palestine. Since Israel's independence in 1948, its Jewish population grew rapidly to one million in 1949, two million in 1962, three million in 1976, four million in 1991, five million in 2001, and 5.9 million in 2012. Mass immigration was the major determinant of growth until the 1960s, when natural increase commenced to predominate as the engine of growth. The total size of Diaspora Jewry declined from 10.8 million in 1948 to 10.1 million in 1970, 9.1 million in 1990, and 7.8 million in 2011. Until about 1970, in spite of mass immigration to Israel (*aliyah*), the number of Jews in the Diaspora did not diminish. World Jewish population, after some postwar recovery, was approaching zero population growth. Natural increase in Israel only slightly overcame the natural decrease in the Diaspora.

70.3.3 Current Size of World Jewish Population

At the onset of the twenty first century, world society, and world Jewry within it, witnessed intensive transformations. Far reaching political, economic, and cultural *globalization* processes involved contraction of time and space, greater interdependence among different and distant components of world society, such as political-military interventions, industrial competition, international trade, and most significantly media and communication networks. Continuing gaps in standards of living and human opportunities stimulated large waves of geographical mobility and generated growing ethnocultural heterogeneity in local societies. While, as part of the continuing drive of modernization, society became increasingly secularized, large masses of people, including Jews, were more than ever involved in a keen quest for spiritual meaning and sought gratification in religious values and ethnic identities.

In 2011, of a world total of about 13.7 million Jews, 95 % lived in nine countries with 100,000 Jews or more: Israel (5,803,000), United States (5,425,000), France (482,000), Canada (375,000), UK (292,000), Russia (199,000), Argentina (182,000), Germany (119,000), and Australia (108,000). Other significant communities were in Brazil (95,000), South Africa (70,000), Ukraine (69,000), Hungary (48,000), Mexico (39,000), and Belgium and the Netherlands (30,000 each) (Table 70.1).

Table 70.1 Jewish population estimates, by country, 1970 and 2011

Country	1970	2011	% Change
WORLD TOTAL	12,665,200	13,657,800	8
AMERICAS TOTAL	6,219,800	6,185,800	0
Canada	286,000	375,000	31
United States	5,420,000	5,425,000	0
Central America, Caribbean	46,800	54,300	16
South America	467,000	331,500	–29
EUROPE TOTAL	3,231,900	1,437,700	–56
European Union 27[a]	1,333,250	1,112,400	–17
Former Soviet Republics	1,831,100	284,600	–84
Other West Europe	21,450	19,400	–10
FSU in Europe[b]	[1,896,700]	[295,800]	[–84]
ASIA TOTAL	2,936,400	5,843,000	99
Total Israel	2,582,000	5,802,900	125
Israel	2,581,000	5,498,700	113
West Bank	1,000	304,200	30,320
FSU in Asia	254,100	20,500	–92
Other Asia	100,300	19,600	–80
AFRICA	207,100	75,700	–63
Northern Africa	82,600	3,700	–96
Sub-Saharan Africa	124,500	72,000	–42
OCEANIA TOTAL	70,000	115,600	–100

Source: DellaPergola (2013: 225)

[a]Including former Republics now members of the European Union

[b]Including areas in Asia

Smaller Jewish communities were found in another 80 countries with at least 100 Jews. Parallel socio-demographic trends prevailed across the Diaspora: intense concentration in major metropolitan areas, suburbanization, high educational levels, specialization in liberal and managerial professions, low fertility, frequent intermarriage, and aging. Recent international migrants became significantly absorbed and acculturated in the context of the receiving countries. The changing distribution of world Jewish population by major regions is displayed in Fig. 70.1 and Table 70.1. Note that the latest estimates of Jewish population worldwide can be found at www.jewishdatabank.org.

About 45 % of the world's Jews in 2011 resided in the Americas, with more than 42 % in North America. About 43 % lived in Asia – mostly in Israel – including the Asian republics of the FSU, but not the Asian parts of the Russian Federation and Turkey. Europe, including the Asian territories of the Russian Federation and Turkey, accounted for about 11 % of the total. Fewer than 2 % of the world's Jews lived in Africa and Oceania.

70.3.4 World Jewish Population Distribution

Reflecting global Jewish population stagnation along with an increasing concentration in few countries, in 2011 97.9 % of world Jewry lived in the largest 16 communities, and excluding Israel from the count, 96.3 % of world Jewry lives in the 15 largest communities of the Diaspora, including 69.1 % who lived in the U.S. (Table 70.2). Besides the two major Jewish populations (Israel and the U.S.) each comprising well over five million persons, another seven countries each had more than 100,000 Jews, and another seven had about 30,000 Jews or more. Of the larger seven, three were in the European Union (EU) (France, the UK, and Germany), one in Eastern Europe (the Russian Federation), one in North America (Canada), one in South America (Argentina), and one in Oceania (Australia). Of the smaller seven, three were in the EU (Hungary, Belgium and the Netherlands), one in Eastern Europe (Ukraine), two in Central and Southern America (Brazil and Mexico), and one in Africa (South Africa). The dominance of Western countries in global Jewish population distribution is a relatively recent phenomenon and reflects the West's relatively more hospitable socioeconomic and political circumstances *vis-à-vis* the Jewish presence.

The growth, or at least the slower decrease, of Jewish population in the more developed Western countries is accompanied by a higher share of Jews in a country's total population. Indeed, the share of Jews in a country's total population tends to be strongly related to the country's level of development (Table 70.2). For 2011, the share of Jews out of the total population was 744.7 per 1,000 in Israel (including Jews in East Jerusalem, the West Bank, and the Golan Heights, but excluding Palestinians in the West Bank and Gaza) which obviously is a special case, yet also a developed country; 17.4 per 1,000 in the U.S.; 3.9 per 1,000 on average in the other seven countries with more than 100,000 Jews; 0.8 per 1,000 on average in the other seven countries with more than 30,000 Jews; and virtually nil in the complex of the many other countries.

To better illustrate the increasing convergence and dependency between the Jewish presence and the level of socioeconomic development of a country, Table 70.2

Table 70.2 Countries with largest core Jewish populations, 2011

Rank	Country	Jewish population	% of Total Jewish Population				Total population	Jews per 1,000 population	HDI rank[a]
			In the World		In the Diaspora				
			%	Cumulative	%	Cumulative			
1	Israel[b]	5,802,900	42.5	42.5	=	=	7,694,900	744.7	15
2	United States	5,425,000	39.7	82.2	69.1	69.1	311,592,000	17.4	4
3	France	482,000	3.5	85.7	6.1	75.2	63,340,000	7.6	14
4	Canada	375,000	2.7	88.5	4.8	80.0	34,500,000	10.9	8
5	United Kingdom	291,500	2.1	90.6	3.7	83.7	62,920,000	4.6	26
6	Russia	199,000	1.5	92.1	2.5	86.2	142,800,000	1.4	65
7	Argentina	182,000	1.3	93.4	2.3	88.5	40,500,000	4.5	46
8	Germany	119,000	0.9	94.3	1.5	90.1	81,800,000	1.5	10
9	Australia	108,000	0.8	95.1	1.4	91.4	22,700,000	4.8	2
Total 3–9		1,756,500	12.8	95.1	22.3	91.4	448,560,000	3.9	24.4
10	Brazil	95,400	0.7	95.8	1.2	92.6	196,700,000	0.5	73
11	South Africa	70,500	0.5	96.8	0.9	94.4	50,500,000	1.4	110
12	Ukraine	69,000	0.5	96.3	0.9	93.5	45,700,000	1.5	69
13	Hungary	48,300	0.4	97.1	0.6	95.0	10,000,000	4.8	36
14	Mexico	39,300	0.3	97.4	0.5	95.5	114,800,000	0.3	56
15	Belgium	30,100	0.2	97.7	0.4	95.9	11,000,000	2.7	18
16	Netherlands	29,900	0.2	97.9	0.4	96.3	16,700,000	1.8	7
Total 10–16		382,500	2.8	97.9	4.9	96.3	445,400,000	0.9	38.6
Rest of world		297,000	2.1	100.0	3.7	100.0	5,773,047,100	0.0	ca. 100

Source: DellaPergola (2013: 225)

[a]The Human Development Index, a synthetic measure of health, education, and income (in terms of U.S. Dollar purchase power parity) among the country's total population. See: United Nations (2011)

[b]Israel's Jewish population includes residents in East Jerusalem, the West Bank, and the Golan Heights. The respective total population includes non-Jews in Israel, including East Jerusalem and the Golan Heights, but does not include Palestinians in the West Bank and Gaza

also reports the Human Development Index (HDI) for each country. The HDI – a composite measure of a society's education, health, and income – provides a general sense of the context in which Jewish communities operate, although it does not necessarily reflect the actual characteristics of the members of those Jewish communities. Of the top 16 Jewish communities, four (Australia, the U.S., Canada, and the Netherlands) live in countries with the ten best HDIs among nearly 200 countries, another four (Germany, France, Israel, and Belgium) are ranked better than 25th, three (UK, Hungary, and Argentina) are better than 50th, two (Russian Republic and Ukraine) are better than 100th, and only one (South Africa, 110th) exhibits a lower HDI. Of course, one should be aware that Jewish communities in these countries may display social and economic data significantly better than the average population of their respective countries.

The increasing overlap of a Jewish presence with higher levels of socioeconomic development in a country, and at the same time the diminution or gradual disappearance of a Jewish presence in less developed areas, is a conspicuous development of the twentieth and early twenty-first centuries. The emerging geographical configuration carries advantages concerning the material and legal conditions of the life of Jews, but it may be associated with growing difficulties at maintaining a sufficient sense of separate Jewish identity in an open and dynamic society; and it also may generate a lack of recognition of, or estrangement toward, Jews on the part of societies in less developed countries that constitute the overwhelming majority of the world's total population.

70.4 Migration of Jews to the U.S.

Most likely, the first settlement of Jews in the New World resulted from the Spanish Inquisition (1492) and the Portuguese Inquisition (1497) during which time, faced with the decision to leave, convert to Christianity, or be burned at the stake, many Jews fled to other lands, including the Mexican territory that now is part of the western U.S. But the first American Jewish community was created when 23 Jews left Pernambuco (Recife), Brazil (as a result of the Portuguese reconquest of the area from the Dutch) and arrived at New Amsterdam in 1654.

Although Peter Stuyvesant, governor of New Amsterdam, was loathe to permit the Jews to remain, they were eventually permitted to stay. When permission was granted for Jews to conduct religious services in private, the agreement stipulated that they must "exercise in all quietness their religion within their houses, for which end they must without doubt endeavor to build their houses close together in a convenient place on one side or the other of New Amsterdam" (Hertzberg 1989: 24). Thus was established the first Jewish neighborhood in America, although Jews in New York were not permitted to build a synagogue until 1728.

Historic estimates of the U.S. Jewish population have been collated by Marcus (1990) and by the American Jewish Historical Society (www.ajhs.org) and form the basis for estimates in this chapter prior to 1990 (see Fig. 70.1).

Four periods of Jewish immigration to the U.S. (see Fig. 70.2) may be identified (Dimont 1978):

1. The Sephardic Migration (1654–1820);
2. The German Migration (1820–1880);
3. The Eastern European Migration (1880–1920s); and
4. The Modern Period of Migration (from the 1930s to the present day).

70.4.1 The Sephardic Immigration Period

Sephardic Jews are those whose ancestors derive from Spain and were expelled in 1492 as a result of the Spanish Inquisition. Many Sephardic Jews migrated to parts of the Ottoman Empire, as the Ottoman Sultan welcomed Jews who were expelled from Spain. During this period, the U.S. Jewish population only increased to about 5,000. These Jews were mostly shopkeepers and merchants. Not having been allowed to own land in most European countries, Jews did not develop farming skills. They were involved in retail activity in Europe and brought those skills to the U.S. During colonial times, while 80 % of Americans in general were farmers, the vast majority of Jews were urbanites. The earliest synagogues were found in New Amsterdam (New York), Newport (Rhode Island), Savannah, Philadelphia, and Charleston.

70.4.2 The German Immigration Period

The second period of Jewish migration, from 1820 to 1880, marks the era of German Jewish migration, with most Jews coming from Bavaria (Hertzberg 1989). While Napoleon's message of liberty, equality, and fraternity had improved conditions for Jews in Europe and had freed them from the confines of the ghetto in many areas (resulting in the *Haskala*, or Enlightenment movement in Jewish history), the end of this era, with the end of the Napoleonic era, made life difficult for Jews in many areas, particularly in Germany. During the first wave of immigration from Germany, from 1820 to 1860, as many as 100,000 German Jews migrated to the U.S. (Hertzberg 1989: 106). Many of these German immigrants were involved in retail trade, particularly in the garment industry. Some began peddling goods from push carts and gradually developed retail outlets. These retail outlets evolved into major department stores, including Abraham & Strauss, Gimbels, Bloomingdale's, Lazarus, Macy's, Lord & Taylor, and others. In New York, by 1880, 80 % of all retail clothing establishments and 90 % of wholesale clothing establishments were owned by Jews (Sachar 1992: 86). In other cities, such as Columbus, Ohio, all retail clothing stores were owned by Jews. In the 1870s, nine out of ten Jews in smaller towns were self-employed businessmen (Hertzberg 1989: 137). When the Gold Rush of 1849 began,

Jewish merchants left the East and became storekeepers in the West, but they did not become gold miners. Levi Strauss, however, was responsible for the design and manufacture of pants for gold miners that have now diffused throughout the world as "jeans."

By 1880, about 280,000 Jews lived in the U.S. Two hundred new synagogues were established, although they were more important in providing immigrant Jews a familiar milieu than they were in providing a location for piety. B'nai B'rith (www.bnaibrith.org), now the largest Jewish organization in the world, began as well as a (nonreligious) group designed to maintain some aspects of Jewishness and to provide self-help. These German Jews also brought with them a new innovation in Jewish worship, Reform Judaism, which had flourished in Germany. Economically, many German Jews prospered and as they moved into the better neighborhoods, non-Jews moved out, leading to "gilded" ghettos. Other German Jews remained poor. The nature of this migration is fully explored in Stephen Birmingham's 1967 classic *Our Crowd, The Great Jewish Families of New York*. This German migration changed the American Jewish community from one in which most Jews were American born, to one in which most Jews were foreign born.

70.4.3 The Eastern European Immigration Period

The third period of Jewish migration began with the fall of Czar Alexander II in Russia in 1881. Following this change in leadership, *pogroms* (anti-Jewish riots) occurred in Russia in 1881 and in Kishinev (in Bessarabia, now Moldova) in 1903 and 1905 (Pasachoff and Littman 1995: 218–221 and 236–239). This led to a significant migration of Jews from Eastern Europe to the U.S., Palestine, and other locations. Often this was a *stage migration* process, with Jews first moving out of small villages and towns into Eastern European cities, and then moving from these cities to the U.S. Jews began to arrive in significant numbers to New York, Baltimore, Philadelphia, Boston, and Chicago, all prominent ports of entry (Sanders 1988: 167).

This migration was to change the face of American Jewry from one that was dominated by German Jews, who by 1880 were, because of very high levels of assimilation, well on their way to becoming another Protestant denomination, to one dominated by the Eastern European Jewish migrants who came between 1880 and 1940. This large scale migration increased the U.S. Jewish population to almost five million by 1937. More than 90 % of Jewish migrants during this period were from Russia. In total, 3,715,000 Jews entered the U.S. between 1880 and 1929. During this period, 8 % of migrants to the U.S. were Jewish (Barnavi 1992: 194–195) and 15 % of all European Jewry moved to the U.S. The Jewish immigrants came to the U.S. to stay. The rate of reverse migration was only 5 % for the Jewish population, compared to 35 % for the general immigrant population (Sherman 1965: 61). This difference is probably related to the fact that while "economic opportunity" was a "pull" factor to the U.S. for all immigrant groups, the "push" factors (persecution) for Jews to leave Europe were clearly more significant than for most, if not all, other ethnic groups.

At first, the German Jews distanced themselves somewhat from the Eastern European group. The German Jews had "made it" and had become quite "Americanized" in the process and had largely adopted Reform Jewish practices. The Eastern European arrivals came with an Eastern European *shtetl* (small town) mentality, were poor, spoke Yiddish, and those that were religious followed more traditional practices. Because the rabbis of Eastern Europe had opposed emigration to the U.S., those with higher incomes and those who were more religious remained in Europe (Hertzberg 1989: 156–157). In both "look and feel," despite the common heritage, Eastern European Jews differed significantly from the German Jews that preceded them. The eventual success of these Eastern European immigrants is fully described in Stephen Birmingham's (1984) *The Rest of Us: The Rise of America's Eastern European Jews* and in Howe's (1976) *World of Our Fathers, the Journey of East European Jews to America and the Life They Found and Made.*

At first, the German Jews wanted to spread the new Jewish immigrants throughout the country. The concept was that if the Jewish population became too geographically clustered, a reaction would occur among non-Jews, resulting in anti-Semitism. This led to the Galveston Plan in 1907, which was to divert some of the immigrants headed for northeastern cities, particularly New York, to Galveston, Texas (Sanders 1988: 235–240). This plan failed, as any plan would that is at odds with an understanding of the chain migration process: very few Jews could be convinced to move to Galveston. Jews wanted to move to the large northeastern cities that already had large Jewish populations, where they could find a *landsmannschaftan* or *landsleite*, a cultural society with membership from their former country, or even their former city (Shamir and Shavit 1986).

With the eventual suburbanization of the Jewish population following World War II, the passing of a generation and significant intermarriage between German and Eastern European Jews, the distinction between the two groups has left the American Jewish psyche.

An important distinction between the German Jews and the Eastern European Jews should be noted. German Jews did not dislike the land they had left. They remained proud of their German heritage and maintained varying degrees of loyalty to Germany until the U.S. entered World War I against the Germans. In today's language, the German Jews were "Jews by religion" or in the language of Germany in the 1800s "Germans of the Mosaic persuasion." The Eastern European Jews, on the other hand, thought of themselves as a people, as an ethnic group, and did not have fond memories of the old country and its government (although they might have had fond memories of their little *shtetl* or Jewish village). The coming of the Eastern European Jews in large numbers did lead to an increase in anti-Semitism. It also led to an initial friction between the groups, although eventually the German Jews did come to the aid of their poor brethren.

With this aid, the Eastern European Jews quickly rose to be one of the most successful American ethnic groups. As early as 1910, about one-fourth of those enrolled in American medical schools were Jewish. By the 1960s, about 20 % of the faculty at non-Church related American universities were Jewish. By the beginning of World War I, the Jews from Eastern Europe had replaced the German Jews as the leaders of the American Jewish community. And the door to further immigration was significantly closed.

70.4.4 The Modern Immigration Period

The fourth period of Jewish migration to the U.S. began in the 1930s. By this time, the First (1921) and Second (1924) Johnson Acts (Sanders 1988: 386–387) had been passed by Congress, practically halting Jewish (and other Eastern and Southern European) immigration (Friesel 1990: 132). Immigration would never return to the high levels that persisted between 1880 and the early 1920s. Unfortunately, this closing of the door to immigration occurred at the worst time for European Jews, as the next two decades saw the rise of Hitler and the extermination of six million Jews in the Shoah. Those Jews who came to the U.S. during World War II clearly came as refugees, not merely as immigrants. Between 1933 and 1937 fewer than 40,000 Jews were permitted to enter the U.S. Even under the Immigration Law of 1924 (the Johnson Acts), this was less than 20 % of the quota. Jewish immigration from Europe saw a temporary surge from Germany following *kristallnacht* in 1939, the event most historians now agree was the initial act of the Shoah. In total, about 110,000 Jews were permitted entry to the U.S. from 1938 to 1941. Wyman's (1984) *The Abandonment of the Jews* provides significant detail on this period.

At the end of World War II, as the full extent of the Shoah became evident, American Jews came to the realization that they were now the largest Diaspora Jewish community, but it soon became evident that the size of the community would no longer be increasing due to large scale immigration. First, as described above, U.S. immigration law no longer permitted unfettered immigration. Second, and perhaps more important, the establishment of Israel in 1948 opened a new haven for Jews throughout the world who were in economic or political distress or had been displaced by World War II. By 1953, Israel had welcomed more than 700,000 refugees. In total, from 1948 to 2011, Israel has settled 3,150,000 new immigrants (www.cbs.gov.il), mostly Jews, but also their non-Jewish family members in the framework of the Law of Return.

Yet, Jewish migrants have continued to enter the U.S. A surge of migrants was seen just after World War II, comprised mostly of 160,000 Holocaust survivors from the displaced persons camps established just after the Holocaust (Shapiro 1992: 26). Since the mid-1960s, more than 400,000 Jews have immigrated to the U.S. from the FSU (Kotler-Berkowitz et al. 2003), with most settling in New York City (particularly Brighton Beach, Bensonhurst, Borough Park, and Bay Ridge). Other important destinations included Boston, Chicago, Los Angeles, and San

Francisco. About 140,000 Israeli Jews now live in the U.S., based on an analysis of the Public Use Microdata Sample (PUMS) from the American Community Survey (Sheskin 2010a), with most residing in New York, California, Florida, and New Jersey.

Smaller numbers of Jews have come to the U.S. from a variety of other locations. More than 10,000 Hungarian Jews moved to the U.S. just after the Hungarian revolution of 1956. A few thousand Cuban Jewish migrants came to Miami in the late 1950s and early 1960s (Liebman 1977). Starting in the 1970s and continuing to the present day, Jews from a number of Middle American and South American countries, particularly from Argentina, Colombia, and Venezuela, have moved to Miami (Sheskin 2005: Table 4–13). After the fall of the Shah of Iran in 1979, Jews moved to the U.S. from Iran (particularly to Los Angeles and New York), resulting in a community of perhaps 60,000–80,000 by 2006 according to an Associated Press report ("Iranian Jews Living in U.S. Have Complex Feelings About Mideast Crisis" Fox News, August 7, 2006). Jewish migrants also came from the Arab world, particularly after the establishment of Israel in 1948.

70.5 Migration of Jews to Israel

70.5.1 General Patterns

The Jewish population of Israel is composed almost entirely of immigrants and descendants of immigrants. In 1951, after the mass immigration following Israel's independence in 1948, three-quarters of the Jewish population were foreign-born, and about half of all Israelis had lived in the country for 5 years or less. Over the years, the share of Israeli-born in the Jewish population rose from 47 % in 1972 to 63 % in 1988 and 71 % in 2010 – after absorbing more than 1.3 million new immigrants (mostly from the FSU) since 1990.

Following an early numerical predominance of the European-born Jews, the share of the Asian/African-born rose markedly as a percentage of all foreign-born Israelis, though the European/American-born remained the larger group. The children's generation shows a different trend: since the early 1970s, Israeli-born Jews of Middle Eastern (Asian)/North African extraction outnumbered those of European/American extraction. Overall, the division between the two main origin groups (comprising foreign- and Israeli-born nationals, including members of the third generation) was quite close. In 1988, the percentage of Israeli Jews of Asian/African extraction stood at 52 % of all Israeli Jews. This percentage would have been slightly higher if it included FSU immigrants from the Caucasus and Central Asian republics, who, like all FSU immigrants, were customarily classified as European-born by the Central Bureau of Statistics (CBS). With the large FSU immigration in recent years, the number of Israelis of European origin rose again. By 2010, 34 % of all Jews were of European/American origin, 28 % were of Asian/African origin, and 38 % were Israelis of third and higher generation.

A sequence of periods of large- and moderate-scale immigration most strongly influenced demographic and socioeconomic change in Israeli society. Immigration directly affected the growth rates and composition of the population in several respects, such as the absolute and relative share of population groups according to religion and ethnic affiliation, country of origin, and duration of stay in Israel. However, immigration also affected other significant variables such as the geographic distribution of the population, its composition by age groups and marital status, and social stratification.

Immigration reached Israel at a very unsteady pace, in the shape of subsequent waves (DellaPergola 1998). Early immigration played an important role in establishing the primary cultural, political and institutional infrastructure of Palestine and, later, in Israel. Roughly 60,000–70,000 Jews immigrated to Palestine toward the end of Ottoman rule (1880–1918), and about 483,000 arrived during the British Mandate (1919–1948). Until 1948, official immigration policies restricted the size of Jewish immigration and directly or indirectly stimulated selectivity of immigration by countries of origin and the social and demographic profile of migrants (Bachi 1977; Friedlander and Goldscheider 1979). Migration opportunities radically changed with Israel's independence in 1948 and its adoption of nearly unrestricted admission policies for Jewish immigrants epitomized by the 1950 Law of Return.

70.5.2 Geographical Correlates of Immigration to Israel

Since 1948, the rhythm of immigration was quite different in the various Diaspora countries. Four principal models provide a basic typology that can be extended to all possible countries of origin of immigrants (DellaPergola 2009):

1. total transfer to Israel between 1949 and 1951 of a country's existing Jewish population – as in Iraq, Yemen, Bulgaria, and to a lesser extent in several other countries in the Middle East and the Balkans;
2. large-scale and unselective migration spread over longer time spans of the absolute majority of pre-existing Jewish population, the pace of migration being mostly determined by the shifting opportunities to leave the country of origin. These, in turn, largely reflected changing migration policies of the respective governments, typically in the FSU, but also in Romania and most other countries in Eastern Europe, Morocco, and Ethiopia;
3. more selective Jewish emigration, uninterrupted over time but of variable intensity reflecting periodic economic and political crises and their consequences for the Jewish community in the respective countries of origin, as in Argentina and most other Latin American countries and South Africa;
4. comparatively scant and highly selective Jewish emigration, with the exception of a more intense period following the 1967 Six-Day war – as in the U.S. and most Western developed countries.

These different models clearly indicate that the principal determinants of the timing and volume of migration to Israel reflected the push of changing and often negative situations in the countries of origin more than Israel's pull. At the same time, the role of cultural-ideological determinants in migration decisions among Diaspora Jews cannot be dismissed. The influence of Zionism and of a broader Jewish identity significantly determined choosing Israel over alternative targets such as the U.S., Canada, Australia, or France, once migration driving factors emerged in the countries of origin.

Following mass immigration, Israel became a contemporary society with the seventh highest percentage of foreign-born population in the world. Yet by 2010, 78 % of Israel's total population, and 71 % of its Jewish population were born in the country. As immigration to Israel reflected greatly different demographic, socioeconomic, and cultural modernization attained across Diaspora communities, many of the typical global East-west or North-south cleavages were introduced through immigration in the Israeli context. In a sweeping generalization, immigrants from Muslim countries comprised a comparatively more traditional population in terms of cultural orientation and socioeconomic development, relative to immigrants from European countries and the relatively small component of American origin.

Concomitant Jewish migration streams from the same countries of origin to Western countries – particularly from North Africa to France and from Eastern Europe to the U.S. – comprised smaller proportions of children and elders relative to migrants to Israel, and featured better levels of socioeconomic training and specialization. In other words, Israel had to absorb and integrate a Jewish population with unusually onerous initial demographic and socioeconomic profiles, which became more frustrating because of the integration elsewhere of a large share of the respective social and intellectual elites.

70.5.3 Emigration from Israel

Out-migration (also called *yerida*, descent) from Israel was a quantitatively far lesser factor. Emigration has been comparatively stable over time. Excluding occasional fluctuations, the number of Israeli residents who left Israel and spent more than 4 years abroad usually ranged between 10,000 and 20,000 per year. As against an overall population in rapid growth, the annual emigration rate tended to decline and ranged around 3–4 per thousand inhabitants. A recent survey by the Pew Research Center (Global Religion and Migration Data Base) suggests that only 230,000 Jews born in Israel are living in other parts of the world. The 4 % of Israelis who live abroad is half the 8 % average figure for the rest of the world.

70.6 The Changing Spatial Distribution of American Jews

Next, we examine the changing spatial distribution of American Jews over the past four decades (Figs. 70.3, 70.4 and 70.5). What the aforementioned Galveston Plan was unsuccessful at in 1907 (distributing the Jewish population throughout the nation), American Jews have accomplished on their own in the past half century.

American Jews are not geographically distributed as are all Americans: In 1970, 46 % of American Jews would have had to move to another state so that the geographic distribution of American Jews would match that of the American population in general. In 2012, the 46 % decreased to 40 %, indicating that the significant difference in geographic distribution between Jews and all Americans, although lessening a bit, has maintained from 1970 to 2012.

In 1970, five states (New York-42 %, California-12 %, Pennsylvania-8 %, New Jersey-7 %, and Illinois-5 %) contained 73 % of American Jews. By 2012, Jews were *somewhat* less geographically concentrated with seven states (New York-26 %, California-18 %, Florida-10 %, New Jersey-8 %, Illinois-4 %, Pennsylvania-4 %, and Massachusetts-4 %) containing 74 % of American Jews. Note that New York's share of American Jews decreased from 42 % in 1970 to 26 % in 2012, while Florida's share increased from 4 % in 1970 to 10 % in 2012 (Sheskin and Dashefsky 2012).

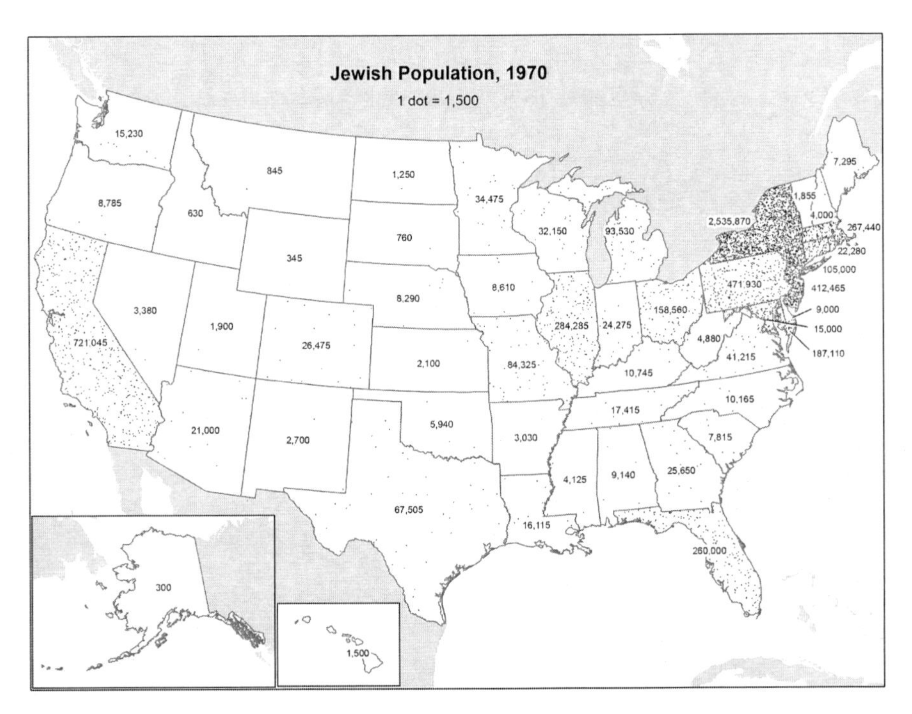

Fig. 70.3 U.S. Jewish Population, 1970 (Map from Sheskin and Dashefsky 2013: 160, used with permission)

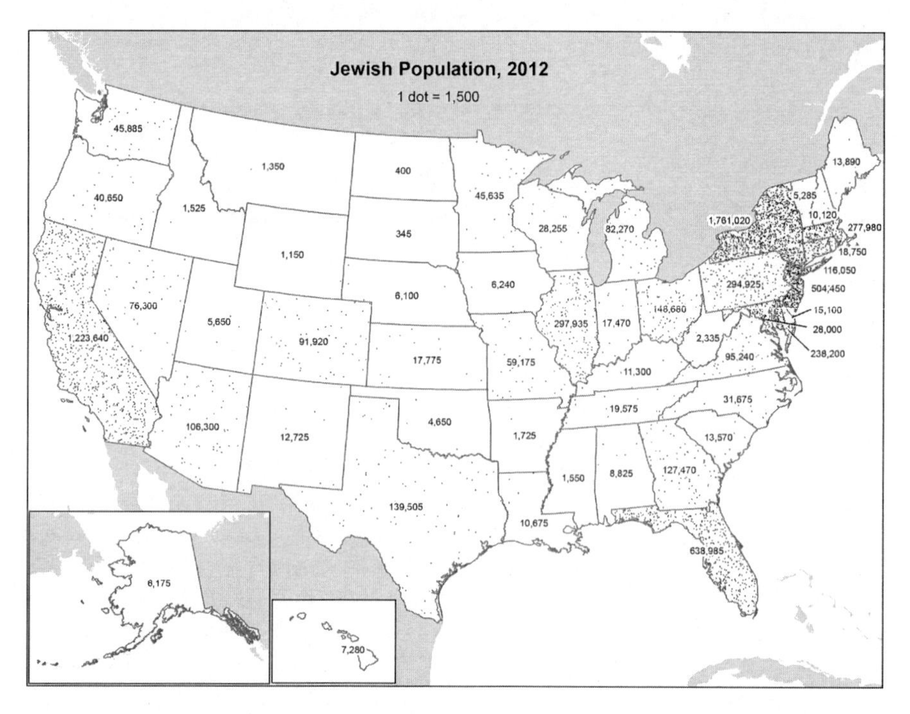

Fig. 70.4 U.S. Jewish Population, 2012 (Map from Sheskin and Dashefsky 2013: 160, used with permission)

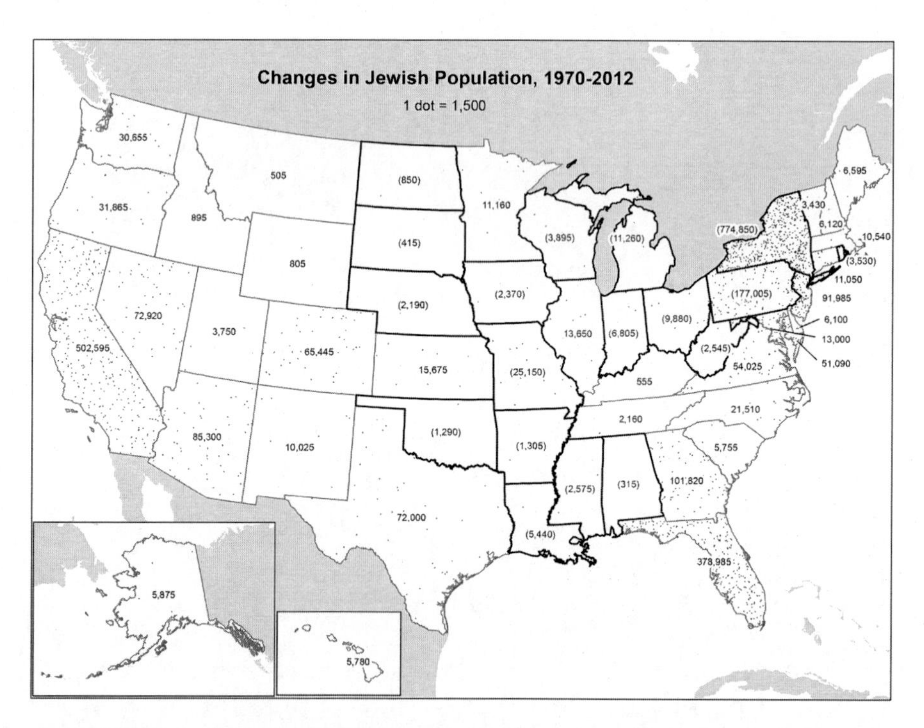

Fig. 70.5 Change in U.S. Jewish population, 1970–2012 (Map from Sheskin and Dashefsky 2013: 161, used with permission)

Table 70.3 Jewish population in the United States, by census region and census division, 2012

	Jewish population		Total population	
Census region/division	Number	Distribution (%)	Number	Distribution
NORTHEAST	3,002,470	44.7	55,521,598	17.8
Middle Atlantic	2,560,395	38.1	41,029,238	13.2
New England	442,075	6.6	14,492,360	4.7
MIDWEST	710,280	10.6	67,158,835	21.6
East North Central	574,610	8.5	46,519,084	14.9
West North Central	135,670	2.0	20,639,751	6.6
SOUTH	1,388,380	20.7	116,046,736	37.2
East South Central	41,250	0.6	18,553,961	6.0
South Atlantic	1,190,575	17.7	60,513,771	19.4
West South Central	156,555	2.3	36,979,004	11.9
WEST	1,620,550	24.1	72,864,748	23.4
Mountain	296,920	4.4	22,373,411	7.2
Pacific	1,323,630	19.7	50,491,337	16.2
TOTAL	6,721,680	100.0	311,591,917	100.0

Source: Sheskin and Dashefsky (2013: 162)

While remaining geographically concentrated in 5–7 states, the geographic distribution among the four census regions has changed significantly away from the Northeast and Midwest and toward the South and West (Table 70.3). From 1970 to 2012, the percentage of Jews in the Northeast decreased from 67 to 45 %, with this decrease being concentrated in the Middle Atlantic. The percentage of Jews in the Midwest decreased slightly from 12 to 11 %, with this decrease being predominantly in the East North Central. The percentage of Jews in the South increased from 12 to 21 %, with this increase being predominantly in the South Atlantic. Finally, the percentage of Jews in the West increased from 13 to 24 %, predominantly in the Pacific. Note that the latest table of Jews by state is available at www.jewishdatabank.org.

Four issues related to the changing spatial distribution of American Jews may be identified. First, as Jews have moved from the immigrant cities of the Northeast and Midwest to the South and West, the difficulty of creating a sense of community in locations where few have roots in the area, leads to a lack of support for local institutions (Tobin 1984). An extreme example is the Jewish community in Palm Beach County, Florida. According to a 2005 survey, 0 % of Jewish adults in the County were born in the County. This is manifested in very low rates of synagogue and Jewish Community Center membership (Sheskin 2006).

Second, the movement to the South and West has meant that enormous monetary resources and volunteer time have been expended recreating Jewish institutions (synagogues, Jewish Community Centers, Jewish day schools, Jewish federations, etc.) in these new communities, rather than using resources for the development of programs that would enhance the lives of American Jews and the general communities in which they reside.

Third, a significant geographic separation of families has occurred. In some cases, grandparents have moved away from their children and grandchildren to retire in Florida and other Sunbelt states. In other cases, children have migrated out of the traditional areas of Jewish settlement in the Northeast and Midwest to seek job opportunities elsewhere. While improvements in communication, such as unlimited long distance telephones, Skype, etc. have somewhat ameliorated the impact of the separation of families, the impact is still significant, particularly as the elderly age and become frail and vulnerable.

Fourth, the changing geographic distribution of American Jews has resulted in increased political power for the Jewish community in presidential elections. The top ten states for Jewish population have 244 electoral votes, with 270 needed to secure the presidency. Although Jews are only about 2 % of all Americans, they are a somewhat larger percentage of all American adults. About 90 % of Jews are registered to vote, compared to about two-thirds of all Americans. For the 2008 presidential election, polls showed that 96 % of registered Jews actually voted, compared to 77 % of all Americans (www.electionstudies.org). Thus, Jews are always a much larger percentage of *voters* than of *the total population*. Some of the states with the largest increases in Jewish population over the past half century also have significant increases in electoral votes, California's electoral votes increased from 40 in 1970 to 55 in 2012; Florida from 14 to 29; Georgia from 12 to 16, Arizona, from 5 to 11; and Texas, from 25 to 38. While Jews are but 1.3–3.4 % of the population of most of these states (Sheskin and Dashefsky 2012), they are a larger percentage of voters in these states. With most elections won by 2–6 percentage points, presidential candidates do pay attention to the concerns of the Jewish community.

70.7 Population Distribution in Israel

The population in Israel is chiefly urban – at least under current definitions that do not necessarily reflect changes in lifestyle or in the social makeup of localities. In Israel all localities with more than 2,000 residents are defined as urban. Among Jews, the share of town dwellers exceeded 90 % since the early 1970s, and was 91 % in 2010. The Arab population, too, has been characterized by a significant transition to a large urban majority: from 63 % in 1961 to 95 % in 2010. The process most responsible for the urbanization of Israeli Arabs was the growth of small rural or semi-urban localities into larger, urban-type localities, as a result of local population growth. The Bedouins have recently shown greater inclination to move to permanent settlements in the south, as opposed to the temporary settlements they occupied in the past.

Another important element for the appraisal of the process of immigrants absorption relates to the preoccupation of Israeli governments with strategic population dispersal over the national territory. Despite its small size, Israel features significant regional climatic and socioeconomic differences. The incoming population naturally tended to concentrate in the more developed and geographically more accessible areas. The veteran population, mostly of European origin, thus tended to

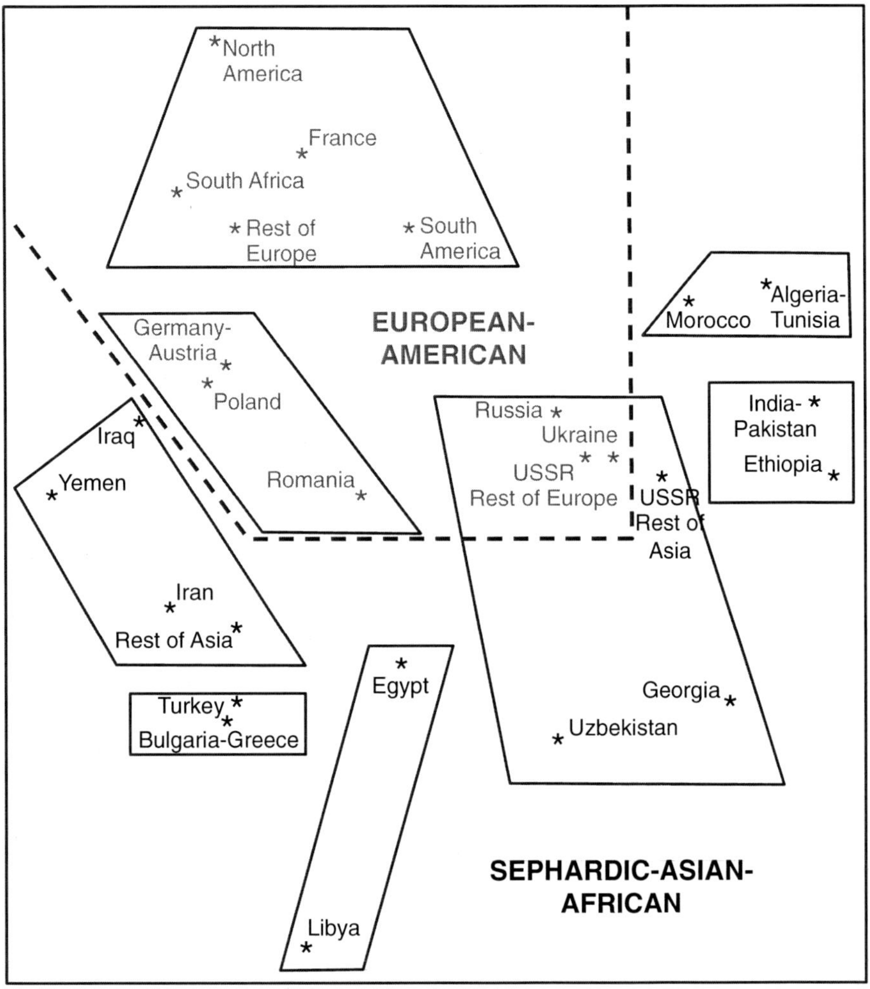

Fig. 70.6 Smallest space analysis of Jewish population distribution in Israel, by country of birth and sub-district, 1995 (Source: DellaPergola 2007)

occupy geographical locations closer to the central areas in the country, in and around the Tel Aviv-Yafo Metropolitan Area, along the Mediterranean coast, and in the Haifa urban area. Subsequent immigrants from Asia and Africa in the 1950s and 1960s, but also more recently from the FSU were largely settled in northern and southern areas which besides being more peripheral were also less economically developed and more often affected by security problems. Looking at the country-wide distribution of Jewish population by countries of origin, by residential districts, one obtains a synthetic picture of similarity and dissimilarity across different origin groups. Figure 70.6 obtains from a Smallest Space Analysis (SSA) which graphically elaborates percentage distributions of each group, by district, as of the

1995 Israel population census (DellaPergola 2007). While each country is represented in each district, there are definite patterns of concentration and convergence that outline the complexity, dynamics and resilience of Israel as an immigrants' absorption country.

70.8 The Metropolitan Scale

70.8.1 The Interurban Scale

Changes in the geographic distribution of Jews have affected their distribution not only among countries, but also within countries. Historically, reflecting legal and economic constraints imposed upon them, Jews tended to be concentrated in cities more than other population groups. However, especially in Eastern Europe, in Middle Eastern countries like Yemen, or in North African countries like Morocco, a large share to an absolute majority of Jews were found in small towns and villages. Jews often constituted a high percentage or even the majority of the total population in those locales. Especially since the mid-nineteenth century, vast masses of Jews relocated from those smaller and relatively peripheral towns and villages to major urban centers. The 15 cities that hosted the largest Jewish communities in 1925 had a combined Jewish population of 4.3 million that constituted nearly 30 % of the total world Jewish population, whereas in 1850 the same 15 cities had a combined Jewish population of less than 150,000, equal to 3 % of world Jewry (DellaPergola 1983). The geographic, socioeconomic, and cultural change involved was huge.

The continuing tendency of Jews to migrate to large metropolitan areas is shown by the overwhelmingly urban concentration of Jewish populations in 2010 (Table 70.4). More than half (53.0 %) of world Jewry lived in only five metropolitan areas. These areas – including the main cities and vast urbanized territories around them – were Tel Aviv, New York, Jerusalem, Los Angeles, and Haifa. Over two-thirds (67.5 %) of world Jewry lived in the five previous areas plus the South Florida (Miami, Fort Lauderdale, West Palm Beach), San Francisco, Be'er Sheva, Washington/Baltimore, and Boston metropolitan areas. The 24 largest metropolitan concentrations of Jewish population encompassed 74.4 % of all Jews worldwide.

Of the 15 largest metropolitan areas of Jewish residence, eight were located in the U.S., four in Israel, and one each in France, the UK, and Canada. Nearly all of the major areas of settlement of contemporary Jewish populations share distinct features, such as being national or regional capital cities, with a high standard of living, a highly-developed infrastructure for higher education, and strong transnational connections.

American Jews have tended to reside in major urban areas. Because Jews were traditionally barred from owning land in many parts of Europe, few became farmers when they came to the U.S. In fact, of the 6.7 million American Jews enumerated in Sheskin & Dashefsky, Table 70.5 shows the 5.3 million (79 %) live in the top 20

Table 70.4 Metropolitan areas with largest core Jewish populations, 2012

Rank	Metropolitan area[a]	Country	Jewish population	Share of world's Jews	
				%	Cumulative %
1	Tel Aviv[b]	Israel	3,070,800	22.2	22.2
2	New York[c]	U.S.	2,099,000	15.3	37.6
3	Jerusalem[d]	Israel	850,900	6.2	43.8
4	Los Angeles[e]	U.S.	688,600	5.0	48.8
5	Haifa[f]	Israel	686,300	5.0	53.8
6	South Florida[g]	U.S.	485,850	3.5	57.3
7	Be'er Sheva[h]	Israel	377,700	2.7	60.1
8	San Francisco[i]	U.S.	345,700	2.5	62.6
9	Washington/Baltimore[j]	U.S.	332,900	2.4	65.0
10	Boston[k]	U.S.	295,700	2.2	67.2
11	Chicago[l]	U.S.	294,700	2.1	69.3
12	Paris[m]	France	284,000	2.1	71.4
13	Philadelphia[n]	U.S.	280,000	2.0	73.4
14	London[o]	United Kingdom	195,000	1.4	74.8
15	Toronto[p]	Canada	180,000	1.3	76.1

Source: DellaPergola 2013, p. 226. For definitions see United States, Executive Office of the President (2008)

[a]Most metropolitan areas include extended inhabited territory and several municipal authorities around the central city. Definitions vary by country. Some of the U.S. estimates may include non-core Jews

[b]Includes Tel Aviv District, Central District, and Ashdod Subdistrict. Principal cities: Tel Aviv, Ramat Gan, Bene Beraq, Petach Tikwa, Bat Yam, Holon, Rishon LeZiyon, Rehovot, Netanya, and Ashdod, all with Jewish populations over 100,000

[c]New York-Northern New Jersey-Long Island, NY-NJ-CT-PA Metropolitan Statistical Area. Principal Cities: New York, NY; White Plains, NY; Newark, NJ; Edison, NJ; Union, NJ; Wayne, NJ; and New Brunswick, NJ

[d]Includes Jerusalem District and parts of Judea and Samaria District

[e]Includes Los Angeles, Orange, Ventura, Riverside, and San Bernardino Counties. Not including 5,000 part-time residents

[f]Includes Haifa District and parts of Northern District

[g]Includes Miami-Dade, Broward, and Palm Beach Counties. Not including 69,275 part-time residents

[h]Includes Be'er Sheva Subdistrict and other parts of Southern District

[i]Our adjustment of original data. Includes the San Francisco area (San Francisco County, San Mateo County, Marin County, and Sonoma County), as well as Alameda County, Contra Costa County, and Silicon Valley. Assumes the San Francisco area currently comprises 60 % of the total Bay area Jewish population, the same as in the 1986 demographic study of that area

[j]Includes DC, Montgomery and Prince Georges Counties in Maryland, and Fairfax, Loudoun, and Prince William Counties in Virginia

[k]Includes North Shore

[l]Includes Clark County, DuPage County, and parts of Lake County

[m]Departments 75, 77, 78, 91, 92, 93, 94, 95

[n]Includes the Cherry Hill, NJ area

[o]Greater London and contiguous postcode areas

[p]Census Metropolitan Area

Table 70.5 Jewish population for the top 20 U.S. Metropolitan Statistical Areas, 2012

MSA Rank	MSA Name	Population Total[a]	Jewish	Percent Jewish (%)
1	New York-Northern New Jersey-Long Island, NY-NJ-PA	19,015,900	2,064,300	10.9
2	Los Angeles-Long Beach-Santa Ana, CA	12,944,801	617,480	4.8
3	Chicago-Joliet-Naperville, IL-IN-WI	9,504,753	294,280	3.1
4	Dallas-Fort Worth-Arlington, TX	6,526,548	55,005	0.8
5	Houston-Sugar Land-Baytown, TX	6,086,538	45,640	0.8
6	Philadelphia-Camden-Wilmington, PA-NJ-DE-MD	5,992,414	275,850	4.6
7	Washington-Arlington-Alexandria, DC-VA-MD-WV	5,703,948	217,390	3.8
8	Miami-Fort Lauderdale-Pompano Beach, FL	5,670,125	555,125	9.8
9	Atlanta-Sandy Springs-Marietta, GA	5,359,205	119,800	2.2
10	Boston-Cambridge-Quincy, MA-NH	4,591,112	251,360	5.5
11	San Francisco-Oakland-Fremont, CA	4,391,037	304,700	6.9
12	Riverside-San Bernardino-Ontario, CA	4,304,997	22,625	0.5
13	Detroit-Warren-Livonia, MI	4,285,832	67,000	1.6
14	Phoenix-Mesa-Glendale, AZ	4,262,236	82,900	1.9
15	Seattle-Tacoma-Bellevue, WA	3,500,026	39,700	1.1
16	Minneapolis-St. Paul-Bloomington, MN-WI	3,318,486	44,500	1.3
17	San Diego-Carlsbad-San Marcos, CA	3,140,069	89,000	2.8
18	Tampa-St. Petersburg-Clearwater, FL	2,824,724	58,350	2.1
19	St. Louis, MO-IL	2,817,355	54,200	1.9
20	Baltimore-Towson, MD	2,729,110	115,400	4.2
Total population in top 20 MSAs		116,969,216	5,298,730	4.5
Total U.S. population		311,591,917	6,721,680[b]	2.2
Percentage of population in top 20 MSAs[c]		37.5 %	78.8 %	

Source: Sheskin and Dashefsky (2013: 156)

[a]Total population is for 2011, Jewish population is for 2012

[b]Total Jewish population of 5,298,730 excludes 75,875 part-year residents who are included in MSAs 8, 12, and 18

[c]See www.census.gov/population/metro/files/lists/2009/List1.txt for a list of counties included in each MSA

U.S. Metropolitan Statistical Areas. (Note that Table 70.5 shows data for Metropolitan Statistical Areas while Table 70.4, to be more consistent with data from other countries, uses data for the Combined Metropolitan Statistical Areas.)

70.8.2 *The Intraurban Scale*

Some of the basic classical models of neighborhood change (Jordan et al. 1996; Boal 1978; for example) do not apply as well to Jews as they might to other groups. Arthur Hertzberg (1989) shows that between 1945 and 1965, one third of Jews left the city for the suburbs and spent at least one billion dollars on new synagogue buildings in the 1950s and 1960s. Thus, the need for capital expenditures noted at the national scale above, also applies at the intraurban scale. Moore (1997) suggests that the suburbanization of Jews, as is true for all Americans, was mostly due to a scarcity of adequate housing in the city as Jewish incomes increased post World War II, the modestly priced single family homes available in the suburbs with easy mortgage rates, and highway construction that allowed the working spouse to work in the city while living in the suburbs.

Nathan Glazer (1957) posited that Jewish suburbanization occurred in three phases since the end of World War I. At first Jews, mostly the Eastern European Jews who arrived via Ellis Island, lived in the areas of *first settlement* (in New York, the Lower East Side neighborhood) and many would define themselves as Orthodox. As the area of first settlement rapidly emptied, second generation Jews, mostly Conservative and Reform, with higher socioeconomic status than the first generation, moved to areas of *second settlement*, also called *gilded ghettos*. Finally, by the 1950s and 1960, Jews were moving to areas of third settlement in suburbs located further from the city. Thus, as one moved further from the city, generation in the U.S. increased, as did socioeconomic status, and assimilation into American society.

Sheskin (1993) notes that this suburbanization differed from some other ethnic groups in that Jews tended to move in large numbers to some suburbs, but not to others. (Recently, Li (2009) has noted a similar phenomenon for Asians suburbs in Los Angeles, which she called "ethnoburbs.") In the case of Jews, while restrictive covenants were mostly passé after World War II, real estate practices that restricted Jews from some neighborhoods were in effect. Sometimes non-Jews moved out of neighborhoods as large numbers of Jews moved in (*gentile flight*). Also, Jewish incomes were generally higher than non-Jews, limiting the neighborhoods into which Jews wanted to move. Finally, Jews wanted to live near other Jews and Jewish institutions, such as synagogues and Jewish schools.

Thus, even with suburbanization, Jews remain a spatially clustered population at the intraurban scale. American Jews are also a highly mobile population. In 2000, 35 % of adult Jews lived in a different house than in 1995 (Kotler-Berkowitz et al 2003). Thus, the changing intraurban distribution is due not just due to intraurban migration but to interurban migration, and international migration. For example, Sheskin (2010b) shows that most of the change in the geographic distribution of

Jews in Milwaukee and Detroit is due to interurban migration. Both intraurban and international migration has contributed to the changing spatial distribution of Jews in New York and the Twin Cities. Interurban migration explains the significant changes in Broward County (FL) and Las Vegas. All three migration types have combined to change the geography of Jews in Miami.

Finally, at the other end of the geographic scale, many small Jewish communities, particularly in the south, have disappeared due to changes in employment, the effect of anti-Semitism, and assimilation (Sheskin 2000).

70.8.3 The Changing Urban Profile in Israel and in Other Countries

Rapid Jewish population growth in Israel has resulted in the formation and development of a number of large metropolitan areas. The largest Jewish urban concentration in Israel, and eventually the largest worldwide, is the Tel Aviv-Yafo Metropolitan Area stretching from Netanya north in Israel's Central District to Ashdod south in the Southern District. The total population passed the three million mark in the first decade of the 2000s. The second largest metropolitan area is Jerusalem, which also includes portions of Palestinian Authority territory in the West Bank. Haifa and the Northern District is the third largest, and Beersheba and surrounding areas in the Southern District is fourth.

As noted above, immigration came in waves, each dominated by different countries of origin. Therefore, the early environments of newcomers tended to be quite homogeneous. Many new immigrants were directed to small rural localities (especially *moshav*-type cooperative villages), to newly built development towns, or to developing residential quarters in older cities where new housing blocks (*shikuním*) were built. While these policies produced reasonably quick and extensive solutions to impelling problems of housing in a situation of unusually rapid population growth, the resulting geographic and social distribution of Jewish population in Israel could well be described during the late 1960s as a mosaic of segregated groups (Klaff 1980).

Compared with the prevailing residential homogeneity of primary settlement boosted by immigration waves, powerful streams of internal geographical mobility operated to dilute residential gaps both between the different administrative divisions at the country level, and within individual localities. Social intermingling and integration occurred particularly in newly developed urban areas that attracted secondary movers who previously experienced geographical mobility and whose socioeconomic characteristics were compatible with the new neighborhoods. A more heterogeneous geographic origin profile developed within each residential area, while the differences between the various areas tended to diminish. At the same time, pockets of ethnic homogeneity persisted among the less mobile, many of whom also shared a lower socioeconomic status. At the same time, high residential segregation continued to prevail between Israeli Jews and Arabs (Schmelz et al. 1991).

Also in other countries, outside of the U.S. and Israel, the residential configuration of Jews in metropolitan areas has certain definite peculiarities worthy of noting. One trait frequently shared by Jewish communities in European countries is that the original location of the ghetto – whether legally or socially determined – was in the Eastern parts of the central city. Such was the case in London's East End, in Paris' Le Marais, or in Rome's Rione St. Angelo, as well as in several other major cities. Later generations tended to move out of the ghetto, resolutely choosing locations mostly to the southwest, west, northwest, or north of the city center. Such movements were of course also a function of a city's physical shape, as delimited by seashores, rivers, and hills. Some of these patterns will become better understood when considering that the value of real estate is related to environmental conditions. In the northern hemisphere, the jet stream tends to flow predominantly from the northwest to the southeast, thus creating a better and cleaner environment in the windward northwestern urban sections. The logic would be, therefore, that other things being equal, one would prefer a Northern-Western to a Southern-Eastern location, which in turn would generate higher land and housing costs in the northwest. In turn, access to real estate's cost is related to the socioeconomic characteristics of households. An additional factor related to physical geography could be the presence of rivers. Clean water would come into the city neighborhoods located upstream. The city would add industrial pollution and sewerage to the river, making home sites downstream far less desirable. That is why Jewish suburbs (and in general higher income suburbs) may often be found upstream, also enjoying the benefit of higher altitude (Fig. 70.7).

Upward mobility processes like those that quite massively characterized most Jewish communities in Western countries during the twentieth century could thus result in residential relocation of significant percentages of a given Jewish community. Indeed, Jewish communities in large cities like London, Paris, Berlin, Vienna,

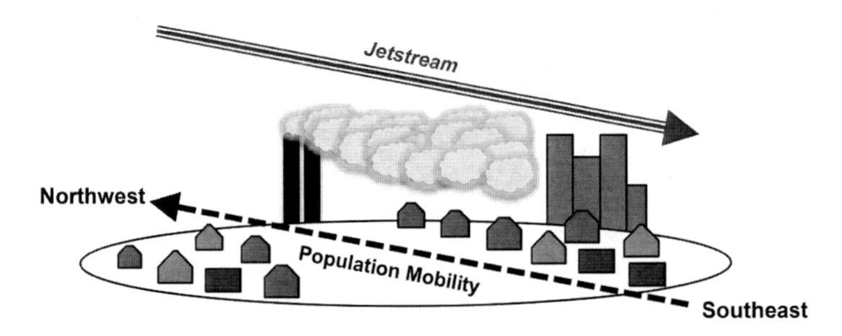

Fig. 70.7 Schematic representation of the possible relationship between physical environment and socioeconomic opportunities as a background to residential choices of Jews in major metropolitan areas in the Northern hemisphere (Source: Sergio DellaPergola and Ira M. Sheskin)

Rome, Milan, Manchester, but also Mexico City and Buenos Aires, all tended over time to move to a more western and/or northern configuration. Jewish urban settlement in several large North American cities followed similar patterns, for example, in Atlanta, Baltimore, Boston, Chicago, Detroit, Indianapolis, Los Angeles, Miami, Minneapolis, Milwaukee, St. Louis, Washington DC, Montreal, and Toronto.

Similar patterns were visible in major cities in North Africa and the Middle East inasmuch as upward social mobility of Jews was at work. In some Jewish communities in the southern hemisphere, like Sydney, São Paolo, or Caracas, the different wind patterns and land value differences prevailing there led to Jewish relocation patterns somewhat symmetric to those observed in the northern hemisphere.

In any case, the significant transformation of Jewish residential patterns implied that the location of institutional buildings and centers of collective activities sometimes remained behind the actual location of the Jewish community. Moreover, in spite of significant acculturation that occurred in the process of socioeconomic promotion, the residential density and concentration of Jews remained substantially high. Urban geography played in the past, and continues to play in the present, an important role in defining the profile of personal and community life among the Jewish population.

70.9 Conclusions

In the course of history, Jewish populations have undergone dramatic geographic transformations that have significantly changed not only the terms of reference and meaning of their relationship with non-Jewish societies, but have fundamentally affected the outlook of the Jews themselves, their existential opportunities, demography, socioeconomic development, security, and quite certainly beliefs. This can be said to be true both in long term historical perspective, in the course of the twentieth century, after the Shoah and the establishment of Israel, and more recently since the demise of the Soviet Union at the beginning of the 1990s. Such dramatic changes concerned movements at interchanges at all possible dimensions of the geographical ladder: across continents, across countries within the same continent, across regions within the same country, across all possible levels of urban hierarchy from peripheral village to capital city and world metropolitan area.

The trends described in this chapter are in no way random, but they reflect consolidated patterns and causal mechanisms that we have outlined: the inner needs of the religio-ethnic group investigated here, its interaction with a social environment on occasion hostile, and broader socioeconomic and cultural transformations shared among Jews and all others. The powerful population dynamics affecting world, regional and national Jewish communities are not set to stop suddenly. It can be postulated that further changes are ahead, following the broader transformative drivers that we have outlined. However, it would be presumptuous to suggest a clear prognosis for the future. Jewish history has taught the analyst to be cautious. Indeed, the geographer and the demographer that, say 100 years ago in 1912, would have ventured a projection of the future of the Jewish people, would not have been able to predict two world wars, the Holocaust of the European Jews, the rebirth of world Jewry thanks to the foundation of Israel, and its younger population. Nor would it have been possible to accurately predict changes in international migration policies and their dramatic consequences, the rise and fall of empires and nations, the exodus from the Islamic countries, the FSU, and Ethiopia, or the demographic consequences of acculturation and integration in the more hospitable societies. The main result of the trends of the past decades was the rise of the Jewish community of Israel to global prominence and primacy, and judging from the existing evidence this quantitative strengthening is expected to continue, while the prospects for Jewish communities in the rest of the world rather point to stability or decline.

References

Bachi, R. (1977). *The population of Israel*. Paris/Jerusalem: CICRED/The Hebrew University.

Barnavi, E. (Ed.). (1992). *A historical atlas of the Jewish people*. New York: Knopf.

Birmingham, S. (1967). *Our crowd, the greater Jewish families of New York*. New York: Harper & Row.

Birmingham, S. (1984). *"The rest of us:" the rise of America's Eastern European Jews*. Boston: Little, Brown.

Boal, F. W. (1978). Ethnic residential segregation. In D. T. Herbert & R. J. Johnston (Eds.), *Social areas in cities* (pp. 57–95). London: Wiley.

DellaPergola, S. (1983). *La trasformazione demografica della Diaspora ebraica*. Torino: Loescher.

DellaPergola, S. (1998). The global context of migration to Israel. In E. Leshem & J. Shuval (Eds.), *Immigration to Israel: Sociological perspectives* (pp. 51–92). New Brunswick/London: Transaction Books.

DellaPergola, S. (2005). Was it the demography: A reassessment of U.S. Jewish population estimates, 1945–2001. *Contemporary Jewry, 25*, 85–131.

DellaPergola, S. (2006). *Demography, encyclopedia Judaica* (Vol. 5, pp. 553–572). Jerusalem: Keter.

DellaPergola, S. (2007). "Sephardi and Oriental" migrations to Israel: Migration, social change and identification. In P. Medding (Ed.), *Studies in contemporary Jewry* (pp. 3–43). New York: Oxford University Press.

DellaPergola, S. (2009). International migration of Jews. In E. Ben-Rafel & Y. Sternberg (Eds.), *Transnationalism: Diasporas and the advent of a new (dis)order* (pp. 213–236). Leiden/Boston: Brill.

DellaPergola, S. (2010). World Jewish population. In *Current Jewish Population Reports, Number 2–2010*. Storrs: Mandell Berman Institute, North American Jewish Data Bank.

DellaPergola, S. (2012). How many Jews in the U.S.? *Contemporary Jewry, 33*, 1–2, 15–42.

DellaPergola, S. (2013). World Jewish Population, 2012. In A. Dashefsky & I. M. Sheskin (Eds.), *The American Jewish year book*. Dordrecht: Springer.

Dimont, M. I. (1978). *The Jews in America, the roots, history, and destiny of American Jews*. New York: Simon and Schuster.

Friedlander, D., & Goldscheider, C. (1979). *The population of Israel*. New York: Columbia University Press.

Friesel, E. (1990). *Atlas of modern Jewish history*. New York: Oxford University Press.

Glazer, N. (1957). *American Judaism*. Chicago: University of Chicago Press.

Hertzberg, A. (1989). *The Jews in America, four centuries of an uneasy encounter*. New York: Simon and Schuster.

Howe, I. (1976). *World of our fathers, the journey of East European Jews to America and the life they found and made*. New York: Galahad Books.
Israel, Central Bureau of Statistics. (2011). *Statistical Abstract of Israel*, table 2–16. Jerusalem.
Jordan, T. G., Domosh, M., & Rowntree, L. (1996). *The human mosaic. A thematic introduction to cultural geography*. New York: Addison Wesley Publishing Company.
Klaff, V. Z. (1980). Residence and integration in Israel: A mosaic of integrated groups. In E. Krausz (Ed.), *Studies of Israeli society: Migration, ethnicity and community* (pp. 53–71). New Brunswick: Transaction.
Kosmin, B. A., & Keysar, A. (2009). *American religious identification survey (ARIS 2008)*. Hartford: Trinity College.
Kosmin, B. A., Goldstein, S., Waksberg, J., Lerer, N., Keysar, A., & Scheckner, J. (1991). *Highlights of the CJF 1990 National Jewish Population Survey*. New York: Council of Jewish Federations.
Kotler-Berkowitz, L., et al. (2003). *Strength, challenge and diversity in the American Jewish population*. New York: United Jewish Communities.
Li, W. (2009). *Ethnoburb: The new ethnic community in urban America*. Honolulu: University of Hawaii Press.
Liebman, S. (1977). Cuban Jewish community in south Florida. In A. D. Lavender (Ed.), *A coat of many colors, Jewish subcommunities in the United States* (pp. 296–304). Westport: Greenwood Press.
Marcus, J. R. (1990). *To count a people: American Jewish population data 1585–1984*. Lanham: University Press of America.
Mayer, E., Kosmin, B. A., & Keysar, A. (2001). *American religious identification survey*. New York: The Center for Jewish Studies: The Graduate School and University Center, City University of New York (CUNY).
Moore, D. D. (1997). Jewish migration in postwar America: The case of Miami and Los Angeles. In J. D. Sarna (Ed.), *The American Jewish experience* (pp. 314–327). New York: Holmes and Meier.
Norland, J., & Freedman, H. (1977). Jewish demographic studies in the context of the census of Canada. In U. O. Schmelz, P. Glikson, & S. DellaPergola (Eds.), *Jewish Population Studies 1973 (Papers in Jewish Demography)* (pp. 59–78). Jerusalem: The Hebrew University, The Avraham Harman Institute of Contemporary Jewry.
Pasachoff, N., & Littman, R. J. (1995). *Jewish history in 100 nutshells*. Northvale: Jason Aronson.
Pew Research Center. (2013). *A portrait of Jewish Americans*. Washiington, DC: Pew Research Center at www.pewforum.org
Rosenwaike, I. (1980). A synthetic estimate of American Jewish population movement over the last three decades. In U. O. Schmelz & S. DellaPergola (Eds.), *Papers in Jewish Demography 1977* (pp. 83–102). Jerusalem: The Hebrew University, The Institute of Contemporary Jewry.
Sachar, H. M. (1992). *A history of the Jews in America*. New York: Knopf.
Sanders, R. (1988). *Shores of refuge. A hundred years of Jewish immigration*. New York: Henry Holt.
Schmelz, U. O., DellaPergola, S., & Avner, U. (1991). *Ethnic differences among Israeli Jews: A new look*. Jerusalem/New York: The Hebrew University/The American Jewish Committee.
Shamir, I., & Shavit, S. (1986). *Encyclopedia of Jewish history*. New York: Facts on File Publications.
Shapiro, E. S. (1992). *A time for healing. American Jewry since World War II*. Baltimore: John Hopkins University Press.
Sherman, C. B. (1965). *The Jew within American society*. Detroit: Wayne State University Press.
Sheskin, I. M. (1993). Jewish ethnic homelands in the United States. *Journal of Cultural Geography, 13*(2), 119–132.
Sheskin, I. M. (2000). The dixie diaspora: The 'loss' of the small southern Jewish community. *The Southeastern Geographer, 40*(1), 52–74.
Sheskin, I. M. (2001). *How Jewish communities differ: Variations in the findings of local Jewish demographic studies*. New York: City University of New York, North American Jewish Data Bank.

Sheskin, I. M. (2005). *The greater Miami Jewish community study*. Miami: The Greater Miami Jewish Federation.

Sheskin, I. M. (2006). *The Jewish community study of Palm Beach County*. West Palm Beach: The Jewish Federation of Palm Beach County.

Sheskin, I. M. (2010a). *Jewish Israelis in the United States*. Paper presented at the International Geographic Union Regional Conference, Tel Aviv.

Sheskin, I. M. (2010b) *The suburbanization of American Jewry*. Paper presented at the Race, Ethnicity and Place V Conference, Binghamton, NY.

Sheskin, I. M., & Dashefsky, A. (2006). Jewish population in the United States, 2006. In D. Singer & L. Grossman (Eds.), *American Jewish year book* (Vol. 106, pp. 134–139). New York: American Jewish Committee.

Sheskin, I. M., & Dashefsky, A. (2012). Jewish population in the United States, 2012. In A. Dashefsky & I. Sheskin (Eds.), *American Jewish year book* (Vol. 106, pp. 134–139). New York: American Jewish Committee.

Sheskin, I., & Dashefsky, A. (2013). A. Jewish population in the United States, 2012. In A. Dashefsky & I. M. Sheskin (Eds.), *The American Jewish year book*. Dordrecht: Springer.

Tighe, E., et al. (2013). *American Jewish population estimates: 2012*. Waltham: Steinhardt Social Research Institute.

Tobin, H. (1984). The myth of community in the Sunbelt. *Journal of Jewish Communal Service, 61*(1), 44–49.

United Nations Development Programme. (2011). *Human development report 2011*. New York: Palgrave-McMillan.

United States, Executive Office of the President, Office of Management and Budget. (2008). *Update of statistical area definitions and guidance on their uses* (OMB Bulletin 09-91). Washington, DC: Office of Management and Budget.

Waksberg, J. (1978). Sampling methods for random digit dialing. *Journal of the American Statistical Association, 73*(361), 40–46.

Wyman, D. (1984). *The abandonment of the Jews: America and the Holocaust, 1941–1945*. New York: Pantheon Books.

Chapter 71
The Narration of Space: Diaspora Church as a Comfort Zone in the Resettlement Process for Post-communist Bulgarians in Toronto

Mariana Mastagar

> *The major stirrings and creative drives of revolutionary events have totality impact. For a brief moment they bring the elements of totality together.... (Lefebvre, 1991: 4)*

71.1 Introduction

The immigration and resettlement phenomenon is a private revolution and disruption of the flow of personal history since tradition and identity are intrinsically linked with location. The habitual emotional and geographical boundaries of home, customs, and embedded values are upset. Diasporic life is a creative process of building new personal and social space and forming new meanings to reharmonize the fragments. Communal space with its relativity and reflectivity offers layered connectedness for the individual and the group in their adaptive journeys.

This study investigates the problematic area of reconstruction and redefinition of religious space and what implications they pose at the personal and communal levels in the diasporic context. The perplexity intensifies by the fact that the post-communist wave of immigrants to Toronto are raised in a non-religious environment, yet include the religious institution as a part of their adaptive space. It evokes a dilemma: how does a conservative tradition such as Eastern Orthodoxy meet with religiously uncommitted settlers and negotiate space? Does the space of a diasporic church function truly as a faith-nurturing institution and, hence, a reconstructive image of traditional religious space? Or, is it a locality that recollects ethnicity and historical rootedness and "reinforces a sense of belonging and community" (Georgiou 2006: 107).

My study focuses on the diasporic church as an organization of physical and social space—architectural and interior design, functions, and activities—to observe how the religious institution accommodates a life of resettlement with its ecclesiastical and secular streams. Comparison with the native Eastern Orthodox Church clarifies

M. Mastagar (✉)
Department of Theology, Trinity College, University of Toronto,
6 Hoskin Ave., Toronto, Canada
e-mail: mariana.dobreva@mail.utoronto.ca

S.D. Brunn (ed.), *The Changing World Religion Map*,
DOI 10.1007/978-94-017-9376-6_71

the role and meaning of the religious institution in the adopted homeland and its new socioeconomic realities. This study is a step towards my larger investigation of south Slavonic diasporas in Canada, looking closely at the post-1990-Bulgarian community in Toronto.

Methodologically, data gathering and a theoretical frame are employed in the study. The research relies on the standard anthropological fieldwork practices of archival records, participant observation, and interviews. The dataset includes a random selection of 23 lay persons from the post-communist wave of immigrants and seven priests who were approached for interviews. The site location consists of four diasporic/ethnic churches in the Greater Toronto area, established between 1910 and 2004 (Fig. 71.1). Three of the churches are designated as "Macedono-Bulgarian Eastern Orthodox" and attendees identify themselves either as Bulgarians, Bulgarians from Macedonian areas, Macedono-Bulgarians, or Macedonians.[1] The newest church (established in 2004) is designated as "Bulgarian Eastern Orthodox" and attended mainly by the post-communist wave of settlers from Bulgaria.

The interdisciplinary theoretical framework involves the conceptual lenses of Eastern Orthodox theology, religion, the sacred, space, diaspora, and identity. Specifically, I presuppose the understanding of Orthodoxy and art (Ouspensky and Lossky 1982; Bulgakov 1988; Meyendorff 1974) and the notion of space as a social construct (Lefebvre 1991). Implicitly included are the concepts of sacred and profane space (Eliade 1959) diaspora (Cohen 1997), and debates about secularization (Bruce 2002; Norris and Inglehart 2004).

The focus group of 23 interviewees are post-communist immigrants from Bulgaria between the ages of 35 and 55. In Bulgaria 79 % of the population identify themselves as Christians but only 8 % are habitual churchgoers, with 0.8 % active and 2 % non-active members of religious organizations and churches.[2] The level of religious practice and commitment to the church is low as it is in the most secularized Western European societies (Norris and Inglehart 2004). The diasporic focus group showed a marked similarity: 80 % did not have a personal connection with religion in their homeland; 40 % had never been to a church; 50 % had attended church occasionally; and 40 % are firm atheists. Soon after arriving in Toronto all began attending church locations relatively frequently. After the second year, 30 % ceased doing so. To provide a benchmark for the discussion that follows I begin by

[1] Though the study group came from Bulgaria and identified themselves as Bulgarians, the diaspora as a whole is Macedono-Bulgarian and consists today of a mix of identity factions. The territorial redistributions of Bulgaria and Balkans resulted in a massive emigration flow from Macedonian areas with two main resettlement points—Bulgaria and North America. This redistributions were a consequence of the Russian-Turkish war (1877–1878) and the liberation of Bulgaria from the Ottoman rule (1878), the Treaty of San Stefano (1878) followed by the Treaty of Berlin (1878), the Unification of Bulgaria (1885), the Independence of Bulgaria (1908), Balkan Wars (1912–1913), and WWI. The immigrants established the first three churches in Toronto (1910, 1941, and 1973), attended today by their children. The Macedonian Orthodox Church was granted autonomy in 1959 by the Serbian Orthodox Church, and the first Macedonian Church in Toronto was established in 1965. Those identifying themselves only as Macedonians formed a separate diaspora group.

[2] EURO 2000 and BBSS Gallup and Balkan British Social Surveys, as cited in Kanev 2002.

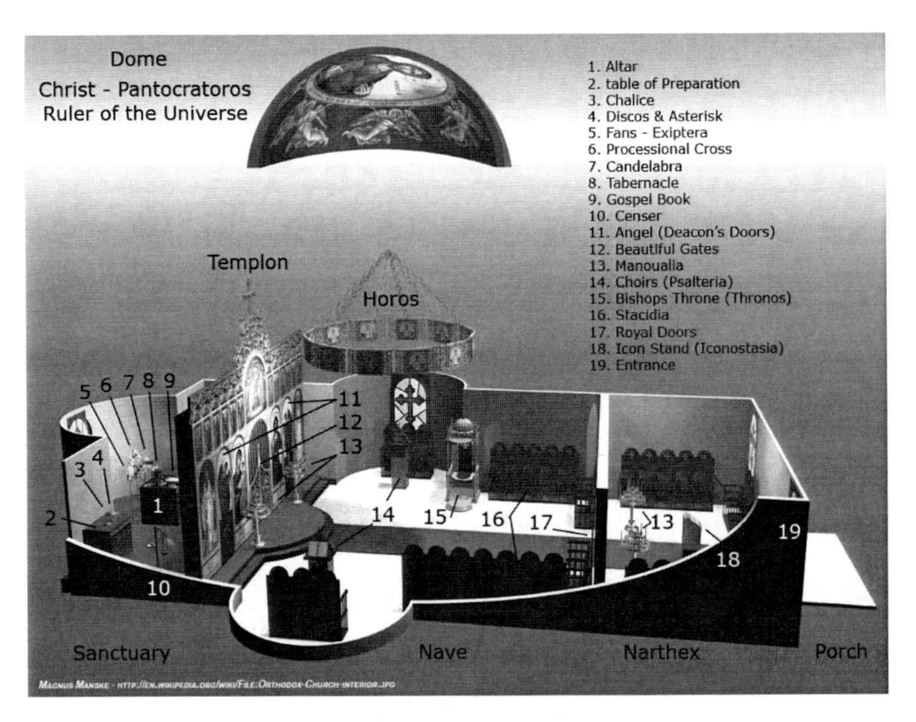

Fig. 71.1 Interior organization of the church space (Photo by Phiddipus, http://en.wikipedia.org/wiki/File:Orthodox-Church-interior.jpg)

reviewing the physical, theological, and functional parameters of space in a Traditional Bulgarian Eastern orthodox church.

71.2 Sacred Space in an Eastern Orthodox Church in Bulgaria

Space—buildings, iconography, and rituals—is a fundamental dimension of Eastern Orthodoxy. While adhering to the belief that "church" is a signifier of people and less of buildings, the Orthodox tradition has developed deep respect and intrinsic attachment to the church space as a home of God and place for his glorification.[3] Thus, space, structure, decoration, and function are configured to inculcate feelings of piety, devotion, awe at transcendence, and a sense of the overwhelming presence of the divine. The entire visual ensemble is designed to convey the believer to the divine

[3]The idea of the church as a sacred space in opposition to the profane was developed by Pseudo-Dionysius the Areopagite, fifth to sixth century CE (see The Ecclesiastical Hierarchy, 1955). "Church" in the first centuries is referred to by the collection of believers as the "temple of the Holy Spirit" (Acts 7:48, 1 Cor 3:16, 6:19; 2Cor 6:16).

threshold. The language of architecture and art is a theologically symbolic narrative (Bulgakov 1988; Meyendorff 1974). It is a complex visualization of the ideas of eternity, time, divinity, good, and evil that educate and elevate spiritually. Icons and mural paintings are a visual credo and confession of faith, not merely decorative elements in the church (Meyendorff 1974; Ouspensky and Lossky 1982; Studite 1981). The creation of church ornamentation requires artistic talent as well as theological understanding of the dogmas every image conveys—the saints' lives, biblical plots, and the history of the local church. Byzantine artistic style crystallized after the iconoclastic period of eighth century and by the thirteenth century it was fully developed and unified as a dogma (Goshev 1952).[4] It remains unchanged today. The form and content of space is controlled by an imagination of the divine hierarchy rather than a response to the needs and sensibilities of the attendees. In this sense, space provides a template for ordering and schooling the emotions of the faithful.

The physicality of the Eastern Orthodox Church is organized and defined by four spatial elements: narthex, nave, iconostasis, and altar (Fig. 71.2). Each relates to the liturgical order and reflects the tabernacle and the organization of the Solomonic

Fig. 71.2 Iconostasis of the Nativity of Christ Church, Arbanasi, Bulgaria, sixteenth to seventeenth century (Photo by Svilen Enev, http://en.wikipedia.org/wiki/File:Arbanasi_Architectural_Preserve.jpg)

[4]The rule for depicting and placing icons and paintings is called Ermenia (from Greek ἑρμηνεία, ἑρμηνεύω, 'to interpret').

Temple. I will describe each of these architectural features briefly before comparing and contrasting them with their diasporic contexts.

The *narthex*, the vestibule, serves as transitional space between the secular and profane worlds. Originally in the early years of Christianity, it was used for the catechumens, penitents, and non-Christians. Liturgically, the space is used in some services, such as the Little Hours during Holy Week and parts of the wedding ceremony. Over the narthex is the usual place for a bell tower. The narthex decoration includes images of the church founder(s) or donor(s) and often the Day of Judgment, which serves to teach, recollect, and heighten one's consciousness of transcendence.

The *nave* is the gathering area symbolizing the earthly. It forms for attendees a praying cosmos through spatial arrangement, rituals, and decoration. In contrast to Western churches, the attendees stand; there are no pews and kneelers because standing is the proper posture before God. Only a few chairs, called *stasidia*, are positioned against the walls and used by people with physical conditions or the very elderly. Those present are free to stroll around at any point in the service. Usually engulfed in their own prayer, they watch the liturgical performance or meditate before icons. The worship may appear unstructured to outsiders, but seemingly passive participants are, in fact, in a self-absorbed meditative journey to build personal bridges to the spiritual. Rarely do attendees sing along or pray in liturgical unison. Priests and deacons, located in front of the iconostasis, go in and out of the altar area and pray, mostly facing the altar. Hence there is the inkling that individual religiosity takes precedence over congregational expression.

The nave's iconographic decoration with its overwhelming abundance of images, serves both veneration and the goal of religious education. The images reflect themes of the Christian Bible, including Christ's birth, Presentation,[5] baptism, transfiguration, Lazarus's resurrection, the descent to hell, the entrance into Jerusalem, the Assumption, the Holy Spirit's descent, etc. Depictions of local saints and historical church events are positioned in designated places. The dome, symbolizing heaven, is decorated with Christ Pantokrator (*Vsederjitel*) surrounded by angels and symbols of the four Evangelists at the base.

Other commemorative decorations include stands with icons of Mother Mary, Christ, or a popular saint. Tall brass candelabras and sand boxes hold candles lit for living and deceased loved ones. Some churches keep relics of saints in a special feretory (reliquary) displayed in either the nave or a basement room for public veneration. Veneration of relics is reminiscent of the gatherings around the tombs of martyrs in early Christianity.

The *iconostasis,* or *templon*, visually divides the nave and altar, the earthly and the heavenly. Symbolically, it is a transitional demarcation between the two. A wooden frame, elaborately carved and sometimes gilded, serves as more than just an icon holder. It has three doors and up to six rows (*yaruses*) of images. For Bulgaria, the importance of the iconostasis frame is underscored further by a tradition of carving that resulted in three distinct styles by the nineteenth century. Historically the altar was separated by a low row of columns with stone plates

[5] On the fortieth day Mary took her son Jesus to the temple, Luke 2:22–40.

between them, but following the Second Nicean Council's endorsement of icons as visual dogma of faith, the iconostasis started to add new rows.[6] Icons are organized as follows. The first row has depictions of the Jewish Bible as a foundation of the Christianity. The next, main row, called the royal or *tsarski*, has the largest images (usually a half life-size torso). At the center of the third, or festive, row is the Last Supper scene with narratives of events from the lives of Mother Mary and Jesus. Apostles, saints, prophets, and patriarchs, are depicted on the remaining rows. The second row is mandatory. Its center is the door known as the Beautiful Gate; to its right are Christ and John the Baptist, and to its left are Mother Mary and the saint for whom the church is named. The two side doors are adorned with icons of the archangels Gabriel and Michael or the first martyr, Stephan. The rest of the images on this row are optional and depend on the popularity of certain saints in the area. For example, admired icons are that of St. Nikolas, St. Mina, and St. George.

The *altar*, the holiest place in the church and symbol of heaven, is where the clergy are secluded for the sacred performance of the *proscomedia* (preparation) and eucharistic transformation of the bread and wine. Here are kept the sacred utensils, books, and garments. Every church must have a miniature particle of a saint's relic embedded in the altar table to sanctify the place and reinforce spiritual strength.[7] As the most sacred area of a church, only priests, deacons, and male persons with blessed permission have access to it.[8] The altar's iconographic themes are closely related to the Eucharist: Mother Mary *Oranta* (hands raised, with Jesus shown in a circle in her womb) or *Odigitria* (holding the child Jesus), Christ taking communion or administrating Holy Communion with the apostles, and sometimes Jewish Bible themes such as the sacrifice of Isaac or Jacob's ladder. Attendees catch a glimpse of the altar a few times during the liturgy when the main door, the Beautiful Gate, is open.

Church architecture has three typical styles: cruciform, elongated rectangular, and less frequently, a rotunda or multi-angled shapes (for example, an octagon).

[6]The iconoclasts anathematized the images and sculptures on the basis of the Decalogue, rejecting every likeness of images made from any material. The second council in Nicea in 787 endorsed the usage of images: "The more frequently they are seen in representation of art, the more are those who see them drawn to remember and long for those who serve as models, and to pay these images the tribute of salutation and respectful veneration. Certainly this is not the full adoration in accordance with faith, which is properly paid only to the divine nature, but it resembles that given to the figure of the honored and life-giving cross, and also to the holy books of the gospels to other sacred cult objects" (Tanner, et al. 1990:132–136).

[7]The holiness of the saint is considered to be what sanctifies the place. It comes from the early tradition when Christians gathered around the tombs of saints, to connect the people with the heroism of the saint, to reinforce the spiritual strength, and to sanctify the believers. In some monasteries and churches the bones of deceases monks or priests are stored in a small space under the altar. Such churches are Boianska (twelfth century), Asenova citadel church (ninth century), and the Alexander Nevski cathedral in Sofia (Goshev 1952).

[8]Women are prohibited from entry into the altar except by special permission for the housecleaning lady, who must be a post-menopausal widow.

Buildings are conservative and it is rather an exception to have a modern design.[9] The preservation of tradition takes priority, for in Bulgaria a new church project (architectural and interior design) would first need to be approved by a censorship committee functioning under the Holy Synod and the Church's historical and archeological museum. A few church buildings include a small narrow adjacent room used for quick distribution of ritual food after funerals or memorials. In sharp contrast with diaspora churches, there is no space for social activities. Interior and architectural designs in the diaspora are a result of the combined decision of the board members, donors, and priests, with participation of the community. Indeed, the preservation of the tradition is a concern. But in no way limiting. In the homeland building of a church is a spiritual act that starts with spiritual devotion; in the diaspora, it is a communal act that starts with fostering care of the community and the building is a communal asset.

Windows are of great architectural significance for the interior of Eastern Orthodox churches. Few, small, and evenly spread, they are positioned high so that the walls remain free for icons, which are considered windows into the spiritual realm. Daylight filtering in from above, flickering candelabras, and lit candles create *sfumato*, or smokey effects that soften the lines, thus forming an encapsulated spiritual space into which attendees withdraw from the outside reality.[10] All elements of the space, including two-dimensional images, incense, and chanting lead to a visual—and also emotional—*sfumato*.

The social institution of the Bulgarian Orthodox Church has been unquestionably respected as a formative and sustainable factor in the 1,300 years of Bulgarian national history. Today, however, its role and influence in public and personal domains is marginal. Yet, what remains popular is the appreciation of the churches and monasteries as historical sites and monuments of art. All interviewees shared this appreciation, and one reported that "going to a church was like going to a museum." The interviewees were not familiar with the concept of Church as communal life. The shortest answer received was, "Community? No." Diaspora priests confirmed that in the homeland the church functioned mainly to celebrate the divine elements and reinforce the religious commitments of the people; their parishes had no allocated space for social activities.

In summary, Orthodoxy is a tradition of beauty—in architecture, garments, images, lyrics, music, and rituals—beauty, not as a superficial earthly goal but as a sign of divine infiltration into the world. As Gochev notes, the "Eastern Orthodox church, as a sacred building is intended for prayer gatherings of believers, and is suited to certain conditions connected with the liturgical purpose of the temple … an artistic building devoted to God, comfortable for services, delighting the eye of the worshipers and ascending their souls" (1952: 320). Accommodating the daily

[9]The Greek Orthodox Church in Milwaukee, Wisconsin, contemporary design by Frank Lloyd Wright and Vanga's church in Petrich, Bulgaria, decorated with modern unorthodox frescos.

[10]Sfumato, a renaissance technique of painting referred mostly to DaVinci, is a style of drawing without clear lines that creates an unfocused smokey effect.

life-needs of people or starting where believers are at, is clearly not a priority for the Bulgarian Eastern Orthodox church.

71.3 Diaspora Churches in Toronto

Diasporic churches stand in sharp contrast with homeland churches. Entering a diaspora church in multicultural Toronto is an exciting experience of familiarity and discovery where sameness is no longer the same. Collective identity factors such as language, home, encounters, politics, culture, and buildings, are replaced with unfamiliarity (Shneer and Aviv 2010). Larger and older diasporic communities (such as Greek and Italian) have, in addition to churches, several public meeting and mediating places: schools, cultural centers, clubs, TV and radio programs, neighborhood cafes, and their members live in areas with high concentrations of the same ethnic population (Georgiou 2006). Those places reconfirm and reconstruct the lost sense of hominess, thereby forming a new responsive space to compensate in a layered way for the loss of collective identity factors.

Relatively smaller and/or newer communities on the diasporic map, however, often have only one point of familiarity and togetherness—their religious institution. Such is the case of the Macedono-Bulgarian group. Church, community, and individuals all reflect the disruption associated with their relocation. While the church credo remains unchanged, the new locality promotes development towards a close interrelationship between the institution and its members, greatly enriched in comparison with the unitary orientation of the native church.

Hence, I turn now to offer an account of the diasporic church building space: its configuration, decoration, and functional facets, along with the ways that clergy and lay individuals understand this space.

71.4 Buildings, Decoration and Halls

In Greater Toronto area there are four diasporic churches which are under the jurisdiction of the Bulgarian synod. Sts. Cyril and Methody and St. Dimitar reflect the traditional style (Figs. 71.3 and 71.4). The former was built in 1948 and is considered the largest cathedral in the dioceses abroad, and the latter, in 2004. Both St. George (built in 1941) and Holy Trinity (established in 1973) are eclectic in style, bearing similarities to Protestant churches (Figs. 71.5 and 71.6). The four buildings possess the spatial elements of the traditional organization discussed. Yet, two new structural elements that appear in all four are enlarged windows and community halls. Decorations mark the space without overwhelming it.

The narthexes of the churches welcome attendees with a mixture of icons and historical pictures, whereas a traditional church would have icons and murals recognizing its donors. In Sts. Cyril and Methody and St. George, black and white photographs

Fig. 71.3 Sts. Cyril and Methody Church, Toronto (Photo by Nina Stratieva Kaloferova, used with permission)

Fig. 71.4 St. Dimitar Church, Brampton (Photo by Nina Stratieva Kaloferova, used with permission)

of all priests who have served are displayed. Instead of *The Day of Judgment*, those portraits witness continuity and stability of the religious organization in a respectful, contemporary manner. As well, they recognize the role of every clergyman who helped sustain the diasporic institution.

Fig. 71.5 St. George Church, Toronto (Photo by Nina Stratieva Kaloferova, used with permission)

The naves have an ambience that is markedly different. An infiltration of light and sense of bodily comfort are noticeable. Tall windows cover most of the naves' wall surface. Dim light with the *sfumato* effect is replaced by abundant light, which suggests spatial connectedness with the outside realm (Figs. 71.7 and 71.8). The pale stained or tinted glass windows have no images. Benches and kneelers also structure the space. People sitting next to each other convey a visual impression of community. As mentioned above, traditional space accommodates spiritual necessities and seems to dismiss the physical comfort. Benches and kneelers can also be observed in Russian and Greek Orthodox churches in Toronto, known to uphold their religious tradition. For example, St. Nikolas Greek Orthodox church, which indeed creates a "piece of heaven on earth," as its website announces, has benches. As Henri Lefebvre comments, "the adoption of another people's gods always entails the adoption of their space and system of measurements" (1991: 111). In our case, the reverse is true as well; the adoption of a new home space may not change the gods/deities, but certainly gives a new perspective in approaching them.

Fig. 71.6 Holy Trinity Church, Toronto (Photo by Nina Stratieva Kaloferova, used with permission)

Fig. 71.7 Interior of St. Dimitar Church, Brampton (Photo by Nina Stratieva Kaloferova, used with permission)

Fig. 71.8 Interior of St. George Church, south wall. The first four icons are of Sts. Sofronii Vrachanski, Patriarh Evtimii Tarnovski, Theophylact Ohridski, and Ivan Rilski (Photo by Nina Stratieva Kaloferova, used with permission)

The iconostasis frames are plain and hold one to three rows of canonically-correct icons. An exception is the new church, St. Dimitar, which has a wood-carved iconostasis. St. Dimitar and St. George churches exhibit collections of icons, suiting both canonic and aesthetic requirements. The icons of the other two churches are of modest artfulness. The priests agree that a minimalistic approach has been applied by arranging the decoration with only the most necessary elements. One of them noted: "Visual means do not have influence the way it is in Bulgaria. The altar rather marks the space and is not an exuberant expression of veneration." Interestingly, priests face the attendees considerably longer during liturgy, spending less time looking toward the iconostasis.

The altars preserve their holy status and architectural elements, but are sparsely decorated and furnished. Every altar has the required saint's relic. The utensils, one priest pointed out, are basic and have not been changed since the 1950s, which reaffirms the practical approach towards the functionality of the altar space and sacred space in general.

On the whole, the decorations and their selection serve as simple reminders rather than conveying the attendees closer to the divine threshold. St. Dimitar has on its dome mural paintings of Christ Pantocrator and the 12 apostles. Sts. Cyril and Methody church has three full-length wall-engraved paintings of Saints John, Paul, and Peter. St. George church exhibits an excellent collection of 17 gilded icons. The nave of Holy Trinity shows a small number of framed pictures of icons. Usually the inscriptions on the icons are in Old Church Slavonic language, but in diasporic churches inscriptions are either in Old Church Slavonic, modern Bulgarian or English with Old Church Slavonic script. However, the sparse decoration makes the image selection a fascinating element. A case in point is St. George's collection of icons. It presents canonized persons who are deeply connected with turning points in Bulgarian history from about the ninth century onwards, including St. Equal to Apostles Cyril, St. Equal to Apostles Methody, Saints Kliment Ohridski, Naum Ohridski, Ioan Kukuzel, Georgi Sofiiski, Pimen Zografski, Paisii Hilendarski, Boyan Enravota, Tsar Boris, Tsar Perter, Ivan Rilski, Ilarion Muglenski, Theophylact Ohridski, Patriarch Evtimii Tarnovski, and Sofronii Vrachanski. The only biblical image among them is Christ the Great Bishop. Holy Trinity church follows a similar pattern, having more than half of these motifs. Indeed, the images represent honored persons of faith. Yet the narrative conveys a succinct historically charged review, tracing critical religious, cultural, and political developments. Every one of the names is familiar to the attendees. This "wall of fame" inspires national pride and reinforces ethnic belonging. The collection of saints acts as a theophoric guardian of ethno-religious roots.

Halls are an architectural intervention in the conceptualization of traditional church space that functions in the North American context as centers for the gravitation of community. Three of the churches have their banquet halls developed on the basement level and are indirectly connected with the liturgical area. Sts. Cyril and Methody, the oldest of the four churches, has its banquet hall situated on the same level as the sanctuary, and is directly connected to the church nave by two doors. The access between both spaces is natural and convenient. Doors are open except during Sunday services. The hall, with its 650 sq. m. (7,000 sq. ft.), high ceilings, sizable stage, kitchen, bar, and sound station attracts a variety of visitors to its desirable downtown location. The community gathers here for celebrations, concerts, and meetings. Two church halls are decorated with pictures of freedom-fighters for Macedonian areas at the turn of the twentieth century, as well as pictures of the community since the mid-1950s. The decoration of the new church has a modern and informative approach. The corridors leading to the hall are a gallery with framed pictures, posters, and portraits of prominent cultural figures, archeological monuments from the Thracian period and the medieval Bulgarian Empire, a history-at-a-glance chart, folk customs and costumes, and traditional foods.

71.5 Use of Church Space

The diasporic church building functions as an instrumental communal meeting and mediational space. Space, particularly social space, is never indifferent and/or independent. Physical space has no reality "without the energy that is deployed within it" (Lefebvre 1991: 13), but contains "a great diversity of objects, both natural and social, including networks and pathways which facilitate the exchange of material things and information. Such ‘objects’ are thus not only things but also relations” (1991: 77). Every action performed within the diaspora church changes its spatial characteristics. While in homeland the Eastern Orthodox religious space focuses on the transcendental relation, diasporic churches consider communal relationships to be within its realm.

Based on observations and email records spanning 7 years, my study analyzes some of the activities that have occurred. Aside from the Sunday service, the church buildings accommodate different types of gatherings. The church of Sts. Cyril and Methody organizes Sunday school, celebration of the patron’s day, the day of the Slavic Alphabet, and co-hosts a New Year’s Eve party. In 2011 it held three concerts of popular folk music. Holy Trinity supports a folklore customs study group for children and adults. St. George is strictly focused on the religious calendar of events, but in 2010 introduced matinee classical music concerts. By way of comparison, the events list of the new church, St. Dimitar, shows striking diversity. The 2009 year list includes a grandmother’s day celebration, Valentine-Trifon Zarezan day, March 3—the national day of Bulgaria, native movie nights with catechism course, the first day of spring, May 24th—the day of the Slavic alphabet, the formation of a choir, a piano recital, a children’s talent competition, charity golf tournaments, Carabram—Brampton’s multinational festival, annual Ilinden picnic, a tennis tournament, Sunday school, and a New Year’s Party.

The remaining three downtown churches, however, cannot be called quiet—their halls host a significant part of the cultural and social events happening in the community. Table 71.1 summarizes the number of events that occurred between 2005 and 2011, organized only by the community members and groups. Those organized by the churches or by individuals and associations outside of the community are not included. The table shows that a total of 98 secular events were organized in a 7-year period, 11 of which were outdoor activities. Hence, of 87 indoor gatherings, 53 (61 %) were held in church halls. One priest disappointedly referred to the church building space as a social club where “halls have become the altar.” Another priest expressed a more balanced opinion that “while to me the spiritual aspect is most important, for people it is everything in the church—faith, language, exchange of information, concerts, and parties. The new immigrants need a place to acclimatize

Table 71.1 Events organized by Macedono-Bulgarian community members and groups

Year	Events held in all venues	Events held in church halls	St. Cyril & Methody	St. George	Holy Trinity	St. Dimitar
2005	8	2	1		1	
2006	4	3	1		2	
2007	13	6	3			3
2008	12	7	5		1	1
2009	15	10	7		2	1
2010	25	13	8		2	3
2011	21	12	6	3	1	2
Total	98	53	31	3	9	10

Source: Mariana Mastagar

and to integrate easily. Nostalgia also … but there is a spark that not everything is corporeal." The clergy still consider the church to be 'God's' home. They admit, nevertheless, that "church here is more worldly minded" and "a place for consolation, healing, hope, nostalgia, and homesickness."

A jazz concert in Sts. Cyril and Methody banquet hall conveyed a sense of confluence between the sacred church space and the gathering zone. People were seated at tables. A buffet with national dishes was situated by the open nave door closer to the iconostasis. Through the door one could see the candelabras with lit candles, and a few people praying. The second open door allowed access to the church area. The jazz concert concluded with horo—a national folklore dance—and many people joined the dance. A newcomer commented in surprise that "one has everything [from home] here—jazz and horo, meat balls and ljutenitsa, icons and candles—everything in the church." This was her first visit to the building.

In fact, the above description is typical of many community gatherings (see Fig. 71.9).[11] Due to a busy pace of life and far distances, not many people can attend services, but while in the building for a social event, they take the opportunity to light a candle. A person who has never set foot in a church in Bulgaria shared: "I have very positive opinion about the church—people have need to meet their ethnic culture. Or maybe it is nostalgia. I am so busy with my job now but I used to go for

Fig. 71.9 Sts. Cyril and Methody Church, Children's Christmas Party (Photo by Dimitar Karaboychev, used with permission)

[11] The picture was taken during a children's Christmas party on Dec 16, 2011. The buffet tables were situated right in front of the first nave door (closer to the iconostasis), which was open. Through the second nave door a male individual is seen sitting on the bench, looking at his smart phone.

meetings of the Engineer Association and some other gatherings." Another individual, who came immediately after the fall of the communist regime in 1989 reported: "I knew nobody and this was the place to exchange a few words in my native language. It was my bridge towards what I left behind, and towards the new home."

The December 24th event of *Budni vecher* (Evening of the Future), Christmas Eve, is quintessential for the perception of the church building as a comfort zone. St. George church established a long-standing tradition of inviting the community for a Divine Liturgy followed by Lenten dinner in the parish hall. The liturgy is performed together with the priests from the two other downtown churches. In Bulgaria the custom associated with *Budni vecher* is a Christianized pagan celebration of the cult of the Sun (the birth of the new Sun), and is not church-related (Georgieva 1993). Rather, it is a dinner for the immediate family, the most family-oriented evening of all yearly celebrations, and one is supposed to stay home on this day. Seven or nine Lenten fare dishes are served—cabbage rolls, beans, stuffed peppers, compote, nuts, wheat, etc. The heart of the custom is a freshly-baked, round, unleavened bread with a silver coin in it; everybody receives a piece and looks to find the coin, which is a sign of abundance and luck in the year ahead.

To find this homey festive event in the diasporic church hall is a pleasurable surprise. The 2011 dinner started with a short prayer and solemn singing of the *Mnogaia Leta* (For Many Years) hymn, followed by the Bulgarian national anthem. A child in a traditional costume sang the folksong "Kazji mi oblache le bialo" ("Tell me White Cloud," called the immigrant's anthem) on a stage adorned with both Canadian and Bulgarian flags. The audience of around 150 people stood and sang along, faces moved, and eyes glistening. The priests broke the round bread. After them, the round bread on each table was broken and shared, and people looked for the lucky coin in their pieces. The main dishes were the same as those in their native land. That evening attracts people from different generations and waves of immigrants, Canadian-born, single people, and families. One of the interviewees mentioned, "We prefer coming here with the kids for Budni vecher, it feels homier." A sense of togetherness links those present in a family-like fashion. In repeating an intimate domestic ritual, the church space is transformed into a symbolic home, a place where cherished memories are reachable, conjoining past and present. Thus, geography of feelings is mapped onto the church hall.

71.6 Conclusion

This study makes two points. First, the church space in a new location does not function as a reconstructive image of the traditional, where the determination of space, function, and decoration reflects only the liturgical dimension. By introducing new structural elements, the church building becomes instrumental as a communal and mediational space to the resettlement process. The church hall appears to be a liminal space between sacred and secular, and is the place where the institutional church and community meet to negotiate new relations with each other; for example, social and quasi-religious activities such as catechetical instruction and Christmas eve both combine religious and secular components. Another element is the interior design. Its minimalistic and selective approach with respect to images allows for more prominence and importance to be assigned to the presence of people than to the credo and mystics conveyed by the art. Hence, the strongest appeal of the church space is that people are drawn to the buildings for many reasons—practical (promoting business or networking), emotional (socializing, nostalgia), and spiritual—and still feel that it is the right place to be. The sacred offers existential security. The stability associated with a religious institution is even more attractive during moments of instability and intense dynamics associated with resettlement.

Second, the traditional *faith-nurturing* institution and its space have evolved into the most socially important institution for the immigrant community in Toronto. It responds to people's needs by assisting them to gain a sense of belonging, ethnic rootedness, and nurtures identity formation in a land that honors multiculturalism. Multiculturalism creates an imperative for resettlement from the disruption of personal history. In Toronto, as one interviewee commented, "everyone needs some kind of belonging." The church building domesticates the new locality, helping the immigrants to bridge their transition to the adoptive homeland because "knowing where we are is as important as knowing who we are," two streams of consciousness that are essentially connected (Silverston 1999: 86). Faith nurturing is enriched by ethno-cultural nurturing, thereby making the church genuinely the theophoric guardian of ethno-religious roots and a symbol of belonging.

References

Bruce, S. (2002). *God is dead*. Oxford: Blackwell Publishers.

Bulgakov, S. (1988). *The Orthodox church*. New York: St. Vladimir's Seminary Press.

Cohen, R. (1997). *Global diasporas: An introduction*. Seattle: University of Washington Press.

Eliade, M. (1959). *The sacred and the profane: The nature of religion*. New York: Harcourt, Brace.

Georgieva, I. (1993). *Balgarska narodna mitologia* [Bulgarian folklore mythology]. Sofia: Nauka & Izkustvo.

Georgiou, M. (2006). *Diaspora, identity and the media*. Cresskill: Hampton Press, Inc.

Goshev, I. (1952). Belejki za troeja i zografisvaneto na pravoslavnite hramove u nas [Notes on building and depicting our orthodox churches]. In *Annual of theology Academy St. Clement Ohridski* (Vol. II, pp. 317–334). Sofia: Clement Ohridski Press.

Kanev, P. (2002). Religion in Bulgaria after 1989: Historical and socio-cultural aspects. *South-East European Review, 1*, 75–95.

Lefebvre, H. (1991). *The production of space*. Oxford: Blackwell.

Meyendorff, J. (1974). *Byzantine theology: Historical trends and doctrinal themes*. New York: Fordham University Press.

Norris, P., & Inglehart, R. (2004). *Sacred and secular*. Cambridge: Cambridge University Press.

Ouspensky, L., & Lossky, V. (1982). *The meaning of icons*. New York: St. Vladimir's Seminary Press.

Shneer, D., & Aviv, C. (2010). Jewish as rooted cosmopolitans: The end of diaspora? In K. Knott & S. McLoughlin (Eds.), *Diasporas: Concepts, intersections, identities* (pp. 263–268). London/New York: Zed Books.

Silverston, R. (1999). *Why study the media*. London: Sage.

Studite, St. Theodore the. (1981). *On the holy icon*. Crestwood: St. Vladimir's Seminary Press.

Tanner, N. P., Alberigo, G., Dossetti, J. A., Joannou, P. P., Leonardi, C., & Prodi, P. (1990). *Decrees of the ecumenical councils. Volume One: Nicaea I to Lateran V*. London/Washington, DC: Sheed & Ward/Georgetown University Press.

Chapter 72
Temples in Diaspora: From Moral Landscapes to Therapeutic Religiosity and the Construction of Consilience in Tamil Toronto

Mark Whitaker

72.1 Introduction

Sri Lanka's recently concluded civil war has not only sent many members of that country's largely Hindu, Tamil minority into diaspora but their religious institutions, their *kooyil* or temples, as well. In Sri Lanka Hindu Temples tend to be tied to the landscape by cultural practices that makes them, at once, centers of local, moral landscapes, and, during the civil war, alternative public spheres where critiques and enquiries otherwise dangerous might be given and asked with relative impunity. Indeed, it seems only in the wake of the Sri Lankan government's defeat of militant Tamil separatism that these last refuges – these geographies of confirmation and expression – are being increasingly penetrated by a victorious state. In Toronto Canada, on the other hand, where perhaps 200,000 Sri Lankan Tamils now reside, most Hindu temples float free from the urban landscapes where they have been constructed, and are created, moved, and adjusted more according to a logic of retail religiosity than by some transformation of the strategic moral cosmologies employed in the homeland. And yet this new practice seems to be at once congenial with the kinds of individualized therapies and tactics the traumas of displacement and of disappointed nationalism seem to require, as well as with the ethos of late-modern capitalism that suffuses Canadian daily activity, while yet, in seeming contradiction, still protecting and promoting Tamil identity. How can Toronto's temples reconcile such seemingly paradoxical activities?

In this chapter I will draw on research in both locations, Tamil speaking Sri Lanka and Toronto, Canada, to map this transformation, the dislocations that occasioned it, and the strategic incoherencies that, I think, allow it to work in both locations. Ultimately, I shall argue that the malleability such temples display is not a

M. Whitaker (✉)
Department of Anthropology, University of Kentucky, Lexington, KY 40506, USA
e-mail: mark.whitaker@uky.edu

S.D. Brunn (ed.), *The Changing World Religion Map*,
DOI 10.1007/978-94-017-9376-6_72

consequence of their gradual assimilation into Canadian life but a subtle kind of cultural persistence, as well as a by-product of the state of exclusion many Tamils in Canada and Sri Lanka alike now find themselves experiencing. My thesis, ultimately, is two-fold. First, that Tamil Hindu identity in Sri Lanka is partly a product of a convergence of enacted confirmations, a *consilience*, that is, perceived as emanating from the same landscape Hindu temples anchor. Second, that practices of strategic incoherence that Sri Lankan Tamil local elites used – at least in the eastern province – to manipulate, outlast, and, in some ways, benefit from the incoherently different practices deployed by the colonial and, later, the nation state to absorb them, are being used again, in Canada, to resist and respond to current conditions in the diaspora. To show why I think this is so, I will proceed, first, to discuss the civil war and the diaspora it produced; turn next to Toronto's Sri Lankan Tamil temples and the compromises and innovations with which they have met the challenge of diaspora[1]; and turn finally to the Hindu temples I know best – the Batticaloa District temples I have been researching since 1981 – to show both how they work and how people used them during the civil war to survive the conflict's outrages.

72.2 Sri Lanka's Civil War and the Diaspora in Toronto

From 1983 to 2009 a bloody civil war propelled between 700,000 and 900,000 Tamil speaking Sri Lankans into exile (Fegulrud 1999; Crisis Group Asia 2010). Roughly 250,000 members of this diaspora, over 80 % now Canadian citizens, currently reside in Canada, most in the greater Toronto area – something now apparent to anyone riding through Mississauga, Markham, Bromley, Scarborough, or other Tamil populated neighborhoods thick with the scatter-shot signs of Toronto's 2000 or so Tamil owned businesses (Canadian Tamil Chamber of Commerce 2006). All of these neighborhoods, of course, are also *shared* landscapes. Slightly over 52 % of the people in the greater Toronto Area are immigrants according to (*Statistics Canada* 2012; Ryerson 2004). In every case, hence, Tamil shops and people are surrounded by signs and people speaking other languages: English, of course, always, but also Chinese, Punjabi, Gujarati, Italian, Portuguese, Tagalog, Arabic, Spanish, Russian, Armenian, Haitian Creole, Hindi, Jamaican English, Urdu, and so on, a

[1]This chapter constitutes a rather preliminary report on research that I see as only partly completed. It is based on participant observation in Sri Lanka, Toronto and London between 2001 and 2012. Much of the data for this work are derived from notes of unstructured interviews conducted during that research with Sri Lankan Tamil people I ran into in shops, meetings, temples and during the large protests and political meetings that occurred after the May 2009 defeat of the Liberation Tigers of Tamil Eelam (LTTE). I have also been following the fortunes of seven families: three in Sri Lanka, two in London, four in Toronto. I also conducted, roughly, forty-seven taped interviews during this period. I say 'roughly' because these interviews often occurred with relatives or friends in the room who sometimes contributed as much to the conversation as the person being, officially, interviewed. I should mention, also, that many of the people I interviewed were either people I have known since the 1980s or their relatives.

proliferating colloquy bespeaking the thorough-going ethnic complexity for which Toronto is so justly famous. Still, by careful navigation, one can always steer a reliably Tamil course and be surprised by who you find there.

This became more apparent to me one day in 2001 when I went out to a family party in Markham and promptly ran into a man I had last seen in Batticaloa, Sri Lanka over 8 years before. Arumugan[2] had been a government accountant and amateur philosopher, well regarded by the local intellectuals of the town, when I first met him there in 1983. But by 1993 I knew he had gone into hiding from the police (who were allegedly suspicious of him for his nationalist writings), and, since no one had seen him or knew anything about his fate, I assumed like most of his friends that he had been arrested and "disappeared." But he had, in fact, left the country illegally, to spend the next 7 years in transit, mostly in the Philippines, awaiting official (I believe UNHCR) classification as a political refugee. Only in 2001 was he finally able to come to Canada. "Amazing place," he said, as I remember it, shaking my hand vigorously, his emaciated frame trembling a bit, his eyes constantly tracking, glittering with the newness, "So...what to say?...fast!"

I was very happy to see him, but not overly incredulous at his sudden emergence from my research past. For by 2007 many of the people I had first met in the early 1980s while doing research in Sri Lanka's Tamil dominated eastern province (or else during trips to Sri Lanka's capital, Colombo, or to Sri Lanka's largest Tamil city, Jaffna) were now in Canada. I would run into them in Tamil stores, or walking along the street, but see them mostly at gatherings – family parties, birthdays, graduation celebrations, the annual community sponsored "Tamil Studies Conference" or at get-togethers arranged to celebrate someone passing through, like me, or just arriving, like Arumukan. Indeed, by 2012, so many Sri Lankan Tamils had come to Canada that many there now joke that Toronto is the largest *Sri Lankan* Tamil city in the world.

But this is a rather grim joke. For Sri Lankan Tamils in diaspora, like other people cast adrift by the civil travails of our troubled century, were thrust into movement by a long series of extraordinarily violent displacements; and these have left their shadows. In the late 1970s and early 1980s, after a postcolonial period of increasing mistreatment, many Sri Lankan Tamils sought to create a separate Tamil nation, an "Eelam," as an answer to the problem of living in a state dominated by a vastly larger Sinhalese ethnic majority, and one animated by its own brand of majoritarian nationalist ardor, that, increasingly viewed Tamils with resentment and fear (Wilson 2000; Tambiah 1986; Whitaker 2007). Eventually, as should have been expected, this communal stand-off between rival ethnic-nationalisms precipitated a series of civic disasters. There were nation-wide, government-sponsored, anti-Tamil riots in 1983 that burned thousands of Tamil homes (Tambiah 1986: 22), and set off 27 years of civil war. There were decades of anti-insurgency warfare, conducted by Sri Lankan government "security forces" often trained or funded by global powers

[2]Arumukan is a pseudonym as are the names of all those interviewed for this chapter. In this chapter I will also occasionally change minor biographical details to further hide identities. This is common ethnographic practice to protect the identities of the people we work with.

(the US, Britain, Israel, Pakistan and India[3] among others, interested in Sri Lanka for their own contradictory reasons) that exposed thousands of young Tamil men and woman, most innocent civilians, to summary justice, torture, political rape, indiscriminant attack and an almost constant atmosphere of danger, epistemological murk, and distrust (Whitaker 1997, 2007; Somasundaram 1998; Trawick 2007). And there were the often paranoid activities of the Liberation Tigers of Tamil Eelam (or LTTE), the primary Tamil nationalist army pursuing the war, which, by the early 1990s, had proclaimed itself the sole legitimate representatives of the Tamil people, and had ruthlessly crushed its competitors (Wilson 2000: 126[4]). The LTTE's tactics, at once efficient and vicious on the battlefield, nonetheless targeted Tamils deemed enemies almost as often as they did the Sri Lankan army or Sinhalese civilians; and its increasingly desperate recruitment drives frequently forced other Tamils (including children) into its ranks, or into exile (See Thiranagama 2011; UNICEF 2012; Save the Children 2012; Trawick 2007: 147–184).[5]

So it was, in short, a horrible war prosecuted with grisly vigor and little regard for human rights by all sides; and it pushed those Tamils possessing the means to flee out into the diaspora, and left those who, for reasons of poverty or conviction, had to stay in Sri Lanka, perilously poised between titanic, grinding forces. Leaving aside, for later, consideration of the dreadful circumstances of Tamils in Sri Lanka during the war, and how they sometimes used their temples to deal with them, it is important to see now how all this geographically distant, but emotionally near, disaster informed the consciousness of Toronto's Tamils.

Take, first, the matter of loyalties. Most exterior views of the Tamil diaspora present it as a hotbed of LTTE-style militancy, and do so even now that the conflict is over (Crisis Group Asia 2010; United Nations 2011). But it is important to note that although, in exile, most diasporic Tamils have, publicly and financially, supported the LTTE – a classic case of "long-distance nationalism" of the Andersonian

[3] India trained and supported various militant Tamil separatist groups, including the LTTE, prior to 1987 and the Indo-Sri Lankan Accord. After the war between the Indian Peace Keeping Force (IPKF) and the LTTE (1987–1990) and the assassination of Rajiv Gandhi in 1991 the Government of India started providing aid to Sri Lankan security forces.

[4] Wilson notes that there were 35 distinct nationalist militant groups in 1983 but only five were "… of significance: the LTTE under the leadership of Velupillai Prabakaran; the People's Liberation Organization of Thamil Eelam (PLOTE, headed by Uma Mahewwaran; the Thamil Eelam Liberation Organization (TELO), controlled by Sri Sabaratnam; the Eelam People's Revolutionary Liberation Front (EPRLF), led by K.S. Padmanabha; the Eelam Revolutionary Organization of Students (EROS), commanded by Velupillai Balakumar." (2002: 126). All of these groups were once supported by India's security service, RAW (Research and Analysis Wing), in the early 1980s; all were effectively destroyed or absorbed into the LTTE by 1990. In 1987 India and Sri Lanka signed an accord which brought India, in the form of an Indian Peace Keeping Force (or IPKF), into Sri Lanka, and set up a war between India and the LTTE, which India had formerly supported.

[5] My own informants in Sri Lanka universally report feeling terrified that their children would be forced into the LTTE. However, many even so supported the LTTE, or a break-away group from the LTTE that also practiced child recruitment. See Trawick (2007) for a discussion of some of these complexities, and see Thiranagama (2011) critique of Trawick's discussion.

variety according to some commentators (Anderson 1992; Fugelrud 1999) – my own research in Toronto and London between 2001 and 2012 reveals that this apparent unity hid a more complex variety of positions and rationales. Everyone I talked to supported, with various degrees of ardor, Tamil nationalism, which they generally called "the Eelam Struggle" But the aims of that struggle could elicit a range of opinions, from flat avowals of the need for a completely separate state to a yearning for something more like federalism. People's attitudes toward the LTTE were even more complex. It is important to remember, in this regard, that some members of the Tamil diaspora were affiliated with militant groups other than the LTTE in Sri Lanka, or were forced to flee by the LTTE, and, hence, sometimes expressed anger at the LTTE equal to, and occasionally greater than, what they felt toward the Sri Lankan government. So the people I interviewed often spoke to me with nuanced ambiguity. They would tell me of their genuine fear of criticizing the LTTE in Toronto (and London), but also of their greater fear of what the Sri Lankan government might do to their relatives and friends back in Sri Lanka were the LTTE to lose. Or they would speak of their pride in the LTTE's fearsomeness in battle ("our boys," most called them) even as they expressed disquiet at the LTTEs totalitarianism and intolerance, or even, in two cases, as they also told me how the LTTE was responsible for the deaths of people they loved. Finally, most often, people revealed their guilt at having escaped the war even as they rehearsed their gladness at having done so. Hence, fear and faith, horror and happiness, approval and distaste, guilt and relief – a tippy seesaw of attitudes rather than a singular, coherent, view – was the complex emotional balancing act necessitated by the war. And all of these uncertain ambiguities became amplified for most Tamils in Toronto because of how the war ended.

This it did, in May 2009, when the Sri Lankan army, pumped up by (quietly distributed) international military aid and political support, surrounded the remnant LTTE along with over 300,000 civilians, and bombed them into oblivion at a cost, according to some UN estimates, of over 30,000 Tamil civilian lives (UN 2011). Much of this was watched in horror by the diaspora via the worldwide communication network of radios and TV stations set up by the LTTE. I remember sitting on a sofa with people in Toronto that May and listening to them identify the towns and temples and, worse, relatives and friends being destroyed in the grainy videos coming in over the net or Tamil TV. One man I interviewed, a former LTTE cadre, told me the images so shook him that, for the first time in his life, he gave in to open, loud, and unpracticed weeping. His family was so unprepared for his uncharacteristic sobbing they thought he was having a heart attack. Moreover, throughout Toronto, images of the May defeat reached across the "generational" divide between Tamils who remembered Sri Lanka and Tamils born or mostly raised in Canada. Suddenly, the Sri Lankan Tamil community was being given the dubious gift of a singular outrage commonly experienced – albeit largely from a distance, via the media. Ultimately it was this communal "trauma" – for lack of a better term – that precipitated waves of intense public protest by Tamil diaspora communities all over the world, much of it ignored by the international press, and a period of intense mourning, disappointed nationalism and communal introspection. I have written elsewhere about how this communal defeat, surprisingly, elicited a more reasonable collective response by the world wide Tamil diaspora than much modern theorizing about diaspora might suggest (Whitaker in press).[6] But it is undeniable that many Toronto Tamils who watched Sri Lanka's spiral of violence from afar displayed thereafter, with increased intensity, some of the sadder markers of post-defeat exiles: remembered trauma, nostalgia for a homeland violently lost, various forms of long-distance nationalism, and expanded worry over far-flung transnational family ties threatened by increasingly well-policed, post-9/11 national borders. The end result was a generalized, existential anxiety about the continued existence of the Sri Lankan Tamil community as such.

But the Sri Lankan Tamils I talked to in Toronto were, I think, as concerned about losing their Tamil identity to assimilation and the increased pace of life in Toronto as they were about the long term fate of their world-wide community and the grim politics of postwar Sri Lanka – though most people also linked these two concerns. Worries about assimilation and remembrance were most intensely felt and acted upon by Sri Lankan Tamils who came to Canada as adults, as can be seen in the kinds of organizations they set up in the last 25 years to ensure Tamil continuity. In the 2007 Canadian Tamil Business Directory (also called Sethi's Tamils' Guide), for example, are listed a host of "old boys" organizations and "sangams" (groups) that hark back to a lost Sri Lanka: clubs for former Bank of Ceylon Employees; still others for former CTB (or Sri Lankan government) bus drivers and for University graduates; even an association for former Sri Lankan (Tamil) police officers. There are also groups dedicated to the memory of particular villages, and 46 "old boy" (or alumni) associations commemorating particular secondary schools. These are, one might say, *organizations of remembrance*, and their job is surely to allow the retention in memory of a lost, communitarian, conviviality. At the same time there are other organizations – such as the Canada Baratha Natya Teachers Association and various Tamil language schools – that are mostly dedicated to conveying Tamil 'heritage' (as people generally put it) to Canadian born Tamils. One might call these groups *organizations of persistence*, and their job is clearly that of preserving a scaffolding of knowledges and practices upon which some kind of future, *Canadian-Tamil* identity might, practically, be built. But these two tasks, remembrance and persistence, are hardly contradictory; and, thus, there are organizations that do both. The most numerous and influential organizations of both remembrance *and* persistence are, obviously, Toronto's Hindu temples.

[6] Several organizations were set up by people in the diaspora in the wake of the May 2009 defeat, most notably the Global Tamil Forum and the Transnational Government of Tamil Eelam. Both organizations have been criticized for mission statements that reaffirm the 1976 Vaddukoddai Resolution, a manifesto proclaiming the necessity of a fully independent Tamil state; and some within the diaspora have suggested this reaffirmation of radical separatism has sidelined them as actors on a world stage still dominated by fears of international terrorism. Interestingly, however, both organizations have embraced non-violence; a move my interviews suggest was widely demanded.

72.3 Identity, Daily Life and Consilience in Tamil Canada

But to grasp why temples play this role (among others) it is important to understand how many older Toronto Tamils feel the sheer pace of life in Canada threatens Tamil identity. The most frequent tropes they used with me had to do with to speed, rootlessness and anonymity. People frequently talked about how the rapidity of their daily lives cut them off from other people and left them feeling unmoored from the people and places that, in Sri Lanka, used to anchor their lives. They spoke of the velocity of daily life, the anonymity of the shifting sights and sounds of the city, of lives suddenly urbanized and dominated by the demands and schedules of (capitalist) work in Canada – all themes that would have made sense to Simmel meditating on the experiential transformations wrought by capitalist modernity, and that have been too often ignored, as Aihwa Ong has noted, in some of the more optimistic musings about diaspora as a kind of beneficial 'critical cosmopolitanism' (1999: 14). Hence, as one young man, originally from Batticaloa told me in 2001, while people there had time to stop and talk on the road, this was not possible in Canada. In Toronto, he claimed, life can only be glimpsed 'through a windshield' as you go by on your way to work. Similarly, a middle aged woman from Jaffna with two children, again in 2001, spoke of how taking a full time job in Canada meant having no time for anything or anybody; for, she insisted, her job as a sorter for the post office had such long hours she rarely got to see her children awake. And when I asked her about this again in 2012 she said, laughing ruefully, "Same job. Still no time!" This shortage of time due to work was something that also struck all eight of the priests or owners of the Sri Lankan Hindu temples I interviewed in 2011 and 2012. "This is why," one priest told me, "I make sure every puja (worship service) happens exactly on time. People can't fool around. They can't be late so *we* can't be late." Several of the priests also noted that in Toronto people were lonely. Many priests pointed out that, *except during festivals,* people rarely came to temples for long, they often came alone or with just one of their children, and that they complained frequently of never getting to see anyone.

I could not understand this at first. Although many Tamils came to Sri Lanka alone in the early 1980s, relatives often followed soon after until the post 9/11 crackdown on immigration and claiming asylum in Canada. Most Tamils I spoke with, hence, live within dense networks of kin. This is why, whenever visiting Toronto, I have generally been taken on several rounds of family visits, and people there have as often complained to me about being caught up in cloying webs of family and community gossip as about being alone. But I gradually came to understand that the "loneliness" and fear of losing "Tamilness" often expressed by older Tamils relates to a more subtle perception that the texture of daily life in Canada is too loosely woven to provide the kind of constant reassurance that maintaining one's identity requires. For example, Rajan, a middle aged man I interviewed in 2001, once tried to explain to me why, regardless of the presence of his family and friends, he nonetheless felt rootless and in danger of losing what he called his "Tamilness."

> Ok. You have been to Ceylon, right? In Sri Lanka the family and extended family is very close knit. You know, for everything the family is there. And when you go off – say that, what you have done in your life, it carries you. People … recognize you, what you have done, what you have… You are a known person. A *nulla teringka al* [a well known person]. A *teringka manithan* [a known man]. A *kaurava manithan* [an honored man]. But here I am living in a city, a big city, one of the largest in the world … I am one of a million.

A *kaurava manithan* or "honored person," in Sri Lankan Tamil, is a person fixed in place by a web of knowledge and activity, sometimes stretching into the karmic past, that is anchored to the family and community and the moral landscape they occupy (Whitaker 1998: 91–95). It is, in other words, a form of identity dependent upon being able to "place" oneself within a specific, every day, "moral geography" (Kingsolver 2010; De Certaeau 1984; Thomas 2002; Birdsall 1996). In eastern Sri Lanka, for example, Sabathambi, in 1981 a middle aged school teacher and Hindu temple board member, once explained to me that he was known" and "honored" because of his jobs, his education, his command of Tamil, and his current behavior; but also, as importantly, because of how he was received by people as he walked about the market or the fields surrounding his village. There, he claimed, people had to take into account his hereditary rights (or *urimai*) in his village of origin (his *uur*) and in temples – rights which also defined his caste, clan and lineage statuses within the Batticaloa district, as well as (he argued) his cosmological standing.[7] Such rights, he often pointed out, were deeply rooted in place and history; his own family's right to a position on the board of the Sri Kanticuvami kooyil stemmed from the role assigned to his lineage by the god, Murukan, at the moment of the temple's creation as laid out in its written history (*kaLvuttu* – literally, 'stone cutting'). In this way, one might say, Sabathambi's sense of himself, his "knownness," was fixed in place by multiple forms of geographically sited and practically enacted assurance – much as some scientific propositions are held to be "true" due to converging, multiple lines of evidence. This is what some historians of science call consilience.[8]

---

[7] Thiranagama has written very trenchantly about the wrenching experience of Sri Lankan Muslims who lost their connection to a particular villages and landscapes after being displaced by the LTTE during the civil war (2011: 172–177).

[8] Now I am using "consilience" here metaphorically rather than literally. I am simply trying to point to the way identity, whether personal or collective, is always a process requiring constant assurance for its continuance rather than a fixed quality of groups or individuals. In this way I see my position as being closer to Foucault's notion that subjectivities are maintained by their embedding in multiple forms of knowledge and practice than to any sort of "realist" claim about individual or collective personhood. Further, I am trying to point here to something vaguely "propositional" and "theoretical" about individual and collective identity claims. They involve assertion as well as disposition. And since, in certain circumstances, assertion can be problematic, even dangerous – as, for example, it is now for Sri Lankan Tamils in Sri Lanka and, sometimes, after 9/11, even in Canada – the circumstances required to make assertion possible must engage our attention. Hence, the term 'consilience' is useful, for me, when trying to describe people struggling to maintain those multiple conditions or vectors of assurance necessary for the comfortable assertion of identity claims. I realize, of course, that my use of consilience in this way is somewhat perverse. The term consilience was originally coined by the nineteenth century polymath philosopher William Whewell to account for the way multiple forms of sometimes weak evidence (or inductions) may, nevertheless, compel adherence to particular scientific theories or conclusions

Hence, Rajan, even when traveling around Sri Lanka, felt that a fabric of place, people and practices ("what you have done") kept him "known," and "close knit." But family alone could not do this, only his conjoint participation in practices that occurred in various shared locations – only, that is, this partly geographic *consilience.*[9] A constant, gentle pressure of confirmation from all sides, he seemed to suggest, was what kept people in Sri Lanka what they were, and insured their individual and collective identities, their "Tamilness." And so Ranjan went on to speak wistfully of how, in Sri Lanka, he would meet people in the market, or at a temple, or on a visit to his sister's house – and everyone knew him and what he was about and that he belonged. Away from all that, however, in the free-floating landscape of urban Toronto, Ranjan, like so many Canadian Tamils who still remember being "known" in Sri Lanka, felt his communal identity, his "Tamilness" diminish, diluted by Toronto's size and numbers, and its absence, I think, of consilience. "I am," as he put it, "one in a million."

These commonly expressed concerns about communal continuity and the lack of what I am calling 'consilience' – of multiple forms of sited and enacted assurance – were often described, however, by older Toronto Tamils, more intimately, in the generational fear that their children were "losing their Tamilness" People worried that children were failing to learn the Tamil language, showing little interest in Tamil cultural practices such as classical dance (Bharatha Natya), spending too much time with Canadian friends in Canadian places (Malls, online, at school) and, thus, were too easily absorbing Canadian personal, political and sexual mores through the "unknown" landscape they were too frequently traversing. One older Tamil man, the CEO of a small firm in Scarborough, hence, gloomily predicted that in 10 years there would no longer be any Tamils in Canada anymore; just Canadians with Tamil names. In Sri Lanka, he claimed, Tamils would be "finished" by the Sinhalese while "here we will just be Canadian." It is this sense of "loss," I suspect, that is precisely what many Toronto Tamils are striving against with their Tamil language and dance schools and of course, with their Hindu temples – a sense of loss become all that much more acute and general in the wake of the May 2009 defeat of the LTTE, and the end of any practical path to an independent Tamil state.

(Fisch 1985). The term was popularized most recently, and rather misleadingly, by E.O. Wilson in his 1998 book, Consilience: The Unity of Knowledge. There Wilson stretches Whewell's concept out of its originally useful smaller shape to include the world as a whole so that he can argue that a unified configuration of truth or knowledge undergirds and confirms the universe as science describes it, a neo-realist (and metaphysical) claim, needless to say, that I am not making here. Indeed, quite the reverse, I am claiming, locally and provisionally, that Sri Lankan Tamils in Toronto are working fiercely to construct new kinds of consilience in Canada to replace the more land-bound forms they lost in Sri Lanka; and that constructing new organizations of remembrance and persistence such as Hindu temples is part of that effort.

[9] 'Consilience', here, is very similar too (and inspired by) Ann Kingsolver's concept of "placing." See Kingsolver 2010: 13. The difference, here, is mostly just a matter of emphasis: I am interested in how a landscape is constructed to provide the signs of assurance that people are invoking in the cultural practice of "placing" Another way of saying this is to point out that consilience is what people have lost when they can no longer comfortably place either themselves or others. This is inevitably a problem of diaspora.

So for many Sri Lankan Tamils it was temples (or other organizations of remembrance and persistence) that stood most effectively against this erosion of identity.[10] It was in Hindu temples, after all, where one found most frequently the Tamil language schools, dance classes, meeting places, marriage classes, and sponsored "cultural events" in which worried parents could enlist their children in order to preserve their "Tamilness" and in which older Tamil men and woman could at least try, once again, to be "known." So it is perhaps, in one way, unsurprising that between 1983 and 2012, the number of Hindu temples in the greater Toronto area grew from 1 to, at least, 40 (Chamber of Commerce 2006; Dr. E. Balasundaram personal communication)[11]; and although only eight of these temples are officially listed under revenue Canada as 'Tamil' temples, many others have influential Tamil members or founders.

But how could temples in a postmodern city-scape characterized by a constant roiling of people, practices and locations provide the kind of anchorage, and ultimately the "consilience," for which Toronto's older Tamils, at least, seemed to yearn? To address that question, I think we must look now at some of Toronto's new Hindu temples.

72.4 Toronto's New Hindu Temples

In May 2011 and June 2012 the folklorist Dr. Balasundaram Elayathamby and I conducted interviews with the founders of seven Tamil (or Tamil dominated) Hindu temples. Most of these temples were started in the last 10 years, and most are now located in buildings relatively featureless outside while impressively, even overwhelmingly, decorated inside. These temples were, to name them: the *NaakapuusaNi ammon* temple, located in a sign-less store front next to a Tamil take-out shop; the *Meenakshi ammon* temple, a converted warehouse in an industrial section of Scarborough; the popular *Iyappan cuvami* temple, with its huge, new, elaborately appointed temple hall built next to the completely plain, off-white storage unit that

[10] Although Christians make up only a small percentage of Tamils in the Toronto area – <5 % – nevertheless roughly half the people I interviewed were Christians or had Christians in their families. This is because one of the larger families I have been following since the 1980s is Christian. Christians Tamils mostly did not agree that Hindu temples were important to preserving "Tamilness" (though several, interestingly, did). Some assigned this role to their Tamil churches; others to Tamil language and dance schools or even political organizations such as the Canadian Tamil Congress. Others, mostly Pentecostals, claimed that Christianity demanded of them that they shift their priorities away from community preservation toward remaking themselves "for Christ." I am conducting interviews at Toronto Tamil Pentecostal churches as well as Hindu temples to see what significance this difference may have.

[11] Or perhaps many more, Revenue Canada lists 84 Hindu temples in the greater Toronto Area. It is unclear whether all these listed entities are temples per. se. however, since many seem to be organizations associated with but not themselves temples. Hence, the 2007 Tamil's Guide lists 22 temples, but many of these are not yet listed with revenue Canada. It also lists 33 Sri Sathya Sai Baba Organizations (1775–6).

was its original home; the *Periya Bhadrakaalii*[12] *ammaL* temple, another converted warehouse with a large, round, above ground swimming pool in its parking lot for he conducting of the sin-cleansing annual ceremony of *thiirtham* or "water cutting"; the *Thirumurkan* temple, likewise found in a warehouse that, alas, was discovered to be facing the wrong way (south instead of east or west, I believe) by an imported temple architect, and thus was having a new entrance door cut into its wall by two pneumatic drills even as we tried to speak to its chief board member; and the *Cuparimalai Iyappan* temple, a warren of multiple, large, rooms and basements, all tucked in together, in seeming geometrical impossibility, behind the grey reflective glass doors of a single strip mall store front.

The impetus for these interviews was a long term research ambition we share to talk to the founders of all 40 or so of the Hindu temples in the Toronto area frequented by Canadian Tamils. But we started with these seven, I must admit, mostly because we could easily get to them by car from Dr. Balasundaram's condominium in Scarborough. But even with so small and initial a sample, it was easy to note patterns that seem likely to hold true for Toronto's newer Hindu temples in general.

First, obviously, a fact of location: all of these temples are currently located either in former warehouses or in storefronts of old strip malls, locations, in other words, where land or rents are cheap and access to parking and major roads are good.[13] Toronto, of course, boasts some elaborate temples, such as the Richmond Hill Hindu temple, 'the largest in North America' according to its website, complete with elaborate walls and a huge gopuram or decorative tower, all obvious from a distance (see www.thehindutemple.ca/). But most of Toronto's Hindu temples are like the seven above: anonymous outside; elaborate inside: that is, seemingly discrete sacred worlds that, like Rajan's unknown Tamil refugees, merge into Toronto's vast urban landscape with seeming acquiescence. Second, all of the temples were created either because there was local demand for a temple devoted to a particular deity or because someone, usually a founder, thought there was. Hence, four of the temples were founded by priests convinced of their being local demand for a particular kind of deity, two by wealthy men and woman interested in a particular deity, and one by a group of devotees who became interested in a South Indian deity, Iyappan, because they felt his attributes matched their challenging circumstances as Sri Lankan Tamils. The point, here, is that unlike in Sri Lanka where all temples are, in theory, created by the direct intervention of the god and attach devotees and communities to particular places, temples in Toronto, according to their founders, were built, and distributed across the landscape as a consequence of a mix of popular demand and real estate practicality as determined by a kind of 'retail' logic. Third, most of the temples, like most temples in Toronto in general, were formed in the last 10 years. Only two of the seven, the *Thurkeswaram Sri Durka* temple and the

[12]The name "Periya Bhadrakaalii" is a pseudonym used here to protect the identites of its priests discussed below.

[13]The one exception to this proves the rule. The Cuparimalai Iyappan temple is located in a storefront of a new strip mall but one with few tenants that is trying to attract more. In short, they got a good deal.

Iyappan Cuvami temple, were founded in the 1980s by members of the diaspora who arrived first; and both of these have been dramatically expanded to answer new demand. This implies, given the above, I think, that the *need* for such temples is increasing rather than decreasing. Finally, all of the temples we visited exhibited practices that, to my mind, combined forms of cultural *remembrance* and *persistence* with an incongruous embrace, however incoherently, of those aspects of Canadian life that people also told me were most threatening to Tamil identity: capitalism, individuality, disappointed nationalism and cosmopolitan urbanity. In other words, just as these temples hid their elaborate religiosity within a cloak of anonymous cityscape – for what looks like a store is not always a store – so they also arranged to at once preserve and promote Tamil identity even while using and sometimes celebrating Canada. They provided Tamil people with consilience, in other words, by somehow subverting some of the very forces threatening it. To see what I mean, here, it is best to turn to a specific example.

In May 2011 I talked to two Sri Lankan Tamil brother priests, whom I will call Chellappa and Chandran,[14] who told me how they came to own and control a Hindu Kali temple in Scarborough, a part of Toronto, Canada, locally famous for its many Tamil shops, groceries and restaurants. We spoke at their relatively new temple, the *Periya Bhadrakaalii ammaL theevashthaaLam,*[15] over milky tea brought by Chellappa's new wife, Jeeva, herself a new arrival from Sri Lanka. Newlyweds, they smiled at each other with genuine warmth, but Jeeva paused only long enough to observe how funny she was finding Toronto – "because no one ever stops." The temple in which we sat, started only 6 years previously, was impressively ornamented, especially inside, where gold leaf walls and silvered *puja* instruments glittered before the flames of many oil lamps and the glow of even more phosphorescent tube lights. But it was nonetheless obviously constructed – as so many diasporic temples are – out of what was originally a plain warehouse in a part of Scarborough zoned for light industry.

Although Canadian citizens now, the brothers explained that they came originally from Karaitivu, a village in Sri Lanka's eastern, coastal, Ampara district, a part of Sri Lanka I know well from over 30 years of research. The brothers told me their father and (maternal) uncles had been priests or *Kurukal* at a large *Pilaiyar* (that is, Ganesha) temple there. Like so many east coast Hindu temples, the Karaitivu *Pilaiya*r temple was, they continued, owned and controlled by a specific named community – in this case the Sinkala matriclan (or *kuTi*) of the VeLLaaLar caste people of Karaitivu village (or *uur*) – and was, thus, a *"kuTi uur* temple," a designation that connects the temple inextricably to both a network of vaguely matrilineal relatives from a geographically specific place and the exact acres of rice-growing (or "paddy") land that the clan's ancestors originally contributed to support the temple. Chellappa and Chandran, in the normal course of things, being themselves

[14]As is conventional in ethnographic accounts the names of people interviewed for this project have been changed to protect their identities.

[15]The word *theevashthaaLam*, like the word *aalayam*, are alternative, more Sanskritic terms for kooyil. Thus, *aalayam* is the term temples are listed under in Senthi's Tamils' Guide 2007.

members of a priestly matrilineage, and thus supported by this heritage, would have followed their uncles, father (although a member of a different lineage), and an elder brother, whom I will call Elanganayagam, into the priesthood. But by 1996 that predictable path had been kicked over by the Sri Lankan Civil War.

That war, as I have already described, was a bitter, sanguinary struggle that cost over 100,000, mostly civilian, lives. In the east, by 1996, government anti-insurgency forces had made daily life for young men and women exceedingly dangerous, and the brothers were frightened. Most roads in the Batticaloa and Ampara districts were blocked every few miles by grim checkpoints manned by suspicious army or security police soldiers, usually accompanied by silent, hooded prisoners there for the anonymous fingering of Tigers "suspects." Young people arrested at such checkpoints were rarely seen alive again (Whitaker 1997).

So the brothers began to leave, following the lead of many other Sri Lankan Tamils. And as the bulk of this exodus ended up in Europe and North America, particularly Canada, it is perhaps no surprise that they did too. Hence, in 1996, Elanganayagam, their elder brother journeyed to the United Kingdom to work in a large London Hindu temple. Soon he was able to save enough money to pay for Chandran, his youngest brother, to attend a school in London; and, finally, for Chellappa, in 2001, to come and work as an assistant priest in the large Kali temple he had by then established in London.

At this juncture the brothers hatched a plan to establish another Kali temple in Toronto. Chandran, the next eldest brother, told me that Kali was particularly important to them because various members of their lineage had been periodically possessed by Kali and, hence, they "belonged" (*paLLaiyam*)[16] to her. So, with the Kali plan in mind, Elanganayagam sent, first, Chandran in 2001, to scout the lay of the land; and then, in 2004, Chellappa, to set the temple up. In Toronto, Challappa told me, he took Chandran's observations (one might almost say his market research) into account, scanned a map of the metropolitan area, and concluded that there was both great demand and cheap land available for a Kali temple in the part of Scarborough zoned for light industry. And so they built their temple there.

Chellappa was aware, however, that this way of establishing a temple constituted a radical departure from Sri Lankan practice, and it appeared to trouble him a bit. Most temples in Sri Lanka, he explained, were built where a god or goddess dictated. There would be a "miracle," usually an appearance by the god in an altered form, or a dream in which the god would come directly to speak to the chosen vessel of action. In either case the eventual founder or founding group would be commanded to build a temple at a particular, precisely defined, location. Alternatively, someone might find an appropriately shaped piece of rock under a tree or near a road suggesting, iconographically, that a particular god or goddess was regularly appearing and so wanted to be worshiped *there*. In either case the resulting temple

[16] 'Belonged' was the gloss they offered for puLLaiyam. A better translation might be "offered," as in saying that their family was often offered to Kali through possession. See Winslow'sA Comprehensive Tamil and English dictionary (1987: 746).

or shrine – always, regardless of size, called *kooyil* – would be attached to the landscape by the god's command.

I was well aware of this from my earlier work in Sri Lanka. In 1982, for example, temple officials at Mandur's *Sri nullapompu* temple, told me that their temple was founded in 1981 because a cobra (the '*nullapumpu* or "good snake" – an incarnation of God) landed on the face of a member of their *veLLaaLar* community while he was napping under a shady tree. Instead of biting him, the cobra reared up imperiously, commanded the building of a temple, and then slithered off with its albino mate. The napping man's bone fides were later attested to, I was told, by a dream that came to a more senior *veLLaLar* man. Similarly the anthropologist Dr. Sasikumar Balasundaram recently told me that Tamil children on Tea Plantations in Sri Lanka's mountainous interior often "plant" temples by worshiping likely looking, often triangular[17] rocks they find in the forest. Frequently, he pointed out, such impromptu shrines later became flourishing temples, sometimes within months. Diane Mines, another anthropologist, has aptly described a similar temple planting processes in Tamil Nadu villages resulting from wandering gods marking and shaping the earth with their sacred, geo-generative power or *cakti*, and then throwing up "shoots" of themselves out of sheer, divine, fecundity (2005: 130).[18]

But in Canada neither a location-fixing dream, nor a directive miracle, nor a divine shape found on a Toronto street was available to help Chandran decide where he should locate their temple. Moreover, given the kaleidoscopic, multicultural and commercial landscape of Toronto, any such location-fixing commandments would have been impractical, even impossible. Toronto is, after all, a city-scape constructed, as de Certeau might say, according to the strategies of alien institutions over which Tamil immigrants have no control (1984: 53); and thus a place that, at best, can be *shared*, and only where fortune dictates. So, Chandran said, "…circumstances demanded that I come up with my own idea." Their kali temple would be, thus, a "kaaraNNakkooyil," a "temple for a reason," that is, a temple arrived at though Chandran's own idea, informed by his brother's research, of where a Kali temple would best work. Here, clearly, something of cultural *persistence*, the notion that Chandran's lineage "belongs" to Kali, was incoherently conjoined with practices of pragmatic and quasi-commercial "retail" religiosity to settle the issue of where the

[17] Triangles are one of the key, sacred, geometric forms or *yantra* so important to the representation of Hindu cosmology and are also used, through-out South Asia, as "cosmo-geographical groundplans for temple architecture" (Lannoy 1971: 21). Mandur's largest temple, the Sri Kantacuvami Kooyil, for example, has triangles (*koonum*) built into its walls as a sign of its geometrical centrality.

[18] Mines, speaking specifically of village goddesses, says: "Long before the humans settled the village, the goddess wandered, and as she wandered, she marked the earth with her substance, her cakti, thus establishing her real presence in places. She is rooted in the soil, and her image might spring up anywhere like a tree's root might throw up shoots far from its trunk. Her power is a quality that shapes the village's topography: it creates zones of greater and lesser energy (*cakti*), and discovering those places links human beings to that power"(205: 130). I am extending this point to include gods in general.

temple was to be built. With this, Chandran smiled, and extended his arm to their obviously successful temple.

The temple is successful, apparently, for a variety of reasons. First, because the brothers made sure that their *pujas* were frequent and on time to appeal to schedule-pressed Toronto Tamils – as this did, for example, to the young, accountant mother I spoke with who was pulling her somewhat unhappy 6 year old (he wished, loudly, to go home and watch mutant ninja turtles *right now!*) through the temple's various stations. Second, because the brothers provided a range of services and festivals designed to aid and publicly reassure Canadian Tamils: ceremonies celebrating school scholarships, civil servants, local Tamil literary scholars (my co-researcher, the Folklorist Dr. Balasundaram Elayathamby, was "garlanded" at one such ceremony) and, on "Canada Day" (a Canadian national holiday), senior citizens. Third, because the brothers had adjusted to what they claimed was the "increased individuality" of Canadian Tamil life by keeping a ledger in which all temple members were listed by their names and addresses so each could be appealed to individually for support and volunteer labor – something, they said, one never had to do in Sri Lanka where such obligations were communal and geographic (distributed by caste and lineage but also by unit, village and so forth) as spelled out in the temple's official history. Finally, the temple was successful because the brothers provided counseling for unhappy couples, music and language classes for children, and, most importantly, annual festivals, like the temple's "Milk Pot Carrying ceremony" and its annual founding celebration or *tiruviLaa*, in which hundreds of Canadian Tamils often participate. Here, surrounded by like devotees, many Toronto Tamils found the enveloping communal reassurance, the sensory, enacted consilience, otherwise so lacking in Toronto daily life.

Such temple festivals, indeed, are probably the most important practices of cultural remembrance and persistence conducted by Hindu temples in Toronto; and yet they betray the same in-twisting of contradictory activities so often found in Toronto Hindu life. At the *Periya Bhadrakaalii* temple, for example, the "milk pot carrying" festival (or *aaTippuuram paaRthiTam*) frequently involves hundreds of devotees walking down a busy main road to the *celvi channathi Murukan* temple (a temple I have not yet visited) 3 mi (4.8. km) away. The brothers assured me that they always had permits for this procession, and hence the cooperative help of the city police in stopping traffic and controlling crowds of curious Canadians. In other words, the temple has managed to get Toronto city government involved in helping them take over the street, at least temporarily, for their moment of collective and individual consilience.

The annual festival, commemorating the temple's recent founding, provides an even better example of the complex and tricky commingling of practices of remembrance and persistence with incoherent features of Toronto daily life that is required by being a temple in diaspora. This festival or *tiruviLaa* was, I was told, generally larger than the "milk pot carrying" festival, and usually lasts for 15 days, culminating in the cathartic *thiirtham* ceremony involving the god and as many people as possible circling the temple and then "water cutting" by submersing themselves in the large, blue, above ground pool in the parking lot. Much of this was modeled on how *tiruviLaa* were conducted at large, important temples in Sri Lanka. Hence, like many large temples in Sri Lanka, the *Periya Bhadrakaalii ammaL* temple allows each of its 15 festival days to be "sponsored" (paid for) by different, named groups. But this similarity reveals, at the same time, key differences between the situation of temples in diaspora and in Sri Lanka.

In Sri Lanka the business of groups sponsoring festival days is, in a way, a culminating moment affixing particular people, kin groups and communities to the moral (and, also, political and historical) landscape for which a temple is, symbolically and geographically, the center. Symbolically this is quite clear. A *tiruviLaa*, or "royal celebration," involves taking a god as a sovereign on a circle around the temple; but since the temple is constructed as a geometric representation of the universe as a whole, and since to circle something is to conquer it according to the well-practiced language of South Asian ritual, then the god and all who accompany him or her are similarly lodged at the heart of a universal moral cosmology.[19] But festivals also make more concrete and local statements about particular moral landscapes. At Mandur's Sri Kantacuvaami temple, for example, as I described in detail in my 1998 book, the 21 days of the festival were sponsored by the villages, caste groups, and lineages that, theoretically, were assigned to the specific jobs, tasks and status positions needed within the social system mapped out by the god, Murukan, at the moment of his miraculous appearance to command its creation. And I am using "mapped out" here quite literally since Murukan's command, as recounted in the temple's history, involved delegating tasks and social positions to eight villages, five castes, five matriclans, and one matrilineage spread out over a delineated geographical segment or *theesam* of the Batticaloa district. Nor was sponsorship there, in the 1980s, merely a matter of the present confirmation of an agreed upon past. Over the course of the year important men, *kaurava manithan,* in the various temple- connected villages and castes competed in many ways to become the leaders, the most important men, the *periya kauravakamana alkal*, of their respective groups, and thus entitled to lead the procession on their group's special day. Nor was this merely a merely a matter of prestigious display. The leaders of four of the villages involved in the temples were called "accountants" (*kaNakapiLLai*), that is, members of the four-person board that controlled the temple and the land that supported it. Its temple festival was also, thus, a reconfirmation of their rights (*urimai)* to hereditary power within the temple social system. In this way, Mandur's temple's festival was, in a way, (among many other things) a festival of consilience that recognized the identities and worth of individuals and groups occupying, with the god, the geographic and cosmological center of an intensely felt, temple-anchored, moral landscape.

But this could not be the case in Toronto, at least not in the same way. Consider, for example, *Periya Bhadrakaalli* temple's own list of festival day sponsors – a list

[19] I say "South Asian" because this principle of using an encircling procession to demonstrate sovereignty is found throughout South Asian symbolic practice. Nor is it particularly Hindu. H.L. Seneiviratne's (1978: 89–147) meticulous description of the Kandy Perahära makes this clear, as well as that this principle of conquest continues to be applied in contemporary Sri Lankan politics.

that in its variety is, I think, fairly characteristic of sponsor lists at other Toronto temples.

Periya Bhadrakaalii Festival Day Sponsors by day

1. The town of Trincomalee, Sri Lanka
2. The village of Odaipattu, Tamil Nadu, India
3. The village of Pangudutivu (an island near Jaffna, Sri Lanka)
4. The coastal town of Karavetty, Jaffna District, Sri Lanka
5. The village of Valvettithurai, on the Jaffna peninsula, Sri Lanka
6. The goldsmith caste
7. The Visvabrama caste
8. Nedunthееvu, a village and island in the Palk Straights between India and Sri Lanka.
9. The temple's own musical 'teams' (students in its classical music classes)
10. The temple's own dance 'teams' (students from its classical dance classes)
11. Toronto merchants
12. The Hindu Society of Canada
13. A business family (that owns several businesses)
14. The temple administration
15. "Water Cutting" conducted by the temple administration

Here Sri Lankan Tamil sponsors are mixed up with Indian Tamil ones, organizations and practices of remembrance and persistence (such as the Hindu Society of Canada, and the dance and music classes) with reminders of an urban, capitalist present ('Toronto merchants'). Instead of the temple festival confirming, as in Sri Lanka, relationships between people, a landscape, and a moral order, here the sponsor list is "heteroglossic," bespeaking not one territory but a scattershot spread of points of origin, and not one remembered and legitimizing past but a selection of lost, remembered *ur* and caste positions, and a complex, capitalist, Canadian present. Moreover, in Toronto, unlike Sri Lanka, the festival confirms no rights of influence, no *urimai*, within the temple itself.

That is so because while the *Periya Bhadrakaalli ammaL* temple technically has a board of temple trustees, its board holds no real power. All decisions are made by the brother priests – the religious entrepreneurs who founded the temple. Again, this displacement of communal rights (*urimai*) in favor of "ownership" (with most using the English word here) and, hence, of Sri Lankan Tamil forms of authority and belonging with Canadian ones where, as in a business, control rests with the owners of the temple in a straight- forward manner, appears to be the rule in Toronto. With the exception of the oldest institution in my small sample, the *Iyappan cuvami* temple, all the other temples shared this new, Canadian, command model.

So once again we find paradoxes and incoherencies. Priests acting like CEOs, anonymous Canadian streets used as momentary ritual avenues, generally indifferent Canadian police working as ritual enablers, castes and remembered villages mixed up with new businesses and wholly Canadian holidays, and on and on. Always it is Tamil Canada careening into Tamil Sri Lanka, the temple as a locus of the god's power vying with the warehouse as a real estate deal, Tamil cultural remembrance and persistence crashing into the very venues and realities that would seem to be calling "Tamil" into question altogether. And yet out of this apparently hopeless mishmash of locations and influences emerges, it would seem, in Toronto's Tamil Temples, the required consilience – at least momentarily. How can this be?

Perhaps Sri Lankan temples can do this because they actually spring from a long history of dealing with this kind of 'postmodern' incoherence.

72.5 Temples in Sri Lanka and in Diaspora and the Preservation of Consilience

One problem, I think, with discussions of diaspora is that the lost homeland is too often presented as a stable entity prior to the trauma that precipitated the dispersal of its people. But this was certainly not the case in Sri Lanka, and perhaps never is anywhere. Sri Lanka experienced close to 400 years of colonial conflict and domination before its civil war and this, obviously, constituted as great a challenge to consilience, to the confident assertion of local identities, as more recent catastrophes. In colonial Ceylon, thus, though the land remained the same, the practices and markers of assurance that attached people to it – the temples and shrines, the monastic and priestly forms of life, the agricultural fields and irrigation tanks, the mountains and lowlands (transformed by the needs of capitalist production into "plantations" and "cash crop" farming areas) – were cast into doubt by the new relationships of power and knowledge that the Portuguese, Dutch and finally the British imposed. And perhaps these things were called into question most grievously because they changed, as it were, while *underfoot*. This is why Richard G. Fox once suggested in a lecture before the American Anthropological Association that the colonized world was "postmodern" long before the colonizing world was; and why the historian, Kumari Jayawardena, has been able to show, so firmly, how often and how unintentionally the flow of influence was reversed (1995). All this incoherence was disorienting because it was so unexpected, at least by people at the time; in diaspora, on the other hand, most people assumed things were going to be different.

In any case, Sri Lanka's complex colonial history has been written of in sufficient length to not require rehashing here. But it is important to realize that this barrage of disorienting new colonial activities and understandings was always resisted there in various ways: by violence at first, then, variously and simultaneously, by political intrigue, the assumption of some European cultural practices (such as Anglicanism, Catholicism, cricket, tea-drinking, and racism), the construction of local (and, alas, contradictory) ethnic nationalisms, and, also, more quietly, through a subtle kind of cultural sleight of hand. This last form of resistance, which I have called, elsewhere, "amiable incoherence," generally involved subverting European practices by using them against themselves (Whitaker 1998). As Anne Balckburn has shown in her excellent work *Locations of Buddhism: Colonialism and Modernity in Sri Lanka* (2010), such minglings of European and local cultural practices for strategic effect was an important tactic used by even major players in

Sri Lanka's colonial drama to trick colonial governance into supporting what it was trying to suppress or replace. But I found "amiable incoherence" most easily identifiable in the circumscribed settings of temple disputes.

Hence in my 1998 study, *Amiable Incoherence: Manipulating Histories and Modernities in a Batticaloa Hindu Temple*, I described how local temples elites involved in Mandur's *Sri Kantacuvami* temple used British laws that were intended to dissolve the temple's independent legal existence to, instead, pursue their own temple-oriented political goals, and managed to get colonial police and legal help in doing so. Later, in the 1990s and in conversations I had with temple officials, I found out they were able to engage in similar incoherent yet strategic interchanges with state security forces and the LTTE, in both cases managing to use the mutually contradictory powers of these forces to support their own Temple-based activities. This did not make life for such elites safer. I know that at least one was killed, another captured and imprisoned, and another forced to pay a huge, ruining "tax" to the LTTE. But the temple and the kind of enveloping consilience it offers – one among many forms of consilience, of course, being offered – was able to survive the war.[20]

Nor is this just a power found in the largest Sri Lankan temples or during human disasters such as colonialism or civil war. In 2004, when a tsunami crashed against Sri Lanka's eastern coast killing 30,000 people, it smashed also into the village of Kallady, just outside of Batticaloa town, and partially destroyed its small, relatively new, Murukan temple, undermining the *muulaStaanam*, the hall where the god resides, and leaving it canted at an odd angle toward the sea – as if, in some strange way, bowing before the force that almost destroyed it. Or so the priest suggested when I met him while out on a walk one night with friends. He pointed out that it was possible that the tsunami was a message from the god. The temple had been founded, after all, by an old, childless man in 1982 who was looking for a place to use his money. One day while sitting on Kallady beach under a tree, the old man was approached by a handsome younger man who extolled the beauty of the day and of this particular spot. The younger man being, obviously, in retrospect, Murukan in human form, the older man decided to indeed build a temple on that spot, both to honor the god who spoke to him, and because he was worried about the increased influence of Christianity in Kallady, a village with a substantial Catholic population. But the temple was built facing inland, and, now, had been twice destroyed by the sea: the first time in 1978, by a cyclone, and then, again, in 2004, by the tsunami.

[20] This, of course, was at the kind of large Hindu temple that people in the Batticaloa district call a *teecattakooyil* or 'national' temple. One might expect some sort of effective resistance there. Yet, Patricia Lawrence (2000), working during a period of intense fighting in the east, was able to show how a village temple devoted to the goddess Kali was able to create a kind of protected "public sphere" within which devotees could ask after loved ones caught in the conflict without being destroyed, as they might be otherwise, by the warring sides. They could do this because, first, they were addressing their questions to the goddess herself, speaking through her chosen oracle (her *vakku colluRatu*, her speaking oracles) rather than to an implacable human; and, second, because the ecstatic religiosity of these events attracted Sinhalese army soldiers too.

Obviously, the priest concluded, the god wanted his *muulaStaanam* turned around to face the sea. It helps, I suppose, to know what you are facing.

My point, here, is that Sri Lankan Hindu temples, as forms of action, are well suited for producing consilience out of the intermingling of mutually incoherent events and practices. And I believe this skill, at least, has proven transportable. Hence, even though Temples in diaspora do not have the kinds of relationships to particular moral landscapes that Sri Lankan Hindu temples generally presuppose, they have retained a subversive ability to use unfamiliar, and "un-Tamil," practices and locations to support, often in spite of themselves, the identities those very things would seem to threaten. In this way temples in diaspora are helping Sri Lankan Tamils find themselves regardless of where they are.

Acknowledgements I would like to thanks Dr. Balasundaram Elayathamby, the prominent folklorist, for his aid in conducting the research for this chapter. He is, of course, not responsible for any errors I may have made in the writing if it.

References

Anderson, B. (1992). *Long-distance nationalism: World capitalism and the rise of identity politics.* Amsterdam: CASA—Centre for Asian Studies, Amsterdam. http://213.207.98.211/asia/wertheim/lectures/WL_Anderson.pdf. Accessed 14 Oct 2012.

Birdsall, S. S. (1996). Regard, respect, and responsibility: Sketches for a moral geography of the everyday. *Annals of the Association of American Geographers, 86*(4), 619–629.

Canadian Tamil Chamber of Commerce. (2006). *The emergence of the Tamil Community in the GTA. Facts and Figures*. www.tamilcanadian.com/Canada/report.pdf. Accessed 31 Aug 2012.

Crisis Group Asia Report No 186. (2010). *The Sri Lankan Tamil diaspora after the LTTE*. www.crisisgroup.org/-/media/Files/south-asia/sri-lanka/186%20The%20Sri%. Accessed 31 Aug 2012.

De Certaeau, M. (1984). *The practice of everyday life*. Berkeley: University of California Press.

Fisch, M. (1985). Whewell's consilience of inductions – An evaluation. *Philosophy of Science, 52*, 239–255.

Fugelrud, O. (1999). *Life on the outside: The Tamil diaspora and long-distance nationalism.* London: Pluto Press.

Jayawardena, K. (1995). *The white woman's other burden: Western women and South Asia during British rule*. New York: Routledge.

Kingsolver, A. (2010). *Tobacco town futures: Global encounters in rural Kentucky*. Long Grove: Waveland Press.

Lannoy, R. (1971). *The speaking tree*. Oxford: Oxford University Press.

Lawrence, P. (2000). Violence, suffering, Ammon: The work of oracles in Sri Lanka's eastern war zone. In V. Das, A. KLeinman, M. Ramphole, & P. Reynolds (Eds.), *Violence and subjectivity* (pp. 171–204). Berkeley: University of California Press.

Mines, D. P. (2005). *Fierce gods: Inequality, ritual, and the politics of dignity in a south Indian village*. Bloomington: University of Indiana Press.

Ong, A. (1999). *Flexible citizenship: The cultural logic of transnationality*. Durham: Duke University Press.

Richmond Hill temple. www.thehindutemple.ca/. Accessed 29 Sept 2012.

Ryerson School of Journalism. (2004). Diversity Watch: Tamils. http://www.diversitywatch.ryerson.ca/backgrounds/tamils.htm. Accessed 31 Aug 2012.

Save the Children. (2012) *Children's situation in Sri Lanka*. http://resourcecentre.savethechildren.se/print/941. Accessed 24 Oct 2012.

Seneviratne, H. L. (1978). *Rituals of the Kandyan state*. Cambridge: Cambridge University Press.

Senthi's Tamil Guide. (2007). *Canadian Tamil business directory*. Toronto: Athavan Publications Inc.

Somasundaram, D. (1998). *Scared minds: Psychological impacts of war on Sri Lankan Tamils*. New Delhi: Sage Publications.

Statistics Canada. (2012) *Community profiles from the 2006 Census, Statistics Canada*. www12.statcan.gc.ca/census-recensement/2006/dp-pd/prof…. Assessed 9 Oct 2012.

Tambiah, S. J. (1986). *Sri Lanka: Ethnic fratricide and the dismantling of democracy*. Chicago: University of Chicago Press.

Thiranagama, S. (2011). *In my mother's house: Civil war in Sri Lanka*. Philadelphia: University of Pennsylvania Press.

Thomas, P. (2002). The river, the road, and the rural-urban divide: A postcolonial moral geography from southeast Madagascar. *American Ethnologist, 29*(2), 366–391.

Trawick, M. (2007). *Enemy lines: Warfare, childhood, and play in Batticaloa*. Berkeley: University of California Press.

UNICEF. (2012) *Once a child soldier now an ice-crème truck driver*. http://www.unicef.org/srilanka/reallives_7924.htm. Accessed 24 Oct 2012.

United Nations. (2011) *Report of the Secretary-General's panel of experts on accountability in Sri Lanka*. New York: United Nations. http://www.UN.org/News/dh/i. Accessed 14 Oct 2012.

Whitaker, M. (1997). Tigers and temples: The politics of nationalist and non-modern violence in Sri Lanka. *Journal of South Asia Studies (Special Issue, Conflict and Community in Contemporary Sri Lanka), 20*, 201–214.

Whitaker, M. (1998). *Amiable incoherence: Manipulating histories and modernities in a Batticaloa Hindu temple*. Amsterdam: VU University Press.

Whitaker, M. (2007). *Learning politics from Sivaram: The life and death of a revolutionary Tamil journalist*. London: Pluto Press.

Whitaker, M. (in press). The Sri Lankan Tamil diaspora and time-bomb versus pragmatic theories of post-defeat ethnicity. In D. Rampton & B. Kapferer (Eds.), *Critical interventions: After the war in Sri Lanka*. Oxford: Burgham Press.

Winslow, M. (1987). *Winslow's: A comprehensive Tamil and English dictionary*. New Delhi: Asian Educational Services.

Wilson, E. O. (1998). *Consilience: The unity of knowledge*. New York: Knopf.

Wilson, A. J. (2000). *Sri Lankan Tamil nationalism: Its origins and development in the 19th and 20th centuries*. Vancouver: UBC Press.

Chapter 73
Golden States of Mind: A Geography of California Consciousness

Erik Davis and Jonathan Taylor

73.1 Introduction

California holds a unique position in the vast and complex cartography of American religion—a position that can seem, depending on the angle of approach, at once central and marginal. On the one hand, California—from its economic opportunities to its world-changing media and culture industries to its quasi-mythological status as a site of personal and collective transformation—has played a dominant role in developing and broadcasting American culture and identity, including diverse forms of American religious culture and identity. At the same time, this influence has oftentimes proceeded at the margins and edges of American culture—and nowhere as obviously as in matters of the spirit. For though mainline religious traditions have played crucial roles in the development of California's religious landscape, and though Los Angeles alone is arguably the most religiously diverse city in the planet (Orr 1999), what stands out as the most influential and globally significant of California's many religious currents is that restless, intense, faddish, and often heterodox religiosity—or "spirituality"—that compels both mockery and fascination. We call this current "California consciousness" (Davis 2006): an imaginative, experimental, eclectic, heretical and sometimes hedonistic quest for human transformation that, while principally rooted in Anglo-American sensibility, has manifested as a highly diverse and recombinant set of sects, "cults," lifestyle movements, cultural practices, ontological beliefs, psychological systems, and personal attitudes. In invariably broad brushstrokes, this paper will attempt to map five of the

E. Davis (✉)
Department of Religious Studies, Rice University, Houston, TX 77005, USA
e-mail: ed8@rice.edu

J. Taylor
Department of Geography, California State University, Fullerton, CA 92834, USA
e-mail: jstaylor@fullerton.edu

S.D. Brunn (ed.), *The Changing World Religion Map*,
DOI 10.1007/978-94-017-9376-6_73

major strands of California consciousness: *nature religion, esotericism, counterculture, east-west hybridity*, and *human potential.*

Defining California consciousness is no easier than defining the New Age, with which it shares a great deal. Though world faiths like Buddhism and Christianity have marked the West Coast's alternative spirituality in fundamental ways, many of the paths that criss-cross California fall into that increasingly popular claim of being "spiritual, but not religious" (Fuller 2001). But even the word *spiritual* barely suffices, since some expressions of California consciousness fuse and confuse sacred and profane and are so embodied as to appear indistinguishable from exercise routines or hedonistic revelry. The loss of boundaries sought by some seekers marks the object itself. To take Los Angeles as an example, as early as 1913—well before the occult boom of the 1920s—the writer Willard Huntington Wright was already claiming that:

> No other city in the United States possesses so large a number of metaphysical charlatans in proportion to its population. Whole buildings are devoted to occult and outlandish orders—mazdaznan clubs, yogi sects, homes of truth, cults of cosmic fluidists, astral planers, Emmanuel movers, Rosicrucians and other boozy transcendentalists. (Davis 2006: 107)

One way to generalize about the heterogeneous phenomenon of California consciousness is to underscore its profoundly phenomenological dimension. California seekers could be said to have taken the bait that William James dangled in *The Varieties of Religious Experience*, where he famously defined religion as "the feelings, acts, and experiences of individual men in their solitude, so far as they apprehend themselves to stand in relation to whatever they may consider the divine" (1961: 42). For James, personal experience was the cornerstone of the religious life, rather than dogma or institution or even belief. Prophesying modes of religious experimentation that would characterize California, James opened up the *wunderkammer* of consciousness and embraced mysticism, the occult, and psychoactive substances as valid points of departure. Experimenting with peyote and nitrous oxide, James argued that exalted states of consciousness had to be integrated into any philosophy worth its salt. Though James' approach hardly exhausts our understanding of religion, it certainly helps illuminate the centrality of cognitive and perceptual experience to California consciousness. His key word "solitude" is also telling, for the very informality and restlessness of California consciousness underscores the singularity of the individual seeker, whose sectarian commitments are often transitory and who inhabits what Colin Campbell calls a "cultic milieu" out of which more sectarian identities crystallize and dissolve (Campbell 2002: 14). The emphasis on phenomenology articulates what amounts to one of the few explicitly ideological strands of California consciousness: the insistence on personal experience as fundamental to religious perspective. But it also creates a basis for the sort of comparative methodology found in the history of religions school, where shared experiences allow commonalities—dominant themes, practices, and controlling images and "myths"—to be traced and constructed through fields of strong diversity of scale and expression. Our five strands arise from such an application.

What is particularly significant for a geography of this religious imagination is how rooted California consciousness is in a certain cartographic self-awareness, a geopolitical "position" that might be approached by way of Walt Whitman's poem "Facing West from California's Shores." A patriarch of America's unchurched spirituality, Whitman never actually visited California, at least if you leave aside whatever phantasm appeared to Allen Ginsberg in "A Supermarket in California." But Whitman intuited that the state's identity, and perhaps its spiritual destiny, lay in space. Speaking with his most expansive poetic "I," Whitman imagines a single being that moves westward from Asia until finally arriving at the Pacific (Whitman 1897: 95):

Facing west from California's shores,
Inquiring, tireless, seeking what is yet unfound,
I, a child, very old, over waves, toward the house of maternity, the land of
  migrations, look afar,
Look off the shores of my Western sea, the circle almost circled.

Here Whitman announces two geographic myths that collide. In one, the West heads west, as Euro-American civilization expands until it metaphorically runs aground in California. At the same time, perhaps sensing California's location on the bubbling Pacific Rim, Whitman folds this finality back towards Asia and its ancient traditions. The circularity of the planetary sphere unveils the utopian possibility of return and reunion at the very apogee of escape. Whitman ends the poem by asking a question that goes to the heart of the region's restless spiritual imagination: "But where is what I started for so long ago?/And why is it yet unfound?"

The seeker orientation that lies at the center of California consciousness is bound up with the economics of the state, including postmodern capitalism—a new socio-economic order that, as the example of Silicon Valley alone proves, California played a privileged role in creating. California's robust "spiritual marketplace" led to a diverse production of social spaces which furthered the flourishing of new and hybridized religious offerings in ever-evolving and profoundly mediated forms. These spaces vary tremendously; from the bohemian enclaves of the Haight-Ashbury or Venice Beach, to the religious landscapes of Tassajara Zen Mountain Center or San Jose's Rosicrucian Park, to the innumerable psychic fairs, desert full-moon raves, online communities and other more transient spaces erected to cater to the needs of the restless and recombinant California spirit. Nature itself—in the form of mountains, woodlands, deserts, the coast and hot springs—is revered, but also transformed into new socially produced sacred (and profane) spaces, which along with their urban counterparts serve to both reflect and reinforce the significance attached to them, in an unending dialectic of social processes and individual experience. In this way California becomes not just the place where currents come together to form a new and globally visible socio-spiritual template, but a landscape of modern religious innovation itself, a "visionary state" (Davis 2006).

73.2 Nature

One wellspring for California's unusual spiritual culture lies in the state's diverse and inspiring natural environment—its dramatic topographies, bioregional heterogeneity, beauty, and often ideal climate. As California grew into a national myth in the minds of nineteenth century Americans, the state's epic wilderness catalyzed an imagination of place infused with religious forces. The very notion of wilderness was, of course, already a feature of the American religious imagination, an ambivalent dimension marked at once by Puritan demonization and the Transcendentalist reverence voiced by Emerson and Thoreau, who unveiled what Catherine Albanese calls "an environmental religion of nature" (Albanese 1990: 80–105). In the case of California, it is this organic Transcendentalism that came to the fore. One example is the reverent awe with which early Anglo-American visitors regarded the Sierra Nevada's groves of giant sequoias, the first and most charismatic symbol of the California wilderness, and one which has routinely inspired religious language—especially comparisons to Gothic cathedrals.

In the Sierras, this ambient nature religion found its voice in John Muir, California's first great prophet of the wild. In his writings, Muir combined Transcendentalist intensity, naturalist detail, the prophetic rhetoric of the Bible, and a strong vein of pantheism. Revising the language of the Gospel of John, for example, Muir once famously wrote "Now we are fairly into the mountains, and they into us" (Albanese 1990: 99). Based in part on his own extraordinary experiences, including a famous climb up Mt. Ritter where Muir felt a preternatural spirit aid him as he negotiated a particularly dangerous passage, Muir's writings were in turn inspirational. They encouraged Americans to retreat to nature as a respite from the "galling harness" of civilization and to discover for themselves a naturalized framework for unchurched and sublime experience. Tracing the specifically religious dimension of Muir's legacy is not easy. Given its informality, interiority, and naturalistic leanings, the religion of nature is sometimes difficult to isolate from hiking, surfing, and other practices of outdoor leisure. At the same time, earth-based and "pagan" sensibilities are also found throughout California counterculture, whose "back to the land" communes, wilderness vision quests, and natural food practices are clearly imbued with green spirituality.

Though the founding of the Sierra Club and Muir's battles to preserve wild places were achievements of governmentality rather than religion, Albanese reminds us that "it was the presence of the religion of nature that gave to preservationism its vital force" (Albanese 1990: 95). Throughout the twentieth century, the religion of nature continued to inform American environmentalism, much of whose theory and practice—from Greenpeace to eco-psychology—was developed up and down the West Coast. Movement texts surrounding the aggressive and militant Earth First! campaigns in the 1980s and 1990s, which included the defense of old growth redwoods in Northern California, were rich with Neo-paganism, and the rallying cry of "Back to the Pleistocene" suggested a pantheistic anti-modernity (Taylor 2009: 71–102). The influential Californian witch Starhawk embraced anti-nuclear activism in the 1980s as a holistic element of her pagan spirituality. Perhaps the most charismatic and media-genic exemplar of Muir's legacy in recent decades was Julia Butterfly Hill, a young activist from Arkansas who, in 1997, climbed 180 ft up a Humboldt County redwood named Luna and remained there until the Pacific Lumber corporation agreed to save the tree. Like Muir, Hill had found her calling after a traumatic, life-changing accident, and though she insisted that she was no saint, her self-performance and language were deeply shaped by American religion, and by the American—and Californian—quest to leave religion behind and discover spirit in the world at hand.

California's environment marks the state's spiritual culture in more grounded ways as well. In Southern California, especially, the mild Mediterranean climate encouraged sunshine living and a measure of ease that stirred up fantasies of a paganish "return to Greece." Indeed, the first of the nation's outdoor Greek Theaters was built in 1901 by the Theosophical Society at Point Loma near San Diego, where Shakespeare and Theosophical dramas were performed in flowing white robes. Meanwhile the burgeoning agricultural industry cemented images of a California Eden of outsized fruits and vegetables, deepening the association of California with nature even as rapid industrialization marked the state. Particularly significant for spiritual culture were the tens of thousands of consumptives, asthmatics, and other invalids who came to Southern California in the late nineteenth and early twentieth centuries, drawn by the promise of health-restoring climate and lifestyle (Frankiel 1988: 59–61). In 1895, one census suggested that Los Angeles featured the largest number of doctors per capita of any city in the world (Baur 1959: 87). Many of these doctors did not practice allopathic medicine, which led one medical examiner at the time to dub Los Angeles County "the Mecca of the quack" (Baur 1959: 81). While alternative medicine was a complex, dynamic, and frequently religious force throughout the United States, Southern California developed, at the dawn of the century, a back-to-nature culture of health food, vegetarianism, raw foodism, and inventive mind-body practices that remain strong to this day (Kennedy 1998). The strong spiritual dimension to healing in Southern California not only sets up the later rise of "holistic health," but also helps explain the earlier runaway success of Christian Science and New Thought in the Southland and eventually throughout the state. It may also lie behind the subtle but significant differences between the spiritual temperament of Northern and Southern California, with the latter arguably more renowned for its embodied expression and commercial exuberance.

California's culture of the spiritualized but hedonic body defies easy divisions into sacred and profane. A good example of this ambiguity is modern yoga, a world and national phenomenon that fuses spirituality and physical culture, and was partly engendered by the California experiment (Syman 2010). In the 1950s, for example, the innovative bodybuilder Walt Baptiste opened the first comprehensive yoga studio with his wife Magana, who had studied with the legendary Hatha yogi Indra Devi in Hollywood; beginning from an ethic of fitness, Baptiste yoga gradually transformed into a more explicitly spiritual expression. Another hedonic spirituality, even less explicitly religious but certainly more grounded in place, is surfing. Though the modern resurrection of the sport began with Hawaii's Duke Kahanamoku, it spread early to Southern California. In Los Angeles, the legendary Tom Blake not only revolutionized the sport with lighter, hollow boards but articulated an explicitly spiritual approach to the practice, one informed by his creed "Nature = God." Despite a sometimes aggressive and adolescent culture, the rituals, ethos, and fluid phenomenology of surfing have made it, in a variety of ways, a particularly informal but influential expression of what Bron Taylor describes as "dark green religion," a pagan pantheism in which sensual experiences in nature constitute a "sacred center" (Taylor 2009: 103–126). And the legacy of Muir remains; Taylor explains that for many surfers their experiences of pleasure, challenge, and flow lead directly to "ethical action in which Mother Nature, and especially its manifestation as Mother Ocean, is considered sacred and worthy of reverent care" (Taylor 2009: 104).

73.3 Esotericism

Over the last few decades, scholars of Western religion have begun to pay increasing attention to esotericism, a frequently disguised and counternormative stream of religious and occult ideas and practices associated variously with gnosticism, hermeticism, alchemy, freemasonry, and ceremonial magic. While definitions of esotericism differ, Antoine Faivre established an influential check-list that includes the idea of enchanted nature; the notion of mystic correspondences (as in the cosmic connections that underlie astrology); the mediating power of the imagination; and the experience of transmutation, or, in more contemporary terms, transformation (Faivre 2010).

Esoteric currents coursed into America from the colonial period onwards (Horowitz 2009). Perhaps the most influential esoteric body in America was the Theosophical Society, an organization cofounded in New York in 1875 by Colonel Henry Olcott and the Russian writer and seer Madame Helena Blavatsky. Weaving together occult Neoplatonism, parapsychological practices, and often rather bowdlerized philosophical ideas from the East, Theosophy helped created the template for today's New Age, which Wouter Hanegraaff characterizes precisely as a secularization of esotericism (Hanegraaff 1996). Theosophy played a role in introducing Buddhism, Hinduism and yoga to the West, and also supported Buddhist and Hindu revival movements in India and Sri Lanka. At the same time, Blavatsky's writings attempted to reframe the supposedly universal truths of mysticism in the light of contemporary scientific advances, including Darwin's account of evolution and the discovery of the electromagnetic spectrum, whose "vibrations" provided a ready model for Theosophical accounts of spiritual energy. Blavatsky believed that humanity was beginning to mutate into a new and superior "sixth race," a transformation that she believed would occur in America (Blavatsky 1888). Annie Besant, another Theosophical leader, believed this would happen specifically in Southern California; Besant claimed that the finest magnetic vibrations in the world were to be found in Pasadena (Davis 2006). In the 1960s, the notion of Theosophical mutation continued into the hippie counterculture with speculation about the birth of a new Aquarian generation (Lachman 2001).

In some ways both women were right: California would become a crucial site in the history of Theosophy, and a principal stage in its transformation into New Age culture. In 1897, Katherine Tingley founded a Theosophical community on San Diego's Point Loma; nicknamed "Lomaland" by the locals, it lasted for over 30 years. The Temple of the People set up its headquarters in Halcyon, near Pismo Beach, while a Theosophical group led by Albert Powell Warrington—and allied with Besant, Tingley's rival—founded the Krotona colony in the Hollywood Hills in the early 1910s. In 1924 Warrington moved Krotona to the remote Ojai Valley near Santa Barbara, a dry green oasis whose name is Chumash for "nest" or "moon." Besant loved Ojai and made it the home of her Happy Valley School—an alternative high school, designed to nurture the coming sixth race, still thriving today.

Ojai would also play a transformative role in the life of Jiddhu Krishnamurti, the most significant teacher to emerge from Theosophy in the twentieth century. As a youth, Krishnamurti was recognized by the Theosophical leader Charles Leadbeater as the human vessel for the coming World Teacher—the Theosophical equivalent of the Messiah. In the early 1920s, Krishnamurti and his brother Nitya moved to Ojai, drawn, like so many incoming Californians, by the promise of health, as Nitya had tuberculosis. In the summer of 1922, beneath a pepper tree in Ojai, Krishnamurti had an overwhelming experience of "God-intoxication." He began to question Theosophy, and in 1929, Krishnamurti left the society. Rejecting the whole notion of mystical schools and proscribed spiritual practices, as well as the very concept of the guru, Krishnamutri proclaimed that the spiritual search cannot be organized, that "truth is a pathless land" (Lutyens 1975: 272). Personal friends with a wide range of Southern California intellectuals, artists, and celebrities, from Aldous Huxley to Charlie Chaplin, his astringent message introduced a powerful existential dimension to the emerging New Age, though Krishnamurti, who kept Ojai as his home base throughout his life, remained a popular and influential anti-guru guru.

Other offshoots of Theosophy took a more recognizably cultic form. In 1930, a mining engineer, Theosophy student, and onetime American Nazi named Guy Ballard claimed to have met a mysterious Ascended Master named Count Saint-Germain in the environs of Mount Shasta. Ballard and his wife went on to found the Mighty I AM Religious Activity in Los Angeles, a controversial sect that popularized many of the mystic themes later bandied about by New Ager leaders: a color-coded chakra system, a fringe science of electronic vibrations, and an insistence that individuals create their own reality. For the followers of I AM, this also meant manifesting material prosperity, a theme that also appears in many of branches of the New Thought movement—a parallel development, descending from Emersonian transcendentalism, that believed in the creative powers of the mind.

In the postwar period, Theosophical esotericism also fed into the modern UFO contactee movement, which began in California in 1952 when George Adamski, who worked at a café near Palomar observatory, claimed to have encountered a golden-haired Venusian named Orthon in the desolate Mojave Desert. As in the 1951 Robert Wise movie *The Day the Earth Stood Still*, the space being chastised humans for waging war with nuclear weapons and poisoning the planet. Unlike those who believed that UFOs were nuts-and-bolts crafts, contactees like Adamski, who had once belonged to a Southern California esoteric group called the Royal Order of Tibet, transformed the Theosophical notion of Ascended Masters into the quasi-scientific idea of the Space Brothers. Saucer sects remained a part of the California spiritual landscape, though not always so optimistically. In 1997, 39 members of the group Heaven's Gate dispatched themselves near San Diego, believing their souls would escape the illusory nightmare of material reality by hitching a ride on a spaceship riding the tail of comet Hale-Bopp (Davis 2006).

Theosophy was not the only source of California esotericism. Perhaps the most important "Rosicrucian" order in the United States, the Ancient Mystical Order of the Rosae Cross, moved to San Jose in 1927 and built, in addition to their temples, a public park and Egyptian museum that exist to this today (Fig. 73.1). Another local source was the work of Manly P. Hall, a Canadian who moved to Los Angeles as a young man, where he soon authored a remarkable, influential, and beautiful compendium of esoteric lore called *The Secret Teachings of All Ages* and subsequently founded the Philosophical Research Society. Hall's copious books, many lectures, and significant library, including many rare alchemical texts, helped seed Southern California with a myriad of esoteric signs, symbols, and information. The Southland

Fig. 73.1 Rosicrucian Park, headquarters of Ancient and Mystical Order Rosæ Crucis (AMORC), San Jose (Photo by Michael Rauner, used with permission)

would also play a pivotal role in the legacy of Aleister Crowley, a ceremonial magician and esoteric writer who would eventually be recognized, along with Blavatsky, as one of the most influential occult figures to emerge from the Victorian period. When Crowley died in obscurity in 1947, Southern California's Agape Lodge #2 was the only still functioning center of Crowley's initiatory society the Ordo Templi Orientis. Led for a time by the remarkable Jack Parsons, an innovative rocket scientist who co-founded Pasadena's Jet Propulsion Laboratory, the Agape Lodge helped keep Crowley's mystical religion of Thelema alive until the occult boom of the 1960s (Carter 1999). Though Crowley was unfairly maligned during his life and posthumously as a "Satanist," California did eventually play host to an overtly diabolic initiatory order when Anton LaVey founded the Church of Satan in San Francisco in 1966. Though espousing more of a hedonic and social Darwinist philosophy than a metaphysical one, the Church of Satan did later spawn the Temple of Set when a disaffected member—the U.S. Army Lieutenant Colonel and PSYOPS (psychological operations) specialist Michael Aquino—founded a left-hand order with far deeper ties to esoteric ideas and practices. So though the lens of California very much helped intensify the "white light" of the New Age's secularized esoterica, the state also served as a home-grown matrix for what another native Californian, George Lucas, would later dub "the dark side of the force."

73.4 East-West Hybridity

Asian religious traditions have played a large role in the development of California's religious landscape and spiritual supermarket. While introduced early on, when Chinese immigrants to Gold Rush California built the first Joss Houses (Taoist temples), it was mainly from the 1920s through the 1970s that Eastern religious and philosophical thought established its pivotal and lasting influence on California consciousness. In addition to a steady stream of teachers from Asia, Theosophists, intellectuals, and countercultural figures were central conduits for this religious dissemination. While east-west hybridization and orientalism far predate this era, at no other time were ideas so utterly non-Western in origin fused so thoroughly into popular culture and thus mass consciousness. While New York and Boston functioned as the primary academic centers for Buddhist and other Asian philosophies in the United States, the cultural and countercultural epicenter of this movement was most broadly established in California.

Orientalism, as discussed by Edward Said, is frequently thought of as a one-sided affair, in which Western depictions of "the Orient" are used to assert the supremacy of Western culture, traditions, and ways of seeing the word; this assertion in turn helps naturalize political, economic, and cultural domination (Said 1978). However, more recent scholarship on Orientalism stresses its role in opening up the West to new heterodox currents of thought that often led to a profound questioning of politics, worldviews, cultural practices, and identity formation in the West (Clarke 1997). This was undoubtedly the case in Orientalist influences on European

Romanticism, in the establishment of American Transcendentalism, and in countercultural California. The most important importations have been the non-dual spiritual philosophies found in Advaita Vedanta; in Taoism; and in Therevada, Zen and Tibetan Buddhism; likewise the more applied techniques of self-transformation found in Tantra, Yoga, Vipassana and other meditative traditions. With a move towards charismatic and spiritual leaders (gurus) in the late 1960s, a variety of sectarian movements based around neo- or pseudo-Asian traditions also emerged, most with major California chapters. Yogi Bhajan's 3HO (Healthy, Happy, Holy Organization), Transcendental Meditation, and the Hare Krishna movements are all well-known examples.

A number of ideas from these traditions strongly impacted California consciousness. One central concept is the non-duality of *Brahman*—the transcendent impersonal reality of unity that lies beyond the world of appearances, found in the Upanishads and most subsequent Hindu philosophy, and in a markedly different form, in China as the Tao. This concept was largely introduced into California through the Advaita Vedanta school of Hinduism, especially in the guise of the Vedanta Society, which founded major centers in San Francisco in 1905 and Los Angeles in 1930 (Fig. 73.2). These centers bear a direct lineage from the celebrated Hindu leader Swami Vivekananda, who modernized the teachings of his guru, the nineteenth century Hindu saint Ramakrishna, founding a Western-friendly "practical Vedanta." The Hollywood Vedanta Society made a particularly important mark on the British expatriate writers Gerald Heard, Christopher Isherwood, and Aldous Huxley. In addition to his influential religious study *The Perennial Philosophy* (1946), Huxley would also express his Vedantist and perennialist concerns with the experiential unity underlying the world's religions in the insightful essays on his mescaline experiments published as *The Doors of Perception* (1954) and *Heaven and Hell* (1956).

Another significant Hindu influence on California consciousness came from the Indian religious leader Paramahansa Yogananda, who established the Self-Realization Fellowship in California in 1920. Still headquartered in Los Angeles, with temples in Hollywood, Encinitas, Pacific Palisades, and around the world, the SRF espoused an ecumenical philosophy of religion, centered on creating a spirit of understanding and goodwill among the world's faiths. Yogananda was a practitioner of kriya yoga, which seeks to attain states of inner peace and attunement with Brahman through advanced techniques of pranayama and meditation. These techniques were widely disseminated by the SRF, and were also promulgated in Yogananda's best-selling *Autobiography of a Yogi* (Yogananda 1959), which owes some of its popularity to its author's account of extraordinary spiritual experiences.

Though Japanese Zen priests were already teaching in San Francisco in the first decade of the twentieth century, Buddhism was not widely embraced by Anglos until the postwar period, when influential North Beach Beats like Gary Snyder, Allen Ginsberg and Jack Kerouac turned towards the dharma (Seagar 2000). They, and the many who followed them, were especially attracted to Zen, which had become popular across the country through the writings of D.T. Suzuki and Bay Area-based scholar, writer, and radio presence Alan Watts. Important formal

Fig. 73.2 Old Vendanta Temple, San Francisco (Photo by Michael Rauner, used with permission)

exchanges between Asian teachers and Western students occurred during this era. Robert Aitken, later one of the most celebrated teachers of Western Buddhism, studied with Nyogen Senzaki at the "Floating Zendo" he set up in Los Angeles following his wartime internment. Another significant transition occurred in 1962, when Zen teacher Shunryu Suzuki left his Japanese congregation in San Francisco to found a meditation center specifically aimed at teaching Westerners (Chadwick 1999). Transplanting the Asian traditions of placing monasteries in remote upland settings, Suzuki's San Francisco Zen Center also built Tassajara—the first Zen monastery built outside of Japan—in a designated wilderness area of the Los Padres National Forest near Carmel. Nowadays, major Zen centers can be found in the San Jacinto mountains near Idylwild (Yokoji-Zen Mountain Center), in the Santa Cruz mountains south of San Francisco (Jikoji), in the San Gabriel Mountains north of Claremont (Mt. Baldy Zen Center), in the hills of Sonoma County near Santa Rosa (Sonoma Mountain Zen Center), and in the urban centers of Los Angeles, San Francisco and San Diego. Chinese Zen temples include the City of Ten Thousand Buddhas in Mendocino County near Ukiah, which competes with the Hsi Lai Temple of Hacienda Height in east Los Angeles County for the honor of the largest

Buddhist temple in the Western Hemisphere. In these various implantations of Buddhist temples into the cultural landscape, we can recognize two regional strands that remain largely distinct: ethnic and often immigrant Buddhist communities with direct ties to Southeast and East Asia, and American Buddhists, often led by Anglos.

Though an increasingly important influence on Western spiritual thought, psychology, and most recently, neuroscience, Tibetan Buddhism was slower to make inroads into California. The first Tibetan site in California, the Tibetan Nyingma Meditation Center, was established by Tarthang Tulku in Berkeley in 1969; later he would build the enormous (and inaccessible) Odiyan Buddhist retreat center, a ritual landscape in coastal Sonoma county. In popular culture, the imagery and basic concepts of Tibetan Buddhism gained exposure after Timothy Leary and his colleagues adapted the so-called *Tibetan Book of the Dead* into an LSD trip guide renamed *The Psychedelic Experience* (Leary et al. 1964), which became a de rigour manual for acid-worshipping acolytes like Laguna Beach's infamous drug smuggling organization the Brotherhood of Eternal Love (Schou 2010). By the early 1970s, a growing number of Buddhist, Taoist, and Hindu texts also became available to the counterculture. As the 1960s drew to a close, many hippies jaded by the degeneration of the scene followed the lead of the Beatles and Richard Alpert (Ram Dass) away from psychedelic Dionysianism towards the other apparent "enlightenments" offered through meditation and guru worship. From the 1970s on new Eastern-themed fads swept the state, which continued to set many countercultural trends, including a rather distorted and sexually-obsessed version of Tantra that some dubbed "California tantra" (Fuerstein 1998). A more broadly significant form of West Coast neo-tantra was identified by the religious scholar Jeffrey Kripal, who argued that the human potential movement centered at Esalen drew its cultural power from the transformations of an essentially erotic energy (Kripal 2007). After ebbs and flows of popularity over the decades, an Orientalized but still body-focused culture of various postural yoga regimens has now become deeply installed in the cultural fabric of the state and the nation.

73.5 Counterculture

In the popular imagination, California has long been linked with hedonism, an association that can partly be traced to promotional boosterism and media portrayals. However, the stereotypes of hedonism and the pursuit of liberated pleasure have their origins in a variety of bohemian and countercultural movements that either emerged or took strong root in California. The roots of these movements are also bound up with the unchurched religion of "spirituality," as well as with the new understanding of the human psyche that took hold after the popularization of the Freudian and Jungian models of the unconscious. Motivated by the tireless quest for intense experience and a creative social life, California counterculture drew together eroticism, psychological (and psychoactive) experimentation, and religious practice into a rich set of exploratory currents that embraced both the sensual body and more subtle altered states that could be tapped within.

California counterculture was seeded early in the twentieth century by a number of transplants from Germany who brought with them a romantic proto-hippie philosophy of natural living and mystical seeking associated with the Wandervogel movement (Kennedy 1998). California's spatial and cultural openness was also recognized by Henry Miller, most famous for his erotically charged semi-autobiographical novels written in the 1930s. In 1940, Miller settled in a remote cottage in Big Sur and remained in California until his death in 1980. Miller's work, whose unabashed sexuality and experimentation was complemented by a growing interest in Eastern mysticism, was in turn a pivotal influence for the writers of the Beat Generation, most of whom ended up in San Francisco in the mid-1950s. Exploring new literary forms and lifestyles, the Beats took their cues not only from the frenetic improvisation of bebop, but also from their experimentation with drugs, sexuality, and Zen, the latter of which was a primary interest of the West Coast poet Gary Snyder, who would formally study the religion in Japan in the 1960s and later co-found a Zen center in the Sierra Nevada.

Beat writing was based on the primacy of lived experience, a crucial theme that laid the groundwork for the rapid growth of the hippie counterculture in the 1960s. A crucial figure in this transition was author and countercultural icon Ken Kesey. Given psychedelic drugs at Stanford as part of a CIA-sponsored research program (Lee and Shlain 1992), Kesey helped catalyze the counterculture through the antics of his group, the Merry Pranksters, who spread an anarchic gospel of personal and social experimentation through a series of multi-media psychedelic "happenings" called the Acid Tests (Fig. 73.3). Experimenting with lighting, electronics, music, and "expanded cinema," these events facilitated immersive experiences that broke down social norms and conventional cognitive frames. While the Acid Tests were short-lived, their influence lived on through their house band, the Grateful Dead, who would grow from their Bay Area roots into de facto psychedelic evangelists who inspired a cultlike following of Deadheads across the U.S. and the world.

The counterculture's embrace of psychedelic drugs, particularly LSD and marijuana, meant that an enormous number of young people were experiencing non-ordinary states of consciousness. This dovetailed with the experiential orientation that many believed lay at the heart of both Asian religion and occultism. The result was an "Aquarian" generation who believed that they were creating a new kind of humanity through the embrace of mystical or magical practices within an ethos of human potential and resistance to the "system," represented by consumer culture and the military-industrial complex. An early "spiritual supermarket" emerged in San Francisco's Haight Ashbury, which was saturated with images, ideas, practices, and teachers drawn from Theosophy, Buddhism, Hinduism, Jungian psychology, Sufi mysticism, shamanism, magick and the occult, astrology, parapsychology, and Christianity (Roszak 1976). This "freaky" mystic brew was also communicated to the rest of the world through San Francisco's popular and influential psychedelic music and art. Even as the Haight became an object of media exposure, its very existence provided a space produced by and for the counterculture, a brief that attracted flocks of young people from across the country, reinforcing California consciousness as well as a cultural politics of liberation. New forms of communal

Fig. 73.3 The Onion, Sepulveda Unitarian Universalist Society, site of a 1966 "Acid Test," Northridge (Photo by Michael Rauner, used with permission)

living, alternative sexualities, street theater, and anarchist-communitarian politics formed, held in uneasy balance with the more militant wings of the political counterculture typified by Berkeley. By late 1967, however, the Haight had begun to degenerate, filled with homeless teen runaways and a growing street trade in heroin and speed. This fueled a movement away from the urban milieu "back to the land" and to rural communes like Black Bear Ranch, founded in 1968 in Siskiyou County.

From the perspective of mainstream America, the darkest legacy of California's hippie counterculture was Charles Manson, who was arrested near Death Valley in October 1969, in association with a series of brutal murders in Los Angeles. Initially based in the Haight-Ashbury, Manson maintained control over his young and largely female "family" through a mix of sex, drugs, and a charismatic authoritarianism based on a rhetoric of fear and psychedelicized non-dualism. Along with the killing of a concertgoer at Altamont at the hands of an allegedly LSD-dosed Hell's Angel (Schou 2010), Manson marked the end of the hippie era. However, by this time, the counterculture had itself been heavily mediated through consumer culture, and as its own "counterculture industry" took off in the 1970s—especially rock music—many of its less radical aspects were incorporated into the cultural mainstream.

The erosion of the 60s dream had a direct influence on American religion by clearing the ground for a variety of new religious movements to take root, many of which proved controversial. Burned out by drugs, hurt by the sexual revolution, and confused by the clash of non-ordinary worldviews, many young people gravitated towards gurus, sects, and sometimes authoritarian groups like the Jesus Movement (its followers referred to as "Jesus Freaks"), Hare Krishna, the Unification Church ("Moonies"), est, Rajneesh, and Scientology. While the fears of dangerous cults and their "brainwashing" techniques loomed large in the public mind, and were intensified with the mass suicide that Jim Jones commanded in 1977 after he moved his People's Temple from San Francisco to Guyana, these new religious movements also must be seen in the context of a larger and more informal "cultic milieu" of spiritual seeking, occult studies, and human potential psychology that characterizes the New Age.

One non-authoritarian new religious movement that strongly took hold in the 1970s was Wicca. Inspired by the largely vanished legacy of Europe's own earth wisdom, British aspirants started constructing modern witchcraft in the 1940s, and its polytheistic beliefs and rich magical practices blossomed alongside the counterculture in England and America, where it frequently was reconceived as part of a larger Neo-pagan current. An obvious site for Neo-pagan developments, California particularly stands out for its feminist focus, as figures like Victor Anderson and especially the Los Angeles-based witch Z. Budapest increasingly focused attention on the Goddess (Hutton 1999). In 1977, the Californian witch Starhawk published *The Spiral Dance*, a best-seller that widely disseminated her feminist, ecological, and deeply engaged form of witchcraft (Starhawk 1979).

The continued countercultural interest in psychedelics set the stage for the arrival of one of the most famous shamans and teachers of the era: the almost certainly fictitious Don Juan, whose gnomic wisdom and challenging practices were disseminated through an enormously popular series of books written by onetime UCLA anthropology graduate student Carlos Castaneda. Despite Castaneda's academic trickery and his eventual formation of an exploitative sect, his first books beautifully evoked a supernatural world only thinly separated from this one, and helped rekindle a romance of the desert and an imaginative engagement with the visionary traditions of the Americas. Another notable Californian "freak" who studied indigenous shamanism was Terence McKenna. A long-standing figure at Esalen, McKenna later became a spokesperson for the psychedelic revival associated with the rave scene in the early 1990s, encouraging listeners to take "heroic doses" of mushrooms or the intense, short-acting substance DMT in order to visit other dimensions and communicate with entities who dwelled therein. While the rave scene was a global phenomenon, it received a particularly hippyish (and Acid Test-worthy) remix in the vital and influential San Francisco and Los Angeles scenes, many of which also held their events at beaches and deserts, manifesting what McKenna called "the archaic revival" (McKenna 1991).

73.6 Mind Science

As Wouter Hanegraaff and others have made clear, one of the definitive elements of the New Age is its embrace of (and parasitism upon) the discourse of science (Hammer 2004). One way to effect this shift is to recode and interpret divine powers as largely untapped forces inherent in our own evolving psycho-physiology—as "human potential" that can be catalyzed through various "scientific" techniques and practices. This concept had already been set in motion within the Theosophical and New Thought movements that preceded the New Age. In the latter, religious or esoteric notions of prayer and sacred power were transformed into the increasingly secular forms of positive thinking. Paradigmatic here is the "the science of mind" announced in 1927 by Ernest Holmes, who founded the Institute for the Religious Science and School of Philosophy in Los Angeles. This loosely "psychological" fusion of religious and scientific aspirations also led to new institutional forms that do not fit comfortably into the familiar sociological triumvirate of church, sect, and cult. Holmes, for example, did not want his Religious Science to take the form of "churches," but rather of teaching institutions, or "centers." By reframing religious aspirations and spiritual practices within essentially secular institutional forms (the center, the institute, the seminar), the broad current of the New Age was able to provide a novel and influential sociological ground for its reconfiguring of the relationship between spirit and science.

Many of these developments occurred in California. Certainly the most influential was the Esalen Institute, which was formed on the stunning Big Sur coast in 1961 by two Stanford graduate students named Michael Murphy and Richard Price. Though Murphy had spent 16 months at the philosopher and mystic Sri Aurobindo's ashram in India, and the bohemian Price had practiced Zen, the two intellectuals were committed to radical psychological development outside of traditional spiritual models of guru-like teachers or religious conformity. In this they were influenced by Gerald Heard and the short-lived Trabuco College that Heard had founded in the 1940s in a remote canyon in the Santa Ana mountains of Orange County. In his own thinking, Heard embraced the possibility of what Jeffrey Kripal calls "evolutionary mysticism," and wanted Trabuco to be what Heard called a "gymnasia of the mind" (Davis 2006: 164).

Esalen became that gymnasia, although it placed at least as much emphasis on the body as the mind, while, of course, insisting on the holistic connections between them. Esalen became ground zero for the human potential movement, an eclectic and influential blend of psychological therapies and secularized spiritual practices that transformed the American image of the self. Once again, individual phenomenology lay at the core of the work. The psychologist Abraham Maslow's conception of self-actualization and the "peak experiences" that helped catalyze such transformation were crucial. By the late 1960s, Esalen had become a petri dish in which an enormous number of techniques were explored: Gestalt therapy, meditation, tai chi chuan, psychedelics, Rolphing, primal scream therapy, holotropic breathwork, hatha yoga, biofeedback, Tantra, massage, and the encounter group. At times,

Esalen participants must have felt like they were surfing the edge of human evolution, as if a new kind of person was being birthed. By the end of the 1970s, Esalen's self-developmental therapies and holistic ideas had spread around the world, even as the institute became the flashpoint for attacks on American narcissism and the Me Generation.

Here an evolutionary embrace with human potential inevitably was engaged with parapsychological topics, another privileged site of the crossover of science and the sacred (Kripal 2010). Just how far such mind sciences could penetrate more orthodox research institutes in California can be seen at the Stanford Research Institute, a facility in Palo Alto, sponsored by Stanford University, that has performed high-end research-and-development work for the government and private corporations for decades. In the early 1970s, Russell Targ and Harold Puthoff, two SRI researchers who specialized in laser technology, became fascinated by the hidden powers of the mind, and convinced the CIA to support an in-depth program of remote viewing—the paranormal ability to mentally visualize and describe a distant place or object, like a toy hidden in a box a mile away (or a submarine beneath the ice). Puthoff and Targ focused on so-called "gifted individuals" like the New York artist Ingo Swann, who, like Puthoff and other research subjects in the program, was a Scientologist. L. Ron Hubbard's Church of Scientology, of course, also has a significant foothold in California, and though its institutional form is hierarchical and sectarian, this explicitly "religious" form grew out of the secular and relatively informal practice and theory of Hubbard's Dianetics, which was itself first announced in the pages of a science fiction magazine (Urban 2011).

73.7 Conclusion

In this brief chapter, we have attempted to identify some of the main currents and subsidiary streams of what we call "California consciousness:" a diverse, inventive, and frequently eclectic culture of spiritual seeking that straddles the borderlines between religion and lifestyle, science and culture, hedonism and mysticism, mind and body. The restless diversity and even contradictions of this major current of American spiritual quest—what Kripal calls the "religion of no religion" (Kripal 2007: 9)—makes it elusive to define, but that very diversity—and even the contradictions it entails—is part of its substance, Campbell's amorphous "cultic milieu" writ large. The "spiritual supermarket," which has become an important concept for studies of modern spirituality and one which California singularly embodies, is not only a sociological index of the incursion of market values into religion (Carrette and King 2005). The spiritual supermarket is also a reflection of the diverse knowledges and practices that characterize global modernity as well as the utopian possibilities of a global, radically integral, even planetary view. Arising in a twentieth century melting pot, California consciousness can thus only be approached in the global spirit of religious comparativism.

At the core of this approach is the attention to human experience over and above the specific sociological or conceptual structures of any given religious or spiritual form. Indeed, it is only by understanding the call of subjective experience that many of the specific characteristics of California consciousness—its restlessness, embrace of altered states, and vacillation between hedonism and asceticism—can be understood. For well over a century, countless Californians have embraced the nonrational as if it were a portal into a deeper freedom, opening their arms to everything from nature mysticism to psychedelic rapture, from Pentecostal fire to the "choiceless awareness" of the deeply meditating mind. In many ways, California consciousness sought to question and transcend the quotidian ego, dissolving the small self into the larger frameworks of mind and body and the wilderness within and without.

California consciousness can also be understood in terms of its geographic mythology. A liminal landmass, whose location over some of the most significant faultlines on the planet lend the territory its geological instability, California and the West Coast stages an encounter between East and West. The bounty and beauty of the state have inspired a contradictory vision of Eden and transformative, often technological futurity. This is why, in the American imagination, California's shores have staged both the fulfillment and decline of the West; its final shot at paradise and its precipitous fall into the sea. That is why the "California dream" encompasses both Arcadian frontier and apocalyptic collapse, particularly in filmic and literary representations of Los Angeles (Davis 2000).

California has long played a unique role in the global imagination, and has served as a major cultural and symbolic node in the networks of unchurched spirituality that have spread across the United States and around the world. But the influences of California consciousness, which already questions the distinction between sacred and secular, also point beyond religion. One secular zone that it has marked is its globally influential information technology industry. In the late 1960s, the leading edge of human-computer interaction was taking place at the Augmentation Research Center at Menlo Park's SRI (mentioned above). There the visionary researcher Douglas Engelbart developed the mouse, the hyperlink, the graphical user interface, e-mail, and video conferencing; in 1969, the first transmission was sent from UCLA to SRI on the Arpanet, the direct ancestor to today's Internet. Like many other California mind scientists, Engelbart's vision was guided by a sense of human potential rooted in the co-evolution of technology and human consciousness, ideas that would also infuse the counterculture (Turner 2006). While at SRI, he also became devoted to est, the Erhard "training seminars" that fused Zen, Scientology, psychocybernetics, and Dale Carnegie and that debuted at San Francisco's Jack Tar Hotel in late 1971. Some of ARC's fundamental research was eventually reworked by Steve Jobs and Steve Woczniak into the first Apple computer. When Jobs passed away in early 2012, his debt to California consciousness became clear. Jobs was a practicing Buddhist, explored alternative medicine, and counted LSD and a youthful wander through India as some of his most formative experiences.

In this sense, California's enormous role as a global powerhouse of biotechnology, IT, and media culture cannot be separated from the California consciousness that in many ways resists and challenges the engines of modernity. In fact, they are integrally related. Today we are in the midst of one of the most turbulent and disturbing periods of transformation humans have ever known. The biosphere we depend on is passing through a severe and possibly disastrous shuddering dubbed the Anthropocene by despondent geologists, while molecular engineering, cognitive science, artificial intelligence, life extension technology, and media technology are staging the emergence of a "posthuman" human being. These developments threaten some of our deepest assurances—about progress, about reason, about gender, about the boundaries between nature and technology, about mortality and immortality, about the very definition of human being—and they demand a spiritual response. California consciousness can thus be seen as a prophetic and paradoxical reflection of the global crisis of our times, at once engaging its transformative potential and providing, or at least attempting to provide, visionary alternatives. The inventive rootlessness of California is a call to inventive rootlessness, the state's shattering and reassembling of traditions a reflection of the loss of tradition and the collectively improvised and emergent tactics that inevitably follow. This process is the moving core of California consciousness.

References

Albanese, C. (1990). *Nature religion in America: From the Algonkian Indians to the New Age*. Chicago: University of Chicago Press.

Baur, J. (1959). *The health seekers of southern California, 1870–1900*. San Marino: Huntington Library Press.

Blavatsky, H. P. (1888). *The secret doctrine*. London: Theosophical Publ. Co.

Campbell, C. (2002). The cut, the cultic milieu, and secularization. In J. Kaplan & H. Lööw (Eds.), *The Cultic Milieu* (pp. 119–136). Walnut Creek: AltaMira Press.

Carrette, J., & King, R. (2005). *Selling spirituality: The silent takeover of religion*. London: Routledge.

Carter, J. (1999). *Sex and rockets: The occult world of Jack Parsons*. Venice: Feral House.

Chadwick, D. (1999). *Crooked cucumber: The life and Zen teaching of Shunryu Suzuki*. New York: Broadway Books.

Clarke, J. J. (1997). *Oriental enlightenment: The encounter between Asian and Western thought*. London: Routledge.

Davis, M. (2000). *City of quartz: Excavating the future in Los Angeles*. New York: Verso.

Davis, E. (2006). *The visionary state: A journey through California's spiritual landscape*. San Francisco: Chronicle Books.

Faivre, A. (2010). *Western esotericism*. Albany: State University of New York Press.

Frankiel, S. (1988). *California's spiritual frontiers: Religious alternatives in Anglo-Protestantism, 1850–1910*. Berkeley: University of California Press.

Fuerstein, G. (1998). *Tantra: The path of ecstasy*. Boston: Shambhala.

Fuller, R. (2001). *Spiritual but not religious: Understanding unchurched America*. New York: Oxford University Press.

Hammer, O. (2004). *Claiming knowledge: Strategies of epistemology from theosophy to the New Age*. Leiden: Brill.

Hanegraaff, W. (1996). *New Age religion and western culture: Esotericism in the mirror of secular thought*. Leiden/New York: Brill.

Horowitz, M. (2009). *Occult America: The secret history of how mysticism shaped our nation*. New York: Bantam Books.
Hutton, R. (1999). *The triumph of the moon: A history of modern pagan witchcraft*. Oxford: Oxford University Press.
Huxley, A. (1946). *The perennial philosophy*. London: Chatto & Windus.
Huxley, A. (1954). *The doors of perception*. London: Chatto & Windus.
Huxley, A. (1956). *Heaven and hell*. London: Chatto & Windus.
James, W. (1961). *The varieties of religious experience: A study in human nature*. New York: Macmillan.
Kennedy, G. (1998). *Children of the sun: A pictorial anthology, from Germany to California 1883–1949*. Ojai: Nirvana Press.
Kripal, J. (2007). *Esalen: America and the religion of no religion*. Chicago: University of Chicago Press.
Kripal, J. (2010). *Authors of the Impossible: The paranormal and the sacred*. Chicago: University of Chicago Press.
Lachman, G. (2001). *Turn off your mind: The mystic sixties and the dark side of the age of Aquarius*. New York: Disinformation.
Leary, T., Alpert, R., & Metzner, R. (1964). *The psychedelic experience: a manual based on the Tibetan book of the dead*. Secaucus: Citadel.
Lee, M., & Shlain, B. (1992). *Acid dreams: The complete social history of LSD: the CIA, the sixties, and beyond*. New York: Grove.
Lutyens, M. (1975). *Krishnamurti: The years of awakening*. Boston: Shambhala.
McKenna, T. (1991). *The archaic revival*. New York: Harper Collins.
Orr, J. (1999) *Religion and multiethnicity in Los Angeles*. Los Angeles: Center for Religion and Civic Culture, University of Southern California. www.prolades.com/glama/CRCC%20demographics%20%20Los%20Angeles.htm
Roszak, T. (1976). *Unfinished anima: The aquarian frontier and the evolution of consciousness*. New York: Harper & Row.
Said, E. (1978). *Orientalism*. New York: Pantheon Books.
Schou, N. (2010). *Orange sunshine: The Brotherhood of Eternal Love and its quest to spread peace, love, and acid to the world*. New York: Thomas Dunne Books.
Seagar, R. H. (2000). *Buddhism in America*. New York: Columbia University Press.
Starhawk. (1979). *The spiral dance: A rebirth of the ancient religion of the great goddess*. San Francisco: Harper & Row.
Syman, S. (2010). *The subtle body: The story of yoga in America*. New York: Farrar, Straus and Giroux.
Taylor, B. (2009). *Dark green religion: Nature spirituality and the planetary future*. Berkeley: University of California Press.
Turner, F. (2006). *From counterculture to cyberculture: Stewart Brand, the Whole Earth Network, and the rise of digital utopianism*. Chicago: University of Chicago Press.
Urban, H. (2011). *The church of scientology: A history of a new religion*. Princeton: Princeton University Press.
Whitman, W. (1897). *Leaves of grass*. Boston: Small, Maynard & Company.
Yogananda, P. (1959). *Autobiography of a Yogi*. Los Angeles: Self-Realization Fellowship.

Chapter 74
Lived Experience of Religion: Hindu Americans in Southern California

Shampa Mazumdar and Sanjoy Mazumdar

74.1 Introduction

Contrary to predictions by theorists that with secularization religion would fade from the public realm, in recent years there has been widespread acknowledgement that far from disappearing religion is continuing to play a significant role in the lives of people (Berger 2007; Leonard et al. 2005). This realization has led to a renewed interest in scholarly research on the topic of religion (Cadge and Ecklund 2007; Leonard et al. 2005; Carnes and Yang 2004; Eck 2001; Min and Kim 2002; Warner and Wittner 1998; Mazumdar and Mazumdar 2006, 2009a, b; among others). Much of this scholarship has focused on understanding the role of religion in the lives of "new" (post 1965) immigrants (to the United States) from Asia, Latin America, Middle East, and Africa.

Several important themes have emerged from this literature (for a detailed review see Cadge and Ecklund 2007; Stepick 2005). Studies have pointed to the multifaceted functions of churches, temples and mosques, including provision of economic assistance, social and psychological support, fostering of transnational ties and preservation and transmission of culture and ethnic traditions (Ebaugh and Chafetz 2000; Guest 2005; Levitt 2001; Lin 1999). From these we learn that temples in New York's Chinatown facilitate employment opportunities by connecting the newly arrived with a network of restaurants owned by their members (Guest 2005), churches in Long Beach target Cambodian youth as potential converts and provide them with economic, social, and educational support (Douglas 2005),

S. Mazumdar
Department of Sociology, University of California, Irvine, CA 92697, USA
e-mail: s2mazumd@uci.edu

S. Mazumdar (✉)
Department of Planning, Policy and Design, University of California, Irvine, CA 92697, USA
e-mail: mazumdar@uci.edu

S.D. Brunn (ed.), *The Changing World Religion Map*,
DOI 10.1007/978-94-017-9376-6_74

and that Chinese Buddhist temples sponsor classes aimed at teaching art, calligraphy and language (Lin 1999). Other studies have focused on understanding change, such as the adoption of the congregational model (Yang and Ebaugh 2001; Warner and Wittner 1998), as well as on documenting the role of women, their greater participation, involvement, fundraising, volunteerism and activism (Ebaugh and Chafetz 2000; Abusharaf 1998; Douglas 2005; Leonard 1997). Some of this research has been limited in scope, focusing primarily on the congregational and institutional structure of organized religion in public space (Berger 2007). This is unfortunate because it gives a partial and incomplete understanding of a complex social phenomenon (Berger 2007). Although religious organizations, institutions and structures continue to be significant, as Berger (2007: iii) points out, "much of religious life takes place outside these institutional locales." But only a few studies (notably Ammerman 2007; Williams 2010; Joshi 2006; McDannell 1995; Mazumdar and Mazumdar 2003, 2009b) have taken a more nuanced approach to the experiences of lived religion, examined the ordinary spatial settings of everyday religion (such as home spaces, dormitory rooms, cars, etc.), expression of religious values in lifestyle choices, such as food, or the ways people engage with the sacred through the physical environment and landscape ecology.

74.1.1 Indians and Hindus in America and Southern California

This study is about Hindu immigrants in Southern California, so a brief introduction. The first significant migration from India occurred primarily from the state of Punjab in the early 1900s. Of these early migrants, only a small number were Hindus, the vast majority being Sikhs (Kitano and Daniels 1995). The immigration act of 1965, with its "detailed preference system based on professional or occupational skills needed in the United States and on family reunification" (Williams 2000: 214), dramatically affected immigration from India. According to the 2010 census there are 3.18 million Asian Indians in the United States. Indians have tended to settle more in urban areas and on the coasts. California, with its premier universities, medical centers, high tech industries and temperate climate, has been a major destination. According to the 2010 census there are 528,176 Indian Americans in California (360,392 in 2000). Approximately 167,800 Indians reside in Southern California's Metropolitan Statistical Areas (MSA).

There does not seem to be an accurate count of the number of Hindus in the U.S., and estimates vary between 40 and 80 % of the total Indian population, from 1 to 1.5 million (Joshi 2006: 19; www.ask.com/wiki/Hinduism_in_the_United_States). The number of Hindus in Southern California MSAs could be estimated to be between 67,100 (40 %) and 134,200 (80 %).

74.1.2 Research Questions of Interest

How do Hindu Americans conduct their religious lives? What are their experiences and in what manner and where do they create meaningful religious lives for themselves? This paper examines the everyday reality of lived religion among Hindu Americans in Southern California in public and private spaces, in homes and temples, through food, music, art and commerce. We will focus primarily on three spatial settings, namely the home, temple, and ethnic enclave.

74.2 Method

To answer the above research questions a Naturalistic Field Research method was employed. The primary objectives were to study the social group in its natural setting, to learn in an in-depth way about it and its practices, and to understand as much as possible from an emic point of view. Scholars have used a number of labels for this kind of research, for example, "naturalistic social research" (Blumer 1969; Lofland 1967), and "qualitative social research" (Taylor and Bogdan 1984; Schwartz and Jacobs 1979). A major difference in this research is an intentional examination of space, buildings, and physical objects and representations, mostly ignored in the social science literature.

Several strategies were used to collect data. The focus was on spatial settings, buildings, artifacts, merchandise, particularly those considered sacred, and on religious rituals, music, and art. Naturalistic data were collected through fieldwork. Detailed observations were made of homes, several Hindu temples as well as the ethnic enclave commonly known as Little India located in Southern California.

Interviews were conducted to obtain answers to questions and to enable people to provide descriptions, verbalize their thoughts, views, aspirations and feelings. This included short, casual unstructured interviews, and intensive in-depth unstructured interviews as well as multiple interviews over several sessions with knowledgeable members.

Ethnic newspapers, newsletters, and other publications of the immigrant community were also examined. The use of multiple data collection techniques was designed to capture a wide variety of information, to cross-check when needed and to help increase the validity of the data.

The research, which was part of a larger on-going team project on ethnic and immigrant groups, was conducted in the Indian immigrant community in Southern California. For the home portion, 30 Hindu immigrant families were studied. Initial contact with a few families led to introduction to others. This "snowball" technique helped provide entry to study settings such as homes, and interior sacred spaces, artifacts, and gardens. Quotes from interview data are included to enable description and expression of the participants. The codes at the end of data quotes refer to their location in the data (much like citations), but the names are pseudonyms.

74.3 Experiences and Practice of Diasporic Hinduism in Southern California

74.3.1 Religion in Homes

Home plays a very important role in the religious lives of Hindu Americans. Within their homes, they express, engage and experience religion through art, artifacts, music, landscape ecology and food. This is evident in both interior spaces as well as in outside décor and landscaping. Sacred artifacts are placed at the front entrance, on the threshold, or on the door/doorpost. These provide a blessing for the dwelling, its occupants and visitors. In one home, for example, a miniature *Ganesha* is strategically located above the entry doorway – anyone entering and leaving then walks under the protective gaze of *Ganesha*, the remover of obstacles. In others, the threshold is decorated with flowers, mango leaves, *torans* (above-door hangings of or resembling mango leaves and/or particular flowers), sacred sandalwood paste, sacred symbols such as the ॐ (*Om* or *Aům*) (Mazumdar and Mazumdar 2009b), and pictures of deities, as exemplified in the following description:

> Outside the front door, we have a piece of red felt symbolizing good fortune. We also have three pictures of Lord Venkateswara, Goddess Lakshmi Devi, and Bhagwan Sri Satya Sai Baba. (Intrv: HIR/Sita/000590)

On important days of the Hindu calendar, such as *Diwāli*, the front entryway receives special attention. On this day, it is auspicious to invite Goddess *Lakshmi* into the home. Decorative patterns (*rangoli* or *alpona*) executed by women adorn her pathway and the warm glow of candles and clay lamps light her way, creating an ambiance that is both ethereal and inviting especially when juxtaposed against the dark, moonless autumn night.

Inside the home, religious artifacts add to the home aesthetics, and also act as reminders of religion (Mazumdar and Mazumdar 2009b). They facilitate the practice of devotional Hinduism, which mostly is a multisensorial experience (Eck 1981). As Eck (1981) points out, this involves "seeing," "hearing," "touching," "tasting" and "smelling" the sacred. One or more senses may be engaged from time to time or all together.

On entering the home one sees the sacred artifacts that adorn the walls, hallways, doorways, shelves, tables, and alcoves. These could be pictures, paintings, sculptures and other forms of religious art (Mazumdar and Mazumdar 2009b) (Fig. 74.1). Maya describes her home in the following way:

> Religious artifacts are ubiquitous yet subtle in my home from the moment you step in. Upon entering the front door is a large three-foot tall representation of the Goddess Lakshmi. In Hinduism, Lakshmi is the Goddess of prosperity, light, generosity, wisdom, and fortune. The statue is hand carved from wood and was made in India … Continuing through this formal living room you encounter a small corner table with a statue of the Lord Ganesha on it, who is widely revered as the remover of obstacles. This statue was handmade in India upon my mother's request, from fiberglass and stands at around two feet … Moving into the kitchen, there is a small nook with a counter and a cabinet above it, which houses a small prayer area. (Intrv: HIR/Maya/000012). Most significant is the seeing of one's deities in

Fig. 74.1 Kitchen altar (Source: Cultural Ecology Research Team)

one's home *mandir* (temple). Every day a devotee seeks *darsan* (visitation) a time for one-on-one personalized devotion and communion (see also Eck 1981). The following is a description of the *mandir* in Kanchana's home:

> The home temple is [located] in a room facing east—the direction of the rising sun. The temple holds representations of [several] Hindu deities, each aligned against double glass doors, enabling sunlight to enter. A main theme color of the room is red ... Every morning and evening, a diya (lamp) is lit and burns throughout the day, illuminating the deities along with the sun. (Intrv: HIR/Kanchana/000012W)

Sounds of devotional music, which may be played at specific times during the day or at any time, may greet visitors. Some families consider it auspicious to be awakened by the sounds of sacred music.

> Religious music is played daily in Kanchana's home ... From the time a person wakes up until 10 p.m. bhajans (devotional music) are playing in the background. The same CD is placed on repeat in the home temple, and the sound softly fills the downstairs. On Saturdays, the CD is switched to that of the Hanumana Chalisa, since Tuesdays and Saturdays are widely revered as days to celebrate Lord Hanumana. Hanumana is central to the Indian epic Ramayana, and is a symbol of strength and devotion. (Intrv: HIR/Kanchana/000012S).

There may be live music performances at other times. In Maya's home, her mother and grandmother (when she visits) often sing during daily *pooja*. This too, is seen as an offering to the gods and deities. Some families host collective *bhajan* (devotional songs in praise of the gods and goddesses) or *kirtan* singing sessions wherein families get together on significant days or weekends and collectively sing *bhajan*s and/or *kirtan*s. There is a lead singer accompanied by traditional Indian musical instruments. The following is a description of a *bhajan* session.

> In our family we host bhajans once a month. Forty to fifty people are invited to participate. It usually takes place in the living room. Furniture is rearranged so that guests can sit on the

floor. Everybody joins in the singing …There are Indian drums (tablā), and Indian keyboard (harmonium) and cymbals. One person usually leads and the congregation follows. (Intrv: HIR/Sita/000390). The sounds of religion can include the chanting of *mantra*s and *shloka*s (verses from the holy texts), the ringing of bells and the blowing of *shankha* (conch shells) for special occasions. And, it can also include (the sounds of) silence to facilitate the tranquility and quiet of meditation.

Sacred smells—the scent of fresh flowers—roses, jasmines, gardenias comingling with the fragrance of sandalwood paste, incense, and camphor help create a sacred microcosm. Home gardens and landscape ecology further contribute to this ambiance. Maya explains:

In my garden, we have a plethora of flowers and fruits that are used during poojas (prayers). Each plant that we have was chosen either for its connection back to India, or something that my mother saw as beautiful and good to offer during prayer. Some of the flowers we have are roses, calla lilies, irises, jasmines, hibiscus … We also have fruits, that are offered during pooja—such as the sacred mango…mango leaves are used in décor and during most Hindu ceremonies … some other fruits that are offered from our garden are oranges, strawberries, guavas, passion fruits and loquats. (Intrv: HIR/Maya/000012W). Similarly, Kanchana says:

Flowers are used in prayers or pooja as an offering for the gods. Many of the flowers are grown at home so that they can be cut and offered fresh … Significant are jasmines, roses, and small flowers that look like elephants, representative of Lord Ganesha. (Intrv: HIR/Kanchana/000012S)

Immigrant Hindu families express their devotion through *pooja*, as described by Radha:

Pooja is a daily process. In our home pooja begins with prayer to Lord Ganesha. Before prayers, we put kumkum (vermillion) on the base of the feet of all the deities. Next, we place fresh flowers or garlands, on the pictures of all the deities. We light incense. After that we offer water, this can be coconut water, plain water with karpuram (camphor) or plain water with tulasi (holy basil) to the gods and goddesses and then we drink it three times. We offer fresh fruit to be blessed by the gods and goddesses. After all of the above is done we prostrate in devotion called ashtanganamaskaram (body prostration). (Intrv: HIR/Radha/000095)

Engagement with the sacred also involves food, the preparation of celebratory/ritual food, eating consecrated food, fasting, and the enforcement of food taboos. For Kanchana's grandmother, each of her daily meals is seen as an offering to the Gods; she meticulously prepares them in the kitchen, refraining from tasting or consciously smelling any of the cooked items; the first taste and smell are reserved for the deities; it is only after the offering is made that she serves herself and sits down to eat. For most families however serving cooked food at the altar is only done on special holidays and/or *pooja*s. On such occasions a vegetarian meal may be prepared or *payasam* (rice pudding) may be offered which when transformed into *prasaad* (consecrated food) through offering, prayer and blessing, is distributed and consumed by family members and invited guests. For holidays such as *Ganesha* *chaturthi*, ritual food is prepared, particularly those that are believed to be favored by *Ganesha*. Families take special care to prepare *modakam*, steamed balls of rice stuffed with coconut and dipped in brown sugar syrup. Ritual foods for other holidays, include *pongal* rice, which is rice cooked with milk and jaggery. While cooking, it is considered auspicious to allow the rice to spill over, symbolizing prosperity for the family. This item is prepared for the *pongal* holiday.

Various food taboos are also enforced in many families. Some of the senior household members exercise more stringent control. Maya's grandmother practices the strictest version of vegetarianism refraining from all meat and meat products (beef, chicken, lamb, goat), fish, eggs, and even onion and garlic. Kanchana's mother and grandmother do not consume any meat. Most Hindus refrain from eating beef, but a few do not follow any food related restrictions.

74.3.2 Religion in Vehicles and on the Body

In transportation vehicles too Hinduism can become evident. Many families take their newly acquired cars to their temple to be sanctified and blessed. This makes them expressive of religion as they place pictures or representations of *Ganesha* in their vehicles to protect the driver, the occupants and the car and to ensure safe journeys.

Amrita explains:

> In each one of my parent's cars, they have a small round Ganesha on the dashboard above the stereo. The Ganeshas in each car are of different colors, but they are smooth in texture, about 2-1/2" in height … They are all stuck using a small white sticky double- sided tape pad. They put the Ganeshas … to protect the driver and any passengers from harm or danger when the car is being used. Also, my parents just like the idea of having something sacred and spiritual not only in their homes but in their cars as well. (Intrv: HIR/Amrita/000012S)

Others may play religious music while driving as described by Maya:

> My mother … keeps a few CD's [of religious music] in her car that she will listen to on her way to work. (Intrv: HIR/Maya/000012S)

Similarly, Savitri's parents when they first enter their car always listen to *bhajans* (devotional music) before they listen to anything else. Furthermore, bumper stickers and car number plates can also be used to express religion.

Additionally, religious jewelry, medallions, amulets and lockets are worn on the body. Of special significance are ॐ ("*Om*" or *Aům*) and *Ganesha*. Made of gold or silver, these may be purchased or acquired during pilgrimage or may be given as gifts, marking important life stage milestones. Individuals also carry pictures of important Hindu deities in their wallets and purses for emotional comfort, blessing, and protection.

74.3.3 Religion in Temples

Temple Hinduism represents another aspect of the religious experience of Hindu Americans. In Southern California, Hindu temples have incorporated different styles and genres of temple design and architectural form.

Some are monumental structures, built from scratch on land purchased by devotees. One example is the Sri *Venkateshwaraswamy* Temple in Calabasas, Southern California, located on a site in close proximity to nature (Fig. 74.2). The architecture follows the South Indian style of temple design with towering *shikharas* (spires), *gopurams* (gateways) and appropriate location of the *garba–griha* (womb chamber), the sacred sanctorum, innermost shrine of the Hindu temple. Inside the temple complex, adjacent to the main shrine dedicated to Lord *Venkateshwara*, are several smaller shrines. There is a spacious courtyard where a large gathering can be accommodated for collective prayers and singing. This space can also be used to create a temporary fire altar, when needed in Hindu ritual (Mazumdar and Mazumdar 2009b).

Pre-built structures may be "modified" to create a place for deities and worship (Mazumdar and Mazumdar 2006; see also Weightman 1993). Such is the case with the *Mandir* in Irvine. Being located in a commercial building, its external décor is devoid of any ornamentation and lacks the appearance of a "traditional" temple. Inside, a large hall has been transformed. Arranged around the wall, on elevated platforms are a number of mini altars or shrines to different Hindu deities such as Lord *Venkateshwara*, *Radha*, *Krishna*, *Jagannath*, *Rama*, *Sita*, *Lakshman*, *Hanumana*, and *Shiva* among others. This modified space functions as individual and as collective prayer space.

The third kind is home temples. Homes or parts are converted into temples with the installation of one or more deities along with space for devotees to pray.

Fig. 74.2 Sri *Venkateshwaraswamy* Temple in Calabasas, North Face (Source: Cultural Ecology Research Team)

Home temples permit people to attend and participate in prayers or pray individually. The *Kālī Mandir* in Laguna Beach is an example. Here the organization of the sanctuary, the arrangement of niches with deities, a courtyard with the sacred *tulasi*, a garden landscaped with sacred trees and placement of outdoor shrines in the midst of flowers and foliage, all provide an intimate and personal ambiance for prayer, meditation, devotion and spirituality.

Although different in architectural style and design, these temples form important settings for individual prayers, collective rituals, and for observance and celebration of Hindu holidays. For example Hindu Americans from the state of Orissa, India, gathered at the *Mandir* in Irvine, to celebrate *Ratha Yatra* (literally translated the journey of the Chariot). On this holiday, in the sacred city of Puri in India, Lord *Jagannath* ritually leaves his temple abode in a glorious chariot pulled by his devotees to symbolically visit the neighborhood. Following this tradition, the diasporic Oriya community, assembled an elaborately decorated *ratha* (chariot). After morning *pooja* (prayers) at the *Mandir*, devotees, some singing and dancing, accompanied by the rhythmic beat of drums and the clashing of cymbals, carefully placed Lord *Jagannath* on the *ratha* (chariot) for a short ride through the streets of Irvine adjacent to the temple (Fig. 74.3). Men, women and children helped pull the chariot in a processional ritual that was colorful, joyous, and spiritually charged.

Besides, collective rituals and celebrations, temples are also settings for performance arts. Singers and dancers "offer" their talent to the deities and provide "live" entertainment for the Gods and Goddesses, many of whom are patron deities for the arts. All classical Indian dance performances begin with an invocation to

Fig. 74.3 Rath Yatra parade on street (Source: Cultural Ecology Research Team)

Lord *Shiva*, who is the Lord of Dance (*Nataraja*) and to *Ganesha* to remove any impediment or glitches in the performance. Goddess *Saraswati* is the patron deity of music and is represented with a *veena* (a stringed instrument) in her hand. In *Kālī Mandir* devotees sing *bhajan*s (devotional songs) on a regular basis after evening prayers. Some lead the assembled devotees in singing while at other times they sing individually. The songs are in praise of the gods or renditions of *shloka*s (sacred verses and mantras). Sometime, traveling musicians from India come to the temple to "offer" their music to the deities. Similarly, exponents of classical Indian dance forms such as *Bharata Natyam* or *Odissi* also perform in temples. For the celebration of *Durga Puja* or *Navaratri* women from the local Guajarati community in Southern California come to *Kālī Mandir* to dance the Garba, a colorful, rhythmic, joyous dance performed with *dandiya*s (sticks).

74.3.4 Religion in Ethnic Enclaves

Religion is practiced and experienced not only in homes, temples, and cars, but also in commercial enterprises and businesses. Little India in Artesia, Southern California, is an example of an ethnic enclave where religion plays an important role especially in the marketing of religious commodities and merchandise. Located on Pioneer Boulevard, this enclave with its many stores and extensive goods, restaurants and markets, caters to the religious needs of Hindu Americans. (Although religious products of other religions are also sold, these are not the focus here and so are not being included here.) Objects and artifacts such as simple and elaborate pre-designed *mandir*s (home temples), large and small *murti*s (carved representations) of Hindu deities: *Shiva*, *Parvati*, *Saraswati*, *Lakshmi*, *Ganesha*, *Krishna* (among others) made out of different materials including gold, silver, brass, marble, wood, and gypsum plaster, some hand carved in India, are sold in shops in the enclave (Fig. 74.4). In addition, ritual objects, used during *pooja*, such as *kalasham* (copper/brass vessels), *deepam* (lamps), pre-made cotton wicks for oil lamps, *karpuram* (camphor), *dhoopam* (incense), *ghanti* (bells), artifacts for creating a *havan* (fire altar), *shānkha* (conch shell), sandalwood, flowers and *toran*s (floral wreath/garlands), decorative items such as religious art (for example *Shiva* as *Nataraja*), wall hangings, paintings depicting narratives from the Hindu Epics the Ramayana and Mahabharata, religious texts religious CD's and videos, Hindu calendars decorated with religious art, as well as protective accessories such as medallions worn as necklaces, bracelets, amulets, religious bumper stickers for automobiles and more are available for purchase.

Businesses in Little India cater to the celebration of Hindu life-cycle rituals of *annapraasana* (ritual feeding of an infant), *upanayan* (thread ceremony for adolescent Brahmin boys), and *vivaha* (wedding). They sell merchandise that include ritual objects, such as silver plates, bowls, cups, spoons, *jayamāla*s (floral garlands) exchanged by bride and groom during wedding ceremony, decorative *mandapam*

Fig. 74.4 Religious icons on sale (Source: Cultural Ecology Research Team)

(altars) for the ceremony, and so on. In addition, for special Hindu holidays stores advertise and market relevant products such as *rakhis* (decorative bracelets exchanged between brother and sister) for *rakshā-bandhan*, *gulal* (colored powder) used in Holi, the festival of colors and *diya*s (clay lamps) lit for *Diwāli*, the festival of lights. Since new clothes are worn on *Diwāli*, clothing stores offer sales and discounts on clothes, such as sarees, *salwaar-kameez*, *ghaghra*, *choli*s, for the occasion. *Diwāli* is also time for families to get together and for friends to visit one another, exchange sweets and host parties. Stores do brisk business in selling varieties of candy and sweets, such as *jalebi*s, *laddoo*s, *pera*s, among others.

Besides products and merchandise on sale, many storeowners have in-store small altars where *pooja* is done every morning before opening the doors for business (Mazumdar and Mazumdar 2005). Flowers and fruits are offered, incense lit and religious music played; some stores have religious art displayed on the walls such as framed and unframed pictures of Hindu deities and statues and sculptures displayed in glass cabinets. Auspicious signs, such as the ॐ (*Om* or *Aům*), are visible as well and floral garlands and mango leaves decorate store entrances on special days of the Hindu calendar. Inauguration of new businesses and opening of new stores often involves religious ceremony, such as *pooja*, with or without the presence and involvement of priests, and the lighting of lamps. In addition, restaurants have incorporated Hindu food taboos in their menu; many will not serve beef or beef products in keeping with the Hindu dietary restrictions and some are strictly vegetarian with no meat, eggs, fish, or chicken on their menu.

74.4 Concluding Discussion

Several important themes emerge from this study. First, this study reinforces the idea that place is important in religious practice, as pointed out by Sopher (1967) (also Weightman 1993; Singh 2011). This study further suggests that the conceptualization of place should be expanded to include temporary, transient, transitory, and unusual places. Religious lives and practices are conducted in various locations and so lived-religion is not confined to one kind of setting. For Hindus, most important, in addition to temples, are homes. Homes are seen as sacred space and set up accordingly. In homes, permanent and fixed altar spaces are where gods and goddesses reside and prayers are offered, but other spaces in the home may be converted temporarily for religious purposes (Mazumdar and Mazumdar 2009b; Mazumdar 2012). Home gardens can serve as spaces for growing flowers and fruits required for worship rituals, and for religious activities (Mazumdar and Mazumdar 2012). For Hindus in America, other locations for religion are natural settings, restaurants, shops, work places (Mazumdar and Mazumdar 2005), ethnic enclaves, streets, and vehicles. Religion and the sacred sphere can be located, fixed, permanent, and formal (as with temples), or transportable and carried along by an individual on the body and in vehicles. Sacredness lies within and without the person.

Second, is the significance of "ordinary" places and spatial settings in the religious lives of immigrants. Attempt is made to transform the "ordinary" into sacred space through the placement of artifacts and performance of rituals. This study documents how homes, vehicles and commercial spaces become important sites of religion. Religion is thus present, visible and relevant in everyday routine activities such as lighting incense in the store/restaurant before opening for business in the morning, or driving to work with sacred music playing in the car.

Third, is the relationship between religion and the arts. In Hinduism, included in lived religion are artifacts (made or purchased), art making, music and dance performance, actions, activities and rituals conducted singly or with others. The making of art, and performance of music and dance are seen as expressions of this intimate personal and devotional connection. Musicians and dancers view their performance not merely as artistic display and of mastery, but also as an act of devotional offering of their talent to the divine in gratitude and self-effaced humility.

Fourth is the symbolic meaning of food and the role of religion in regulating food choices, establishing boundaries through taboos, promoting self-discipline through fasting, and expressing joy through celebratory foods.

Fifth is the experiential aspects of religious behavior and how ordinary everyday activities from landscaping, nurturing fruit trees, selecting flowers for prayer, decorating homes, choosing colors, lighting lamps and candles can help cultivate a deeper sensitivity and engagement with religion and with nature (Mazumdar and Mazumdar 2009b). Experiencing religion does not only take the form of congregational prayer or sermon on a mandated day and time, but can include the nuanced embrace of religion in every day life and activities.

Sixth, individual communion is seen as very important. This implies that many activities and events might be individual and family focused. Co-presence of large numbers of persons, and collective activities may also occur and be significant.

Seventh, this study was about the practice and experience of Hinduism by immigrants in Southern California. Increased focus on lived religion and how religion affects everyday activities and spaces may reveal more about how this happens, or does not happen in other religions. There is evidence that some of the practices and experiences described here may not be peculiar to Hinduism alone but may occur in a few other religions (for example Shintoism and Buddhism) (see also Mazumdar et al. 2000; Mazumdar 2011).

Finally, this study supports Berger's (2007) claim that secularization theorists were not entirely correct and that religion is continuing to play an important role in various settings. However, the lens to study religion should be widened to include at least the settings described here, namely formal temples, home temples, homes, neighborhoods and public spaces, workplaces (Mazumdar and Mazumdar 2005), vehicles, the body, and more. Not doing so results in missing important aspects of religious practice and experience. The geography of religion seems to be much broader than previously thought.

References

Abusharaf, R. M. (1998). Structural adaptations in an immigrant Muslim congregation in New York. In R. S. Warner & J. G. Wittner (Eds.), *Gatherings in diaspora* (pp. 235–261). Philadelphia: Temple University Press.

Ammerman, N. T. (Ed.). (2007). *Everyday religion: Observing modern religious lives*. New York: Oxford University Press.

Berger, P. L. (1967). *The sacred canopy: Elements of a sociological theory of religion*. Garden City: Doubleday.

Berger, P. L. (2007). Foreword. In K. I. Leonard, A. Stepick, M. A. Vasquez, & J. Holdaway (Eds.), *Immigrant faiths: Transforming religious life in America* (pp. iii–viii). New York: AltaMira Press.

Bhardwaj, S. M., & Rao, M. N. (1998). The temple as symbol of Hindu identity in America? *Journal of Cultural Geography, 17*(2), 125–143.

Blumer, H. (1969). *Symbolic interactionism: Perspective and method*. Englewood Cliffs: Prentice Hall.

Cadge, W., & Ecklund, E. H. (2007). Immigration and religion. *Annual Review of Sociology, 33*, 359–379.

Carnes, T., & Yang, F. (Eds.). (2004). *Asian American religions: The making and remaking of borders and boundaries*. New York: New York University Press.

Douglas, T. J. (2005). Changing religious practices among Cambodian immigrants in Long Beach and Seattle. In K. I. Leonard, A. Stepick, M. Vasquez, & J. Holdaway (Eds.), *Immigrant faiths: Transforming religious life in America* (pp. 123–144). Lanham: Altamira Press.

Ebaugh, H. R., & Chafetz, J. S. (2000). *Religion and the new immigrants: Continuities and adaptations in immigrant congregations*. Walnut Creek: Altamira Press.

Eck, D. L. (1981). *Darsan: Seeing the divine image in India*. Chambersburg: Anima Books.

Eck, D. L. (2001). *A new religious America: How a "Christian country" has now become the world's most religiously diverse nation*. New York: HarperSanFrancisco.

Guest, K. J. (2005). Religion and transnational migration in the New Chinatown. In K. I. Leonard, A. Stepick, M. Vasquez, & J. Holdaway (Eds.), *Immigrant faiths: Transforming religious life in America* (pp. 145–163). Lanham: Altamira Press.

Joshi, K. Y. (2006). *New roots in America's sacred ground: Religion, race, and ethnicity in Indian America*. New Brunswick: Rutgers University Press.

Kitano, H. H., & Daniels, R. (1995). *Asian Americans: Emerging minorities*. Englewood Cliffs: Prentice Hall.

Leonard, K. I. (1997). *South Asian Americans*. Westport: Greenwood Publishing Company.

Leonard, K. I., Stepick, A., Vasquez, M., & Holdaway, J. (Eds.). (2005). *Immigrant faiths: Transforming religious life in America*. Lanham: Altamira Press.

Levitt, P. (2001). *Transnational villages*. Berkeley: University of California Press.

Lin, I. (1999). Journey to the far West: Chinese Buddhism in America. In D. K. Yoo (Ed.), *New spiritual homes: Religion and Asian Americans* (pp. 134–166). Honolulu: University of Hawaii Press.

Lofland, J. (1967). Notes on naturalism in sociology. *Kansas Journal of Sociology, 3*, 4–61.

Lofland, J., & Lofland, L. H. (1984). *Analyzing social settings: A guide to qualitative observation and analysis*. Belmont: Wadsworth.

Mazumdar, S. (2011). Yearned crowding, desired closeness: Hakata Gion Yamakasa. *Journal of Asian Urbanism, 4*, 12–15.

Mazumdar, S. (2012). Religion and housing. In A. T. Carswell (Ed.), *The encyclopedia of housing* (2nd ed., Vol. 2, pp. 591–596). Thousand Oaks: Sage Publications.

Mazumdar, S., & Mazumdar, S. (2003). Creating the sacred: Altars in the Hindu American home. In J. N. Iwamura & P. Spickard (Eds.), *Revealing the sacred in Asian and Pacific America* (pp. 143–157). New York: Routledge.

Mazumdar, S., & Mazumdar, S. (2005). How organizations interface with religion: A typology. *Journal of Management, Spirituality, and Religion, 2*(2), 199–220.

Mazumdar, S., & Mazumdar, S. (2006). Hindu temple building in Southern California: A study of Immigrant religion. *Journal of Ritual Studies, 20*(2), 43–57.

Mazumdar, S., & Mazumdar, S. (2009a). Religious placemaking and community building in diaspora. *Environment & Behavior, 41*(2), 307–337.

Mazumdar, S., & Mazumdar, S. (2009b). Religion, immigration, and home making in diaspora: Hindu space in Southern California. *Journal of Environmental Psychology, 29*(2), 256–266.

Mazumdar, S., & Mazumdar, S. (2012). Immigrant home gardens: Places of religion, culture, ecology, and family. *Landscape and Urban Planning, 105*(3), 258–265.

Mazumdar, S., Mazumdar, S., Docuyanan, F., & McLaughlin, C. M. (2000). Creating a sense of place: The Vietnamese and Little Saigon. *Journal of Environmental Psychology, 20*(4), 319–333.

McDannell, C. (1995). *Material Christianity: Religion and popular culture in America*. New Haven: Yale University Press.

Min, P. G. (2005). Religion and the maintenance of ethnicity among immigrants: A comparison of Indian Hindus and Korean Protestants. In K. I. Leonard, A. Stepick, M. Vasquez, & J. Holdaway (Eds.), *Immigrant faiths: Transforming religious life in America* (pp. 99–122). Lanham: Altamira Press.

Min, P. G., & Kim, J. H. (2002). *Building faith communities*. New York: Altamira Press.

Schwartz, H., & Jacobs, J. (1979). *Qualitative methodology: A method to the madness*. New York: Free Press.

Singh, R. P. B. (2011). Sacredscapes and sense of geography: Some reflections. In R. B. Singh (Ed.), *Sacredscapes and pilgrimage landscapes* (Planet Earth & Cultural Understanding Series #7, pp. 5–46). New Delhi: Shubhi Publication.

Sopher, D. E. (1967). *Geography of religions*. Englewood Cliffs: Prentice-Hall.

Stepick, A. (2005). God is apparently not dead: The obvious, the emergent, and the still unknown in immigration and religion. In K. L. Leonard, I. Karen, A. Stepick, M. A. Vasquez, & J. Holdaway (Eds.), *Immigrant faiths: Transforming religious life in America* (pp. 11–37). New York: AltaMira Press.

Taylor, S. J., & Bogdan, R. (1984). *Introduction to qualitative research methods: The search for meanings* (2nd ed.). New York: Wiley.

Warner, R. S., & Wittner, J. G. (Eds.). (1998). *Gatherings in diaspora: Religious communities and the new immigration*. Philadelphia: Temple University Press.

Weightman, B. A. (1993). Changing religious landscapes in Los Angeles. *Journal of Cultural Geography, 14*(1), 1–20.

Williams, R. B. (2000). South Asians in the United States. In H. Coward, J. R. Hinnells, & R. B. Williams (Eds.), *The South Asian religious diaspora in Britain, Canada, and the United States* (pp. 213–217). Albany: State University of New York Press.

Williams, R. R. (2010). Space for god: Lived religion at work, home and play. *Sociology of Religion, 71*(3), 257–279.

Yang, F., & Ebaugh, H. R. (2001). Transformations in new immigrant religions and their global implications. *American Sociological Review, 66*(2), 269–288.

Chapter 75
Multiscalar Analysis of Religious Geography in the United States

Samuel M. Otterstrom

75.1 Studying U.S. Religions at Multiple Scales

In the United States the diversity of religion is large. Although the ranks of the religious are mostly attached to one of hundreds of different Christian denominations, there are significant numbers of people who are associated with non-Christian faiths. The geographic complexity associated with religion is apparent in the clustering of some churches in certain parts of large cities and the preponderance of one religion in certain areas of the country.

This research examines distinct differences in the national geography of major religions in the U.S. as well as local distinctions in the income of neighborhoods where certain churches are located. Both the macro and micro geographies of religions in this country illustrate the history and socioeconomics of the adherents of different churches.

75.2 Religious Geography in the U.S.

There is tremendous geographical diversity in religious practice around the world. Kong (1990, 2001) summarized geographical research in religion into a number of categories including studies that explore cultural religious practices across a number of different scale levels, and others that incorporate historical inquiries into the evolution of specific religious landscapes. In the U.S., religion, race, and ethnicity combined together with the patterns of historical settlement go far in explaining the distribution of current religious practice. In this section I briefly review literature

S.M. Otterstrom (✉)
Department of Geography, Brigham Young University, Provo, UT 84602, USA
e-mail: samuel_otterstrom@byu.edu

S.D. Brunn (ed.), *The Changing World Religion Map*,
DOI 10.1007/978-94-017-9376-6_75

that has focused on overall religious affiliation geography in the U.S. and then I highlight several studies that have considered the changing macro patterns of the adherents of specific churches and how ethnic geographies affect religious geography (Otterstrom 2008). I finish with a recital of studies relevant to the local micro-geography of worship practices and patterns.

America's religious geography is in a constant state of reconfiguration. Many scholars have attempted to capture the "current" state of religion in the nation even while synthesizing models and trends to explain ongoing transformations. Over 50 years ago, Wilbur Zelinsky (1961) used the 1952 census of religion in the U.S. to describe and explain the distribution of the largest religious faiths in the country. Edwin Gaustad's (1962) historical atlas of religion complemented Zelinsky's work with its ambitious sets of maps and commentaries regarding the spatial configurations of this country's diverse religious populace. Over the years, others have updated and built upon these pioneering works in the study of America's religious geography (for example, Halvorson and Newman 1994; Carroll et al. 1979; Shortridge 1976; Newman and Halvorson 1984; Robbins and Anthony 1990; Sherkat 1999; Gaustad and Barlow 2001; Crawford 2005; Warf and Winsberg 2008).

Additional research has approached shifting U.S. geographies from more specialized perspectives. Some studies have limited their scope to regions of the country (for example, Barcus and Brunn (2004) in the Great Plains and Webster (2000) in Georgia) or to factors such as migration and conversion that lead to geographic change in religious landscapes (Stump 1984; Sherkat 2001; Paul 2003). Others have considered the enduring associations that certain ethnic groups or areas of the country have with distinct religious groups. These include considering immigrant Koreans (Min 1992), Hispanic Protestants (Hunt 1999), Jews (Sheskin 2000), Amish (Crowley 1978), Catholics (Shortridge 1978), Muslims, Hindus, Buddhists (Smith 2002), Protestants (Smith and Seokho 2005), and Mormons (LDS) (Meinig 1965; Laing 2002; Otterstrom 2008; Jackson 1978; Jackson and Jackson 2003).

At the more local level, scholars have considered the where and why of meetinghouse locations and religiosity among residents in different cities. One study indicated that most worshipers (88 %) lived within 20 min of the church they attended (Woolever and Bruce 2002) even while a majority actually went to church outside of their own neighborhoods (Diamond 1999). Some have argued that churches are more a gathering of people with common cultural attributes rather than shared neighborhoods (McRoberts 2003), while others have contended that neighborhood and distance to houses of worship still matter (Sinha et al. 2007). Also, Zelinsky (2001) studied the "exceptionalism" of the placement of churches in urban America that does not conform to economic logic, and Tillman and Emmett (1999) revealed how changing urban ethnic characteristics have led to different religious congregations using the same meetinghouse over time.

There are additional studies concerning the evolving geographies of United States religion than what I have summarized here. However, it is sufficient to say that both macro- and micro- studies have emphasized the relevance of geographic factors in religious practice. Still, there is room for a study that takes both a statistical look at the underlying geographic distribution of large religious groups in the U.S. in connection with the overall population, and one that explores the local manifestations of these distinct faiths. The methods I employ will highlight in what way the spatial concentration and differential distribution of the major faiths have grown and changed over the last 60 years, and also to emphasize the distinct geographic niche that each religious group occupies.

75.3 The Approach

75.3.1 Macro Level

Using both macro- and micro- approaches I consider the distinct geographies that each large religious body has within the United States. Simple spatial concentration maps of particular religions' adherents will illustrate their geographic differences. In addition, I quantify the level of differences compared with the underlying national geography and general population.

My approach at the macro- scale consists of deriving descriptive statistics which illustrate geographical variations of major faiths on two scales. The scales are constructed using various agglomerations of counties as the basic unit of analysis. Combining counties creates the "city-systems" (urban counties and their surrounding areas that are economically tied to the core cities) and the nation of the analyses Otterstrom (2003). The two different scales afford useful views of the distributions of the largest faiths and their spatial contrasts with the general population. City-system analyses measures urban-rural relationships in their larger social and political structure.

The national scale is useful to highlight the overall trends in geographical distributions of the various religious populations. While analyzing the recent changes in the population of a particular faith in the country, it also provides a template for comparing growth and change between two different populations within the same area. At the national level I focus on the top 14 religious bodies in the U.S. up to 2000. I use the Glenmary Research Center survey of churches (Jones et al. 2002), which also includes the surveys of 1990, 1980, 1971, and 1952, so I can examine how these Churches geographies have changed since the 1950s. I pattern my analysis procedure after Otterstrom (2008).

I first seek to differentiate each of the top religions by how concentrated their population distributions are compared with the general population. I calculated Hoover indices of church population concentration for the city-systems and nation for each of the data years, and compared those values with the general population (for 1952 and 1971 religion data I used the 1950 and 1970 censuses). The Hoover index (HI) derives the relative population density in a grouping of areas. This measure has been used in many studies of population concentration (for example, Long and Nucci 1997; Otterstrom 2003) and gives a comparable and logical value. Below is a description of how the HI is calculated:

Let P_{it} be the fraction of a nation's [or city-system's] population in sub-area I in year t and a_i be this area's fraction of the nation's [or city-system's] land area, and let there be k sub-areas. Then, the Hoover index (Vining and Strauss 1976), H_t is given by:

$$H_t = \left(\sum_{I=1}^{k} \left| P_{it} - a_i \right| \right) \times 50 \quad (75.1)$$

The HI can theoretically range from 0 to 100. High values indicate high concentration, or a very uneven population distribution. Low concentration values translate into a landscape of more equal population distribution of the particular group in question. Put another way, the HI is the percentage of people across the group of counties in question that would have to be moved to make an even population density in each county. A rise in the HI between decades means that the population growth or decline in the overall population (or religious group's membership) resulted in concentration to fewer counties. At the city-system level, these counties would usually be the core ones. Conversely, a declining HI value between years indicates migration, natural increase, or conversions (in the case of religions) in or to dispersed areas of the city-system or nation. The HI will show the relative geographic distribution of each of the religions and how these vary across various city-regions nationwide, along with the changes over time within individual faiths.

An alternative measure considers relative comparisons of specific religions in county agglomerations with the overall population of those counties not of that faith. When contrasting the HI of the general population with those of a certain religion, we cannot tell where the density distributions of each group are. In other words, two different populations in the U.S. could both have HI values of 50. However, the actual relative locations of these two groups could be quite different. For example, one religious group might have elevated densities in the Northeast, while the other has high densities in the Upper Great Plains, but their HI values could be very similar. In this example, a religion that has large relative numbers in the Northeast would more closely match the general population, which has high densities in the megalopolis of Washington D.C. to Boston. On the other hand, a church that has larger relative numbers in the Upper Great Plains would be very unlike the general population in its geographic distribution, because the Upper Great Plains is not densely settled. The statistic that I use that can show these relative differences in general population versus specific religion's distributions in area and measure shifts over the years is a slightly modified Duncan index (see Funkhouser 2000; Sakoda 1981). In the chapter I call it the "differential geographical index" (DG) at year y that is calculated as follows:

$$DG_y = \sum_{I=1}^{k} \left| \frac{L_c}{L_t} - \frac{G_c}{G_t} \right| * 50 \quad (75.2)$$

Where *I* is the nation (or city-system) having *k* counties, *L* is the religion's population and *G* is the overall population (minus the religion's adherents) in the county (*c*) or region/nation (*t*) respectively at year (*y*). I derived the DG index for the city-systems and nation for the different churches for each study year.

Similar to the HI, the DG has a range from between 0 and 100. Zero indicates equal proportions for both groups of their overall total throughout the city-system or nation. A value of 100 is impossible in this case because it would mean a complete physical separation of the members of one church from all the rest of the nonmembers of that faith (that is, all the church's members are in one set of counties, and all of the nonmembers are in a completely different set of counties), but relatively high values would indicate that the two groups have very different distributions among the counties of the area in question.

One other way to utilize the DG is to figure the ratio between the number of counties within the nation or city-system that have relatively higher proportions of a certain religion compared with those that have relatively fewer numbers (in reference to the overall population in the counties). A value over 1.0 means relatively more counties have greater proportions of the adherents of that religion compared with the general population in the counties, while a value less than 1.0 means that there are more counties that have relatively fewer members. In essence, regions over 1.0 have a more expansive distribution of the religion being analyzed with those under 1.0 that are more concentrated. I computed this ratio for both the city-systems and the nation for each of the 14 largest faiths.

These different statistics– the Hoover index, Differential Geographical index, and the DG ratio– used together deepen our understanding of the contrasting geographies that the major religious bodies have had in the United States since 1952 (the national Hoover indices do not include Alaska and Hawaii, while the DG indices do include those two states). These indicate which churches are most dispersed or concentrated, and which ones most match overall U.S. population patterns or which are trending toward convergence or divergence in their particular geographies, due to internal population shifts and religious affiliation changes.

75.3.2 Micro Level

At the local level, I consider a sample of cities to see where certain faiths place their meetinghouses in comparison with the income levels of the neighborhoods. Data for member addresses are not available; therefore I can only emphasize a general correlation between the two variables. Alternatively, this could answer the question of whether dominant faiths in a city locate in wealthier neighborhoods. Are they able to afford nicer locations based on the concentration of members? In other words, does a higher percentage of people of that faith in the county equate to a generally higher quality of church locations?

My local worship place geography analysis is of several larger religious groups such as Catholics, Southern Baptists, and Jews in three large metropolitan counties, and LDS in some of their centers in the western United States. By comparing income characteristics of neighborhoods that surround the meeting places of these groups in the counties surrounding Dallas, Los Angeles, and Chicago, I highlight how locational choices for meeting places have some consistency among the groups across the country in terms of the relative income levels surrounding the meetinghouses in these areas. In terms of the LDS I use Multnomah County, Oregon, King County, Washington, Orange and Los Angeles counties in California; Maricopa County, Arizona; and Salt Lake County, Utah. For each of the faiths in the different cities, the meetinghouses were first located using public directories and maps, and the census block group median income from the 2000 U.S. Census was obtained. I then used the median point of all these 2000 block group household median incomes that surrounded the meetinghouses of these faiths. With these data I made comparisons with the median incomes of the county in question and with the other sample religions. This method allows me to broach the question of "Does Church neighborhood location relate to the general socio-economic class of adherents?"

75.4 Patterns

75.4.1 Macro Population Concentrations and Distributions

Each of the top 14 religious bodies in the United States has a distinctive geographic distribution (Table 75.1). Jews are the most concentrated of any of the faiths. Jews have traditionally been clustered in the largest cities such as New York and Chicago

Table 75.1 Hoover index of population concentration for the top 14 religions, 1952–2000

Religion	Adherents 2000	1952	1971	1980	1990	2000
Catholic	62,035,042	75.58	76.20	75.28	75.53	74.01
Southern Baptist	19,881,467	56.37	61.90	62.95	64.07	65.72
United Methodist	10,350,629	55.68	56.89	56.57	57.19	58.62
Jewish Estimate	6,141,325	85.69	0.00	0.00	79.65	79.16
Evangelical Lutheran (ELCA)	5,113,418	0.00	0.00	0.00	60.91	61.35
Latter-day Saints	4,224,026	66.84	75.59	64.07	63.66	63.39
Presbyterian	3,141,566	0.00	0.00	0.00	63.68	64.28
Assemblies of God	2,561,998	55.08	0.00	56.89	59.59	61.30
Lutheran (MO Synod)	2,521,062	66.81	65.53	63.80	63.59	64.05
Episcopal	2,314,756	72.27	70.54	70.78	69.72	69.59
American Baptist	1,767,462	63.34	61.08	62.91	63.59	65.12
United Church of Christ	1,698,918	0.00	65.85	64.84	64.63	65.03
Churches of Christ	1,645,584	0.00	0.00	62.03	63.47	62.98
Muslim Estimate	1,559,294	0.00	0.00	0.00	0.00	71.44
US Hoover index for total population				61.95	62.55	62.61

Source: Samuel Otterstrom

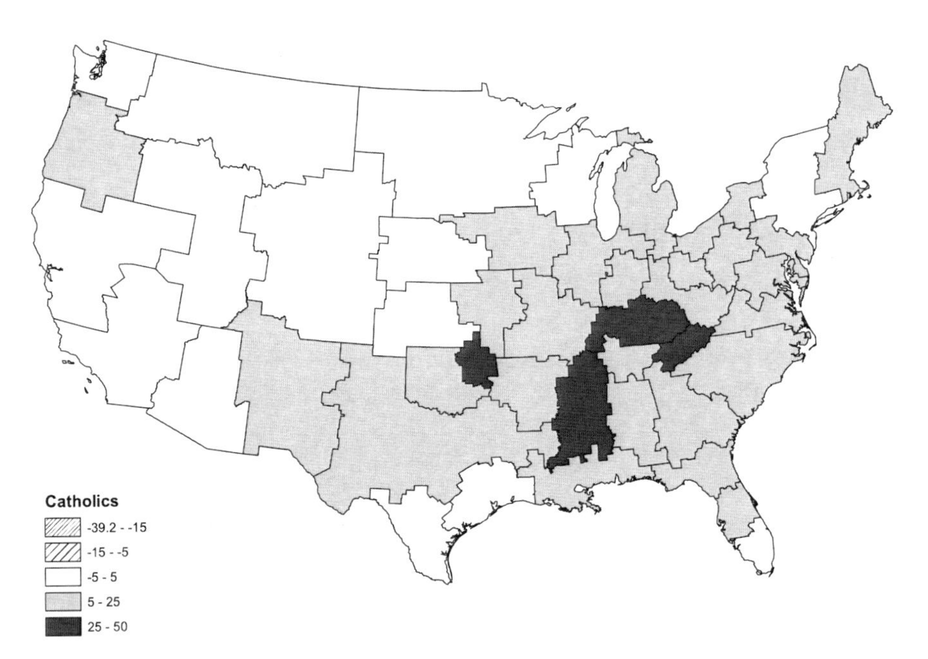

Fig. 75.1 Differences in Hoover Index of Concentration between religion's population (Catholic) and general population, 2000 (Map by Samuel Otterstrom)

and other metropolitan areas. Their Hoover index (HI) of population concentration was a very high 85.7 in 1952. It dropped to a still elevated 79.2 in 2000. The Roman Catholic Church is the largest single faith in the county and it has the second highest level of population concentration. Catholics are also more concentrated than the general population in a majority of the city-systems of the country (Fig. 75.1).

In addition, the Catholic DG index at the national level was 40.3 in 1990 and 36.5 in 2000, which is much lower than many other churches, meaning that even though Catholics have high concentration levels, they are found in areas that also have high population densities of non-Catholics across the country compared with other faiths (Table 75.2).

Southern Baptists and the LDS both have HI of concentration similar to the general population, but Differential Geographical (DG) indices that are high (meaning non-proportional distribution in relation to the general population). This is because they have their core areas of strength that are not in the most densely populated parts of the country. Southern Baptists have become more spatially concentrated since 1952, but their HI values, unlike the Jews and Catholics, are very similar to the concentration levels of the general population. Southern Baptists are very strong in the South, but they are also spread throughout the country and they are not greatly different than the general population in their overall concentration. The north-central U.S. is where they are more concentrated, owing to their smaller presence there (Fig. 75.2).

LDS have generally become more dispersed since 1971, so that their HI in 2000 was only 0.78 higher than the general population at the national level (Table 75.3). In city-systems LDS are more deconcentrated than the Southern Baptists though,

Table 75.2 Differential geographical index (DG), percent of counties with higher (POS) and lower (NEG) shares of the religion's adherents in comparison with the general population, and ratio of number of higher (POS) to lower (NEG) counties, 1990 and 2000

Religion	1990 POS	1990 NEG	1990 DG	1990 RATIO	2000 POS	2000 NEG	2000 DG	2000 RATIO
Catholic	20.25	79.75	40.31	0.2539	21.97	78.03	36.49	0.2815
Southern Baptist	45.18	54.82	63.58	0.8240	45.46	54.54	61.84	0.8336
United Methodist	67.37	32.63	36.10	2.0644	66.89	33.11	36.62	2.0202
Jewish Estimate	2.77	97.23	54.68	0.0285	3.18	96.82	55.54	0.0329
Evangelical Lutheran (ELCA)	27.54	72.46	52.65	0.3801	27.83	72.17	53.58	0.3855
Latter-day Saints	15.50	84.50	60.66	0.1835	16.27	83.73	58.07	0.1943
Presbyterian	37.15	62.85	31.00	0.5912	36.45	63.55	31.53	0.5736
Assemblies of God	42.02	57.98	31.56	0.7249	40.50	59.50	31.25	0.6806
Lutheran (MO Synod)	30.12	69.88	49.69	0.4310	29.80	70.20	50.43	0.4245
Episcopal	22.48	77.52	28.14	0.2899	22.86	77.14	28.55	0.2963
American Baptist	22.92	77.08	48.24	0.2974	22.45	77.55	50.75	0.2894
United Church of Christ	24.74	75.26	55.27	0.3287	23.78	76.22	57.48	0.3120
Churches of Christ	35.02	64.98	52.09	0.5390	35.85	64.15	52.31	0.5588
Muslim Estimate	0.00	100.00	0.00	0.0000	4.52	95.48	51.44	0.0473

Source: Samuel Otterstrom

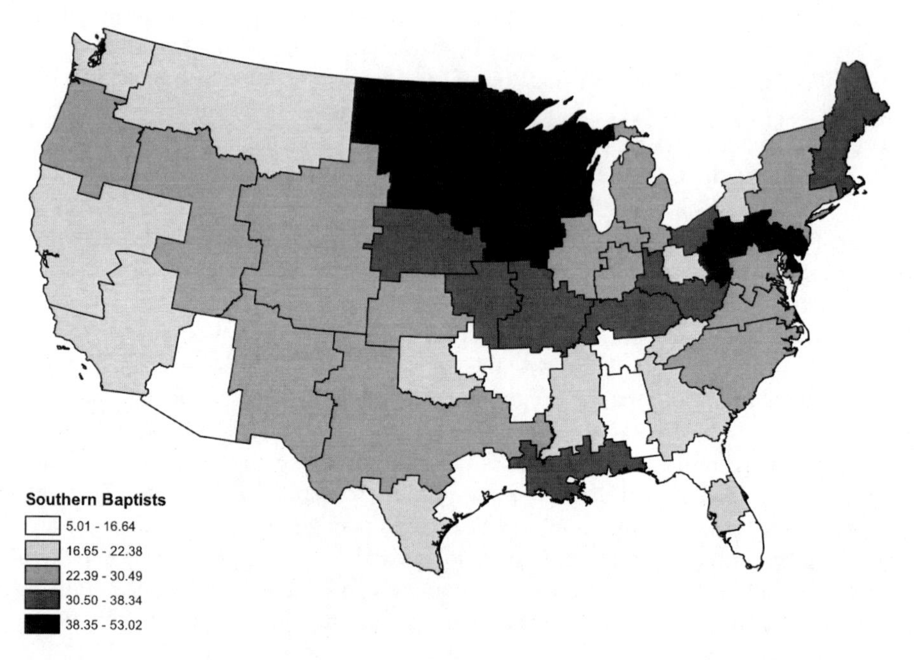

Fig. 75.2 Differential Geographical Index by City-System for Southern Baptists, 2000 (Map by Samuel Otterstrom)

Table 75.3 Difference between Hoover index for the top 14 religions and the overall population Hoover index, 1980–2000

Religion	1980 DIFF	1990 DIFF	2000 DIFF
Catholic	13.33	12.98	11.40
Southern Baptist	1.00	1.52	3.11
United Methodist	−5.38	−5.36	−3.99
Jewish Estimate		17.10	16.55
Evangelical Lutheran (ELCA)		−1.64	−1.26
Latter-day Saints	2.12	1.11	0.78
Presbyterian		1.13	1.67
Assemblies of God	−5.06	−2.96	−1.31
Lutheran (MO Synod)	1.85	1.04	1.44
Episcopal	8.83	7.17	6.98
American Baptist	0.96	1.04	2.51
United Church of Christ	2.89	2.08	2.42
Churches of Christ	0.08	0.92	0.37
Muslim Estimate			8.83
US Hoover index for total population	61.95	62.55	62.61

Source: Samuel Otterstrom

which probably indicates the propensity for the LDS to be found in larger numbers in the suburbs beyond the central cities. The Southern Baptists and the LDS had the two highest DG values in 2000 of the largest faiths, even though those values had dropped somewhat since 1990, pointing to the still regional nature of these churches.

The Methodists and Assemblies of God members are distinctive. They both are more dispersed in their membership structure across the country than the general population. The Methodists were especially dispersed in the northern tier of 23 city-systems in 2000 (Fig. 75.3). In contrast, adherents of the Assemblies of God actually have notably higher city-system concentration levels in 12 city-systems, and are only much more dispersed in four (Fig. 75.4).

At the national level these two faiths are similar in HI of concentration, but at the city-system level they are very different. Furthermore, the Methodists are unique because they are the only faith that has a greater relative proportion of members versus the general population in more than 50 % of the counties. This means that the Methodists are more rural than any of the other large faiths. The Assemblies of God, on the other hand, have one of the lowest DG values so they both have low HI concentration and a population that matches the general distribution of nonmembers of their faith.

Only the Episcopalians had a lower Differential Geographical index than the Assemblies of God in 2000, with the Presbyterians only slightly higher in the index than the Assemblies of God. These three faiths, then, best reflect the overall distribution of the U.S. population. Still, they each have subtle differences in their geography with Episcopalians being more concentrated than the overall U.S. HI and higher than all the churches, except the Jews, Catholics, and Muslims. Furthermore, Episcopalians are less proportionally represented in the population than Presbyterians

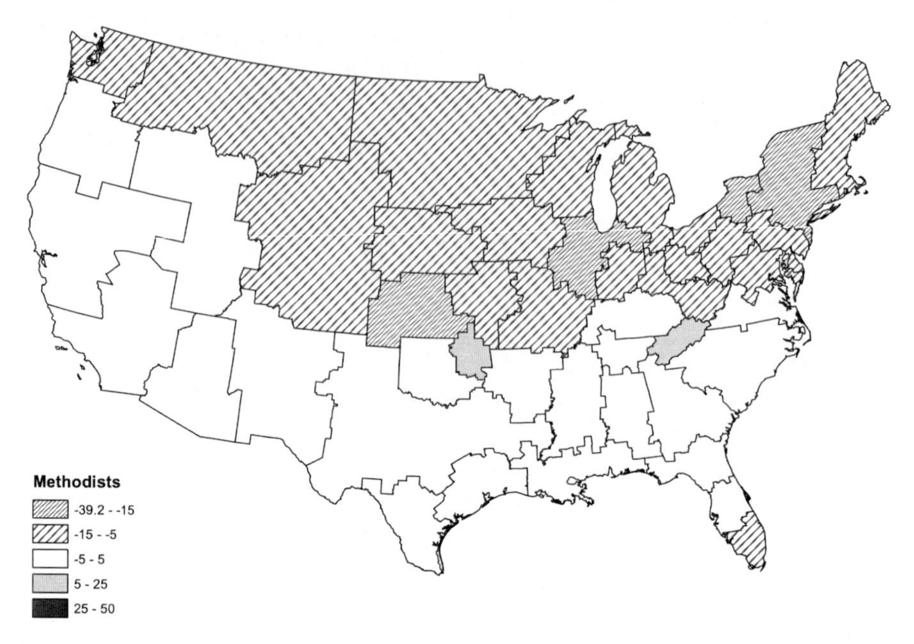

Fig. 75.3 Differences in Hoover Index of Concentration between religion's population (Methodists) and general population for city-systems, 2000 (Map by Samuel Otterstrom)

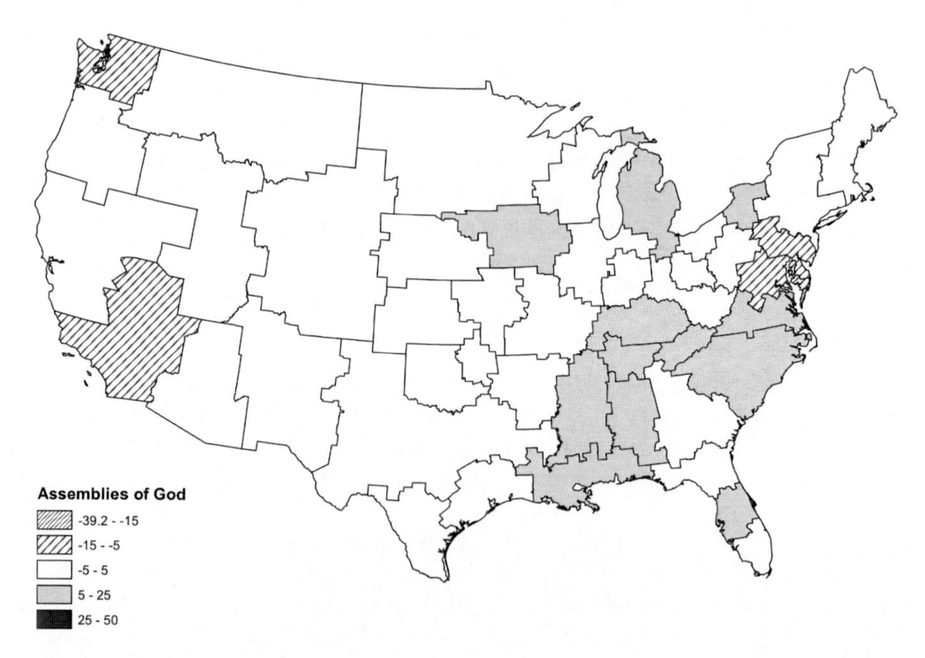

Fig. 75.4 Differences in Hoover Index of Concentration between religion's population (Assemblies of God) and general population for city-systems, 2000 (Map by Samuel Otterstrom)

and Assemblies of God members (77.6 % of U.S. counties have a lower representation of Presbyterians than the general non-Presbyterian population, while similar statistics for Presbyterians were at 64.6 % and Assemblies of God at 59.5 %).

Each of the other top 14 faiths has their comparative similarities and differences with the other churches. Evangelical Lutherans (ELCA), the fifth largest church in 2000, had lower HI than the national population (like the Methodist and Assemblies of God), yet had high DG values or spatial dissimilarity similar to the other major Lutheran church (Missouri Synod). The Muslims (14th largest faith in 2000) were most similar in their geography to the Jews, being spatially concentrated in the HI, very dissimilar with the DG, and highly localized in a relatively small number of counties (only 3.8 % and 4.5 % of the counties, for Jews and Muslims respectively, had relatively higher population densities of those religions than the general population). Finally, the 11th, 12th, and 13th largest churches, the American Baptists, United Church of Christ, and Churches of Christ, were fairly similar to each other with somewhat higher HI than the national, and DG indices in the 50s, indicating concentration in areas different than where the general population was concentrated in 2000, similar to the Southern Baptists and LDS.

75.4.2 Micro Church Locations and Household Income Distributions

Time and space only allow a handful of case studies about the relationship of church location and religious affiliation. Therefore, the three counties of Los Angeles, California; Cook, Illinois; and Dallas, Texas, yield useful samples, because of their large size and location in different U.S. regions. In addition, the LDS-specific multi-county study illustrates how one religion varies across the West in its church building placement.

In Cook County, six faiths were considered for 2000. The Catholics had the largest share of the church-going population. With some 298 church locations located, the median of all the block group median household incomes was $44,765, which was very similar to the county's median household income of $45,922. The Lutheran (MO-Synod) churches had a somewhat higher block group median at $47,452. The 14 Muslim mosques were in the poorest neighborhoods of any of the faiths being in areas of only an average 70 % of the county median income. The Southern Baptist churches were also in below median-level neighborhood income areas. In contrast, the 110 Jewish synagogues were placed in much higher income areas. The median block group incomes were 135 % of the county median income (Fig. 75.5).

Only three faiths were analyzed in Los Angeles County. Here the Catholics' 269 church buildings were in block groups with median incomes of only 86 % of the household median income of the county. This lower relative neighborhood income level compared with Chicago's Catholic churches could be partly related to Hispanics being traditionally Catholic and that many Hispanic areas in Los Angeles County were in lower income zones. The Baptist churches (432 of them) had an even lower block group median household income level at just

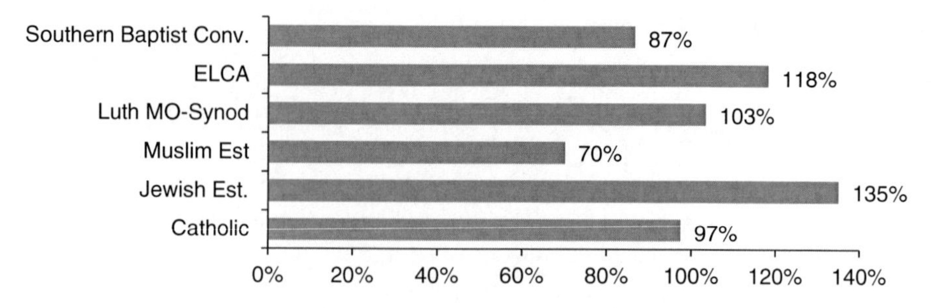

Fig. 75.5 Block group income as percent of county median income for locations of houses of worship for selected religious bodies, Cook County, Illinois, 1999 (Source: Samuel Otterstrom and Bryce Peterson)

83 % of the $43,518 county median. Once again the Jewish synagogues were generally located in above median- income areas at over 115 % of the county median income.

More religions were considered for Dallas County. Here the Catholic churches overall were located in block groups that were right at the county median household income, and so were the Assembly of God churches. Here too, the Baptist churches were located in relatively lower income block groups. Consistent with Los Angeles and Chicago, the 19 Jewish synagogues were in neighborhoods with median incomes at about 167 % of the county median income, which is notably high. The Episcopalian churches were also in high-income neighborhoods, while the meeting-houses for the Presbyterians and the United Methodists were in above average income neighborhoods (Fig. 75.6).

Membership in the LDS church is mostly concentrated in the western states. On average, LDS churches in five of the six counties considered were located in higher income neighborhoods. Only in Multnomah County, Oregon (Portland area) did the LDS locate a greater share of its buildings in lower than median income neighborhoods. However, this lower median is not statistically significant because of the small sample of buildings, and neither is Orange County's higher median block group income level. What is most noteworthy, then, is that the median block group incomes of LDS meetinghouses locations were significantly higher for the other four counties (Table 75.4). Of particular interest is Salt Lake County, because it is central to the greatest numbers of LDS of any county in the U.S. It contains 467 meetinghouses. So, even with buildings throughout the Salt Lake valley, there was a tendency of the neighborhoods around these buildings to have higher household incomes than median incomes in the county overall. This is evidence that the LDS church either intentionally builds chapels in higher income areas, and/or that members of that church in Salt Lake have somewhat higher median incomes. Additionally,

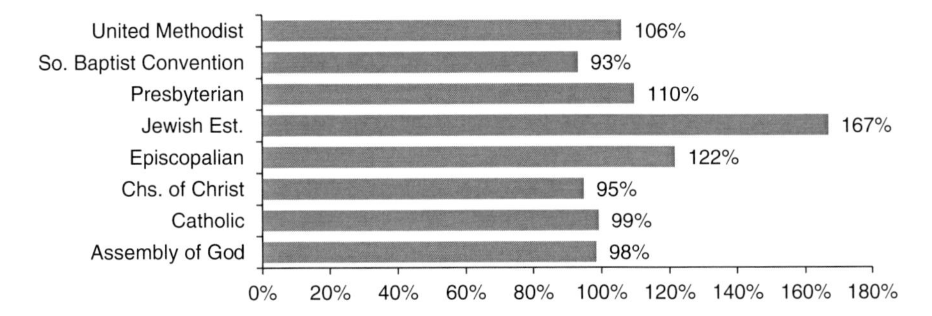

Fig. 75.6 Block group income as percent of county median income for locations of houses of worship for selected religious bodies, Dallas County, Texas, 1999 (Source: Samuel Otterstrom and Bryce Peterson)

the LDS Church is similar to the Catholic Church in that its members are assigned a congregation to meet with according to geographic boundaries, so that the buildings in an area of high density of LDS adherents (such as Salt Lake County) should be located close to the church-going population.

75.5 Putting It Together

The combined macro and micro analysis illustrates the distinct geographic niches that various religions fill in the U.S. Although I have focused on only the 14 largest faiths, the diversities in geographic distribution and concentrations illustrated are tremendous. The United Methodists are the most dispersed throughout the country, while the Jews and Catholics are the most concentrated. The Presbyterians and adherents to the Assemblies of God most closely match the overall distribution of the U.S. population. The Jews and Muslims show relatively higher numbers (compared with the general population) in only 2–4.5 % of the counties, meaning that they are very urban in nature, while the United Methodists have greater relative numbers in nearly 67 % of the counties, illustrating how rural their membership is. At the micro level the Jewish synagogues were located in significantly higher income areas in the three case study counties, while the Catholics were nearer the median, except for Los Angeles County. The Episcopalian, LDS, and Evangelical Lutheran churches also tended to be located in significantly higher income areas, while the Southern Baptist churches in Chicago and Los Angeles, and the Muslim mosques in Chicago were in neighborhoods with incomes significantly below the median for the respective counties (Table 75.5).

Table 75.4 Church of Jesus Christ of Latter-day Saints: Median block group income levels of church locations compared with county median household incomes, 1999

County	Adherents as % of population	Adherents	Churches	Adherents per church	CBG median income 1999	County median household income 1999	CBG income as % of county median	County population, 2000	Standard deviation	Z Score[a]	Two-tailed P-value
Multnomah	1.57 %	10,389	13	799.15	$39,934	$41,278	96.74 %	660,486	10,179	–0.48	0.6426
King	2.33 %	40,525	45	900.56	$64,306	$53,157	120.97 %	1,737,034	19,224	3.89	0.0003
Los Angeles	1.02 %	97,347	122	797.93	$50,436	$42,189	119.55 %	9,519,338	22,762	4.00	0.0001
Maricopa	5.01 %	153,980	139	1,107.77	$57,200	$45,358	126.11 %	3,072,149	19,661	7.10	<0.0001
Orange	1.71 %	48,776	46	1,060.35	$62,688	$58,820	106.58 %	2,846,289	25,017	1.05	0.2999
Salt Lake	56.04 %	503,476	467	1,078.11	$52,188	$48,373	107.89 %	898,387	18,686	4.41	<0.0001

Source: Samuel Otterstrom

CBG Church Block Group

[a]Z Score = Sample Median to County Median

Table 75.5 Cook County, IL; Dallas County, TX, and Los Angeles County, CA: Median block group income levels of church locations compared with county median household incomes for selected religious groups, 1999

County	Denomination	Adherents as % of population	Adherents	Churches	Adherents per Church	CBG median income 1999	CBG income as% of county median	Standard deviation	Z Score[a]	Two-tailed P value
Dallas	Assembly of God	1.39 %	30,891	85	363.42	$42,667	98.5 %	18,512	−0.33	0.7443
Dallas	Catholic	21.66 %	480,510	50	9,610.20	$42,970	99.2 %	23,784	−0.11	0.9165
Los Angeles	Catholic	39.99 %	3,806,377	269	14,150.10	$37,283	85.7 %	23,446	−4.36	<0.0001
Cook	Catholic	39.93 %	2,146,961	298	7,204.57	$44,765	97.5 %	25,614	−0.78	0.436
Dallas	Churches of Christ	2.03 %	45,037	103	437.25	$41,094	94.9 %	31,366	−0.72	0.4722
Cook	ELCA	1.18 %	63,421	143	443.50	$59,603	129.8 %	27,572	5.93	<0.0001
Dallas	Episcopalian	1.25 %	27,740	38	730.00	$52,696	121.6 %	30,909	1.87	0.0695
Los Angeles	Jewish Est	5.93 %	564,700	143	3,948.95	$50,263	115.5 %	26,594	3.03	0.0029
Dallas	Jewish Est.	1.72 %	38,250	19	2,013.16	$72,328	166.9 %	39,869	3.17	0.0053
Cook	Jewish Est.	4.36 %	234,400	110	2,130.91	$62,045	135.1 %	45,156	3.74	0.0003
Cook	Luth MO-Synod	1.19 %	63,814	101	631.82	$47,452	103.3 %	18,512	0.83	0.4082
Cook	Muslim Est	1.78 %	95,623	14	6,830.21	$32,188	70.1 %	21,170	−2.43	0.0305
Dallas	Presbyterian	1.26 %	27,923	36	775.64	$47,543	109.7 %	38,646	0.65	0.5168
Los Angeles	Southern Baptist Conv.	1.17 %	456,702	432	258.41	$36,250	83.3 %	28,271	−5.34	<0.0001
Dallas	Southern Baptist Conv.	12.71 %	281,984	445	633.67	$40,423	93.3 %	38,909	−1.57	0.1165
Cook	Southern Baptist Conv.	1.17 %	63,084	191	330.28	$39,889	86.9 %	22,096	−3.77	0.0002
Dallas	United Methodist	4.78 %	106,100	86	1,233.72	$45,953	106.1 %	39,066	0.62	0.5343

Source: Samuel Otterstrom
CBG Church Block Group
[a]Z Score = Sample Median to County Median

Each faith's membership in the U.S. has a distinct geographical presence. Even as the population of the country has made significant shifts, religious patterns have a tendency to retain their character over time. For instance, the largest faith, the Catholics, have had high HI concentrations at nearly the same level over the past 50 years since 1952. The second largest faith, the Southern Baptists, also have had a long period of relative stability in their concentration levels. Even smaller churches, such as the American Baptists, have had remarkable consistency in the population distribution. Additionally, local level placement of locations of worship has shown fairly consistent neighborhood income levels among faiths. Thus, macro and micro geographical snapshots underscore the unique place niche of each faith, which is tied to history and culture, while emphasizing a noticeable consistency and stability within each religion. Freedom of religion in the U.S. will continue to encourage and foster these important elements of distinctive spatial character by religion in this exceptional diverse national cultural geography.

Acknowledgements Thanks to Bryce Peterson for his work compiling the "micro" income and church location data for the various large cities of the study, and, thanks to Briona Derrick for converting the map figures to black and white.

References

Barcus, H., & Brunn, S. D. (2004). Mapping changes in denominational membership in the Great Plains, 1952–2000. *Great Plains Research, 14*(1), 19–48.

Carroll, J. W., Johnson, D. W., & Marty, M. E. (1979). *Religion in America 1950 to the present*. San Francisco: Harper & Row.

Crawford, T. W. (2005). Stability and change on the American religious landscape: A centrographic analysis of major U.S. religious groups. *Journal of Cultural Geography, 22*(2), 51–86.

Crowley, W. K. (1978). Old order Amish settlement: Diffusion and growth. *Annals of the Association of American Geographers, 68*(2), 249–265.

Diamond, E. (1999). Religion and mobility in 20th century Indianapolis. *Research Notes from the Project on Religion and Urban Culture, 2*(5), 1–7.

Funkhouser, E. (2000). Changes in the geographic concentration and location of residence of residents. *International Migration Review, 34*(2), 489–510.

Gaustad, E. S. (1962). *Historical atlas of religion in America*. New York: Harper & Row.

Gaustad, E. S., & Barlow, P. L. (2001). *New historical atlas of religion in America*. New York: Oxford University Press.

Halvorson, P. L., & Newman, W. M. (1994). *Atlas of religions change in America, 1952–1990*. Atlanta: Glenmary Research Center.

Hunt, L. (1999). Hispanic Protestantism in the United States: Trends by decade generation. *Social Forces, 77*(4), 1601–1625.

Jackson, R. H. (Ed.). (1978). *The Mormon role in the settlement of the west*. Provo: Brigham Young University Press.

Jackson, R. H., & Jackson, M. W. (Eds.). (2003). *Geography, culture and change in the Mormon west, 1847–2003*. Jacksonville: National Council for Geographic Education.

Jones, D. E., Doty, S., Grammich, C., Horsch, J. E., Houseal, R., Lynn, M., Marcum, J. P., Sanchagrin, K. M., & Taylor, R. H. (2002). *Religious congregations & membership in the United States 2000: An enumeration by region, state and county based on data reported for 149 Religious Bodies*. Nashville: Glenmary Research Center.

Kong, L. (1990). Geography and religion: Trend and prospects. *Progress in Human Geography, 14*, 355–371.

Kong, L. (2001). Mapping 'new' geographies of religion: Politics and poetics in modernity. *Progress in Human Geography, 25*(2), 211–234.

Laing, C. R. (2002). The Latter-day Saint diaspora in the United States and the South. *Southeastern Geographer, 42*(2), 228–247.

Long, L., & Nucci, A. (1997). The Hoover index of population concentration: A correction and update. *The Professional Geographer, 49*, 431–440.

McRoberts, O. M. (2003). *Streets of glory: Church and community in a black urban neighborhood*. Chicago: University of Chicago Press.

Meinig, D. W. (1965). The Mormon culture region: Strategies patterns in the geography of the American west, 1847–1964. *Annals of the Association of American Geographers, 55*(2), 191–220.

Min, G. P. (1992). The structure and social functions of Korean immigrant churches in the United States. *International Migration Review, 26*(4), 1370–1394.

Newman, W. M., & Halvorson, P. L. (1984). Religion and regional culture: Patterns of concentration and change among American religious denominations, 1952–1980. *Journal for the Scientific Study of Religion, 23*(3), 304–331.

Otterstrom, S. M. (2003). Population concentration in United States city-systems from 1790 to 2000: Historical trends and current phases. *Tijdschrift voor Economische en Sociale Geografie, 94*(4), 492–510.

Otterstrom, S. M. (2008). Divergent growth of the Church of Jesus Christ of Latter-day Saints in the United States, 1990–2004: Diaspora, gathering, and the east-west divide. *Population, Space and Place, 14*(3), 231–252.

Paul, P. (2003). Religious identity and mobility. *American Demographics, 25*(2), 20–21.

Robbins, T., & Anthony, D. (1990). *In Gods we trust: New patterns of religious pluralism in America*. New Brunswick: Transaction Publishers.

Sakoda, J. M. (1981). A generalized index of dissimilarity. *Demography, 18*(2), 245–250.

Sherkat, D. E. (1999). Tracking the 'other': Dynamics and composition of 'other' religions in the general social survey, 1973–1996. *Journal for the Scientific Study of Religion, 38*(4), 551–561.

Sherkat, D. E. (2001). Tracking the restructuring of American religion: Religious affiliation and patterns of religious mobility, 1973–1998. *Social Forces, 79*(4), 1459–1493.

Sheskin, I. M. (2000). The Dixie diaspora: The 'loss' of the small southern Jewish community. *Southeastern Geographer, 40*(1), 52–74.

Shortridge, J. R. (1976). Patterns of religion in the United States. *Geographical Review, 66*, 420–434.

Shortridge, J. R. (1978). The pattern of American Catholicism 1971. *Journal of Geography, 77*, 56–60.

Sinha, J. W., Hillier, A., Cnaan, R. A., & McGrew, C. C. (2007). Proximity matters: Exploring relationships among neighborhoods, congregations, and the residential patterns of members. *Journal for the Scientific Study of Religion, 46*(2), 245–260.

Smith, T. W. (2002). Religious diversity in America: The emergence of Muslims, Buddhists, Hindus, and others. *Journal for the Scientific Study of Religion, 41*(3), 577–585.

Smith, T. W., & Seokho, K. (2005). The vanishing Protestant majority. *Journal for the Scientific Study of Religion, 44*(2), 211–223.

Stump, R. W. (1984). Regional migration and religious commitment in the United States. *Journal for the Scientific Study of Religion, 23*(3), 292–303.

Tillman, B. F., & Emmett, C. F. (1999). Spatial succession of sacred space in Chicago. *Journal of Cultural Geography, 18*(2), 79–108.

Vining, D. R., & Strauss, A. (1976). *A demonstration that current deconcentrations trends are a clean break with past trends*. Philadelphia: Regional Science Research Institute.

Warf, B., & Winsberg, M. (2008). The religious diversity in the United States. *The Professional Geographer, 60*(3), 413–424.

Webster, G. R. (2000). Geographical patterns of religious denomination affiliation in Georgia, 1970–1990: Population change and growing urban diversity. *Southeastern Geographer, 40*(1), 25–51.

Woolever, C., & Bruce, D. (2002). *A field guide to U.S. congregations: Who's going where and why*. Louisville: Westminster John Knox.

Zelinsky, W. (1961). An approach to the religious geography of the United States: Patterns of church membership in 1952. *Annals of the Association of American Geographers, 51*, 139–193.

Zelinsky, W. (2001). The uniqueness of the American religious landscape. *Geographical Review, 91*(3), 565–585.

Chapter 76
Bible Belt Membership Patterns, Correlates and Landscapes

Gerald R. Webster, Robert H. Watrel, J. Clark Archer, and Stanley D. Brunn

76.1 Introduction

The literal interpretation of the Bible is a dominant and distinguishing feature of the American religious landscape. According to a 2007 Gallup Poll, one-third of Americans "believe the Bible is accurate and should be taken literally word for word" (Newport 2007). Another 47 % consider the Bible to be "inspired by the word of God" and 54 % of those who are regular church attendees view the Bible "as the absolute word of God." Regular church attendance in the U.S. is highest in the South, the same region where 41 % accept the Bible literally. A Gallup Poll in December 2010 reported that 40 % of Americans "believe that God created humans in something like their present form within the past 10,000 years" (Burke 2010). The Pew Forum on Religion and Public Life (2008), a group regularly issuing a barometer on the country's religious outlook, reported that 33 % believe the word of God should be taken literally. While among Evangelicals the percentage was 59 %, it was only 22 % for mainline church members (Jordan 2007: 9).

G.R. Webster (✉)
Department of Geography, University of Wyoming, Laramie, WY 82071, USA
e-mail: gwebste1@uwyo.edu

R.H. Watrel
Department of Geography, South Dakota State University, Brookings, SD 57006, USA
e-mail: robert.watrel@sdstate.edu

J.C. Archer
Department of Geography, School of Natural Resources, University of Nebraska-Lincoln, Lincoln, NE 68588, USA
e-mail: jarcher@unlserve.unl.edu

S.D. Brunn
Department of Geography, University of Kentucky, Lexington, KY 40506, USA
e-mail: brunn@uky.edu

S.D. Brunn (ed.), *The Changing World Religion Map*,
DOI 10.1007/978-94-017-9376-6_76

Evangelical groups are strongest in the U.S. South, which would lead many observers and scholars to label the region a Bible Belt, a term coined by H. L. Mencken in the 1920s following his coverage of the Scopes "monkey trial" in Dayton, Tennessee (Mencken 1927, 1947, 1949). Mencken did not provide a specific location, but believed it included all or parts of Mississippi, Arkansas, Alabama, Georgia, South Carolina, and "darkest Tennessee" (Tweedie 1978: 865). Wilson (1989: 1312) felt that Mencken was also referring to "rural areas of the Midwest," but especially the "Baptist backwaters of the South." Today the groups with the highest percentages supporting the literal interpretation of the Bible are the Church of Jesus Christ of Latter Day Saints (Mormons) (35 %), Jehovah's Witnesses (48 %), Muslims (50 %) and members of historically African American churches (62 %). About 60 % of evangelicals live in the South as do 34 % of mainline church adherents.

While these high percentages may astound some scholars outside the U.S., they are not shocking for those who study and monitor the role that religion plays in daily and community life, especially in the South (Webster 1997). Religion is part of the daily fabric of a large percentage of households, many organizations and also those in the public arena. In presidential elections religion enters discussions and debates regarding abortion, gay rights, same sex marriage, stem cell research, alcohol availability, capital punishment, euthanasia as well as support for the state of Israel. Religion also enters the fray in debates pertaining to the teaching of creationism as science in the public schools, the placement of the Ten Commandments in public spaces, Christian nativity scenes assembled at government cost, and prayers in public schools before sanctioned events. The legal landscape regarding religion in education, the workplace and larger culture is one where state and local governments are the scene of some contentious religious and secular battles in the U.S. South and elsewhere. For these reasons alone, continued study of the contemporary geographic character and social implications of the Bible Belt, or more accurately Bible Belts, merits further inquiry. These are the focus of this study.

76.2 Background Literature

The Bible Belt is a subject of interest to scholars from a number of different disciplines, many who recognize the importance of studying this phenomenon in regards to place and region, while mapping identities, affiliations and distinguishing landscape features. The contributions by historians and sociologists are numerous and include the work by Reed (1972), Halvorson and Newman (1978), Hill (1983, 1990), Marty (1998), Gaustad and Barlow (2000), Eck (2001), Hill and Lippy (2005), Neitz (2005), Williams (2005) and Conser and Payne (2008). Geographers, as one would expect, have been fascinated with this topic and region; see, for example, the classic work by Zelinsky (1961) on patterns of Christian denomination membership; see also Shortridge (1976, 1977), Tweedie (1978), Vogeler (1979),

Webster (1997, 2000), Crawford (2005), Jordan (2007), Warf and Winsberg (2008) and Shelton et al. (2012).

For those in the cultural geography community, the pioneering study specifically on the Bible Belt was done by Heatwole (1978) which delineated the Bible Belt by the geographic patterns of affiliation of adherents to 24 Christian denominations that believe in a literal interpretation of the Bible. His methodology was used by Brunn et al. (2011a) to update and map county membership changes in Bible Belt denominations during a three-decade (1970–2000) period. Their study addressed questions about classifying Bible Belt denominations (a topic not discussed here) and includes references to numerous studies about defining the term, delineating the region, and the potential location of the Belt's "buckle" or "buckles" in the U. S. (candidates include Jackson, Mississippi; Columbia, South Carolina; Memphis, Nashville and Knoxville, Tennessee; Lubbock, Texas; Springfield, Missouri, and Oklahoma City, Oklahoma). Scholars have used different criteria to map the Bible Belt (see Smith 1987; Corbett 1993; Shibley 1998; Steensland et al. 2000) and have focused on how the Belt overlaps with other cultural and social variables and indexes. Authors have linked the Bible Belt to a "hookworm belt," a "divorce belt," a "beauty pageant belt," "lynching belt," and higher rates of corporal punishment. The reader is referred to Brunn et al. (2011a) for further discussion of these correlates and additional references. More than one of the recent studies has noted the shift (without maps) in the core of the Bible Belt from its original focus in western North Carolina and eastern Tennessee to the Texas and Oklahoma borderlands, patterns illustrated in Brunn et al. (2011a).

This study builds on Brunn et al. (2011a) and has three purposes: (1) to map, describe and discuss the current (2010) pattern of adherents to Bible Belt denominations on a county basis, (2) to seek explanations for the patterns by looking at correlates of county population change, rural-urban mix, age and income structures, among others, and (3) to apply and examine the relevance of Stump's (2008: 222) concept of religious territoriality to the landscape of the Bible Belt by looking at signage (outdoor advertising, public and private) in the region. Each of these three perspectives—cartographic, spatial analytic, and photographic—will shed light on the current state of the U.S. Bible Belt or Belts.

76.3 Data Sources and Overall Patterns

In this study we use a recent update to the Glenmary Research Center database which was used by Heatwole (1978) to map Bible Belt counties using 1971 data, and also by Brunn et al. (2011a) who added data from 1980, 1990, and 2000. The 2010 data used in this study were drawn from the *2010 U.S. Religion Census: Religious Congregations and Membership Study* which was conducted by the Association of Statisticians of American Religious Bodies (Grammich et al. 2012). The methods and denominational coding employed by the Association generally followed those established in the earlier Glenmary Research Center works, to ensure cross-temporal comparability of the 2010 census with the earlier reports for 1971, 1980, 1990 and 2000.

Heatwole defined the Bible Belt on the basis of 24 denominations that believe in a literal interpretation of the Bible. Brunn et al. (2011a) used the same denominations as Heatwole (1978), except that General Baptists were excluded because no data were reported for this denomination in the 2000 religious census. In the discussion below we use the same set of denominations to investigate the Bible Belt landscape in 2010 as used in the Brunn, Webster and Archer study to examine the region between 1971 and 2000, except that the Presbyterian Church in America has been added to the Bible Belt denomination set for both 2000 and 2010. Also no data were reported for the Baptist Missionary Association of America and the Mennonite Church for 2010. To achieve consistency between values for 2010 and 2000, the numbers of adherents of the Presbyterian Church in America were obtained for 2000 as well. Figures for two other denominations which had been included in Heatwole's (1978) Bible Belt denominations set—Evangelical Congregation Church, and Seventh Day Baptist General Conference—also were collected for 2010 though these remained missing for 2000.

76.4 The Bible Belt in 2010

Table 76.1 identifies the set of Bible Belt denominations studied here; it displays the number of adherents in each of the included denominations observed for both 2000 and for 2010. The symbol "NA" is used to indicate missing data. By far the largest denomination is the Southern Baptist Convention with nearly 20 million adherents in both years. Notably, the percentage increase in Southern Baptist adherents from 2000 to 2010 was less than one-tenth of 1 %, considerably less than the nearly 10 % increase in the overall total population of the United States during the decade. Other denominations with over one million adherents reported for 2010 include the Lutheran Church—Missouri Synod (2.3 million), Christian Churches and Churches of Christ (1.5 million), Seventh-Day Adventist Church (1.2 million), and Church of God (Cleveland, Tennessee) (1.1 million).

Summing over the entire set of Bible Belt denominations for 2010 yielded a total of 29.4 million adherents in the U.S. The corresponding sum for 2000 was nearly 29.6 million adherents, indicating a decrease of –0.62 % or a decline of about 183,000 adherents to Bible Belt denominations from 2000 to 2010. The total of all adherents to religious divisions in the United States increased nearly 6.6 %, from nearly 141 million in 2000 to nearly 150.7 million in 2010. But the 6.6 % increase in total adherents for all faiths lagged behind the 9.7 % increase in the total population of the U.S. (from 281 million in 2000 to 309 million in 2010).

Several Bible Belt denominations reported considerable growth from 2000 to 2010. Specific denominations with 10 % or greater increases included the Seventh Day Adventist Church, Church of God (Cleveland, Tennessee), International Pentecostal Holiness Church, Free Methodist Church of North America, North

Table 76.1 Bible-Belt denominations, adherents, and percent change, 2000–2010

Denomination	Number of adherents 2010	Number of adherents 2000	Percent change 2000–2010
Southern Baptist Convention	19,896,279	19,881,467	0.07
Lutheran Church—Missouri Synod	2,270,921	2,521,162	−9.93
Christian Churches & Churches of Christ	1,453,160	1,439,253	0.97
Seventh-day Adventist Church	1,194,996	923,067	29.46
Church of God (Cleveland, Tennessee)	1,109,992	974,198	13.94
Church of the Nazarene	893,649	907,331	−1.51
Wisconsin Evangelical Lutheran Synod	382,883	405,078	−5.48
Presbyterian Church in America	341,431	315,293	8.29
Reformed Church in America	295,120	335,677	−12.08
International Pentecostal Holiness Church	289,475	241,828	19.7
Converge Worldwide/Baptist General Conference	260,100	238,920	8.86
Church of God (Anderson, Indiana)	225,753	238,609	−5.39
Christian Reformed Church in North America	224,003	248,938	−10.02
National Association of Free Will Baptists	217,560	254,170	−14.4
Mennonite Church USA	127,363	156,345	−18.54
Free Methodist Church of North America	107,271	96,237	11.47
North American Baptist Conference	57,219	49,762	14.99
Orthodox Presbyterian Church	28,559	26,346	8.4
Evangelical Congregational Church	20,592	N/A	N/A
Seventh Day Baptist General Conference, USA and Canada	5,168	N/A	N/A
Church of God General Conference	3,832	4,925	−22.19
Fellowship of Evangelical Bible Churches	2,121	1,811	17.12
Baptist Missionary Association of America	N/A	295,239	N/A
Mennonite Church	N/A	34,617	N/A
Total Bible-Belt Denominations Adherents	29,407,447	29,590,273	−0.62
Total Adherents of all Reporting Denominations in US	150,686,156	141,371,963	6.59
Total Population of US including AK & HI	308,745,538	281,421,906	9.71
Bible-Belt Denominations Adherents percent of US population	9.52	10.51	−9.42
Bible-Belt Denominations Adherents percent of Total Reported Adherents	19.52	20.93	−6.74
Total Reported Adherents percent of US Population	48.81	50.23	−2.83

Source: Calculated by J.C. Archer from Grammich et al. (2012)

American Baptist Conference, and the Fellowship of Evangelical Bible Churches. Several other Bible Belt denominations witnessed declines of 10 % or more in numbers of adherents from 2000 to 2010, including the Reformed Church in America, Christian Reformed Church in North America, National Association of Free Will Baptists, Mennonite Church USA, and the Church of God General Conference.

Comparatively speaking, adherents to Bible Belt denominations declined from 20.9 % of adherents to all faiths reporting and 10.5 % of total U.S. population in 2000, to 20 % of all adherents to reporting faiths and 9.5 % of the total U.S. population in 2010. However, these changes were not geographically uniform across the United States, so that an investigation focusing on regional and sub-regional variations is warranted. As a result it is worthwhile to consider the geographic pattern of adherents to Bible Belt denominations with the general population, and in terms of comparisons of the numbers of adherents to Bible Belt denominations from one region or residential setting to another. Whereas the numbers of adherents can be regarded as indicating the likely numbers and sizes of church facilities needed to provide adequate ministry to believers, population percentages can be regarded as indicating the relative strength of Bible Belt denominations in comparison with other denominations across different regions or sub-regions of the United States.

Brunn et al. (2011a) closely followed Heatwole (1978) in distinguishing and mapping counties using 3 intervals including below 25 %, 25–50 % and above 50 % of a county's population reported as adherents to Bible Belt denominations. Our preliminary analysis of the 2010 data, however, demonstrated that somewhat lower thresholds including three categories of less than 20 %, 20–40 % and above 40 % of a county's population being adherents to Bible Belt denominations are more suitable to display cartographic patterns for the later time period.

Using these somewhat lower thresholds, counties with 20 % or higher Bible Belt denomination adherents, a total of 1,075 counties, were identified as belonging to the Bible Belt geographical area in 2010. These counties show the general outline of the Bible Belt, with counties having over 40 % Bible Belt populations considered its core area(s) (Fig. 76.1).

As would be expected, the majority of U.S. counties or 2,034 jurisdictions (65 % of all counties) had less than 20 % of their populations adhering to Bible Belt denominations (Table 76.2). These counties account for 54 % or 15.8 million of all Bible Belt adherents in the U.S. The two remaining classes for Bible Belt counties accounted for a combined 34.6 % of all U.S. counties, and jointly have nearly the same number of Bible Belt adherents, at 13.4 million or 46 % of the total, as the first class.

However, when looking at the spatial distribution of these counties a more subtly detailed geographic picture of the Bible Belt emerges (see Fig. 76.1). When studying this figure, it becomes apparent that the Bible Belt is largely a feature of the U.S. South. Indeed, of the 1,075 Bible Belt counties identified, only 109 or 10.1 % are not found in the southern states extending from Virginia to Missouri to Oklahoma and Texas. However, even though Bible Belt counties are concentrated in the South, their geographic pattern can hardly be considered uniform or solid. Rather it is a

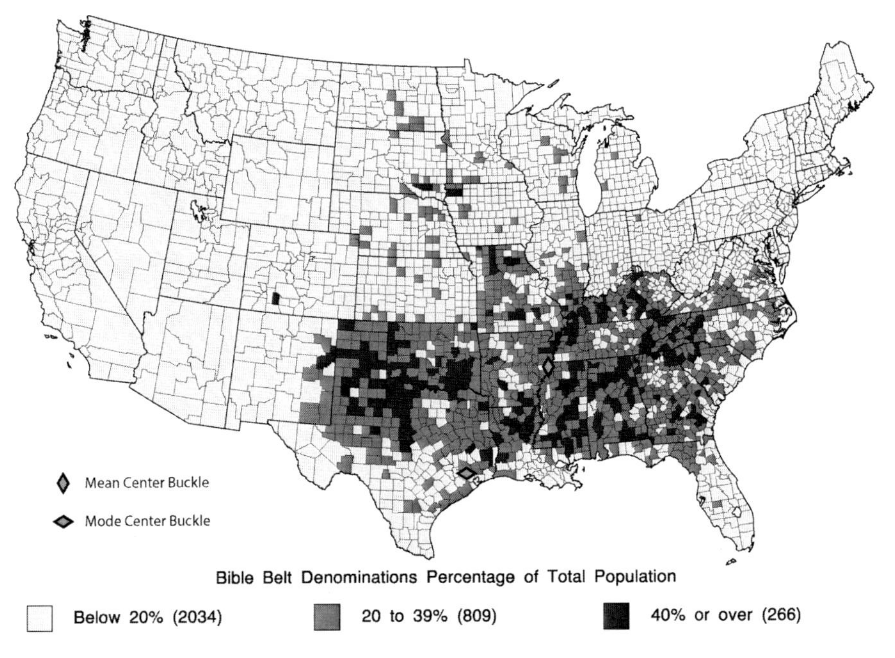

Fig. 76.1 Percent Bible Belt population, by county, 2010 (Map by the authors)

Table 76.2 Bible Belt denomination adherents as percent of population, 2010, by county

Bible Belt denomination adherents as a percent of population 2010	Number of counties[a]	Percent of counties[a]	Number of Bible Belt denomination adherents 2010	Percent of total Bible Belt denomination adherents in contiguous United States
0–19 %	2,034	65.4 %	15,876,875	54.13 %
20–39 %	809	26.0 %	10,124,301	34.52 %
40–121 %	266	8.6 %	3,329,241	11.35 %
Total	3,109	100.0 %	29,330,417	100.0 %

Source: Calculated by J.C. Archer from Grammich et al. (2012)
[a]Includes only counties in the 48 contiguous states

pattern of "several belts" with one running from northeastern Arkansas through Kentucky and another from northern Louisiana, through Mississippi and Alabama, to eastern North Carolina. There are also some "buckles" or concentrations, the most noticeable being located in northern Texas and southern Oklahoma. There are also some "notches" where there are few adherents to Bible Belt denominations, including southern Texas, southern Louisiana and southern Florida, these three locations being heavily Catholic. However, many of the notches in the Bible Belt are urban areas including the cities of Atlanta, Nashville, Memphis, Dallas and Fort Worth. The Bible Belt also contains some counties that are geographic outliers to

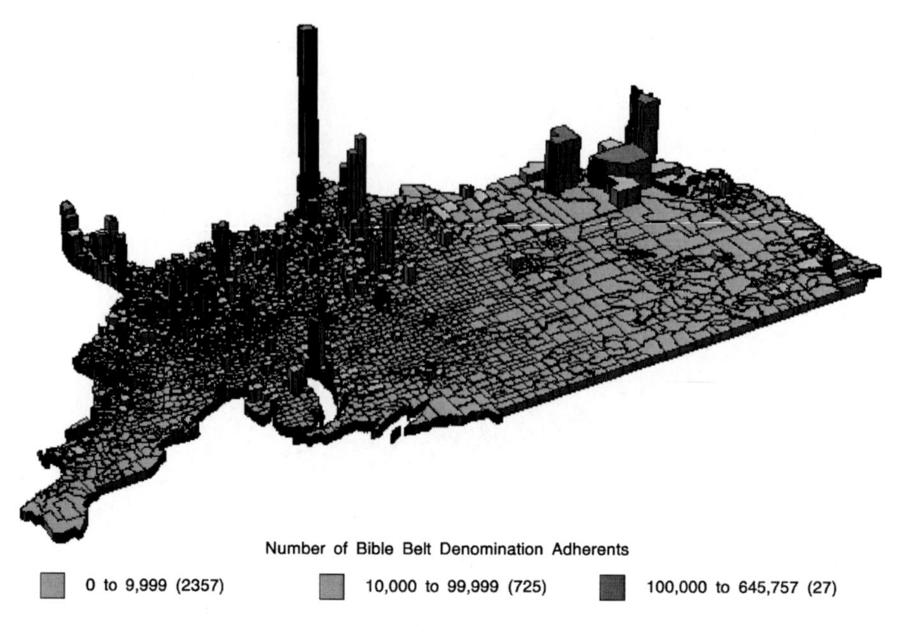

Fig. 76.2 Bible Belt denomination adherents, by county, 2010 (Map by the authors)

the core of the region. These extensions include some counties located in eastern New Mexico and southern Illinois, as well as others found in the central and northern Great Plains and Midwest states.

From visual inspection of the map of adherents to Bible Belt denominations as a percentage of population in 2010, it appears that many counties with large proportions of Bible Belt adherents tend to be situated in less geographically accessible and more lightly populated rural settings (Fig. 76.2). This pattern can be further explored by visually examining a map which represents the total numbers of adherents of Bible Belt denominations by county across the contiguous or "Lower-48" states. Although the distribution of total Bible Belt adherents by county is still concentrated in the U.S. South, counties along the West Coast and northern Atlantic Coast, and in the Southwest, Midwest, and in southern Florida have large numbers of adherents to Bible Belt denominations. The three-dimensional map of numbers of adherents to Bible Belt denominations also reveals that many highly urban or metropolitan counties in the South and elsewhere can be regarded as "spikes" with conspicuous elevations above their surrounding areas. Several of the most conspicuous pinnacles are associated with urban or metropolitan counties outside the South.

Important aspects of the geographical distribution of adherents to Bible Belt denominations in 2010 can be further explored by identifying the 15 counties having the highest population percentages and absolute numbers of adherents to Bible Belt denominations (Table 76.3). The lower threshold for inclusion among the 15 counties with the highest percentages of adherents to Bible Belt denominations in 2010 is slightly more than 67 % of population, the proportion found for Chickasaw

Table 76.3 Leading Bible Belt denomination counties, 2010

Part A. Counties with highest proportions of Bible-Belt denominations adherents percentages of population, 2010

County, State	County's largest settlement	Bible-Belt denominations percent of population 2010	Bible-Belt denominations number of adherents 2010	County population 2010 Census of Population
Harmon, OK	Hollis	120.64	3,525	2,922
Yalobusha, MS	Water Valley	81.24	10,299	12,678
Hickman, KY	Columbus	79.99	3,921	4,902
Hancock, KY	Lewisport	76.15	6,522	8,565
Caldwell, KY	Princeton	73.37	9,526	12,984
Hancock, TN	Sneedville	72.81	4,965	6,819
Collingsworth, TX	Wellington	72.33	2,211	3,057
Haskell, TX	Haskell	72.32	4,266	5,899
Clay, NC	Hayesville	71.84	7,606	10,587
Borden, TX	Gail	70.05	449	641
Calhoun, MS	Pittsboro	69.54	10,405	14,962
Green, KY	Greensburg	68.59	7,722	11,258
Cottle, TX	Paducah	67.64	1,018	1,505
Cotton, OK	Walters	67.27	4,166	6,193
Chickasaw, MS	Houston	66.78	11,615	17,392

Part B. Counties with highest numbers of Bible Belt denominations adherents, 2010

County, State	County's largest settlement	Bible-Belt denominations percent of population 2010	Bible-Belt denominations number of adherents 2010	County population 2010 Census of Population
Harris, TX	Houston	15.78	645,736	4,092,459
Tarrant, TX	Ft. Worth	19.18	346,917	1,809,034
Dallas, TX	Dallas	12.86	304,531	2,368,139
Los Angeles, CA	Los Angeles	2.64	258,836	9,818,605
Jefferson, AL	Birmingham	33.51	220,659	658,466
Oklahoma, OK	Oklahoma City	26.08	187,437	718,633
Maricopa, AZ	Phoenix	4.62	176,449	3,817,117
Cook, IL	Chicago	3.12	162,206	5,194,675
Jefferson, KY	Louisville	21.65	160,456	741,096
Duval, FL	Jacksonville	18.08	156,261	864,263
Shelby, TN	Memphis	16.66	154,534	927,644
Greenville, SC	Greenville	32.13	144,987	451,225
Orange, CA	Anaheim	4.81	144,836	3,010,232
Knox, TN	Knoxville	32.98	142,566	432,226
Bexar, TX	San Antonio	8.27	141,883	1,714,773

Source: Calculated by J.C. Archer from Grammich et al. (2012)

County, Mississippi. At the highest end, Harmon County, Oklahoma had 3,525 adherents to Bible Belt denominations in 2010, which is actually a greater number than the population figure of 2,922 people reported by the *2010 Census of Population* for the county. Such seemingly impossible figures can occur in thinly populated locations experiencing out-migration because some churches maintain out-migrating members of extended families on their membership rolls including the Southern Baptist Convention. An examination of the figures reported for particular denominations by county in the *2010 U.S. Religion Census* (Grammich et al. 2012: 467) indicates that there were 3,487 adherents to the Southern Baptist Convention in Harmon County. Notably, no other county in the contiguous 48 states was found to have a percentage of adherents to Bible Belt denominations of the population greater than 81 % for 2010.

None of the top 15 counties in terms of percentage of their total populations being adherents to Bible Belt denominations had a total 2010 population greater than 18,000 people. The average 2010 Census population of these 15 counties is just 8,024 people, and the smallest (Borden County, Texas) contained merely 641 people in 2010. Thus, the highest population proportions of adherents to Bible Belt denominations tend to be found in less populous, more rural settings. Bible Belt denominations may be quite locally dominant in such settings, where they can constitute large majorities of the local population.

What about counties with the largest number of adherents to Bible Belt denominations? Table 76.3 also lists the 15 counties in the contiguous lower-48 states containing the absolute largest number of adherents to Bible Belt denominations. It is obvious that the greatest numbers of adherents to Bible Belt denominations are found, perhaps not surprisingly, in large metropolitan centers, especially in the South but also in some other regions of the country as well. Three counties in Texas—Harris County, Tarrant County and Dallas County—each contain more than 300,000 adherents to Bible Belt denominations, with nearly 646,000 in Harris County, Texas alone. These three counties contain the million scale plus metropolitan centers of Houston, Fort Worth and Dallas, TX. Other notable counties among the 15 containing the largest number of adherents to Bible Belt denominations include Los Angeles County and Orange County, CA; Maricopa County, AZ; and Cook County, IL, which include the large metropolitan centers of Los Angeles and Anaheim, CA; Phoenix, AZ; and Chicago, IL. Los Angeles and Chicago both received many migrants from the South during the twentieth century.

Although quite significant numbers of adherents to Bible Belt denominations are concentrated in such metropolitan settings, they do not tend to dominate these settings to the same extent that smaller numbers of adherents to Bible Belt denominations do in smaller, more rural settings. Although adherents to Bible Belt denominations are found to comprise more than one-quarter of the total population of several of the top 15 counties which contain the cities of Birmingham, AL, Oklahoma City, OK, Greenville, SC, and Knoxville, TN, they comprise barely 2.6 % of the population of Los Angeles, CA. While Los Angeles County's 259,000 adherents to Bible Belt denominations ranks 4th in the U.S. in terms of total adherents to Bible Belt denominations, this is out of a total population of 9.8 million people. Clearly, metropolitan areas tend to exhibit greater religious diversity than do more rural settings.

Computerized Geographic Information System (GIS) techniques were used to further examine the aggregate geographical patterns of adherents to Bible Belt denominations within the contiguous lower-48 state area of the United States (Strategic Mapping 1990; ESRI 2004). Although it has become quite evident that region and residential setting have a great deal to do with the overall geographical pattern of adherents to Bible Belt denominations, more specificity can be added to this discussion.

If counties are classified in terms of "region and settlement groups," there are some very striking distinctions in the patterns of religious adherence. The "North," "South" and "West" division of the U.S. into regions applied in this statistical exercise was previously derived through quantitative electoral geography research which concluded that citizens in each of these three regions have historically exhibited distinct patterns of voter support for major party presidential candidates from one election to the next over many decades (Shelley et al. 1996; Brunn et al. 2011b). Many observers believe that religious views and political views tend to exhibit coincident cleavages within the United States, so applying a regionalization based on political behavior is appropriate.

The "Rural," "Micropolitan" and "Metropolitan" residential settings division applied in this statistical analysis was drawn from the U.S. Census distinction between various types of Core-Based Statistical Areas (CBSA) for 2006. "Metropolitan" areas encompass metropolitan central cities of 50,000 population or more, plus adjacent counties of metropolitan character. "Micropolitan" areas encompass smaller cities with populations of 10,000–49,999, plus adjacent counties of urban character. Other counties which are outside the boundaries of CBSA's are deemed to be "Rural."

Table 76.4 summarizes the numerical values which were derived after partitioning the 3,109 counties and county-equivalents (for example parishes in Louisiana) into various groupings based on regional setting and settlement class. The numbers in the table are not averages, but are rather values derived through GIS methods as aggregates for all counties within a region's settlement class considered collectively. For example, the aggregate total population of all counties in the South region in 2010 was 105 million and the aggregate total number of adherents to Bible Belt denominations in the South region in 2010 was 20 million. Hence, adherents to Bible Belt denominations comprised 19 % of total population in the South region in 2010. The similarly derived proportions for the other two regions were 5 % of total population for the North region, and 4 % for the West region in 2010. Thus, adherents to Bible Belt denominations are roughly four times more prevalent in the South than in the North or West regions of the contiguous United States.

Distinctions by settlement class are similarly striking. In the 1,334 counties outside CBSA boundaries, nearly one out of five residents were an adherent to a Bible Belt denomination in 2010. More specifically, there were 3.7 million adherents to Bible Belt denominations out of an overall rural population of 20 million, with adherents to Bible Belt denominations comprising 19 % of the rural population in

Table 76.4 Change in Bible-Belt denomination adherents, by region and settlement class, 2000–2010

Region & settlement class	Number of counties 2010	Census population 2010	Bible-Belt denominations adherents 2010	Bible-Belt denominations percent of population 2010	Number of counties 2000	Census population 2000	Bible-Belt denominations adherents 2000	Bible-Belt denominations percent of population 2000	Percent change in Bible-Belt denominations adherents 2000–2010
Contiguous 48 States	3,109	306,675,006	29,330,417	9.56	3,109	279,583,437	29,517,373	10.56	–0.63
South Region	1,340	105,429,541	20,003,075	18.97	1,341	91,776,331	19,904,161	21.69	0.5
North Region	1,038	125,204,214	6,220,069	4.97	1,038	120,650,918	6,461,773	5.36	–3.74
West Region	731	76,041,251	3,107,273	4.09	730	67,156,188	3,151,439	4.69	–1.4
Rural Settlement	1,334	19,488,316	3,693,755	18.95	1,335	19,123,498	3,961,654	20.72	–6.76
Micropolitan Settlement	687	30,620,632	4,764,446	15.56	687	28,953,736	5,023,945	17.35	–5.17
Metropolitan Settlement	1,088	256,566,058	20,872,216	8.14	1,087	231,506,203	20,531,774	8.87	1.66
South-Rural	544	8,807,730	2,653,827	30.13	545	8,516,497	2,802,929	32.91	–5.32
South-Micropolitan	285	12,395,629	3,415,857	27.56	285	11,534,851	3,554,087	30.81	–3.89
South-Metropolitan	511	84,226,182	13,933,391	16.54	511	71,724,983	13,547,145	18.89	2.85
North-Rural	379	7,295,792	800,880	10.98	379	7,275,860	886,804	12.19	–9.69
North-Micropolitan	247	12,495,975	966,223	7.73	247	12,153,455	1,039,065	8.55	–7.01
North-Metropolitan	412	105,412,447	4,452,966	4.22	412	101,221,603	4,535,904	4.48	–1.83
West-Rural	411	3,384,794	239,048	7.06	411	3,331,141	271,921	8.16	–12.09
West-Micropolitan	155	5,729,028	382,366	6.67	155	5,265,430	430,793	8.18	–11.24
West-Metropolitan	165	66,927,429	2,485,859	3.71	164	58,559,617	2,448,725	4.18	1.52

Source: Calculated by J.C. Archer from Grammich et al. (2012)

2010. Nearly 4.7 million adherents to Bible Belt denominations out of a total population of 31 million were found in Micropolitan counties, with adherents to Bible Belt denominations comprising 15.6 % of the Micropolitan population in 2010. And nearly 21 million adherents to Bible Belt denominations out of a total population of 257 million residents were reported in Metropolitan counties, with adherents to Bible Belt denominations comprising a somewhat lower 8 % of the population in 2010. Hence, although Rural and Micropolitan settings contain only about one-sixth of the total population of the contiguous United States, adherents to Bible Belt denominations are about twice as prevalent in Rural or Micropolitan settings as they are in Metropolitan settings.

Even sharper distinctions emerge when both region and residential setting are considered jointly, though some details are better left to the reader to identify in Table 76.4 because discussion of a (3×3) or 9 class division can become quite tedious. By far the highest proportion of adherents to Bible Belt denominations are found in rural counties in the South, where nearly one-third of the population subscribed to Bible Belt denominational beliefs in 2010. The proportions for South-Micropolitan (28 %) and South-Metropolitan (17 %) both exceed the overall U.S. proportion (10 %) of adherents to Bible Belt denominations by a substantial margin. The lowest value of merely 3.7 % of population is found for West-Metropolitan counties, which echoes the pattern noted above for Los Angeles, California. But, in all three regions of the United States—South, North and West—the same pattern is repeated, with considerably higher proportions of adherents to Bible Belt denominations found in Rural or Micropolitan locales than in Metropolitan places.

Comparable values were calculated for 2000 and these are also included in Table 76.4. Broadly speaking, the same relative patterns described for 2010 also can be identified for 2000. For example, both the relative regional scale and the relative settlement scale rankings are quite similar for both years. However, it is notable that the specific proportions are generally higher for 2000 than for 2010. Something which is especially worthy of note is that when changes in adherents to Bible Belt denominations numbers and proportions of the population are closely compared by region and settlement class for the decade between 2000 and 2010 (in the right-most column of Table 76.4), that greatest percentage declines in adherents to Bible Belt denominations are found in rural settings in each of the three regions. To a large extent this probably reflects the national pattern of much larger overall population growth in Metropolitan settings, but it also identifies a notable diminishing trend of the relative influence among Bible Belt denominations in their traditional rural core settings.

76.5 Correlates of Bible Belt Membership

Additional insights into the salient geographical patterns of adherence to Bible Belt denominations across the United States at the county level can be identified by using the multivariate statistical method of multiple regression analysis (Rogerson

2010; SYSTAT 2004). Although overall patterns and trends indeed emerge from the GIS-based analysis presented in Table 76.4, such aggregate results derived by summing over larger scale region and residential setting groupings can obscure the statistical variance discernible only at a more local county-by-county scale. In the regression results reported in Table 76.5, each of the 3,108 counties in the contiguous United States makes an individual contribution to the analysis (one county was dropped due to missing data).

The "dependent variable" chosen for the regression modeling is the proportion of adherents to Bible Belt denominations, calculated by dividing the number of adherents reported for each county in the *2010 U.S. Religion Census* (Grammich et al. 2012) by the population of each county reported in the *2010 Census of Population* (U.S. Bureau of Census 2011). A set of demographic and economic "independent variables" also measured at the county level was selected from the *2010 Census of Population* for 2010, and from U.S. Census *American Community Survey* reports for the 2005–2009 time period (U.S. Bureau of Census 2011, 2012). An additional political "independent variable" consisting of the percent of popular vote cast for the Democratic presidential candidate in each county in 2008 was obtained from previous research by the authors (Brunn et al. 2011b). Succinct definitions of each variable and the numerical results of the multiple regression modeling effort are shown in Table 76.5.

In statistical terms, a Multiple-R value of 0.730, a Multiple-R-Square value of 0.532, and an overall F-ratio of 220.026 (with 16 and 3,091 degrees of freedom) indicate a successful and highly statistically significant modeling effort, with slightly more than half of the statistical variance in the proportion of the total county population adhering to a Bible Belt denomination explained by the independent variables.

Very briefly, the multiple regression results can be interpreted as follows. Substantially greater proportions of adherents to Bible Belt denominations are associated with the South region (S independent variable). Somewhat greater proportions of adherents to Bible Belt denominations are associated with percent of population under age 16 (PCTLT16) and percent of population age 65 or older (PCTGTE65), as well as percent of adults with college degrees or more education (PCT_BA_OVR). In contrast, lower population proportions of adherents to Bible Belt denominations are associated with Metropolitan settlements (MET), the foreign born percent of the population (FBORN10), the percent of adults with a high school or more education (PCT_HS_OVR), the percent of workers employed in agriculture or other primary sector employment (PCT_AG), median family income (MEDFAMINC), support for the Democratic candidate in the 2008 presidential election (DEM08), and population growth between 2000 and 2010 (PCNG2000T10). Notably none of the ethnic composition measures indicating percent of population White, Black, Native American, Asian, or Hispanic (PCTWHITE, PCTBLACK, PCTAMIND, PCTASIAN, and PCTHISP) were found to systematically covary with population proportions of adherents to Bible Belt denominations to a statistically significant extent. The overall impression which emerges from the multiple regression modeling effort is that higher population proportions of adherents to

Table 76.5 Multiple regression analysis of Bible Belt denomination percentage of the population

Dependent variable:	
PCTAD2010	2010 Bible-Belt Denominations Adherents percent of 2010 Census Population by county
Independent variables:	
S	South Region
MET	Metropolitan Settlement County (CBSA 2006 Definition)
PCNG2000T10	Percent population change Census 2000 to Census 2010
PCTWHITE	Percent non-Hispanic White 2010
PCTBLACK	Percent Black 2010
PCTASIAN	Percent Asian 2010
PCTHISP	Percent Hispanic 2010
FBORN10	Percent Foreign Born 2010
PCTLT16	Percent Under age 16 2010
PCTGTE65	Percent 65 or older 2010
PCT_HS_OVR	Percent High School graduate or more education 2005–2009
PCT_BA_OVR	Percent Bachelors Degree or more education 2005–2009
PCT_AG	Percent employed in primary industries 2005–2009
MEDFAMINC	Median Family income 2005–2009
DEM09	Percent Democratic Presidential Vote 2008
Number of observations:	3,108 counties in contiguous United States (Excluding AK and HI)
Multiple-R:	0.730
Multiple-R-Square:	0.532

Regression analysis of variance:

Source	Sum-of-squares	df	Mean-square	F-ratio	P
Regression	362,472.945	16	22,654.559	220.026	0.000*
Residual	318,258.450		3,091	102.963	

Regression coefficients

Effect	Coefficient	Std error	Std coef	Tolerance	t	P(2 Tail)
CONSTANT	31.816	7.555	0.000	–	4.211	0.000*
S	15.925	0.557	0.533	0.435	28.571	0.000*
MET	−1.355	0.490	−0.044	0.606	−2.764	0.006*
PCNG2000T10	−0.200	0.018	−0.178	0.581	−11.024	0.000*
PCTWHITE	−0.038	0.060	−0.042	0.034	−0.631	0.528
PCTBLACK	−0.045	0.061	−0.044	0.042	−0.728	0.466
PCTAMIND	−0.068	0.069	−0.031	0.157	−0.985	0.325
PCTASIAN	−0.117	0.126	−0.019	0.377	−0.928	0.354
PCTHISP	−0.026	0.021	−0.025	0.393	−1.263	0.207
FBORN10	−0.278	0.048	−0.102	0.495	−5.815	0.000*
PCTLT16	0.655	0.083	0.134	0.528	7.930	0.000*
PCTGTE65	0.267	0.068	0.075	0.415	3.908	0.000*
PCT_HS_OVR	−0.203	0.050	−0.101	0.251	−4.092	0.000*
PCT_BA_OVR	0.173	0.039	0.101	0.285	4.386	0.000*
PCT_AG	−0.293	0.031	−0.151	0.578	−9.317	0.000*
MEDFAMINC	−0.001	0.000	−0.079	0.342	−3.768	0.000*
DEM08	−0.311	0.017	−0.291	0.578	−17.957	0.000*

Source: Calculated by J.C. Archer from Grammich et al. (2012)
*Significance level .01 or greater

Bible Belt denominations at county-level tend to be associated with greater economic uncertainty (negative sign for Median Family Income), outmigration (negative sign for Percent Population Change from 2000 to 2010), and geographical locations in more rural (negative sign for Metropolitan status) or more Southern settings (positive sign for South indicator variable).

Further reflection on the various cartographic and statistical results reported here suggest it is worth keeping in mind that, on the one side, higher population proportions of adherents to Bible Belt denominations are typically encountered in counties located in the rural South, but that, on the other hand, greater absolute numbers of adherents to Bible Belt denominations are typically encountered in large metropolitan centers located in the South and even in other regions of the United States. These two important aspects of the geography of adherence to Bible Belt denominations in the United States are cartographically symbolized on maps presented in this essay. The vertical diamond symbol on the map of Percent Bible Belt Population by County (see Fig. 76.1) represents the location of the "Mean Center Buckle" for 2010, which was calculated as the geographical coordinate pair which corresponds to the mean latitude and the mean longitude of the 266 counties where Bible Belt denominations adherents comprised 40 % or more of population in 2010. This center, at 34.7169° north latitude and 90.4444° west longitude, is located in rural northwestern Mississippi, a bit east of the Mississippi River and near the town of Tunica, Mississippi. The horizontal diamond on the map of Percent Bible Belt Population by County (see Fig. 76.1) represents the location of the "Mode Center Buckle," which corresponds to the geographic centroid of Harris County, Texas, with 645,736 adherents to Bible Belt denominations in 2010 and home to the sprawling city of Houston, Texas. On the 3-dimensional map of "Bible Belt Denomination Adherents by County 2010" (see Fig. 76.2), Harris County, Texas is represented by the very conspicuous polygon near the Texas coastline which towers over and visually dominates the entire map. Dallas-Fort Worth, Los Angeles, Chicago, and Miami are also conspicuous on the 3-dimensional map.

76.6 The Bible Belt Visual Landscape

Anyone familiar with the Bible Belt (and its "buckles" and "notches") in the U. S. South and lower Midwest is aware that there are distinctive visible features in the cultural landscape. These include not only the dominance or predominance of churches and church denominations that adhere to literal interpretations of the Bible, but also promotional signs outside church buildings with messages such as "Are You Saved or Going To Hell?" or "Where Will You Spend Eternity?" Additional features are outdoor advertising (billboards) one sees along major highways, Calvary crosses on hillsides, and homemade signs for revivals, camp meetings and other church related events. And there are advertisements against abortion, the sale of alcohol, and promotions for religious theme parks and museums. The densities of these "place" and landscape features are especially high on major highways in the Bible Belt in both rural and urban settings.

Religious signage illustrates and underscores the often passionate religiosity of Bible Belt residents about social issues in public and private life. While controversies over religious issues occur in virtually every region of the United States, it could easily be argued that their numbers and intensities in the Bible Belt contribute to making the region distinctive. For example, in October 2012, 15 members of the middle and high school cheerleading squads in Kountze, Texas, 25 miles north of Beaumont, sued the public school district when it prohibited them from putting Bible verses on the banners displayed during athletic events. The controversy stirred emotions to the extent that a police officer with an assault rifle and binoculars was stationed on the Hardin County Courthouse during court hearings (Fernandez 2012). In mid-October 2012 District Judge Steve Thomas granted an injunction against the school's prohibition, allowing the cheerleaders to put Bible verses on their banners until the case goes to trial in June 2013 (Tomlinson 2012).

In December 2012, a letter to the editor published in the *Tuscaloosa News*, Tuscaloosa, Alabama, caused controversy when it claimed President Obama had changed his opinion on same-sex marriage because he needed the gay and lesbian vote to win the 2012 election. The letter's author also went on to say that "When you get to the pearly gates, Saint Peter will say, 'Homosexuals, I don't support that.' He will say, 'I know how you voted. I am sorry but you are at the wrong gate. You must go downstairs.' Wake up before it's too late. Hell is a real place" (Price 2012). One commentator responded to the dismay expressed by other readers by reminding them "you are in the bible belt" (Flanigan 2012).

Bible Belt religiosity and controversy also spills out on to the landscape in the placement of religious symbols, billboards and signage. During the past few years various agnostic and atheist groups including the Madison, Wisconsin based Freedom from Religion Foundation have been placing anti-religion billboards along freeways around the country, with many being located in Bible Belt states. The nonprofit group has a membership of 19,000 in all 50 states (http://ffrf.org/about). Starting in Wisconsin in 2010, the group has placed signs in Ohio, Oregon, Colorado, Pennsylvania, Arizona, North Carolina, Oklahoma and Texas, among other states (Nunez 2011). In reaction to the Catholic Church's resistance to the contraception coverage mandates in President Obama's health care legislation, a billboard between Dallas and Ft. Worth stated "Quit the church: Put women's rights over bishops' wrongs" (Llorens 2012). The Washington, D.C. based United Coalition of Reason erected signs which read "DON'T BELIEVE IN GOD? Join the club" in 31 states from 2010 through 2012 (http://unitedcor.org/national/page/about-us). While the signs generated controversy elsewhere, reactions were particularly heated in Bible Belt locations such as Wichita, KS, San Antonio, TX, Richmond, VA and Lexington, KY (http://unitedcor.org/national/news; Richard 2012). In response, Christian groups began erecting "Think God" billboards including at least four in metropolitan San Antonio (Levy 2012).

The placement of religious signage and symbols on the landscape is common across the U.S. Thus, while it is quite possible to see a "Jesus Saves" sign in San

Diego, Denver, Minneapolis, or Buffalo, such signs are much more likely to be found in Bible Belt cities and towns in Alabama, Mississippi, Tennessee, Texas and Oklahoma. As stated by *The Economist* (2012: 25), "For many people in the world, Hell still exists; not just as a concept, but as a place on the map. 'Hell is Real,' declare the billboards across the American South: as real as the next town." While the "American South" and the Bible Belt are not geographically the same, they have substantial overlap and the writer could have used "American Bible Belt" and increased the accuracy of the statement (Brunn et al. 2011a).

The placement and interpretation of religious icons and billboards on the landscape can be examined from a number of different perspectives. Among these are sense/experience of place (Inge 2003; MacDonald 2003a; Tuan 2005, 2009), visual rhetoric (Olson et al. 2008; see also Rose 2003, 2007), or religious territoriality (Stump 2008). While "sense of place" has varied meanings, in the context of the Bible Belt we might ask what makes the region distinct? Are the many landscape markers of religiosity "experienced" similarly by residents and those traveling through? How is the identity of residents affected by religious landscape markers? Humans give places emotive meaning, and a comparatively large number of religious billboards can clearly make our sense of place different in Amarillo than Seattle (MacDonald 2003b). Visual rhetoric studies emphasize the cultural meanings of images, and the perspective has been used to examine memorials, photography, books, films, stamps and billboards, among other topics (Olson et al. 2008). It has been argued that societies in the West during the past two centuries have become "ocularcentric," eye or visually centered due to the growth in visual media (Gronbeck 2008). Many Bible Belt denominations, including the Southern Baptist Convention and its emphasis on evangelism, target non-believers. The placement of large billboards with religious messages along freeways blends ocularity and missionary work together (Southern Baptist Convention N.D.a).

While both experience of place and visual rhetoric are appropriate perspectives for examining religious billboards, we find Stump's concept of *religious territoriality* particularly useful when considering religious landscape iconography such as crosses and billboards with religious messages. Stump (2008: 222) states that territoriality is not

> an atavistic biological instinct but … a form of cultural strategy through which individuals and groups seek to exert control over the meanings and uses of particular portions of geographical space. … Territoriality in sum represents an intrinsic product of the spatiality of human activities. In reference to the relationships between religious systems and secular space, expressions of territoriality serve as the means by which adherents integrate their religious beliefs and practices into the spatial structure of their daily lives.

While this territoriality is clearly critical to the preservation and protection of sacred spaces such as churchscapes, it has relevance to secular spaces as well since the extension of religious territoriality into these spaces "represents the application of the ultimate truths accepted by believers to the spaces of their everyday world" (Stump 2008: 222). Stump (2008: 222) further argues that religious territoriality links the adherents' "understanding of the indisputable truth of their religious system" to the "organization and control of space." He identifies multiple expressions for religious territoriality including reproducing the "religious group's system of beliefs and practices, both by reinforcing the faith of existing adherents and by communicating that faith to children and, among proselytic religions, to converts" (Stump 2008: 223). Thus, it can be argued that religious signage in the Bible Belt is not only to inculcate the articles of faith among the region's residents including the children, but also to "witness" to potential converts traversing the region, thereby also satisfying the common doctrinal duty to "endeavor to make disciples of all nations" (Southern Baptist Convention N.D.b).

An excellent example of religious territoriality extending itself into what might be viewed as public space is the "Jesus Christ is Lord Travel Center" off of Interstate 40 in Amarillo, TX (www.jesuschristislordtravelcenter.com). The truck stop's walls and signage include religious messages such as "The Lord is good and his mercy endureth forever" (Fig. 76.3). While a private business, the location of the travel center off of Interstate 40 insures that thousands of motorists will see the religious messaging on the site.

Crosses, whether singular or in a row of three to reflect the crucifixion of Jesus at Calvary (Golgotha), are common throughout the Bible Belt. The scene in Fig. 76.4 is near Refuge, MS off of state highway 82 near the Arkansas border. The three crosses representing Calvary sit on private land with a corn field behind. This is, therefore, a public expression of personal faith on private land, common throughout the Bible Belt. A United States flag is also included in the yard, underscoring that patriotism and religious faith are frequently mixed in the Bible Belt, and the strict separation of church from politics largely rejected. Large crosses are less common on the landscape of the Bible Belt, likely due to substantial cost when compared to more modest displays. One of the most substantial crosses in the Bible Belt is shown in Fig. 76.5. This cross is off of Interstate 40 near Groom, TX and was erected by

Fig. 76.3 Jesus Travel Center, Interstate 40, Amarillo, Texas (Photo by Gerald R. Webster)

Fig. 76.4 Three-cross Calvary display on Highway 82 near Refuge, Mississippi (Photo by Gerald R. Webster)

Fig. 76.5 Cross Ministries Christian cross on Interstate 40 in Groom, Texas (Photo by Gerald R. Webster)

Cross Ministries (www.crossministries.com). The cross is 190 ft (58 m) high, was completed in 1995, and can be seen from at least a dozen miles away. The site also has bronze statues depicting the 14 stations of the cross. Another large cross is located just off Interstate 35 near Edmond, OK on the campus of LifeChurch and is 163 ft (49 m) in height (www.LifeChurch.tv). LifeChurch, headquartered in Edmond, OK, is associated with the Evangelical Convenant Church (www.life-church.tv/who-we-are/beliefs), and has 15 campuses with 12 being located in Oklahoma and Texas. In terms of scale, the large crosses located at Groom, TX and Edmond, OK are in stark contrast to the three cross Calvary display in Mississippi. While the latter serves as a personal display of faith, the crosses erected by LifeChurch and Cross Ministries are parallel efforts at a large institutional scale. Secondly, while the traffic across Mississippi on state highway 82 is significant, it is only a small fraction of the traffic passing by the two larger crosses on Interstates 40 and 35. Thus, the latter are less about private expressions of faith and more about institutional proselytizing. Finally, the visibility of the larger crosses aids in the normalization of adding religious iconography to what are otherwise largely public viewscapes, emphasizing if not confirming the dominance of Christian theology to Bible Belt denomination adherents.

Religious signs along Bible Belt roads and highways can differ substantially in terms of size, visibility and cost. Figures 76.6 and 76.7 are along the Interstate 40 corridor in Oklahoma and highly visible from the freeway. Figure 76.6 is a huge billboard near Dewar, OK erected by the Silver Spring Baptist Church including a phone number to contact the church directly. The billboard has a traditional image

Fig. 76.6 Jesus billboard on Interstate 40 near Dewar, Oklahoma (Photo by Gerald R. Webster)

Fig. 76.7 Bible verse billboard on Interstate 40 near Roland, Oklahoma (Photo by Gerald R. Webster)

of Jesus (light complexion, blue eyes, brown hair, a beard) overlaid with a verse from Isaiah ("Fear not…I am with you"). Figure 76.7 is also a highly visible billboard along Interstate 40 near Roland, OK. It includes Bible verses from both Proverbs and John. The Proverbs verse underscores the traditional conservatism of many Bible Belt denominations, highlighting "Use the rod on your children and save their life." Notably, Wiehe (1990) found greater support for the use of corporal punishment in the Bible Belt than in other portions of the United States. This is arguably due to the belief that the Bible is infallible and the direct work of God. The second verse from the book of John underscores the importance attached to being "born again" among most Bible Belt denominations. A billboard near Muldrow, OK along Interstate 40 lists the Ten Commandments (Exodus 20), central articles of faith emphasized by Bible Belt denominations, and Christians generally.

Not all religious signage in the Bible Belt involves high visibility or substantial initial costs. An excellent example of this is Fig. 76.8 placed off to the side of Alabama highway 43 near Samantha, AL. The sign is handmade and states "Thank God For Jesus." Such signs are common in much of the rural Deep South, and represent very personal statements of faith. Notably, they are inexpensive to make and display, and allow individuals to fulfill their theological obligation to witness to others.

Only somewhat more substantial is Fig. 76.9, a small billboard along Highway 82 near Gordo, AL. The sign poses the question, "Who Holds the Only Key to Heaven?" answering it with "It's Jesus." The sign also includes both a Christian

Fig. 76.8 "Thanks God" sign in trees on Highway 43 near Samantha, Alabama (Photo by Gerald R. Webster)

Fig. 76.9 "Key to Heaven" billboard on Highway 43 near Gordo, Alabama (Photo by Gerald R. Webster)

Fig. 76.10 Handmade "Jesus is Lord" billboard on Interstate 70 near Russell, Kansas (Photo by Gerald R. Webster)

cross and a Jewish star of David. It is notable that no church is listed on the small billboard, in contrast to Fig. 76.6. An additional example of this less expensive and arguably more personal signage is displayed in Fig. 76.10 located in central Kansas along Interstate 70, near Russell. It is handmade, located just inside a pasture along the freeway and simply states "Jesus is Lord." Abortion has long been a defining issue among Bible Belt denominations, and one homemade sign states that "The cost of an abortion: A human life." While much smaller, nearly always less expensive and frequently less visible, these lesser scale signs are also much more plentiful on the landscape of the Bible Belt.

This examination of the religious signage and Christian iconography on the Bible Belt's landscape has been limited to selected examples. But arguably it demonstrates that a larger and more systematic study using Stump's concept of "religious territoriality" could produce both interesting and valuable results. This is particularly the case when considering religious versus secular landscapes which become overlapping when highway viewscapes are considered. While there exists to our knowledge no systematic count of the density of such landscape markers of faith, we suspect the density of such signage is greatest along highways in Oklahoma, Texas and Kansas. Another interesting question pertains to the likely different reactions to such signage and symbols by local residents and those passing through to other destinations. We suspect that those involved in through-transit notice such signs more than locals due to their comparative rarity elsewhere, with local residents accepting such signage as normal and not particularly distinct.

76.7 Summary and Future Research

The description and analysis above clearly illustrates several points. First, the Bible Belt is not a simple and singular belt, but a combination of "belts, buckles and notches" mostly, but not entirely concentrated in the South. It is not a label which can be or should be applied to all parts of the rural and urban South. When we compare the spatial distribution of Bible Belt counties from 2000 to 2010, the patterns remain very similar (see Brunn et al. 2011a). As in 2000, the greatest concentration of Bible Belt counties remains in northern Texas and southern Oklahoma, however the number of counties with over 40 % adherents to Bible Belt denominations has decreased. Similarly, minor belts within the South identified above remain similar to the 2000 pattern, but have been somewhat diluted.

Table 76.4 includes data on the changes in the Bible Belt's population from 2000 to 2010, for the U.S. as a whole and by region, and by Metropolitan, Micropolitan, and Rural counties. Second, there are some internal changes taking place within Bible Belt counties. That is, there is both consistency and also fluidity. Four changes that are most evident are (a) that the number of adherents to Bible Belt denominations in the U.S. declined 0.65 % between 2000 and 2010, losing over 180,000 adherents in total, (b) that the majority of Bible Belt counties in the U.S. generally, and the South in particular, are Rural or Micropolitan, (c) that the population of these counties has declined between 2000 and 2010, and (d) that increases in the numbers of adherents to Bible Belt denominations between 2000 and 2010 occurred primarily in Metropolitan counties. The number of adherents to Bible Belt denominations in these counties increased by 1.7 % between 2000 and 2010. Regionally, Metropolitan counties in the South and West saw an increase in the number of adherents to Bible Belt denominations, while Metropolitan counties in the North had a decline in adherents to Bible Belt denominations.

We have also observed that the visible Bible Belt landscape is a distinctive feature of much of rural and urban South. The advertisements include publicity for individual churches, images or words of Jesus, familiar biblical verses and positions on specific religious/secular controversies. Some of these advertisements are part of a national effort and are professionally designed and expensive, but many signs are individually constructed with only very small budgets. Promoting the theology and practices of Bible Belt denominations and believers will almost certainly continue to be part of the traditions of the rural and urban South.

In conclusion, we offer the following topics for future study into Bible Belt landscapes, theology and culture in the U.S. *First*, we think it would be worthwhile to examine in greater detail the individual membership patterns of these 24 denominations, specifically the largest ones: their numerical changes (growths and declines) and where (county level) these changes have occurred. As we have discussed above, it would be worth discussing these changes with respect to county population changes, the rural-urban mix, age, education and income categories. Analyzing changes involving the Southern Baptist Convention, Lutheran Church-Missouri Synod, and Christian Churches and Churches of Christ would be especially interesting, but also perhaps those that have less than one-half million members. *Second*, it would be useful to examine closely the growth of Bible Belt churches in metropolitan areas both inside the South, but especially outside the South where the growth is now occurring. For example, what denominations have built churches or satellite churches in Los Angeles, Anaheim, Phoenix and Chicago? Did these new churches have "sister and brother" supporting churches in the urban South? *Third*, the publications of these Bible Belt denominations are deserved of greater study, not only the educational/promotional materials used on Sunday mornings for children and adults, but also the texts used in religion and science classes (geology, biology, anthropology and even geography) in colleges and universities, but also the content of outdoor advertising, denominational and seminary websites (Brunn and Long 2000). *Fourth*, since many of these denominations have their own colleges, it would be worthwhile to look at their faculty recruiting, specifically to what extent are there "exchanges" of faculty (do they send and receive faculty from other Bible Belt colleges and universities?) and are the faculty hiring searches regional (as opposed to national) in scope? Answering these questions would reveal something about the inclusiveness and academic networking of Bible Belt colleges. *Fifth*, and finally, the Bible Belt phenomenon would be worth exploring in other countries where there are active fundamentalist and evangelical groups that comprise a small but important part of the religious landscape. Examples might include groups in Germany, the Netherlands, Honduras, Russia and Canada, among others. These topics are important for disciplinary and interdisciplinary scholars who investigate not only adherents to Bible Belt denominations and the correlates to their geographic patterns, but also their images on human landscapes and their role in the science, education and policy arenas.

References

Brunn, S. D., & Long, E. (2000). The worldviews of southern seminaries: Images, mission statements and curricula. *Southeastern Geographer, 40*, 1–22.

Brunn, S. D., Webster, G. R., & Archer, J. C. (2011a). The Bible Belt in a changing South: Shrinking, relocating, and multiple buckles. *Southeastern Geographer, 51*, 513–549.

Brunn, S. D., Webster, G. R., Morrill, R. M., Shelley, F. M., Lavin, S. J., & Archer, J. C. (2011b). *Atlas of the 2008 elections* (pp. 229–240 on religion). Lanham: Rowman and Littlefield.

Burke, D. (2010). Creationism holds steady at four in ten Americans, Gallup says. http://www.religionnews.com/index. Accessed 21 Dec 2011.

Conser, W. H., & Payne, R. M. (2008). *Southern crossroads: Perspectives on religion and culture*. Lexington: University Press of Kentucky.

Corbett, J. M. (1993). Religion in the United States: Notes toward a new classification. *Religion and American Culture: A Journal of Interpretation, 3*, 91–112.

Crawford, T. W. (2005). Stability and change on the American religious landscape: A centrographic analysis of major U.S. religious groups. *Journal of Cultural Geography, 22*, 51–86.

Eck, D. (2001). *A new religious America: How a "Christian country" has become the world most religiously diverse nation*. San Francisco: Harper San Francisco.

Economist. (2012, December 22). Hell: Into everlasting fire. *The Economist*, pp. 25–28.

ESRI. (2004). *ArcGIS 9: Using ArcMap*. Redlands: ESRI Inc.

Fernandez, M. (2012, October 4). Cheerleaders with bible verses set off a debate. *New York Times*. http://www.nytimes.com/2012/10/05/us/in-texas-cheerleaders-signs-of-faith-at-issue.html?pagewanted=all&_r=0. Accessed 31 Dec 2012.
Flanigan, D. (2012, December 6). Reply to Price letter. *Tuscaloosa News*. www.tuscaloosanews.com/article/20121206/NEWS/121209864?tc=obinsite. Accessed 7 Dec 2012.
Gaustad, E. S., & Bralow, P. L. (2000). *New historical atlas of religion in America*. New York: Oxford University Press.
Grammich, C., Hadaway, K., Houseal, R., Jones, D. E., Krindatch, A., Stanley, R., & Taylor, R. H. (2012). *2010 U.S. religion census: Religious congregations and membership study*. Kansas City: Association of Statisticians of American Religious Bodies, Nazarene Publishing House.
Gronbeck, B. E. (2008). Forward: Visual rhetoric studies: Traces through time and space. In L. C. Olson, C. A. Finnegan, & E. S. Hope (Eds.), *Visual rhetoric* (pp. xxi–xxvi). Los Angeles: Sage.
Halvorson, P. L., & Newman, W. M. (1978). *Atlas of religious change in America, 1952–1971*. Washington, DC: Glenmary Research Center.
Heatwole, C. (1978). The Bible belt: A problem in regional definition. *Journal of Geography, 77*, 50–55.
Hill, S. S. (1983). *Religion in southern studies: A historical study*. Macon: Mercer University Press.
Hill, S. S. (Ed.). (1990). *Handbook of denominations in the United States*. Nashville: Abington Press.
Hill, S. S., & Lippy, C. H. (Eds.). (2005). *Encyclopedia of religion in the South*. Macon: Mercer University Press.
Inge, J. (2003). *A Christian theology of place*. Burlington: Ashgate.
Jordan, L. M. (2007). Religious adherence and diversity in the United States: A geographic analysis. *Geographies of Religions and Belief Systems, 2*, 3–20.
Levy, A. (2012, May 9). Billboard sends out call for atheists to shed fear. *San Antonio Express-News*. http://www.mysanantonio.com/news/local_news/article/Billboard-sends-out-call-for-atheists-to-shed-fear-3544003.php. Accessed 4 Dec 2012.
Llorens, I. (2012, June 30). Atheist billboard in Texas targets Catholics with phrase 'put women's rights over bishops' wrongs. *Huffington Post*. www.huffingtonpost.com2012/02/13/athiest-billboard-in-texas-catholics-womens-rights. Accessed 31 Dec 2012.
MacDonald, M. N. (Ed.). (2003a). *Experience of place*. Cambridge, MA: Harvard University.
MacDonald, M. N. (2003b). Introduction: Place and the study of religion. In M. N. MacDonald (Ed.), *Experience of place* (pp. 1–20). Cambridge, MA: Harvard University Press.
Marty, M. E. (1998). Revising the map of American religion. *Annals of the American Academy of Political and Social Science, 558*, 13–27.
Mencken, H. L. (1927). *Prejudices: Sixth series*. New York: Knopf.
Mencken, H. L. (1947). *The days of H.L. Mencken*. New York: Knopf.
Mencken, H. L. (1949). Modern chrestomathy (pp. 76–77). New York: Knopf. (Reprinted from *American Mercury* (1925), pp. 268–269)
Neitz, M. J. (2005). Reflections on religion and place: Rural churches and American religion. *Journal for the Scientific Study of Religion, 44*(3), 243–248.
Newport, F. (2007). One-third of Americans believe the Bible to be literally true. Gallup Polls. http://www.gallup.com/poll/27682. Accessed 1 Jan 2011.
Nunez, D. M. (2011, December 10). Billboard campaign hopes to dispel atheist stereotypes. *USAToday.com*. http://www.azcentral.com/news/articles/2011/12/01/20111201billboard-atheist-stereotypes.html. Accessed 1 Jan 2013.
Olson, L. C., Finnegan, C. A., & Hope, D. S. (Eds.). (2008). *Visual rhetoric: A reader in communication and American culture*. Los Angeles: Sage.
Pew Forum on Religion and Public Life. (2008). *U.S. religious landscape survey: Religious affiliations*. Washington, DC: Pew Forum on Religion and Public Life.
Price, J. (2012, December 6). Letters: Teach children that homosexuality is sin. *Tuscaloosa News*. www.tuscaloosanews.com/article/20121206/NEWS/121209864?tc=obinsite. Accessed 7 Dec 2012

Reed, J. S. (1972). *The enduring South*. Chapel Hill: University of North Carolina Press.
Richard, J. (2012, February 13). Atheism billboard planned for Dallas sparks controversy. *Huffington Post*. http://www.huffingtonpost.com/2012/02/13/atheism-billboard-dallas_n_1274823.html. Accessed 31 Dec 2012.
Rogerson, P. A. (2010). *Statistical methods for geography* (3rd ed.). Thousand Oaks: Sage.
Rose, G. (2003). Intervention roundtable: On the need to ask how, exactly, is geography 'visual'? *Antipode, 35*, 212–221.
Rose, G. (2007). *Visual methodologies*. Thousand Oaks: Sage.
Shelley, F. M., Archer, J. C., Davidson, F. M., & Brunn, S. D. (1996). *Political geography of the United States*. New York: Guilford Press.
Shelton, T., Zook, M., & Graham, M. (2012). The technology of religion: Mapping religious cyberspaces. *The Professional Geographer, 64*, 602–617.
Shibley, M. A. (1998). Contemporary evangelicals: Born-again and world affecting. *Annals of the American Academy of Political and Social Science, 558*, 667–687.
Shortridge, J. R. (1976). Patterns of religion in the United States. *Geographical Review, 66*, 420–434.
Shortridge, J. R. (1977). A new regionalization of American religion. *Journal for the Scientific Study of Religion, 16*, 143–153.
Smith, T. W. (1987). Classifying Protestant denominations. *Review of Religious Research, 31*, 225–245. http://www.icpsr.umich.edu/GSS99/report/m-report/meth43,htm. Accessed 1 Jan 2011.
Southern Baptist Convention. (N.D.a). Evangelism. North American Mission Board. www.namb.net/evangelism/. Accessed 11 Dec 2012.
Southern Baptist Convention. (N.D.b). Missions work, Article XI. www.sbc.net/missionswork.asp. Accessed 11 Dec 2012.
Steensland, B., Park, J. Z., Regnerus, M. D., Robinson, L. D., Wilcox, W. B., & Woodberry, R. D. (2000). The measure of American religion: Toward improving the state of the art. *Social Forces, 79*, 291–318.
Strategic Mapping Inc. (1990). *Atlas*GIS: Desktop Geographic Information System*. San Jose: Strategic Mapping Inc.
Stump, R. W. (2008). *The geography of religion: Faith, place and space*. Boulder: Rowman and Littlefield.
SYSTAT. (2004). *SYSTAT 11*. Richmond: SYSTAT Software, Inc.
Tomlinson, C. (2012, October 18). Kountze high school bible banner case: Judge rules for cheerleaders. *Huffington Post*. http://www.huffingtonpost.com/2012/10/18/judge-rules-for-cheerlead_0_n_1981865.html. Accessed 6 Jan 2013.
Tuan, Y. T. (2005). *Space and place: The perspective of experience*. Minneapolis: University of Minnesota Press.
Tuan, Y. T. (2009). *Religion: From place to placelessness*. Chicago: Center for American Places.
Tweedie, S. W. (1978). Viewing the Bible Belt. *Journal of Popular Culture, 11*, 865–876.
U.S. Bureau of Census. (2011). *2010 Census of Population*. Washington, DC: U.S. Bureau of Census.
U.S. Bureau of Census. (2012). *American Community Survey, 2005–2009*. Washington, DC: U.S. Bureau of Census.
Vogeler, I. (1979). The Roman Catholic culture region of central Minnesota. *Pioneer America Society Transactions, 8*, 71–83.
Warf, B., & Winsberg, M. (2008). The geography of religious diversity in the United States. *The Professional Geographer, 60*, 413–424.
Webster, G. R. (1997). Religion and politics in the American South. *Pennsylvania Geographer, 35*, 151–172.
Webster, G. R. (2000). Geographical patterns of religious denomination affiliation in Georgia, 1970–1990: Population change and growing urban diversity. *Southeastern Geographer, 40*, 25–51.

Wiehe, V. (1990). Religious influence on parental attitudes towards the use of corporal punishment. *Journal of Family Violence, 5*(2), 173–186.
Williams, R. H. (2005). Introduction to a forum on religion and place. *Journal for the Scientific Study of Religion, 44*, 239–242.
Wilson, C. R. (1989). Bible belt. In C. R. Wilson & W. Ferris (Eds.), *Encyclopedia of Southern culture* (pp. 1312–1313). Chapel Hill: University of North Carolina Press.
Zelinsky, W. (1961). An approach to the religious geography of the United States: Patterns of church membership in 1952. *Annals of the Association of American Geographers, 51*, 139–193.

Chapter 77
Transnationalism and the Sôka Gakkai: Perspective and Representation Outside and Inside Japan

Alexandre Benod

77.1 Introduction

Sôka Gakkai dogma stresses the importance of the Lotus Sutra and, as a mahayanist school, emphasizes the missionary diffusion.[1] Makiguchi with his disciple Toda Josei (1900–1958) converted to Nichiren Shoshû Buddhism in 1928. In 1930 he and Toda created an association called Sôka Kyoiku Gakkai, the "Educational Society for the Creation of Values," the very first step toward making Sôka Gakkai a lay organization (Inoue et al. 1990: 903). Due to their opposition to the Shintô State religion during the context of WWII, Makiguchi and his colleagues were incarcerated. He died in jail in 1944. His protégé Josei Toda freed from prison after the war was determined to recreate and expand the group. Toda renamed the association Sôka Gakkai "Value-Creation Society" as a lay Organization affiliated with the Nichiren Shoshû. He assumed the presidency of the movement with high expectations on recruiting; official Sôka Gakkai figures say he reached the goal of converting approximately one million persons before his death in 1958 (Inoue et al. 1990: 194; see SGI.org).[2] Nevertheless, the most important change in the ideology of what they

[1] Buddhism has three main branches: (1) Theravada, "the Teaching of the Elders," is mostly present in Southeast Asia; it is generally accepted by scholars that the Hînayâna "Inferior Vehicle" is equivalent. (2) Mahâyâna, "the Great Vehicle," is the largest tradition and mostly present in East Asia. It differs from Theravada on the importance given to Bodhisattva, who are characters that attained Buddhahood for the benefit of all sentient beings, and, therefore, explains the expansionist nature of the mayahanism. (3) Vajrayâna, "Diamond Vehicle," or Tantrayâna, "Tantric Vehicle," is mostly present in Tibet. As one of the specificities, the Vajrayâna is considered the "path of the Fruit" and emphasizes esoteric practices, whereas sutras-based vehicles are seen as the "path of the Cause." On the propagation of Buddhism in general, see: Learman 2005.

[2] According to the Shinshûkyô Jiten, official statistics were transmitted from Sôka Gakkai to the government only since 1975.

A. Benod (✉)
Research Division, Department of Japanese Studies, Université de Lyon, Lyon, France
e-mail: alexandre.benod@gmail.com

S.D. Brunn (ed.), *The Changing World Religion Map*,
DOI 10.1007/978-94-017-9376-6_77

called the "human revolution" occurred under the third presidency of Sôka Gakkai, Ikeda Daisaku (1928–) (Bornholdt 2009). The globalization of Sôka Gakkai started in earnest with him and constituted the movement's primary source of appeal, "not only among the Japanese public, but also to individuals worldwide who were experiencing the process of social change taking place in the late twentieth century." (David Machacek and Bryan Wilson 2001: 3). Ikeda has been traveling all over the world to support even small communities of followers. In 1960 he went to North and South America, in 1961 Southeast Asia, India, then Europe (Ronan Alves Pereira 2008: 95–113). Since this period, Sôka Gakkai has entered in a phase to strengthen its legitimacy inside and outside Japan. For instance, in 1963, the movement was legally recognized in the United States as a non-profit Organization. On January 26, 1975, Ikeda Daisaku rebuilt the movement on the island of Guam and named it Sôka Gakkai International (SGI).[3] Ikeda has also founded diverse institutes, such as the Institute of Oriental Philosophy (1961), the Boston Research Center for the Twenty First Century (1993) and the Toda Institute for Global Peace and Policy Research (1996). Several institutions were founded aiming to promote "mutual respect and understanding" (Matsudô 2000: 59) such as the Min-On Concert Association (1963), the Tokyo Fuji Art Museum (1991) and La Maison Littéraire de Victor Hugo in Bièvre, France (1992) (Bornholdt 2009). SGI was registered as a Non-Governmental Organization (NGO) with the United Nations High Commission for Refugees and the Department of Public Information in 1981 and registered with the UN's Economic and Social Council 2 years later (Bornholdt 2009). 1994 also saw the opening of the Soka University of America in California, which declares on its website that their mission is to "foster a steady stream of global citizens committed to living a contributive life" (see www.soka.edu/about_soka/mission_and_values.aspx) This initiative makes SGI one of the most efficient new Japanese religious movements in terms of socio-political development, but also simply in term of followers. The first Sôka university as opened in Tokyo in 1969; there are currently enrolled about 8,500 students. Today, Sôka Gakkai claims more than 12 million members in 192 countries. Witnessing this success, new religions such as Agonshû (Agama Sûtra Sect), or Kôfuku no Kagaku (Happy Science) founded at the end of the 1970s in Japan, envied these results and tried to integrate Sôka Gakkai's path as model of success to influence their own politics.[4] For instance, since the end of the twentieth century, Agonshû has extended its activities abroad to pray for World Peace and performed numerous Goma ceremonies (fire rite to liberate souls of Dead) outside Japan. The choice of the place of the ceremony is led by its importance during World War 2, like the Goma held in the Pacific Ocean in 2012 or at Khabarovsk (Siberia) in 2007 where many WW2 Japanese soldiers had been buried. All these rituals were broadcasted into Japan via satellite. The study revealed

[3]The current president of the "original" Sôka Gakkai is Harada Minoru (1941–) who is the 6th generation.

[4]For example, Agonshû's leader, Kiriyama Seiyû, in his most autobiographical book, wrote that during the 1960s he used to frequent a circle of Sôka Gakkai's students in Tôda (Saitama prefecture). Apparently, Kiriyama was asked to take the lead of their group and he ironically declared that if he had accepted back then he might be now leading the current SGI instead of Ikeda Daisaku (Kiriyama 1983: 99).

that the recent acceleration of Goma's performances conducted out of Japan (Auschwitz, 2006; Khabarovsk, 2007; Jerusalem, 2008; Guadalcanal, 2009; Pacific Ocean, 2012), emphasizes the eagerness of Agonshû to develop the movement inside Japan rather to expand its overseas activities.

This chapter shows how the development and propagation of the SGI had an impact on the orientation of more recently founded religions in Japan. I will, therefore, first link SGI with transnationalism and focus on the criteria of the development of its reputation.[5] To do so, this analysis has to be seen in the context of the history of Japanese religions abroad, whose significant first steps outside the archipelago are linked with the political shift of the Meiji Restoration (1868). Indeed, this development is the result of labor migrations, military expansion and colonization (Inoue et al. 1990: 608–672; Jaffe 2010: 1–7). It is also important to keep in mind that of the currently existing 2,000 Japanese new religions, only 100 have tried to expand overseas (Sakashita 1998: 6). Among these experiences, I will emphasize the example of the Japanese community in Brazil, then I will analyze how New Religions such as Agonshû (Agama Sûtra Sect), or Kôfuku no Kagaku (Happy Science) founded at the end of the 1970s in Japan integrated Sôka Gakkai's path as a model of success to influence their own politics in Japan.[6]

77.2 Sôka Gakkai as a New Religion in Brazil

77.2.1 History of the Japanese Diaspora in Brazil

As François Laplantine (2001) stated, the plurality of the composition, structure and history of the population in Brazil, at the crossroads of indigenity, colonization and migrations, fuels the understanding of *métissage*, that is, "a kind of polyphony completely original in culture."[7] During its modern history, Brazil has received a large number of migrants. At the turn of the twentieth century, agriculture in Brazil focused on the export of coffee and required many workers since the abolition of slavery in the 1850s. The government made an announcement to attract European immigrants to work in their plantations with the result that many Italians settled in Brazil. Because working conditions and financial rewards were extremely poor, most of the Italian workers left the country. Meanwhile, industrialization in Japan caused massive poverty in the countryside. Many searched for an improvement in

[5] Transnationalism is defined as "the processes by which immigrants forge and sustain multi-stranded social relations that link together their societies of origin and settlement. We call theses processes transnationalism to emphasize that many immigrants today build social fields that cross-geographic, cultural, and political borders. An essential elements is the multiplicity of involvements that transmigrants sustain in both home and host societies" (Basch et al. 1994: 6).

[6] On the social shift from New Religions to the so-called New New Religions in the Japanese context, see Susumu 1992.

[7] In the context of social sciences, the French word "métissage" describes the hybridization of socio-cultural influences (Laplantine 2001).

Fig. 77.1 Japanese Workers in Coffee Plantation, 1930 (Photo from http://commons.wikimedia.org/wiki/File:Japanese_Workers_in_Coffee_Plantation.jpg)

their living conditions.[8] In 1907 Brazil and Japan signed a treaty to allow Japanese to migrate to Brazil. *Nikkei Burajiru jin* or *nipo-brasileiro* is the name of the descendants of those who took part in the Japanese mass migrations that began in 1908; the first migrants arrived at the port of Santos in the state of São Paulo (Rocha 2000: 31–55). By 1941 historians estimate that 234,000 Japanese emigrants had arrived in Brazil (Clarke 2005: 129). Today, Brazil has the largest Japanese diaspora of any country in the world (Fig. 77.1). The Brazilian Institute of Geography and Statistics (IBGE) stated that more than 1.4 million citizens have Japanese roots (Ministério do Planejamento 2008).

77.2.2 Religions and the Japanese Diaspora in Brazil in Prewar Society

The first generations of migrants paid little attention to religious matters, because they mostly hoped they could return to their homeland and perform the *senzo kuyô* "ancestor cult," which is at the core of the Japanese religiousness.[9] They imagined that if they

[8] "By 1963 around one million had emigrated. The largest number of emigrants eventually settled in Manchuria, followed by Brazil, Hawai'i, mainland U.S., Southeast Asia, Australasia, Canada, Peru, Mexico, and Colombia" (Clarke 2005: 129). This kind of economic migration was not the first one. For instance, in 1868, a huge number of laborers left for Hawai'i to work in sugarcane exploitations.

[9] In addition to the genze riyaku "worldly benefits" that can be found either in Buddhist or Shintô; the funeral service is generally restricted to Buddhism. On the ancestors' worship see Kunio 1945: 9, and on the genze riyaku see Ian and Geores 2004.

died abroad, their soul would naturally go back to the ancestral motherland (Clarke 2005: 129). At the beginning, migrants did not learn much about religions at school; they followed the Imperial Ordinance. This law proclaimed filial piety to the Emperor since 1890. However, even if in the 1930s there was evidence of the presence of new religious movements in Japan (founded in the mid-nineteenth to early twentieth century), including Tenrikyô, Ômotokyô and historical Buddhists sects such as Jôdo Shinshû, Zen, Shingon; all have substantially developed in postwar Japanese society. In Brazil's case, the formal propagation of the religions of the Japanese community did not increase until after World War II. Before 1941, the Japanese government thought that the introduction of Buddhism or Shintô in a mostly Christian country would only exacerbate the anti-Japanese climate. Because of this the Japanese Ministry of Foreign Affairs only permitted the migration of Catholic missionaries and imposed a self-imposed restraint on all other missionary activity.[10] Indeed for some Japanese who had converted to Catholicism, it was a means of integration into Brazilian society, though most of them settled into exclusively Japanese communities. By the middle of the twentieth century, Japanese migrants had begun to consider Brazil as their new home and while their integration into Brazilian society was far from being complete, returning to the motherland was unthinkable for most.

77.3 New Religions in Japan

After the defeat of Japan, the Allies, led by General MacArthur, occupied the ravaged country, and proclaimed several policies about religions and especially Shintô. Firstly, the American administration cleaned up the State Shintô of its most nationalistic and military symbols, such as the shared lineage of the Emperor with deities. This policy eventually led to the separation of Church and State in Article 20 of the Constitution promulgated in 1947, which guaranteed freedom of religion for the Japanese people.

The direct consequence of this action was the rapid emergence and propagation of new religions which gained thousands of followers, especially those affiliated with Buddhism. At that time these new religions were heavily criticized by established religions, as the latter were increasingly incapable of satisfying the spiritual needs of the Japanese and were consequently losing members. Ancient Buddhism portrayed new religions as being dogmatically poor, toned down and taking advantage of the misery witnessed in the aftermath of WWII. Indeed, established Buddhism was limited to a funeral role mostly because of its strong ties with secularization by the Tokugawa.[11] During the Meiji Period, the social role of the Buddhist

[10] "In this respect, the Brazilian case contrasts with the case of both Hawai'i and the United States, where temples and shrines were built for existing Buddhist sects shortly after Japanese migration began" (Watanabe 2008: 115).

[11] During the Tokugawa period (1603–1868), authorization had been given to some sects to eat meat and to get married (nikujikisaitai act). Most of monks were also involved in teaching and holding civil cadasters (Jaffe 2001).

clergy was minimized, due to a nationalist policy (as an Indian/Chinese heritage) and the monks were restricted to their funereal functions, leading to the expression Funeral Buddhism (*sôshiki bukkyô*). During this time, Shintô was sponsored by the State and the "first" New Religions appeared. Many contemporary Japanese sociologists agree that the new religious boom started in the middle of the nineteenth century and mostly concerned movements related to the Shintô (like Ômotokyô or Tenrikyô) which collapsed after Japan's surrender. Buddhist new religions and, in a way, Judeo-Christian new religions (few were founded in Japan while others were imported by Americans during the occupation, such as Jehovah's Witnesses) gained power (*taitô*) in the aftermath of WWII. The new religions like Sôka Gakkai successfully attracted the Japanese trying to escape the poverty, illness, loneliness and alienation that were consequences of war. Antagonism grew between established and new religions, since any new religion was labeled "new religious movement:" religions either voluntarily stood out of established religions or were rejected by them.[12] At the end of the 1950s Sôka Gakkai counted almost 800,000 members. Following the missionary pattern of propagation of the Mahayanist school, the movement was exported abroad. The religious scholar Inoue Nobutaka refers to this as an overseas dispatch model (*kaigai schuchô kata*) (Inoue et al. 1990: 189). In the very beginning, the proselytizing mission paid little attention to attracting non-Japanese followers, but rather on providing spiritual services to the expatriate community. I will next focus on the Brazilian example through the SGI experience.

77.4 Sôka Gakkai in Brazil

Following WWII a great shift occurred among the Japanese Community in Brazil. Inherent to diasporic issues (knowledge of the mother tongue among the descendants, return to a changed homeland, misrepresentations of the original culture), a great number of the migrants started to think they would definitely settle in Brazil. By extension they began to express interest in religions, developing and adapting, for instance, their own protecting deities (*ujigami*) to the local geography. The clearest example Clarke (2005) provides is the Great Shinto Shrine of Brazil, the Dwelling of the Myriad Deities (Kaminoya yaoyorozu kyo) established in 1966, which stresses the universal character of the Japanese sun Goddess Amaterasu, who has the most important position in the Japanese mythology. According to the words of the founders Clarke interviewed, "in the afterlife, there will be no nationalities, only people." (Clarke 2005: 129–130). Clarke noted that this view does not differ from other Shintô shrines in Brazil in that it includes Catholic, African-Brazilian

[12] The success of these new religious movements was such that a field of study was developed in the 1960s to describe them. The study of new religions is a direct heritage of the fieldwork methods established by folklorists fifty years earlier. It is intended to develop several criteria to analyze this huge social change. Nevertheless, defining the scope of the field of New Religious Movements here is complicated.

and Amerindian iconography in the outer shrine and the inner sanctuary is dedicated to the Sun Goddess Amaterasu (Clarke 2005: 130). This new approach in the transformation of Japanese religions in a foreign context drew the attention of Sôka Gakkai. The first Sôka Gakkai believers arrived in Brazil in the late 1950s (Pereira 2008: 98). In 1958, Sôka Gakkai landed for the first time in the State of Bahia, northeast Brazil (Clarke 2005: 130).[13] Among those founding members, The Kominatsus performed an intense propagation among the Japanese community (discussion meeting, etc. restricted to the family circle, neighbors and friends), acted like what Pereira called a "pioneering period" (Pereira 2008: 98). Then came a period of structuring and legalization. Sôka Gakkai established a branch in 1960 on the occasion of Ikeda Daisaku's trip to Brazil. Ikeda's role was crucial in the transnationalization of Sôka Gakkai,

> for instance, there is no SGI event, inauguration, anniversary of the group [...] without a note from Master Ikeda written for the special event. [...] Member sees Ikeda as a great leader and mentor who is tireless to spread the teachings of Nichiren and advance the cause of world peace. (Pereira 2009: 174)

The religious group has been open to non-Japanese only since the late 1960s, despite the fact that ceremonies and rites remain in Japanese language at this time (Fig. 77.2). Jay Sakashita mentioned in his thesis that:

> The strength of the Japanese economy during this [postwar] period provided the backbone for expansion abroad and, to some extent, lent an air of legitimation to Japanese new religions. (Sakashita 1998: 4–5)

According to Clarke, SGI proclaimed to be the most genuine and authentic form of Buddhism under the Toda presidency, and with a canon that clearly denied other forms of religions, even within Buddhism (Clarke 2005: 127). In contrast, under Ikeda's presidency, a great shift was made to open Sôka Gakkai to non-Japanese, to transform the Brazilian experience as a propagation scheme that fits with the multinational membership model (*takokuseki kata*) as Inoue described it (Inoue 1996: 189). Indeed, a shift was made after a fierce proselytism (SGI method of recruitment is called *shakubuku* and sometimes controversial due to its intensity) to cultural activities since second visit of Ikeda in 1966 (Pereira 2008: 98). The regular venues of several parliamentarians from the political wing of SGI, the Kômeitô (Clean Government Party) alarmed Brazil authorities about the potential impacts about the official religious purpose of SGI.[14] The Brazilian Department of Political and Social Order put SGI activities under surveillance: in 1974, Daisaku Ikeda was refused entry into the country when he was supposed to visit the country for a third time

[13] Inoue stressed the existence of a different but close model of propagation: in the years preceding formal missionary work in California, most Sôka Gakkai members were Japanese women who moved to the USA after their marriages to American soldiers. The wife introduced her beliefs and practices to her friends, husband's friends and neighbors. Once the community was big enough in America, they were able to organize, meet, and lay the foundation for successful missionary activity (Inoue 1983: 102).

[14] Similar critics rose in Japan and the Komeito and SGI officially split in 1970.

Fig. 77.2 BSGI web page, 2013 (Source: www.bsgi.org.br/, used with permission)

(Pereira 2008: 96). Sôka Gakkai decided to make a move in its own orientation by giving up on proselyting and emphasizing involvement in cultural activities, such as government-sponsored activities (like the Independence Day) or festivals related to Japanese migration memory (Pereira 2008: 99). From the mid-1980s onwards the rate of non-Japanese members quickly increase and the *dekasegi* policy (Japanese going back to Japan under special conditions) in Japan provided opportunities for Brazilian to assume higher positions in the organization (Pereira 2008: 99). Today, BSGI (Brazilian SGI) is in a period of stabilization and estimates the number of its members at 150,000, 90 % are non-Japanese who live in the State of São Paulo (Watanabe 2008).[15] From this perspective one can admit that SGI has succeeded in its goal in propagating its religion even among non-Buddhist households. This is what supporters claim as universalism: Shimazono underlines

> in a capitalistic competitive society, one's legitimacy is graphically brought home on the basis of success in expanding numbers. What is more, when the following of one's teaching

[15] The state of São Paulo is approximately the same size as Japan.

> by people of other cultures is felt to be proof of your religion's universal adequacy, missionary activity to people of other cultures overseas can stir up stronger impulses than propagation among one's own compatriots. (Shimazono 1993: 282)

This success led several other new religions to follow in SGI's path. I next want to move this discussion into a different direction, not to analyze how a movement like Agonshû developed outside Japan, but rather to understand how it has influenced its own politics inside the archipelago. One point Sôka Gakkai brings with it legitimacy. Many adherents, many branches, many socially engaged association (at least on the surface) and raise and solve the question of legitimation (in terms of brand equity. i.e. an becoming a recognized institution). In his analysis of the NR Shinyoen, Jay Sakashita stated that even if new religions are seeking financial profit at the very end, the first goal is

> an affirmation of the truth of their beliefs and practices and the validity of their universal nature, that is, things which can be gained by winning of followers abroad. The recruitment of even a few non-Japanese members from the local population lends legitimacy to a group. [...] The perceived international acceptance of a new Japanese religion's teachings will lend self-confidence to its members. This in turn may lead to deeper commitment and more enthusiasm, which, in the end, may very well result in financial gain. (Sakashita 1998: 26–27)

This shift to a potential universalism of Japanese religion spotlights the repercussions SGI has drawn in Japan and as several Agonshû's followers told me during interviews "they want Agonshû to influence the World as much as SGI is doing." For rival new religions, the case of SGI serves to show the path to attain a global reach in terms of space and universalism in terms of diversity of the population and also in terms of policy. As Sakashita stresses in his study of the new religion Shinyoen, the shift from esoteric practice to ethic is recurrent in Japan (Sakashita 1998: 102). SGI diversities have expressed itself as a long evolution in the trend of NGO and the promotion of World Peace (Fig. 77.3).

77.5 Consequences on New New Religions

77.5.1 Situation in Japan After the Economic Boom

After the Japanese post-war economic miracle in the 1960s and 1970s, the socio-historical evolution of Japan revealed many changes taking place in the population, and "new new religious movements" rebuilt the Japanese religious landscape. Established Buddhist sects were still declining and marginalized in their role, and many young people who were seeking alternatives to the modern concept of freedom and rationality found satisfaction in what many Japanese sociologists call, "new new religions." The youth begun to place their hopes on religious precepts, ascetic morality, meditation, and spiritual and mystical experiences. The new new religions represented a second religious boom that mainly attracted urban and cultivated young people. Moreover, these movements pointed to a more general dissatisfaction and an increasing lack of faith in modern scientific rationalism, the

Fig. 77.3 Soka Gakkai International web page, 2013 (Source: www.sgi.org/resource-center/ngo-resources/peace-disarmament/transforming-the-human-spirit.html, used with permission)

Japanese educational system and the business world.[16] In the 1970s and 1980s, young people found redemption in movements like Agonshû (circa 1978), an esoteric media-centered new religion which promotes individual happiness and World Peace through fire rituals.[17]

77.5.2 Agonshû and Internationalization

Agonshû was founded in 1978 by Kiriyama Seiyû (1921–). According to a video I saw in during an initiation, Agonshû estimated the number of its followers to be about 500,000. The video is 10 years old. At present, there are about 30 centers in Japan and branches in Taiwan, Brazil and the United Kingdom. Kiriyama claims to have discovered the Original Buddhism thanks to Agama sûtras. According to the group's books and followers I interviewed, the leader first mastered the techniques of Shingon esoteric Buddhism and then created his original systems incorporating Indian yoga, Taoism elements and a particular form of Shugendô.[18] In addition, the practice includes various voluntary activities, such as cleaning the religious centers, help the visitors, etc.

[16] In the extreme case of Aum Shinrikyo, the cult founded in 1984 by Asahara Shoko which perpetrated the sarin gas attack in the Tokyo metro on March 1995. Most of the followers were in their twenties and a considerable number were young graduates from Japan's elite universities, itself a sign of a huge social crisis.

[17] My data on Agonshû are based on fieldwork done in Agonshû's facilities in January-February 2008 in Tôkyô and Chiba and from September 2008 through July 2009 and also in August 2012 in Tôkyô.

[18] Shugendô are ascetics exercises ingrained in Japanese folklore and mountain oriented cults.

Fig. 77.4 Hoshi Matsuri 2009: the two fires (Photo by Alexandre Benod)

The teachings of Agonshû are based on the belief that all of life's problems and misfortunes are the result of spiritual and karmic hindrances. This debt is inherited from the past and from ancestors. Everyone and every household inherit karmic debt. This belief is rooted in Japanese folk religious traditions; unhappy spirits of the dead afflicting the living and causing spiritual hindrances and pollution (*tatari*). The leader focuses on this point in his lectures, explaining to the members the importance of positive thinking. He presents himself as a model of a happy, positive (and wealthy) life.

The monthly rituals include the *tsuitachi goma* a fire rite held on the first day of each month at the Tokyo headquarters, the *meitokusai* a fire rite for the ancestors performed on the 16th of each month at the Kyoto headquarters. And also the *ressai,* a fire rite performed on one of the last weekend days of each month, reflects Agonshû's both preoccupations which are (1) the removal of karmic hindrances and (2) the realization of wishes through positive thinking as seen in worldly benefits (*genze riyaku*).[19]

The main festival of the group is the "Star Festival" (*Hoshi matsuri*) held each Founding of the Nation Day (11 February) at the Yamashina headquarters near Kyoto. During the *matsuri* two huge fires represents the two esoteric Buddhist mandalas. Prayers written on millions of wooden sticks (*gomagi*) are throw into the flames. The two fires are interpreted in Agonshû as having different functions. Both come from esoteric Buddhism, the "wombworld" (*taizokai*) and "diamond world" (*kongokai*). The *taizokai* is for the liberation of the souls of the dead which might otherwise cause spiritual hindrances; whereas the *kongokai* is for transforming peoples' inner wishes into reality (Fig. 77.4). This is a spectacular

[19] Worldly benefits are the result of, for example, a ritual or the power of an amulet that will provide the purchaser: education success, traffic safety, business prosperity, protection from illness, etc.

public rite (it is said that the festival attracts over half a million visitors annually) is usually held for benefit of World Peace.[20]

The Hoshi Matsuri is symptomatic of the need for legitimacy or acceptance. The event was clearly prepared to welcome foreigners: dozens of English-translators available, most of the advertising was translated into English. Despite the huge number of visitors I was one of the few foreigners who attended to the Hoshi Matsuri. All the preparation provided a general atmosphere that Agonshû was truly international in scale. The (almost) annually World Peace rituals abroad emphasize this same idea.

77.5.3 Agonshû and the Promotion of World Peace

When Agonshû performed Goma ceremonies abroad for promotion of World Peace, it is always for the *gedatsujôbutsu: becoming a Buddha*. According to followers, "goma rites purpose is to provide spiritual aid and comfort to the souls of the departed, to lead them to nirvana (*jôbutsu*) and avoid to disturb us in our daily life." This is how Agonshû justify to perform Goma in places deeply marked by WW2. For example, on June 13, 2006, a Fire Rite Ceremony was held on the land adjacent to the Auschwitz concentration camp in dedication to world peace and in commemoration for holocaust victims. There was also a fire rite in 2007 at Khabarovsk in Siberia, where many Japanese Soldiers died. In 2009, they took a fire rite in Guadalcanal in the name of World Peace and appeasing numerous victims of War in the Pacific Ocean (Fig. 77.5).

According to a follower I interviewed in March, "we must hurry up to liberate all the souls of departed soldiers if we want to avoid the third Nuclear Impact." The suffering against war is double: (1) the feeling that Japanese army acted badly during WWII by killing millions of people and (2) the fear of War happening again.

To cure this Agonshû developed its activities abroad perceived as social engagement by performing Goma rites in the name of World Peace. The method of action is twofold: (1) active participation to rites in real life by gaining psycho-emotional effects through praying, such as catharsis. Catharsis refers to any cleansing of emotion experienced by an audience in relation to drama. (2) passive participation by talking, listening, etc. about World Peace to perform a particular memory work, which means a process of engaging with the past which has both an ethical and historical dimension. By talking, explaining, purging moral, historical facts, it becomes easier to assimilate and people can go forward.

Agonshû has developed the use of its satellite broadcasting to cover two types of events: indoor rituals and outdoor rituals. The video and audio are transmitted to other Agonshû centers from abroad. In the Kanto Betsu in, members are gathered in front of large video screens. There are four 3-m screens and one 8 m. Followers participate in the rite in Japan while it is being performed on the other side of the

[20] However, in 2009, the main theme was substituted for the "national economic healthiness."

Fig. 77.5 Guadalcanal event advertisement in the Tôkyô Metro (Photo by Alexandre Benod)

planet. Live rituals make the religious experience more vivid, more realistic. Members watching on the screen responded in very much the same way as those attending the "live" performance. The ritual seemed to be extended by the media. The members "participate" in the ritual, even though they do so at a distance and are actively involved. The synchronism of satellite communications creates a "sense of unity" and the "feeling of solidarity."

> Even if the place is distant it is wonderful sensation of sharing the same prayer and the same feelings. In the traditional pilgrimage the sense of unity comes from being in the same place, even if at different time. With the satellite communications it is the contrary: being in a different place at the same time. (Baffelli 2008)

Thus they have created what the anthropologist Erica Baffelli defined as a Virtual Nationwide Temple and in case of Goma rites performed abroad, I will call a Virtual Worldwide Temple (Baffelli 2008). In the case of a Virtual World Wide Temple, the ceremony is always held for the promotion of World Peace. I think that while historical Buddhism treats social suffering through secularized social engagements activities, NRs like Agonshû brings a religious response to Japanese people suffering by performing religious ceremonies.

In fall 2009, Agonshû went to Guadalcanal to liberate souls of dead soldiers. After the ritual, Agonshû communication service produced a tv program broadcast on cable TV and also DVD. It is highly interesting in the way the video stresses the ambiguous position of Agonshû concerning the Memory Work. While they performed a ceremony to promote World Peace between nations, especially Japan and the U.S. in the case of the Pacific War, the video itself shows the three tv hosts enjoying Japanese Army huge machine guns for massive murders and 5 min later visiting a primary school sponsored with Japanese funds. This duality of the rituals actually questions the impact of the Memory Work and the path that some Japanese new religions are walking on.

77.6 Conclusion

In conclusion, I want to underline some interpretative hypotheses on Modernization of Buddhism which are not only based on technology, but and also through changes within society itself. Theories of religious modernization, originating in the monotheistic context of Western societies (including Max Weber and his theory of the "elimination of magic" and "inner isolation" in the Protestant Ethic and the Spirit of Capitalism), have focused on two crucial issues which are beyond the challenging case of their transposition in non-Western settings. First, religions in Japanese society have proved to follow other historical paths than the model of a modern "disenchantment" or "secularization." Second, recent theoretical developments in sociology have refashioned old conceptions of a single universal "modernity" and put emphasis on "multiple modernities" (Arnason 1997, 2002 and Shmuel Eisenstadt 1996). The relevance of the models of "modernization" remains subsequently unsolved, unless one examines local dynamics of change. The Japanese situation is very interesting especially in the adaptation of the New Religious Movements. As Suzuki Sadami (2005) showed with the emergence of a cultural nationalism, it can be particularly relevant in the religious studies field. Indeed, Tsukada Hotaka (2009) stressed the co-existence of universalism and nationalism in two other new new religions: Mahikari and Kôfuku no Kagaku. Mahikaris nationalism consists of Japan-centrism, the theory that the birthplace of human beings is Japan and that the emperor of Japan is the ruler of the world. In the early 1990s, Mahikaris nationalism was inspired by the consciousness that Japan is an advanced nation and should, therefore, be diplomatically more present through religious activities. Koufuku-no-Kagaku's nationalism is based on the superiority of the Japanese economy and the logic that many higher spiritual presences exist in modern Japan. In the early 1990s, it reacted to the "bubble economy" and international affairs by insisting that Japan should export its superior religious thought to the world against the background of its economic superiority. Through comparative studies such as this, we can observe that Agonshû is actually moving in the same way. By acting exactly like the promotion of the peace model of the Japanese State, Agonshû is developing its own religious group brand/soft power. Beyond completing a Memory Work, it is also a way to gain authority and prestige outside, but also inside Japan. Nationalism is on the core of these rituals. As followers told me during interviews: only Japan can accomplish this, only Agonshû, in the very same way SGI claims its legitimacy (also originated in the fierce predications of Nichiren). These kinds of affirmations stress the edge where, in our case studies, Sôka Gakkai and Agonshû, sit. On the one hand there is the promotion of universalism among human being and on the other hand the absolute necessity that only Japanese religions can help the World.

References

Arnason, J. P. (1997). *Social theory and Japanese experience: The Dual Civilization*. London: Kegan Paul International.

Arnason, J. P. (2002). *The peripheral centre: Essays on Japanese history and civilization*. Melbourne: Pacific Press.

Baffelli, E. (2008, January 15–29). *Media and religion in Japan: The Aum affair as a turning point*. Paper presented during the EASA Media Anthropology Network & EASA Religion Network joint e-seminar.

Basch, L., Glick Schiller, N., & Szanton, B. C. (1994). *Nations unbound: Transnationalized projects and the deterritorialized nation-state* (p. 6). New York: Gordon and Breach.

Bornholdt, S. C. (2009, June 11–14). *Social action and ambiguities of* Soka Gakkai *in Brazil*. Paper presented at the 2009 Congress of the Latin American Studies Association, Rio de Janeiro, Brazil.

Clarke, P. B. (2005). Sôka Gakkai in Brazil. In L. Learman (Ed.), *Buddhist missionaries in the era of globalization* (pp. 123–139). Hawaii: University of Hawai'i Press.

Eisenstadt, S. N. (1996). *Japanese civilization: A comparative view*. Chicago: University of Chicago Press.

Ian, R., & Georges, T. (2004). *Practically religious and the common religion of Japan*. Honolulu: University of Hawaii Press.

Inoue, N. (1983). NSA and non-Japanese members in California. In Y. Keichi (Ed.), *Japanese religions in California: A report on research within and without the Japanese-American community* (pp. 99–161). Tokyo: University of Tokyo Press.
Inoue, N. (1996). *Shinshûkyô no kaidoku*. Tokyo: Chikuma Library.
Inoue, N. et al. (Eds.). (1990). *Shinshûkyô Jiten* [Dictionary of Japanese new religions]. Tôkyô: Kôbundô.
Jaffe, R. (2001). *Neither monk nor layman: Clerical marriage in modern Japanese*. Honolulu: University of Hawaii Press.
Jaffe, R. M. (2010). Religion and the Japanese empire. *Japanese Journal of Religious Studies, 37*, 1–7.
Kiriyama, S. (1983). *Gense jôbutsu: Waga jinsei, waga shûkyô* [Transformation of the Buddha: My life, my religion]. Tôkyô: Rikitomi Shobô.
Kunio, Y. (1945). Senzo no Hanashi [About our ancestors]. In Yanagita Kunio Zenshû [Complete works of Yanagita Kunio], 13. Tokyo: Chikuma Bunkô, 9–209.
Laplantine, F. (2001). Brésil. In François Laplantine & Alexis Nouss (Eds.), *Métissage* (pp. 121–125). Paris: Fayard.
Learman, L. (Ed.). (2005). *Buddhist missionaries in the ear of globalization*. Honolulu: University of Hawai'i Press.
Machacek, D., & Wilson, B. (2001). Introduction. In D. Machacek & B. Wilson (Eds.), *Global citizens: The Soka Gakkai Buddhist movement in the world* (pp. 1–12). New York: Oxford University Press.
Matsudô, Y. (2000). Protestant character of modern Buddhist movements. *Buddhist-Christian Studies, 20*, 59–69.
Ministério do Planejamento, Orçamento e Gestão, Instituto Brasileiro de Geografia e Estatística-IBGE, Centro de Documentação e Disseminação de Informações. (2008). Resistance & Integration: 100 years of Japanese immigration in Brazil.
Pereira, R. A. (2008). The transplantation of Soka Gakkai to Brazil: Building "the Closest Organization to the Heart of Ikeda-Sensei". *Japanese Journal of Religious Studies, 35*, 95–113.
Pereira, R. A. (2009, August 11–14). Transnationalization of Japanese religions in a globalized world: Some considerations based on case studies in Brazil. Proceedings of a paper presented at the 2009 International symposium on *management and marketing of globalizing Asian religion* (pp. 163–179), Museum of Ethnology, Ôsaka, Japan.
Rocha, C. (2000). Zen Buddhism in Brazil: Japanese or Brazilian? *Journal of Global Buddhism, 1*, 31–55.
Sakashita, J. (1998). *Shinnyoen and the transmission of Japanese new religions abroad*. Stirling: University of Stirling. Ph.D. Supervisor: Ian Reader.
Shimazono, S. (1993). The expansion of Japan's new religions in foreign cultures. In M. Mullins, P. Swanson, & S. Susumu (Eds.), *Religion and society in modern Japan* (pp. 273–299). Berkeley: Asian Humanities Press.
Susumu, S. (1992). *Shin-shinshûkyô to Shûkyô Bumu*. Tôkyô, Iwanami Booklets n°.237
Suzuki, S. (2005). *Nihon no bunka nashonarizumu* [Cultural Nationalism in Japan]. Tokyo: Heibonsha.
Tsukada, H. (2009). *Shinshinshûkyô ni okeru bunka teki nashonarizumu no shosô: Mahikari to Kôfuku no Kagaku ni okeru nihon, nihonjinkan no* ronri to hensen [Diverse aspects of cultural nationalism in Japanese 'new-new religions': Origins and changes in perspective toward Japan or Japanese in Mahikari and Koufuku-no-Kagak]. *Religion & Society, 15*, 67–90.
Watanabe, M. (2008). The development of Japanese new religions in Brazil and their propagation in foreign culture. *Japanese Journal of Religious Studies, 35*, 115–144.

Chapter 78
The Place and Role of Alternative Forms of Religiousness in Contemporary Russia

Demyan Belyaev

78.1 Introduction

It is perhaps the history of the twentieth century that in the first place underpins the "typical" vision of Russia as a definitely "non-Western" country, which is common in the West both in political discourse and in academic research. Indeed, Russia has for decades been "the enemy" of the West and its "avant-garde," the United States of America. But the divide obviously goes further than the economic and political differences of the last hundred years. Indeed, throughout the Middle Ages and thereafter Russia had also consistently stayed apart from the "advanced humankind," or Europe, in another highly important area of social life – religion.

Russia had never been Muslim so as to enter a direct religiously motivated conflict with the West, such as the ones that took place during the Crusades or on the part of the Ottoman Empire were not part of Russia's history. However, while Russia was also Christian, it adopted its particular form of Christianity, which denied any subordination to Western religious authorities and always maintained a "cautious distance" to the latter. And later, in the Soviet times, Russia became notorious for seeking to "wipe out" the name of God from the face of Earth at the level of officially declared state policy, thus a land-to-be of dedicated communists and atheists.

In 1991, the Soviet Union collapsed. What was to come in the field of religious beliefs in this situation? A revival of "pagan superstitions"? A new Christianization? Or a reign of blatant atheism? Hardly anyone would dare to predict, but it was meant to be something *different* from the West, just because Russia has always been so different. This chapter seeks to provide a view on what really happened.

D. Belyaev (✉)
Collegium de Lyon/Institute of Advanced Studies, 15 Parvis René Descartes, 69342 Lyon, France
e-mail: demyan.belyaev@ens-lyon.fr

S.D. Brunn (ed.), *The Changing World Religion Map*,
DOI 10.1007/978-94-017-9376-6_78

78.2 An Overview of Existing Approaches to the Religious Situation in Russia in 1990s–2000s

The first and most widespread approach postulates an irreversible advancement of the *secularization* process in contemporary Russia (Greeley 1994; Furman 1997; Kääriäinen 1999; Borowik 2002; Agadjanian 2006). Russians have increasingly demonstrated an "amorphous," "eclectic" religious consciousness that does not depend on denomination: "It is neither Christian nor anti-Christian" (Furman 1997: 299), a private kind of religiosity, where the most important criterion is the ability to construct one's own view of the world (Borowik 2002) drawing on such diverse sources as science, "parascience," theosophy, magic, occultism and Far Eastern religions as well as the traditional Christian doctrine. Some call such situation "triumphant neo-paganism" (Nikonov 1997: 307); others speak of a "popular religion that encompasses many different elements from different areas" (Lewis 2000: 295).

According to Kääriäinen (1999), the presence of an eclectic mixture of beliefs in things like astrology should be understood as a sign of an absence of any kind of strong religiousness. Hence secularization theorists imply that only "official" religiousness, with corresponding institutions, public rituals and state recognition should count as "real." People who cannot say clearly what they believe are considered non-religious. This line of thought is continued, for example, in Agadjanian (2000), according to whom the public consciousness in Russia is dominated by a "spiritual entropy." He argues that even the Orthodox Church is not a source of belief and values anymore, but just a "public religion," a source of national ideology and identity. He further argues that Russia´s religious life suffers from a "syndrome of missed modernity" and the absence of experience in independent public conduct with religious questions in a pluralist context (Agadjanian 2006: 179–180).

Indeed, there are substantial reasons as to why the Orthodox Church might have lost its spiritual credibility. Apart from the hypothesis that it never had a monopoly on faith in Russia (see Plaggenborg 1997), it cannot be forgotten that the most violent repressions against the Church, including destruction of buildings and murdering of priests, were only conducted in 1930s under Joseph Stalin. In 1943, in a critical situation during the World War Two, he, however, changed his attitude towards the church and grasped to it as a useful ideological resource for additional mobilization of population to resistance against the enemy invasion. The Church has been again allowed to function, but priests were obliged to corroborate with the secret security service, later known as KGB. Among other things, they were reporting to KGB about anti-Soviet views of the people who came to confess to them. This, however, is not a purely Soviet practice, but was common in Russia at least since the rule of the emperor Peter I who subordinated the church to the state bureaucrats rather than to an independent (at least formally) Patriarch. Thus, in the early 1990s there were good reasons for the Church to be "spiritually bankrupt" (Pankhurst 1996: 355).

On the other hand, the number of functioning Orthodox parishes in Russia grew tremendously between 1990 and 2010, from 7,000 to about 30,000, not least with the explicit state support. While only under 50 % newly born children were baptized in 1980s, nowadays most of them are. Thus Russians might continue to be religious, or at least wish to look as such. Some authors also point out to the fact that Soviet communism had quite many traits of religion and thus in a way was fulfilling religious needs of Russians (McFarland 1998; Wunder 2005). According to Furman (1997), it is a mistake to assume that the period of communist rule was a time of secularization and the 1990s accordingly was that of "de-secularization." He argues that a dogmatic and "quasi-religious" system of atheism does not have anything to do with secularization and only the 1990s finally brought into Russia a secularization, even if in a somewhat "post-rational" or "adogmatic-eclectic" form.

Another strand of thought draws on the studies on alternative forms of religiousness in the West (Wassner 1991; Helsper 1992; Bochinger 1994; Boy 2002; Heelas and Woodhead 2005). These researchers represent the view that the "eclectic" beliefs identified above, in particular beliefs in supposedly supernatural phenomena not recognized either by science or by Christian church – for example, astrology, spiritualism, telepathy – constitute an important novel form of religiousness, which may be complementary to its more traditional forms or "fill the gap" left by the latter in the course of the secularization of Western societies.

78.3 Alternative Forms of Religiousness in Russia: Views and Controversies

While there is nearly a consensus among scholars that "alternative," that is, non-Christian and not institutionally associated with any other major established confession, forms of religiousness exist and have become widely spread in contemporary Russia, that is also where the consensus ends. Scholars have rather different opinions as regards the terminology to be used, as well as the scope and influence of this "alternative" religiousness. In my view, this mainly depends on a scholar´s own spiritual sympathies and/or affiliations as well as on the nature of his or her contacts used during field work in Russia (if any).

Almost no one uses anymore the term "superstition" which was quite common in the 1970s (Jarvis (1980: 288) uses superstition for a "non-institutionalized kind of belief"). Instead, many prefer to speak of "cults," "sects," or "new religious movements" which are not much better because, in the common language they carry a negative connotation (Shterin 2000). Thus some scholars have suggested replacing these terms with "alternative religion" (Barker 2001). Others opt for "occult(ism)" (Rosenthal 1997; Menzel 2007) which in my view implies, through its Latin root, a certain measure of secretiveness, which is not the case for most forms of alternative religiousness in Russia today.

A term that in my opinion suits the Russian context somewhat better is "esoteric knowledge" which Stenger (1989) describes as a kind of "inner and spiritual knowledge" one gains only through experiences in one's life. The "esotericism" term also finds favor with Bochinger (2005) who sees it as a modern phenomenon, a collective answer to Enlightenment and a 'non-dogmatic' approach to (the dogmatized) religiosity (pp. 68–69). On the contrary, "New Age" seems to be rather adequate only for specific alternative religious movements which flourished in the West in the 1960s–1980s (York 2001).

Elsewhere (Belyaev 2009) I have proposed my own terminology. I argue that all these alternative religious worldviews shared the fact that they did not develop within publicly recognized systems of knowledge which in Russia's case are scientific materialism and the Orthodox Christianity. Hence these worldviews stand in a conflict relationship with the norms and values taught by public institutions such as the education system and state-supported churches, which allows them to interpret themselves as providers of heterodox knowledge, while the education system and state-recognized churches are providers of publicly authorized, and hence an orthodox knowledge about the world. In this model, knowledge is not separated from belief, because belief, insofar it is taken as truth and guides human action, is functionally not different from the knowledge that had been obtained experimentally. While I still hold to this theoretical model, in this chapter I speak of "alternative religiousness" so as to avoid misunderstandings, which could appear in the absence of detailed explanations of my own model.

Researchers have extensively speculated on possible reasons for the growth of alternative religiousness in Russia today. Behind this growth they have seen, among other things, the need for a "spiritual medicine" in difficult times (Monas 1999: 518), the reaction to the 'old atheistic pseudoscientific view of the world' (Heino 2000: 304), the "consoling" alternative to the rational view of the world propagated by scientific materialism (DeNio Stephens 1997), the loss of faith in science (Vizgin 2001), the complexity and uncertainties of life, which science is not able to make sense of (Balagushkin 2001) and the spiritual search by people considering the traditional religious worldviews "primitive" for worldviews that would seem more rational or scientific (Fesenkova 2001). While most see alternative religiousness as a *reply* to something, some scholars also regard it as rooted in the utopian ideas of socialism, and hence their natural companion (Potrata 2004). Still, the question about reasons remains open and the subject of debate.

78.4 Alternative Religiousness in Russia: A New or an Old Phenomenon?

This issue definitely belongs to those who emerged first when considering our topic. Those who would argue for a connection between socialism and the following upsurge in alternative religiousness would definitely expect the latter to be a new phenomenon. This would perfectly fit their "suppression [under socialism] – liberation [under capitalism]" paradigm and also the theory of "spiritual vacuum" which now needed to be filled.

The hypothetical contrary point of view would point out that alternative forms of religiousness have always existed in Russia to some extent, be it during the Empire, or during the Soviet period, und what we observe now is simply their much greater emergence in the public field due to the growth of mass media and the internet both which serve as much more powerful vehicles for spreading various teachings and worldviews than those that were available during the previous historical epochs.

In order to shed some light on solving this dilemma, below I will make a brief and very subjective (as is always the case with qualitative studies) overview of alternative religiousness in Russia before and after the breakup of the Soviet Union.

In 1875 Dmitri Mendeleyev, the world famous chemist and inventor of the periodic table of elements, was appointed by the Academy of Sciences to lead a scientific commission for the investigation of the phenomena of spiritualism (Vinokurov 2005). This fact reflected nothing else than the extent to which spiritualism (contacting ghosts) had become popular in the Russian educated society: several spiritualist magazines were published; famous foreign mediums were invited to parlors of noble families; several reputable scientists such as A. Aksakov and A. Butlerov argued for the existence of ghosts.

In the very same year, a Russian émigrée to the United States known as Helena Blavatsky founded the Theosophical Society in New York, becoming a world famous authority in alternative religiousness. She claimed to have united various esoteric teachings in her doctrine known as "theosophy" (Fuller 1988; Cranston 1993; Carlson 1993). Blavatsky influenced such famous western thinkers and occultists as R. Steiner and found many followers in Russia as well (von Maydell 2005). Blavatsky and Steiner left a profound influence on Russian poets and artists of the early twentieth century, including A. Bely, M. Tsvetayeva, V. Kandinsky and K. Malevich (Parton 1995; Bogomolov 2000; Seidel-Dreffke 2004; Hansen-Löve 2004). Esoteric influences can also be found in the works of Russian philosophers and scientists of that time, such as V. Solov`yev, N. Berdyayev, K. Tsiolkovsky, V. Vernadsky, A. Losev and P. Florensky (Hagemeister 1997; Kravchenko 1997; Menzel 2007).

Somewhat later, almost as famous in the West as Helena Blavatsky, was George Gurdjieff whose alternative religious doctrine also originated in Russia and was based on the teachings of Sufis and Tibetan Buddhists (Lefort 1967; Bennett 1973; Webb 1980; Rovner 2002).

Further Russian alternative religious authors whose teachings have attained high prominence throughout the twentieth century include the artist Nicholas Roerich who lived in India and developed there his "Agni-Yoga" (Williams 1980; Shishkin 2003) which is still important in Russia today. In many cities there exist so-called 'Roerich clubs' with a membership drawn mostly from the intelligentsia (Lunkin and Filatov 2000); the visionary Daniil Andreyev whose self-published esoteric work *Roza mira*, written in prison in 1950–1959 on the basis of his own revelations and contacts with invisible non-material beings describes the invisible world supposedly surrounding us, and the future of humankind (Epstein 1997); the writer Yuri Mamleyev who led an occult group of the Moscow intelligentsia in the 1960s (Polikovskaya 1997; Menzel 2007); Aleksandr Aseyev who edited Russian-language journal *Okkul'tizm i ioga* published in Paraguay and widely read by Russian émigrés in all parts of the world in the 1960–1970s; Vladimir

Shuktomov who in the late 1970s founded an occult group "*Kunta-Yoga*" which sought to hide from state persecution in the Armenian mountains (Belyayev 2004); the village healer Porfiri Ivanov who preached a "natural way of life." spent years in psychiatric hospitals and finally wrote a letter to the Central Committee of the Communist Party, following which the party-affiliated magazine *Ogonyok* published about him a praisingworthy article (Ogonyok 1982), making his teachings widely popular.

In general, there is evidence that alternative religious groups continued their activities in the Soviet Russia despite the official suppression of all kinds of religion. Templars ('*tampliyery*'), freemasons and other secret societies are reported to have existed even in the 1930s Moscow (Nikitin 2000). Stalin himself allegedly once consulted the fortune-teller Wolf Messing, a German Jew who fled to the Soviet Union and was allowed to work there as a stage magician (Nepomnyashchy 1999; Küppers 2002). Since 1970s, wide circles of the intelligentsia had been interested in Christianity, Buddhism and Yoga (Menzel 2007); by 1980s, numerous charismatic individuals filled the Russian alternative religious "underground" and gathered many followers (Lebedko 2000). In 1980, Leonid Brezhnev, the general secretary of the Soviet Communist Party, received treatment by the Georgia-born faith healer Dzhuna (Daily Telegraph 1984).

The Soviet Union conducted substantial research on "parapsychology" (Huneeus 1990; Ostrander and Schroeder 1997), that is, on the paranormal phenomena and abilities whose existence has been postulated by many alternative religious worldviews, but denied by the official Soviet and Western science and the Russian Orthodox Church. At the University of Leningrad, there was a department on the study of telepathic phenomena led by Professor Leonid Vasil`yev (Wassiliew 1965); an international conference on parapsychology was held in Moscow in 1968 (Ebon 1977). In the 1980s, a special laboratory was set up at the Moscow Institute of Normal Physiology to scientifically study the abilities of Dzhuna and the "telekinetist" Ninel´ Kulagina (Torin 1997). The phenomenon of Yeti was discussed in popular scientific magazines as early as 1958 (Menzel 2007) and an expedition consisting of scientists and journalists was financed in 1982 to search for it in the Pamir mountains (Yeltsin 2001). In 1980, the Geographical Society of the Soviet Academy of Sciences established a commission for the investigation of abnormal phenomena whose task was to trace UFOs. It has been reported that research in these areas still continues at some leading scientific institutions in Moscow (Dubrov and Lee 1998; Boldyreva and Sotina 2002; Yamamoto 2003).

From this overview we can conclude that alternative religious worldviews in Russia unambiguously predate the breakup of the Soviet Union. Some even argue that 'heathen' and 'mystic-utopian' beliefs are far more deeply rooted in Russian "popular" religion than in the West (Ryan 1999). As much stereotypical as it may sound, it should be recognized that Russia, despite its socialist and anti-religious experience, has been at least as much susceptible to alternative forms of religiousness as Western countries, if not more.

78.5 Alternative Religious Teachings in Russia of the 1990s and 2000s: How Religious Are They?

Alternative religious teachings which have notable numbers of followers in today´s Russia are numerous and very diverse in their nature. On the one hand, still alive are the long-standing alternative religious traditions such as those drawing on Blavatsky, Roerich or Gurdjieff (Menzel 2007); shamanism and animism (Balzer 1997; Tchervonnaya 1997); Slavic neopaganism (Kochanek 1998; Laruelle 2004); or "new religious movements" such as Hare Krishna (Burdo and Filatov 2006). On the other hand, there have emerged many new teachings which are not directly related to any established traditions, whether orthodox or heterodox, have been developed by contemporary figures in the last two decades, do not seek to withdraw their followers from everyday life as is common in many "new religious movements," nor worship the founder of their doctrine. With these distinction criteria in mind, I will give some examples below.

The teachers of alternative religious worldviews focus on very diverse abilities and techniques which are meant to improve the life of individuals who apply them. We may consider these teachings as religious, because their content is not scientifically confirmed (and often contradicts views of scientists) and thus needs to be accepted by belief. We could also simply call them heterodox science, but since the mainstream science, both natural and social, widely denies whatsoever validity of their teachings, in a social scientific context, they need to be considered religious.

Examples of the content of such alternative religious teachings are: existence of and the possibility to use "bio-energy," existence of and possibility to develop extrasensory perception, or self-help and life management methods based on the assumption of the existence of certain supernatural laws and/or powers. Such teachings are most often spread through books and workshops on "personal development."

Some of the influential teachers of the last two decades include Boris Zolotov, an "expert on complex systems;" Gennadi Malakhov whose four-volume work on alternative self-healing techniques *Tselitel'nyye sily* (*Healing Powers*) sold over four million copies in the 1990s; Mirzakarim Norbekov who teaches how to recover from myopia and to earn money through intuition; Andrej Levshinov who teaches yoga, qi-gong and organizes spiritual training retreats for VIPs; Vyacheslav Bronnikov who claims to have learned "direct vision" through the brain with closed eyes and even allowed his disciples to undergo a scientific study at the Institute for Brain Research of the Russian Academy of Sciences (Bekhtereva et al. 2002); Valeri Sinel`nikov who teaches that illnesses happen for people's good; Aleksandr Sviyash and Vadim Zeland who address a wide variety of topics (Fig. 78.1).

All these people earn money from their teachings, especially from workshops, which leads some critics to accuse them of recurring to alternative religious teachings as an easy source of money during the difficult period of transition from socialism to (supposedly) democracy and market economy. It will be pointed to such persons as Grigorii Grabovoi, an alternative religious teacher of a more radical kind, who

Fig. 78.1 A vast book supply on alternative religious topics in a typical Russian bookshop (Photo by Demyan Belyaev)

allegedly even offered to resurrect children killed during the terrorist attack on a school in Beslan in 2004. In 2008 he was sentenced to 11 years in prison for fraud. However, it was also speculated that his imprisonment was related to his intention to run for presidency in the same year's elections, and he was eventually released in 2010. It should further be noted that most of the people listed above had knowingly started their involvement with alternative religiousness way before the fall of socialism.

Beyond these individual teachers and their books, there are also media outlets specializing in the propaganda of alterative religious beliefs. The first newspaper devoted exclusively to such matters appeared in 1990 under the title *Anomaliya* and the slogan "Miracles do not contradict Nature itself, but only what we know about Nature." Already in 1991, its print run reached 250,000 per issue. Today the highest print run in this sector is 550,000 per issue manifested by monthly esoteric newspaper *Orakul*, which unlike the pioneering publication mentioned above is full of commercial advertising.

Television was responsible for the instantaneous fame of the hypnotic healer Anatoli Kashpirovsky whose sessions were broadcast nationwide in 1989; the astrologist Pavel Globa; or the magician Yuri Longo who apparently levitated and even resurrected dead people in a sort of reality show. In recent years, a TV competition show called "The battle of psychics" has become very popular and undergone several editions. The number of healers, magicians and astrologists working openly has grown dramatically as well (DeNio Stephens 1997; Lindquist 2006). Most of them provide specific one-time services of a counseling (astrologers, card readers), magical or healing nature (psychics, "extrasensitives" (*ekstrasens*) and other types of healers). As they are not organized in associations (unlike it is the case, for example, in Germany), it is very difficult to estimate their real numbers.

We could hence arrive at the twofold conclusion: first, alternative religiousness has long been present in Russia and attracted the interest of the population; second, it is also true that the liberalization of mass media *in conjunction with* the extension of opportunities for commercialization of this kind of religiousness have led to the rise both in the intensity of interest in alternative religiousness among the population and in the scale of its proliferation in the latter during 1990s and 2000s.

78.6 How Many Russians Hold Alternative Religious Views?

This is a very intriguing question that until recently could not be answered satisfactorily due to the complete lack of data. To close this gap, I have designed and carried out the first comprehensive representative population survey in Russia specifically devoted to the measurement of various forms of alternative religiousness (see Belyaev (2011) for more details). This survey was commissioned by the University of Heidelberg, with financial support from the *Deutsche Forschungsgemeinschaft*, and administered by the leading Russian polling organization *Levada-Tsentr* in September 2006. The survey collected data from 1,601 people in 128 communities (cities, towns and villages) in 46 regions across the entire Russia (Fig. 78.2).

Table 78.1 shows the share of population which believes in various traditional and alternative religious concepts. It can be inferred from these results that alternative religious concepts, although denied by both scientific and Orthodox Christian mainstream, are at least as widely believed as traditional religious concepts.

The questions in the table are about respondents' *beliefs*. However, it would be even more interesting to assess to what extent these beliefs guide individuals' *actions*, for example, urging them to come in contact with active carriers of alternative

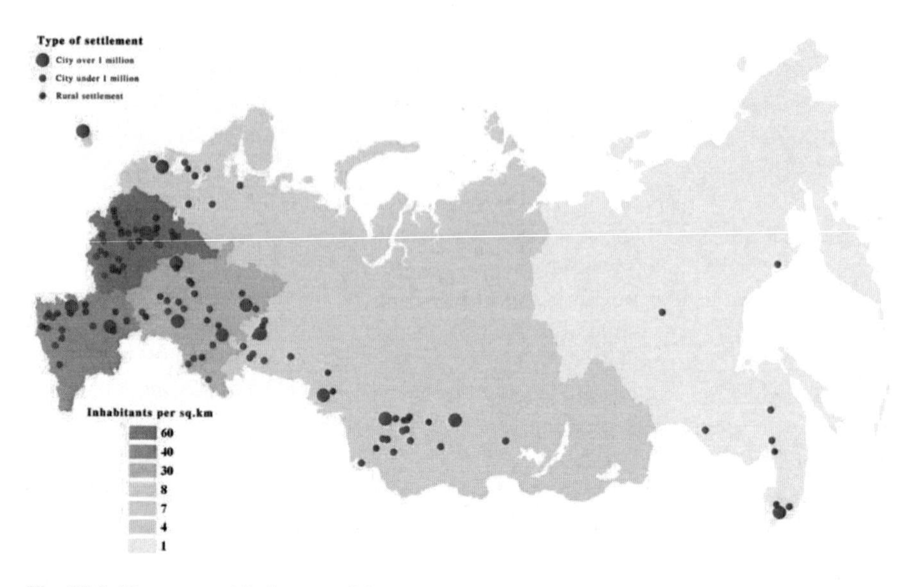

Fig. 78.2 The geographical scope of the survey (Map by Demyan Belyaev)

Table 78.1 Attitudes toward various religious concepts

Concepts present in traditional Orthodox Christian teachings		Concepts present only in alternative religious teachings	
The existence of…	Is believed in by:	The existence of…	Is believed in by:
A higher power like God	69 %	Magic that can be used to influence people's lives	53 %
A higher justice that rewards or punishes people	61 %	Healers who can heal from sicknesses incurable through Western medicine	52 %
Angels, demons and ghosts	44 %	A connection between Zodiac signs and a person's character	49 %
Soul that can exist independently of the human body	51 %	Talismans that can influence real-life events	43 %
Jesus Christ who really was the Son of God	53 %	Predictions of the future	58 %
		Telepathy (extrasensory thought transmission)	44 %
		Telekinesis (moving physical objects by thought)	24 %
		The possibility to contact spirits of the dead	19 %

Source: Author's survey

Table 78.2 Frequency of practical experiences with alternative religiousness. The gray background shows a potential influence area of alternative religiousness

	Percent of total sample who have at least once:				
	Read alternative religious literature	Practiced yoga, meditation etc.	Seen an astrologist, magician, fortune-teller	Seen a faith healer, shaman	Used alternative healing methods
Have had this kind of experience and gained from it	16	5	5	12	19
Have had this kind of experience and NOT gained from it	19	5	9	10	12
Have NOT had this kind of experience, but would like to try it	16	15	12	19	22
Have NOT had this kind of experience and would NOT like to try it	49	74	74	59	48

Source: Author's survey

religiosity. We asked people about their *practical* experiences with alternative religiousness and its practitioners, distinguishing between five different kinds of experiences. The gray-shaded area shows the total share of population which is actually or potentially involved in some of these activities (Table 78.2).

Almost 35 % of the adult population has read some kind of esoteric literature and about half of those said that they had profited from the information it contained. Furthermore, over 22 % of people in Russia have had experience with a faith healer, with a subjective success rate of almost 56 %. For comparison, in Germany only 11 % have been treated by a faith healer, with a subjective success rate of 58 % (Andritzky 1998). In France, 13 % of the population has seen an astrologist (Boy and Michelat 1986). These results suggest that alternative religiousness may be wider spread in today's Russia than in the West.

78.7 Spatial Aspects of Alternative Religiousness in Russia

Geographers would surely be interested in the question as to whether and how the spatial dimension is related to the spread of alternative religiousness in Russia. Contrary to what could be expected, the survey could not find any meaningful statistically significant differences in alternative religious beliefs or practices in connection with the varying size of settlement or geographical macro-region within Russia. Alternative religiousness seems to be spread fairly equally in physical space.

However, a few, although not obvious, kinds of interconnectedness between alternative religiousness and spatiality have still been discovered in Russia. The first of them is the link between geographical mobility and alternative religious beliefs. The idea to look for such a link goes back to Nelson (1975) who found that people who were geographically mobile believed less in traditional Christian ideas and more often in heterodox ideas like ghosts, predicting the future and faith healing. More recently, a similar connection was also found in Canada (Orenstein 2002).

Hence we asked the people surveyed how often they had changed their place (town) of residence in the last 10 years. In our sample, 77 % did not change it, 14 % changed it once and 8 % more than once. We found no difference between geographically mobile and non-mobile people concerning their belief in traditional religious concepts. However, people with higher mobility believe significantly more often in such alternative religious concepts as reincarnation, telekinesis, (90 % significance) and contact with spirits (95 % significance). An impressively high and statistically significant was the difference between mobile and non-mobile people regarding the presence of own experiences with alternative religiousness (measured through participation in the activities listed in Table 78.2).

In my view, the explanation for this finding lies in the fact that moving relatively often implies a certain psychological flexibility and ability to adapt to new environments. People who are mobile are also more open to meeting new people and hence more often faced with new ideas, which also renders them more open and receptive to new spiritual currents.

Furthermore, it has been speculated that there may be a connection between religiousness and the bond with place of birth, or "localism" (Roof 1974; Eisinga et al. 1990) who found localism to be connected to religious (Christian) beliefs. Following the assumption that a stronger bond to places also leads to a bigger impact exercised by various spatial contexts on one's behavior, our survey included a question on the degree of emotional connectedness with the place of origin. Forty percent of the people questioned felt a strong bond to the place or region they grew up in, as compared to the 57 % who said that they felt only a weak bond.

We have found that those who feel a strong bond to their native place tend to believe in numerous concepts of both traditional and alternative religiousness more often than those who feel only a loose such bond. The highest statistically significant differences were reached with regard to the beliefs in predicting the future, telepathy, telekinesis and contact with spirits, followed by the beliefs in faith healing and magic. As an explanation of this finding I would suggest that people who feel a stronger bond have a more emotional personality overall, which also leads them to believe in concepts which have low level of conformity to the rational thinking.

A further important discovery related to the role of spatial dimension is related to the notion of "social spaces." The results of our survey convincingly demonstrate that social spaces of contact are a decisive factor in turning to alternative religiousness. The effect of social networks on the recruitment of followers of "traditional" religions is well known (Stark and Bainbridge 1985; Sherkat and Wilson 1995). Regarding alternative religiousness, Mears and Ellison (2000) found, that in the United States the demand for printed matter of corresponding contents was distributed quite evenly among different socio-demographic groups, and the only variable with a strong predictive power turned out to be the integration in personal networks with people who themselves had had contact with alternative religiousness: the people whose friends and relatives had acquired "New Age" publications were more likely than others to do the same.

In our sample, 34 % had in their social space no people with the experience of previous contact with alternative religiousness, be it in the form of reading literature on esoteric knowledge or spiritual growth, the use of healing methods dismissed by conventional medicine, the consultation of faith healers, astrologists, magicians, card readers or through the practice of Yoga or meditation. Seventeen percent said they had one such person, 33 % had two to four such people and 16 % even had five or more!

We found that there was a definitive link between people's beliefs in alternative religious concepts and the presence of individuals with alternative religious experiences within their social spaces. Those who have no close relationships to people with such experiences tend towards an especially low frequency of beliefs. Conversely, people who often spend time together with people with such experiences, tend to believe more often themselves. For some alternative religious concepts, the percentage of believers increases proportionately to the increase in the number of close people with corresponding experiences.

This finding seems to us very important because networks on a spatial microscale contribute decisively to the so called spatial and social context, which influences people. The people who are socially close to the individual and who have had own active contact with alternative religiousness function as important *intermediaries* on the way of that individual to alternative religious *belief*. They do not necessarily constitute the *main source* of information on alternative religious worldviews, but their presence as such is shaping a social-spatial context which significantly raises the *receptiveness* of the people living in that context to relevant information from whatever source it might come.

Equally important is the social space of an individual for his/her propensity to be involved in *actions* related to alternative religiousness. In our sample, the individuals whose close acquaintances had alternative religious experiences were more likely to have some form of such experience as well, although not necessarily of exactly the same kind (except for yoga and meditation). This implies that what happens is not a mere "mimicry" of behavior, but rather a general influence of the social space on one´s interest towards activities with alternative religious motivation.

78.8 Conclusion

This chapter has discussed a number of interesting findings on alternative forms of religiousness in contemporary Russia which can be briefly summed up in the following way. Russia seems to be far less different from Western countries in the evolution of religiosity than in its political and economic history. Alternative religious views had since long occupied a firm place in the Russian society, which has expanded further with the post-socialist growth of mass media and of the opportunities to raise money from alternative religious activities just as the Christian Church used to do from traditional ones. Today at least as many Russians share at least some alternative religious views as traditional Christian ones – sometimes in combination with the latter, sometimes in opposition to them. Although alternative religiousness seems to be spread fairly homogenously over the country from the spatial point of view, there are still a few important interconnections between it and spatiality. More specifically, people with a higher geographical mobility significantly often share certain alternative religious views and usually have significantly more personal experience with this kind of religiousness. People who feel strongly connected with their place of origin, tend to believe more in both alternative and traditional religious concepts. Finally, the presence of people with an experience of direct contact with alternative religiousness within the social interaction space of an individual shapes a spatial context that substantially increases the receptiveness of the latter to alternative religious ideas from many different sources.

Acknowledgments The author gratefully acknowledges the financial support of the *Deutsche Forschungsgemeinschaft* to the research project which led to the findings presented in this chapter.

References

Agadjanian, A. (2000). Russian religion in media discourse: Entropy interlude in ideocratic tradition. In M. Kotiranta (Ed.), *Religious transition in Russia* (pp. 251–288). Helsinki: Kikimora.

Agadjanian, A. (2006). The search for privacy and the return of a grand narrative: Religion in a post-communist society. *Social Compass, 53*(2), 169–184.

Andritzky, W. (1998). Spiritualität, Psychotherapie, Gesundheitsverhalten. *Grenzgebiete der Wissenschaft, 47*(1), 25–65.

Balagushkin, E. (2001). Ezoterika v novykh religioznykh dvizheniyakh. In L. Fesenkova (Ed.), *Diskursy ezoteriki (filosofsky analiz)* (pp. 214–239). Moscow: URSS.

Balzer, M. (1997). *Shamanic worlds: Rituals and lore of Siberia and Central Asia*. New York: North Castle Books.

Barker, E. (2001). New religious movements. In N. Smelser & P. Baltes (Eds.), *International encyclopedia of the social and behavioral sciences* (pp. 10631–10634). Amsterdam: Elsevier.

Bekhtereva, N. P., Danko, S. G., Melyucheva, L. A., Medvedev, S. V., & Davitaya, S. Z. (2002). O tak nazyvayemom alternativnom zrenii ili fenomene pryamogo videniya. *Fiziologiya cheloveka, 28*(1), 23–34.

Belyaev, D. (2009). Zur Rolle des heterodoxen Wissens in Russland nach dem Zusammenbruch des kommunistischen Systems. *Aries, 9*(1), 1–35.

Belyaev, D. (2011). "Heterodox religiousness" in today's Russia: Results of an empirical study. *Social Compass, 58*(3), 1–20.
Belyayev, I. (2004). *Ostriye kunty: put` russkogo mistika*. Moscow: Profit Stail.
Bennett, J. (1973). *Gurdjieff: Making a new world*. New York: Harper Colophone.
Bochinger, C. (1994). *New Age und moderne Religion: Religionswissenschaftliche Analysen*. Gütersloh: Gütersloher Verlagshaus.
Bochinger, C. (2005). The invisible inside the visible – The visible inside the invisible: Theoretical and methodological aspects of research on New Age and contemporary esotericism. *Journal of Alternative Spiritualities and New Age Studies, 1*, 59–73.
Bogomolov, N. (2000). *Russkaya literatura nachala XX veka i okkultizm*. Moscow: NLO.
Boldyreva, L., & Sotina, N. (2002). *Physicists in parapsychology*. Moscow: Hatrol.
Borowik, I. (2002). Between Orthodoxy and eclecticism: On the religious transformations of Russia, Belarus and Ukraine. *Social Compass, 49*(4), 497–508.
Boy, D. (2002). Les Français et les para-sciences: Vingt ans de mesures. *Revue Française de Sociologie, 43*(1), 35–45.
Boy, D., & Michelat, G. (1986). Croyances aux parasciences: dimensions sociales et culturelles. *Revue Française de Sociologie, 27*(2), 175–204.
Burdo, M., & Filatov, S. (Eds.). (2006). *Sovremennaya religioznaya žizn Rossii: opyt sistematicheskogo opisaniya* (Vols. 3 & 4). Moscow: Logos.
Carlson, M. (1993). *No religion higher than truth: A history of the Theosophical Movement in Russia, 1875–1922*. Princeton: Princeton University Press.
Cranston, S. (1993). *HPB: The extraordinary life and influence of Helene Blavatsky*. New York: Putnam.
Daily Telegraph. (1984, August 8). Faith healers thrive in land of Rasputin. *Daily Telegraph*, 15.
DeNio Stephens, H. (1997). The occult in Russia today. In B. Rosenthal (Ed.), *The occult in Russian and Soviet culture* (pp. 357–376). Ithaca: Cornell University Press.
Dubrov, A., & Lee, A. (1998). *Contemporary problems of parapsychology: Parapsychological research on the verge of the twenty-first century*. Moscow: Fond parapsikhologii im. Vasilyeva.
Ebon, M. (Ed.). (1977). *PSI in der UDSSR: Religion ohne Kreuz*. München: Langen-Müller.
Eisinga, R., Lammers, J., & Peters, J. (1990). Localism and religiosity in the Netherlands. *Journal for the Scientific Study of Religion, 29*(4), 496–504.
Epstein, M. (1997). Daniil Andreev and the mysticism of femininity. In B. Rosenthal (Ed.), *The occult in Russian and Soviet culture* (pp. 325–356). Ithaca: Cornell University Press.
Fesenkova, L. (2001). Teosofiya segodnya. In L. Fesenkova (Ed.), *Diskursy ezoteriki (filosofsky analiz)* (pp. 10–35). Moscow: URSS.
Fuller, J. (1988). *Blavatsky and her teachers: An investigative biography*. London: East-West Publications.
Furman, D. (1997). Religion and politics in mass consciousness in contemporary Russia. In H. Lehmann (Ed.), *Säkularisierung, Dechristianisierung, Rechristianisierung im neuzeitlichen Europa* (pp. 291–303). Göttingen: Vandenhoeck und Ruprecht.
Greeley, A. (1994). A religious revival in Russia? *Journal for the Scientific Study of Religion, 33*(3), 253–272.
Hagemeister, M. (1997). Russian cosmism in the 1920s and today. In B. Rosenthal (Ed.), *The occult in Russian and Soviet culture* (pp. 185–202). Ithaca: Cornell University Press.
Hansen-Löve, A. (2004). *Kazimir Malevich: Gott ist nicht gestürzt! Schriften zu Kunst, Kirche, Fabrik*. Munich: Carl Hanser.
Heelas, P., & Woodhead, L. (2005). *The spiritual revolution: Why religion is giving way to spirituality*. Oxford: Blackwell.
Heino, H. (2000). What is unique in Russian religious life? In M. Kotiranta (Ed.), *Religious transition in Russia* (pp. 289–304). Helsinki: Kikimora.
Helsper, W. (1992). *Okkultismus – die neue Jugendreligion?* Opladen: Leske+Budrich.
Huneeus, A. (1990). *Study guide to UFOs, psychic and paranormal phenomena in the USSR*. New York: Inner Light.

Jarvis, P. (1980). Towards a sociological understanding of superstition. *Social Compass, 27*(2–3), 285–295.

Kääriäinen, K. (1999). Religiousness in Russia after the collapse of communism. *Social Compass, 46*(1), 35–46.

Kochanek, H. (1998). Die Ethnienlehre Lev N. Gumilevs: zu den Anfängen neurechter Ideologie-Entwicklung im spätkommunistischen Russland. *Osteuropa, 48*(11–12), 84–97.

Kravchenko, V. V. (1997). *Vestniki russkogo mistitsizma*. Moscow: Izdattsentr.

Küppers, T. (2002). *Wolf Messing: Hellseher und Magier*. Munich: Langen Müller.

Laruelle, M. (2004). The two faces of contemporary Eurasianism: An imperial version of Russian nationalism. *Nationalities Papers, 32*(1), 115–136.

Lebedko, V. (2000). *Khroniki rossiiskoi Sanyasy: iz zhizni rossiiskikh mistikov 1960kh–1990kh godov*. Moscow: Institut obshchegumanitarnyh issledovanii.

Lefort, R. (1967). *The teachers of Gurdjieff*. London: Victor Gollancz.

Lewis, D. (2000). *After atheism: Religion and ethnicity in Russia and Central Asia*. London: Lynne Rienner.

Lindquist, G. (2006). *Conjuring hope: Healing and magic in contemporary Russia*. London: Berghahn Books.

Lunkin, R., & Filatov, S. (2000). The Rerikh movement: A homegrown Russian "new religious movement". *Religion, State & Society, 28*(1), 135–148.

McFarland, S. (1998). Communism as religion. *International Journal for the Psychology of Religion, 8*(1), 33–48.

Mears, D., & Ellison, C. (2000). Who buys New Age materials? Exploring sociodemographic, religious, network, and contextual correlates of New Age consumption. *Sociology of Religion, 61*(3), 289–313.

Menzel, B. (2007). The occult revival in Russia today and its impact on literature. *The Harriman Review, 16*(1), 2–14.

Monas, S. (1999). Review of the occult in Russian and Soviet culture by B. Rosenthal. *The Journal of Modern History, 71*(2), 517–518.

Nelson, G. (1975). Towards a sociology of the psychic. *Review of Religious Research, 16*(3), 166–173.

Nepomnyashchy, N. (1999). *Volf Messing*. Moscow: Olimp.

Nikitin, A. (2000). *Mistiki, rozenkreitsery i tampliyery v sovetskoi Rossii: issledovaniya i materialy*. Moscow: Agraf.

Nikonov, K. (1997). Beobachtungen zur Religiosität im nachkommunistischen Russland. In H. Lehmann (Ed.), *Säkularisierung, Dechristianisierung, Rechristianisierung im neuzeitlichen Europa* (pp. 304–307). Göttingen: Vandenhoeck & Ruprecht.

Ogonyok. (1982, February 20). Eksperiment dlinoyu v polveka. *Ogonyok*, 30–31.

Orenstein, A. (2002). Religion and paranormal belief. *Journal for the Scientific Study of Religion, 41*(2), 301–311.

Ostrander, S., & Schroeder, L. (1997). *Psychic discoveries*. New York: Marlowe.

Pankhurst, J. (1996). Religion in Russia today. In D. Shalin (Ed.), *Russian culture at the crossroads: Paradoxes of postcommunist consciousness* (pp. 127–156). Boulder: Westview Press.

Parton, A. (1995). Avantgarde und mystische Tradition in Russland 1900–1915. In W. Hagen (Ed.), *Okkultismus und Avantgarde* (pp. 193–237). Ostfildern: Temmen.

Plaggenborg, S. (1997). Säkularisierung und Konversion in Russland und der Sowjetunion. In H. Lehmann (Ed.), *Säkularisierung, Dechristianisierung, Rechristianisierung im neuzeitlichen Europa* (pp. 275–292). Göttingen: Vandenhoeck & Ruprecht.

Polikovskaya, L. (1997). *Ploschad Mayakovskogo 1958–1965*. Moskva: Zvenya.

Potrata, B. (2004). New age, socialism and other millenarianisms: Affirming and struggling with (post)socialism. *Religion, State & Society, 32*(4), 365–379.

Roof, W. (1974). Religious Orthodoxy and minority prejudice: Casual relationship or reflection of localistic world view? *American Journal of Sociology, 80*(3), 643–664.

Rosenthal, B. (1997). Political implications of the early twentieth-century occult revival. In B. Rosenthal (Ed.), *The occult in Russian and Soviet culture* (pp. 379–418). Ithaca: Cornell University Press.
Rovner, A. (2002). *Gurdzhiyev i Uspensky*. Moscow: Sofiya.
Ryan, W. (1999). *The bathhouse at midnight: Magic in Russia*. University Park: Pennsylvania State UP.
Seidel-Dreffke, B. (2004). *Die russische Literatur Ende des 19. und zu Beginn des 20. Jahrhunderts und die Theosophie E.P.Blavatskajas*. Frankfurt am Main: Haag+Herchen.
Sherkat, D., & Wilson, J. (1995). Preferences, constraints, and choices in religious markets: An examination of religious switching and apostasy. *Social Forces, 73*, 993–1026.
Shishkin, O. (2003). *Bitva za Gimalai: NKVD: magiya i shpionazh*. Moscow: Eksmo.
Shterin, M. (2000). New religious movements in Russia in the 1990s. In M. Kotiranta (Ed.), *Religious transition in Russia* (pp. 185–207). Helsinki: Kikimora.
Stark, R., & Bainbridge, W. (1985). *The future of religion*. Berkeley: University of California Press.
Stenger, H. (1989). Der "okkulte Alltag": Beschreibungen und wissenssoziologische Deutungen des "New Age". *Zeitschrift für Soziologie, 18*, 119–135.
Tchervonnaya, S. (1997). The revival of animistic religion in the Mari El Republic. In I. Borowik & G. Babinski (Eds.), *New religious phenomena in Central and Eastern Europe* (pp. 369–382). Kraków: Nomos.
Torin, A. (1997). *Die wahre Geschichte der Extrasensologie in Russland*. Unpublished manuscript.
Vinokurov, I. (2005). *Dukhi i mediumy*. Moscow: AST.
Vizgin, V. (2001). Nauka, religiya i ezotericheskaya traditsiya: ot moderna k postmodernu. In L. Fesenkova (Ed.), *Diskursy ezoteriki (filosofsky analiz)* (pp. 79–99). Moscow: URSS.
von Maydell, R. (2005). *Vor dem Thore: ein Vierteljahrhundert Anthroposophie in Russland*. Bochum: Projekt.
Wassiliew, L. (1965). *Experimentelle Untersuchungen zur Mentalsuggestion*. Munich: Francke.
Wassner, R. (1991). Neue Religiöse Bewegungen in Deutschland: Ein soziologischer Bericht (EZW-Informationen, 113).
Webb, J. (1980). *The harmonious circle: The lives and work of G.I. Gurdjieff, P. D. Ouspensky, and their followers*. London: Thames & Hudson.
Williams, R. (1980). *Russian art and American money, 1900–1940*. Cambridge: Cambridge University Press.
Wunder, E. (2005). *Religion in der postkonfessionellen Gesellschaft: ein Beitrag zur sozialwissenschaftlichen Theorieentwicklung in der Religionsgeographie*. Stuttgart: Steiner.
Yamamoto, M. (2003). Report in 2002 for Russian research. *Journal of International Society of Life Information Science, 21*(1), 188–192.
Yeltsin, M. (2001). *Vizit nesushchego drakona*. Moscow: Institut obshchegumanitarnyh issledovanii.
York, M. (2001). New age commodification and appropriation of spirituality. *Journal of Contemporary Religion, 16*(3), 361–372.

Chapter 79
The Cow and the Cross: South Asians in Russia and the Russian Christian Orthodox Church

Igor Kotin

79.1 Introduction

Russia remains the most important successor state of the Soviet Union, while the Soviet Union had as its predecessors the Russian Empire, the Moscow Rus and the Kyev Rus. Indians constitute a small proportion in the population of present day Russia, but they can claim a history of at least four centuries of having a South Asian presence there. Indian traders used to visit Kyev even before the Mongol invasion of Rus. In the twelfth century the Mongol invasion interrupted Indian-Russian trade contacts and the Golden Horde, one of the successor states of the Mongol Empire in the Russian steppes and the Volga region; it blocked direct movement of Indian merchants to Rus. Instead the Indians went to Sarai, the Tatar capital city on the Volga River rather than Kyev and Vladimir, but not further westwards. Russian merchants also frequently visited the capital of the Golden Horde of Tatars, who blocked them from coming further east than the Horde capital. The collapse of the Golden Horde resulted in the emergence of the Tatar Khanate in Astrakhan, not far from the ruins of Sarai, where both Indians and Russians had their trading sites.

The invasion in the sixteenth century of the Astrakhan Khanate by Moscow Tsar Ivan the Terrible resulted in the incorporation into the Russian state the existing trading colony of Indians in Astrakhan. (former Haji Tarkhan). By then, the Indian enterprise went as far as Isfaghan in Persia and Kizlyar in the North Caucasus. The Archives in Astrakhan, Moscow and St. Petersburg contain significant information about the activity of the Indian merchants and artisans in this Russian city at the mouth of the Volga river and near the Caspian Sea (See: Palmov 1934; Goldberg 1949, 1958; Gopal 1989). From the archives we know that in 1624 a special trading court (Gostiniy Dvor) for Indian merchants was erected in Astrakhan along with

I. Kotin (✉)
Museum of Anthropology and Ethnography, Russian Academy of Sciences,
St. Petersburg, Russia
e-mail: igorkotin@mail.ru

S.D. Brunn (ed.), *The Changing World Religion Map*,
DOI 10.1007/978-94-017-9376-6_79

separate courts for Armenian and Persian merchants. More than 100 Indian merchants with their servants lived there. They traded jewelry and medicines in Astrakhan. Since 1632 clerks in Astrakhan and in the Posolskiy Prikaz (Foreign Ministry) in Moscow kept a special file on India and Indians. In 1645 an Indian merchant dared to go as far as Kazan and Moscow and while there sold his goods with much success. As a result of his success, 25 more Indian traders came to Astrakhan from Persia. In 1650 Indian merchants sold their goods in Yaroslavl on the Volga River, not far from Moscow. They often traveled from Astrakhan to Moscow, to Kazan and to Nizhniy Novgorod and to the site of the famous Makarievskaya Fair nearby. By that time Indians mostly sold cloths and precious stones and bought furs. They also established themselves as important money-lenders in Astrakhan. The Russian Tsar Alexei Mikhaylovich also invited Indian artisans to Moscow to introduce the textile industry there.

Those Indians who came were mostly Hindus, but also the Sikhs (Vorobiov-Desyatovski 1955: 48) and Muslims. They were generally traders and in various services, including missionary activities. Russian documents from the seventeenth century mentioned that the colony of Indians there included not only merchants, but several Hindu Sadhus, a Brahmin priest, a Brahmin cook, beggars and servants. English traveler George Forster mentions Indians who went to Astrakhan especially to teach their religion, probably Hinduism, Jainism or Sikhism. Reports by Russian merchants and comments by Russian officers of an Armenian merchant referred to these Indians as cow-worshippers. These remarks suggest that among Indians there were several strict Hindus (Sanatana Hindus) who worshipped cows and avoided beef-eating. An eighteenth century account of Hindu life in Astrakhan by Peter Simon Pallas tells of mutton-eating by some Hindus; he also stresses that beef-eating was strongly prohibited (Nikolskaya 2003: 135).

Among Hindu *Murtis* (idols) mentioned in the Indian colony in Astrakhan are those of Rama, Sita and the *Thakur* (probably Krishna, but may be also saints of the Krishna cult: Chaitanya or Nityanand). Reports of Indian Muslims and Sikhs are also present in the archives. The Russian archives contain information that Indians had their own living quarters and a temple along with a trading center in Astrakhan.

The international trade of the local Indian colony was interrupted with the worsening of political relations between Moghul India and the Persian Empire. Indian traders also survived political turmoil of the establishment of new political power in what is now known as the Southern Russia. They kept control of the East-west trade, but their links with India itself became irregular. This resulted in the concentration of Indians on the trade route with Persia via Kizlyar and Derbent and also the growing of importance of money-lending as their main occupation. They also tried to enter the trade of horses from the Volga region to Moscow. The lack of constant contacts with India also resulted in marriages of Astrakhan Indians with Tatar women as well.

The growing isolation of the Astrakhan Hindus from their home country and intermarriages with Tatars and Russians led them to convert to Christianity and Islam. By the time then this isolation became nearly complete the Hindu community in Astrakhan had at least 200 years of existence.

The first decades of Russian authority over Astrakhan seemed to be a time of relative religious freedom in the city that had thriving Muslim, Armenian and Hindu communities. Later, however, Russian merchants tended to show less tolerance to believers of other religions, particularly or Hindus whom they considered as pagans. Religious intolerance combined with the competition in trade resulted in a series of reports by Russian merchants to authorities condemning pagans and seeking their removal from the city. It is interesting that Russian merchants did not demand forced conversion of these Hindus, but rather asked to expel them from Astrakhan. The reason was clear. Hindus who turned to Christianity obtained the same rights as Russian traders; their loyalty to the Tsar and religious belonging were considered main features of Russian identity then. At the same time, as foreign born persons these Indian merchants did not bear the obligations and duties of Russian traders. The petitions initiated by the Russian merchants, however, did not receive positive reactions of the authorities and did not result in any formal ban on Hindu religion from the side of Russian Tsar or any religious figure. Tsar Alexey Mikhailovich allowed Indian traders to follow their own religious rites, including that of cremating dead bodies; also the Hindu temple remained in the city of Astrakhan. This tolerance to Hindus was unprecedented in the seventeenth century Russia. The Moscow state was the country where the Orthodox Church, which since the sixteenth century had been independent from the Constantinople Patriarchate, enjoyed the status of the state religion. The Orthodox Church was in charge of education and culture (Kotin and Krindatch 2005: 147). However, under Tsar Alexey Mikhailovich and the then Patriarch Nikon state-sanctioned attempts to modify the Christian liturgy led to Church schism and generated social unrest. Struggle with the dissenters (the so called Old Believers) led to religious tension and to a rise of intolerance towards all non-Orthodox Christians as well. Yet Hindus were given a freedom of religious worship. This fact can be partly explained by the small size of their community and their importance as suppliers of precious goods to the Tsar.

Under Alexey Mikhailovich the Hindu community in Astrakhan lived as an isolated society with its own sacred objects and a sacred place. Hindu Brahmins used poured water from the sacred Indian river of the Ganges and the Volga River in Russia. They also started to consider the Volga as their local Ganges in ceremonies. They freely prayed to their gods and made cremation of their dead despite gossip and hostility from Muslims and Christians who considered them as pagans. Several new petitions had been sent to Moscow aiming at removing the Murtis from the Indian trading center. Yet, the local administrative head (*voevoda*) was given instructions from Moscow to allow Hindus to follow their rites of passage.

The establishment of the Russian Empire by Peter the Great slightly changed the place of Russian Orthodoxy which now became more strictly controlled by the state. The position of the Patriarch had been cancelled and the Synod or the Ministry in Charge of the Russian Christian Church was established in 1721. The Church of the majority fell under the direct control of the state. Other religious minorities controlled by the state had more freedom in their religious life. In 1722 new Tsar Peter the Great issued a special decree on tolerance towards Indians and their beliefs. The decree was included in to the Collection of the Laws of the Russian

Empire. After the death of Peter the Great and a short reign of Catherine the First, Peter's niece Anna Ioannovna issued a decree confirming religious freedom of Indians. The decree was also confirmed by Peter's daughter Elisabeth who ascended to the throne soon after Anna's death in the mid-eighteenth century. It is interesting that these laws were issued decades before religious rights were given by Catherine the Great to the much more numerous Muslim subjects of the Empire. In 1773, Czarina Catherine the Great also issued the law on Religious tolerance which gave formal rights to religious minorities, as they felt less protected outside their core areas. For Hindus, Astrakhan remained their fortress.

In the early eighteenth century Indian merchants lived not only in Astrakhan, but also in Moscow. It is reported in Russian chronicles that in 1723 a group of Hindu traders left Moscow in protest of the local administration's objection to the *sati* (sacrifice of the widow) of a rich Hindu merchant. This record also indicates a presence of Hindu traders in Moscow in the eighteenth century. Thus, we observe that in Moscow locals were less aware of Hindu traditions than in Astrakhan. There was also the custom of *sati* which often alienated locals from Hindus, as this custom did not appear to be very dignified.

79.2 Conversion

With the expansion of their trade to central Russia and to a capital city of St. Petersburg many Hindu traders converted to the Orthodox Christianity. As usual, they received Russian Christian names and surnames. Since the 1740s we have several records of "Russian Indians" with surnames like Ivanov and Feodorov. Many Russian families in Astrakhan also have an oral history of their Hindu forefathers. The first conversions, however, took place a century ago. In the seventeenth century there were recorded cases of Hindus converting to the Christian Orthodox religion. One of new converts named Jukka used the conversion as the way of escape from other Hindus. From other records we know that this Jukka reported to the Tsar that these Hindus did not pay taxes in full (Goldberg 1958: 248–249). Another Indian convert is mentioned by his new Christian name Bogdan (a name means by the God); a precise translation of Sanskrit name is *Devadatta*. Other converts had typical Russian names and surnames without any indication of their foreign origin. The reason for this naming is simple. Converts were given the name of the saint of the day of their conversion, while their godfather gave them their surname. Thus Indian converts often had names like Ivan Petrov or Petr Ivanov unlike a son of a Muslim convert from Sikhism or Hinduism with a name of Suleiman Amardasov, that is, Suleiman, a son of Amardas. We can only guess that some Russians from Astrakhan with typical Russian surnames and Oriental appearances may have Indian ancestors.

From the Astrakhan records we can assume that the main reason any conversion was to gain some economic privileges and freedom of travel in the Moscow state (Russian Empire). These were not mass conversions, but individual ones. One of the archives in St. Petersburg records the conversion of a Hindu name Netoram into the Russian Orthodox Christianity. Netoram was given an exceptional welcome by the Russian Church. After a temporary stay at the highest rank in the Alexandro-Nevskiy Monastery in St. Petersburg as a probationer, he was means special approval of the *Svjashennij Synod* (the Ministry in charge of the state control over the Orthodox Church) and became a Christian. The state even gave Netoram financial help after his conversion into Christianity. This case however was unique; no other Indian was recorded as a probationer in that monastery.

With declines in the trade with India and the conversion of Astrakhan Indians into Christianity and Islam, the Indian community in Astrakhan lost its significance, By 1799 the famous Indian trading lines had drastically shrunk and the governor of Astrakhan ordered their demolition. This decline also had a productive side. By the early nineteenth century Russians had re- started attempts to establish their own companies for trade with India. The Russian Indians had been invited as advisers to Orenburg in the Urals region, where such agreements were made. But as a separate ethnic community or trading colony, Indians by the early nineteenth century had disappeared in Russia. They had been assimilated into the Russian and Tatar populations. Many Astrakhan inhabitants today have a family story of their Indian ancestors being baptized to Christianity. However, in the nineteenth century few locals there could claim Indian heritage. One of few famous cases of a conversion was that of a Count of Vazipur whose family for a half a century was listed among the Russian nobility.

79.3 The Twentieth Century

The Indian diamond trade was known then in Moscow and St. Petersburg. Later Moscow and St. Petersburg had several visiting Indian adventurers, including the Sikh hero Dalip Singh (1838–1893). In 1887 Dalip Sing, disguised as an Irishman (sic!), visited Moscow where he met a popular Russian publisher and politician Katkov. Indians from Central Asia, where their numbers by the early twentieth century were estimated to be 6,000–8,000 (Dmitriev 1972; Levi 2002). They visited Moscow and participated at the famous Nizhniy Novgorod Makarievskiy Fair (market) there. Yet, it is impossible to speak of a continuous Indian presence from the previous time. Astrakhan Indians dispersed to Kazan, Moscow, St. Petersburg and Astrakhan and their descendents had assimilated, although it seems possible to assume that some families of "Russian Indians" still keep memories of their South Asian ancestors.

The Soviet era witnessed the emergence of an Indian Communist community in Moscow and Leningrad in 1920s–1930s. From the mid-1950s onwards, significant numbers of Indian students came to major educational institutions in Moscow, Leningrad, Sverdlovsk, Kursk and other cities. Yet, their presence was not expected to be long. A few managed to remain in Russia after completing their education. They did not form a diaspora; the temporary presence of Indians in major Russian cities was not questioned because of strict immigration and residence rules. Their religious life remained a domestic affair and they not attract public attention.

79.4 Post-Soviet Russia

The Indian-Russian situation changed with the collapse of the Soviet Union. Although economic hardships for a while made post-Soviet space unattractive for foreign students depending on their stipend, those better off, and adventurers, found Russian conditions suitable, even though Russian immigration and residence rules had been neglected for a decade by Russian authorities themselves. As the result, a new wave of Indians came to Russia that consisted mostly of students, but of them only half (mostly medical students) made study abroad their main aim. The rest found it a good opportunity to combine a study abroad with starting a small business, often in retailing which they continued after graduating or left the universities as failed students. As the result, Indians in Russian Federation are now nearly as numerous as students. Some are successful and rich; others are just petty traders. Rich Indian business owners are involved in tea and garments trade, construction industry, and most recently they invested in St. Petersburg and Moscow breweries. A new brand of Indian beer "Kobra" ("Nag") of Sun Breweries is winning place in the prosperous beer market of Russia. Another area of success is the diamond industry. Moscow is experiencing the establishment of Indian firms that have a wide clientele in India and western Europe. New large projects of Indian business leaders in the Russian Federation are involved in the Indian investment in Russian oil fields, particularly in Sakhalin Island in the Russian Far East and in steel production where Mr. Mitthal, an East African Indian, is viewed as a competitor or a partner for a steel empire based in London. Cheaper jewelry, garments, tea and spices are also sold by many Indians in Russia.

The 2002 census of Russian population listed 2,000 Indians in Russia. The Report of the High Level Committee on Indian diaspora provides a figure of 16,000 Non-Resident Indians in the country. Some experts estimate a figure of 40,000, a figure that seems reasonable. At least, 10,000 Indians study in Moscow. A similar number are residents in St. Petersburg, Nizhniy Novgorod, Kazan, Kursk and other Russian cities. There are at least 10,000 illegal migrants from South Asia also living in Moscow. Ukraine has at least 10,000 Indian students and also a small number of Indian professionals there. The importance of Russia and Ukraine, however, is not in the numbers of South Asians currently present, but in the importance of these countries as transit regions on the main inland routes of those migrating from South Asia to Western Europe. The economic potential of these countries for the Indian business leaders is also high, and the political turmoil and administrative chaos, which often complicates the business atmosphere, give South Asians a competing edge in their would-be struggle for the markets there with western investors. And last, but not least, is the importance of Russia and Ukraine as successor states of the Soviet Union, an old friend of India. Russian society was in love with Indian movies and music for a long time, it is more likely to accept Indians as a part of its emergent multicultural mosaic replacing Soviet uniformity. Porous borders of the Russian Federation with both Asian countries of the former Soviet Union and with new member states of the European Community make the Russian Federation and Ukraine ideal stepping stones for Indian transit migrants moving westwards.

Thus, Russian Federation and Ukraine are both emerging as an important transit territories as well as the final destination of Indian immigrants.

The first theme of present day Russian life, which often retains an Indian connection, relates to transit migration. The second important theme is the rise of racism in Russia. Many Russians treat Indians like a racial group, the Asians or even "Blacks." The third important theme relates to Indian cultural institutions, and particularly, of a Hindu temple which was initially planned to be constructed on the Hodinskoye Pole in Moscow. The announcement of construction plans faced strong opposition from the Russian Orthodox Church. The process of globalization involved Russia, especially after the breakup of the Soviet Union and the emergence of the Russian Federation as the stepping point of transit migration from China and South Asia to Western Europe. The result of the South Asian penetration in Russia is that no less than 40,000 Indians, Pakistanis and Bangladeshis have settled in Russia, especially in the main cities of Moscow and St. Petersburg. While Pakistanis and Bangladeshis can use local mosques, migrants from India belonging to Hindu, Jain, Parsee and Sikh communities until recently lacked any possibility to formally express their faith and perform their public religious rituals in Russia. While Jain and Parsee communities are quite small in Russia, Hindu and Sikh groups are larger.

79.5 Proposed Hindu Temple

The announcement of plans for the construction of a Hindu temple in Moscow gave rise to heated discussion in Russian society. On October 17, 2003 Mr. Sanjit Kumar Jha, a leader of the Association of Indians in Russia, announced the foundation of the Center of Vedic Culture (Fig. 79.1).

The Foundation was supposed to be a sort of continuation of the success which provided the Indian culture with the establishment of the Swami Narayan Temple in Neasden in London. The driving force behind the Moscow project, however, was different. While rich members of the Gujarati community in London and Swaminarayan Sect leaders both in India and in the diaspora were among the main sponsors and organizers of the temple construction in London, in Moscow the project was promoted by local Indian business leaders and Hare Krishna activists. Famous Alfred Ford of the Ford automobile empire also promised to support the project. Thus, although it was welcomed by Indians as a part of their culture, the project was viewed by the Russian Orthodox clerics as another attack on Russia from the West. Pressure was applied on the Moscow city mayor, Mr. Luzhkov, not to allow the planned construction. Also an old Hare Krishna temple in Moscow was under threat of demolition due to the construction plans for the planned residential complex on the site. By accepting the Hare Krishna movement in Russia as the representative of the Vedic culture the Indians in Moscow found themselves in an extremely hostile environment in present-day Russian capital, once to be known as the "city of forties of forties of churches."

Fig. 79.1 The initial plan model of the Vedic Center (Photo by Igor Kotin, http://www.veda.ru/gallery/market/)

The rise of the Hare Krishna movement took place in the last Soviet years when the impact of the Soviet state was weakened and while the attractiveness of both Soviet atheist and the Orthodox religions faded due to their connectedness with the state. The decade after 1991 witnessed a period of religious revival due to the collapse of the Soviet state and ideology. The August 1991 "revolution" led to the collapse of the stagnating community ideology and to the rise of freedom, including personal freedom and religious freedom (Kotin and Krindatch 2005: 150). The multinational and multi-confessional composition of Russia's population does not support the choice of Orthodox Christianity as the only ideology of the Russian state. Thus, the Russian Federation was proclaimed a secular state. Yet the state supports four main churches as traditional or historical churches, viz., Russian Orthodox Christianity, Islam, Judaism and Buddhism as opposed to other churches, particularly New Protestant Churches and New Religious Movements of Neo-Hinduism type. The latter saw the number of their organizations increase from 9 in 1991 to 96 in 2003. It is against these "new cults" that the 1997 Law on the Freedom of Conscience and Religious Organizations was passed in addition to the rather neutral Religious Freedom Act of 1991.

The "Hare Krishna Temple issue" highlighted the importance of Indian business elite in Russia who are no more interested in being an "invisible minority." The allocation of a territory to the future temple was lobbied by the Asian Business Club and the Association of Indians in Russian Federation. Concern was expressed by the demolition of the previous temple, delay in the construction of new one, and public dissatisfaction with the proposed project. Mr. Jha said that "this is our sacred thing. The demolition of the temple will be a hard blow for the Indian community of Moscow and a direct insult of the feelings of Hindus all over the world, the insult, that can have far-reaching consequences." Mr. Sanjiv Kumar Jha was among the organizers of several high –profile visits to Moscow including that of then-Speaker of the Indian Parliament. Mr. Joshi offered a prayer in the old Krishna temple on Begovaya street in Moscow and the then- Prime-Minister of India, Mr. Atal Bihari Vajpeyee, promised to inaugurate the new mandir and vedic center.

The defeat of the Bharatiya Janata Party and its National Democratic Alliance in 2004 in the 13th general elections in India and the return to power of the Indian National Congress heading the secularist United Progressive Alliance downplayed the importance of a new Hindu temple in Moscow. In Russia itself, however, the importance of the issue increased with the involvement of high level clerics in the discussion of the nature of the Hare Krishna Movement.

In December 2005 senior Russian Orthodox authority Archbishop Nickon of Bashkortastan sent an open letter to the Moscow mayor Yuri Luzhkov. In his letter Bashkortostan archbishop made comments on the cult of Krishna in general and on the Hare Krishna movement in the Russian Federation in particular. The letter became known to the general public in days of the state visit by Indian Prime Minister Mr. Manmohan Singh. It triggered the concerns of the Indian media in India and the Indian community in Russia. The letter of the Archbishop referred to Krishna as the evil "deity." Not surprisingly, the President of the Association of Indians in the Russian Federation, Mr. Sanjeev Kumar Jha sent a letter of protest to the Head of Russian Orthodox Church Patriarch Alexiy II. The letter questioned the tolerance of the Russian Orthodox Church in general and its tolerance towards devotees of Hinduism in particular. A few days earlier the Prosecutors Body of the Russian Federation made a decision and declared that the Moscow government's application for disallowing a grant for land for the Hindu temple was invalid (Figs. 79.2 and 79.3). The Moscow government agreed with the decision.

Fig. 79.2 New model of the Vedic Center (Photo by Igor Kotin, http://www.veda.ru/gallery/market/)

Fig. 79.3 Vedic Center at night (Photo by Igor Kotin, http://www.veda.ru/gallery/market/)

79.6 The Current Status

The failure to construct the Hindu temple coincided with the electoral defeat of the Vajpayee government. However, yet another important supporter of the project, the Congress Chief Minister of Delhi, Sheila Dikshit, remains in power in the Indian capital. The incumbent Indian government remains interested in promoting India's culture in Russia though not stressing the Hindu religion. It is Mr. Sanjiv Kumar Jha who lobbied the foundation of Indian Cultural centers in Saratov in middle Volga, in Nizhniy Novgorod and in Yaroslavl. The IndoRussia business network, which is closely tied to the Association of Indians in the Russian Federation, supports the collection of money for the Shiva temple initiated by the Indian cultural group "Rishikesh" as well as the establishment of Sikh temple Gurdwara Guru Nanak Darbar, initiated by the Sikh Cultural Center in Moscow.

It should be added that the establishment of a Sikh temple and cultural center in Moscow did not cause any trouble. The Inter-Confessional Council of Russia on November 28, 2003 approved this step by providing a normal religious life to nearly 300 Sikhs in Moscow (Interfax Agency, Moscow, 28 November 2003). Although it was dubbed as a branch of "traditional Hinduism," Sikhism was considered a "traditional" religion and supported by clerics of the Russian Orthodox, Muslim, Buddhist and Judaist Churches. In this context, opposition to the construction of a Hare Krishna temple in Moscow appears not due to the Indian origin of the religion, but because of its western interpretation and the missionary activity of the Hare Krishna movement among persons of Russian Orthodox Christian culture.

Having considered more than three centuries of Hindu-Christian relations in Russia, we observe that the main attitude of the Russian Orthodox Church towards Hinduism was to allow the Hindus to live their way, but not to allow any propaganda of their religion in Russia. While cases of conversion of Hindus to Christianity can be explained by economical or matrimonial reasons, the conversion of Russians to Hinduism or a Hinduism-named new religion leads to a conflict between the Indians and the Russian Orthodox Christian Church. Recently the Vedic Center lobbyists obtained approval for its temple construction near the village of Vereskino, north of Moscow, thus accommodating the decision acceptable to the Moscow Patriarchate and for the most vigilant Russian Orthodox activists. As for those Indians who came from India and from families with Christian heritage, their numbers in Russia are miniscule. They belong either to the Roman Catholic or to the Anglican or Methodist Churches. These few Christian Indians remain the most invisible even among rather invisible Indian community of Russia.

Acknowledgements This research was funded by the Special Projects Office, Special and Extension Programs of the Central European University Foundation (CEUBPF). The thesis explained represents the ideas of the author, and not necessarily the opinion of CEUBPF.

References

Dmitriev, G. L. (1972). Iz Istorii indiiskih kolonii v Srednei Azii (2-aia polovina 19-nachalo 20 v.) [On history of Indian colonies in Central Asia. 2nd hald 19th c.-early 20th century]. In D. A. Olderogge (Ed.), *Strani I Narodi Vostoka*. India. Strana I Narod. 2 (pp. 234–247). Moscow: Nauka, Vostochnaya Literatura (in Russian).

Goldberg, N. M. (1949). *Russko-indiyskie otnosheniya v 17m veke* [Russian-Indian relations in the 17th century] (pp. 127–148). In *Uchenie Zapiski Tikhookeanskogo Instituta*. Moscow: Vostochnaya Literatura.

Goldberg, N. M. (Ed.). (1958). *Russko-indiyskie otnosheniya v 17m veke. Sbornik dokumentov* [Russian-Indian relations in the 17th century. Published Documents]. Moscow. Vostochnaya Literatura (in Russian).

Gopal, S. (1989). *Indians in Russia in the 17 & 18 centuries*. New York: South Asia Book.

Kotin, I. (with A. Krindatch). (2005). Religious revival in a multicultural landscape. In H. Knippenberg (Ed.), The changing religious landscape of Europe. Amsterdam: Het Spinhus.

Levi, S. C. (2002). Hindus beyond the Hindu Kush: Indians in the Central Asian lands. *Journal of the Royal Asiatic Society, 12*(Part 3), 277–288.

Nikolskaya, K. P. (2003). Simon Pallas on the Astrakhan Indians. In A. Monanty (Ed.), *India – Russia: Dialogue of civilizations* (pp. 135–153). Moscow: Nauka (in Russian).

Palmov, N. N. (1934). Astrakhan Archives (pp. 162–182). In *Zapiski Instituta Vostokovedeniya AN USSR, 2*(44), 161–182 (in Russian).

Vedic Center. (2011). http://krishna.ru/index.php?option=com_content&task=view&id=126&Itemid=341; illustrations www.veda.ru/gallery/maket/ and www.veda.ru/gallery/appearance/. Accessed 15 Dec 2011.

Vorobyev-Desyatovski, V. S. (1955). Concerning early contacts between India and Russia. *Journal of Indo-Soviet Cultural Society, 2*, 48.

Chapter 80
Islam and Buddhism in the Changing Post-Soviet Religious Landscape

Edward C. Holland and Meagan Todd

80.1 Introduction

Since the collapse of the Soviet Union and the state policy of scientific atheism, public religiosity in Russia has blossomed. Accordingly, this volume features a wide range of scholarship on the changing geographies of religion in Russia, including the dynamics of paganism within the Krina community in St. Petersburg (Aitamurto), the suburbanization of Russian Orthodoxy in Moscow (Sidorov), and the controversial construction of mosques and Buddhist temples in St. Petersburg and Moscow (Kotin). This chapter supplements these discussions through an examination of the changing landscapes of religion in two of Russia's southern regions, Astrakhan Oblast and the Republic of Kalmykia. Its emphasis is on Islam and Buddhism, two of the four faiths legislatively recognized as "traditional" by the 1997 Law on Freedom of Conscience and Religious Associations (along with Orthodoxy and Judaism).[1] Each religion has experienced a substantial revival in the post-Soviet period, both institutionally and among individual practitioners.

The chapter opens with a brief review of how changes in the religious landscape reflect the continued relevance of religion as publically practiced. We then turn to legislative approaches to Islam and Buddhism in contemporary, secular Russia. In

---

Both authors contributed equally to the writing of this chapter.

[1] Though these four religions often referred to as traditional, this term does not appear in the 1997 legislation. We place the term in quotations marks throughout the chapter to emphasize the distinction between the label "traditional" and the content of the law (see also Fagan 2013).

E.C. Holland (✉)
Havighurst Center for Russian and Post-Soviet Studies, Miami University, Oxford, OH 45056, USA
e-mail: hollanec@miamioh.edu

M. Todd
Department of Geography, University of Colorado, Boulder, CO 80309, USA
e-mail: meagan.todd@colorado.edu

S.D. Brunn (ed.), *The Changing World Religion Map*,
DOI 10.1007/978-94-017-9376-6_80

order to examine the effects of the legal institutionalization of Islam and Buddhism, we next consider how religion is changing the urban landscape in two cities in Russia's south. In Astrakhan, a multiethnic and multi-confessional city located on the banks of the Volga River near the Caspian Sea, the restoration of historical mosques is celebrated as a sign of friendship between state and religious authorities. However, opposition to new mosque construction challenges Astrakhan's reputation as a city of tolerance. In Elista, Kalmykia's capital, the construction of religious buildings has substantially altered the city's post-communist appearance. The building of temples (referred to locally as *khuruls*) has reaffirmed the role of Buddhism in the national identity of the Kalmyks, while communist symbols have been displaced from positions of prominence by other religious markers. Together, these two cases illustrate that the negotiation of spatial and social identities is an on-going project in post-Soviet Russia for a wide range of national and confessional groups.

80.2 Changes in the Landscape: The Public Face of Religion in Contemporary Russia

Secularization, understood here not as the relegation of religion to the private sphere but as the practice of shaping religion as a category by modern state power, is not a universal concept (Asad 2003). This chapter examines how secularism is historically and spatially situated in Russia, drawing in particular on the work of Jose Casanova (1994). In challenging thinking on secularization Casanova identifies a "deprivatization" of religion, whereby "religious traditions throughout the world are refusing to accept the marginal and privatized role which theories of modernity as well as theories of secularization have reserved for them" (Casanova 1994: 5). More broadly, this debate hinges on the consequences of differentiation as a result of modernization (Wilford 2010). The deprivatization of religion currently underway is occurring at the intermediate scale; as religions no longer maintain their position as "grand legitimators" due to the separation of church and state, they "can become movements and pressure groups that vie with rivals in the public sphere" (Gorski and Altınordu 2008: 58). For individuals, the erosion of the central place of religious actors in the political, economic, and social spheres results in greater freedom in terms of religious choice; differentiation, in turn, "serve[s] as the social context within which religious organizations in civil society must adapt" (Wilford 2010: 336).

This deprivatization of religion is apparent in changes to the religious landscape in contemporary Russia. Most religious organizations in the country have unequivocally accepted the principle of differentiation in part because of the climate of secular tolerance that now exists, a divergence from the Soviet period. The number of registered religious organizations has increased substantially in the past two decades, as have outward, public markers of faith through the construction of religious buildings and symbols. This growth in registration and construction has notable political consequences, and is itself an outcome of Russia's particular variety of secularism. For example, Sidorov's (2000) work on the Cathedral of Christ the

Savior as a religious *qua* national monument reflects the mismatch between the constructed object, which has been used for the purpose of political legitimation, and the local and national scales at which such legitimation occurs. The nation is itself a political, social, and religious construct, one that has required a particular form of (re-)legitimation in the post-Soviet period. Sidorov argues that the reconstruction of the Cathedral has privileged the Russian Orthodox Church in comparison to other confessions, though the rebuilding has been less relevant to regions beyond Moscow; "it is only a slight exaggeration to conclude that, in many respects, the scale of the national state has become local" (Sidorov 2000: 564).

In this chapter, we are interested in the increasingly public face of religion as practiced by Russia's ethnic minorities in geographic contexts far removed from the debate over Moscow's "new" Orthodox Church. We explore the extent to which the principals of deprivatization and differentiation vis-à-vis the prominence of the Orthodox Church have disciplined post-Soviet religious identities, using landscape as an analytical lens. Our approach in this chapter is informed by past and contemporary work on religion in the discipline of geography.

Following Carl Sauer, cultural geographers have long considered the effect of religion on the landscape. Isaac (1961–62: 12, quoted in Kong 2004), provided an early definition of the geography of religion. He suggested that it is "the study of the part played by the religious motive in man's [sic] transformation of the landscape." The new cultural geography turned away from physical artifacts as markers of the landscape and indicators of cultural distinctiveness, and instead focused on the semantic and semiotic nature of landscape (Cosgrove and Jackson 1987). The production of landscape by religious actors and institutions underscores the contentious nature of space as conditioned by the politics of the sacred. Lily Kong (2001) has suggested that geographers consider both the politics and poetics of place; for many, religion is the "search for the immanent and transcendent" (Kong 2001: 218; see also Kong 2004) Yet, this search is regulated heavily by state intervention, as seen in the relegation of religious practice to the private sphere in the Soviet period and the closure or repurposing of churches, mosques, and temples throughout the USSR. The cultural geography of religion and the poetics of the religious motive in the transformation of the landscape are, in other words, highly political.

As such, we are influenced not only by work on the geographies of religion, but also by scholarship from religious studies influenced by the "spatial turn" in social theory. Kim Knott (2009), a religious studies scholar, explores how spatial methodologies can further the study of religions through a reconceptualization of locality. She argues that the recent debates over the Western construction of world religions have treated localities as containers rather than as dynamic scales of study, and emphasizes the importance of localized approaches in the academic study of religion. Influenced by geographers and spatial theorists such as Massey, Harvey, and Lefebvre, she conceptualizes locality not as a context in which religious activities take place, but as influenced by multiple scales and processes. In other words, localities are a lens through which to examine the interconnections and dynamics of religious activities with large-scale social, political, and economic processes, including nation-building, the spread of democracy, and secularization in the social and political spheres. Local level research provides insight into how these processes are both the means of and modes for the production of religious spaces and identities. Both of the case studies in this chapter are based on fieldwork in the respective regions of Astrakhan and Kalmykia, and consider religious landscapes not as markers or texts, but as products and producers of political and religious debates occurring at multiple scales.

80.3 Religious Policy in the Soviet Union and Russian Federation

Before turning to the discussion of our case studies, we summarize religious policy as enacted in the Russian Federation. The Russian Federation was the primary successor state to the Soviet Union; it inherited many of the legal and political structures of the communist state, including the 1990 Law on Freedom of Religions, which guaranteed freedom of religious choice, legal equality for all religions, and the separation of church and state (Shterin and Richardson 1998). The 1993 Russian constitution reaffirmed the provisions of the Soviet law, guaranteeing freedom of conscience and freedom of worship (Article 28), proscribing the restriction of rights on religious grounds (Article 19.2), and disavowing the endorsement of any one religion as the official religion of the state (Article 14.1).

Primus inter pares—at least in terms of the historical "hierarchy of religion"—is the Russian Orthodox Church (ROC). Prior work on religion in the Russian Federation has focused on the ROC, with the Moscow Patriarchate as the key political actor (Knox 2004; Daniel 2006; Papkova 2011a). This is unsurprising, not only given the historical importance of Orthodoxy in Russian history, but also the high levels of self-identification by ethnic Russians with Orthodoxy and the close relationship that the Patriarchate has attempted to cultivate with the post-communist Russian state; on the last point, as Papkova (2011a: 3) writes: "The church's outwardly cozy relationship with the state and consequent presumed political influence have become stable assumptions in the discourse on Russian politics, particularly since the advent of the Putin era."

One of the important outcomes of this special relationship was the 1997 Law on Freedom of Conscience and Religious Associations.[2] The product of close collaboration between the Church and the federal government (particularly with the Duma, the lower house of Russia's parliament), the 1997 law endorsed freedom of religion and establishment as guaranteed in the constitution, while also avowing the favored position of Russia's historical religions—Orthodoxy, but also Islam, Buddhism and Judaism—by differentiating religious organizations from religious groups. The former were accorded the right to full legal status, with the stipulation that they have confirmation from local authorities of their existence in a given territory for 15 years.

[2]An English translation of the 1997 law is available at: www2.stetson.edu/~psteeves/relnews/freedomofconscienceeng.html

The rights of religious groups were more circumscribed, as they functioned without full legal recognition by the state, itself impossible to obtain without the 15-year presence (Chapter II of the 1997 Law). The 1997 law was the culmination of lobbying by the ROC and other leaders of "traditional" religions against the perceived threat to Russian society from new religious movements (NRMs) and other non-"traditional" faiths, including Mormonism, Jehovah's Witnesses, Scientology, and Aum Shinrikyo. A number of authors have suggested that the Russian Orthodox Church was motivated by the perceived threat to its standing in Russian society that could result from widespread conversion; "as early as 1993 the ROC was actively supporting a campaign for radical restrictions on the activities of foreign missionaries, seen by the ROC as "soulhunters" trespassing on its canonical territory" (Verkhovsky 2002: 334).

The law resolved some of the legal questions that had arisen in the previous decade, specifically by creating a national law that brought coherence to the legislative hodgepodge that had developed on the regional level in response to the diversification of religion (Daniel and Marsh 2007). More detailed analysis has questioned the effectiveness with which the Moscow Patriarchate has influenced Russia's political course in the past two decades (see Papkova 2011a). Although Vladimir Putin's first stint as president was characterized by a political course distinct from the preferences of the Russian Orthodox Church on a range of issues, including religious education and the presence of clergy in the armed forces, Dmitry Medvedev's presidency was marked by increased collaboration between the ROC and the Russian government, such as the decision by the ROC to allow its clergy to enter politics, educational reforms requiring students to take the course "Fundamentals of Religious Culture and Secular Ethics," and the transferal of pre-Revolutionary confiscated church property to the ROC (Anderson 2007; Papkova 2011b).

Of secondary importance in the literature on religion in post-Soviet Russia are the minority religions of Islam, Buddhism and Judaism. When discussed, it is frequently in reference to the 1997 Law on Freedom of Conscience and Religious Associations; to reiterate, for adherents to some minority religions in the Russian Federation, the 1997 legislation accorded *de facto* legal recognition and status as "traditional" religions. In the preamble, Orthodoxy is cited for its special role in contributing to the Russian state system, while Islam, Buddhism, Judaism and Christianity are recognized as "constituting an integral part of the historical heritage of the peoples of Russia."[3]

Islam and Buddhism, along with Judaism, are viewed as religious traditions with a long historical presence on the present-day territory of the Russian Federation. Russia is home to a diversity of national groups, and these minorities frequently practice the "traditional" religions with which they are historically associated. In the North Caucasus, the peoples of Dagestan, in addition to the Chechens, Ingush, Karachays, Balkars and the Circassian populations have practiced Islam since at least the

[3] It should be noted that the preamble has no legal force backing it.

eighteenth century.[4] Pockets of Islam-practicing nations are also found along the banks of the Volga River and many Muslims living in Russia's capital of Moscow. Buddhism is "traditional"ly associated with three ethno-national minorities—the Buryats, the Kalmyks, and the Tuvans—though syncretism with shamanism is common, particularly among the latter group. During the Soviet period, Russia's Jewish population was granted its own autonomous oblast in the Russian Far East (the Jewish Autonomous Oblast), though it never served, as intended, as the "homeland" for the country's Jews; rather, during the late Soviet period and into the 1990s most Jews emigrated to Israel and the United States following the liberalization of travel restrictions.

In post-Soviet Russia, religion has served as a touchstone for national identity in a period of transition and upheaval. This is true for both minority religions, like Buddhism and Islam, as well as Russian Orthodoxy. Religion has consolidated its position as a mechanism for legitimating, and in certain instances mobilizing, political identities; as Agadjanian (2001: 478) writes: "religious identity [provides] cohesive symbolic networks both on the level of an "imagined" public religion … and on the level of private everyday religiosity." In an example of the former instance, Knox (2004) argues that Russian Orthodoxy has been co-opted by national chauvinists to legitimate an exclusivist national identity, one which connects to the discourses of xenophobia and racism present in the current Russian political discourse. The role of religion as a foundation for national identity—its public function as a basis for the imagined community (see also Brubaker 2012)—will be discussed below in detail with reference to Russia's Muslim and Buddhist populations. Our main interest going forward in this chapter concerns the place of religion in the public sphere; specifically, we consider how the contestations over public space and changes in the religious landscape reflect the public nature of religion in contemporary Russia.

80.4 Islam's Revival in Post-Soviet Russia: The Case of Astrakhan

This section explores the differing relationships of two mosque communities in Astrakhan, Russia to local, regional, and national governments and official Islamic structures. Based on interviews, participant observation, and archival research conducted in 2009, this local project revealed how mosque construction in Astrakhan is related to questions on the right to build mosques in public spaces in secular Russia. Notably, Astrakhan has a reputation within Russia for being tolerant despite its complex ethnic and religious makeup. Astrakhan is located on the Volga River, near the Caspian Sea. The city's proximity to Sarai, the capital of the Golden Horde and

[4] Islam has a long-standing historical presence in the North Caucasus. Muslim Arabs first reached the Dagestani city of Derbent—in the republic's south and sited on the Caspian Sea—in the 640s, not long after Muhammad's death (Reynolds 2005). The region's mountainous geography hindered the spread of Islam into the region's interior, however; it was not until late eighteenth and early nineteenth century that Islam gained a substantial and widespread following in the North Caucasus. Reynolds (2005: 36) suggests that the spread of Islam was associated with the establishment of legal norms: "concurrent with the spread of Islam was the diffusion of the notions of formalized law and the state."

the Silk Road, as well as the location on Russia's main waterway, made Astrakhan a center of trade and commerce for many peoples, leading to a history of ethnic and religious diversity that is still present today (Riasanovsky and Steinberg 2005).

Although Russians are a majority at 73 % of Astrakhan's urban population, members from over 130 ethnic groups live in the city. Besides being multiethnic, the city is also multi-confessional. Within this city of over 520,000 people, the public landscape of religion reflects its confessional diversity: there are 34 Orthodox churches, 34 mosques, a Buddhist temple, a synagogue, and a Catholic cathedral. Only St. Petersburg and Moscow contain such a diverse ensemble of religious buildings in Russia. The Islamic community within Astrakhan is ethnically diverse, consisting of members of Avar, Azeri, Tatar, Kazakh, and Russian ethnic groups. Because no census data on religion are collected, and the estimate of Muslims in Russia is unreliable and depends on political use, there is no agreed upon figure for the number of Muslims residing in Astrakhan (Heleniak 2006; Walker 2005).

Fieldwork in the region included visits to several mosques within Astrakhan, including Astrakhan's Central Mosque (Fig. 80.1). This mosque is located in a Tatar

Fig. 80.1 Astrakhan's central mosque (Photo by M. Todd, July 2009)

neighborhood in downtown Astrakhan and is also a historical monument. It was built in 1898 and funded by a Tatar mullah, Abd al-Vakhlab Aliev, and a Tatar merchant, Shakir Kazakov (Riasanovsky and Steinberg 2005). Each mosque during this time was also surrounded by a marketplace. Over a dozen mosques and markets were located in this neighborhood at the time of construction. Although the mosque was closed at the height of Stalin's Great Terror, it re-opened in 1950 and was the only functioning mosque in the city for the remainder of the Soviet period. Astrakhan's Central Mosque serves as the headquarters of the Astrakhan Regional Spiritual Board of Muslims (ARDUM), which is affiliated with the Central Spiritual Directorate of Muslims of Russia (TsDUM), one of three centralized Islamic boards in Russia.

Along with the two other Islamic boards in Russia (the Council of Muftis of Russia and the Coordinating Council of the Muslims of the Northern Caucasus), the chairman of TsDUM, Talgat Tajuddin, meets annually with the Russian president. These three groups often share opinions with the Kremlin on matters related to Islam and politics, such as denouncing the insurgency of Islamic fundamentalists in the second Chechen War, and are at the center of relations between Islam and the state in Russia. Importantly, the leadership of TsDUM believes that Astrakhan's Muslim community is an example of religious tolerance. During a meeting with the Astrakhan governor, Tajuddin stated, rather colorfully, that regional leadership has "managed to find the golden rod of fraternal relations," in reference to the mutual support of Astrakhan's local government and Muslim communities (Office of Press Service 2006).

The leadership of Astrakhan's official Islamic community is proud of this concordance between local Islamic communities and the Russian government, as evidenced by the prominent display in the Central Mosque's office of an image of Putin shaking hands with Nazymbek Ilyasov, the mufti of ARDUM. This photograph also hangs in an Islamic shop adjacent to the mosque. Putin's handshake with Ilyasov displays good relations between the nation-state and the local Islamic communities affiliated with ARDUM. The photograph is a visual reminder that Astrakhan's Central Mosque retains and celebrates its close ties with the Russian government not only at the local level, but also at the level of the nation-state.

The leaders of Astrakhan's Central Mosque believe that Putin and Medvedev are both strong supporters of the development of "traditional" Islam within Russia's borders. In interviews with mosque volunteers, a number of interlocutors expressed the sentiment that the Russian government tries to promote the good reputation of "traditional" Islam, which is associated with the Tatar and Bashkir ethnic groups of the Volga-Urals region. In an interview with M. Todd (Central Mosque Astrakhan, 10 July 2009), one official stated that "we tolerate and respect all confessions recognized by the government." This sentiment implies that laws of religious tolerance do not apply to Islamic peoples at war with the Russian state. The mosque leaders listed Wahhabism, which in Russia is associated with Caucasian rural Islamic communities, as a threat to the development of "traditional" Islam within Russia.

Fig. 80.2 Mosque 34, sitting half-way between construction and demolition (Photo by M. Todd, July 2009)

Despite this discourse of tolerance of "traditional" Islam and Astrakhan's discovery of the "golden rod of fraternal relations," a different mosque community in Astrakhan voices concerns over Islamaphobia in the local and national governments (Fagan 2006). Mosque 34 (Fig. 80.2) is located along the highway to Astrakhan's airport. Mosque 34 is not associated with Wahhabism, and many of its patrons also visit other mosques in the city. However, this mosque is in dispute with the local government because it has been ordered to relocate and demolish its existing structure by the city government. The mosque received building permits in 2003; however, they were revoked in 2005 and the community was ordered by the local government to demolish what they had already built.

Mosque 34's community has fought the orders to demolish through a series of lawsuits against the Astrakhan government. The discourses surrounding the demolition orders and lawsuits of Mosque 34 provide insight into how secularism can work as a tool of the state to aestheticize religious identities, and also into how minority religious communities draw upon secular claims to promote their rights. This debate illustrates the principles of deprivatization of religion—the way religions play public and frequently political roles in secular societies.

Astrakhan's local government framed its orders for demolition in purely pragmatic terms, over a concern with safety issues involved in construction. The regional court cited failure to pay rent, inappropriate land use, deviations from the proposed architectural plan, and failure to obtain consent from the power company for building near an electric grid as reasons for demolition. Importantly, the minaret of the mosque was located 7 m (23 ft) from power lines, when safety standards call for a minimum distance of 20 m (65.6 ft). The state presents itself as rational and unbiased in the orders to demolish the mosque; it is concerned with safety hazards and threats to citizens, not with the existence of the mosque itself or the nature of Islamic teaching within the mosque. In turn, the Islamic community affiliated with Mosque 34 is portrayed as irrational because they ignored these pragmatic concerns over planning issues, refused to move, and subsequently countersued the government.

However, the local community believed that identity politics rather than legitimate safety violations underpin the demolition request. The airport highway, as a gateway into the city, is one of the first places visitors see when arriving to Astrakhan. Thus, the Muslim congregations believe that concern over Astrakhan's religious identity as expressed in this landscape is the reason behind the ordering of the demolition. The mosque's supporters cite Putin's visit to the city in 2005 as the turning point in the decision to demolish the mosque, as well as the election of a new mayor. According to the mosque's supporters, in August 2005 President Putin stated to the regional governor and mayor that "they had not chosen a good place for a mosque" (Bureau of Democracy 2006). Following the failure of their countersuit to stop demolition, a mosque leader stated to the local press, "The current mayor of the city of Astrakhan Sergei Bazhenov and Governor of Astrakhan Region Alexander Zhilikin do not like the fact that a mosque and not a church is located at the entrance to the town" (Sova Center 2006). In other words, the proximity of the mosque to the airport, instead of the minaret to the power line, is the underlying motivation for demolishing the structure.

Since the Islamic community could not get a permit to protest in Astrakhan, they framed their case as a national-level issue rather than as a local battle between a mosque and city government. A letter from Islamic leaders to Putin states that they view the demolition as akin to political persecution (Open Letter 2007). The letter was signed by more than 3,000 Russian Muslims, including two co–chairmen of Russia's Mufti Council, Nafigulla Ashirov and Muqaddas Bibarsov, and was published in Izvestia, a national newspaper. This letter appealed to Putin, as guarantor of the Russian Constitution, to protect Islamic citizens from abuses from regional and city governments.

However, Putin did not reply directly to the letter. Rather, a low-ranking member of the Presidential Administration released a statement online, informing the signees that the Russian courts are subject to the Constitution, and that if they have problems with the judicial branch they need to communicate with that branch. Although Mosque 34's community, with support of Russia's Islamic community, employed discourses of Islamic rights, the state at all levels responded with pragmatic discourses. Now, Mosque 34 continues to sit on the highway, halfway between demolition and construction. The mosque community brought their case against demolition to the European Court of Human Rights (ECHR) in Strasbourg, France (Application no. 40482/06); the Court decided it would hear the case in November 2006 (Fagan 2007). Although the case application subsequently received prioritization from the ECHR in July 2007, a verdict is still pending.

In Astrakhan, the local government, supported by regional and national courts, has played a large role in regulating the production of the mosque landscape. The close ties and mutual support between the local and national government is on display in the Central Mosque. This mosque is located in a historic Tatar neighborhood in downtown Astrakhan, while Mosque 34 stands at the new gateway to the city, near the airport on the outskirts of town. The controversy over Mosque 34 shows that the aesthetic identity of Astrakhan is at stake when investigating the right to build mosques. Astrakhan has a reputation as a multi-confessional city with a diverse religious landscape, at once reflected by Putin's photographic presence in the Central Mosque but refracted by the unfinished project of Mosque 34.

80.5 The Public Place of Buddhism in the Republic of Kalmykia

In this section, we contrast Astrakhan multi-confessional religious landscape to that of its neighbor to the west, the predominantly Buddhist republic of Kalmykia. The first Buddhist organization was registered in Kalmykia in 1988, during Gorbachev's period of glasnost. The next year, the first prayer house opened in the republic, headed by the Buryat lama Tuvan Dordzh. Two years later (1991), the Buddhist Union of Kalmykia (OBK) was set up, uniting the religious organizations that had been established in the larger towns and villages throughout the republic. In 1992, Telo Tulku Rinpoche, an American citizen of Kalmyk descent, was appointed to head the association. The OBK was the primary force behind the opening, in 1996, of the Buddhist temple "Syakyusn-Syume" in the republic's capital of Elista (Sinclair 2008; Badmaev and Ulanov 2010). Frequently described as the "largest Buddhist temple in Europe," it was constructed with substantial monetary support, approximately $500,000 USD, from the republic's government (Fagan 2003). State assistance also extended to construction, as of 2003, of 20 more Buddhist temples (known locally as *khuruls*) beyond the capital. Agadjanian (2009: 244) underscores the importance of state funding for the construction of religious buildings: "In Kalmykia (more so than in the other Buddhist republics—Buryatia and Tuva) Buddhism is actively and openly supported by the local government," although religious pluralism is still respected and Buddhism is not legally recognized as the "state" religion. Filatov (2003: 290) points to the election of Kirsan Ilyumzhinov as President of the republic in 1993 as the starting point for the "genuine resurrection of Buddhism in Kalmykia," thanks to the resultant government support (though, as Humphrey [2002] notes, Ilyumzhinov's tenure has led to other points of political tension). A prominent example of Buddhism's public presence is the Pagoda of Seven Days, situated in the square at the heart of Elista, near the republic's parliament and main university (Fig. 80.3). The Pagoda displaced the city's Lenin statue to the square's periphery.

One of the important events in the past decade in Kalmykia was the opening of the Golden Temple complex in Elista in December 2005. The Golden Temple is a source of pride for the nation. After the transition from communism, attendance at the makeshift Buddhist temple near Elista's center was sparse, at less than ten persons per day—this increased to 50 or so on religious holidays. The impetus for further construction was provided by the Dalai Lama, as well as the promise by Ilyumzhinov that a grand temple would be constructed. With a reported cost of $25 million USD, the Golden Temple is certainly grand; it houses the largest

Fig. 80.3 The newly-built pagoda in downtown Elista (Photo by E. C. Holland, February 2010)

statue of Buddha in Europe (Parfitt 2006; Fig. 80.4). The Temple was constructed quickly, in roughly a year between November 2004 and December 2005. In an interview, Telo Tulku Rinpoche (head of the OBK) emphasized the importance of the temple to Kalmyk society—it established the republic as a center for Buddhism in Europe and united the Kalmyk people behind a common cause (Interview with E.C. Holland, Elista, Kalmykia, 3 March 2010.) Continued harmony is essential, as financially the upkeep of the temple is supported by private donations and Buddhism is one of the foundational elements of national identity: This is achievable because Buddhism is commonly viewed as one of the foundational elements of Kalmyk national identity; according to Telo Tulku Rinpoche.

Beyond the construction of new religious facilities and the close collaboration between religious and political authorities, Buddhism's prominence in contemporary Kalmykia is primarily as a marker of national identity. Numerous authors, in addition to the interlocutors discussed above, have noted this importance. To paraphrase Bakaeva (1994: 20), religion is now one of the key components in defining Kalmyk national consciousness. Buddhism is one of a set of elements—also including the national epic, the Dzangar, and the legacies of shamanism as a pre-Buddhist belief system—that defines the spirituality of the Kalmyk nation and religious life in the republic today (Filatov 2009). Similarly, other commentators have suggested that Buddhism is more important as a component of cultural, as opposed to religious, identity (Badmaev and Ulanov 2010).

However, serious challenges remain for Buddhism and Buddhist practice in the republic. Badmaev and Ulanov (2010) identify a number of issues complicating the religion's revival: a lack of understanding among self-identified Buddhists of the

Fig. 80.4 Elista's Golden Temple, foregrounded by Beliy Starets (Photo by E. C. Holland, February 2010)

foundations of the religion, problems surrounding the Dalai Lama's visits to the region, continued debates between "traditional"ists and modernizers (see also Ulanov 2009), and the need to import religious leaders from other "traditional"ly Buddhist regions, including Buryatia and Tibet, due to the closure of *khuruls* and institutions for religious training during the Soviet period (Bakaeva 2008).

Also questioned by some religious elites is the depth of Buddhism's revival in the republic. Sinclair (2008) discusses the nature of Kalmykia's Buddhist revival, contrasting this localized reawakening with discourses of reform advocated for by Tibetan Buddhists now serving in the republic. Prior to 1917 Buddhism in Kalmykia was closely tied to Tibet, as monks frequently traveled east to study Buddhist practice before returning to Kalmykia to serve as local religious teachers. In discussing the situation currently in Kalmykia, Sinclair (2008: 242) distinguishes between revival and reform, "two types of social movement that may be conceptually distinct but are entwined in practice, with each enabling the emergence of the other." These competing discourses play out within the Buddhist community, rather than as a rivalry between Buddhists and non-Buddhists. The outer markers of a religious revival, as viewed through changes in Elista's urban landscape and the construction of new sites of worship, underscore the internal tension within this community; Buddhism's public position has been strengthened, but the private role of the religion remains less clear.

80.6 Conclusion

In this chapter, we first examined religious law in Russia, and in turn how these laws affect the urban landscapes in Elista and Astrakhan in order to understand the spatialization of religion and secularism in Russia. We operationalized Knott's spatial theory of locality to examine how (some of) Russia's religious minority groups work within these legal frameworks to negotiate their identities in localized public spheres, using religious landscapes as analytical lenses to understand the spatial differentiation of de-privatized Islamic and Buddhist religious identities. Harmony and conflict over mosque construction in multi-confessional Astrakhan show that Russia's promotion of religious tolerance is underlined by a sometimes exclusionary identity politics, while Elista's Buddhist temples and their continued maintenance show that the co-production of nationalism and religious identity is an ongoing project.

As religion and politics work together continue to re-spatialize and reshape the lives of Russia's peoples, research agendas in the region will adapt accordingly. No longer is Russia inaccessible and prohibitive of place-based field research—thanks to funding agencies and changing visa regimes, scholars are able to engage in the region for extended periods of time. As Russian scholars gain access and familiarity with Western academic scholarship on geography and the academic study of religion, and as Western scholars gain access to the region and see how valorized concepts of religion, Orientalism, and space translate in the post-Soviet Russian milieu, projects will proliferate. We view continued work on Russia's religious minorities as particularly important. The future of research on the political and cultural geographies of religion in Russia is glimpsed through the "Religion, Politics and Policy-making in Russia: Domestic and International Dimensions" conference, held in June 2012 at the University of Tartu, Estonia. At this conference, scholars from universities in Russia, the United States, and the European Union presented on diverse projects and project proposals from education and religion to the politics of the Russian Orthodox Church to the role of nationalism and religion in separatist movements in the Caucasus. This conference shows that there is an intensifying focus on the role of religion in Russia's politics from multiple places, disciplines, and scholarly approaches—the concordance of this diversity illustrates that the interweaving of religion and political life transcends Russia and is increasingly part of a more widespread research agenda.

References

Agadjanian, A. (2001). Revising Pandora's gifts: Religious and national identity in the post-Soviet societal fabric. *Europe-Asia Studies, 53*(3), 473–488.
Agadjanian, A. (2009). Buddizm v sovremennom mire: Myagkaia al'ternativa globalismu. In A. Malashenko & S. Filatov (Eds.), *Religiia i gloablizatsiia na prostorakh Evrazii* (pp. 222–255). Moscow: ROSSPEN.
Anderson, J. (2007). Putin and the Russian Orthodox Church: Asymmetric symphonia? *Journal of International Affairs, 61*(1), 185–201.
Asad, T. (2003). *Formations of the secular: Christianity, Islam, modernity*. Stanford: Stanford University Press.
Badmaev, V., & Ulanov, M. (2010). Buddizm v Kalmuikii: Problemy vozrozhdeniia i perspektivy razvitiia. In *Yug Rossii v Pervom Desyatiletii XXI Veka: Itogi, Problemi i Perspektivi* (pp. 74–86). Rostov-on-Don: Izdatel'stvo SKNTs VSh YUFU.
Bakaeva, E. P. (1994). *Buddizm v Kalmykii: istoriko-etnograficheskie ocherki*. Elista: Kalmytskoe Knizhnoe Izd-vo.
Bakaeva, E. P. (2008). Kalmytskii Buddizm: Istoriia i covremennost. In N. L. Zhukovskaya (Ed.), *Religiia v istorii i kul'ture Mongoloiazychnykh narodov Rossii* (pp. 161–200). Moscow: RAN.
Brubaker, R. (2012). Religion and nationalism: Four approaches. *Nations and Nationalism, 18*(1), 2–20.
Bureau of Democracy, Human Rights, and Labor. (2006). *Russian Country International Religious Freedom Report*. Retrieved June 18, 2012, from http://moscow.usembassy.gov/policy/reports-on-russia/russia-country-international-religious-freedom-report-2006.htmlz
Casanova, J. (1994). *Public religions in the modern world*. Chicago: University of Chicago Press.
Cosgrove, D., & Jackson, P. (1987). New directions in cultural geography. *Area, 19*(2), 95–101.
Daniel, W. L. (2006). *The Orthodox church and civil society in Russia*. College Station: Texas A&M University Press.
Daniel, W. L., & Marsh, C. (2007). Russia's 1997 law on freedom of conscience in context and retrospect. *Journal of Church and State, 49*(1), 5–17.
Fagan, G. (2003, April 11). Few complaints over Kalmykia's state support for Buddhism. *Forum 18 News Service*. Oslo, Norway. Retrieved May 3, 2012, from www.forum18.org/Archive.php?article_id=29
Fagan, G. (2006, March 20). Russia: Blocks to acquiring places of worship. *Forum 18 News Service*. Oslo, Norway. Retrieved June 19, 2012, from www.forum18.org/Archive.php?article_id=746
Fagan, G. (2007, October 30). Russia: Threats to demolish churches and mosques continue. *Forum 18 News Service*. Oslo, Norway. Retrieved July 16, 2012, from www.forum18.org/Archive.php?article_id=1040
Fagan, G. (2013). *Believing in Russia – religious policy after communism*. New York: Routledge.
Filatov, S. (2003). Buddizm v Respublike Kalmykiia. In M. Bourdeaux & S. Filatov (Eds.), *Sovremennaia religioznaia zhizn' Rossii: opyt sistematicheskogo opisaniia* (pp. 284–297). Moscow: Logos.
Filatov, S. (2009, June–July 2009). Buddisty Kalmykii: Mezhdu Tibetom, zavetami predkov i evrobuddizmom. *Russian Review, 38*. Retrieved May 4, 2012, from www.keston.org.uk/_russianreview/edition38/02-kalmik-budd-from-filatov.html
Gorski, P., & Altınordu, A. (2008). After secularization? *Annual Review of Sociology, 34*, 55–85.
Heleniak, T. (2006). Regional distribution of the Muslim population of Russia. *Eurasian Geography and Economics, 47*(4), 426–448.
Humphrey, C. (2002). Eurasia: Ideology and the political imagination in provincial Russia. In C. M. Hahn (Ed.), *Postsocialism: Ideals, ideologies and practices in Eurasia* (pp. 258–276). London: Routledge.
Isaac, E. (1961–62). The act and the covenant: The impact of religion on the landscape. *Landscape, 11*, 12–17.

Knott, K. (2009). From locality to location and back again: A spatial journey in the study of religion. *Religion, 39*(2), 154–160.

Knox, Z. (2004). *Russian society and the Orthodox church: Religion in Russia after communism*. New York: RoutledgeCurzon.

Kong, L. (2001). Mapping 'new' geographies of religion: Politics and poetics in modernity. *Progress in Human Geography, 25*(2), 211–233.

Kong, L. (2004). Religious landscapes. In J. S. Duncan, N. C. Johnson, & R. H. Schein (Eds.), *A companion to cultural geography* (pp. 365–381). Oxford: Blackwell.

Office of Press Service and Information Administration of the Governor of Astrakhan. (2006, April 24). *Supreme mufti of Russia Talgat Tajuddin visited the Astrakhan region*. Retrieved June 20, 2012, from http://astrakhan.net/index.php?ai=11656

Open letter of Muslim community to President V.V. Putin. (2007, March 7). Retrieved June 19, 2012, from www.worldbulletin.net/news_detail.php?id=2658

Papkova, I. (2011a). *The Orthodox Church and Russian politics*. Washington, DC: Woodrow Wilson Center Press.

Papkova, I. (2011b). Russian Orthodox concordat? Church and state under Medvedev. *Nationalities Papers, 39*(5), 667–683.

Parfitt, T. (2006, September 20). King of Kalmykia. *The Guardian*. Retrieved July 16, 2012, from www.guardian.co.uk/world/2006/sep/21/russia.chess

Reynolds, M. (2005). Myths and mysticism: A longitudinal perspective on Islam and conflict in the North Caucasus. *Middle Eastern Studies, 41*(1), 31–54.

Riasanovsky, N., & Steinberg, M. (2005). *A history of Russia* (7th ed.). Oxford: Oxford University Press.

Shterin, M. S., & Richardson, J. T. (1998). Local laws restricting religion in Russia: Precursors of Russia's new national law. *Journal of Church and State, 40*(2), 319–341.

Sidorov, D. (2000). National monumentalization and the politics of scale: The resurrections of the Cathedral of Christ the Savior in Moscow. *Annals of the Association of American Geographers, 90*(3), 548–572.

Sinclair, T. (2008). Tibetan reform and the Kalmyk revival of Buddhism. *Inner Asia, 10*(2), 241–259.

Sova Center. (2006, September 2). The Supreme Court upheld the decision to demolish the mosque in Astrakhan. *Sova Center: Religion in Secular Society*. Retrieved July 16, 2012, from www.sova-center.ru/religion/news/harassment/places-for-prayer/2006/02/d7231

Ulanov, M. S. (2009). *Buddizm v sotsiokul'turnom prostranstve Rossii*. Elista: Izdat. Kalmuitskogo Univ.

Verkhovsky, A. (2002). The role of the Russian Orthodox Church in nationalist, xenophobic and antiwestern tendencies in Russia today: Not nationalism, but fundamentalism. *Religion, State and Society, 30*(4), 333–345.

Walker, E. (2005). Islam, territory, and contested space in post-Soviet Russia. *Eurasian Geography and Economics, 46*(4), 247–271.

Wilford, J. (2010). Sacred archipelagos: Geographies of secularization. *Progress in Human Geography, 16*(4), 328–348.

Chapter 81
Back to the Future: Popular Belief in Russia Today

Jeanmarie Rouhier-Willoughby and Tatiana V. Filosofova

81.1 Introduction

A complex multilayered system of popular belief lies at the heart of any culture's folklore. It is a core element of mythological folk consciousness and imagination. Popular belief finds practical implementation through religious rituals and everyday customs, the folk calendar and folk medicine. Popular belief also stimulates an abstract folk thought and imagination that responds to our desire for creativity, an understanding of the world around us and the universal cycle of life. The nature and complexity of the concept of "popular belief" allows us to use this term in a very broad cultural context. The popular belief systems of different peoples have been the subject of humanitarian sciences for more than a century. However, even present day scholars cannot agree either on the definition of the concept of "popular belief" itself or on the use of a unified terminology concerning this subject. Today, the term "popular belief" along with similar terms, such as "folk religion," "vernacular religion," "popular religion," "lived religion" and "religious folk imagination" are frequently used by historians, specialists in religious studies, art historians, folklorists, anthropologists, ethnographers and philosophers when describing various aspects of folklore, for example, folk rituals, popular culture, folk art, folk poetry and narratives, popular customs and superstitions (for examples of the various definitions of popular belief as well as the range of terminology in current scholarship, see Balzer 2010; Bernshtam 1989; Levin 1993; Gromyko 1994, 1995; Primiano 1995;

J. Rouhier-Willoughby (✉)
Department of Modern and Classical Languages, Literatures and Cultures,
University of Kentucky, Lexington, KY 40506, USA
e-mail: j.rouhier@uky.edu

T.V. Filosofova
Department of World Languages, Literatures, and Cultures, University of North Texas,
Denton, TX 76203, USA
e-mail: tatiana.filosofova@unt.edu

S.D. Brunn (ed.), *The Changing World Religion Map*,
DOI 10.1007/978-94-017-9376-6_81

Panchenko 1999a; Levin 2004; Himka and Zayarnyuk 2006; Rock 2007; Wanner 2012). For many years popular belief was conceived of as both religious and everyday rituals and customs rooted in polytheism. Recently, the term "popular Orthodoxy," concerning vernacular religious beliefs of the Eastern Slavs, was introduced to define contemporary folk rituals and customs that are widely practiced in Russia, Belarus and Ukraine (Kononenko 2006). Bearing in mind the complexity of the concept of "popular belief," is it feasible to define this phenomenon at all? It is too simplistic to define a system of popular belief as a kind of alternative religion in opposition to a dominant, "official" religion often supported by the state. Moreover, in a system of national popular belief, we can find traces from various world religions and world mythologies. Therefore, it is more accurate to define "popular belief" as a worldview based on the natural cycle of birth, life, death and the afterlife. In this natural cycle of physical and spiritual life humankind and god are inseparable and essential elements of the natural cycle. Therefore, the stages of the life cycle remain at the heart of every system of popular belief, but changes occur in the attendant practices. Therefore, the practical application of popular belief through rituals, customs and interpretation of the supernatural is often determined by changing political regimes, state ideology, the domination of religious institutions, social and cultural circumstances and even the development of new technologies.

81.2 Periods in the Development of Popular Orthodoxy

In the academic world, for instance, it has become a broadly accepted fact that after Christianity had spread across Europe, polytheistic beliefs and rituals did not disappear without trace from folk customs, everyday rituals and traditions, poetry and narratives, but were rather assimilated into the newly adopted religious format. For example, the Christian celebration of the life cycle of Christ corresponds to polytheistic rituals that revolved around the birth of summer and the death of winter. This was the main reason why, in the early days of Christianity, for almost 300 years, the Roman Church decided against the celebration of Christ's birth at the winter solstice, so as to avoid such an obvious reminder of the agrarian festival of the birth of the Unconquerable Sun. The necessity of communicating to a deity a person's needs and desires remained unchanged after Christianity became the dominant religion, but the means of communication changed, from blood sacrifices in polytheistic society to monetary donations and commissioned services of the Church. Many ancient pre-Christian rituals concerning the cults of water and earth were transformed into a new format of everyday customs and rituals determined by the Church. For example, polytheistic communities would collect the harvest together and then celebrate it by sharing food, dancing and singing, performing rites to ensure an abundant harvest the next year as well. After Christianization, a church would organize a collective harvest and its celebration in much the same way, providing assistance to the needy under its auspices. This tradition survived in many European rural communities throughout the last century.

An examination of vernacular religion in the context of recent Russian history provides a great deal of evidence on how popular belief reflects dramatic changes in political and social systems. For example, after the Russian Revolution a state policy of atheism was imposed by the Bolsheviks with the aim of establishing a new morality based on scientific materialism. That policy had a dramatic impact on popular culture on the surface, for example, the introduction of new Soviet holidays. It also damaged the Orthodox Church as an institution, by significantly reducing the number of parishes and congregations. However, this step neither significantly weakened the position of Orthodoxy as a cultural phenomenon nor eradicated the existence of popular Orthodoxy in Soviet society. This situation was accommodated by the Soviet government who actually made use of the Orthodox Church during critical periods in World War II (discussed in more detail below) and allowed the blessing of the troops, for example, by a priest to boost morale. The process of change in popular belief is continuous, of course; even today in Russia when the Orthodox Church enjoys popularity and holds a very strong position in society, new technology has become an important means of advertising, spreading information and communicating on religious matters. For example, the Internet has become a means of communicating with the saints and a virtual space for sharing one's experiences with other believers (http://www.hristianstvo.ru/internet/dialogue/).

Interest in folklore and popular belief increases as a result of important political changes, when a nation needs to re-discover its national identity and find a way to move forward after a crisis. Not unexpectedly, the collapse of the Soviet political system and dramatic social and economic changes in Russia have given rise to a new wave of interest in Russian history, the Orthodox Church, Russian self-perception and identity, and in Russian culture, art, literature and folklore. In the last two decades, in particular, the system of popular belief in Russian folklore has become one of the most discussed and studied topics among scholars in Russia, Europe and America (for an overview of the state of scholarship, see Panchenko 1999a, 2006).

81.3 Characteristics of Popular Russian Orthodoxy

Contemporary Russian vernacular religion is no different from any other belief system in its essence and its aims. However, at the same time, the system of Russian popular belief can be distinguished from other similar systems in a number of significant ways:

- First, it is notable for its completeness – an excellent preservation, quality and quantity of the material culture (folk icons and artifacts) and a body of folk literature that has survived to the present day.
- Second, Russian folklore has a very large body of religious Christian folk poetry, songs, music, legends and penitential lyrical songs that were created under the strong influence of the theological, ethical and aesthetical values of Orthodox Christianity.

Finally, dramatic events in Russian history and in the history of the Russian Orthodox Church have determined that Christian folklore, especially poetry and narratives, has always been and remains an area for popular creativity. In present day Russia, after the collapse of the Soviet system, the Russian Orthodox Church has flourished with state and popular support, the growth of churchgoers, and a great interest in Orthodox culture. This has stimulated a re-discovery of religious folklore and a desire for a continuation of this tradition in a more sophisticated way.

81.3.1 Dvoeverie 'Dual Faith' in the Russian Religious Tradition

From the nineteenth century on, the concept of *dvoeverie* 'dual faith' was held up as a hallmark of Russian Orthodoxy (Levin 2004; Rock 2007). Theorists often contended that the Russian believer, particularly the peasant one, worked with a dual system of beliefs (polytheistic and Christian) and held onto the polytheistic system much more overtly than other Christian peoples. For example, they pointed to the pre-Christian features of much of the yearly, religious cycle, e.g., the burning of an effigy of winter as part of the pre-Lenten carnival week *Maslenitsa*; mumming rites, when people dressed as animals or monstrous creatures between Christmas and New Year's; and the ancestral cult as depicted by feeding the dead during visits to graves around Easter or leaving food out and heating the bathhouse for departed souls at Christmas. There were also many traditions associated with the land, conceived of as a feminine entity, for example, Moist Mother Earth (*Mat' syra zemlia*) or the motherland (*rodina*), that were absorbed into Christianity. Therefore, before spring work on the land started each year, peasant farmers would ask forgiveness from the land for "ripping open her breast." Similarly, in the Russian Orthodox Church, during confession, it was customary to publicly ask the land and community (*mir*) for forgiveness before confessing to a priest. A literary example of this phenomenon may be found Dostoevsky's novel (1945: 508–510), *Crime and Punishment*, when Raskolnikov decides to confess his crime publicly in front of the crowd in the square, after asking for forgiveness from the motherland and community. Another interesting agrarian survival is taking a bit of earth from home to carry during a long journey. If a person dies and is to be buried away from his/her home, then s/he was buried with a small piece of earth from the homeland.

The theory of "dual faith" was particularly advantageous within the context of the state policy of atheism. Soviet analysts of folk culture thus could assert that the peasantry did not actually believe in Christianity and argue that the pre-Christian remnants showed the people's disdain for religion, in particular their dissatisfaction with Church institutions and the social hierarchy (Sokolov 1950: 174–179; Sukhanov 1976: 68). By the late Soviet period, theorists such as Vlasov (1992) and Bernshtam (1992) were casting doubts on this approach to Orthodox belief among the peasantry. Levin (2004) has also convincingly challenged these assumptions about the strength of *dvoeverie*, illustrating that the Russian believer was no more likely to exhibit dual faith than any Christian believer in nineteenth century Europe.

That is not to say that syncretic practices were not in evidence; they were, but they did not represent an active alternation between two belief systems. Rather, believers viewed the pre-Christian system as part of a larger whole that was incorporated into a unified Orthodox belief (albeit often non-doctrinal) system. Therefore, while the peasantry may have performed rites that may be traced to agrarian practices, these rites were actually viewed from the context of their conceptions of Christianity. A typical example is when a person would recite a charm to bring good fortune or ward off negative consequences of other actions. Charms typically include formulas such as, "I stand up, having blessed myself, I go out, having crossed myself," which was followed by bowing to all four directions, a remnant of pre-Christian belief, but seen through the lens of Christian faith, for example, crossing oneself (Peskov 1996: 13). While these actions may not have conformed to established church doctrine (and indeed the Orthodox Church did battle against many folk practices, as discussed below), they illustrate more about the conception of the world and otherworld in the folk belief system than attest to a lack of faith in Christianity. The peasantry certainly did also evince dissatisfaction with institutional norms, and agrarian practices may have been part of their means of resistance, but these behaviors relate more to institutional and social power relations than to faith, in our view.[1]

81.4 Popular Belief in the Soviet Era

The situation did not change overnight once the official policy of atheism was introduced after the Bolshevik Revolution in 1917. Until the 1920s, the goal of the revolutionary government and its theorists was to totally remake society and to eliminate any vestiges of the "backwardness" that characterized the Russian patriarchal and religious systems. Remaking the family (and its rites) and eliminating religion and religious institutions generally was a primary goal (Stites 1989). All that changed in the 1930s under Stalin, who realized that public celebrations and rituals (and to some extent, even the Orthodox Church)[2] had an important role to play within

[1] The literature on resistance supports this claim, for example, Abu Lughod (1986) and Scott (1985) demonstrate how codified resistance in actually reifies hierarchical social systems by allowing for dissent in an established way, so as to assuage resentment. At the same time, the powerless in the hierarchy are not attempting to overthrow the system in these rites. Rather, after having expressed their dissent with the existing structure, they actually actively participate in it and reaffirm its pattern.

[2] For example, in 1918 the Decree of Separation of Church and State eliminated the Orthodox Church as a legal entity (Mazyrin 2009: 28). However, Stalin reinstituted the Moscow Patriarchate in 1943. This act was certainly motivated by political ends, both internal and external; Stalin realized that the Church could help the citizenry cope with wartime devastation and also put him in a good light while working with his World War II allies in Britain and the United States (Pospielovsky 1997). Nevertheless, it did allow for the existence of religious institutions, albeit on a restricted basis, through much of the Soviet period.

Soviet society. As Rouhier-Willoughby (2008) has shown, the Soviet ritual theorists attempting to create new life-cycle rites for birth, weddings and funerals, faced a difficult dilemma. They knew that the populace had retained some religious beliefs and needed to create rites that did not reinforce them. Nevertheless, their only models for formal rites were indeed Orthodox Christian ones. It is not surprising, then, that the rites (and spaces where they took place) look remarkably similar to christenings, religious weddings and funerals, even down to keeping the phrase from the Orthodox hymn, *Vechnaia pamiat'* 'eternal memory' in the latter rite. In their writings, ritual theorists (Sukahnov 1976: 54, 201; Ugrinovich 1975: 137) strived to explain that these rites may have seemed similar to religious ones, but that they were not; they were designed to inculcate *dukhovnye tsennosti* "spiritual values" born of the Soviet ethos, not of religion. Thus, the two sponsors who attended a Soviet-era naming rite for a newborn may have appeared to function as godparents, but they did not. Rather, they were illustrating their dedication to support the family as they raised a child within Soviet society.

Over time the populace came to accept Soviet values as well, such that the twentieth century system of vernacular belief drew on three strands of its history: folk practices, Orthodoxy and socialist tenets. The combination of these three strands was particularly evident in life-cycle rituals, which featured elements derived from all these traditions (see Rouhier-Willoughby 2008 for a detailed discussion). For example, in the Soviet-era funeral, many people retained aspects of religious practices as best they could without priests. They arranged for women to read the psalter over the body, put religious items in the casket, and even ordered requiem masses when possible. Thus, even if a priest was not present at the gravesite, proper acts to ensure the soul's peace were performed in a church. However, people also retained folk beliefs about feeding the dead by leaving food on graves at various times of the year and on the anniversary of a person's death, a practice both the Orthodox Church and the Soviet authorities frowned upon at various times, if for different reasons. The Church was wary of the non-doctrinal view of the return of the soul of the dead to the earthly plane that such practices assumed, while the secular government was concerned that these were religious practices. That is one reason that new cemeteries during the Soviet period were placed in distant locations, to make visitations on *Roditel'skie subboty* "Parents' Saturdays" more difficult (Merridale 2000: 279). While a funeral may have preserved aspects of both doctrinal and non-doctrinal religious practice, it also incorporated socialist tenets. For example, citizens accepted the role of the state in funerals under the auspices of the workplace. Employers named a person responsible for organization of transportation (using company vehicles) and the commemoration at the graveside, which emphasized the deceased's professional life and contributions to socialist values.

This combination of socialist, folk and religious values also existed in the oral tradition. Folk legends serve as a particularly rich source of the intersection between the three systems, because they were a popular genre in Russia throughout the twentieth century. In Russian folklore scholarship, legends are defined as retellings of biblical content or apocryphal stories featuring Biblical figures or saints. These legends reflect the vernacular understanding of Orthodoxy and incorporate established folk patterns and motifs, for example, in the cycle about Noah (attested from the nineteenth century onward (Afanas'ev 1859/1990; Matveeva 2005)), the devil tricks Noah's wife, so that he can get onto the ark and survive the flood. Similar plots are attested throughout the Russian tale tradition, in which a minor demon or imp fools a peasant (for additional examples, see Haney 2009 and Ivanits 1992). In these plots, devils are often conflated with or replaced by forest spirits or similar malicious otherworldly figures from Russian mythology, which reflects the intersection of folk and religious traditions. One particular variant of a legend recorded in Siberia will illustrate how all three strands may be combined. Christ (or God) and St. Peter are on Earth visiting Russia (Novikov 1941: 75; Soboleva and Kargapolova 1992: 162; Vlasova and Zhekulina 2001: 197). During their visit, Peter asks Christ to make women superior to men (either because they are better Christians or because Peter feels sorry for them, because men have rights and they do not).[3] Once women have gained power, they begin to act in violent and unchristian ways. During the course of the story, the wife housing Peter and Christ ends up giving Peter a beating twice (she intends to beat Christ as well, but he tricks Peter into taking both beatings). He, in turn, asks Christ to restore men to power, and since then women have had no rights. However, the Soboleva and Kargapolova (1992) version ends with the narrator saying that this situation held true until the time of the Bolshevik Revolution, after which men and women had become equal. Thus, the beliefs determined by God expressed by the Siberian woman who told the story were fluidly combined with Soviet reality, which established equal rights for women in its constitution.

81.5 Religious Revival in the Post-socialist Period

The post-socialist context is even more complicated with regard to the interaction of the various strands in vernacular religion. In the 20 years since the fall of the USSR, religion has flourished. The Russian Orthodox Church has assumed a preeminent role in society and indeed has joined with the government to spread its influence in many ways. Church lands have been returned; in 1997 laws to complicate registration of religions "foreign" to Russia were passed;[4] and in 2009, a class on religious instruction was established in all public schools for 4th graders, textbooks for which must be approved by Church authorities as of 2012 (Knox 2005; Shakhanovich 2013). Parents have the option to choose from six tracks for the religious instruction component: Orthodoxy, Judaism, Islam, Buddhism, World Religions or Secular Ethics. These courses were chosen to reflect the religious heritage of the country (see note 4). Across the country, parents, 80 % of whom state they are believers, largely opt for the Secular Ethics class for their children (42.7 % of school children

[3] In the Vlasova and Zhekulina version, women are superior from the start, but their misbehavior toward Peter prompts his request for Christ to demote them and put men in charge.

[4] A term referring to certain Protestant sects, especially Pentecostals, and to the Church of Latter Day Saints.

attend that class; 31.7 % attend Orthodoxy; 21.2 % World Religions; 4 % Islam and less than 1 % Buddhism or Judaism (Shakhanovich 2013)).[5] These data suggest not only that parents prefer to have their children get their religious education at home or in religious schools affiliated with a church or temple. In addition, it appears that the heritage of secularism inherited from the USSR maintains its currency even during the religious revival. At present, the Russian Orthodox Church has over 23,000 parishes and 620 monasteries in the country. Priests are now educated in five theological academies and 32 seminaries. Additionally, the Church supports a number of church music and icon-painting schools and Sunday schools for children in every reasonably size parish (http://www.orthodoxworld.ru/en/segodnia/1/index.htm). Believers have embraced the return of their churches, even if they are not active members of a congregation (attendance at church services, even on high holy days, is still relatively low in Russia (Titarenko 2008: 246–247)). Christening of children is now the norm, and even adults, especially women, have opted to be baptized in the post-socialist era. In addition, people display icons in their homes and workplaces and carry them while travelling; it is rare to find a bus, taxi or van without an icon hanging from the rear view mirror. As part of the post-wedding ceremony tour of the city, nearly every couple makes a stop at a holy site, even if they do not intend to have a religious wedding (which most still do not) (Rouhier-Willoughby 2008: 257–258).

81.5.1 The Russian Orthodox Church as Institution and Cultural Symbol

Before we begin a detailed examination of the current system of popular belief in Russia and its interaction with the Orthodox Church, it is necessary to make clear the following regarding the state of the Church in Russia historically and today. First, the Orthodox Church has assumed a role of a religious institution that presents itself as an arbiter of social life. As a result, the Church faces resistance from citizens to its attempts to maintain control over belief, that is, to ensure that it is consistent with doctrine. Conflicts between the Church and its flock, as mentioned above, have been a longstanding issue in Russian Orthodoxy. Second, Orthodoxy as a cultural phenomenon has long served as a core element of Russian national identity and in the post-socialist frame is particularly vibrant. Third, popular Orthodoxy is characterized by a discordant and diverse body of oral religious literature that demonstrates

[5] Note that Shakhnovich's research does indicate some regional differences in areas traditionally associated with a particular religion, e.g., in Chechnia and Ingushetiia 100 % of parents chose the class on Islam; however, in Tatarstan (also a Muslim area), 61.3 % chose World Religions and 38.7 % chose Secular Ethics. Given the history of armed conflict and political resistance in the Chechen-Ingush region, as compared to the largely peaceful relations with the federal government in Tatarstan, the choice likely conveys a political and nationalist stance for parents in these regions. In Buryatia, the heart of the Buddhist confession, only 8 % chose the class on Buddhism; 56 % preferred Secular Ethics and 29.2 % opted for Orthodoxy.

the life of vernacular religion as a system. In addition, Christian moral, ethical and aesthetical values became the essential source for folk creativity in all oral genres (tales, song, legends, poetry). Fourth, as with other Christian faiths, we find survivals of rituals, customs, omens, and folk mythology based on a polytheistic worldview and ancient pre-Christian agrarian belief (as discussed earlier in the chapter). And finally, the Church has always been required to respond to socio-political history, from the acceptance of Christianity into Russia through the Bolshevik Revolution and into post-Soviet reality. These responses have been as diverse as the events themselves, but the Church has always negotiated, to varying degrees, a balance between the needs of the faithful and official teachings.

While historically the peasantry before 1861 may have conceived of itself as Orthodox, it stubbornly held onto its traditions, many of which the Church condemned (Rock 2007: 30, 42–44; Vlasov 1992: 25). As Bernshtam (1992: 37–40) shows, there was a wide gulf between practice in the urban centers, where doctrine and rite were much more controlled, than in the vast stretches of the countryside, where itinerant priests used every means at their disposal, including apocryphal and other non-canonical materials, to convey the Gospel. Similarly, without a priest on hand at all times, rural churches developed their own local traditions (often incorporating pre-Christian agrarian practices) and interpretations of doctrine. The Orthodox Church began its battle against the "'false' faith of the masses" in the fifteenth and sixteenth centuries, but their edicts were largely ignored by the people and their priests (Bernshtam 1992: 40). As a result, these congregations were characterized by an independence from the central Church hierarchy throughout the history of Russian Orthodoxy until the Bolshevik Revolution (and beyond).

81.6 The Miraculous Icon of Snopot: The Development of Popular Religion from the Nineteenth Century to the Present Day

In the Soviet context, the official policy of atheism in the post-Revolutionary period did not much affect rural practice or belief. While certainly religious ceremonies were curtailed, in particular after the purge of priests and other religious in the 1920s–1930s or by the destruction of churches, a tradition for independent worship already existed among the populace, which often maintained former practices. For example, E. and S. Minyonok (1997) have documented an annual procession dedicated to a miraculous icon of the Virgin Mary in the village of Snopot in the Briansk region of Russia. This annual holiday was celebrated each year from the nineteenth century throughout the Soviet era, from 1917 to its fall in 1991, and has continued in the post-socialist era. Traditionally unwanted icons could not be destroyed, but were often placed in the river when no longer needed. One such icon of the Virgin Mary washed up on the shores of this village. Thereafter, the spring in this town became blessed, resulting in miraculous cures and a corresponding legend cycle describing these events. In the Soviet period, priests were not available for the ceremony, so that local women took the task upon themselves, preserving the rite (and the legends themselves) for fellow believers.

Near the spring, a church dedicated to the icon had been built. However, Russian forces destroyed it during World War II, fearing that the Germans would use the church to their strategic advantage. A monument to locals who died in the Second World War now stands on the site of the former church. The ceremony once involved circling the church three times with the icon before going to the sacred spring. Once the memorial was erected, villagers began to circle the monument.[6] We see evidence here of the interconnections between two seemingly disparate belief systems (faith in Orthodoxy and honor of Soviet war dead). In this case, official Soviet institutions intersected with a local legend and Christian ritual practice related to a miraculous icon and spring.

Despite the shift to a post-socialist state, the Soviet-era World War II memorial still plays a central role in the ritual to this day. In the post-Soviet period, priests have taken the place of the village women who preserved the rite during the Soviet era. Villagers note that the priests do not perform the ritual properly, rushing people paying homage to the icon and the spring. They also no longer allow women to take holy water from the spring, reserving this honor for men. This interaction with the religious institution that was absent for 70 years illustrates the long-standing dissent between official Orthodoxy and the local version of belief as well as its interaction with both Soviet and post-Soviet institutions.

81.7 Popular Religion and the Russian Orthodox Church

In the present day we can find a wide range of responses to popular Orthodoxy by the Church. In some cases, the Orthodox Church, remarkably, apparently takes a tolerant and supportive approach to popular responses to the revival of popular Orthodox beliefs and culture in society (Figs. 81.1 and 81.2). For example, the Church has helped to restore an important aspect of everyday popular Orthodox culture, namely religious tourism or pilgrimage to holy places (monasteries such as Valaam and Solovki) and venerated sacred objects – miraculous icons, graves of saints, and sacred springs, etc. Pilgrimage to holy places in Russia and abroad has a thousand-year history in Orthodox culture and has always been a way for believers to demonstrate their devotion. On the other hand, pilgrimage was always an important source for folk creativity in the creation of legends, travel stories and stories about miracles related to sacred objects. In 1999 the Pilgrimage Center in Moscow, which aims to develop religious-related trips, was established under the supervision of the Department for external church relations of the Moscow

[6]Tumarkin (1994) has studied the development of the World War II cult in Russia. This cult parallels many of the features of the attitudes toward deceased ancestors in Russian folk religion (Rouhier-Willoughby 2008: 156, 180).

Fig. 81.1 Kreshchenie: "Epiphany" procession by the parish priest and members of the congregation of the Life-Bearing Spring (established in 2003) from the church of a nearby holy spring, Iskitim, Russia, 2011 (Photo by Jeanmarie Rouhier-Willoughby)

Fig. 81.2 Epiphany. Blessing of the waters at the holy spring near Iskitim, 2011 (Photo by Jeanmarie Rouhier-Willoughby)

Fig. 81.3 Bathing in the holy lake near the Vazheorzerskii Monastery, Leningrad Region, July 2012 (Photo by Tatiana Filosofova)

Fig. 81.4 Collecting holy water near the Vvedeno-Oiatsky Convent, Leningrad Region, July 2012 (Photo by Tatiana Filosofova)

Patriarchate. For the last few decades, organized pilgrimage has become a norm of everyday life and is supported by the Church (Figs. 81.3 and 81.4). However, surprisingly, the Church has also assumed a tolerant approach to non-doctrinal phenomena of popular belief, for example, correspondence to the saints, as in the Center for St. Xenia (discussed below in more detail) in St. Petersburg. The

Church endorses correspondence and publication of the letters (Kormina and Shtyrkov 2012). Believers can also correspond with the saints through websites (www.woman.ru/health/medley7/thread/4038348/ and www.woman.ru/health/medley7/thread/3703406/)

81.7.1 Matrona Moskovskaia: The "Soviet" Saint

However, the contemporary Orthodox Church must contend with the tradition of independent worship and a positive view of the Soviet past among the populace. The legend of Saint Matrona Moskovskaia, a woman born on the eve of the Bolshevik Revolution in 1885 and who died in 1952, has been, strangely enough, intricately intertwined with her contemporary Joseph Stalin (b. 1878, d. 1953), who ruled the USSR for much of her life. Her story encapsulates the dilemma faced by the post-Soviet Orthodox Church in its attempts to reconstruct itself after the fall of the USSR. The Russian Orthodox Church has essentially taken the position that anyone who died at the hands of the Soviet government, regardless of his/her personal beliefs, suffered to some degree for the Orthodox faith. The foremost example of this position was the canonization of Nicholas II and his family in 1996. The former monarch (and hundreds of others who died at the hands of Soviet authorities) were named Russian New Martyrs and Confessors in the Role of Saints.

Matrona Moskovskaia (born Matrona Dmitrievna Nikonova) is one of the so-called "people's saints." From her birth in 1885 in the village of Sebeno in the Tula region until her death on May 3, 1952 in Moscow, she maintained her Orthodox faith and used its power to perform miracles that have led to her canonization. Her path to sainthood was not an easy one and was primarily the result of pressure from believers who lobbied in her favor. In essence, because of the absence of doctrinal or priestly control during the Soviet period, people began to name saints of their own volition. The Church has resisted the people's demands to canonize most of these folk saints (Sedanov et al. 2008: 48–49). It has been able to do so primarily because their popularity is usually limited to rural villages or to smaller, regional cities. The larger community of believers thus does not bring enough pressure to force the Church to move forward on canonization, despite fervent local belief in their saintly qualities.

Matrona's case presented a very different and more complicated situation for the Russian Orthodox Church. While she was indeed born in a village, in 1925, at the age of 40, she left Sebeno for Moscow. She did so because her brothers were die-hard Bolsheviks who were afraid that her presence could threaten their professional careers and, indeed, their lives. She spent most of the rest of her life in Moscow itself, only leaving on occasion for the suburbs to escape arrest. As a result, at her request, she was buried in Moscow at the cemetery of the Danilovsky Monastery. According to her saint's life (Khudoshin 2009: 20), from the moment of her interment in the cemetery, her gravesite has produced miracles. The word of

Matrona's holiness spread among believers in Moscow and beyond, and her grave has become the site of pilgrimage by believers from across the country. Due to the fact that she had been buried in the largest city in Russia, in a prominent cemetery often visited by tourists, the Church authorities could not ignore her (or the large portion of the faithful demanding canonization) as easily as it could a religious figure from a small town. In fact, this popularity led the nuns at the Intercession Convent in Moscow, who had cared for her grave, to petition for her remains to be moved to the convent in 1998. The relics were moved on March 8 of that year and now rest in an elaborate shrine there. The situation was complicated by the fact that the fervent belief in Matrona's saintliness threatened Church authority. In an environment when the Church was trying to rebuild itself after 70 years of stagnation, it could ill afford to have Matrona become more popular than either the official saints or than the Church itself. As a result, not surprisingly, the Expert Commission established by the Orthodox Church in 1999, which has debunked the majority of the miracles performed by peoples' saints, accepted that Matrona was worthy of sainthood that year (Sedanov et al. 2008: 48; Khudoshin 2009: 3).

Certainly the fact that Matrona's miracles are perceived as more powerful than typical folk healing and that she lived an ascetic life enhance the average believer's faith in her holiness. However, the main reason these facets of her life are underscored in the Church's official hagiography is because of the doctrine on folk religion. Church officials roundly criticize non-doctrinal Orthodoxy (particularly as attested during the Soviet period, when the Church was unable to systematically control belief in everyday life). The preeminent role of women as representatives of the faith during the Soviet era poses a serious threat to the Orthodox hierarchy, because the average believer had no qualms with accepting the wisdom and holiness of elderly women healers. The Church is striving to claim a place in the lives of the ordinary faithful, while at the same time criticizing some essential tenets of those believers' faith: that elderly (holy yet secular) women represent God's power; that these same women (or ones very much like them) kept Orthodoxy alive during 70 years of socialism and were (and remain) at the heart of the transmission of the belief system; and that interpretation of doctrine and spirituality on the individual level, undiluted by priestly control, has become the norm.[7]

The most famous story about Saint Matrona is not attested in her saint's life, but does appear on sites dedicated to her on the Internet and is widespread among the laity (http://bgforum.ru/flame/16694/). Stalin asked Matrona to take the Kazan icon of the Virgin Mary, known for its protection of Russia against Tatar invasions, in an airplane around the borders of the Soviet Union during the Second World War.[8] Because of Matrona's prayers (and the icon's protection), the nation was saved from being overrun by the German forces. A similar story about Matrona and Stalin

[7] Russian Orthodoxy has a tradition of female monasticism as well. However, the nuns do not play a central role in establishing doctrine. Nor did they perform rites in the way that secular women did during the Soviet era.

[8] In some versions of this legend, Matrona took along not the Kazan icon, but an icon of the Virgin Mary commissioned by Matrona herself as a child.

Fig. 81.5 Painting of Matrona blessing Stalin in Saint Olga's Cathedral in St. Petersburg (Photo by Jeanmarie Rouhier-Willoughby)

(also not in the official saint's life) tells of how the Soviet leader came to see Matrona in 1941 to ask her about the outcome of the war. She replied, "the red rooster will conquer the black rooster," indicating that the USSR's "red" troops would triumph over the black-clad Nazis. In another variant of the story, St. Matrona replied, "Everyone will flee Moscow, but you [Stalin] will remain." According to the newspapers *Moscow Komsomol* and *Pravda,* this event led to the painting of an icon depicting Matrona blessing Stalin (Fig. 81.5), which was "discovered" in Saint Olga's Cathedral in Saint Petersburg in 2008 (although the parish priest reported that it had long been known of in the area) (http://cprfspb.ru/5538.html; http://www.pravda.ru/faith/religions/orthodoxy/25-10-2010/1054781-stalin-0/; http://kprf.ru/rus_soc/83916.html; Kormina 2010: 12).

81.8 Attitudes of the Populace Toward the Soviet Era

These two legends illustrate the complexity of post-Soviet Orthodoxy and the dilemma faced by the Church in this context. As discussed, the Church has roundly condemned the Soviet era. It cannot but do so, given that a central tenet of the government at the time was the elimination of religion and the Church as an institution. However, in doing so, the Church threatens its standing among a faithful who view the era with nostalgia. Among a significant portion of the populace, the Soviet era, especially during Stalin's reign when the USSR became a superpower, stands as a symbol for the greatness of Russia itself. Its fall not only threatened Russia's preeminence on the world stage, but led to economic collapse and severe personal hardships. The country that billed itself as the savior of Europe in World War II no longer merited fear or respect from its former allies. The average middle-aged person, even those who have thrived in the new Russia, still views the Soviet Union of his/her childhood as a better society than the one s/he experiences today. People often cite the argument that it was safer, easier to live, and more stable overall for the population. Certainly the nation had a stronger place in the world and commanded a degree of respect that it has lost, a factor that plays an important role in conceptions of the past.

81.8.1 Attitudes Toward the Soviet Era in the Russian Orthodox Church

The glorification of the Stalin era and of the Soviet Union broadly has emerged from the populace at a time when more people are returning to the Church. The Church has made its position on the "demonic" nature of the Soviet period clear. But a large portion of its flock does not (and cannot) share its opinion. Rather, they must somehow reconcile their faith in both systems, the religious and the political, by connecting the two. This saint, particularly the popular stories about her interactions with Stalin, illustrates this process. While in the official saint's life, Matrona roundly condemns the Soviet Union and its representatives for the harm they did to the Church and the nation, that portion of her hagiography is overshadowed by folk legends that thrive among her adherents. In these legends, as the icon depicts, she not only acceded to Stalin's wishes, but anointed him as the leader and savior of Russia when he visited her in 1941. Stalin, then, was actually a true believer leading an Orthodox nation, another factor in his favor in the contemporary world. Many of the faithful create an alternate vision of the past on the basis of legends such as these. Matrona's story allows them to take the position that the Soviet Union was indeed an Orthodox state and represented a better society than the one they live in now. In addition, the defense of the legitimacy of the icon by the parish priest at St. Olga's, where the Matrona-Stalin icon hangs, also demonstrates a similar position among local clergy, who understand the situation on the ground. If they hope to keep the flock's trust and respect, they too may (genuinely or not) advocate for a more tolerant view of the Soviet period. Vernacular beliefs about Matrona incorporate all these attitudes and allow for the average believer to forge a non-doctrinal view of the Soviet past and of religious faith therein, a past that the Church finds troubling. Matrona's induction as a saint thus demonstrates how the Church is attempting to minimize these threats to its control over faith by reframing her life from the point of view of official doctrine, but also is trying to come to terms with its own Soviet past and its role in post-Soviet Russia.

81.9 Saint Xenia: The "Populist" Saint

Unlike the popular saint Matrona, the protector of Saint Petersburg, St. Xenia is a people's saint recognized by the Church in all her aspects. Her cult also provides us with an excellent example of a new level of an interaction between "official" and "popular" Orthodoxy in Russian contemporary society. In 1988 Xenia the Blessed was officially canonized by the Orthodox Church, and her Life of Saint Xenia of Petersburg largely based on folk legends and stories, was approved and published. Within the popular imagination she functions as an understanding, approachable protector, helper and savior. Popular religious legends describe St. Xenia of Petersburg (Fig. 81.6) as a holy fool (or a fool for Christ), who voluntarily chose a

Fig. 81.6 Painting of St. Xenia of Petersburg (Photo by Tatiana Filosofova)

life of suffering and prayer (Panchenko 1999b). In Russian Orthodox culture, the act of suffering and tolerance has always been especially valued, and, as a result, Orthodox Russia has produced many holy fools approved by the Church and venerated in the popular religious imagination. By humiliating herself and practicing self-denial of material goods, Xenia reached the highest level of spiritual perfection and gained the ability to perform miracles and foresee the future. Xenia, as a result, has become the primary symbol of charity, hope, support and consolation for the people. According to the legend, Xenia was a real person, a 26-year-old widow who lived in St Petersburg between 1719 and 1800. After the unexpected death of her beloved husband, who passed away without confession and last rites, she gave all her possessions to the poor and assumed the role of a holy fool, praying for his salvation. Everyday she wandered around the city and spent the nights in fields praying until daybreak to provide relief to his soul. In her legendary life Xenia has functioned as a savior, who cured disease and protected people from death or misfortune; a helper to the needy; and a seer.

As we saw in our discussion of the legend of Christ and Peter, Russian saints may take on corporeal form and help people in need. Therefore, legends from the eighteenth century to the present day tell of how she appears in the streets of Saint Petersburg and elsewhere as a simply dressed old lady. For example, one legend states that Xenia helped a person to regain her place in the bread line during the German Blockade of Leningrad and by doing so saved the woman's life (Sindalovsky 2010: 146). A contemporary legend describes Xenia helping a soldier during recent conflicts in Chechnya (Sindalovsky 2010: 147). In the legend a soldier, originally from Saint Petersburg, was on patrol with a group of soldiers. At the last moment, when they were just about to leave, the soldier was called to the main office as his mother had come to visit him. Unfortunately, all the soldiers in the patrol, except him, perished. Naturally, he discovered that his mother had not visited him that day and he firmly believes that it was St. Xenia who saved his life.

Nevertheless, like other beloved Russian saints, she also could be quite severe in her punishment of sin. According to one legend from the twentieth century, a young night watchman at the Smolensk Cemetery in Saint Petersburg got very drunk. He bet with his friends that he "would sleep with Xenia herself!" He spent a night in Xenia's chapel and the next morning he discovered, to his horror, that traces of mold had covered his clothes and his body. Apparently, the man displayed symptoms of the so-called "pharaoh's disease," an illness that struck anyone who dared to disturb the tombs of Egyptian pharaohs (Sindalovsky 2010: 145).

Many stories and legends about Xenia concern the tiny chapel built above her grave in Saint Peterburg's Smolensk Cemetery in the early nineteenth century, which is a place of popular pilgrimage. During the Soviet era, it served as a studio for making park sculptures. However, people who worked there said that sculptures made during the day were destroyed each night. The workers believed that this was the work of Xenia, who strongly disapproved of the closing of her chapel. In the 1960s–1970s after the chapel has been closed and fenced off, a new way of communicating with Xenia was established. The faithful started sticking notes with prayers and requests for help into the fence and gates (Filicheva 2013). This tradition is still very much alive and, according to popular belief, if one leaves a note under the candle stand or in the chapel wall and then walks around the chapel three times, one's wish will come true. Today a large number of pilgrims going to pray to Xenia in her chapel usually stands in a long queue to enter. This experience is a perfect opportunity to communicate with other believers, hear other peoples' stories of the help given by St. Xenia and to gain advice from more experienced church-goers.

81.10 Conclusion

The Russian Orthodox Church is grappling with the socialist legacy and the independence of its flock from the institution of the Church historically. As a result, it has staked its future on ties with the federal government building a relationship termed a "symphony." Under Putin and Medvedev, this strategy has been a fruitful one for the Church. Laws favoring the Church, such as the institution of the religious instruction controlled by the Church in public schools and against "foreign" confessions, have been passed, largely at the behest of the Church hierarchy. In addition, the Church has been able to prosecute various cases against artists and musicians whose work expressed anti-Church attitudes, for example, the cases against the Beware, Religion and Forbidden Art exhibits at the Sakharov Museum in Moscow in 2003 and 2006 and the punk group Pussy Riot in 2012 (http://www.nytimes.com/2010/07/13/arts/design/13curators.html?_r=0; http://themediaproject.org/article/art-bears-brunt-russias-religious-defamation-debates?page=0,1; http://www.guardian.co.uk/music/2012/aug/17/pussy-riot-sentenced-prison-putin). The most vocal protests against these events emerge from strongly nationalist, neo-fascist groups, for example, the *Pravoslavnaia druzhina* "Orthodox Defenders," an alarming trend with which secular and religious institutions will likely have to contend in the future.

In many ways, the post-socialist states are thus more secular than those in western Europe, for example, lower membership in churches, less frequent attendance at church services, and a clear desire to separate religion from the state. However, Titarenko (2008: 241) cites data indicating that between 1991 and 2005 the percentage of those declaring themselves to be believers in some confession grew from 23 to 53 %, while the number of self-declared atheists fell from 35 to 6 %. This trend, unlike in western Europe, is true of younger generations as well (Titarenko 2008: 240). In addition, as Titarenko (2008: 242–243) discusses, Orthodox believers among those surveyed overwhelmingly admit to holding beliefs that may be classified as non-Christian or non-doctrinal. As we have discussed, the Church has often been tolerant of these deviations from the traditional faith. However, they are also made uneasy by certain of them, and not surprisingly, they have pushed for control over the school curricula related to religion. Priests create and manage Internet sites and question and answer forums designed to correct beliefs that emerge from vernacular Christian belief or alternative, non-traditional religions, for example, neopaganism and New Age movements. In response to the pre-eminence of the Church in the Russian state, universities are also opening Departments of Theology, mainly to train teachers for the public school religion class or to prepare them to take holy orders. These departments have caused a great deal of controversy, because they are separate from the religious studies and philosophy programs that exist at major universities, and their curricula are designed to conform to Orthodox doctrine. They are often viewed as propaganda arms for the Church that do not merit inclusion in an academic institution of higher learning (http://rbth.ru/articles/2012/09/05/russia_to_produce_qualified_specialists_of_divinity_17941.html).

While we have focussed on Orthodoxy here, as the dominant religion in the country, similar historical developments, patterns of belief, and institutional concerns could be identified in the other major confessions, for example, Judaism, Buddhism and Islam, all of which are experiencing similar revivals since the fall of the USSR in 1992. The nature and limits of dissent against religious institutions, the range and interaction of secular and religious world views and the wisdom of cooperation between governmental and religious institutions will be at the center of public discussions in Russia for the foreseeable future. Consideration of these issues by scholars (and by various institutions in Russia itself) is complicated by a vibrant vernacular belief system among the populace. As we have seen, popular belief, in particular popular Orthodoxy, is a complex and variable system in its practice. One must consider the historical, cultural and institutional contexts of the practices at various times to gain any sort of complete understanding of the system. We have demonstrated that there are multiple strains in the development of the belief system itself, from pre-Christian polytheism to Christian doctrine to non-canonical sources and even socialist tenets, all of which intersect at different cultural levels and in coherent ways. Vernacular Orthodoxy has become an inseparable part of popular culture in the broader context of a revived public interest in the Orthodox Church. In addition, it spreads Orthodox culture through the oral tradition and even in contemporary media and technology, including the Internet. Despite the challenges it presents, scholars must deal with the fragmented, flexible and fast changing phenomenon of popular belief in order to conceptualize contemporary socio-cultural mores of post-socialist Russia.

References

Abu Lughod, L. (1986). *Veiled sentiments. Honor and poetry in a Bedouin society*. Berkeley: University of California Press.

Afanas'ev, A. N. (1990). *Narodnye russkie legendy A.N. Afanas'eva*. Foreward, compilation and commentary V. S. Kuznetsova (based on the 1859 ed.). Novosibirsk: Nauka.

Balzer, M. M. (2010). *Religion and politics in Russia: A reader*. Armonk: M.E. Sharpe.
Bernshtam, T. A. (1989). Russkaia narodnaia religiia. *Sovetskaya etnografiia, No 1*, 91–100.
Bernshtam, T. A. (1992). Russian folk culture and folk religion. In M. M. Balzer (Ed.), *Russian traditional culture. Religion, gender and customary law* (pp. 34–47). Armonk: M. E. Sharpe.
Bogest (the forum editor). (2006). Bogorodskii gorodskoi forum. http://bgforum.ru/flame/16694/. Accessed 23 Feb 2013.
Dostoevsky, F. (1945). *Crime and punishment* (C. Garnett, Trans.). New York: Modern Library.
Elder, M. (2010, August 17). Pussy Riot Sentenced to two years in prison colony over anti-Putin protest. The Guardian. http://www.guardian.co.uk/music/2012/aug/17/pussy-riot-sentenced-prison-putin. Accessed 2 Mar 2013.
Filicheva, O. N. (2013). Zapiski dlia Ksenii Blazhennoi: pozitsiia tserkovnosluzhitelei i narodnyi obuchai. http://www.ruthenia.ru/folklore/filicheva2.htm. Accessed 20 Feb 2013.
Ganeev, T. (2012, September 12). Russian to produce divinity specialists. http://rbth.ru/articles/2012/09/05/russia_to_produce_qualified_specialists_of_divinity_17941.html. Accessed 2 Mar 2013.
Gromyko, M. M. (1994). Pravoslavie v zhizni russkogo krest'ianina. *Zhivaia starina, No 3*, 3–5.
Gromyko, M. M. (1995). Etnograficheskoe izuchenie religioznosti naroda: zamentki o predmete, podkhodakh i osobennostiakh sovremennogo etapa issledovanii. *Etnograficheskoe obozrenie, No 5*, 77–83.
Gumanova, O. (2010, October 25). Chto delaet Stalin na ikone? http://www.pravda.ru/faith/religions/orthodoxy/25-10-2010/1054781-stalin-0/. Accessed 23 Feb 2013.
Haney, J. (2009). *An anthology of Russian folktales*. Armonk: M.E. Sharpe.
Himka, J.-P., & Zayarnyuk, A. (Eds.). (2006). *Letters from heaven. Popular religion in Russia and Ukraine*. Toronto: University of Toronto Press.
Ivanits, L. (1992). *Russian folk belief*. Armonk: M.E. Sharpe.
Khudoshin, A. (2009). *Zhitie svyatoi blazhennoi staritsy Matrony Moskovskoi i ee chudotvoreniia XX–XXI vv*. Moscow: Idel Press.
Kishkovsky, S. (2010, July 12). Organizers of art show convicted in Moscow. *The New York Times*. http://www.nytimes.com/2010/07/13/arts/design/13curators.html?_r=0. Accessed 2 Mar 2013.
Knox, Z. (2005). *Russian society and the Orthodox church. Religion in Russia after communism*. Oxford: Routledge.
Kononenko, N. (2006). Folk orthodoxy: Popular religion in contemporary Ukraine. In J.-P. Himka & A. Zayarnyuk (Eds.), *Letters from heaven. Popular religion in Russia and Ukraine* (pp. 46–75). Toronto/Buffalo/London: University of Toronto Press.
Kormina, Z. (2010) Politicheskie personazhi v sovremmennoi agiografii: kak Matrona Stalina blagoslovila. *Russkii politicheskii fol'klor, No 12*, 1–28. http://anthropologie.kunstkamera.ru/files/pdf/012online/12_online_kormina.pdf. Accessed 23 Feb 2013.
Kormina, J., & Shtyrkov, S. (2012). Believers' letters as advertising. St Xenia of Petersburg's 'National Reception Centre'. In A. Baburin, C. Kelly, & N. Vakhtin (Eds.), *Russian cultural anthropology after the collapse of communism* (pp. 155–182). London/New York: Routledge.
Levin, E. (1993). *Dvoeverie* and popular religion. In S. K. Batalden (Ed.), *Seeking God: The recovery of religious identity in Orthodox Russia, Ukraine, and Georgia* (pp. 31–52). DeKalb: Northern Illinois University Press.
Levin, E. (2004). *Dvoeverie i narodnaia religia v istorii Rossii* (A. L. Toporkova & Z. N. Izidorovoi, Trans.). Moscow: Indrik.
Matushka Matrona. Komu ona pomogla? (2006, March 18). http://www.woman.ru/health/medley7/thread/3703406/. Accessed 15 Feb 2013.
Matveeva, R. P. (2005). *Narodno-poeticheskoe tvorchestvo starobriadtsev Zabaikal'ia (semeiskikh)*. Ulan-Ude: BNTS.
Mazyrin, A. (2009). Legalizing the Moscow patriarchate in 1927: The secret aims of the authorities. *Social Sciences, 1*, 28–42.
Merridale, C. (2000). *Night of stone: Death and memory in 20th century Russia*. New York: Viking.

Novikov, N. V. (Ed.). (1941). *Skazki Fillipa Pavlovicha Gospodareva*. Petrozavodsk: s.n.

Panchenko, A. A. (1999a). Religioznye praktiki: K izucheniiu 'narodnoi religii'. In K. A. Bogdanov & A. A. Panchenko (Eds.), *Mifologiia i povsednevnost'. Vypusk 2. Materialy nauchnoi konferentsii 24–26 fevralia 1999 goda* (pp. 198–218). St. Petersburg: Institut Russkoi Literatury (Pushkinskii Dom) Rossiiskoi Akademii Nauk. Otdel Folklora.

Panchenko, A. M. (1999b). Iurodivye na Rusi. In A. M. Panchenko (Ed.), *Russkaya istoriia i kul'tura: Raboty raznykh let* (pp. 392–407). St. Petersburg: Iuna.

Panchenko, A. A. (2006). Novye religioznye dvizheniia i rabota foklorista. In Z. V. Kormina, A. A. Panchenko, & S. A. Shytkova (Eds.), *Sny bogoroditsy. Issledovanniia po antropologii religii* (pp. 119–130). St. Petersburg: Izdatel'stvo Evropeiskogo universiteta.

Peskov, A. M. (1996). *Mezhdu angelom i domovym*. In Peskov and Peskov, *Oberegi i zaklinaniia russkogo naroda* (pp. 3–20). Moscow: Kron-Press.

Pospielovsky, D. (1997). The 'best years' of Stalin's church policy (1942–1948) in light of archival documents. *Religion, State and Society, 25*, 139–162.

Pravoslanoe khristianstvo: Pravoslavnii internet: Obshchenie pravoslavnykh. (2014). http://www.hristianstvo.ru/internet/dialogue/. Accessed 24 Sept 2014.

Primiano, L. N. (1995). Vernacular religion and the search for method in religious folklife. *Western Folklore, 54*(No 1, Special issue), 37–56.

Rock, S. (2007). *Popular religion in Russia. 'Double belief' and the making of an academic myth*. London: Routledge.

Rouhier-Willoughby, J. (2008). *Village values: Negotiating identity, gender and resistance in urban Russian life-cycle rituals*. Bloomington: Slavica.

Scott, J. C. (1985). *Weapons of the weak: Everyday forms of peasant resistance*. New Haven: Yale University Press.

Sedanov, P., Shevchenko, D., & Prokop'ev, I. (2008). *Khot' sviatikh zanosi. Newsweek* (Russian version) 18:2, 2/24/08 (pp. 46–49).

Shakhanovich, M. (2013, February). *Religion and public education in the Russian federation: From experiment to regular practice*. Presentation at the conference: Post-Atheism: Religion and Society in Post-Communist Eastern Europe and Eurasia, Tempe, Arizona State University.

Sindalovsky, N. (2010). *Prizraki Severnoi stolitsy. Legendy i mify piterskogo zazerkal'ia*. Moscow: Tsentrpoligraf. Retrieved February 15, 2013, from http://ariom.ru/forum/t38566.html&sid=073f8f920825b54ce2a3fdd09f0739f9

Snopot. (1997). Minyonok, S., dir., Russian folklore expedition. Film.

Soboleva, N. V., & Kargapolova, N. A. (Eds.). (1992). *Russkie skazki Sibiri i Dal'nego Vostoka. Legendarnye i bytovye*. Novosibirsk: Nauka.

Sokolov, Y. M. (1950). *Russian folklore* (C. R. Smith, Trans.). New York: Macmillan.

Stites, R. (1989). *Revolutionary dreams: Utopian vision and experimental life in the Russian revolution*. New York: Oxford University Press.

Sukhanov, I. V. (1976). *Obychai, traditsii i preemstvennost' pokolenii*. Moscow: Politizdat.

Sviataia Matrona Moskovskaia skazala Stalinu v 1941-m: Krasnyi petukh pobedit hernogo petukha. (2010). http://cprfspb.ru/5538.html. Accessed 23 Feb 2013.

Sviataia Matrona Moskovskaia skazala Stalinu v 1941-m: Krasnyi petukh pobedit hernogo petukha. (2010). http://kprf.ru/rus_soc/83916.html. Accessed 23 Feb 2013.

Titarenko, L. (2008). On the shifting nature of religion during the ongoing post-communist transformation in Russia, Belarus and Ukraine. *Social Compass, 55*, 237–254.

Tumarkin, N. (1994). *The living and the dead*. New York: Basic.

Ugrinovich, D. M. (1975). *Obriady. Za i protiv*. Moscow: Politizdat.

Vlasov, V. G. (1992). The Christianization of the Russian peasants. In M. M. Balzer (Ed.), *Russian traditional culture: Religion, gender and customary law* (pp. 16–33). Armonk: Sharpe.

Vlasova, M. N., & Zhekulina, V. I. (Eds.). (2001). *Traditsionnyi fol'klor novgorodskoi oblasti. Skazki. Legendy. Predaniia. Bylichki. Zagovory (po zapisiam 1963–1994)*. Saint Petersburg: Aleteiia.

Wanner, C. (Ed.). (2012). *State secularism and lived religion in Soviet Russia and Ukraine*. New York: Oxford University Press.

Zapiski Ksenii Peterburgskoi. Komu nado peredat'-pishite!! (2011, February 8). http://www.woman.ru/health/medley7/thread/4038348/. Accessed 15 Feb 2013.

Zolotov, A. (2010, December 16). 'Offensive' art abuses religious symbols. http://themediaproject.org/article/art-bears-brunt-russias-religious-defamation-debates?page=0,1. Accessed 2 Mar 2013.

Chapter 82
The Changing Religious Mosaic of Ukraine

Esther Long Ratajeski

82.1 Introduction

Ukraine, the largest country in eastern Europe after Russia, is home to the region's most religiously diverse population, and is a country in which organized religious life has flourished in the past two decades. This paper examines key aspects of Ukraine's religious change since its independence in 1991, focusing on its largest religious groups: Orthodox Christian, Catholic Christian, Protestant, and Muslim. Each of these groups has a complex makeup and history as well as a unique path to the present, but several common themes also rise to the surface. Firstly, all four of these religious groups have been impacted to a greater or lesser extent by international connections: Orthodoxy experiences the tension between Russia and Ukraine; Catholicism is situated in a unique position between Rome and Eastern Orthodoxy; and especially noteworthy is how Protestantism and Islam have been transformed through interactions with people in other countries. These interactions include deepening social relationships with religious congregations and individuals in other countries, and changes to the built environment as foreign money has helped pay for new church buildings and mosques. A second important issue that has affected religious groups across Ukraine has been the reclamation and reconstruction of religious sites, most of them Orthodox but also including various other religions, and sometimes with legal conflict over ownership. Finally are changes that religious groups are bringing to civic life in Ukraine, through the establishment of institutions for religious education and through humanitarian efforts such as soup kitchens, drug rehabilitation centers, and medical clinics. These types of privately-run outreach efforts had been banned prior to Ukraine's independence, and have sprung up in an atmosphere with little such history. The advent of religious freedom that arrived

E.L. Ratajeski (✉)
Independent Scholar, Lexington, KY 40504, USA
e-mail: esther.long@gmail.com

S.D. Brunn (ed.), *The Changing World Religion Map*,
DOI 10.1007/978-94-017-9376-6_82

with independence in 1991 laid the foundation for deep changes in Ukraine's religious life, as well as changes to Ukrainian society at large.

82.2 Overview of Religious Growth Since 1991

Ukraine has historically been more religiously diverse than its neighboring countries, with strong representations of multiple strains of Orthodox and Catholic Christianity, Judaism, and also viable Protestant and Islamic communities. The various religious groups have ebbed and flowed over the centuries, and were severely reduced in number and size by the Second World War (Judaism) and by the Soviet regime. By the end of the Soviet era Ukraine's official religious life had been weakened, but believers were poised for a comeback. The multiplication and growth of religious communities in Ukraine since 1991 has been dramatic. Dozens of new religious communities have begun in Ukraine over the past two decades, while at the same time traditional religious groups have also prospered. This growth has been aided by Ukraine's permissive policy towards religious organizations and openness towards foreign missionaries working in Ukraine, a policy that contrasts sharply with that of former Soviet countries like Russia and Belarus which put strong restrictions on foreign religious organizations. In Belarus, for example, it is nearly impossible to legally register new religious organizations, and missionary activity there has been severely curtailed in recent years.

The Soviet Union passed a religious freedom law in 1990, the year before its demise, when it adopted the Law on the Freedom of Conscience. For the first time since the Revolution, religious organizations were permitted the right to own property and stand in court, and preaching was permitted in public (Druzenko 2010: 726). The next year, Ukraine passed a similar law, the Law of Ukraine on Freedom of Conscience and Religious Organizations.[1] This law, along with Article 35 of Ukraine's constitution,[2] established Ukraine as a secular state and lifted control over religious organizations from the shoulders of the government. While the Ukrainian government does not completely follow everything laid out in the Freedom of Conscience law, the U.S. State Department concluded that Ukraine experiences the "generally free practice of religion," and that the Ukrainian government "generally respect[s] religious freedom in practice" (U.S. State Department 2010).

While the Russian Orthodox Church dominates the religious sphere in much of the former Soviet Union, because of historical legacies Ukraine has three major Orthodox churches, each with its own government and structure. Moreover, western

[1] An English translation of Ukraine's law on freedom of conscience is available at www.iupdp.org/index.php?option=com_content&view=article&id=68:30e-11-law-of-ukraine-on-the-freedom-of-conscience-and-religious-organizations-april-23-1991&catid=40:effective-lawDOUBLEHYPHENnd-subordinate-legislation&Itemid=75

[2] An English translation of Ukraine's constitution can be found at http://gska2.rada.gov.ua/site/const_eng/constitution_eng.htm

Ukraine is home to a large Catholic presence. This means that there is not one obvious church for the Ukrainian state to endorse (as in Russia, for example), and Ukraine is considered a "model of religious pluralism among formerly socialist societies" (Wanner 2004: 736). Ukraine's history of religious diversity has set the stage for a degree of religious freedom in the current era that has led to dramatic growth of all religious faiths represented in the country, including three main divisions of Orthodox Christianity, two forms of Catholicism, historical and new Protestant churches, and the Islamic community, among others (Table 82.1). By 2009 the Ukrainian government reported that 42 religious faiths, plus unspecified "others,"

Table 82.1 Registered religious communities in 1994 and 2009

Name of religious denomination	1994	2009	Percent change
Ukrainian Orthodox Church – Moscow Patriarchate	5,998	11,444	91 %
Ukrainian Orthodox Church – Kiev Patriarchate	1,932	4,093	112 %
Ukrainian Autocephalous Orthodox Church	289	1,183	309 %
Russian Orthodox Old Believers	80	63	–21 %
Ukrainian Greek Catholic Church	2,932	3,566	22 %
Roman Catholic Church in Ukraine	588	901	53 %
Armenian Apostolic Church	9	26	189 %
All-Ukraine Union of the Association of Evangelical Christian Baptists	1,364	2,516	84 %
Other Unions of Evangelical Christians-Baptists	204	241	18 %
All-Ukraine Union of Christians of the Evangelical Faith – Pentecostals	798	1,455	82 %
Full Gospel Churches	33	601	1,721 %
Other Charismatic communities	44	396	800 %
Ukrainian Union Conference of Seventh-day Adventists Church	380	1,005	164 %
Transcarpathian Reformed Church	91	114	25 %
Lutheran	13	96	638 %
Other Protestant communities	30	1,435	4,683 %
Church of Jesus Christ of Latter-day Saints (Mormons)	0	34	n/a
Jehovah's Witnesses	458	685	50 %
Jewish communities	66	280	324 %
Muslim communities	77	521	577 %
Society of Krishna Consciousness	24	21	–13 %
Other Eastern Religious Communities	18	83	361 %
Pagan communities	19	105	453 %
Other Religious Organizations	12	387	3,125 %
Total	15,459	31,251	202.15 %

Data source: Religious Information Service of Ukraine and the State Committee for Religious Affairs

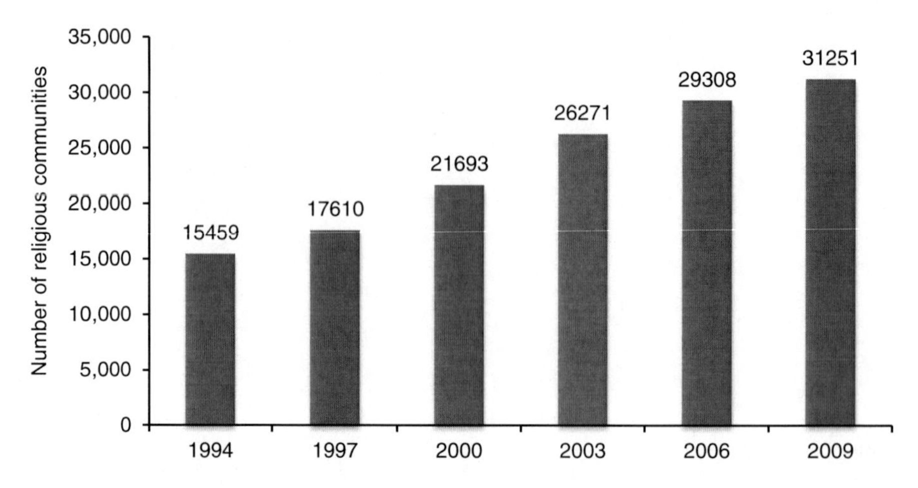

Fig. 82.1 Number of registered religious communities in Ukraine, 1994–2009 (Data source: Religious Information Service of Ukraine and the State Committee for Religious Affairs)

were officially registered in the country, comprising 31,251 distinct religious communities (RISU 2009).[3,4] As shown in Fig. 82.1, this was an increase of 102 % from 1994, when only 15,459 religious communities were registered in the country (State Committee for Religious Affairs 1994).

By 2009 there were over 33,000 registered and unregistered religious communities in Ukraine. Of these, about half were Orthodox Christian, one fourth were Protestant Christian, and 14 % were Catholic Christian. The rest of the religious communities were divided between Muslim, Jewish, pagan, Eastern, and other religions including Jehovah's Witnesses and the Church of Jesus Christ of Latter-Day Saints (Fig. 82.2). Ukraine's Jewish population, which had been large and vibrant prior to being decimated during the Second World War and later emigration to Israel, has used the past two decades to rebuild its institutions and community life. By 2009 there were 280 registered Jewish communities in Ukraine (RISU 2009), although Jews made up only about 0.2 % of the country's population (CIA 2012). Ukraine also is home to a growing number of pagans, up to 105 communities by 2009, including the Native Ukrainian National Faith, who turn to Ukraine's pre-Christian pagan roots for their spiritual inspiration.

Information that is not readily available, however, includes the number of practitioners in each religious community. While a quarter of religious communities are Protestant, most experts estimate that Protestants make up only 2–3 % of the country's 45 million people (Central Intelligence Agency 2012). Certain places in Ukraine have even fewer Protestants than that: Lyubashchenko (2010: 273) reports

[3]The Religious Information Service of Ukraine reported a total of 31,257 registered religious communities in 2009, but independently adding up their data only finds 31,251 registered communities, for a total of 33,077 registered and unregistered communities. This is a difference of only six communities, which is a minor error.

[4]"Religious communities" include places of regular worship, administrative centers of religious organizations, monasteries, missions and theological educational institutions.

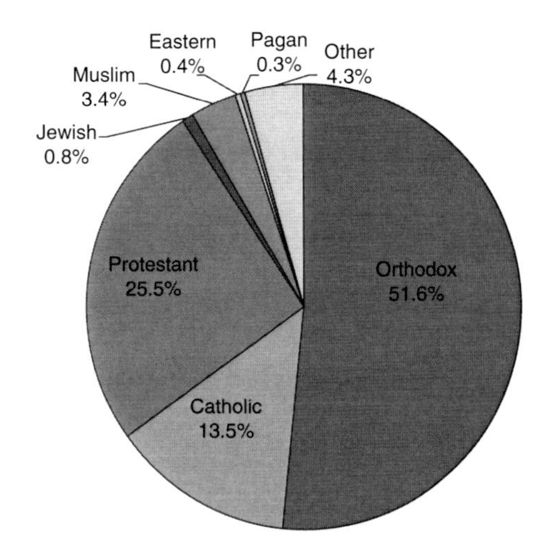

Fig. 82.2 Registered and unregistered religious communities in Ukraine in 2009 as a percentage of the total (Data source: Religious Information Service of Ukraine)

that a 2008 religiosity survey found only 1.5 % of Kyiv residents were affiliated with Protestantism. Meanwhile, about 84 % of Ukraine's population is Orthodox and about 10 % is Catholic (Central Intelligence Agency 2012).

In 2008 two-thirds of Ukrainians self-identified as extremely religious, very religious, or somewhat religious. However, the degree of religiosity in Ukraine varies greatly by region, with western Ukraine being by far the most religious. The average resident of western Ukraine attends church 22–23 times annually, while residents of eastern and southern Ukraine only visit church 7–8 times each year (Yelensky 2010: 215). This puts western Ukrainians at a par with Italians, Portuguese, and Slovaks in terms of church attendance, while those in eastern Ukraine attend church at the level of residents of France, but more frequently than Russians, Czechs or East Germans (222). Even among those 36.5 % of Ukrainians who considered themselves to be believers of a particular religious group in 2007, a relatively small portion (about 25 %) attended church at least once per month (U.S. State Department 2008). The difference between the high levels of religiosity and the low levels of church attendance indicate that for many Ukrainians, religion is a private matter, or at least one that does not require frequent public affirmation. This could also be a sign of what Mitrokhin calls a "low level of religious culture" among Ukrainians, several generations of which lived largely without places of worship. So although they have a sense of religious belief, they do not have the habit of regular church attendance (2001: 179).

82.3 Orthodoxy in Ukraine

During the Soviet period all Orthodox churches were part of the Ukrainian Exarchate of the Russian Orthodox Church, with its headquarters in Moscow. In 1990, at the dawn of the new era, the Russian church, in an attempt to satisfy the desire for a national church in Ukraine, formed the Ukrainian Orthodox Church

(UOC) – Moscow Patriarchate (Davis 1995: 74). This church has the most number of communities of any religious organization in Ukraine; by 2009 they had 11,444 registered communities, or 37 % of all religious communities in the country (RISU 2009). However, there are two other Orthodox churches that also vie for national prominence. The largest of these is the Ukrainian Orthodox Church – Kyiv Patriarchate, which was formed in 1992 to challenge the supremacy of Moscow, and has over 4,000 communities, or about 13 % of all religious communities in the country (Davis 1995: 99; RISU 2009). The other is the Ukrainian Autocephalous Orthodox Church (UAOC), formed in the 1920s to challenge the Russian Orthodox Church, suppressed by Moscow, and then resurrected in 1989 in Western Ukraine (Davis 1995: 71). The UAOC has close to 1,200 communities, or below 4 % of Ukrainian religious communities. Despite the smaller number of parishes in these alternate Orthodox churches, together they have more than twice as many adherents as the Moscow-based church. The Ukrainian Orthodox Church – Kyiv Patriarchate claims about 50 % of the Ukrainian population, the Autocephalous Church claims about 7.2 % of the population, and the Ukrainian Orthodox Church – Moscow Patriarchate has about 26 % of Ukraine's population (CIA 2012). Similarly, in a 2006 religiosity survey 40 % of respondents self-identified with the Kyiv Patriarchate church, while only 30 % self-identified with the Moscow Patriarchate church, even though the Moscow-based church has more than twice as many parishes as the Kyiv-based church (Yelensky 2010: 217). This discrepancy evidently reflects a desire to affirm a Ukrainian national identity. Yelensky writes that religious decisions in Ukraine are often "motivated by a desire to affirm national and cultural identity" and "reflect political affiliations" (222). The Ukrainian Orthodox Church – Kyiv Patriarchate benefits from this sentiment.

These three Orthodox churches (and besides these are some much smaller ones, such as the Russian Orthodox Old Believers' Church), highlight a dominant theme of Orthodox Christianity in Ukraine, viz., the evolving relationship between Ukraine and Russia, and between Ukraine and the international community. The Moscow-based Orthodox church would dearly love all of the Orthodox churches in Ukraine to return to its fold, but that would be an extraordinarily unlikely development. The Kyiv-based Orthodox church, for its part, would like the chief patriarch of all Orthodox churches, Patriarch Bartholomew of Constantinople, to recognize its legitimacy and grant it status as an autocephalous Orthodox church. These desires came to the forefront during a 2008 visit to Ukraine of both Patriarch Bartholomew of Constantinople and Patriarch Aleksi of Moscow, and the following year during a visit from Patriarch Kirill of Moscow, Aleksi's successor. The Ukrainian president at the time, Viktor Yushchenko, personally fetched Bartholomew from the airport during his visit, but was disappointed that the Patriarch declined to recognize the Kyiv Patriarchate as Ukraine's national Orthodox church. The visits of Aleksi and Kirill also failed to produce demonstrable progress for the Russian side. Although Kirill was warmly welcomed in the ethnically Russian parts of eastern Ukraine, his visit to more nationalist western Ukraine was controversial. The patriarch of the UOC-Kyiv Patriarchate, Patriarch Filaret, accused Kirill of coming to see "his flock in order to expand his influence beyond the walls of the church to Ukrainian state and society in pursuit of Russian state interests" (Tonoyan and Payne 2010: 261). Filaret himself experienced opposition when he traveled to eastern Ukraine in 1999, to Maryupil in the Donetsk oblast. He went to consecrate land for a new UOC-Kyiv Patriarchate church, and was physically assaulted by supporters of the UOC-Moscow Patriarchate, which is the dominant church in the mostly ethnically Russian part of Ukraine. Protesters poured a bucket of holy water over the patriarch's head, then beat him on his head with it (Mitrokhin 2001).

Regional conflict over Orthodoxy continues to the present. The most recent Religious Freedom Report published by the U.S. State Department gives examples of problems the Ukrainian Orthodox Church – Moscow Patriarchate (UOC-MP) faces in western Ukraine. The UOC-MP complained that officials in four towns in L'viv oblast denied their request for land allocation for new church buildings. They further stated that officials in the province of Ivano-Frankivsk refused to issue a building permit for constructing a church to Saint John the Baptist (U.S. State Department 2010).

President Yushchenko's involvement in church affairs, mentioned previously, is typical of all Ukrainian presidents since independence in 1991. Although Ukraine is officially a secular state with religious freedom and legal rights for all religious organizations, figures in the central government have openly backed Orthodoxy and stated numerous times their desire for all Orthodox churches in Ukraine to be united into one national church. This government involvement in Orthodox church affairs has led to complaints from some other religious leaders, such as those from major Protestant denominations (U.S. State Department 2008).

In addition to lobbying for the establishment of a unified, national Orthodox church, the Ukrainian government has publicly funded the renovation and reconstruction of numerous Orthodox church buildings. Some of the most famous were historic structures that the previous government had destroyed, such as the Dormition Cathedral of Kyiv Caves Monastery and the St. Michael's Golden-Domed Monastery in Kyiv (Fig. 82.3). Others, however, were new churches.

Among these is a church paid for by the national railroad company, an Orthodox church for St. George the Victory-Bearer, constructed near Kyiv's main railway station (Fig. 82.4). Another is the St. Nikolas church at the National University of the State Tax Administration. A third example is the Church of St. George the Victory-Bearer at the Central Hospital of the Ministry of Internal Affairs (Druzenko 2010: 733).

A third way that the government has been involved with religious affairs, and one that also tilts towards Orthodoxy, has been in the advocacy of teaching Christian ethics in public schools. In 2005 the president declared a plan to introduce an optional course on Christian ethics in public schools. This was done with the support of both Orthodox and Greek Catholic leaders, and was not intended to teach specific church doctrine. However, because of concerns over the separation of church and state, and because some non-Christian religious leaders objected to the state supporting Christianity over other religions, the program has not made much headway. From time to time Orthodox religious leaders will also announce plans to initiate religious education in public schools, such as an announcement by an

Fig. 82.3 St. Michael's Golden-Domed Monastery complex in Kyiv. Nearly the entire complex was demolished by the Soviets in the 1930s and rebuilt at the expense of the city government of Kyiv in the 1990s. It now belongs to the Ukrainian Orthodox Church-Kyiv Patriarchate (Photo by Esther Long Ratajeski)

official in the UOC-Moscow Patriarchate in late 2011 (RISU 2011a). Although no broad national-scale religious education project was underway as of early 2012, courses in "spiritual and moral direction" are taught in some provinces, such as L'viv, but not country-wide (Institute for Religious Freedom 2012).

82.4 Catholic Christianity in Ukraine

Approximately 10 % of Ukraine's population is affiliated with the Catholic Church. Of these, most (about 8 % of Ukrainians) belong to the Greek Catholic, or Uniate, church (CIA 2012). This church falls under the jurisdiction of Rome, but follows tenets of Byzantine orthodoxy, such as worship style and rules such as those permitting priests to marry. The Ukrainian Greek Catholic church is the largest Eastern Rite Catholic Church in the world, with roots in the late sixteenth century Union of Brest, when local bishops in what is now Western Ukraine made a strategic partnership with the Roman Catholic church, supported by the Roman Catholic political leadership in the Polish-Lithuanian Commonwealth at the time (Subtelny 2000: 100).

Fig. 82.4 The church of St. George the Victory-Bearer, constructed in 2001 near Kyiv's main train station, was funded by the national train corporation (Photo by Austin Crane, used with permission)

Despite being outlawed and absorbed by the Russian Orthodox Church during the Soviet period, the Greek Catholic Church dominates the religious sphere in western Ukraine to this day and is the largest religious organization in the provinces of Ivano-Frankivsk, Lviv, and Ternopil, as well as having a strong presence in Transcarpathia, in the Carpathian mountains of far western Ukraine (RISU 2004). Of Ukraine's four million Greek Catholics, 93 % live in the western part of the country (U.S. State Department 2010).

Ukraine's one million Roman Catholics and their 901 registered communities are closely connected to Ukraine's Polish population, and are also concentrated in the western part of the country (U.S. State Department 2010). A website of the Roman Catholic Church in Ukraine, for example, is published in both the Ukrainian and Polish languages (Roman Catholic Church in Ukraine 2012), and the current archbishop of the L'viv Archdiocese in western Ukraine is ethnically Polish, as were all of his predecessors dating back to 1412 (Cheney 2012). This Polish lineage points to western Ukraine's history as part of Poland, when a sizable portion of its population, especially the wealthy landowners and nobility, was ethnically Polish. The Polish population of Ukraine plummeted during and after the Second World War, and stands today at only 0.3 % of the country's population (CIA 2012).

Of the various issues facing Ukrainian Catholicism since 1991, one of the most significant has been the reclamation of church property from the Russian Orthodox (and later, Ukrainian Orthodox – Moscow Patriarchate) Church or the Ukrainian government to both the Roman Catholic and Greek Catholic Churches. All Greek Catholic church buildings had been Russian Orthodox or secular state property during the Soviet era and deciding how to return them to their Greek Catholic status was a difficult headache for religious and civic leaders. As early as 1989, while the USSR was crumbling, Greek Catholics began occupying parish churches across western Ukraine, claiming them for their own. Violent clashes between Greek Catholic and Orthodox demonstrators took place in a number of towns. The Russian Orthodox Church agreed that in towns with a majority Greek Catholic population the church would be returned to the Greek Catholics. If a town had more than one church, then one church would remain Orthodox and the other would become Greek Catholic (Davis 1995: 74). In practice, the transfer of property was a contested affair. Civil authorities also got involved, such as when city officials in L'viv gave the Cathedral of St. Yuri to the Greek Catholics, despite vehement Orthodox protests (75). By 1990, 90 % of Orthodox churches in the western Ukrainian regions of L'viv and Ivano-Frankivsk had been transferred or forcibly seized by the Greek Catholics or the Autocephalist Orthodox Church (76).

Some former Catholic church properties, for example the Roman Catholic St. Nikolas Cathedral in Kyiv and a former bishops' residence, remain in government hands (Fig. 82.5). St. Nikolas, a neo-Gothic structure built in the late nineteenth century, was confiscated by the Soviet government in the 1930s, used by the KGB, and later became the National House of Organ and Chamber Music in Ukraine, remaining so to this day. The Church is permitted to use the structure for religious services, but not to have complete ownership. The Roman Catholic Church also complained about properties that have not been returned to it in Dnipropetrovsk (in eastern Ukraine), L'viv (in western Ukraine), Mykolayiv and Odessa (in southern Ukraine), and in Sevastopol and Simferopol, on the Crimean peninsula. In another property dispute, a local village council refused to implement the regional government's order to return a convent in the Ternopil oblast to the Roman Catholic Church (U.S. State Department 2008).

As of 2010, only one significant property dispute regarding the Greek Catholic Church remained. A building next to St. George's Cathedral in L'viv had not been returned to the church, but local officials said they did not have enough money to purchase new homes for a dozen families who had lived in it since Soviet times (U.S. State Department 2010).

Fig. 82.5 Cathedral of St. Nikolas in Kyiv, formerly a Roman Catholic property but now the National House of Organ and Chamber Music (Photo by Austin Crane, used with permission)

Despite these individual examples to the contrary, most property disputes over local parishes between the various branches of Orthodoxy and Catholicism had been resolved by the late 1990s. This was mostly a result of the construction of new church buildings to accommodate worshippers who did not "win" the right to keep the village church. In a typical town, the majority of the townspeople, or the most powerful townspeople, were able to decide if the church would remain Ukrainian Orthodox, which branch of Orthodoxy it would belong to, or if it would become Greek Catholic. The minority group was then forced to build a new church building, if they had the means to do so. The decision over which group would retain control of the church property occasionally led to violence and the involvement of civil authorities (Mitrokhin 2001: 182).

An area of current difficulty is the resistance the Greek Catholics are finding as they seek to expand outside of their traditional base in western Ukraine. They are facing resistance from municipal and oblast governments to the allocation of land and construction of new church buildings in eastern and southern Ukraine. A recent report listed these types of problems in Kharkiv, Kyiv, Kyiv Oblast, Horokhiv, Luhansk, Krasny Luch, Odessa, Poltava, Shatsk, Sumy, Simferopol, Yalta, and Yevpatoriya (U.S. State Department 2010).

82.5 Protestantism in Ukraine

Although the percentage of Protestant Ukrainians remains much smaller than those in the dominant religious groups, the past 20 years has seen remarkable growth in this community. According to sociological surveys, in 1994 only 0.6 % of Ukraine's population considered themselves to be Protestant (Krindatch 2003: 42). Today in Ukraine Protestants make up 2–3 % of the population – 2.4 % according to one 2006 religiosity survey (Yelensky 2010: 217) – and about one-fourth of all religious communities in the country (RISU 2009). Overall Protestant growth was on par with the growth of other religions from 1994 to 2009 (166 % growth); however, within that figure charismatic and Pentecostal groups exhibited much higher growth rates. For example, Full Gospel churches expanded from 33 to 601 communities; other charismatic communities grew from 44 to 396 communities. Yelensky reports that by the early 2000s Ukraine was the country with the most sizable charismatic, Pentecostal and Baptist communities in all of Central and Eastern Europe (2010: 218).

The broad spectrum of evangelical churches in Ukraine encompasses most Protestant ideologies and practices. The churches hold varying views on issues including alcohol consumption, speaking in tongues, security of salvation, free will and the predestination of believers, church governance, worship style, baptism, communion, and so forth. One way to think of the complex mosaic of Protestant churches in Ukraine is to divide them into those groups founded prior to the 1917 Revolution and those that arrived in Ukraine after Ukraine's independence in 1991. In the first group is the largest Protestant denomination in Ukraine, the All-Ukraine Union of Evangelical Christians-Baptists (ECB), which had 2,516 registered communities in 2009, making it by that measure the fifth-largest denomination of any religion in Ukraine (RISU 2009). The ECB has its roots in the late nineteenth century, when some Ukrainian farm workers converted to Protestant Christianity by way of German Mennonite settlers in southern Ukraine (Rowe 1994). In the second group are myriad other denominations, nearly all of them begun through the work of missionaries from North America and Europe. The establishment of religious freedom in 1990 opened the floodgates for missionaries, who came from foreign denominations such as the Southern Baptist Convention and the Presbyterian Church in America, from non-denominational mission boards such as SEND International (both of these examples are U.S.-based), and from individual congregations abroad. Other missionaries came as individual workers, the most famous of these being Nigerian-born Sunday Adelaja, who started Ukraine's largest charismatic church, the Embassy of the Blessed Kingdom of God for all Nations. Some of the missionaries partnered with existing churches such as the ECB, while others formed new churches of all Protestant traditions, from Charismatic to Baptist, Nazarene, independent evangelical, Presbyterian and Reformed, Methodist, and more. Also active were groups like the Latter-Day Saints and the Jehovah's Witnesses.

All Protestant churches in Ukraine, the historic churches that survived the Soviet era as well as new churches started by foreign missionaries, have been marked by intense interaction with people, ideas, and money from abroad. For instance, Ukrainian clergy and laity have studied at religious colleges and seminaries in Ukraine and abroad funded by international partners; church members have found employment helping construct church buildings, working in religious publishing houses, or administering humanitarian aid paid for by Western partners. Ukrainians have developed international friendships and even families as a result of social interactions with foreign missionaries, and humanitarian aid including medical relief has impacted people involved with Protestant churches and their surrounding communities. In other words, all Ukrainian Protestant churches, whether historic denominations or new endeavors, are highly integrated into transnational religious networks.

Determining the number of foreign religious workers in Ukraine in the early years of independence is nearly impossible, as is calculating how much money the international connections poured into the country. One expert estimated 561 mission agencies, along with over 5,000 foreign religious workers, active in all of the countries of the former Soviet Union in 1997, but missionaries specific to Ukraine were not identified in this report (Elliott 1997: 10). Just a few years later, in 2003, the Ukrainian government reported that nearly 12,000 religious workers entered the country, with more than half of them (6,283) U.S. citizens (U.S. State Department 2004). The amount of international financial support for Protestant churches in Ukraine is also unknowable, but the example of the largest Protestant denomination is instructive. The Evangelical Christians-Baptists (ECB), which have over 2,500 communities across the country, have very close ties with churches in the United States and elsewhere. During a series of interviews in four Baptist churches in different Ukrainian cities in 2002–2003, ECB members and pastors spoke at length about American groups who had worked with them in evangelistic, educational, humanitarian, and building projects (Long 2005). Most Ukrainian cities, and many small towns, now have Baptist church buildings that were paid for by Western supporters and constructed with Ukrainian labor.

One such church is the central Baptist church in Vinnytsia, an oblast capital in west-central Ukraine (Fig. 82.6). It is widely known to be one of the largest ECB churches in Ukraine, with about 1,500 members as of 2003, not including children or unbaptized regular attenders. The impressive worship hall has three balconies, faux-painted concrete columns made to look like marble, gold-painted flourishes throughout, and a large mural of a pastoral scene with an angel flying overhead. The building was primarily financed by churches and individuals in the United States, but constructed by local church members. Raising money for building projects such as the church itself and for an adjacent Bible college was a chief aim of church officials' international fundraising efforts.

Fig. 82.6 Vinnytsia's largest Baptist church (Photo by Esther Long Ratajeski)

Although most Protestant congregations worship in newly-constructed buildings or rented halls, a few have gone the route of Ukraine's Orthodox and Catholic churches and meet in historic structures that had been used for other purposes during the Soviet era. This building strategy was used by the Evangelical Presbyterian Church of Ukraine for their main congregation in Odessa. The EPCU is quite small (fewer than 2,000 members overall), but active in a number of cities across Ukraine. The Evangelical Reformed Presbyterian Church of Odessa meets in a building constructed in the late nineteenth century by Reformed Protestant immigrants from France, Germany, and Switzerland (Fig. 82.7). Odessa was a cosmopolitan port city at the time, with a vibrant international population. The imperial government did not permit services to be held there in Russian, the dominant language of Odessa, so French and German were exclusively used in the liturgy until about 1914. At some point after the 1917 revolution the property was confiscated by the Soviets and eventually converted into a puppet theater and the headquarters of the local actors' union.

Fig. 82.7 Odessa Reformed Evangelical Presbyterian Church, the former building of the late nineteenth century Reformed Church and a Soviet-era puppet theater (Photo by Esther Long Ratajeski)

The process of transferring ownership from the government to the EPCU after Ukraine's independence was a time consuming and expensive process. After the first small Presbyterian congregation was formed in Odessa through the efforts of missionaries from the United States, some of the new Ukrainian Presbyterians made contact with an elderly pastor from western Ukraine. This man had been ordained in the historic Reformed denomination that had originally owned the Odessa property, but which had nearly become defunct in recent decades. A handful of Reformed congregations survived in western Ukraine. This Reformed minister, as one of the last representatives of the earlier group, wrote a letter declaring that the new Evangelical Presbyterian Church of Ukraine (EPCU) should be considered physical and moral heir of the Reformed church in Odessa. This letter permitted the nascent

Evangelical Presbyterian denomination to argue in court that they should be given the Reformed Church building/puppet theater in downtown Odessa. In return, the Ukrainian government expected that the EPCU would renovate the historic structure. Missionaries were able to raise hundreds of thousands of dollars in the United States for the endeavor, and renovations were complete by the mid-2000s. Limitations of the transfer of church property are demonstrated, however, in that the Presbyterian church has thus far been unable to use office space on the first floor of the building, because the actors' guild, which used those rooms during the Soviet period, has refused to move out. The church rents office space for themselves elsewhere. This case highlights a weakness in Ukraine's justice system, in that even though the church won a high court case over the matter, local and regional courts refused to carry out the decision to evict the actors' guild from the church property (U.S. State Department 2008).

A discussion of Protestantism in Ukraine post 1991 would be incomplete without considering the impact of the charismatic movement on Ukraine's religious landscape. Ukraine's charismatic community is emblematic of the rapid growth of charismatic churches around the world (see Jenkins 2011) in that the country's fastest-growing congregations of any religion are charismatic Protestant Christians. According to Ukrainian government data, the number of churches in the "other Charismatic" community increased 800 % between 1994 and 2009, from 44 to 396 communities (RISU 2009). The charismatic community is strongest in east-central Ukraine, where it is said to account for 70 % of all Protestants. Eastern Ukraine is the region of the most rapid growth for charismatic Protestants, as it is the part of Ukraine where traditional religions are the weakest. Charismatic Christianity in Ukraine tends to be youth-oriented and focused in urban areas. It is also the branch of Protestant Christianity in Ukraine where you are most likely to find a female pastor, perhaps but not exclusively part of a husband-wife pastoral team (Lyubashchenko 2010: 275).

The most famous charismatic church in Ukraine is Kyiv's Embassy of the Blessed Kingdom of God for All Nations, founded and led by Nigerian-born Sunday Adelaja. The story of this church is often recounted in books and articles about the growth of charismatic Christianity: a Nigerian student studying in Belarus moved to Kyiv in 1994 and began a small church. The church grew rapidly, attracting people from all walks of life, including the (now former) mayor of Kyiv. The church founded satellite congregations in cities across Ukraine and surrounding countries, and by 2012, according to their website, they had 100,000 members in the entire country (www.godembassy.com), although Wanner estimates that the church has only 25,000 members (Wanner 2010: 12). Some observers have called God's Embassy church the largest in all of Europe (Marsh and Tonoyan 2009: 510). God's Embassy Church is a typical charismatic church in that it believes in the Bible as the inerrant word of God, it practices an emotional worship style, and it teaches Pentecostal theology such as speaking in tongues, faith healing, and prophecy. It is also known as a "prosperity gospel" church, which promotes the teaching that God does not want his people to be poor, and that financial wealth demonstrates God's favor on the believer. As of 2012 God's Embassy Church was constructing a building intended to seat 6,000 worshippers in Kyiv; they have been meeting in temporary facilities since their founding in 1994. Among many programs organized by God's Embassy Church are drug and alcohol rehabilitation programs and humanitarian outreaches to the poor.

In recent years the God's Embassy church has faced serious backlash from the Ukrainian establishment, culminating in a court case in which Pastor Adelaja has been accused of fraud and heading a criminal organization, in effect, a financial Ponzi scheme that was run by some of his church members. Adelaja denies any knowledge of or involvement in the scheme, but a court case is ongoing and could send him to prison (Barton 2012). Orthodox leaders accuse Adelaja of leading a cult and brainwashing church members. Instead of leaving the country, Adelaja is remaining in Ukraine "to clear his name of all kinds of corruption, allegations and accusations" (Embassy 2012).

The attack on God's Embassy church is emblematic of an anti-charismatic atmosphere in Ukraine on the part of secular authorities as well as the traditional religions, including historic Protestant churches. Government officials think that charismatics are "victims of manipulation or even fraud and their leaders imposters who are furthering their own business projects under the guise of religion" (Lyubashchenko 2010: 275–276). Leaders of historic Protestant churches such as the Pentecostal churches believe that charismatics teach a false theology, promote a loose morality, and ignore and dishonor traditions that led to persecution and even martyrdom for Ukrainian Protestant forbearers.

Within Protestant communities of all kinds in Ukraine, there is a marked attention to educating upcoming generations of believers in matters of the faith. This includes weekly Bible teaching of children at Sunday schools or similar programs, the establishment of private Christian elementary and high schools, and the organization of Bible colleges and theological seminaries to prepare men and women for careers in the church. Ukrainian law has thus far forbidden the establishment of theological departments in state universities, or the establishment of religious elementary or high schools recognized by the state (Druzenko 2010: 734). Despite that, evangelical churches have been opening religious schools and colleges across Ukraine since the early 2000s. One solution to the legal problem of the state not recognizing religious education is to register the school or university as both a secular private institution as well as a "spiritual educational establishment," provided that the teachers of the secular institution meet the legal accreditation standards (735). In Ukraine's province of Transcarpathia 62 % of the population is ethnic Hungarian, and 70 % of them belong to the Reformed Church, a historic Protestant denomination. This area had a tradition of church-run high schools, but that practice ended when it became part of the Soviet Union. Since the mid-1990s a number of towns in Transcarpathia have again established religious-based institutions for secondary education sponsored by church-related organizations, most of them Reformed but also Roman Catholic (Molnár 2008).

Protestant churches have also established a number of theological seminaries for training future clergy and laypeople. In fact, the largest Protestant denominations have more educational institutions than the three main Orthodox denominations combined. While as of 2009 there were 42 Orthodox educational institutions, there were 42 educational institutions run by the Evangelical Christians-Baptists alone. The numbers of full-time students at the Protestant-run educational institutions also is impressive, especially when considering how much smaller those denominations are compared to the Orthodox churches. The Evangelical Christians-Baptists and the Pentecostals had over 3,000 full-time students in their institutions of higher learning, while the three large Orthodox denominations had only 2,400 full-time students at Orthodox educational institutions. Meanwhile, there were 1,535 full-time students at 23 Greek and Roman Catholic educational institutions (RISU 2009). The emphasis on Christian education is more prominent among Protestants and Catholics than Orthodox, perhaps because Orthodox churches, "satisfied with their numerical superiority, prefer to direct their financial resources to the restoration of churches and conduct religious education on an *ad hoc* basis" (Mitrokhin 2001: 179).

82.6 Islam in Ukraine

Muslims, the final religious group considered here, have had a presence in the territory that is now Ukraine since the seventh century, when the first Muslims arrived on the Crimean peninsula. A stronger foothold was established by the thirteenth century, when the first known mosque was built (Yakubovich 2010: 292). In the twenty-first century, Crimea remains the locus of Islam in Ukraine, as most Ukrainian Muslims are Crimean Tatars. Although Crimea's Sunni Muslim community survived the Russian Empire, it barely held on after the Second World War (1944), when Stalin deported most Tatars to various distant locales such as the Kazakh and Uzbek Soviet Socialist Republics. Nearly 200,000 Tatars were removed from their homes under pretext of Nazi collaboration, and only about half of them survived the resettlement process (Subtelny 2000: 483). The exile was reversed in the late 1980s and 1990s, as tens of thousands of Tatars were repatriated to Crimea. A trickle of Tatar immigrants to Crimea continues to the present day.

The exact number of Muslims in Ukraine is unknown, and estimates range from a low of 206,000 to an unlikely two million (Yakubovich 2010: 294). One survey estimated 456,000 Muslims living in Ukraine as of 2007, or about 1 % of the country's population (Pew Forum 2009). These Muslims were active in 1,135 registered and unregistered Islamic religious communities, or about 3.4 % of all religious organizations in Ukraine (RISU 2009). More than half of these Muslims are likely Crimean Tatars (who had a 2008 population of about 250,000), and the rest are Volga Tatars (especially in Luhansk and Donetsk regions), Azerbaijanis (mostly in Eastern Ukraine), and immigrants from Islamic countries such as Afghanistan, Iran, Pakistan, and Turkey (Yakubovych 2010: 294).

While Ukraine's relatively small Muslim community does not hold a central position in Ukrainian society, Muslim institutions are active and growing. For example, the Spiritual Administration of Crimean Muslims (SACM) is a religious organization that also seeks the restoration of Crimean Tatar sites, the protection of the Tatar community from hostile groups, and the publishing of material in the Tatar language (Yakubovich 2010: 294). The Spiritual Administration of Muslims in

Ukraine (SAMU) represents fewer people, but is often named as Ukraine's "official" Muslim organization. The duties of its leader, Lebanese-born Sheikh Ahmad Tamim, include receiving official delegations from foreign countries, such as the early (2012) visit of parliamentarians from Iraq (Ministry of Foreign Affairs). Finally, the All-Ukrainian Association of Social Organizations "Alraid" operates free elementary schools in the Arabic language, publishes a Russian language newspaper and website (http://gazeta.arraid.org/), and organizes religious-based conferences, among other outreach activities. A recent conference held in Kiev promoted an active role for Muslim women in society, and brought the (female) ambassador to Ukraine from Indonesia as the guest speaker (Alraid 2012) (Fig. 82.8).

Like Ukraine's other religious groups, Muslim communities have struggled with property issues in recent years. For example, the Simferopol city council had refused to grant land to the Spiritual Directorate of Crimean Muslims for the construction of a mosque, and the high court of Crimea ordered the city to do so in 2010. In other areas the Muslims struggled to have historic mosques returned to them. For example, the U.S. State Department recorded a dispute regarding a 118-year old mosque in Mykolayiv in southern Ukraine, a 150-year old building in Masandra in Crimea, and the ruins of an even older mosque along the Crimean coast (U.S. State Department 2010).

Unlike most other groups, Muslims in Ukraine have suffered incidents of discrimination in the workplace and in society at large. These are difficult to categorize as religious in nature, however, but could also be considered ethnic discrimination, since most Muslims in Ukraine are Tatar, not ethnic Ukrainians or Russians. One example of discrimination occurred in Crimea, where Muslims complained that

Fig. 82.8 Ar-Rahma Mosque in Kyiv was completed in 2000, and provides an example of the new mosques being constructed in the post-Soviet era (Photo by Austin Crane, used with permission)

ethnic Russian officials denied them employment in government offices (U.S. State Department 2010). In December 2011 the Spiritual Board of Muslims of the Crimea complained that a village council in Dobre cancelled its decision to allow Muslims land to build a mosque. They accused the officials of trying to "break the will" of Muslims by "repressions" and by ignoring their constitutional rights (RISU 2011b).

82.7 Conclusions

Religious organizations in Ukraine have faced some common challenges, despite reaping countless benefits from this historically unprecedented era of tolerance for religious expression. For instance, they all face the reality that most Ukrainians do not have a habit of frequent religious practice. Various groups deal with this differently, such as emphasizing religious education and training (Protestants) or focusing on gaining acceptance as the dominant religion in as much of Ukraine as possible (Orthodox). All deal with a legal system that has not always facilitated the property rights of religious organizations. This is especially clear when local governments blatantly ignore rulings from provincial or national courts ordering the return of church property. In practice, it appears that the local officials have the most power in these relationships. All religious groups entered an era of transnational religious networking with little experience and few contacts, and have spent the past two decades seeking their position in an international context, with varying degrees of success. Ukraine's religious groups have to operate in a society with little history of religious organizations being involved in civic life, such as in education, medical care, or anti-poverty programs. Many of them, particularly the Protestant churches, have filled a social vacuum by providing medical services, drug rehabilitation programs, and soup kitchens for needy Ukrainians. Finally, all groups have dealt with Ukraine's complicated regional mosaic of religious preferences and history, which makes some religious organizations more or less accepted depending on their location. Chiefly relevant here is the hostile reception to the Ukrainian Orthodox Church – Moscow Patriarchate in western Ukraine, and the Ukrainian Orthodox Church – Kiev Patriarchate and Greek Catholic church in eastern Ukraine. But also noteworthy is the lack of appreciation within mainstream society for the religious expression of nontraditional groups such as charismatics, especially in parts of the country with a more traditionally religious population. Ukraine's Muslims face an additional set of circumstances, as they live in a society with a level of ethnic/racial tension that also lacks understanding of their religion.

Despite these challenges, Ukraine's religious communities have managed to weave together a vibrant tapestry of religious life that involves not just believers themselves, but a robust network of religious institutions, most of which did not exist 25 years ago. These institutions, and the people who make them up, are leaving their imprint on Ukrainian society, and future generations will inhabit a Ukraine that is quite different from what it would have been without these religious pioneers.

References

All-Ukrainian Association of Social Organizations "Alraid." (2012, April 10). *International female conference in Kiev: Muslim women should be active members of society*. Retrieved April 17, 2012, from www.arraid.org/en/index.php?pagess=main&id=496&butt=1

Barton, J. (2012, February 13). Ukraine's Embassy of God evangelical church struggles with founder's controversy. *The World*. Retrieved May 9, 2012, from www.theworld.org/2012/02/embassy-of-god-ukraine/

Central Intelligence Agency. (2012). *World Factbook*. Retrieved April 27, 2012, from https://www.cia.gov/library/publications/the-world-factbook/geos/up.html

Cheney, D. (2012). Archdiocese of L'viv. *Catholic Hierarchy*. Retrieved May 5, 2012, from www.catholic-hierarchy.org/diocese/dlvla.html

Davis, N. (1995). *A long walk to church: A contemporary history of Russian Orthodoxy*. Boulder: Westview.

Druzenko, G. (2010). Religion and the secular state in Ukraine. In J. Martínez-Torrón & W. C. Durham Jr. (Eds.), *Religion and the secular state: Interim national reports* (pp. 719–736). Provo: International Center for Law and Religion Studies.

Elliott, M. (1997). Updated statistics on the Protestant missionary presence in the former Soviet Union. *East-West Church and Ministry Report, 5*(2), 10.

Embassy of the Blessed Kingdom of God for All Nations. (2012). *Update on the court situation with Pastor Sunday*. Retrieved May 10, 2012, from www.godembassy.com/main/the-truth/item/565-update-on-the-court-situation-with-pastor-sunday.html

Institute for Religious Freedom. (2012, February 15). *Ukrainian state does not fully use the potential of religious education*. Retrieved May 13, 2012, from www.irf.in.ua/eng/index.php?option=com_content&view=article&id=304:1&catid=34:ua&Itemid=61

Jenkins, P. (2011). *The next Christendom: The coming of global Christianity* (3rd ed.). Oxford: Oxford University Press.

Krindatch, A. (2003). Religion in post-Soviet Ukraine as a factor in regional, ethno-cultural and political diversity. *Religion, State and Society, 31*(1), 37–73.

Long, E. (2005). *Identity in evangelical Ukraine: Negotiating regionalism, nationalism, and transnationalism*. Ph.D. dissertation, Department of Geography, University of Kentucky, Lexington.

Lyubashchenko, V. (2010). Protestantism in Ukraine: Achievements and losses. *Religion, State and Society, 38*(3), 265–289.

Marsh, C., & Tonoyan, A. (2009). The Civic, economic, and political consequences of Pentecostalism in Russia and Ukraine. *Society, 46*, 510–516.

Ministry of Foreign Affairs. (2012). *The Iraqi Parliamentary Delegation visits the Religious Administration of Ukrainian Muslims. Embassy of the Republic of Iraq in Kyiv*. Retrieved April 14, 2012, from www.mofamission.gov.iq/UKR/en/articledisplay.aspx?gid=1&id=6076

Mitrokhin, N. (2001). Aspects of the religious situation in Ukraine. *Religion, State and Society, 29*(3), 173–196.

Molnár, E. (2008). The conditions of functioning of denominational educational institutions in Ukraine. In G. Pusztai (Ed.), *Region and education III: Education and church in Central- and Eastern-Europe at first glance* (pp. 85–100). Debrecen: Center for Higher Education Research and Development University of Debrecen and Hungarian Academy of Sciences Board of Educational Sociology.

Pew Forum on Religion in Public Life. (2009). *Mapping the global Muslim population: A report on the size and distribution of the world's Muslim population*. Washington, DC: Pew Research Center.

Religious Information Service in Ukraine (RISU). (2004). *Number of religious organizations as of 1 January, 2004 broken down according to regions of Ukraine*. Retrieved May 1, 2012, from http://old.risu.org.ua/eng/resources/statistics/reg2004/

RISU. (2009). *Religious organizations in Ukraine as of 1 January, 2009*. Retrieved April 24, 2012, from http://old.risu.org.ua/eng/resources/statistics/ukr2009

RISU. (2011a, October 25). *Ukrainian Orthodox Church-Moscow Patriarchate intends to introduce Orthodox education at schools*. Retrieved May 5, 2012, from http://risu.org.ua/en/index/all_news/state/legislation/45057/

RISU. (2011b, December 16). *Spiritual Board of Muslims of the Crimea states that rights of Muslims are violated*. Retrieved May 2, 2012, from http://risu.org.ua/en/index/all_news/community/land_and_property_problems/45963/

Roman Catholic Church in Ukraine. (2012). *The L'viv Archdiocese of the Roman Catholic Church in Ukraine*. Retrieved May 5, 2012, from www.rkc.lviv.ua

Rowe, M. (1994). *Russian resurrection: Strength in suffering, A history of Russia's Evangelical church*. London: Marshall Pickering.

State Committee for Religious Affairs. (1994–2003). Data published in *Liudyna i Svit*. Kyiv, Ukraine.

Subtelny, O. (2000). *Ukraine: A history* (3rd ed.). Toronto: University of Toronto Press.

Tonoyan, L. S., & Payne, D. (2010). The visit of Patriarch Kirill to Ukraine in 2009 and its significance in Ukraine's political and religious life. *Religion State and Society, 38*(3), 253–264.

United States Department of State. (2004). *International religious freedom report 2004*. Retrieved May 25, 2012, from www.state.gov/j/drl/rls/irf/2004/35491.htm

United States Department of State. (2008). *International religious freedom report 2008*. Retrieved May 13, 2012, from www.state.gov/j/drl/rls/irf/2008/108477.htm

United States Department of State. (2010). *International religious freedom report 2010*. Retrieved May 15, 2012, from www.state.gov/j/drl/rls/irf/2010/148993.htm

Wanner, C. (2004). Missionaries of faith and culture: Evangelical encounters in Ukraine. *Slavic Review, 63*(4), 732–755.

Wanner, C. (2010). Social ministry and missions in Ukrainian mega churches: Two case studies. *East-West Church and Ministry Report, 18*(4), 12–14.

Yakubovych, M. (2010). Islam and Muslims in contemporary Ukraine: Common backgrounds, different images. *Religion, State and Society, 38*(3), 291–304.

Yelensky, V. (2010). Religiosity in Ukraine according to sociological surveys. *Religion, State and Society, 38*(3), 213–227.

Chapter 83
An Exception in the Balkans: Albania's Multiconfessional Identity

Peter Jordan

83.1 Introduction

Much in contrast to the other nation states in the Balkans, Albanian national identity is not bound to a denomination. While all the Orthodox nations (Greeks, Bulgarians, Macedonians, Romanians, Serbs, Montenegrins), but also the Roman Catholic Croats and the Muslim Bosniaks define themselves primarily by religion, (see Suttner 1997a, b, 2001; Hösch 1995[3]; Hatschikjan and Troebst 1999) the population of Albania is split into three confessions roughly in the proportion 70 % Muslims, 20 % Orthodox and 10 % Roman Catholics, which makes religion not an appropriate identity marker.

Communist Albania had even declared to be an atheist state and had suppressed religion more than other Communist countries, not the least to overcome this internal division and to enforce national homogeneity. Post-Communist Albania was quick to restore freedom of religious confession and to reintroduce all religious denominations into their religious and societal functions. All of them have indeed used this opportunity very well and substantially gained ground. They are appreciated by the population as institutions not having been involved into Communist affairs and neutral in political terms and today play important roles also in the social field and in education. And although the vast majority of Albanians are Muslims, the state has avoided any special relation to this faith and is also on the international scene not at all acting as a Muslim country or cultivating special relations with other Muslim countries in the wider region (for example, Turkey). It goes even as far as to style the Roman-Catholic nun Mother Tereza, an Albanian from Skopje in Macedonia, as an icon of Albanian identity – stressing in this way the nation's multi-confessional identity.

P. Jordan (✉)
Austrian Academy of Sciences, Institute of Urban and Regional Research,
Postgasse 7/4/2, A-1010 Wien, Austria
e-mail: peter.jordan@oeaw.ac.at

S.D. Brunn (ed.), *The Changing World Religion Map*,
DOI 10.1007/978-94-017-9376-6_83

This chapter will first be embedded into Manfred Büttner's "Bochum Model" of religion geography. Then a survey of the role of religion and religious communities in other Balkan and eastern European countries will follow. Thirdly it will look into the role of religion within Albanian nation-building in the nineteenth and early twentieth centuries. Fourthly, it examines Albania's current religious structure as well as the status of religious communities in public life as well as their relationship to the state. In conclusion I try to answer the question, why it was possible to develop a national identity across religious boundaries under the conditions of the coincidence of religion and nation in the Balkans. As a kind of appendix, I examine whether this Albanian model was transferable to the second essentially multireligious Balkan state, viz., Bosnia and Hercegovina.

The paper is, as regards its historical references, to a larger extent based on the works of the historians Aydin Babuna (2004), Nathalie Clayer (2003) and Oliver Jens Schmitt (2012). Also the findings of Gottfried Hofmann in his diploma work at the University of Vienna (Hofmann 2009) have been a valuable source.[1]

83.2 Büttner's Bochum Model of the Geography of Religion

Manfred Büttner characterizes the relationship between religion and geographical space by a model with three discreet levels (Fig. 83.1, Büttner 1998: 59). The "corpus of religion" ("*Religionskörper*") including the members of a religious community as well the organizational infrastructure of this community are located at the central level integrated with the rest of the society, that is, other religious communities and people not affiliated to a religious community, but also nations as another category of imagined communities. These groups interact at this level with the "corpus of religion" and affect each other. Büttner calls this level the "social level" ("*Sozialebene*").

It shapes the so-called "indicator level" ("*Indikatorebene*"), which comprises tangible features of the geographical space including the cultural landscape. This level is able "to indicate" the presence and effects of communities, but also the effects of a "corpus of religion." The "corpus of religion" manifests itself at this level, for example, by churches and other religious buildings, frequently in a rather symbolic way by establishing landmarks.

[1]The author acknowledges also the works of Brunnbauer (2002), Büttner (1985), Daxner et al. (2005), Deutsch (1989), Doka (2001, 2005), Goehrke and Gilly (2000), Harris (1993), Ivanišević et al. (2002), Kahl et al. (2006), Kahl and Lienau (2009), Kraas and Stadelbauer (2002), Lichtenberger (1976), Lienau and Prinzing (1986), Lienau (2001), Louis (1927), Lucic (2012), Matuz (1994[3]), Milo (2002), Mitchell (2000), Moore (2002), Okuka, (2002), Paulston and Peckham (1998), Pichler (2002), Roth (2005), Rugg (1994), Schermerhorn (1970), Schmidt (1961), Schubert (2002), Seewann (1995), Seewann and Dippold (1998), Smith et al. (1998), Sugar (1980), Sundhaussen (1994), Suppan and Heuberger (1991), Symmons-Symonolewicz (1985), Todorova (1997, 2002) and White (2000) as basic for this chapter and refers to his own findings in Jordan (2007), Jordan et al. (2003, 2006) as well as Jordan and Lukan (1998).

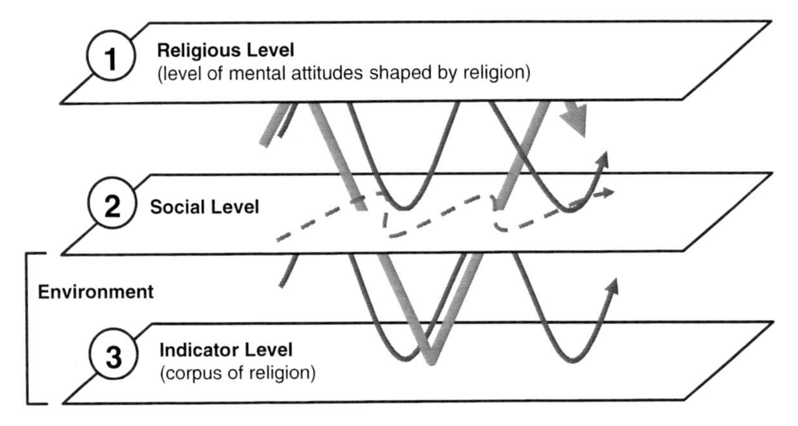

Fig. 83.1 "Bochum Model" of religion geography by Manfred Büttner (Modified from Büttner 1998: 59, used with permission)

To the "social level" the "religion level" ("*Religionsebene*") or the "level of mental attitudes shaped by religion" ("*Ebene der religiös geprägten Geisteshaltungen*") then is superimposed. This level represents the totality of ideas and thoughts motivating people to develop a certain "corpus of religion." But the relationship with the "social level" is a mutual one, since religious and other communities (located at the "social level") produce these ideas and thoughts or indeed modify them. Thus, the emergence of a new national idea and the subsequent formation of a new national community can, of course, modify also mental attitudes in the field of religion.

This chapter will focus on Büttner's "social level" by investigating a case study, in which within a given ethnic and linguistic community the roles and ranks for group identity between religious communities and national community have been exchanged: At the expense of religious identities and their formerly primary role a solid national identity has emerged and occupied the first place. Effects of this process on the "level of mental attitudes shaped by religion" and the "indicator level" will, however, not be neglected.

83.3 The Usual Twinning of Nation/State and Religion/Church in the Balkans

While secularisation has proceeded far in Western Europe and especially in (former) Protestant lands, the situation is quite different in Europe's eastern part.

Here, with the major exception of the Czechs,[2] whose affinity to religion has obviously been expelled by two violent and distracting counter-reformations, and

[2]According to the Czech population census of 2001, 59 % of the population declared themselves not to be affiliated to a religious denomination.

the minor exceptions of Hungarians and Slovenes, who have in more recent times rather distanced themselves from their former significant Roman Catholic identity, religion and churches play still an important role in society and as markers of national identity.

This is even true for East Central Europe with its roots in Western Christianity and its relatively high extent of Westernization. Thus, for Poles Roman Catholicism has certainly remained one of the most prominent characteristics of their nation. This may in more recent history be traced back to its role as a stronghold of the nation for more than a century, when Poland did not exist as a state and the Roman Catholic church's function as a base for all dissidents in the Communist era, not the least also to the contribution of the Polish pope to the fall of Communism in Poland. Catholicism may certainly also be regarded as a means of differentiation from the Poles' traditional rivals in the wider region, the Protestant Prussians and (North) Germans as well as the Orthodox Russians.

Also for Slovaks, their strong Roman Catholic identity is a means of differentiation – in the first line from the linguistically very related, but reformed and today rather secularised Czechs, as well as from the partly reformed Hungarians, with whom the Slovaks lived together as a common state up to 1918.

The Catholic identity of Croatians has ever since the "awakening" of a Croatian national consciousness in the nineteenth century been a means of differentiation from other South Slavonic nations, especially Serbs, Bosniaks and Montenegrins, with whom Croatians share a very similar idiom based on the Štokavian group of dialects. Catholic identification grew in importance after the idea of Yugolavism had lost its momentum already in the interwar period and especially during World War II, when the Croatian Ustaša State instrumentalized Roman Catholicism as an identity marker in contrast to the Orthodox Serbs. The earlier image of self as an "antemurum Christianitatis "against the Muslim Ottoman Empire has at that time at the lastest received the new meaning of a wall of Western Christianity against Byzantine (and more specifically Serbian Orthodox) Christianity. In the wars following and accompanying the dissolution of Yugoslavia, this meaning has again been emphasized and remains vigorous up to the present day.

Much closer still than in the predominantly Western Christian countries of East Central Europe, the Orthodox churches of East and Southeast Europe or the Balkans[3] are related to nation and state, although none of the constitutions grants anyone a privileged status.[4]

This close relationship has its roots in the traditions of the later Roman Empire,[5] which have, much in contrast to the European West, been consequently preserved by East Rome and Byzantium. While in Western Europe, more specifically in the then powerful Frankonian Empire of the eighth and ninth centuries, a parallel existence of ecclesiastical and secular systems was established sharing powers and even competing which each other, in the East Roman, later (from the eighth century onward called) the Byzantine Empire, the Christian church used to be closely affiliated and even subordinated to the secular power. The emperor was the

[3] "Southeast Europe" and "the Balkans" are used here as synonyms.

[4] A last relevant clause was deleted from the Macedonian constitution after the Ohrid Agreement in 2001.

[5] The Roman Empire had adopted Christianity as its exclusive state religion in 391.

supreme authority also of the church and was adequately worshiped as God's representative on Earth.

The intimate identification of church and nation was further been enforced under Ottoman rule, when the Orthodox churches were not only religious institutions, but represented, in absence of a Christian secular power, non-Muslims opposite the Islamic state. Also in political terms the Orthodox churches were in this protective function well accepted by the Ottoman authorities ("millet system"). Their umbrella function was neither confined to a certain ethnic group, nor to the real faithful which may have resulted in a more pronounced political attitude of Orthodox churches in territories once ruled by the Ottoman Empire and also later.

It is, therefore, also not a surprise that the "national awaking" of Christian nations in the nineteenth century in all the lands that had been influenced by Byzantine culture occurred either under the umbrella of an already existing Orthodox church (in the Balkans: Serbs, Bulgarians) or prompted the immediate implementation of a new national Orthodox church. Prominent cases corresponding to the latter variant are Romanians and Greeks.

The emergence of a Romanian national consciousness, based on a Romance and Latin identity, had for the new Romanians the resulting consequence of terminating their membership in the Serbian-Orthodox church, to which Walachians (as Romanians were called earlier) used to be affiliated and which had in the meantime turned into a body of Serbian-Slavonic nation-building. They founded their own national Romanian-Orthodox church, at first actually three of them, in accordance with the different political entities inhabited by Romanians. (They were the patriarchate of Sibiu for Hungarian Transylvania in 1864, the patriarchate of Suceava for Austrian Bucovina in 1873 and the patriarchate of Bucharest [Bucureşti] for Romania 1885 Magocsi 2002: 117).

Greek nation-building could refer to modern Greek as the language of the elites, the Byzantine (Greek) rite and the Greek Koine (ancient Greek) as the sacred language of Byzantine Orthodoxy as well as to the cultural prestige of ancient Greece. Nevertheless the Greeks were in need of a national Greek-Orthodox church, which was established in 1833 and recognized as an autocephalous church by the ecumenical patriarch of Constantinople [İstanbul] in 1850.

With the foundation of nation states accompanying the withdrawal of the Ottoman Empire from the Balkans (starting in 1830 with Greece and ending in 1912), the national Orthodox churches became also the exclusive official churches of their states. This meant their full involvement into state institutions as well as mutual support of church and state. With the exception of Albania, the relationship between Orthodox church and state remained relatively close also in the Communist era, although Communism tried to confine church activities to the private sphere. Occasionally the church was politically instrumentalized by the state. Only rarely did members of the Communist party, even those in higher ranks, separate themselves from the church. For Petru Groza, the first Communist president of Romania, for example, the Romanian-Orthodox patriarch celebrated the event with a state funeral.

After the fall of Communism around 1990, the role of Orthodox churches in society was fully been restored – in spite of trends towards westernization and secularisation. Their role in public life has regained its former importance, not the least due to a widespread and profound distrust in the party system and state institutions. The revival of Orthodox churches brought about impressive traces in the cultural landscape. Many new church buildings were constructed, sometimes in the middle of former Socialist housing quarters, but also at very symbolic places.

Today every nation in the Byzantine cultural sphere conceiving itself as a cultural nation in the full sense, possesses its own national church. This is also true for the newest nations like the Macedonian and the Montengrin, who emerged only after World War II and the dissolution of Yugoslavia, respectively.[6] Their churches have, however, so far not been recognized by the ecumenical patriarch as well as by their older sister churches. It is significant that among the Orthodox nations, only Moldavians and Belarusians lack national churches: their national identity is still faint and oscillating.

Like Orthodoxy, Islam is closely related to the state. However, in contrast to Orthodoxy it is a-national. Thus, the Ottoman Empire was an Islamic state, in which religious law coincided with state law and Muslims were the only ones to enjoy full civic rights. At the same time it disregarded national or ethnic affiliations. Albanians were not treated differently from Turks, when they were Muslims. This situation started to change only in the final stages of the Ottoman Empire's presence in the Balkans, when towards the end of the nineteenth century the movement of the Young Turks propagated a Turkish national identity and advocated the transformation of the supranational Ottoman Empire into a Turkish nation state. This effect gave reason for non-Turkish Muslim communities, who had so far only conceived themselves as ethnic, linguistic or regional, to think about their national identity.

Earlier, in 1878, for the Muslim Bosniaks the unity of religion and state had been broken by the Austro-Hungarian occupation of Bosnia-Hercegovina, that is, by having been taken over by a Christian and predominantly Catholic rule. Later, the Bosniaks in the Sandshak of Novi Pazar as well as Muslim Albanians in what is today Kosovo, Macedonia and Albania followed suit, when they had to acknowledge Austro-Hungarian, later Serbian supremacy, in Albania the rule of a secular state. For all of them except Albanians in Albania, however, religion (Islam), at least in the sense of a cultural background, remained an important, if not the most prominent (Bosniaks) marker of national identity up to the present day.

83.4 Albanian Nation-Building on the Background of Strong Religions Identities

Nation-building in the Balkans as a European periphery not affected by the ideas of French Enlightenment started with about half a century delay. The Albanians were even in this environment certainly not the first to start. Their

[6]A Macedonian nation was immediately after World War II established by the Tito regime as a means of implementing a federative structure and national balancing within the second, federative Yugoslavia. Besides a Macedonian standard language consequently also a Macedonian-Orthodox church was established. Montenegrin nation-building proceeded in the later 1990s, when the government lead by Milo Đukanović progressively dissociated itself from Belgrade. In due consequence also a Montenegrin-Orthodox church was founded.

national awakening was also much more a reaction to other national movements than spontaneous.

This late awakening was mainly due to the fact that the majority of Albanians were Sunni Muslims and as such not only a-national in attitude, but also part of the Ottoman system. They were well-embedded into this system and had little reason to think in new categories. Just the political decline of the Empire and new political directions in it prompted them to react in order to save their interests against centralist forces as well as national and territorial aspirations of neighbors, especially the Serbs and Greeks.

More active in promoting an Albanian national idea were the religious minorities: the Dervish order of the Bektashi,[7] the Orthodox in the South and the Roman-Catholics in the North (Fig. 83.2). The Bektashi became a driving force, although they were Muslims like the majority of Albanians, but felt less loyal to the Sunni Islamic Ottoman state and to represent a true Albanian variant of Islam.

In contrast to the Muslim Bektashi, the Christian communities had to cross a much higher barrier of religious identities towards a common Albanian consciousness. Among them the Orthodox could easily have assumed a Greek rather than an Albanian national identity, if Greek nation-building would have been less exclusive and puristic. But this was not the case as Greek exclusivism pushed them into the arms of an Albanian national identity. The small and isolated group of Roman Catholics had in practice no alternative and became, therefore, in many respects a forerunner in Albanian nation-building.

Albanian nation-building did, however, not start from scratch. The feeling to be an ethnic and linguistic community, that is, a kind of a "proto-national consciousness" (Schmitt 2012: 135), had already existed before 1800. The national movement brought it only to the point and transformed it into political action. This national movement, titled "renaissance" [rilindja] in analogy to the Italian *risorgimento* and the Bulgarian *văzraždane*, was prompted and promoted by the following factors:

- Ottoman reform policy ("Tanzimat") in the middle of the nineteenth century aimed at strengthening the central power and questioning privileges of local and regional Muslim potentates;

[7] While Sunni Muslims were in full concordance with the Empire in religious as well as in political terms, the mystical Dervish order of the Bektashi did and does practice a more liberal, "European" variant of Islam integrating in a syncretistic way elements of Shia as well as Christianity (Clayer 1996). The Bektashi migrated form Anatolia, according to the expansion of the Ottoman Empire, to various parts of the Balkans. From 1780 onwards they developed Gjirokastër in the South of Albania into their stronghold. Their number there grew towards the end of the nineteenth century considerably, and southern Albania turned into their center in the Balkans. Due to this fact and based on their balanced "European" variant of Islam, they conceive themselves as a true Albanian community and engage themselves much more than Sunni Muslims in Albanian nation-building. Certain intellectuals styled Bektashism also as a "bridge between Islam and Christianity" (Clayer 2003: 279).

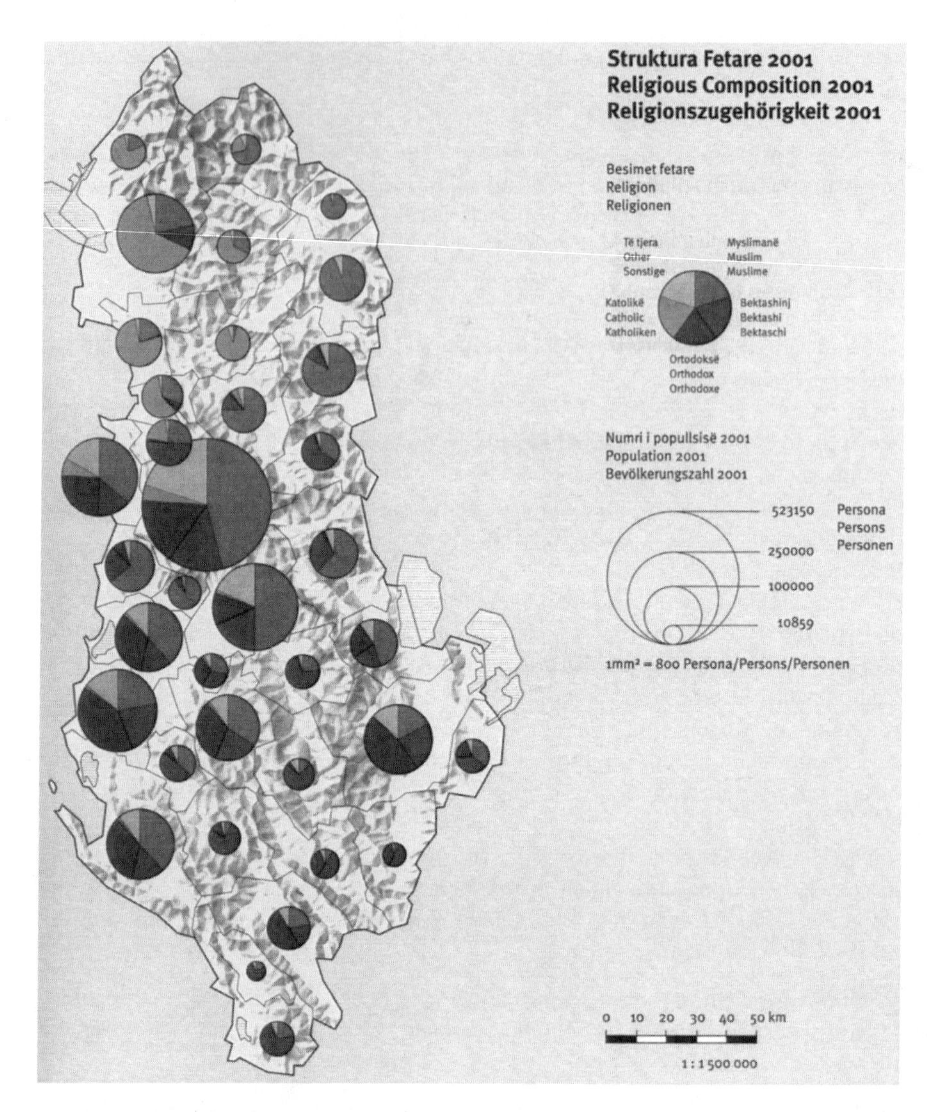

Fig. 83.2 Albania's religious structure 2001 by Arqile Bërxholi (Map from Bërxholi et al. 2003: 101, used with permission)

- heavy territorial losses of the Ottoman Empire (Bulgaria, Romania, Serbia, Montenegro, Bosnia and Hercegovina) after the Eastern Crisis (1875–1878) and the Congress of Berlin (1878) and fear of flight and expulsion; mark it with a dot like "heavy territorial losses" above attempts to nationalize the Orthodox by Serbia and Greece (Schmitt 2012: 136). The Orthodox in southern Albania were especially exposed to an aggressive Greek nationalism from 1878 onward regarding modern Greek not anymore as a supra-national language, but as the Greek national language and abandoning the theory of a common ethnic descent (Theory of the Pelasgians).

- The movement of the Young Turks towards the end of the nineteenth century proclaiming a Turkish national identity and offering Albanian Muslims to opt for this identity.

External supporters of an Albanian national consciousness were Austria-Hungary and Italy in competition with each other and in pursuit of their own interests. Both of them promoted at first Albanian autonomy within the Ottoman Empire, later an independent Albanian state, which would prevent Serbia from reaching the Adriatic coast and thus increase of Russian influence in the Balkans. They regarded Albanian national identity a mandatory prerequisite for such a state and invested many efforts in supporting it, frequently using the Albanian Roman Catholics as a bridgehead.

In the sense of a transfer of ideas Albanian nation-building was from the outside also influenced by the Italian *risorgimento*, essentially communicated to the Albanians in the Balkans by the Arbëresh, the Albanian community in Southern Italy. As a Westernized subgroup of Albanians, they played a similar role for the Albanian national movement as Piemont did for the Italian, the Vojvodina for the Serbian, Transylvania for the Romanian or Galicia for the Ukrainian national movements.

Other external models of nation-building with an impact on the Albanian process were the Greek national and cultural idea as well as (similar to many other cases especially in the Balkans) the French national idea, especially its distinct laicism.

For the rise of a national Albanian movement the following steps may be regarded as characteristic:

- From the beginning of the nineteenth century up to the Crimean War (1853–1856) it was just a cultural movement cultivated by isolated elites.
- The climax of the Tanzimat policy between 1856 and 1876 caused Albanian-minded elites of the various religious communities to join and to emphasize an Albanian identity especially oppositing Greece with its aggressive national policy among Orthodox Albanians.
- The Eastern Crisis of 1875–1878 embarrassed wider ranges of the population and made them aware of the necessity to defend Albanian interests against all neighbors.
- The Young Turks revolution in 1908 disappointed the expectations of Albanian Muslims so far loyal to the Ottoman Empire, since the Young Turks started a policy of Turkish nationalization and centralization.

The Albanian national idea had from the beginning the following main features: autochthony ("one of the oldest peoples in the wider region"), settlement continuity from Illyrian times,[8] specifics and age of the spoken language (not belonging to one

[8] This assumption has, however, been seriously contested by Gottfried Schramm's findings that Albanian toponymy on the territory of modern Albania has adopted many Slavonic names, which means that a layer of Slavonic settlement must have existed before Albanians arrived (Schramm 1999²).

of the larger Indo-European language groups), religious indifference and/or tolerance, a feeling of communality transgressing religious communities.

Among these items the specific language is certainly the most important also from an objective perspective. A language, especially a language very distinct from others in the surroundings, is always very relevant for community building due to its inclusive as well as exclusive function. It is also no question that Albanian-speakers had always felt some kind of communality. But so far language had been regarded as second to other identity markers.

Up to the end of the nineteenth century, however, the Albanian national movement aimed only at more autonomy within the Ottoman Empire. This autonomy should, according to a concept elaborated by the Bektashi and Orthodox in South Albania in 1878, be realized within a province comprising all Albanians and recognizing just a faint suzerainty of the Supreme Porte (Schmitt 2012: 141). The League of Prizren, convened in this town in modern Kosovo in 1878, is frequently regarded as a starting point of national-Albanian political activity. But it remained in fact an interethnic alliance of Sunni Muslims, in which Christian Albanians were present, but had no essential voice.

The most decisive steps towards nation-building were set after 1900 by the Roman Catholics in Shkodër (mainly the clergy) as well as the Orthodox and Bektashi in the South (mainly traders and craftsmen). They elaborated two essential symbols of a common Albanian national identity: a codified standard language and a specific Albanian alphabet (Schmitt 2012: 144). This alphabet in Latin script was adopted by the Congress of Monastir [Bitola] in 1908 against the resistance of Muslim clergy and the Ottoman central power. It replaced three scripts, in which Albanian speakers had (besides Latin, used by the Roman Catholics) written before: Arabic (written by Muslims), Greek (written by the Albanian Orthodox), Cyrillic (written by the Orthodox Aromunians and Slavonic minorities).

The Albanians were in this way the first predominantly Muslim nation to write in Latin script. This was meant as a signal that they conceived themselves as a European nation, as different from the Turks (who converted to Latin script only in 1928), as being not fundamentalist, but European Muslims and able to dismiss a religious symbol, the script of Koran.

Literature in the new standard language and alphabet could, however, until 1912/13 (the foundation of an independent state) only be published and distributed outside the Ottoman Empire or (against the resistance of Ottoman authorities) by foreign schools and other institutions on Ottoman territory. Up to the last moment, that is, the retreat of the Ottoman Empire from "Albanian territories," however, the Albanian national movement presented itself as a heterogeneous conglomerate. It was obviously not easy for Albanians to transcend the boundaries of religion. The Albanian national idea was also rather based on stressing the difference between competing national identities than on developing a proper image of self. Also the formation of an Albanian state and its independence was not so much due to vigorous Albanian efforts, but to competition between the large European powers for influence in the wider region.

It was, however, introduced by an Albanian effort to implement an autonomous province under the umbrella of the Ottoman Empire. This was to include all lands settled by Albanians resembling what was later often called a "Greater Albania." It would have become a country dominated by Sunni Muslims[9] with Skopje as its capital. For this goal Albanian armed activists marched in the summer of 1912 towards Skopje. But this made Serbia, Montenegro and Greece reacting immediately in order to prevent a *fait accompli*. They invaded the "Albanian territories" and occupied them almost completely. The declaration of an independent Albania as of 28 November 1912 in Vlora (at the Adriatic coast) was not much more than the desperate attempt of a disparate group to attract the attention of the large powers. But these had anyway decided to react: For Austria-Hungary and Italy it was unacceptable to see the Serbs (and Russians as their supporters) at the coasts of the Adriatic, and it was in fact sufficient to threaten with war to urge the immediate withdrawal of the occupators. But only on 31 July 1913 the so-called Conference of Ambassadors in London determined the borders of the new country (roughly coinciding with the borders of modern Albania), and it was not until 1922 that the new state was in the position to control most of its territory.

Now the institutions of the state could be used to implement the Albanian national idea profoundly and to disseminate it further. It was also in 1922 that, in accordance with the principle of national-ecclesiastical correspondence and union, an autocephalous Albanian-Orthodox church was founded, which was recognized by the ecumenical patriarch in 1937 (Schmitt 2012: 155).

The new state's borders, however, did not encompass all Albanians. After 1913 the Albanians outside Albania were subject to Serbian domination in Serbia and Montenegro, later in the State of the Serbs, Croats and Slovenes, renamed as Yugoslavia in 1929, as well as to Greek domination in Greece. This action naturally prompted nationalist reactions and an effort for all-Albanian national unity, which was indeed almost achieved during World War II (1941–1944), but again only due to external support and at the expense of external domination, this time by Italy.

Interwar Albania developed into a multi-confessional society with the replacement of the Islamic religious law (Sharia) by civil law in 1928/29 as the essential turning point. Christians, especially clerics, became part of the system and achieved higher state positions. Members of the parliament had to represent proportionally the confessional composition of the population (Clayer 2003: 280). Most of the administration, however, remained in the hands of Muslims, the administrators of the Ottoman era, who needed some time to accept Christians as fellow citizens with equal rights and who frequently behaved as if they still owned the state (Schmitt 2012: 159). The interdiction of private and religious schools affected in principle all religious communities, but in practice it had a stronger effect on religious minorities, for which their ecclesiastical educational system functioned as the backbone of cultural identity. This was especially true for Roman Catholics who had developed such a system to perfection. It was, therefore, not a surprise that they

[9] Most Albanians outside modern Albania, for example, in modern Kosovo and Macedonia, were (and are) Sunni Muslims.

responded by severe protest, asked Italy for support and proposed an administrative subdivision of Albania into three cantons: a Roman Catholic North, a Muslim Center and an Orthodox South.

In general, however, the religious communities remained in interwar Albania important forces in society, well controlled, but also supported by the government under the "royal dictator" Ahmet Zogu, who was appreciating them as a firewall against Bolshevism.

The Communist movement after World War II led by Enver Hoxha was anti-religious from the very beginning and dissociated itself from all religious communities, much more than in the other Communist countries of the Balkans. Religious communities were regarded as a threat to national homogeneity, as a part of the old system and as an obstacle on the path towards a new Communist society.

This attitude culminated in 1967 in the ban of all religious communities and in the proclamation of the "first atheist state." Religious doctrines and rites were replaced by the Communist ideology and a cult around Skanderbeg, the hero in the wars against Ottoman occupation in the fifteenth century. Clerics and underground religious activists were severely persecuted, mosques and churches demolished or used for other purposes. Also a new standard of the Albanian literary language was introduced in 1972 replacing the "clerical" Ghegian northern standard (originating in the Roman Catholic stronghold Shkodër) by a codification based on the Toskian dialect[10] spoken in southern Albania, where Enver Hoxha came from.

83.5 Albania's Current Religious Structure and the Status of Religious Communities

Post-Communist Albania immediately recognized the religious communities and supported their re-consolidation, also as a measure to defuse the general crisis (Clayer 2003: 277). Already in 1990, in the Ethem Bey mosque in Tiranë and in the Orthodox village Derviçan near Gjirokastër as well as in the Roman Catholic cathedral of Shkodër (Fig. 83.3), which had in the era of Communism functioned as a sports hall, the first religious services were celebrated under the attendance of a large number of people (Hofmann 2009: 44; Clayer 2003: 288). All kinds of religious landmarks emerged very quickly all over the country, mainly due to local and individual initiatives. Also organizational structures, caring and educational institutions were established already in the first post-Communist phase, first mainly as grassroot activities, later financed and frequently also conducted by sister communities abroad (Clayer 2003: 288f).

[10] The Ghegs (in the North) and the Tosks (in the South) are the main tribal groups of Albanians speaking specific dialects. Albanians in Kosovo and Macedonia belong in general to the Ghegian group, who migrated to these regions under Ottoman rule, while the Toskian group was a major source of Albanian emigration to abroad.

Fig. 83.3 Roman-Catholic cathedral in Shkodër (Photo by Peter Jordan 2012)

But post-Communist Albania also confirmed already in 1991 by the "Law on the most important constitutional rules" that it is a secular state (Clayer 2003: 285). The new Albanian constitution of 1998 mentions in its Article 10 that "1. In the Republic of Albania there is no official religion. 2. The state is neutral in questions of belief and conscience, and also, it guarantees the freedom of their expression in public life. 3. The state recognizes the equality of religious communities." (Albanian Constitution 2012).

Since the results of the 2011 Albanian population census with reference to religion are not yet published, it is difficult to determine accurately Albania's current religious structure. A map authored by Arqile Bërxholi in the *Demographic Atlas of Albania* (Bërxholi et al. 2003: 101) seems to convey by circle diagrams and districts [rrheti] quite a detailed picture of the structure in 2001 (see Fig. 83.2), but is actually only based on the official religion census of 1942, the last one prior to 2011. Bërxholi used demographic data as recorded by later population censuses to extrapolate the census data of 1942. In this way he estimates the 68.5 % Muslims (60.3 % Sunni Muslims, 8.2 % Bektashi), 21.8 % Orthodox Christians and 9.7 % Roman Catholic Christians for 2001 (Bërxholi 2003: 41) for Albania in total.

Gottfried Hofmann, however, based on expert interviews, more recent publications (Gashi and Steiner 1997: 85; Laksham-Lepain 2000a, b; Peters 2003: 1, 235ff) as well as censuses conducted by the individual religious communities, concludes that the developments after the fall of Communism have resulted in a considerable shift in favour of the Christian communities, first and foremost of the Roman Catholic church (Hofmann 2009: 28f). He estimates the share of Sunni Muslims at between 54 and 56 %, of Bektashi at 7–8 %, of Orthodox Christians at between 21 and 23 %, of Roman Catholics at between 13 and 15 %, of traditional Christian Protestant groups at about 1 % and of New Protestants between 0.3 and 0.5 % (Hofmann 2009: 29).

The reasons for this shift may have been the higher population growth in the Catholic North (Clayer 2003: 282), but also the fact that the Catholic church with its powerful support from Rome had been the preferred target of Communist persecution and received now the most sympathy. Another explanation is that Marxism as a European culture erased the Islamic culture more than other religious cultures (Clayer 2003: 282). The relative growth of the Christian communities may further be attributed to the fact that during recent decades of Communism watching Italian television was a way out of the "prison," which again promoted rather European than Islamic lifestyle (Clayer 2003: 282).

Whatever the exact figures may be, it is clear enough that the religious structure of Albania before the Communist era has been re-established and that 45 years of religious suppression and active and vigorous atheism (since 1967) could not prevent a full recovery of the religious communities. Having, at least compared to some Orthodox churches in other Communist countries, not the least collaborated with the Communist regime, they received immediately full credit by the population. They assumed also very much the role of nodes of the social network at the local level. This function prevails certainly with younger people, who had been educated in the period of atheism and did certainly, at least in the first years, not join religious communities mainly for religious motives.

As already mentioned, Albania presents itself today as a secular state in similar distance or being in equidistance to all religious communities. The distance between state and religion is today certainly wider than it was in the interwar period. In 1995, for example, Albania legalized abortion completely, against heavy protests of all three major religious communities (Schmidt 2007: 101). Also in 1995, state institutions and courts did not intervene at all, when a second Albanian-Orthodox church was founded and the metropolit of the older church urged the state to interfere (Schmidt 2007: 101).[11] National affirmation is now a necessity for all religious communities (Clayer 2003: 283).

Albania cultivates also the image of a tolerant society, in which rather divergent religious communities can co-exist and where also Islam is far from being fundamentalist. This position corresponds to reality insofar as during the Communist period secularization was very much advanced. The majority of the (on the average rather young) population has been educated in 23 years of official state atheism and is emotionally not much affected by religion, but regards joining religious communities as an activity of social networking. It is also true that, partly in contrast to Bosnia-Hercegovina, efforts of Arab and other Islamic countries to provide influence by financial support for the Albanian Muslim communities failed.

[11] This second Albanian-Orthodox church did not gain ground because it was just the undertaking of an elitarian dissident group and did not receive support by a wider part of the population.

This image of tolerance and harmonious co-existence of three larger religious communities is officially demonstrated by styling the Catholic Albanian nun Mother Tereza (1910–1997), who actually descends from Macedonia, an icon of Albanian identity. She is a frequent subject of posters like at the opera house of Tiranë (Fig. 83.4), is celebrated by many monuments like in Shkodër at the main traffic node of the downtown at Bulevard Skenderbeu nearby one of the main mosques (Xhamia e Madhe) (Fig. 83.5) or by naming the recently enlarged and renovated airport of Tiranë, the main airport of the country, after her ("Tirana International Airport Nënë Tereza").

Fig. 83.4 Mother Tereza at the Opera House in Tiranë (Photo from author's collection, taken by Mihai Szabo 2007)

Fig. 83.5 Monument of Mother Tereza at Bulevard Skenderbeu in Shkodër with Xhamia e Madhe in the background (Photo by Peter Jordan 2012)

But although religious practice is not intensive and especially the younger generations are not very much affected by religion emotionally, religion still structures society and one's religious origin is known by everybody. Religion still divides society into "camps," which have today not so much a religious meaning, but are conceived as an affiliation to Western versus Oriental lifestyles (Clayer 2003: 285). At the same time the "camps" of the religious communities have been internally diversified due to the involvement of foreign actors as well as the appearance of New Protestant communities, who embody the "Western World."

It is also a fact that at the political scene the three main religions are not by any law, but implicitly represented by a proportional share in high ranking positions and representative bodies and that in the political game religious communities play a certain role. Thus, the two main political parties, Democratic and Socialist Party, have affinities to different religious groups: while the Democrats are linked with Muslims and Roman Catholics, the Socialists are closer to the Orthodox (Clayer 2003).

Finally, it has also be mentioned that the status of a secular state where all religious communities are considered equal is not accepted by all subgroups of society and that internal political developments, for example, the change of party majorities, as well as external influences may well modify it.

83.6 In Conclusion: Why Were Albanians Able to Develop a National Identity Across Religious Boundaries?

The usual answer of Albanian researchers to this question is that Albanians were in principle indifferent and exceptionally tolerant in religious matters and had, for example, accepted Islam only superficially.

It is true that an official conversion to Islam and a hidden continuation of Christian practices and faith as well as conversion only of the family father, while the rest of the household remained Christian, were widespread phenomena among Albanians. But it is also a fact that these phenomena were not at all confined to Albanians, but equally present also in other parts of the Balkans and the Ottoman Empire, where autochthonous people also converted to Islam, for example, in Bosnia-Hercegovina.

In fact, as historical evidence shows, religion functioned among Albanians not less than with other Balkan groups as the main dividing line in society. Conflict between Albanian religious communities was frequent in history (Clayer 2003: 278f). Assuming an Ottoman or Turkish national identity was at times a realistic and acceptable option for Albanian Muslims, such as to become in national terms a Greek or an Aromunian was an option for Orthodox Albanians. Orthodox Albanians used to have sympathies for Greeks or Romanians, Roman-Catholic Albanians for Italians and Croats rather than for their co-national Muslims. The religious community formed the endogamic group and assuming the partner's language was the usual procedure. Note that not before the twentieth century a multidenominational Albanian nation began to surface and to play a role at the political scene.

It has also to be mentioned that the idea of a multidenominational nation is not unique, even not in the Balkans. Thus, the Greater Serbian ideology as formulated by Ilija Garašanin and Vuk Karadžić in the middle of the nineteenth century also left space for non-Orthodox members such as Catholic and Muslim Serbs. It was, however, not accepted by these groups.

There are four reasons why it was accepted by non-Islamic Albanians:

- A common and very specific language, which contrasted Albanians from all neighbors and gave them a feeling of communality.
- The imagination of a common ethnic descent from the Illyrians.
- Exclusive national ideas and policies of all neighbors (mainly Serbs and Greeks) that offered Albanian-speakers finally no alternative for national affiliation.
- The formation of an independent state, where national consciousness could be further cultivated.

83.7 Albania – A Model for Bosnia-Hercegovina?

Like Albania, the population of Bosnia-Hercegovina is split into three denominations, the Muslim, the Serbian-Orthodox and the Roman-Catholic. The three religious communities fully correspond to the three nations of the state, the (Muslim) Bosniaks, the (Orthodox) Serbs and the (Catholic) Croats. According to estimates (no data on religious or ethnic structure were published after 1991) the proportion between the three groups was in 2003 in the Federation of Bosnia-Hercegovina 72.6 %: 21.8 %: 4.6 %, while 1.0 % remained for smaller minorities (Federalni zavod za statistiku, 2004), in the Serbian Republic 7 %, 91 %, 1 % (Josipovič 2005).[12] So far the three nations have not developed a common national identity. Each conceives their religion (or at least religion as a cultural background) also as the main marker of their national identity.

This significant difference from the Albanian case may be explained by the following facts:

- The languages spoken by the three nations are quite similar and have only been differentiated to some minor extent after the dissolution of Yugoslavia. But they resemble at the same time the languages spoken in the neighborhood, certainly in Croatia, Serbia and Montenegro. In fact, they link two of the three Bosnian nations (Serbs, Croats) with communities abroad rather than with Bosnia and do not contribute to an all-Bosnian communality.
- There is no imagination of a specific ethnic descent different from the wider Slavonic environment.
- National policies of neighboring countries (especially Croatia and Serbia) are rather inclusive in regards to the Croatian and Serbian communities in

[12] The two entities (Federation and Serbian Republic) are almost equal in population number.

Bosnia-Hercegovina. Together with linguistic aspects this supports centrifugal rather than centripetal forces. The Muslim Bosniaks have no real option in this respect. It must, however, also be mentioned that especially the Serbian community in Bosnia-Hercegovina is by some cultural characteristics (for example, the Ijekavian variant of the Serbian language in contrast to the Ekavian variant forming the Serbian standard) different from Serbs in Serbia.

- The Bosnian-Hercegovinian state, following the Dayton Accord in 1995, is not accepted by all Bosnian nations in the same way. The Bosniaks as the largest and (also in history) dominant group wish to maintain and even enforce it, but the Bosnian Serbs aim at strengthen their autonomy and the Bosnian Croats have in fact mentally emigrated (Skočibušić 2005).

In regards to the capacity for nation-building, the positive difference between Bosnia-Hercegovina and Albania is perhaps the political tradition reaching back to the High Middle Ages, when long before any Ottoman occupation a Bosnian Kingdom ruled larger parts of the western Balkans. Also in the Ottoman period (1463–1878) Bosnia-Hercegovina formed always a rather autonomous and high-ranking administrative unit of the Empire. Austro-Hungarian occupation in 1878 and annexation in 1908 preserved this role up to World War I.

References

Albanian Constitution. (2012). Retrieved July 20, 2012, from www.ipls.org/services/kusht/cp1.html

Babuna, A. (2004). The Bosnian Muslims and Albanians: Islam and nationalism. *Nationalities Papers, 32*(2), 287–321.

Bërxholi, A. (2003). Ethnische und konfessionelle Struktur der Bevölkerung Albaniens. In P. Jordan, K. Kaser, W. Lukan, St. Schwandner-Sievers, & H. Sundhaussen (Eds.), *Albanien. Geographie – Historische Anthropologie – Geschichte – Kultur – Postkommunistische Transformation* (pp. 33–41). Wien et al.: Peter Lang.

Bërxholi, A., Doka, Dh., & Asche, H. (Eds.). (2003). *Atlas of Albania. Demographic atlas of Albania.* Tiranë.

Brunnbauer, U. (Ed.). (2002). *Umstrittene Identitäten. Ethnizität und Nationalität in Südosteuropa.* Frankfurt a. M.: Peter Lang.

Büttner, M. (1985). *Grundfragen der Religionsgeographie.* Berlin: Reimer.

Büttner, M. (1998). *Geographie und Theologie. Zur Geschichte einer engen Beziehung.* Frankfurt am Main: Peter Lang.

Clayer, N. (1996). La Bektachiyya. In A. Popovi & G. Veinstein (Eds.), *Les Voies d'Allah. Les ordres mystiques dans le monde musulman des origines à aujourd'hui* (pp. 468–474). Paris: Fayard.

Clayer, N. (2003). God in the "Land of the Mercedes." The religious communities in Albania since 1990. In P. Jordan, K. Kaser, W. Lukan, St. Schwandner-Sievers, & H. Sundhaussen (Eds.), *Albanien. Geographie – Historische Anthropologie – Geschichte – Kultur – Postkommunistische Transformation* (pp. 277–314). Wien/Frankfurt a. M. et al.: Peter Lang.

Daxner, M., Jordan, P., Leifer, P., Roth, K., & Vyslonzil, E. (Eds.). (2005). *Bilanz Balkan.* Wien/München: Verlag für Geschichte und Politik/Oldenbourg.

Deutsch, E. (1989). Statistische Angaben über Albaniens Katholiken im letzten Viertel des 19. Jahrhunderts in k. u. k. Konsulatsberichten. *Österreichische Osthefte, 31*(3), 417–445.

Doka, D. (2001). Wirtschaftsräumliche Entwicklungen in Albanien nach der Wende. In C. Lienau (Ed.), *Raumstrukturen und Grenzen in Südosteuropa* (pp. 333–344). München: Südosteuropa-Gesellschaft.

Doka, D. (2005). *Regionale und lokale Entwicklungen in Albanien – ausgewählte Beispiele*. Potsdam: Institut für Geographie der Universität Potsdam.

Gashi, D., & Steiner, I. (1997). *Albanien: Archaisch, orientalisch, europäisch* (2nd ed.). Wien: Promedia.

Goehrke, C., & Gilly, S. (Eds.). (2000). *Transformation und historisches Erbe in den Staaten des europäischen Ostens*. Bern: Peter Lang.

Harris, C. D. (1993). New European countries and their minorities. *Geographical Review, 83*, 301–320.

Hatschikjan, M., & Troebst, S. (Eds.). (1999). *Südosteuropa. Ein Handbuch. Gesellschaft, Politik, Wirtschaft, Kultur*. München: Beck.

Hofmann, G. (2009). *Die Rolle der römisch-katholischen Kirche im multikonfessionellen Staat Albanien – eine sozialgeographische Untersuchung*. Diploma work, University of Vienna.

Hösch, E. (1995[3]). *Geschichte der Balkanländer. Von der Frühzeit bis zur Gegenwart*. München: Verlag Beck.

Ivanišević, A., Kappeler, A., Lukan, W., & Suppan, A. (Eds.). (2002). *Klio ohne Fesseln? Historiographie im östlichen Europa nach dem Zusammenbruch des Kommunismus*. Wien et al.: Peter Lang.

Jordan, P. (2007). Geopolitical developments in South East Europe. The political-geographical rearrangement of South East Europe. *Europa Regional, 15*(7), 87–98.

Jordan, P., & Lukan, W. (Eds.). (1998). *Makedonien. Geographie – Ethnische Struktur – Geschichte – Sprache und Kultur – Politik – Wirtschaft – Recht*. Wien et al.: Peter Lang.

Jordan, P., Kaser, K., Lukan, W., St. Schwandner-Sievers, & Sundhaussen, H. (Eds.). (2003). *Albanien. Geographie – Historische Anthropologie – Geschichte – Kultur – Postkommunistische Transformation*. Wien/Frankfurt a. M. et al.: Peter Lang.

Jordan, P., & Kocsis, K. et al. (2006). Ethnic consciousness in Central and Southeast Europe around 2000. In Jordan, P. (Ed.), *Atlas of Eastern and Southeastern Europe*, 2.9-G9. Stuttgart: Borntraeger.

Josipovič, D. (2005, November 28–30). *Suitability of the Dayton territorial division for the process of integration of Bosnia-Hercegovina*. Paper, International conference "Dayton – 10 years after: Conflict resolution and cooperation perspectives," Sarajevo.

Kahl, T., & Lienau, C. (Eds.). (2009). *Christen und Muslime. Interethnische Koexistenz in südost-europäischen Peripheriegebieten*. Wien: LIT.

Kahl, T., Maksuti, I., & Ramaj, A. (Eds.). (2006). *Die Albaner in der Republik Makedonien. Fakten, Analysen, Meinungen zur interethnischen Koexistenz*. Wien/Berlin: LIT.

Kraas, F., & Stadelbauer, J. (Eds.). (2002). *Nationalitäten und Minderheiten in Mittel- und Osteuropa*. Wien: Braumüller.

Laksham-Lepain, R. (2000a). *Bektashis of Albania. Report*. Center for Documentation of Minorities in Europe – Southeast Europe (CEDIME-SE).

Laksham-Lepain, R. (2000b). *Catholics of Albania. Report*. Center for Documentation of Minorities in Europe – Southeast Europe (CEDIME-SE).

Lichtenberger, E. (1976). Albanien, der isolierte Staat als gesellschaftspolitisches Modell. *Mitteilungen der Österreichischen Geographischen Gesellschaft, 118*, 109–136.

Lienau, C. (Ed.). (2001). *Raumstrukturen und Grenzen in Südosteuropa*. München: Südosteuropa-Gesellschaft.

Lienau, C., & Prinzing, G. (1986). *Beiträge zur Geographie und Geschichte Albaniens*. Münster: Geographisches Institut, Universität Münster.

Louis, H. (1927). *Albanien. Eine Landeskunde, vornehmlich auf Grund eigener Reisen*. Stuttgart.

Lucic, I. (2012). In the service of the nation: Intellectuals' articulation of the Muslim national identity. *Nationalities Papers, 40*(1), 23–44.

Magocsi, P. R. (Ed.). (2002). *Historical atlas of central Europe*. Toronto: University of Toronto Press.

Matuz, J. (1994³). *Das Osmanische Reich. Grundlinien seiner Geschichte*. Darmstadt: Wissenschaftliche Buchgesellschaft.
Milo, P. (2002). The relations of Albania with its neighbouring countries. *Südosteuropa-Mitteilungen, 42*(5–6), 12–19.
Mitchell, D. (2000). *Cultural geography: A critical introduction*. Oxford/Malden: Blackwell Publishers.
Moore, P. (2002). The international community and Albania. *Südosteuropa-Mitteilungen, 42*(5–6), 20–23.
Okuka, M. (Ed.). (2002). *Lexikon der Sprachen des europäischen Ostens*. Klagenfurt/Celovec: Wieser.
Paulston, C. B., & Peckham, D. (1998). *Linguistic minorities in central and eastern Europe*. Clevedon/Philadelphia/Toronto/Sydney/Johannesburg: Multilingual Matters.
Peters, M. W. E. (2003). *Geschichte der Katholischen Kirche in Albanien. 1919–1993*. Wiesbaden: Harrassowitz.
Pichler, R. (2002). Die albanische Historiographie seit der Wende. In A. Ivanišević, A. Kappeler, W. Lukan, & A. Suppan (Eds.), *Klio ohne Fesseln? Historiographie im östlichen Europa nach dem Zusammenbruch des Kommunismus* (pp. 521–526). Wien et al.: Peter Lang.
Roth, K. (2005). Institutionelles und persönliches Vertrauen. Südosteuropa auf dem schwierigen Weg in die Europäische Union. In M. Daxner, P. Jordan, P. Leifer, K. Roth, & E. Vyslonzil (Eds.), *Bilanz Balkan* (pp. 47–53). Wien/München: Verlag für Geschichte und Politik/ Oldenbourg.
Rugg, D. S. (1994). Communist legacies in the Albanian landscape. *Geographical Review, 84*(1), 59–73.
Schermerhorn, R. A. (1970). *Comparative ethnic relations: A framework for theory and research*. New York: Random House.
Schmidt, G. (1961). Albanien. Ein landeskundlicher Abriss unter Verwertung von Reiseeindrücken. *Geographische Rundschau, 13*(10), 396–407.
Schmidt, F. (2007). Religion in Albanien. *Südosteuropa-Mitteilungen, 47*(5–6), 94–101.
Schmitt, O. J. (2012). *Die Albaner. Eine Geschichte zwischen Orient und Okzident*. München: Beck.
Schramm, G. (1999²). *Anfänge des albanischen Christentums. Die frühe Bekehrung der Bessen und ihre langen Folgen*. Freiburg im Breisgau: Rombach.
Schubert, P. (2002). Albanische Identität und europäische Integration. *Südosteuropa-Mitteilungen, 42*(5–6), 24–33.
Seewann, G. (Ed.). (1995). *Minderheiten als Konfliktpotential in Ostmittel- und Südosteuropa*. München: Oldenbourg, Südosteuropa-Gesellschaft.
Seewann, G., & Dippold, P. (1998). *Bibliographisches Handbuch der ethnischen Gruppen Südosteuropas*. München: Oldenbourg.
Skočibušić, N. (2005). *Identitätsmerkmale der Kroaten in der Herzegowina*. Diploma work, Humboldt University Berlin.
Smith, G., et al. (1998). *Nation-building in the post-Soviet borderlands. The politics of national identities*. Cambridge: Cambridge University Press.
Sugar, P. F. (1980). *Ethnic diversity and conflict in eastern Europe*. Santa Barbara: ABC-Clio.
Sundhaussen, H. (1994). Institutionen und institutioneller Wandel in den Balkanländern aus historischer Perspektive. In H. C. Papalekas (Ed.), *Institutionen und institutioneller Wandel in Südosteuropa* (pp. 35–54). München: Oldenbourg.
Suppan, A., & Heuberger, V. (1991). Staaten und Minderheiten im Donauraum (1945–1990). *Österreichische Osthefte, 33*(2), 41–58.
Suttner, E. C. (1997a). Das religiöse Moment in seiner Bedeutung für Gesellschaft, Nationsbildung und Kultur Südosteuropas. In H. D. Döpmann (Ed.), *Religion und Gesellschaft in Südosteuropa* (pp. 78–96). München: Südosteuropa-Gesellschaft.
Suttner, E. C. (1997b). *Kirche und Nation. Beiträge zur Frage nach dem Verhältnis der Kirche zu den Völkern und der Völker zur Religion*. Würzburg: Augustinus-Verlag.

Suttner, E. C. (2001). Von "Kirchennationen" zu Nationalkirchen. In I. Gabriel, C. Schnabl, & P. M. Zulehner (Eds.), *Einmischungen. Zur politische Relevanz der Theologie* (pp. 76–91). Ostfildern: Schwabenverlag.

Symmons-Symonolewicz, K. (1985). The concept of nationhood: Towards a theoretical clarification. *Canadian Review of Studies in Nationalism, 12*, 215–222.

Todorova, M. (1997). *Imagining the Balkans*. New York: Oxford University Press.

Todorova, M. (2002). Der Balkan als Analysekategorie: Grenzen, Raum, Zeit. *Geschichte und Gesellschaft, 28*, 470–492.

White, G. W. (2000). *Nationalism and territory*. Oxford: Rowman & Littlefield.

Chapter 84
Social and Spatial Visibility of Religion in Question: The Case of Pluricultural and Multiconfessional France

Lionel Obadia

84.1 Introduction

Since the early 2010s, the monthly supplements of the prominent and esteemed French magazine *Le Monde* were dedicated to atlases. Among special issues on "civilizations" "languages" or "minorities," an *Atlas of Religions* was published in 2011: dozens of maps, complex diagrams and tables have been drawn and illustrated the textual analysis of the changing landscape of France (and to a certain extent, other countries). Maps of a multi-confessional France for instance illustrated the distribution and the dynamics of Christianity, Judaism or Islam, and a wide audience –not always familiar with geographic issues relating to religion – could visually appreciate the changes on the location, concentration and dispersion of believers of one tradition or more. Surprisingly, French popular magazines seem much more interested in issues and methods of geography of religion than academic institutions. In stark contrast with the situation in other countries, religious issues, after having been central to the developments of postwar French Academies and Universities, have significantly withdrawn from the forefront of scientific programs of geography. Religious issues and topics, though they had never really disappeared from geographic analysis, have recently been relocated within the scientific agenda of Geography. For example,: the political and social topicality of religion, which has become more and more spatially and socially visible in a French allegedly secular state, nowadays compels scholars and students in religious study to engage, more or less explicitly, in a *spatial turn*. This paper aims at highlighting the ways geographic approaches to religion have been framed in the academic and political context of secular, but pluralistic France.

L. Obadia (✉)
Department of Anthropology, Université de Lyon, Lyon, France
e-mail: lionel.obadia@un-lyon2.fr

S.D. Brunn (ed.), *The Changing World Religion Map*,
DOI 10.1007/978-94-017-9376-6_84

84.2 Religion in the Medias and the Academy in France

Three decades ago, religion was considered as a minor subject-matter in social sciences and humanities in France. The great years of history and sociology of religion were almost over with the late 1960s and early 1970s. Prominent theoretical models of religious life, like those offered by Mircea Eliade, George Dumézil or Claude Lévi-Strauss had a declining influence. France's Christian heritage was fading, as demonstrated by the early works of Hervé Lebras and Chanoine Boulard in the interwar period on demographics of churches in the whole country. The sociology of religion was interested in the changing demographics of adherents, and inclined to conform to mainstream theories of secularization (see Desroche 1968).

In the 1970s, things changed dramatically with the surfacing of a multitude of new religious movements, most of them originating from unexpected (Western) ancient or exotic (Eastern) matrixes. A new generation of scholars, mainly sociologists, quickly seized the topic of religion and launched new researches. At the time, in the 1970s, the research interests of this renewed science of religion started to focus on the bourgeoning and genuine forms of religiosity, rather than the fading of traditional ones (see Hervieu-Léger 1993a). In the 1980s, religion, especially newly fashioned religiosities (New religious movements) became a fashionable item, while the secularization theory was gradually replaced by the "revival of religion" thesis. Besides the current new, spiritual, modern and individually-oriented movements, the surprise came from the parallel expansion of anti-modern, community-oriented and fundamentalist Christian, Jewish and Muslim groups – what political scientists have called the "revenge" of monotheist traditions (Kepel 1991). Since them, and in the following decades, increased attention has logically been drawn to religions, especially when they assumed the "extreme" forms of fundamentalisms and sects; also important research funds have been reserved for such subject matter. More classical religious topics (mythology, texts, rituals, beliefs) of more peaceful traditions did not vanish from the researchers' concerns, though. But each period has witnessed a particular fashion for a specific religious topic in the Academies and the Media. After the post-war first version of the secularization thesis (the decline of religion, in the views of Emile Durkheim and his heirs), the second version of the secularization thesis appeared in the 1970s (the "renewal of religion"); the 1990s and 2000s have been the years of tensions between religious minorities, "sects" and other alternative movements, and the secular state. Studies in secularism have consequently been stimulated by the massive eruption of religious claims in public spaces and requests for alleviations of legal norms of *laïcité*. In parallel, issues in cultural and religious pluralism have regenerated discussions on the role of religion in regulating society. Furthermore, the centenary of the separation between Church and State in 2005 has also been the occasion for scholars to reappraise the social acceptation of the principles of *laïcité* and the prospect of secular politics in a more and more "desecularized" France.

As such, the media-coverage of religious issues has played a crucial role in the relocation of beliefs and confessional loyalty at the forefront of social and political issues. But making religion a popular topic also meant to regard it as a "problem" for society. Press and TV privilege the more spectacular or impressive facts: the radicalization of minorities (like Islam), the trials of sects accused of brainwashing and abuse (like Scientology), or even mass-murders of their members (Ordre du Temple Solaire). In the same time, the Press relates regularly themes like the crisis of the Roman Catholic Church or the political power of ancient esoteric groups like Freemasons. But the topically of religion is also linked with more challenging events. France has indeed been the site of violence in the name of religion: destruction of cinemas by Christians in the late 1980s, subway bombings by Muslims in the 1990s are two well-known illustrations of this phenomena. Conversions of teenagers or young adults to foreign religions (Islam and Buddhism) and the demands for places of worships for foreign religions and sects/cults (expressed by their representatives and adherents) led to an ongoing debate on the tolerance of French "laïque" (secular) society to religions.

It therefore comes as no surprise to observe the rebirth of sciences of religion, whose productions gained visibility and generated a massive interest. Historians and sociologists (Danièle Hervieu-Léger) philosophers (Marcel Gauchet 1999) political scientists (Gilles Kepel) epitomized this generation of scholars emerging in the 1980s–1990s questioning the relationships between religion and modernity. More recent generations of scholars have prolonged but discussed their three theoretical models – respectively, pluralization and individualization, secularization, rebirth of fundamentalism. Other topics, like gender and religion (especially feminist contestations of male domination, Planté 2002), ecology and religion (after a pioneer book edited by Danièle Hervieu-Léger 1993b), the body and religion, and new approaches to belief are quickly emerging as new repertoires for research and reflection (Aubin-Boltansky et al. 2014), alongside more classical ones (religious texts, religions and politics, organizational forms of religions), and more "imperative" ones (religion and migration, religion and ethnicity, religious freedom and secular tolerance, etc.). Today studies on the cultural and social integration of Islam in France represent the primary source of interest for public administrations; it is also a major research theme, especially for sociology, as it has offered copious funding opportunities. It did not, however, replace the "classical" themes of religion in Antiquity (studied by means of archeology, epigraphy or codicology) and in History (by means of historical methods and sources) but the latter is still bounded by geographic limits. History of religion in France is first and foremost the history of religion in France or Europe, that is, the history of Christianity and monotheistic traditions. Among other topical religious items, apocalyptic and messianic groups, alternative spiritual therapies range among interesting topics for public powers – and research has been funded consequently. The focus upon these marginal and "dangerous" spiritual trends and organizations is the result of a two decades control, by State administration, of "sectarian drifts" (Fournier and Monroy 1999) or "prejudicial practice of unconventional medicine." The MIVILUDES, a State-driven committee against sects and sectarian drifts, has published each year since 1994, a report on the current trends in alternative and "threatening" spiritual movements. It recently pointed out the progresses of apocalyptic sects and neo-shamanic therapies in France, after a 20 years presence of New Age and neo-Asian groups. With the broadcasting of an increasing number of reportages blaming the "threats" of such groups for individuals and society, Press and TV shadow the government's attention towards these unconventional expressions of beliefs and religiosity, while Sunday morning TV broadcasts of national TV networks offer a panel of official religions' talk shows.

As a secular state, France is much concerned by the religious issues and public and academic debates on "sects" have generated harsh controversies about religious freedom, the defense or condemnation of alternative religiosities (see Introvigne's 1996 response to the first parliamentary report on "sects" in France), and, in a larger arena, of mainstream religions. In the 2000s, the argument was relating to the declining dominant position of secular knowledge on religion. The difference between North America and other parts of the world is that in France, the sciences of religion and theology were abruptly disconnected in the early twentieth century, and the latter is nowadays somewhat relegated in a handful of universities where curricula in theology have been absorbed into philosophical departments. Most of the university teachers are self-confessed atheists or agnostics, and traditionally, some of them even avow distaste for religion as faith, even while they admire religion as "civilizational fact." This stance is understandable with reference to the history of secularism in France, which started in schools and sciences, and assumed the form of anti-clerical movements and secularism's assertive ideological forces. Yet, a certain connivance still exists between religion and science of religion, at least in the case of the study of monotheist traditions – for in which, indeed, many scholars have been socialized in religious settings and still are practitioners. Extreme atheism (anti-religious attitudes) is not common among scholars in social sciences and humanities studying religions in France, but these sciences of religion (sociology, anthropology, history and psychology) are based upon a postulate of "methodological [rather than ideological] atheism" that has been criticized recently for dissolving the *reality* of the divine (for believers) in the naïve *imagination* of credulous actors (for observers) (Piette 2003).

While theologians and religious leaders do not have a wide social audience, as they used to until the 1950s and 1960s, religion has also been relocated in the heart of public debates and opinion by secular scholars. And the paradox of France maybe lies in the fact that the more secular the country is supposed to be, the more it is interested in religious issues on its own soil and heritage. Religious courses are banished from public primary and secondary schools in France after the process of laicization of schools in between 1882 and 1886, expect in two regions (Alsace-Moselle) whose legal status has not been following the norm of the 1905s separation between Church and State Law. Private confessional schools have been founded for Christian and Jewish communities. But given the demands expressed from religious groups, the government has engaged in a reflection on religious teachings in schools (Debray 2002). Scholars in sociology (Willaime and Beraud 2009), history (Attias and Benbassa 2007), and education have attempted to demonstrate that the prevailing mode of teaching religious issues in universities (secular-based and non-confessional) that could be applied for secondary and high schools. Nevertheless, recent events recorded in high schools (conflicts between pupils and teachers on religious topics) reveal that religion, more than one century after the establishment of the regime of *laïcité*, remains a controversial subject-matter, even in the alleged more secularized sector of French society, i.e., school.

84.3 Academic and Political Causes of the *Spatial Turn* in French Religious Studies

The return of religion in the agenda of geography and the shift of religious studies towards spatial approaches might relate, at first glance, to theoretical and methodological changes, viz., geography rediscovers the importance of the subject-matter of religion (Racine and Walther 2003). Religious studies that have otherwise considered spatial dynamics and especially territorial processes in interdisciplinary frameworks now are accompanied by anthropologically-oriented works (such as Vincent et al. 1995). Two anthropological approaches have received particular attention in the spatial analysis of religion: first, the theory of "non-places" in postmodern societies, coined by French anthropologist Marc Augé (1992); he has generated a debate on the dissolution of "places" as socialized sites. Second, the concept of "deterritorialization," initially coined in the philosophy of Gilles Deleuze and Félix Guattari (1972), has been transposed in the study of dynamics of symbolic and spatial displacements and relocations of religion (Bastian et al. 2001; Roy 2008).

Parallel to the rebirth of geography at the forefront of social sciences, and pertaining to substantial epistemological renewal, is the terminological colonization of social sciences and religious sciences by the conceptual repertoires of geography. There is, as a result, a blurring of the effective contribution of geography can provide an understanding of social and cultural dynamics from a spatial point of view (Lussault 2007). In other words, metaphorical references to the "religious landscape" in sciences of religion, especially in sociology, and the recent methodological emphasis on spatial entities (sites, places, territories, networks, locations, etc.) have, as well, contributed to the interest for geography in religious sciences – or at least for the *objects* of geography, but not necessarily for the *discipline* itself. Recent contributions of French anthropologists and geographers on the spatial dimension of religion or on the fabric of religious territories have drawn attention upon the convergences between geography and other disciplines. While sociologists were prompt to proclaim the renewal of religion on the basis of geography evidence, that is, its relocation on the forefront of public social scene in France, and the pluralization of religious composition of French society (Hervieu-Léger 2001). Though not explicitly labeled as such, French social sciences of religion have thus undertaken a kind of "geographic turn" (even if Doreen Massey's 2005 *For Space* has not been translated in French). It might sound odd, since the masterpiece *Géographie et religion* by Pierre Deffontaines (published in 1948) ranks among the major contributions to a study of French (translocal and comparative) geography of religious *processes* and *dynamics*. Half a century later, this French geography of religion is more focused on a geography of religion *in France* or *of France* for contextual reasons.

84.4 Academic Religious Geography?

While sociologists and historians have thus paid a growing attention to the mutations (secularization, renewal and pluralization) of religion in the country of Descartes' rationalism and anti-religious scientism, interest of French geographers towards religion is still – at least officially – low. Major journals in cultural geography (like *L'espace géographique*) have remained attached to classical objects, but little interested in the development of religious themes. Others, like *Géographie & Cultures* range among the few periodicals that have proved a (slight, but significant) interest in religious issues (for instance, the special issues on "Space" 3/1992 or "Beliefs" 2/2002). In the 2000s, studies in "sacred geographies" have surfaced again after years of scientific neglect. Religious issues, however, focused classically on the ways human activity (symbolism and imagination, practical activities) were shaping landscapes by way of religious ideas and rites (again, following Deffontaines' perspective, see Deffontaines 1948), or on the physical-geographic embeddedness of religions. The beliefs associated with natural landscapes (forests, mountains, rivers, etc.) have been extensively and regularly studied. Geographic-oriented or inspired approaches to religion surface again much more for political and social reasons, that is, the demands of ethnic and confessional groups have been a source of trouble for the norms of *laïcité*, and led to tensions between religious communities in public spaces, and between communities and the whole society.

The spatial or geographic turn in the study of religion is not *only* an academic affair in France. It has become a crucial debate due to the topicality of religion in *public spaces* of a country that has established a principle of relative invisibility in regards to confessional expressions. France is indeed an interesting site for the study of the changing face of religion in Europe, and in the whole world as well. In the country known for having established a political model of separation between Church and State, the *laïcité* (often translated literally as "secularism"), religion has become a major social and political issue. After having imagined that religion (that is, mainstream Christianity) was "in decline," the (recent) recognition of the plurality of traditions and religious expressions in the cultural and religious landscape of France led scholars in religion to revise their views on the *location* of religion in modern societies (Baubérot 1994; Willaime 2009). Previously considered as a "peripheral" matter, religion had turned influential again in society, though not "central," but "crucial." France, a secular country, has witnessed, in the last decades, a series of "affairs" in specific *sites* – the "Islamic veil" in schools (in the 1990s), the prohibition of religious clothes in public spaces (*Burqa*, in 2011), the media-covered controversy about the Muslims' prayers in the streets of large cities in 2010 (due to a lack of places of worship), violence against Jewish people in public transportation or schools, intentional degradations of religious buildings (churches, mosques and synagogues as well) in urban areas, social and physical isolation of members of sects in remote locations are among the significant events which have revealed the need of a *geographic* perspective on religious dynamics. Geographic issues are thus sensitive in the political debates and have impelled for a more socially grounded approach for the understanding of the logics underlying the location, the concentration, the dispersion of religions, and the modes of coexistence of ancient and newly arrived confessional groups with different strategies of visibility in public spaces where they are supposed to have withdrawn from. This empirically-based geography is, moreover, completed by the scrutiny of moral geographies in a kind of multi-planed territoriality, the plan of evidences and the plan of imagination. Generally labeled as a "failure" of the "social and cultural integration" of religious minorities, violence against religious traditions that have lately taken root in France indeed unveils the problematic representation of the "religious Other" in France. Not surprisingly, "foreign" religions are allocated different images and degree of acceptance by French society. While North African Islam, geographically close to French culture and a (monotheistic) religion, is considered as ideologically and ethnically distant (Liogier 2010), Buddhism, an Asian (mainly Tibetan and Japanese) polytheism, on the reverse, is geographically distant, but seen as culturally close to the values of modernity and secularism (Obadia 2007). Only the first one (Islam) is considered as a "problem" for France, whereas both of them are "oriental" religions, but coming from different "Orients."

84.4.1 A "Geographic" Understanding of Religion

The practice of geography of religions in France is hence circumscribed by both academic and ideological issues. But there is one more difficulty. To the contrary of other countries, confessional demographics as well as "ethnic" statistics are banished from French official census. It is, therefore, rather difficult to have reliable information about the number of religious adherents, the location of religious cultures and the spatial distribution of groups in France. Both religious and ethnic statistics are considered as potentially susceptible of bad uses and discriminatory purposes. Non-official censuses however exist, produced by academics (sociographies, for instance) as well as private institutes. Relevant information about the geography of religions is, therefore, mainly provided by non-governmental sources, The IFOP, *Institut Français d'Opinion Publique* (French Institute of Public Opinion), previously demonstrated little interested in religious concerns has now published since the late 2000s several documents on the "sociology and geography" of major religions (Christianity in 2010 and Islam in 2011). It attempted to offer relevant demographics and cartographic attempts to describe the shape and transformations of the religious landscape of France. Other sources, based on quantitative data collected at a wider scale, provide complementary information on the national and regional logics of religious distribution (for instance on minorities, see www.eurel.info/). They supply information where the official institutions do not.

All in all, as goes the sociological slogan, the "religious landscape of France is changing." But the modes of changes also affect in rather different ways the religious traditions in France. Monotheistic traditions, such as Christianity, Judaism and Islam, have been confronting different aspects of modernity. According to the latest estimates (provided by private sector's and non-governmental inquiries), in the mid-2000s, about 65 % of French people declared themselves "Catholics," but the observance of weekly religious practices has decreased: church attendance only amounts 4.5 % of Catholics while 57 % declare themselves "non-practitioners" (against 81 % self-labeled Catholics and 27 % of them attending churches in the early 1950s). The rates of non-religious people are slowly but indubitably increasing (28 % of people in the early 2010s, against 23 % in the early 2000s, 24 % in the 1990s and 21 % in the 1980s). Most important is the fact that besides "mainstream" religions (monotheistic traditions), religious minorities (new religious movements, sects, cults, and other "spiritual" movements) are blossoming. They, however, present the mosaic face of a diffuse and hybrid religiosity, socially and spatially dispersed, rather difficult to map. Françoise Champion, a French sociologist, used an astrophysical metaphor to point at the difficulties to delineate the location and contours of this assortment of alternative, New Age and other esoteric movements or "spiritualities." She described as "a mystical-esoteric nebula" (Champion 1990). Actually, each religious tradition, group or movement raises specific social and academic issues, but in a certain way, all are related to a geographic mode of existence. In the following lines, I will attempt to portray very briefly what I believe to be the salient geographic issue relating to each tradition or family of traditions.

84.4.2 Christianity

In France, like elsewhere, Christianity is losing social and ideological importance. Secularization of France is a fact, and since the separation between Church and State in 1905 it has been installed in the legal framework of French constitution. The disconnection between religion and society has had a critical impact on religious loyalty of Roman Catholics. The demographics of self-definition as "Christians" are decreasing and the beliefs in traditional dogmas of Christianity (God, paradise afterlife) are gradually replaced by alternative (New Age-inspired) ones, or erased by secular ideas. The rate of Christians going to the Sunday service has dramatically decreased (<5 %). The mutations of Christianity have been first studied by means of sociological analysis and demographic accounts, but geographic dimensions of religions are an important key to understanding their mutations. Roman Catholic edifices are less and less in use and a growing number of church structures are physically deteriorating, some of them are even sold. The decay of real estate and material heritage of the Roman Catholic Church in France shadows the withering of Christianity itself, and what French sociologist Danièle Hervieu-Léger calls the end of the "parish civilization" (Hervieu-Léger 2003). Moreover, if secularists have been engaged *ideologically* against Christianity since the nineteenth century, they have won the legal battle. The traditional Catholic Church of France enters the early twenty-first century an unexpected rate of *physical* damage against religious buildings. Considering the numbers of profaned cemeteries and tagged churches each year, a parliamentary group has been established in 2011 with the explicit aim to protect churches and other religious buildings. But not all Christian groups witness a deterioration of their geographic and social conditions. Newly arrived or created Christian groups contributed for instance to the partial but significant reengagement of Christianity in French modern society. Catholic charismatics and Protestant evangelical groups – Jehovah Witnesses, Pentecostalism, Mormons – have established parishes in the suburbs and aim at "recapturing" disaffiliated or unchurched Christians and convert new followers.

84.4.3 Judaism

Judaism has a long history in France, maybe for two millennia and has slowly been integrated in what has become the nation of France. They were "emancipated" in the late eighteenth century French revolution, and have gradually been assimilated despite ongoing persecutions, an enduring anti-Semitism and the destruction of a large part of the community in Germany with the help of collaborationist government of Vichy during WWII. France has one of the more important diasporic Jewish communities: Jews, first from East Europe (*Ashkenazi*) and much later from North African (*Sefaradi*) represent around 400,000 persons, most of them living in urban settings, and as such, have long been highly visible in the specific areas (big cities of France) where they have settled for local reasons. Judaism indeed is both practiced in family context and in synagogues, and has development its own network of religious schools (*Yeshivot*). Geographic political and academic issues for Judaism embrace concentration and dispersion, visibility and invisibility. Indeed, Judaism is primarily concerned by the distance between Jews and Gentiles, and devoted practitioners display specific outfits to disclose publicly their confessions (hat or *kippa,* black suit, beard), especially among the orthodox Ashkenazi (Lubavitch, Haredi, Hasidic). Most of the Jews of France, however, are "traditionalist" (cultural) Jews, well integrated in French society and with a sporadic participation to cult, but given the status of Judaism as "minority," still share a sense of community. The concept of "ghetto," first a community compelled by external force to promiscuity and domination, and later a collective choice against assimilation, is still a symbolic reference (of resistance) but a geographic reality (as neighborhood and for community ties) for French Jews, despite their integration (Benbassa 1997), a sense of community which has been reinforced in the last decades due to the diaspora's solidarity with Israel (Azria 2010).

84.4.4 Islam

Islam is the latest and obviously more "problematic" monotheistic religions which have rooted in France. The history of Islam in France traces back from the 1960s, due to economic migrations. There are nowadays between 4 and 5 million Muslim in France, most of them originating from North-Africa. Islam is in the very heart of academic and political issue, and for geographic issues. Indeed, after the 1960s work migrations of North African Muslims, the 1970s was the decade of "family regrouping" (*regroupement familial*) and Arab temporary migration turned into a durable (or definitive) settlement in France. Workers from Maghreb have been present in the South of France (due to proximity with sea routes to the South), North and North-East regions (industrial sites where Maghreb people were recruited in coal mines or engineering) and, finally, Paris and suburban zones (for other employment facilities). Recent geographic analysis (*Ifop* 2011) demonstrates that these zones still illustrate these patterns of territorial concentration of Muslims. A decade later, in the 1980s, Muslims of France began to lay claims for places of worship (Etienne 1989). The 1990s started the very first "affairs" relating to Islam and especially the famous "veil controversy," which was a way for Muslims to make their body a new site for expressing their religious particularity – veils but also beards, clothes (djellaba) and even T-shirts (with the slogan "Don't Panic: I'm Muslim!") illustrate what has been labeled a "Muslim Pride." In the meantime, Islam expanded rapidly by means of proselytism and conversion, especially among deprived strata living in harsh conditions in the suburbs of large cities. Since the 2000s Islam is increasingly torn between claims for recognition of specific rights and (communitarianism) and attempts to integrate into a French society sometimes little disposed to let mosques flourish throughout the territory.

Hence, we can sum up the attitudes of monotheisms in France as follows. The geographic dynamics of Christianity are relating to historical settlement and cultural absorption, the geographic logics of Judaism, to integration and emancipation, and the geographic issues of Islam to expansion and ethnicity. Moreover, Judaism is in "the Ghetto," Islam and new religious movements (including protestant one) are in "the Suburbs" and traditional Christianity (that is, Roman Catholicism) is "in the city."

84.4.5 The Geography of Non-theistic Traditions

Beside the expected "great" traditions or monotheistic organizations, more interesting are the non-theistic and non-official or less Institutionalized forms of religion. Asian traditions, in particular, have taken root in France by means of migration (from Southeast Asia) and missionary activities (from Tibet and Japan); Buddhism is also practiced by about 850,000 people in France, most of them are of Asian origin. Buddhism and Hinduism are practiced in temples; most are concentrated in urban areas; only the great monastic complexes of Tibetan Buddhism have settled in rural zones (Dordogne and Bourgogne are where the largest monastic complexes are built). Between 80,000 and 150,000 persons are labeled Hindus or practice traditional forms of Hinduism in France, mainly in South Indian and Caribbean islands: La Réunion and La Martinique. Most of them are migrants or descendants. A handful of temples have been built and are mainly attended by native-Hindu practitioners. Buddhism is a more widespread with in between 750,000 and 800,000 followers, including 150,000 converts. Hinduism and Buddhism have in common to be practiced in temples, and, therefore, they also need sites and locations. While Hinduism is bounded by the concentration of migrants and "ethnic" Hindus, the geographic logics of Buddhism are twofold: "Asian" or "native" Buddhists have also made up "ethnic" communities (in big cities), whereas their converts have scattered throughout the country and contribute to the swarming of Buddhist temples and meditation centers. Hinduism and Buddhism have established physical spaces (*sites*) and social spaces where they can perform rituals. Other Asian traditions, like Chinese religion, might be more discreet since they are mainly practiced in domestic spaces, with small altars in honor of spirits and ancestor. The only moment where Chinese religion becomes, for instance, fully visible and clearly unmistakable on the Chinese New Year's day when fake dragons wander the streets of great cities, accompanied with drums and firecrackers.

New religious movements, labeled "*sectes*" in France (similar to *cults* in the United States) have flourished since the 1970s. Despite undersized demographics, sects in France are both located in specific headquarters, and scattered throughout the country, most of the time in hidden sites and places, that is, in rural areas. New Age movements and sects have gradually overflown the urban contexts that have been so propitious to their birth and development, to settle in rural areas. Unexpected in the French religious landscape, Shamanism has recently been established in the soil of Catholic France. Reinvented forms of shamanism under the legacy of ancient animistic pre-Christian substrates (druidism, antique magical cults) is actually imported from the United States and is a consequence of the globalization of religion. Shamanism, like Buddhism, Hinduism, and New Age or sectarian groups, is *bi-located*, that is, located in urban settings, where meditation sessions and ordinary cultic practices are organized in leisure time and in rural settings for spiritual retreats in larger temples and more remote sites. While socially recognized and culturally accepted monotheistic traditions are still subjected to regional variations, this *trans-regional dual territoriality* is a common pattern for all the non-conventional, exported and alternative religions that have taken root in France.

84.5 Visible and Invisible Religions: Public Space as a Battleground for Community Claims

Issues in visibility of religious expressions mobilize political and theoretical issues in France. Indeed, according to the principles of French secularism, or *laïcité*, which can be considered as a very particular form of secularism, religion is not supposed to have disappeared from society itself (which was the first version of secularism in France), but to have withdrawn from "public" spaces and spheres. Indeed, according to the 1905s law on *laïcité*, religion is supposed to have turned "private." The sociological theories of the 1980s and early 1990s followed this intellectual fashion, asserting that religious expressions would have withdrawn from public spaces. In the late 1990s and the 2000-2010s, in France, like in most of the secularized countries, public spaces appropriated by religions in various ways are quite the reverse.

The French views and arrangements of "public spaces," as theorized as a small scaled poetics by Pierre Sansot (1973) in the early 1970s, make them sites subjected to individual appropriations and institutional control. Michel de Certeau had previously evoked public spaces as "arrangements of different modes of life" (de Certeau 1980). Sansot and Certeau did not imagine to what extent, in the 2000s, that the public spaces they described in terms of relationships between the individual and its (mostly urban) environment would have been converted in sites appropriated for ethnic and religious claims. The principles of *laïcité*, and the predictable withdrawal of religion from public spaces (streets, administrations) have been perturbed by an unexpected wave of signs and events heralding the transformation of common social and physical environment as saturated by semiotic elements of religious confessions. The 1990s Muslim's "veil affairs" (the expulsion from schools of young veiled women) was one among the strategies of Muslims to make Islam more visible and the 2010s Islam in France has become problematic not for a question of *belief*, but for a question of *identity* (Dargent 2003).

The issue of visibility is not only relating to the issue of religious symbols in the semiotics of urban landscapes. It is also associated with the matter of the intention, the context and the message underlying the public exhibition of religious symbols. One of the key concepts, thought adjective rather than substantive, of France's last decades' laws on "religious symbols in the public sphere" (Muslim's veil and burka) was the *ostensible* purpose of such an exhibition. Therefore, the debates on *laïcité* assumed a very political dimension: while, on the one side, the 1905s law was mainly aiming at a disarticulating the political and economic link between public administrations and the Church. The late twentieth century's interpretations of *laïcité* turn more socially oriented that the initial spirit was actually (Liogier 2006). The issue of public visibility has only turned more crucial since the 1980s/1990s, when subgroups of French society began to express claims for the recognition of their specificity in a religiously and culturally mosaic landscape. Public spaces have turned into more of a site for performing a newly found, reinvented or accredited ethnicity: Jews, Muslims, Buddhists, Hindus, Sikhs (a very small minority). All reinvest the cities' urban management and in the bodily semiotics of ambulatory spaces (streets). But the acceptance of the semiotic presence of religion in the public sphere is relating to the image of each tradition: while orthodox Judaism (Hassidim and Lubavitch wearing eighteenth century east-European dark clothes) seems exotic, but are tolerated, while Muslims praying by hundreds in the streets of Paris have raised bitter reactions – extreme-right groups have decided to organize "counter-manifestations" to protest against what they call the "Islamization of France."

Otherwise, some ancient esoteric groups, such as the free masons, previously considered as a discreet community of followers, still perform cultic activities in private spaces, only known by the initiated members (to the exception of Lodges Headquarters), but have, in parallel adopted a paradoxical strategy of media visibility. Freemasonry is indeed regularly at the forefront of the press, most of the times described as the "secret rulers of the country." An increasing number of eminent intellectuals or artists nowadays unveil their belonging to this ancient esoteric society. But to the exception of pictures and information publicized by freemasons themselves, Freemasonry remains discreet and only the headquarters of the Grand Loge de France or of the Grand Orient de France are known by the rest of the people. Freemasonry is a good example, but it is far to be the only one, of the subtle interplay between public and private physical (mostly urban) spaces, conditioned by the law on *laïcité*, and the (printed or electronic) media coverage of religious issues.

84.6 Religious Studies: "Back" to Geography?

What are the chances, then, for geography, to get back at the forefront of religious Studies in France? Cutting with essentialist views of Eliade in *Le sacré et le profane* (1950) in which he made "space" a static and the essentialized site for the social dynamics of religion, French geographic approaches to religion have lately, but quickly, jumped in the spatial turn in which all social sciences are engaged. It remains rather challenging to affirm that empirical issues of spatial logics (distribution) and social logics (appropriations and negotiations of space), which rely on a "geographic" approach of religion, could (or not) contribute to the relocation of geography (as a discipline) in the scientific agenda of religious studies. As an anthropologist interested by the new ways of thinking the relationships between culture Arjun Appadurai was praised for a "decisive shift from traits geography to processes geography" (Appadurai 2001). The case of religion in France demonstrates how spatial and territorial processes are of primary importance, for instance, relating to (a) shifts in the patterns of religious territorialization, by means of migration and mission, (b) changes in the relationships between the whole society and the sites or spaces of religion, and (c) changes in the modes of religious expression in their relationship to space – spatial and religious claims of confessional groups. Mapping these new expressions using the "old-fashioned" tools of geography (and even social sciences at large) is not easy. Yet, maps of spatial distribution and patterns of demographic density still remain useful descriptive devices. While "old schools" of French or French-speaking geography of religion have been attached to qualitative methodology beyond the first phenomenological approach of Eliade, new generations of scholars attempt to maintain the "anthropological" legacy of geography of religion (conceptual devices) with the more classical tools of geography (methods and patterns) (see for instance, Hoyez 2012). The social and political issues of struggle for the visibility or invisibility of religious belongings or heritages relate to logics of territorialization, that is, processes that are geographic in shape, sociological in aim, and anthropological in nature. New trends in geography are fertilized by transdisciplinary perspectives.

In the days I am completing this paper (mid-2012), a new large mosque is finally about to be built in Marseille, South of France, where an important Muslim community resides, after long discussions, tensions, and resistances of the local population. A handful of smaller ones have been the target of antiMuslim acts of violence.

In 2011, the representative of the Muslim Council of France declared that in between 100 and 150 mosques were under construction in France, an assertion that reactivated the fear of an "islamization" of France. The same year, however, the first Sikh temple (or *Gurdwara*) of France was finally erected in Bobigny. And the suburbs again, in Bussy-Saint-Georges, the building of an "esplanade of religions" (as it is labeled by the mayor of the town) will be achieved in 2014. The site will include side by side two pagodas, a mosque and a synagogue – a kind of interfaith spot, built with the aim of lowering the social tension in a location where community claims (ethnic and religious) have led confessional groups to reduce physical and social proximity, and build moral and cultural boundaries. Lately, a first Mormon temple might be built in Le Chesnay, in the western suburban area nearby Paris, but here again, local governments resist against the construction – Mormonism being considered as "sect" in France. These few examples confirm the crucial role of space in the dynamics of religion and, not surprisingly, religious minorities and foreign religions, be they exported monotheisms like Islam, sectarian like Mormonism, are the most concerned by, engaged in, and constrained by spatial issues, that is, the negotiation of visibility in public spaces and of legitimacy in society as a whole. Yet again, empirically speaking, the recent historical and religious past and the future of France confirms that religious changes have a direct connection with territorialization issues, and the agency religious actors. French geography of religion is, therefore, definitely bound to the politics and poetics of space, just like in other countries or contexts (Kong 2001), but in the context of secularized public spaces, ruled by the government.

84.7 Conclusion

France, historically the "elder daughter of Roman Catholic Church," has been compelled, during the last decades to recognize its cultural and religion diversity. Religion takes here a mosaic shape as the traditions in the French religious landscape ranges from historical monotheisms (Christianity, Judaism and Islam) to more recent polytheisms (Buddhism, Hinduism), and from organized new religious movements (labeled "sects") to more individually-oriented spiritual techniques. The mapping of these traditions, organizations and practices is a challenge for scholars in religious studies. Indeed, on the one hand, as a secular state, France prohibits religious census and ethnic statistics. Yet scholars' attempts to map the religious expressions have been more accurate, compelling them to invent new approaches to religious geography while adding to "classical" ones. On another hand, and despite the alleged "privatization" of religion according to the principles of 1905s law on *laïcité* (secularism), religious expressions nowadays are gaining visibility in the public sphere, especially for religious-based ethnic claims, Muslims' "veil" affairs and public prayers, Sikhs' turban for instance. One of the paradoxical aspects of this "secular" society is the interplay between the strategies of visibility or of invisibility that different religions adopt in such a context. The geographer's attention, professional or originating from another disciplinary horizon, must be paid to the different scales and sites of observation of such strategies, from the micro to the macro levels: bodily attitudes, clothes and fashions, individual or collective behaviors, religious concentrations in specific zones, buildings and cultic places (the status of which varies depending upon the religion), territorialization processes and proselytizing (concerning mainly new religious movements and "foreign religions"), community and network ties. All these phenomena comprise part of the diversification of the religious landscape of France, but also question the political control of religious pluralism and freedom and of the logics of belongings and confessional / identity claims. Therefore, France offers a very interesting site for the understanding of these geographies of the visible/the invisible in the study of religion, and for the reinvention, *mutatis mutandis*, of the methods and perspectives of geography of religion.

References

Appadurai, A. (2001). Grassroots globalization and the research imagination. In A. Appadurai (Ed.), *Globalisation* (pp. 1–21). Durham/London: Duke University Press.
Attias, J.-C., & Benbassa, E. (Eds.). (2007). *Des cultures et des dieux*. Paris: Fayard.
Aubin-Boltanski, E., Lamine, A.-S., & Luca, N. (Eds.). (2014). *Croire en actes: Distance, intensité ou excès?* Paris: L'Harmattan.
Augé, M. (1992). *Non-Lieux. Introduction à une anthropologie de la surmodernité*. Paris: Le Seuil.
Azria, R. (2010). *Le Judaïsme*. Paris: La Découverte.
Bastian, J.-P., Champion, F., & Rousselet, K. (Eds.). (2001). *La globalisation du religieux*. Paris: L'Harmattan.
Baubérot, J. (1994). *Religions et laïcité dans l'Europe des douze*. Paris: Syros.
Benbassa, E. (1997). *Histoire des Juifs de France*. Paris: Le Seuil.
Béraud, C., & Willaime, J.-P. (Eds.). (2009). *L'école, les jeunes et la religion. L'approche française en perspective européenne*. Paris: Bayard.
Champion, F. (1990). La nébuleuse mystique-ésotérique. Orientations psycho-religieuses des courants mystiques et ésotériques contemporains. In F. Champion & D. Hervieu-Léger (Eds.), *De l'émotion en religion. Renouveaux et traditions* (pp. 17–70). Paris: Le Centurion.
Dargent, C. (2003). Les musulmans déclarés en France: affirmation religieuse, subordination sociale et progressisme politique. *Cahiers du CEVIPOF, Notes et Etudes de l'OIP*, n°39.
de Certeau, M. (1980). *L'invention du quotidien. 1. Arts de Faire*. Paris: Gallimard.
Debray, R. (2002). *L'enseignement du fait religieux dans l'école laïque*. Paris: Ministère de l'Education Nationale.
Deffontaines, P. (1948). *Géographie et religions*. Paris: Presses Universitaires de France.
Deleuze, G., & Guattari, F. (1972). *Capitalisme et schizophrénie*. Paris: Editions de Minuit.
Desroche, H. (1968). *Sociologies religieuses*. Paris: Presses Universitaires de France.
Etienne, B. (1989). *La France et l'islam*. Paris: Hachette.
Fournier, A., & Monroy, M. (1999). *La dérive sectaire*. Paris: Presses Universitaires de France.
Gauchet, M. (1999). *The disenchantment of the world* (O. Burge, Trans.). Princeton: Princeton University Press.
Hervieu-Léger, D. (1993a). *La religion pour mémoire*. Paris: Le Cerf.
Hervieu-Léger, D. (Ed.). (1993b). *Religion et écologie*. Paris: Le Cerf.
Hervieu-Léger, D. (2001). *La religion en miettes ou La question des sectes*. Paris: Calmann-Lévy.
Hervieu-Léger, D. (2003). *Catholicisme, la fin d'un monde*. Paris: Bayard.

Hoyez, A.-C. (2012). *L'espace-monde du yoga. De la santé aux paysages thérapeutiques mondialisés*. Rennes: PUR.
IFOP. (2011). *Enquête sur l'implantation et l'évolution de l'islam en France*. Paris: IFOP. Available at http://www.ifop.com/media/pressdocument/343-1-document_file.pdf.
Introvigne, M. (Ed.). (1996). *Pour en finir avec les sectes*. Milan: CESNUR.
Kepel, G. (1991). *La revanche de Dieu: Chrétiens, Juifs et Musulmans à la reconquête du monde*. Paris: Le Seuil.
Kong, L. (2001). Mapping 'new' geographies of religion: Politics and poetics in modernity. *Progress in Human Geography, 25*(2), 211–233.
Liogier, R. (2006). *Une laïcité "légitime," la France et ses religions d'État*. Paris: Médicis Entrelacs.
Liogier, R. (2010). La distinction sociocognitive et normative entre bonne et mauvaise religion en contexte européen: le cas de l'islam et du bouddhisme. In M. MIlot, P. Portier, & J.-P. Willaime (Eds.), *Pluralisme religieux et citoyenneté* (pp. 99–122). Rennes: PUR.
Lussault, M. (2007). *L'homme spatial*. Paris: Le Seuil.
Massey, D. (2005). *For space*. London: Sage.
Obadia, L. (2007). *Le Bouddhisme en Occident*. Paris: La Découverte.
Piette, A. (2003). *Le fait religieux. Une théorie de la religion ordinaire*. Paris: Economica.
Planté, C. (Ed.). (2002). *Sorcières et sorcellerie*. Lyon: Cahiers Masculin/Féminin/Presses Universitaires de Lyon.
Racine, J.-B., & Walther, O. (2003). Géographie et religions: une approche territoriale du religieux et du sacré. *L'Information géographique, 3*, 193–221.
Roy, O. (2008). *La sainte Ignorance. Le temps de la religion sans culture*. Paris: Le Seuil.
Sansot, P. (1973). *Poétique de la ville*. Paris: Klincksieck.
Vincent, J.-F., Dory, D., & Verdier, R. (Eds.). (1995). *La construction religieuse du territoire*. Paris: L'Harmattan.
Willaime, J.-P. (2009). European Integration, laïcité and religion. *Religion, State and Society, 37*(1–2), i23–i35.

Chapter 85
"A Most Difficult Assignment:" Mapping the Emergence of Jehovah's Witnesses in Ireland

David J. Butler

85.1 Introduction

Jehovah's Witnesses are one of the fastest growing Christian religious denominations, with members and adherents in 236 countries worldwide. In the eyes of many, they are seen as a new religious cult movement, largely due to their comparatively recent arrival in most areas, but their foundation actually dates from 1872. The idea for the foundation of the denomination which eventually became known as Jehovah's Witnesses originated in the mind of Charles Taze Russell (1852–1916), in Pittsburgh, Pennsylvania. In the early years, the movement was known successively as "The International Bible Students Association" and "The Watch Tower Bible and Tract Society"; the final title "Jehovah's Witnesses" was applied from 1931 by J. F. Rutherford (1869–1942), Second President of the Society. This title is based upon the words of Isaiah "Ye are my witnesses, saith Jehovah" (Isaiah 43.10). Jehovah is the ancient title for God, frequently used in ancient Biblical texts, but seldom used nowadays, save in connection with the Witnesses.

Jehovah's Witnesses comprised a worldwide membership of just 50,000 as late as 1938, which had increased to 1.3 million by 1971. Since then, however, the rate of increase has been rapid, with 2.7 million members in 1990, 5.91 million at the end of 1999 and 7.66 million at the end of 2011 (Table 85.1).[1] By that time, the Witnesses had 6,006 members in 115 congregations scattered throughout the island of Ireland.[2] Yet, beyond a few simple facts such as the location of the nearest Kingdom Hall of Jehovah's Witnesses or the home of a member, the majority of people are totally

[1] During 2011 alone, the Witness' numbers grew by 263,131 baptisms to a worldwide total membership of 7,659,019, meeting in a total of 109,403 congregations.

[2] The number in 1999 was 4,600 members in 114 congregations on the island of Ireland.

D.J. Butler (✉)
Department of Geography, University of Ireland, Cork, Ireland
e-mail: d.butler@ucc.ie

S.D. Brunn (ed.), *The Changing World Religion Map*,
DOI 10.1007/978-94-017-9376-6_85

Table 85.1 Worldwide organization of Jehovah's Witnesses, 2011

Number of lands reporting	236
Watch tower branches	98
Branches divide into districts	1–10
Districts divide into circuits	5–20
Circuits divide into congregations	109,403
Worldwide total active publishers	7,659,019

Data source: JW Irish Desk in London

uninformed as to the historical origins, beliefs and practices, distribution and territorial organization of the Witnesses. Most are particularly unaware of the difficult emergence of Jehovah's Witnesses in twentieth century predominantly Roman Catholic Ireland, often barely tolerated by a small and generally declining Protestant community suspicious of their newer reformed denomination. This paper seeks to analyse and map the emergence of Jehovah's Witnesses in Ireland, a place acknowledged by the Witnesses themselves as "a most difficult assignment."

85.2 Irish Protestant Minorities in Context

The study of minorities forms a very large part of research on ethnicity and nationalism and, consequently, Europe's traditional ethnic minorities and the conflicts over their place in the state and nation are the focus of continuing comparative research. For the two centuries that followed the Reformation, states everywhere sought religious homogeneity – the principle of *Cuius regio, eius religio* provided for internal religious unity within a state: the religion of the prince became the religion of the state and all its inhabitants – and marginalized and persecuted their religious minorities. In the nineteenth century the emphasis was on emancipation and integration, but religious identities remained strong and older conflicts frequently emerged in new guise. Today – again with the exception of Northern Ireland – traditional religious oppositions, rivalries and conflicts have faded and religious identities have lost much of their intensity.

Contemporary Europe's geo-religious foundations were set in the crucial two centuries of the wars of religion. This might be seen as a Protestant north European core surrounded on three sides (west, south, and east) by Catholic counter-reformation countries stretching from Ireland in the northwest to Poland in the northeast, with a mixed interface region between them. The island of Ireland, too, reflected this distribution and, prior to 1650, outside of the historic province of Ulster and the larger port towns and cities, Irish Protestantism essentially comprised adherents of the Anglican Established Church. The decade of the Cromwellian protectorate changed this, simultaneously increasing the reformed Christian community island wide, while also splintering it into Presbyterian, Independent/Congregationalist, Baptist and Society of Friends (Quaker) groups, as well as Anglican. From the re-establishment of the Anglican Church in 1660–1662, the dissenters became a declining, though still quite powerful group; at all times, they remained in danger of assimilation into the Anglican community and/or the Roman Catholic majority through intermarriage or contact conversion, as the austerity of their faiths conflicted with the mindset and growing economic prestige of subsequent generations. The Anglican Church in Ireland, for its part, purposefully positioned itself as doctrinally "low church," with widespread success in assimilating much of the southern Irish dissenter community by the later nineteenth century.

The last quarter of the eighteenth and the first quarter of the nineteenth century saw Irish Protestants struggle to stem their declining hegemony against a now emancipated and rising Roman Catholic majority who had never accepted the legitimacy of their defeat. Protestants had two options in this situation. One was to try to lead in the re-construction of the Irish nation and, in the process, secure their own place within it. While a few tried to do this, the overwhelming majority stuck with the traditional option: to align themselves with Britain, the union and the empire in the hope that Britain would defend them against the Catholic threat. Irish Protestants saw themselves as Irish, but in a different way from the Catholic Irish (Ruane and Butler 2007). It was not simply a common national narrative that had to be created, but a common national identification. There was also the matter of Catholic material grievances, particularly in relation to land. Finally, there was geo-politics. Irish Protestants were the ruling class, allied with and traditionally loyal to the British crown: lower class Protestants benefited from this relationship and remained loyal (and deferential) to the upper class. The problem with the traditional option, however, was that British strategic calculations were changing. By the late nineteenth century, placating the increasingly nationalist Irish Catholic majority had a higher priority than protecting the privileges of the traditionally loyal Protestant minority. Demands for Home Rule led on to armed insurrection and, eventually, after 2 years of a nationalist guerrilla campaign (1919–1921), the British government decided to cut its losses and cede the larger part of the island to an independent Irish government. Irish Free State independence placed Southern Protestants in the position that for centuries they had struggled to avoid: becoming a minority in a Catholic-dominated state.

The culmination of the above processes was to make Christian denominations in Ireland all the more possessive of their adherents, whether one is referring to the Roman Catholic Church in the face of Protestant evangelism or to inter-church "sheep-stealing" among the reformed churches. It was against this backdrop that Jehovah's Witnesses commenced their evangelisation of Ireland.

85.3 The Witnesses in Ireland – Origins

Jehovah's Witnesses owe their origins in Ireland to the single visit of their founder in 1891, when en route from the United States to Europe. Charles T. Russell disembarked at the port town of Cobh, in county Cork and worked northwards to Dublin and Belfast, where in both cities he established "ecclesias" or congregations. "Brothers" distributed religious tracts at the doors of Protestant Churches in an informal, unorganized witnessing effort. Protestants alone were evangelized as, given the denominational situation in Ireland historically, they were more receptive to this process and accustomed to evangelical terminology. By 1908, Belfast had 24 Bible Students in membership and Dublin had 40. Outreach continued under the direction of the central organisation until 1919, when Dublin congregation seceded, having exhibited exclusive, independent tendencies. Of a healthy congregation of 100 persons, only four remained in connection with the Bible Students, as the Witnesses were then still called. Preaching and witnessing in Dublin, of necessity, virtually ceased. Belfast congregation was affected to a lesser extent and suffered instead from the proliferation of born-again sects, clergy opposition and paramilitary threats present in the newly-established Northern Ireland, which combined to make it difficult to obtain a preaching hall for more than one engagement (The Watch Tower Society 1988: 65–68).

In the South too, there was unremitting opposition. From 1920 until 1945, only four Witnesses remained in Dublin, and there was but scattered colportage (carriage and dispersal of religious tracts by itinerant preachers) elsewhere. However, missionaries were sent from overseas, mainly England, to the three next largest settlements in the south of Ireland, namely the port cities of Cork, Limerick and Waterford.

In 1926, four members arrived in Waterford, where there were already three members resident. These had been attacked by a Roman Catholic mob, which resulted in a severe curtailment of witnessing among the majority community of that city. The Witnesses obtained lodging in Protestant homes, and largely confined their work to Protestant households. It was the era of mission lectures, slideshows and film reels in Protestantism, and their evangelical outreach in this regard proved popular. However, with the potential of Waterford city severely constrained, the Witnesses began to work through its hinterland. In the face of local opposition, including press warnings, the Witnesses devised a mission strategy aimed at overcoming the geopolitics of the regions, by dividing the witnessing district into zones, working from the outer limits of their chosen territory inward and finishing in the town of their accommodation.

From 1931, five Witnesses worked in Cork, with another five at Limerick and, combined, they also covered county Tipperary. There were setbacks, most notably at Roscrea, in the Irish midlands, where all witnessing literature was stolen in an armed ambush, soaked with petrol and set alight, while local police, clergy and children stood around the bonfire singing "Faith Of Our Fathers" (The Watch Tower Society 1988: 90–95).

In 1937, the present Constitution came into force in Ireland, which guaranteed tolerance and religious freedom to all religious denominations then in existence in the state in the spirit of its predecessor. About this time, Ireland was described in the Jehovah's Witness *Yearbook* as "the darkest place in the British Isles … Priests follow the pioneers from place to place, find where literature has been left and immediately cause its destruction" (see Jehovah's Witnesses in the Divine Purpose). Next to India, where the Hindu's had proved "notoriously difficult" to convert to Christianity in any substantial number, Ireland was considered "the second most difficult assignment" worldwide by the Witnesses. By 1945, about 120 members lived at Belfast and about 20 at Dublin (The Watch Tower Society 1988: 80).

85.3.1 The Witnesses in Ireland – Persecution

In having a converted, rather than an inherited membership, the Witnesses experienced many difficulties. Those converting from one of the mainstream Protestant churches were seen as traitors to their church in an almost totally Roman Catholic environment. Only the very occasional Roman Catholic was converted, particularly in rural Ireland, where the small rural communities were tightly woven social and kinship networks, hostile to the Witnesses, or any other perceived threat to the *status quo*. It is surprising that there were any additions at all, considering rural Ireland was seen as keeper of the Roman Catholic faith, membership of which was equated to the essence of Irishness itself. As many pioneer Witnesses were foreign nationals, particularly British subjects, they also experienced opposition on grounds of nationality. In the post-war period, they were portrayed as Communist foreigners, and there were several instances of mob violence leading to actual injury. Such an incident occurred in 1948 at Cork, and led to a highly publicised court case, where the Witnesses were vindicated with the assistance of evidence from a Roman Catholic policeman. This, and other key incidents, led – especially during the 1950s – to a growing realisation that Jehovah's Witnesses were entitled under the Constitution to freely practice their religion. The Witnesses, for their part, recognized the Irish mission field would make slow progress, for "the people are fearful of letting go of the traditions so long cherished and so the progress is such that it needs much tact and patience".

From the mid-twentieth century, Irish newspapers began to take note of anti-Witness injustices, leading to a religious debate, particularly in Dublin, despite the pre-Vatican II Roman Catholic church position of preaching the excommunication of members who read evangelical literature, while at the same time appealing to republican and nationalist sentiments by preaching "Ireland's heritage of faith, retained through centuries of persecution, is not to be bartered for a mass of Brooklyn [New York] pottage" (The Watch Tower Society 1988: 100). Even within the Protestant community, there was an abhorrence of perceived "losses" to the Jehovah's Witnesses, and this community had its own subtle way of dealing with these losses. One such 1950s instance, recalled in a recently published obituary, noted how the deceased's husband had been:

> a clerk with a Protestant timber firm in Cork city, but that calamity struck, when he was let go after joining the Jehovah's Witnesses. While no reason was given for his lay-off, the couple were convinced it was because of his membership of the Jehovah's Witnesses. A friend advised them to go into the fish-and-chip business and they opened [shop] 54 years ago this month [July 1958] at a derelict cottage in the suburb of Douglas. In an ecumenical gesture, the O'Callaghans, near the North Cathedral, and Jackie Lennox, whose name is over Cork's most famous fish-and-chip shop, brought Ken behind the counter to show him the ropes. The shop was soon denounced from the pulpit by the local priest who discouraged parishioners from going there. (Anon 2012)

By 1950, there were still only a handful of Jehovah's Witness congregations in Ireland, namely two in Cork, and one each in Dublin, Limerick and Waterford. All met in inconvenient upstairs rooms and private houses until 1953, when the Dublin congregation obtained a renovated building, which became the first Kingdom Hall of Ireland. This was a key moment in the Witness' movement in Ireland, as the first component of its "formal positive expression of religion on the landscape" (Sopher 1967: 77). However, such movement of most congregations to decent accommodation was still several decades away. As part of the Irish mission strategy, use was made of the existing small pioneer groups scattered throughout the country, and this system was expanded. As lodging was difficult to obtain in most towns, unfurnished cottages were rented near towns, with furniture provided by the Watchtower Society at a nominal rent. These cottages also served as meeting places.

The Jehovah's Witnesses evangelical methodology of systematically selling religious literature door-to-door had not previously been experienced in rural Ireland, as colportage had principally targeted church meetings and families with evangelical leanings. The difficulties still faced by Christian evangelists in the face of the power exerted by the pre-Vatican II Roman Catholic Church over rural Ireland is well illustrated by an incident of May 13, 1956, when an attack was perpetrated on a pair of Witnesses actively evangelising and selling scripture readings at Doonass, in the Clonlara district of county Clare, on the mid-western seaboard. The resulting case at Limerick District Court – at which the Roman Catholic curate of Clonlara, The Revd Patrick Ryan and nine of his parishioners were charged with maliciously damaging £3 worth of books, bibles and other literature, besides assaulting the two Witnesses who were "preaching the Gospel" on that day – received extensive coverage in the press. The Roman Catholic Church took a great interest in the hearing, which lasted almost 2 h, to the extent that the bishop of Killaloe, The Most Revd Dr Rodgers, personally attended the proceedings, whereupon

> the crowded courtroom rose to their feet as His Lordship entered … also present were some secretaries of the Men's Arch-Confraternity of the Holy Family, attached to the Redemptorist Church, Limerick. (Clare Champion, August 4, 1956.)

The case resulted in all charges being dismissed and, in a centuries-old equation of Roman Catholicism with Irishness, the justice declared the pair

> guilty of blasphemy in the [Roman] Catholic understanding of the word … they come into this village [of Clonlara] and attack and outrage all that these simple Irishmen hold dear. I think the two men were lucky to escape so lightly.

He went on to "ensure there would not be a repetition of the case"' by binding them over to the peace in personal bonds of £100 with each two independent sureties of £100 each (Sopher 1967: 77).

The matter did not rest there. On returning home from the case of *The Attorney General v The Revd Patrick Ryan*, the bishop penned an epistle to the Irish Prime Minister, An Taoiseach, John A. Costello, expressing incredulity how, in the knowledge that the Attorney General had been "fully aware of the pernicious and blasphemous literature distributed and sold in my diocese," he had "proceeded against one of my priests for upholding and defending the fundamental truths of our treasured Catholic faith" (Cassidy 1956: 4). The reply of An Taoiseach fully appreciated "the just indignation aroused among the clergy and the people by the activities of the Jehovah's Witnesses," but emphasized that the law provided the

> means of dealing with persons whose conduct was calculated to lead to a breach of the peace or who uttered blasphemy. It would be incompatible with the duty of those responsible for the maintenance of peace and order to acquiesce in the adoption of other methods. (Letter from An Taoiseach, John A. Costello to Most Revd Dr Joseph Rodgers, bishop of Killaloe, dated August 14, 1956)

A more forceful incident involving Jehovah's Witnesses and zealous clerical defenders of the faith – in April 1960, at Wexford, where two clergy and a companion inflicted actual bodily harm on two Jehovah's Witnesses–generated a memorandum that the clergy and their companion "be prosecuted for assault ... I spoke to the Taoiseach and he agreed to the course being adopted." However, the governments of the day were for clerical appeasement, and the final entry in the Chief State Solicitor's file was of the

> opinion that it would be better to let the matter rest, in the hope that these people [the Jehovah's Witnesses] will leave the town peacefully on the direction of the Principals. But should there be any further trouble ... then I think proceedings should be brought for the purposes of having them bound to be of good behaviour. (Letter from the Chief State Solicitor to Mr. James Coghlan, solicitor, New Ross, dated July 29, 1960)

The past is indeed another country, whose landscapes we gaze at almost without recognition.[3]

85.3.2 The Witnesses in Ireland – Emergence

One of the catalysts in the emergence and acceptance of Jehovah's Witnesses in Ireland was the holding of an International Convention in 1965 at Dublin, in the teeth of sustained opposition from several city councillors who wanted to refuse planning permission. The Irish tourism authority, Bord Fáilte, gave the go-ahead, and thus Ireland was freed, in a very public manner, of the religious persecution of minority groups. Permission was subsequently granted by the Dublin City Council, but set-up and catering difficulties remained. Many hotels maintained they had "no vacancies" on being contacted, so that 3,500 Witnesses had to be housed in homes around the city. Eventually, prejudice dropped and, in 1966, a branch office was opened in Dublin for the whole island of Ireland. At this point in time, there were 268 Witnesses in the Republic and 474 in Northern Ireland, a total of 742 members on the island.

The commencement of the Northern Ireland "troubles" in 1968–1969 created further difficulties in the field ministry, although the strict neutrality of the Witnesses throughout allowed access to the Roman Catholic communities. All over Ireland, pioneers from abroad, predominantly England, remained as congregational elders in Ireland. Individual Witnesses married and began to have children from the later 1960s especially, leading to greater congregational stability.

The Jehovah's Witnesses remained undaunted by these precedents, but the obtaining of meeting premises remained a problem in parts of Ireland as late as the

[3]The release of these State Papers drew a special issue of Kevin Myers in An Irishman's Diary, Irish Times, January 8, 2003.

mid-1970s, such as in Ennis, Co. Clare; here, in 1974 a former bakery premises was made available to the Witnesses, the owner being at pains to mention that he was "not a God-fearing man or he would not be renting to them" (Oral testimony of Mr. John Denning, Jehovah's Witness, August 31, 2005). Gradually, however, throughout the 1980s and 1990s in particular, most congregations obtained a purpose-built Kingdom Hall, many of quick-build construction, as set out in Table 85.2. Up to the mid-1980s, few congregations had proper meeting facilities, with many meeting in small rooms up many flights of stairs or in hotel rooms. Members were reluctant to abandon the field ministry for weeks and months of fundraising and building new halls, but the "quick-build" Kingdom Hall technology changed all this. They could be constructed and finished within 2 or 3 days. Excellent meeting facilities were provided without disrupting preaching for long periods. The first such hall to be constructed in Ireland was at Downpatrick, in June 1985, when over 600 Witness volunteers from all over the British Isles converged and labored together, so that a congregation of 19 Publishers would have their own Hall. These quick-builds were also perfect Witnessing opportunities, as they typically attracted large crowds of onlookers. Soon, Ireland had its own quick-build hall team and many congregations benefited. In 1982, membership numbers surpassed 2,000 for the first time, with 2,021 members recorded on the island of Ireland. Numbers continued to climb; reaching 2,661 in 1987 and 3,012 Witnesses were recorded at the beginning of 1989 (The Watch Tower Society 1993). Since that time, Witness numbers have increased by about a thousand every 7 years so that, by the end of 2011, slightly in excess of 6,000 members resided on the island of Ireland.

At the start of the year 2000, there were 114 congregations meeting on the island of Ireland – 79 in Ireland and 35 in Northern Ireland – and while membership numbers have grown considerably in the decade since and several halls have been improved or rebuilt, only one additional Kingdom Hall has been provided (The Watch Tower Society 2000). In October 2011, it was reported that the Irish headquarters of Jehovah's Witnesses, built in 1996 at Newcastle, near Greystones, Co Wicklow, was to be sold (Fagan 2011). The decision to sell the facility was prompted by a world-wide restructuring of the organization, with future administration of the island of Ireland taken over by the British branch at its London headquarters and the needs of Irish members catered for in the 115 separate congregations already functioning around the island of Ireland (The Watch Tower Society 2012). Nevertheless, the profile of the Witnesses on the island of Ireland continues to increase, with the holding of a recent international 3-day convention in Dublin, the first of its kind in Ireland since 1978, attracting some 1,700 international delegates, and up to 9,000 delegates in attendance each day (Fig. 85.1) (Witnesses 2012).

85.3.3 Organization and Evangelism

How is it, then, that Jehovah's Witnesses have increased their number to such a large degree, particular in the final third of the twentieth century? The key to their success, in my judgment, has been a carefully run organization, tightly controlled in

Table 85.2 Ireland Kingdom Hall history – construction

Year	Month	Congregation	Type
1985	May – June	Downpatrick	New
1986	May	Dun Laoghaire	New
1986	July	Carrick-on-Suir	New
1986	August	Fermoy	New
1986	September	Antrim	New
1986	October	Mullingar	New
1987	May	Athlone	New
1987	August	Port Laoise	New
1987	September	Clondalkin	New
1988	April	Ballymoney	New
1988	May	Tuam	New
1988	June	Clonmel	New
1988	August	Newry	New
1988	September	Londonderry	New
1988	October	Swords	New
1988	October	Dundalk	Remodel existing building
1989	May	Ballyshannon	New
1989	August	Ennis	New
1989	October	Letterkenny	New
1990	February	Arklow	Warehouse/remodel
1990	May	Newtownabbey	Old school
1990	June	Larne	New
1990	August	Boyle	New
1990	September	Strabane	New
1990	October	Cork	New
1990	November	Dunmurry	New
1991	Spring	Bray	Shop/remodel
1991	April	Fortwilliam	
1991	May	Carrigaline	New
1992	April	Roscommon	New
1992	May	Waterford	New
1992	June	Listowel	New
1992	July	Navan	Converted shop
1992	August	Dungarvan	New
1992	October	Portadown	New
1992?	October	Lisburn	
1992	November	Omagh	New
1993	May	Killarney	New
1993	June	New Ross	New
1993	September	Tallaght	New
1993	September	Lurgan	New
1993	October	Blanchardstown	New

(continued)

Table 85.2 (continued)

Year	Month	Congregation	Type
1994	April	Buncrana	New
1994	May	Finglas	New
1994	June	Birr	New
1994	September	Newtowards	New
1994	October	Ballina	New
1995	April	Tipperary	New
1995	May	Castlebar	New
1995	December	Cookstown	House converted to KH
1996	June	Dungannon	New
1996	August	Athy	New
1996	October	Ballymena	Demolish/new
1997	May	Monaghan	New
1997	August	Dungloe	New
1997	September	Cavan	New
1997	Finish August 1998	Donegal	Old school/remodel
1998	September	Bantry	Cottage converted to KH
1998	September	Newcastle West	New
2000	June	Shannon	New
2001	April	Balbriggan	New
2004	May	Tralee	New
2004	August	Ballyhaunis	New
2004	September	Sligo	New
2005	September	Wexford	New
2006	April	Thurles	New
2006	August	Youghal	Demolish/new
2009	January	Belmullet	KH – annex building of Tony Eaton

Data source: JW Irish Desk in London

all its aspects by central headquarters in Brooklyn, New York, and run with all the efficiency of a modern international business-house.

Jehovah's Witnesses are organized in a hierarchical arrangement they call a "theocracy," which they believe is an earthly expression of God's heavenly organization. Based in the Watch Tower Society's headquarters, the organization is headed by the Governing Body of Jehovah's Witnesses, with members of the Body and "helpers" organized into six committees responsible for various administrative functions within the global Witness community, including publication, assembly programs and evangelizing activity.

The Governing Body and its committees supervise operations of 30 global "zones," comprising 98 branch offices worldwide. Each branch office oversees activities of Jehovah's Witnesses in a particular country or region, and may include facilities for the publication and distribution of Watch Tower Society literature.

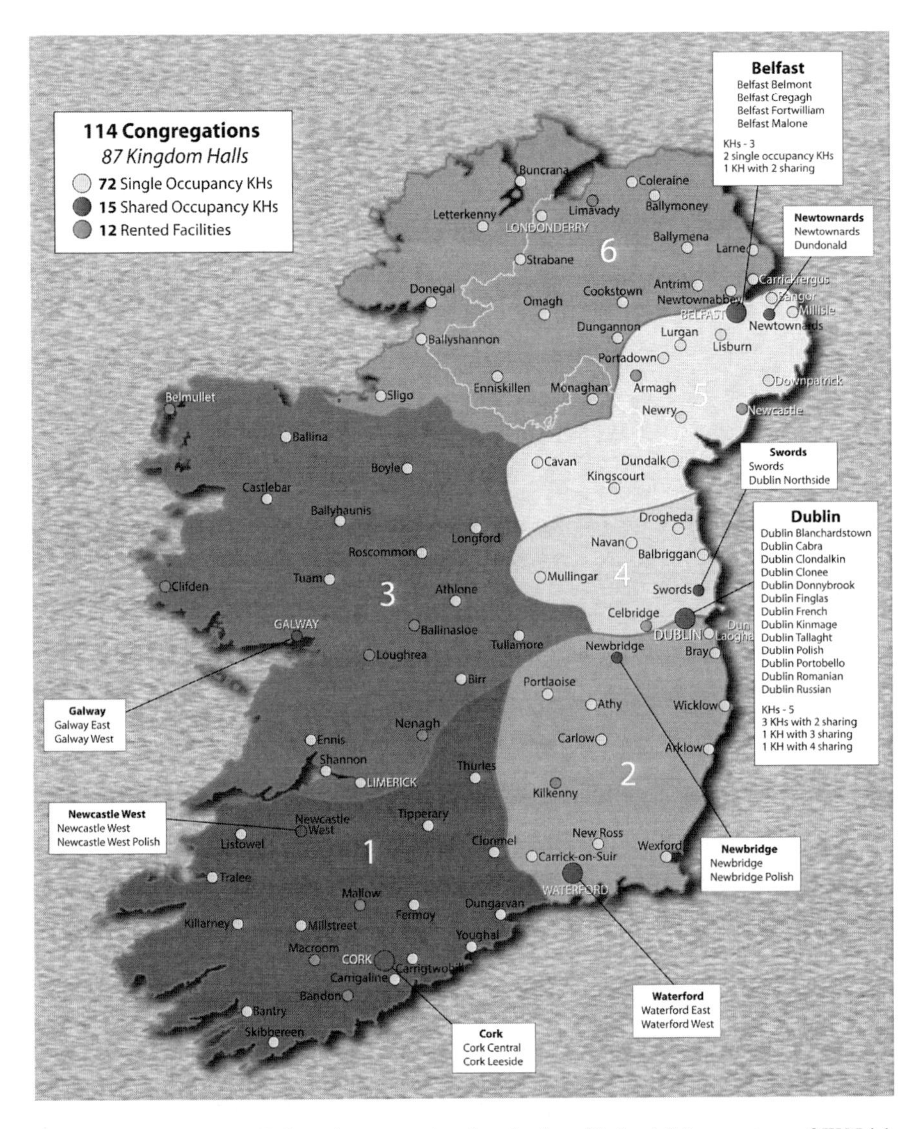

Fig. 85.1 The Kingdom Halls and congregational territories of Ireland (Map courtesy of JW Irish desk, London)

Directly appointed by the Governing Body, branch committees make local appointments and perform other administrative functions for congregations, which are organized within “circuits” and “districts.”

The Governing Body directly appoints “travelling overseers” as its representatives to supervise activities at various levels – “circuit overseers” visit circuits of about 20 congregations; “district overseers” work with the circuit overseers of a number of adjoining circuits; and "zone overseers" who visit groups of branch offices in a particular zone and report back to the Governing Body.

Each congregation is served by a group of locally recommended, branch-appointed male elders and "ministerial servants" (the latter being the Witnesses' term for deacons). Elders take responsibility for congregational governance, pastoral work, setting meeting times, selecting speakers and conducting meetings, directing the public preaching work, and creating judicial committees to investigate and decide disciplinary action for cases that are seen as breaching scriptural laws. Ministerial servants fulfil clerical and attendant duties, but may also teach and conduct meetings.

When a congregation is sufficiently numerous, it obtains a proper place of worship, called "Kingdom Hall." At the head of each congregation is the "Presiding Overseer." (originally Pastor, but translated to "clergyman" in some languages, which is too church in style) who is judged the most spiritually mature, and is responsible to the Branch Servant for the running of the congregation. He is assisted by a "Service Committee," (comprising the Presiding Overseer, Secretary and Field Service Overseer as the three senior elders of each congregation), which takes charge of various activities, particularly of the "back calls," that is, repeated visitation of interested contacts. Each member of the body of elders obeys the orders of his superior without question and, throughout the organization, posts are held by men; women being discouraged from seeking administrative or spiritual office. They are, however, allowed full membership and are leaders in the preaching and publishing mission workers fields (Davies 1954: 72).

The founder of the Witnesses, Charles Taze Russell, had the gift of simplicity of expression in language and the ability to draw illustrations in abundance from everyday life. This tradition continues to this day in the two main Witness booklets *The Watchtower: Announcing God's Kingdom* and *Awake*, both published fortnightly. They are attractively presented, with articles of historical and contemporary interest, which appeal to Witnesses and non-Witnesses alike. In combining these skills with sustained Bible study, Witnesses have an unrivalled knowledge of Scripture and can quote chapter and verse as required at a moment's notice.

From the outset, Jehovah's Witnesses had the foresight to count upon each and every member, male and female, in the promotion of the organization, and supplied them with the most up-to-date aids, such as taped sermons and attractively printed, bound and illustrated volumes. Indeed, the official name given by the organization to its members is "publishers." All baptized members, including women and teenagers, undertake door-to-door missionary work on a weekly basis. Each member was until recently expected to devote at least 10 h per month to this work, but this requirement has been removed, so that more members could enter the part-time preaching ministry, in amounts proportionate to their free time.

Details of all visitations are carefully noted in specially printed service notebooks, which are sent to branch (usually national) office level, and thence to headquarters in New York. In addition to this part-time missionary work, full-time "pioneer" missioners and missionary couples are engaged in evangelical work. The publications dispersed to the public cost some $65 million per annum to produce, using voluntary labor, and production costs are offset by "Book Money" collections in Witness congregations for the worldwide work. A second collection is held for the needs of the local congregation, while other weekly collections are held for relief in areas experiencing famine or natural disasters (The Watchtower, vol. 121).

Each congregation is assigned a territory; members are requested to attend the congregation of the territory in which they reside. Figures 85.2, 85.3, 85.4, 85.5, 85.6, and 85.7 illustrate the congregations and their witnessing territories in the Irish context.

The "pioneers" and part-time mission evangelists share much common heritage with the various evangelical Protestant churches, particularly those established in the eighteenth and nineteenth centuries, with their itinerant preachers travelling from settlement to settlement amid widespread mob violence and hostile confrontation. Denominations which struggled for acceptance and membership in the Christian community in Ireland of this period included the Methodists, Baptists, Brethren and Congregationalists. Of other shared aspects of reformed Christian heritage, one

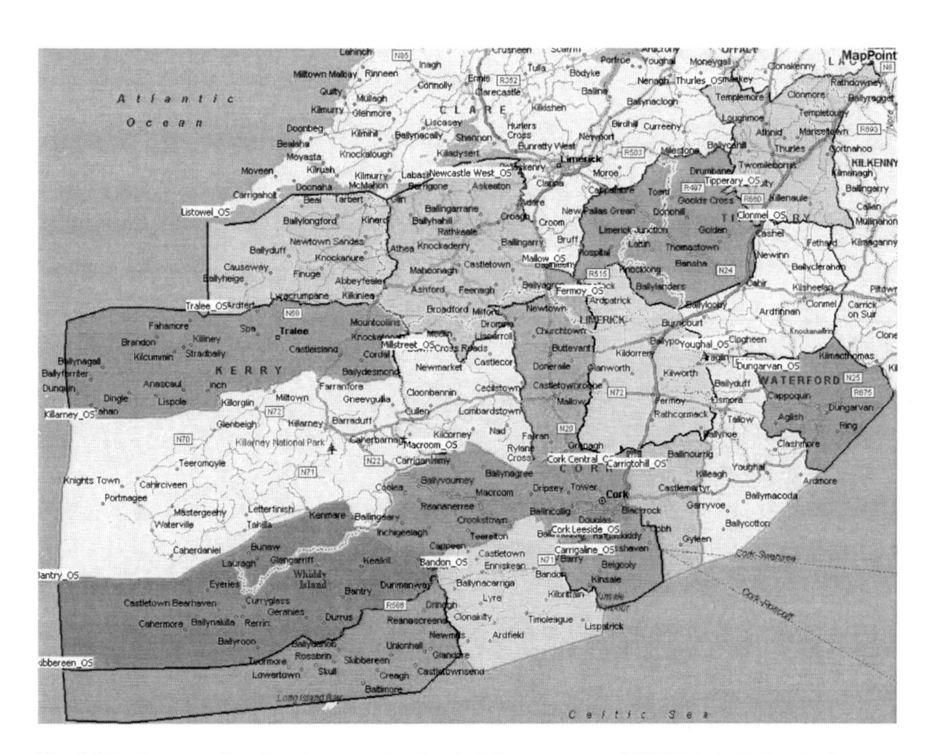

Fig. 85.2 Congregational territory no. 1 – South (Map courtesy of JW Irish desk, London)

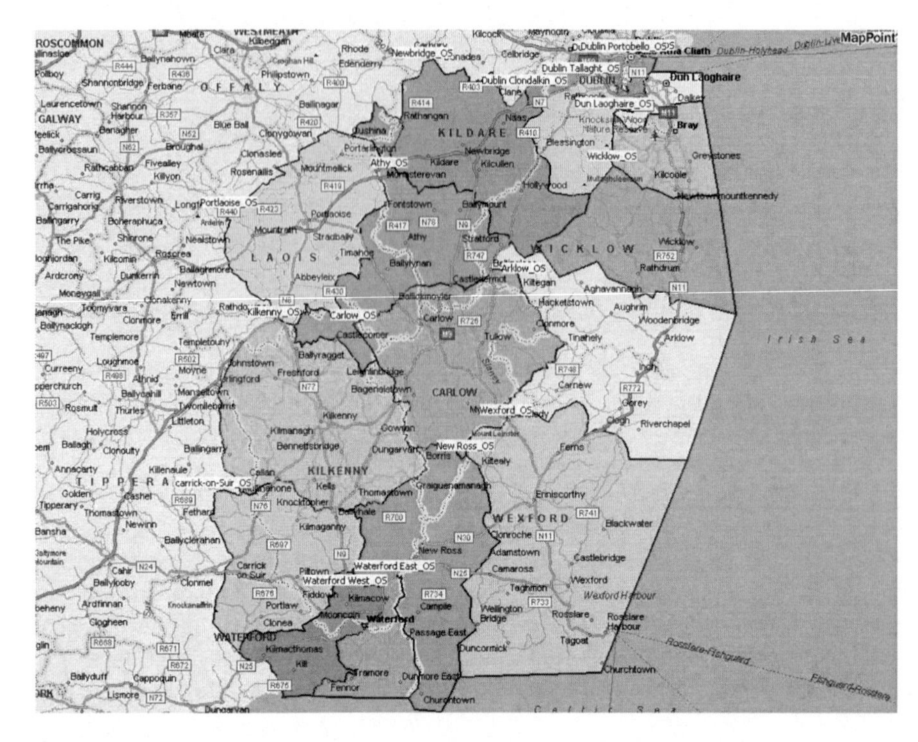

Fig. 85.3 Congregational territory no. 2 – South Eastern (Map courtesy of JW Irish desk, London)

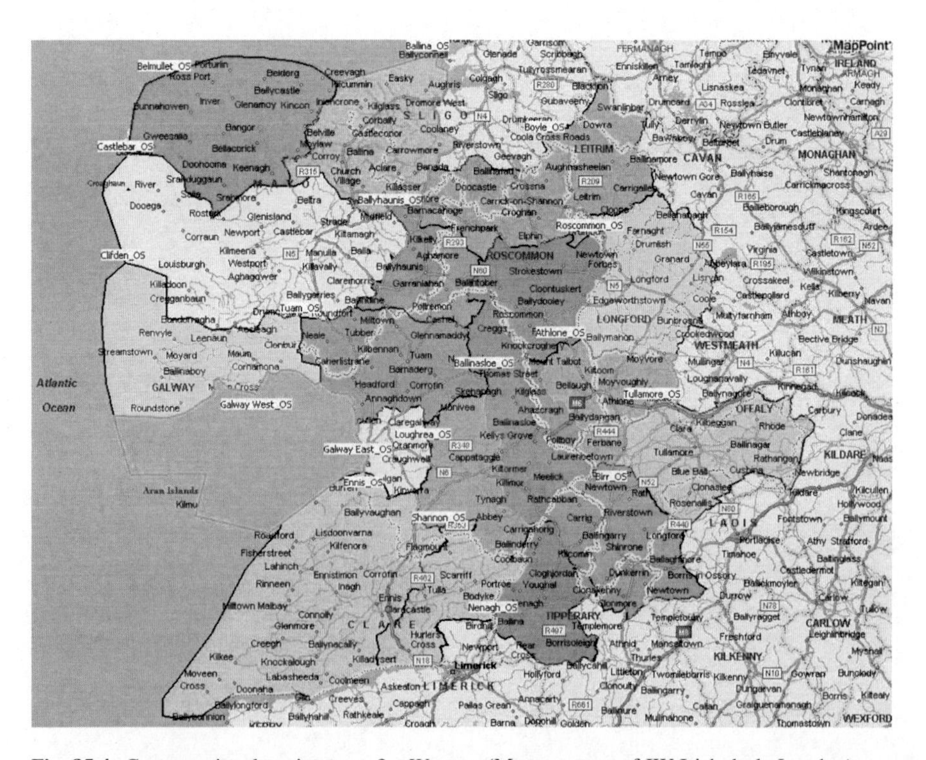

Fig. 85.4 Congregational territory no. 3 – Western (Map courtesy of JW Irish desk, London)

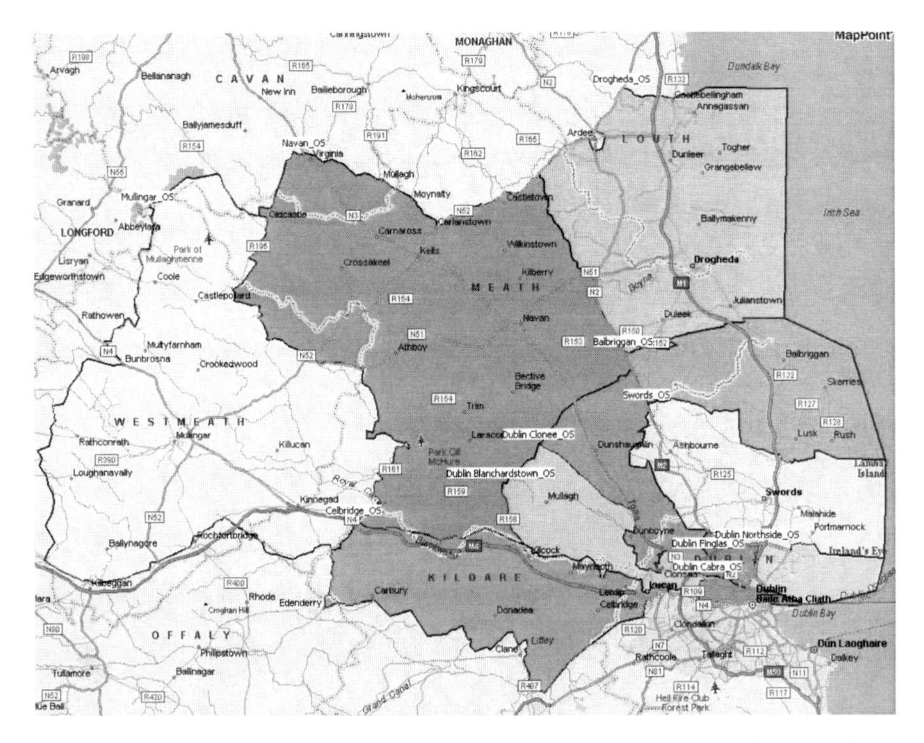

Fig. 85.5 Congregational territory no. 4 – Dublin (Map courtesy of JW Irish desk, London)

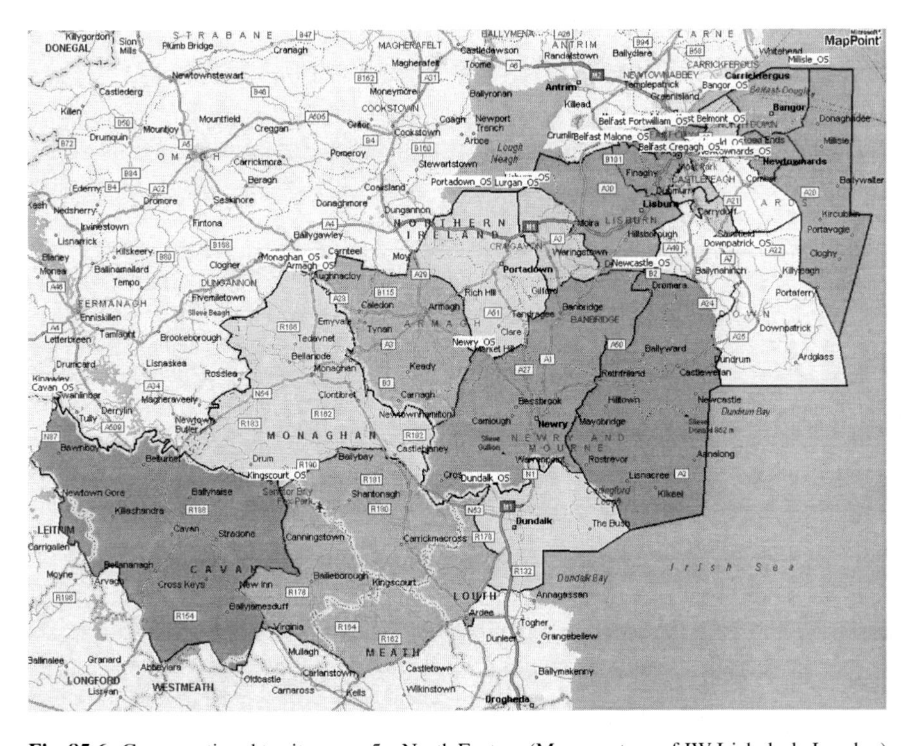

Fig. 85.6 Congregational territory no. 5 – North Eastern (Map courtesy of JW Irish desk, London)

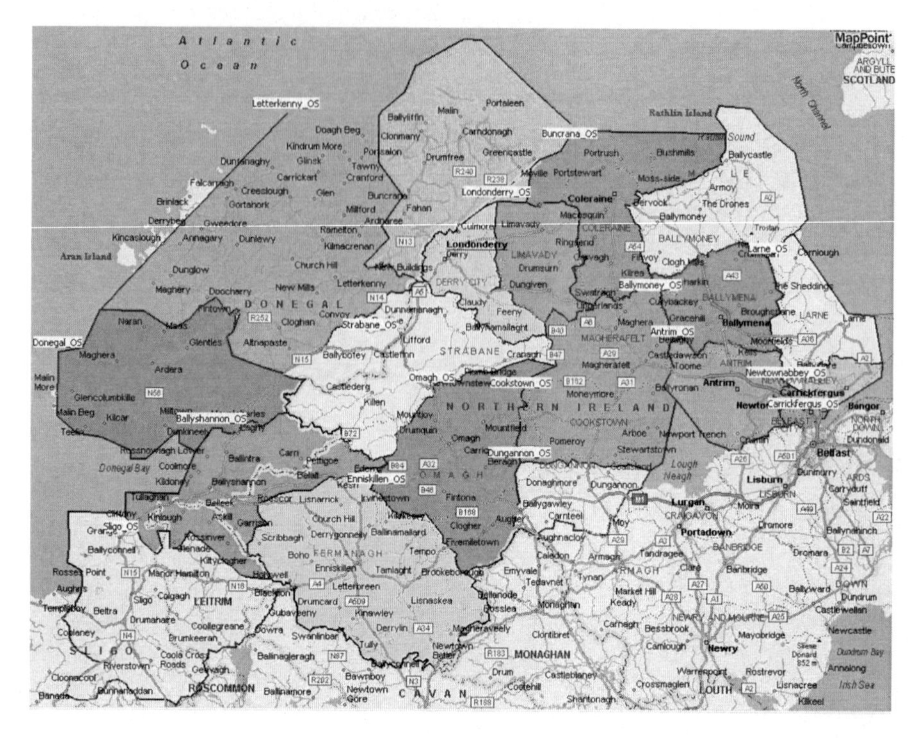

Fig. 85.7 Congregational territory no. 6 – North Western (Map courtesy of JW Irish desk, London)

of the more obvious is the strict emphasis of Jehovah's Witnesses upon a simple layout in their meeting houses. Kingdom Halls are invariably of wood and brick combination, devoid of any unnecessary ornamentation, and extremely functional. Indeed, many meetings continue to take place in rented rooms and converted buildings. However, if a building of historic interest is obtained, as with a 1930s art-deco cinema at Jersey City, U.S.A., everything is conserved and turned to new uses (The Watch Tower Society 1987: 1).

The focus in the interior layout of each Kingdom Hall is the speaker's stand, a wooden lectern or reading desk, typical of the Christian evangelical style, though understated, and not of dramatic height or design. Speakers come from the body of elders, who are assigned 45-min lectures, and other elders are frequently invited to speak, from another part of the circuit, often up to 50 or 60 mi (80–97 km) distant. Thus, elements of the circuit ministry common in the wider reformed Christian community are practiced.

A relic from pre-ecumenical days in most churches is the ruling that Witnesses may not attend service in the church or hall of any other religious denomination, a rule which most reformed denominations, and even the Roman Catholic Church, dropped during the twentieth century. Witnesses may, however, attend weddings in churches of other denominations as observers without a role, and a burial, providing the funeral service is avoided; both situations are left up to the individual conscience of each Witness (Saliba 1995). Jehovah's Witnesses impose a strict moral code upon all members, disallowing drunkenness, malicious gossiping and gambling. Unlike the Mormon Church, alcohol is allowed, for wine used at the Last Supper had been the fruit of the previous year's harvest and, therefore, fermented. The social life of a Witness is largely focused on church activities, reinforced by several meetings for worship to attend each week, all of around 1.5 h duration.

The Witnesses have a further characteristic that is to their benefit and to society as a whole. They are the work ethic personified, are never unemployed and place great emphasis on self-employment and the employment of fellow members in like industrious manner. No employment is seen as menial, but rather as the carrying out of God's work on earth. Members are employed in especially large numbers in building maintenance and construction and the cleaning of domestic and public buildings and areas. Tasks are always carried out efficiently and honestly, and are seen as opportunities for contact building and evangelical witnessing of the Christian message. Members do not especially aim to convert any particular section of society, but converts have traditionally come from amongst the working and middle classes. Education and worldly achievement are not seen as paramount, with lower than average community participation in third-level education, for the concern of members is with "God's Kingdom, shortly to be established on earth," rather than "the world," which members believe to be under the sway of Satan (Barker 1982: 342).

85.3.4 Establishing a Congregation

A typical congregation in provincial Ireland is that at Tipperary town; here, the Witnesses commenced work in about 1970, when two families settled at Pallasgreen, in one house divided into two apartments. Tuesday night house meetings were held there, under the jurisdiction of Limerick congregation, where both families went for Thursday and Sunday meetings.

In 1972, the old library building at O'Brien Street was bought using a loan from a private source. It was very unusual at the time to have such a proper meeting place. Attendance soon comprised 5 families, or 16 individuals, some having moved to the area from the jurisdiction of Limerick Congregation to assist in the mission field. From the early 1980s, local residents commenced in membership, especially from 1984, so that although some of the original "pioneers" had left the area, by 1995, there was a regular attendance of 40–45 Witnesses, as well as some children.

In April 1995, a quick-build Kingdom Hall was built on the Bansha Road, on a comparatively narrow site between the street front, and the railway embankment behind. By 2000, the membership comprised 64 Witnesses, drawn from 27 households, and the number attending including children was 80. By that point, the membership was largely confined to Tipperary town and its suburbs, while the Mission District, 24 by 20 mi (39 by 32 km) covers the settlements and hinterlands of Cashel, Tipperary, Bansha, as far as Cahir, where it meets Clonmel congregation; Dundrum and Hollyford, where it meets Thurles congregation; Cappamore, Cappawhite and

Dromkeen, where it meets Limerick congregation; Knocklong, Mitchelstown and Anglesborough, where it meets Fermoy congregation.

85.4 Conclusion

The period surrounding the independence of the Irish Free State changed many things. It was recently argued to have been a period when "narrow, self-conscious and extravagant national pride was all around … I heard at every turn the espousal of Ireland and everything Irish, but always felt that self praise is no praise" (Semple 2012). Many Protestant families were haunted by the constant spectre of inter-church marriage with a triumphalist Roman Catholic majority backed by the *Ne Temere* papal decree on the one side, and a largely stagnant economy with the prospect of emigration on the other.

The first four decades after independence were, perhaps, the most difficult to be a reformed Christian in the south of Ireland, and it was during this period of fear and suspicion that Jehovah's Witnesses faced their great time of trial. It is testament to their tenacity and tendency to see positive even in the midst of vehement opposition that their growth, albeit tentative by international standards, has been sustained at a fairly constant pace over the second half of the twentieth century, such that a critical mass of population has now been attained, demonstrative of their acceptance as part of Irish society.

85.5 Jehovah's Witnesses – Doctrine and Beliefs

Jehovah's Witnesses, in their doctrine of God, are monotheistic, if not actually Unitarian – that is, they do not subscribe to the Trinity – and it is particularly in relation to their teaching about the person of Christ that relates best to the Arians of the fourth century Anno. Domino, for they assert that the Son of God is a created being or second Adam, as quoted in "the Word was made flesh." They are also committed to the curious belief that before his creation, Jesus was the Archangel Michael, which they believe is taught in the Book of Daniel, Chap. 12. They hold that Jesus gave up his angelic nature in the days of his flesh and was an ordinary fallible mortal and, although he was not divine, paid at his death the ransom necessary to set men free from death, but that his work of Atonement "will be completed with the close of the Millennial Age."

The Witnesses do not believe that the redeeming work of Christ was completed on the Cross or that those who have faith in him are saved from their sins and inherit eternal life. Thus, far from being saved by Christ, each person must work out his or her own salvation. It is, therefore, not surprising that the Witnesses find the doctrine of the Trinity irrational since, in their view, the Son of God was a fallible mortal and they conceive of the Holy Spirit as merely the invisible influence of Jehovah.

Christian theologians argue against the doctrine of the Witnesses in three main ways:

1. Their creed is based upon an arbitrary selection of texts from the scriptures, but the main body of teaching of Jesus and his apostles (New Testament) is largely ignored.
2. Their doctrine is largely based on apocalyptic books such as Daniel and Revelations.
3. The Bible is used as a mechanism of prediction, which is to misunderstand its purpose.

In reply, the Witnesses argue that they equally accept both Old and New Testaments; that Daniel and Revelations are very relevant to the world today; that Jesus himself predicted signs and events; and that most of Christendom has no interest in the topic (Beckford 1986). In addition to having beliefs that are controversial in the wider Christian sense, Jehovah's Witnesses are also controversial in the civic sense, as a result of some of their religious beliefs. As members of God's Kingdom, not of any earthly jurisdiction, the Witnesses consider themselves politically neutral in earthly politics and, while honest and generally law-abiding, will not bear arms or swear allegiance to any state. They refuse to salute any flag, vote or hold political or administrative office (van Baalen 1938). The Witnesses state that this is because they are an international organisation and actions are interpreted differently in many countries where they are working (Rutherford 1938). Their neutral policy means they cannot be accused of being Zionists or Communists by opposing countries and has been of huge benefit to them in countries suffering political tensions, especially in the developing world, and also in zones of religious contention, as in Northern Ireland. In the latter instance, though broadly evangelical Protestant in outlook, the Witnesses did not align themselves politically to either camp, thus allowing equal access to both communities. Indeed, some "B Specials" and "Blanket Protesters", sworn, enemies during the disturbances in Northern Ireland, have become Witnesses, often in the same congregation.

Acknowledgements This chapter benefitted from the oral testimony of many Jehovah's Witnesses, particularly Bob and Charlotte Bramham, Jonathan Moore, Peter Hill, Malcolm Heathcote, Lynton Mason and Arthur Matthews. I am especially grateful to Mark O'Malley of the Irish desk at The Watch Tower Society, London offices, for his cartographic and general assistance with this paper.

References

Anon. (2012, July 7). Dietician behind fish and chip shop in Douglas. *Irish Times*.

Barker, E. (Ed.). (1982). *New religious movements: A perspective for understanding society: Studies in religion and society* (Vol. 3). New York: Edwin Mellen Press.

Beckford, J. A. (Ed.). (1986). *New religious movements and rapid social change*. London: Sage Publications.

Cassidy. (1956). Letter from Most Revd Dr Joseph Rodgers, bishop of Killaloe to an Taoiseach, John A. Costello, dated July 27, 1956, cited in Cassidy, C. Bishop was not amused by Jehovah Witness case. *Irish Times*, p. 4.

Davies, H. (1954). *Christian deviations: Essays in defence of the Christian faith*. London: S.C.M. Press.
Fagan, J. (2011, October 12). Jehovah's Witness to sell Irish HQ. *Irish Times*.
Ruane, J., & Butler, D. J. (2007). Southern Irish protestants: An example of de-ethnicisation? *Nations & Nationalism, 13*(4), 619–635.
Rutherford, J. F. (1938). *Studies in the scriptures*. Pittsburgh: The Watch Tower Bible and Tract Society.
Saliba, J. A. (1995). *Perspectives on new religious movements*. London: Chapman.
Semple. (2012, July 17). The Revd Patrick Semple, I am neither Gaeilgeoir nor Catholic – but I am still Irish. *Irish Times*.
Sopher, D. (1967). *Geography of religions*. Prentice-Hall: Englewood Cliffs.
The Watch Tower Society. (1987). Yearbook of Jehovah's Witnesses, 1987, containing the report for the service year, 1986.
The Watch Tower Society. (1988). Yearbook of Jehovah's Witnesses, 1988, containing the report for the service year, 1987.
The Watch Tower Society. (1993). *Jehovah's Witnesses: Proclaimers of Gods kingdom*. Brooklyn: The Watch Tower Bible and Tract Society, New York.
The Watch Tower Society. (2000). Yearbook of Jehovah's Witnesses, 2000, containing the report for the service year, 1999.
The Watch Tower Society. (2012). Yearbook of Jehovah's Witnesses, 2012 containing the report for the service year, 2011.
The Watchtower: Announcing Jehovah's Kingdom, Vol. 121, No. 1; Oral Testimony.
van Baalen, J. K. (1938). *The chaos of cults*. Grand Rapids: Eerdmans.
Witnesses. (2012, July 16). Jehovah's Witnesses gather in west Dublin to delve into their hearts. *Irish Times*.

Chapter 86
The Multifaith City in an Era of Post-secularism: The Complicated Geographies of Christians, Non-Christians and Non-faithful Across Sydney, Australia

Kevin M. Dunn and Awais Piracha

86.1 A Post-secular Landscape?

There has been a good deal of conceptual debate, especially in western countries, about how best to characterize recent trends in religiosity and belief. In the 1960s and 1970s, the growth in atheism and religious indifference was perceived as secularization, an anticipated long-running consequence of modernism and the decline of Christianity (Cox 1965). But in the 1980s and 1990s there have been returns to faith in many western nations, sometimes as a sustained trend, sometimes haltingly. This gave rise to suggestions that there was a de-secularization occurring. More latterly, the term "post-secular" has been coined to describe an era in which there are co-existent trends towards faith and away from faith. These trends are uneven across time, and also space. Religiosity and secularism vary across the globe (Casanova 2006; Davie 2006); it varies within nations and even across cities (Stevenson et al. 2010). But religiosity and secularism are not necessarily mutually exclusive concepts, and could easily co-exist. Adding to this complexity, there are new trends within religiosity, with the rise of popular religions, hyper-religions and new age faiths (Possamai 2005). And, in countries like Australia, there is a growth of faiths associated with immigration, including non-Christian religions like Islam, Buddhism and Hinduism. Immigration also feeds the growth of some Christian denominations, reviving congregations and generating new infrastructure and practices. Finally, the ancient continent of Australia retains numerous Indigenous faiths associated with the hundreds of Aboriginal language groups. These faiths include cosmologies, art and oral tradition. They also involve ceremony associated with daily life, intercultural relations and cross-species respect. Even the embracing concept of post-secularism struggles to contemplate the above diversity.

K.M. Dunn (✉) • A. Piracha
School of Social Sciences and Psychology, University of Western Sydney, Penrith, NSW 2751, Australia
e-mail: k.dunn@uws.edu.au; a.piracha@uws.edu.au

S.D. Brunn (ed.), *The Changing World Religion Map*,
DOI 10.1007/978-94-017-9376-6_86

In the next section we outline four perspectives on faith in Australia: a land of ancient Indigenous faiths; a Christian country; a secular nation, and; a multi-faith society. These begin to give the fuller sense of the post-secular landscape in Australia, of the complicated geographies of Christians, non-Christians, and the non-faithful. We outline the bases for these four perspectives, followed by a description of important temporal changes in faith. Finally, we focus in on Sydney, Australia's largest and most cosmopolitan city, and look at the variations in faith across that world city. We end by looking in-depth at a suburb within the centre of Sydney. In 2011, Auburn was unique in being split three ways between Christians, non-Christians and the non-religious, while Indigenous tribal faiths were barely present. In that suburb, the proportion of non-Christians is expanding, and the Christians are declining proportionally and absolutely across the census spans. It is a microcosm and an accentuated harbinger of the national trends in faith in some parts in Australia.

86.2 Australian Faiths

86.2.1 A Land of Ancient Dreaming

The faiths indigenous to Australia are those of the Aboriginal custodians of the land. Australia is both a country and an island continent. Humans have been physically dated as being present on the Australian continental land mass for at least 40,000 years. Before the arrival of Europeans and the British conquest 200 years ago, the land had an array of indigenous cultures each with its own cosmology. These were called 'dreaming stories' and they explained the creation of the fauna, flora and landscapes. These Indigenous faiths were likely influenced by the traders and explorers from nearby landmasses, including parts of what is now Indonesia. The Macassan trepang harvesters would stay in Northern Australian for extended periods, and there is evidence of spiritual exchange, including influence upon indigenous cosmologies (Isaacs 1982; Macknight 1976). From the 1600s, these Macassan sailors would likely have been Muslims (Noorduyn 1984), adding to the ancient "dreaming" cosmologies. Since the European invasion, the numerical dominance of these Indigenous faiths across the land mass has faded. By mid-2011, only about 0.03 % of Australians reported to census collectors that they were adherents to an Australian Aboriginal traditional religion (Table 86.1). Nonetheless, the original religions of Australia are these Indigenous faiths.

86.2.2 A Christian Country

Australia can also be described as a Christian country. After European invasion Australia was founded as a set of Christian colonies. Funds and convict labor were deployed to assist with the building of Christian infrastructure like churches and schools.

Table 86.1 Religious affiliation and non-belief, Australia and Sydney, 2006 and 2011

	% of population			
	2011		2006	
Religion	Australia	Sydney	Australia	Sydney
Christianity:				
Anglican	17.11	16.12	18.53	17.68
Baptist	1.64	1.46	1.58	1.47
Catholic	25.29	28.26	25.56	28.77
Lutheran	1.17	0.39	1.25	0.44
Oriental Christian				
Orthodox	2.81	4.71	2.87	4.79
Pentecostal	1.11	1.10	1.10	1.11
Presbyterian and reformed	2.79	2.41	2.97	2.58
Uniting Church	4.96	2.83	5.66	3.31
TOTAL CHRISTIAN	61.14	60.86	63.23	63.32
Buddhism	2.46	4.11	2.09	3.69
Hinduism	1.28	2.58	0.74	1.69
Islam	2.21	4.74	1.70	3.88
Judaism	0.45	0.87	0.44	0.83
Australian aboriginal traditional religions	0.03	0.01	0.03	0.01
Other non-Christian affiliation[a]	0.75	0.90	0.67	0.55
TOTAL NON-CHRISTIAN	7.19	13.20	6.18	11.32
No religion	22.3	17.65	18.48	13.98
Religious affiliation not stated	9.36	8.30	11.09	10.26

Data sources: ABS, 2006 and 2011 census community profiles

[a]This category also covered beliefs that were hitherto listed in the Inadequately described/not stated category: for example, Jeddi-ism and New Age

The salaries of Protestant clergy, catechists and schoolmasters were paid by the state. The same level of colonial state assistance was not initially available to Catholic and other Christian denominations (Ngui 2008), and there was persecution of Catholic religious practice in early colonial Australia (Breward 1993). The Church of England was the effective "state faith" in post-invasion Australia (Mason 2006; Frame 2006). But, by the mid-1800s most Christian denominations were in receipt of state aid. The built environment of cities like Sydney bare the legacy of this state aid, and this is reinforced through contemporary heritage policies and funding (Ngui 2008). Australia is also a "Christian country" insofar as key institutions reflect and reinforce Christian practice (Fozdar 2011; Maddox 2005). This institutional dominance is reflected in the use of Christian prayer in Parliament, and in the dominance of the Christian religious calendar. Public opinion polling suggests that between one-third and two-fifths of Australians believe that being Christian is important to being truly Australian (Australian Survey Social Attitudes 2003; Nelson et al. 2013). Finally, 61 % of Australians reported a Christian affiliation to Census collectors in mid-2011. For these reasons the Australian religious landscape can be described as Christian.

86.2.3 A Secular Nation

In the spirit of modernist toleration the *Australian Constitution* included an overt prohibition on establishmentarianism. Section 116 of the *Australian Constitution*, as proclaimed in 1900, proscribed there being a state religion as well as any state actions that would limit the freedom of religion.

> The Commonwealth shall not make any law for establishing any religion, or for imposing any religious observance, or for prohibiting the free exercise of any religion, and no religious test shall be required as a qualification for any office or public trust under the Commonwealth. (Section 116, Commonwealth not to legislate in respect of religion. Commonwealth of Australia Constitution Act, 9th July 1900)

This secularist injunction served to contain sectarian tensions between Protestant Australians and Roman Catholic communities in Australia. In national surveys, a majority of Australian respondents have stated that religion is a force for division and there is strong support for a separation of church and state, with only 40 % preferring that religion had more rather than less political influence (International Social Survey Programme 1998: V29–V36). At the 2011 Census 22.3 % of respondents stated that they had "No Religion," this included those who stated that they were agnostics, atheists, humanists and rationalists. A further 8.6 % provided no declaration of a religious affiliation. Together these two categories (almost one-third) would be unlikely to support the notion that Australia was a Christian or even religious nation. For the above reasons Australia can be characterized as a secular country.

86.2.4 A Multifaith Country

In the post-war era secularism was supplemented by a national commitment to the Universal Declaration of Human Rights (December 10th 1948). Article 18 set out the universal human right to freedom of thought, conscience and religion. The political discourses of the Second World War in Australia reinforced the principles of democratic and religious freedoms. In 1993 the national Australian government declared that the United Nations Declaration on the Elimination of All Forms of Intolerance and of Discrimination Based on Religion or Belief (25 November 1981) was an international instrument for the purpose of the *Human Rights Act*. It is no surprise then that freedom of religion is a strongly held belief within Australian society (Dunn 2009; Nelson et al. 2013).

Since the 1970s, Australia began to officially acknowledge the cultural diversity among the citizenry (Australian Council on Population and Ethnic Affairs 1982; Grassby 1973). These neophyte statements on diversity deployed useful metaphors that advocated how cultural difference, between "family members" or within "a neighborhood," could easily be accommodated within an identifiable whole. The statements argued that if families and neighborhoods could operate with unreconciled diversity, then why couldn't a nation-state? Multiculturalism was given official standing with the *National Agenda* declaration in 1989 (Office of Multicultural Affairs (OMA)). In the Agenda statement respect for freedom of religion was entrenched as was a cultural right to religious expression (OMA 1989: vii). Recent statements on multiculturalism, from the Rudd/Gillard Government, have made reference to religious diversity. "Multiculturalism is in Australia's national interest and speaks to fairness and inclusion. It enhances respect and support for cultural, religious and linguistic diversity" (*The People of Australia: The Australian Multicultural Advisory Council's Statement on Cultural Diversity and Recommendations to Government* 2010: 2). Those who see Australia as a multi-faith nation can point to these statements on multiculturalism that advocate religious diversity and tolerance. They can also point to the 39 % of Australians who are not Christians by affiliation (see Table 86.1).

Official policy on the status of religion in Australia is as mixed as the census data. In their public inquiry into religion and religious freedoms in Australia, Bouma et al. (2011), found that for some people Australia was a Christian country, to others it was a secular country and to others it was multi-faith country. All of these positions have a demographic basis for their claim. They can also point to policy instruments and political practice which supports their case. There is a haphazard official position and this aligns with the confusion and pastiche of a post-secularist world.

86.3 The Australian Religious Landscape, Over Time

In the 1911 Census, 96 % of Australians reported that they were Christians (Table 86.2). Three percent gave no response or provided an answer that was inadequate for coding. Not even half a percent were prepared to say they were non-religious, in the form of atheism or agnosticism. Almost one-in-one-hundred said they had a non-Christian religious affiliation. This was the White Australia Policy era (promulgated through the *Immigration Restriction Act, 1901*). White Australia thinking was hegemonic and largely unchallenged (Kamp 2010). The extent to which White Australia policy and thinking was bound up with Christian-centrism has not been sufficiently critiqued to date. But the White Australia era was clearly a "Christian Australia" period. Not only were non-Christian religions a minority, but so were the non-religious. This reinforces the argument that Section 116 of the Constitution was more about managing relations across Christian denominations than it was about cross-faith relations or about secularism per se (Nelson and Dunn 2013).

Over the Twentieth Century, the proportion of non-believers ('no religion' plus 'not stated') increased in Australia, from being marginal to later constituting almost one-third of Australians. The first sign of significant non-religiosity in Australians, from the census figures, was the expanding proportion of the religiously indifferent from the 1930s. Stevenson et al. (2010) have referred to those who refuse to answer the question on faith, and those whose answers are un-codable, as the "religiously shy." Similarly, Frame (2009) has characterized them as the religiously disinterested / indifferent.

Table 86.2 Faith in Australia, 1911–2011

Census year	Christian	Non-Christian	No religion	Not stated[a]	Total (thousands)
1911	95.9	0.8	0.4	2.9	4,455.00
1921	96.9	0.7	0.5	1.9	5,435.70
1933	86.4	0.4	0.2	12.3	6,629.80
1947	88.0	0.5	0.3	11.1	7,579.40
1954	89.4	0.6	0.3	9.7	8,986.50
1961	88.3	0.7	0.4	10.7	10,508.20
1966	88.2	0.7	0.8	10.3	11,599.50
1971	86.2	0.8	6.7	6.3	12,755.60
1976	78.6	1.0	8.3	11.8	13,548.40
1981	76.4	1.4	10.8	11.4	14,576.30
1986	73.0	2	12.7	12.3	15,602.20
1991	74.0	2.6	12.9	10.5	16,850.20
1996	70.6	3.5	16.5	9.0	17,753.00
2001	67.3	4.8	15.3	11.5	18,769.25
2006	63.2	6.2	18.5	11.0	19,855.28
2011	61.1	7.19	22.3	9.36	21,507,719

Data sources: ABS, 1991: 1; Bouma, 1997: 13; ABS Online 2012
[a]Includes "Inadequately described"

While there were only 2 % of the religiously shy in 1921, this jumped to 12 % by 1933. The proportion in this category has stayed much the same since. However, there has been a steady increase in the proportion of those who do a make declarative statement in the Census that they are without a religious faith. The proportions went from under 1 % in 1966 to 7 % 5 years later in 1971. From the 1970s onward the proportion of the non-religious has expanded, to be over 22 % of the Australian population in 2011. Some public commentators became excited about the slowed rate of people moving away from religion. Certainly in the 1991 Census there was a sense of a recovery in Christianity, and the proportion of non-Christians had again risen, such that the proportion of those with faith had increased from 75 to 76.5 %. Some had called it a return to faith. But in the 1996 Census the trend of falling proportions of Christians, and the expansion of the non-religious, both returned.

For most of the twentieth century Christianity dominated faith among Australians. Non-Christian belief was confined to <1 % of the population. This changed with the decline of the White Australia Policy after the Second World War, and especially after its formal dissembling from the 1970s. Immigration from East Asia, from Australia's near neighbors, and from Lebanon and other parts of Asia, saw an increase in the proportions of non-Christians in Australia. For example, the expansion of the number of Muslims in the 1970s and 1980s was fed initially by immigration from Lebanon and Turkey. Between the 5-year Census spans of 1971 to 1976, and then to 1981, the number of Muslims in Australia doubled (22,311; 45,206; 76,398) (Dunn 2004). By 2011, there were 476,291 Muslims, constituting 2.2 % of the Australian population. Non-Christians have come to comprise over 7 % of the

Australian population, and Christians only 61 % of the population at the census. Immigration has also fed numbers into some Christian denominations, most especially Roman Catholicism, and without this the decline in Christian numerical dominance would have been more dramatic. Since the 1970s, Australia has become "more than Christian," it is now a multi-faith nation. Over the 110 years since 1901, there has been a change to the Australian religious landscape, such that there are altered demographic claims to the moral and religious base of the nation. However, the preponderance of official policy on religion and national character in the last two decades has edged towards the multi-faith definition. There still remains however a strong legacy of Christian-centrism within political discourse and practice (Fozdar 2011; Maddox 2005). Key forms of Christian privilege remain unchallenged, such as exemptions from anti-discrimination provisions, such that religious bodies (mostly Christians) are able to discriminate against job applicants on the basis of faith (Nelson and Dunn 2013).

86.4 The Changing Australian Religions Map

The post-secularity of the changing Australian religious map is a product of the temporal changes and competing trends described above. However, Australia's post-secularity is also manifest spatially. The trends in faith described above are not consistent across Australia. Table 86.1 shows the variations and similarities in the religious maps of Australia and Sydney. At first glance the proportion of Christians is similar at 61 %. However, this apparent similarity masks some considerable denominational variations. Roman Catholics constitute 28 % of the Sydney population, but only 25 % of the national population. The relative presence of Orthodox Christians is more dramatically different, being almost 5 % of Sydney (one-in-twenty) whereas they constituted less than 3 % of the Australia-wide population. The relative presence in Sydney was almost double that of the national proportion. The higher proportions of Catholic and Orthodox Christians in Sydney is migration related. Post war migration saw the establishment of strong communities and religious infrastructure in Sydney, including communities from eastern, southern and western Europe. These include orthodox Greek, but also Serbian and Russian churches. Catholic numbers and church attendances have benefited from the large immigration waves from Ireland, the United Kingdom, as well as Italy and Croatia. Since the 1970s, Catholics have come to Australia from South America and Asia. While Sydney has a higher proportion of Catholic and Orthodox Christians than does Australia, the same is not true for many of the Protestant faiths. Anglicans are 17 % of the Australian population, yet only 16 % of the Sydney population. The number of Lutherans, Uniting Church, Presbyterian and Reformed Christians is much lesser in Sydney. These Protestant denominations constitute almost 9 % of the Australian population, but they are only 6 % of the Sydney population.

Non-Christians are a much larger proportional presence in Sydney than in the rest of Australia. The collapse of the "Great White Walls" (Price 1974) of the White

Australia Policy can be seen in its religious effects most starkly in Sydney. While non-Christians constitute 7 % of the Australian population they are 13 % of the Sydney population. More than one-in-eight Sydney residents have a non-Christian religious affiliation, for the purposes of the Census, and this is dramatically different to Australia at the turn of the twentieth century. It is also much higher than for Australia generally, reflecting immigration, migrant settlement choices, and the focus of religious and cultural infrastructures described above.

While Sydney has a higher proportion of non-Christians than does the whole of Australia, it has a lower proportion of the non-religious. In other words atheists and agnostics, and others who make a clear statement of non religion in the Census, are less prominent among the Sydney population. While 22.3 % of Australians make this declarative self-affiliation, only 18 % of Sydney residents did so. The proportions of the faith shy are only slightly lesser in Sydney than for Australia. What Sydney lacks in Atheists, compared to Australia, it makes up for in non-Christians. This makes Sydney generally more faithful than the rest of Australia. The Christians and non-Christians in Australia constitute 68 %, whereas in Sydney they reach almost 75 %. Sydney is clearly not a city "from where the Gods have fled" (after Cox 1965), indeed this cosmopolitan world city is more faithful than the country as a whole.

The distribution of faith and atheism, and the trends in affiliation to specific faiths, does not follow a singular pattern across the globe. The only reliable trend is that of diverse trajectories, of a post-secular age. Even across a country like Australia the trends are varied, there are shifts across time, and that changing religious map varies across the country. In the following sections we graphically illustrate the post-secular variations in faith and non-faith across a city like Sydney.

Figure 86.1 shows the change in the proportion of Christians in each Local Government Area (LGA) across the Statistical Division (SD) of Sydney for the latest census period of 2006 to 2011. The overall picture is the declining proportion of Christians. The strongest resistance to the trend, and retention of Christian proportions, is in the perimetropolitan areas like Wollondilly and the Hawkesbury where three-quarters of census respondents identified as Christian. Both areas held their proportion of Christians at 76 and 73 % respectively. The strongest loss of Christian proportions (drops of six or more percentage points) was in the central western Local Government Areas (LGAs) like Holroyd, Parramatta and Strathfield. In these LGAs the proportion Christian was 59, 52 and 50 % respectively. In the nearby LGA of Auburn Christians are no longer the majority (only 34 %). There are also substantial losses of Christians within the southwest (Bankstown, Canterbury, Hurstville, Kogarah) and inner south west (Ashfield and Marrickville). These are areas of strongest immigrant impact, and as we mentioned above, the increase in non-Christians has been strongly influenced by immigration. The proportion Christian in these areas hovers between 50 and 60 %, and at current trends the Christian majorities will soon disappear, as it had in Auburn at an earlier census. In the 2011 Census the additional non-majority Christian LGAs were Marrickville (46 %), Waverley (41 %) and Wollahra (47 %). The latter two are not in western Sydney, nor are they immigration reception zones, and the increases in those LGAs

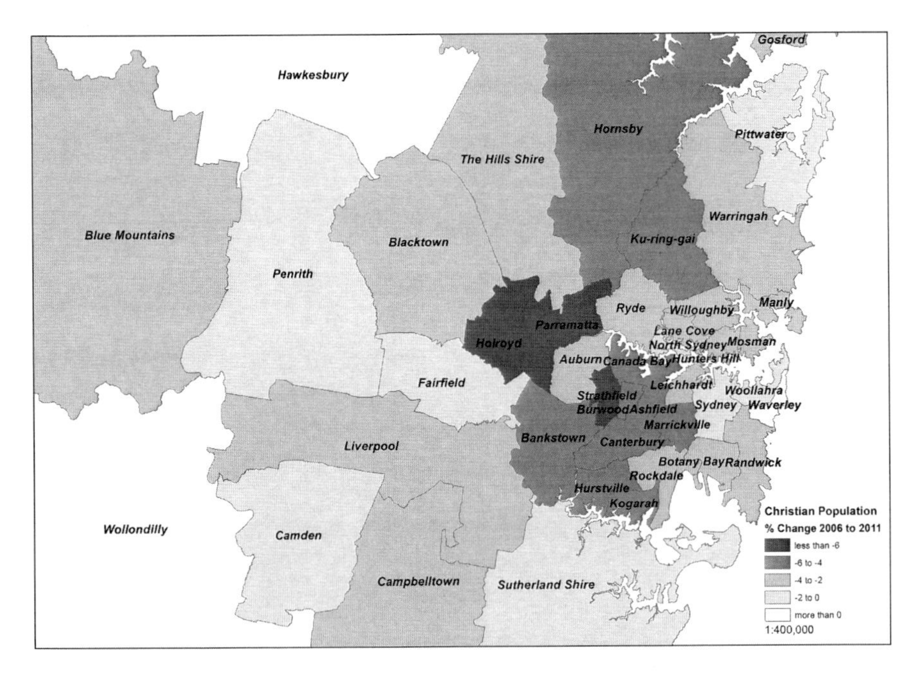

Fig. 86.1 Change in Christian population in Sydney local government areas (2006–2011) (Map by Kevin M. Dunn and Awais Piracha using 2006 and 2011 Australian Bureau of Statistics census data and Australian Standard Geographical Classification 2011 digital boundaries)

are towards the non-religious, as we discuss below. Still other areas are only just majority Christian, such as Strathfield (49.7 %), Leichhardt (50.6 %), Parramatta (52 %) North Sydney (51 %). Almost half (22 plus) of Sydney's LGAs are likely to be non-majority Christian by the 2016 Census. Even parts of the so-called "Bible belt" in Sydney (Connell 2005), such as Hornsby and the Hills Shire lost proportions of Christian (the latter moving from 72.5 % in 2006 to 69.3 % in 2011). But these shifts are varied in their intensity and outcomes across the city.

Immigrant settlement areas have been the principal focus of increase in the proportion of non-Christians. In areas of central western Sydney the proportions who are non-Christian have appreciated by four percentage points (Fig. 86.2). This includes central LGAs like Blacktown in the north through central areas like Holroyd and Parramatta to the southern areas of Bankstown and Liverpool. In amongst these areas is the LGA of Auburn where 41 % of the population is non-Christian. In these surrounding areas of non-Christian growth a quarter of the population now affiliates with a non-Christian faith. Islam, Hinduism and Buddhism are the three principal components of this non-Christian presence. In the northern group of these LGAs Islam and Hinduism are the stronger contributors, while Buddhism and Islam (again) are more prominent among the southernmost grouping of these central LGAs. The increase in non-Christians has been less dramatic in the more affluent northern (Willoughby north to Pittwater) and eastern suburbs (Wollahra, Waverley). The western perimetropolitan fringe areas have had a lesser increase in non-Christian proportions. This includes LGAs like

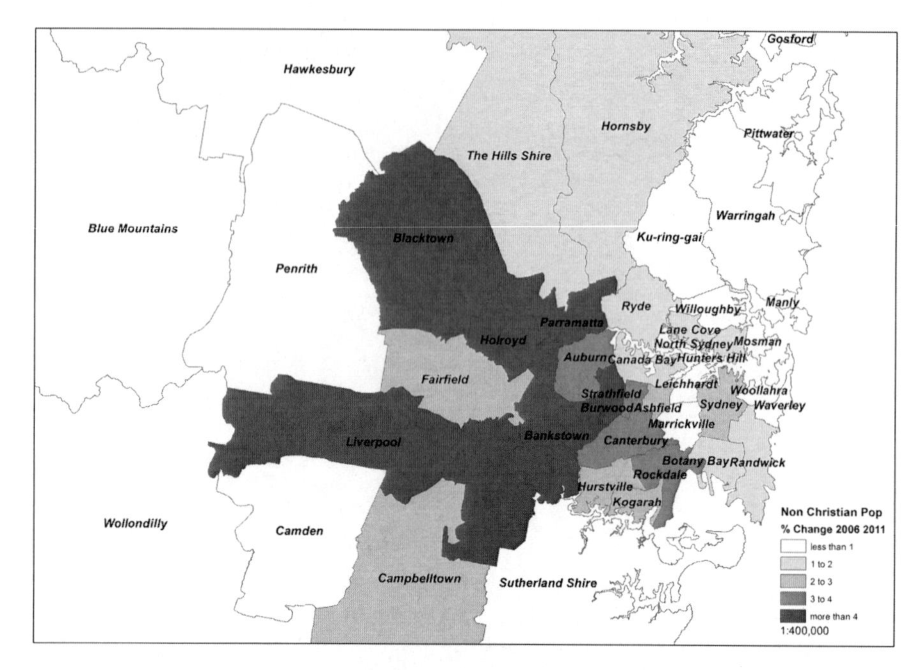

Fig. 86.2 Change in Non-Christian population in Sydney local government areas (2006–2011) (Map by Kevin M. Dunn and Awais Piracha using 2006 and 2011 Australian Bureau of Statistics census data and Australian Standard Geographical Classification 2011 digital boundaries)

Hawkesbury, Penrith, Blue Mountains, Wollondilly and Camden. But even in these places the proportion of non-Christians has still increased. However, the non-Christian presence in these LGAs is modest, such as 1.2 % in Wollondilly, 1.6 % in the Hawkesbury and only 3.1 % in the Blue Mountains. The trend in Sydney is for an increase in non-Christians and a decrease in Christians, but that trend is dramatically uneven across the metropolis.

The proportion of the non-religious, those who selected atheist or agnostic for example in the Census, increased in every LGA across Sydney. This was a consistent Sydney trend in faith(lessness) between 2006 and 2011. The most dramatic increases were in the cosmopolitan inner city areas such as the LGA of Sydney as well as Marrickville and Leichhardt (Fig. 86.3). In these areas as well as in beach-side LGAs of Manly and Waverly the proportion of the non-religious increased by more than six percentage points, in just the 5 year period. In these areas the non-religious constitute one-third of the population. The affluent northern suburbs are also a zone of increasing non-religious presence, by four or more percentage points. The Blue Mountains in the west, an area of small non-Christian presence and little growth, was also a place of increasing faithlessness, indicating that the major shift there is from Christianity to atheism and to other non-religious stances. Meanwhile, in the immigration reception areas, where the increase in non-Christians has been strongest, there was a more modest increase in the non-religious. This suggests that in these middle suburban areas (Blacktown south to Liverpool) the shift in faith is

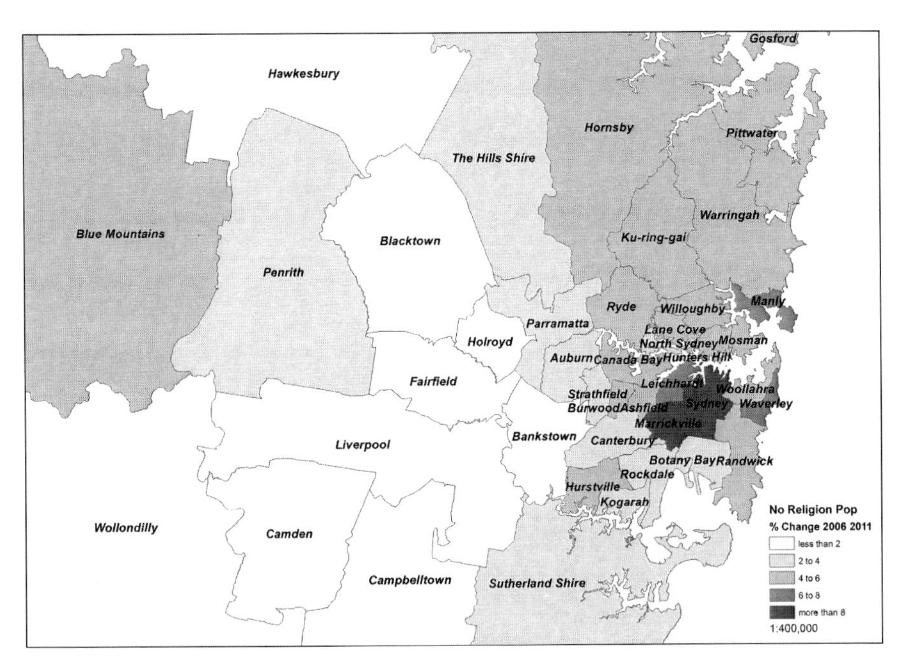

Fig. 86.3 Change in population with "no religion" in Sydney local government areas (2006–2011) (Map by Kevin M. Dunn and Awais Piracha using 2006 and 2011 Australian Bureau of Statistics census data and Australian Standard Geographical Classification 2011 digital boundaries)

from Christian to non-Christian rather than from Christian to faithless as pertains in the cosmopolitan inner city, in the affluent north shore, and in the Blue Mountains in the westernmost part of Sydney.

In the previous 2006 Census there had been a substantial increase in those who were faith shy – who did not answer the question on religious affiliation. The proportion of the faith shy in Australia had increased from 9 % in 1996 to 11.1 % in 2006 (see Table 86.1). In 2011, this proportion has fallen back to 9.4 %. The loss of the faith shy pertains across most of Sydney. However, it is strongest in cosmopolitan inner and eastern Sydney, in areas where the increase in the overtly non-religious has been largest (for example, Sydney, Leichhardt, Marrickville, Woollahra, Waverley) (Fig. 86.4). For example, in Sydney LGA the proportion of the faith shy went down from 29 % in 2006 to 17 % at the 2011 Census, while the non-religious went in the opposite direction from 24 to 34 %. There was almost a direct swap of 10 percentage points between the religion not stated (faith shy) and the definitively non-religious. The two categories together decreased from 53 to 51 %. In the cosmopolitan inner city (Leichhardt, Marrickville) there was a similar exchange between the faith shy and the non-religious, with the combined trend being an appreciation. In those two areas the non-religious and faith shy now constitute 44 and 45 % of the population.

The trends outlined above are complex. The multiplicity of post-secularism is detectable across Sydney. While there are general, city wide, trends they do not

Fig. 86.4 Change in population with "religion not specified" in Sydney local government areas (2006–2011) (Map by Kevin M. Dunn and Awais Piracha using 2006 and 2011 Australian Bureau of Statistics census data and Australian Standard Geographical Classification 2011 digital boundaries)

pertain in all areas. More consistently, the trends are stronger and meeker depending on whether it is the inner city or the perimetropolitan parts of the city, across the affluent or struggle areas, and in the immigration settlement areas or those less affected by immigration. Stevenson et al. (2010: 345–346) were able to detect five distinct geographies of religion in Sydney using the 2006 census. These included those areas were immigration had been a catalyst for growth in non-Christian faiths, the relative decline in Christians, and a maintained level of faith. In these places the Christian numerical majorities are fading. A second set of areas were those places where the proportion of Christians had been maintained, or had decreased much more slowly. Our analysis shows that such areas closer to the city are not retaining Christian proportions, but the perimetropolitan areas are places of slower change. A third geography of this post-secular mix are those areas least effected by change, principally the affluent north shore. But the 2011 figures suggest that change is gathering pace there as well, mostly through a shift of Christians to the non-religious. A fourth geography identified by Stevenson et al. (2010) was the affluent eastern suburbs (Waverley and Woollahra) where the non-religious were expanding and where there was a longstanding though minor non-Christian presence (Jewish Australians). A final and fifth geography was cosmopolitan (secularizing) Sydney where the big story was the steady increase in the faith-less alongside a dramatic decline in the proportion of Christians. This is only one way of characterizing the geographies of faith in the post-secular city of Sydney, there could be others. This geography nonetheless provides a glimpse at the nature of that diversity, and the complicated and contradictory trends in belief and non-belief across the city.

86.5 A Post-secular Australian Suburb: Auburn

The suburb of Auburn is somewhat unique within Sydney in being split three ways between Christians, non-Christians and the non-religious. In the 2006 Census, this three-way split of the Auburn population was almost even, with 36 % Christian, 37 % believers of non-Christian faiths and 26 % non-religious or not stated. However, the presence of non-Christians has continued to expand, and 5 years later in 2011 the non-Christians account for 41 % of the LGA. The reductions have occurred for the Christians (down to 34 %) and to a lesser extent the non-religious and faith shy (25 %). Figure 86.5 illustrates the major changes between the two Censuses in Auburn, and specifically the decrease in the Christian proportion and the increase in the proportion of non-Christians. The figure also shows the extent to which the composition of faith and faithlessness in Auburn varies from Sydney in general. It is in LGAs like Auburn, the first geography described in the previous section, where the increased presence of non-Christians has been most pronounced.

While there has been a decline in the proportion of Christians within Auburn, some denominations have fared better than others. The number of Anglicans had

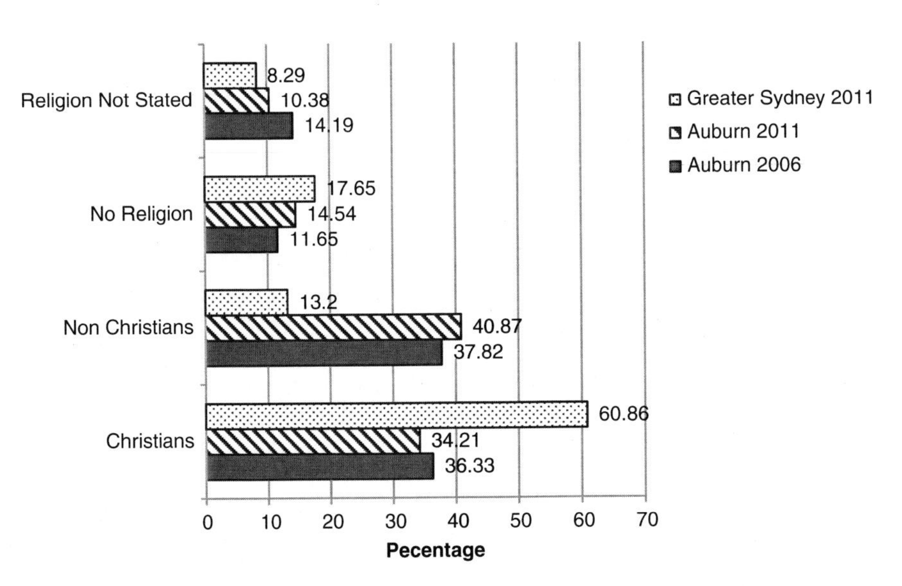

Fig. 86.5 Religious affiliation and non-affiliation, Auburn LGA and Greater Sydney, 2011 (Data source: 2006 and 2011 Australian Bureau of Statistics census)

Table 86.3 Religious affiliations and non-belief, Auburn, 2006 and 2011

Religion	2006	2011
Christianity:		
Anglican	3,399	3,083
Baptist	829	1,032
Catholic	12,794	13,480
Lutheran	131	98
Oriental orthodox	204	229
Eastern orthodox	1,632	1,460
Pentecostal	504	604
Presbyterian and reformed	1,204	1,696
Uniting Church	1,353	1,525
TOTAL CHRISTIAN	23,600	25,227
Buddhism	6,009	6,732
Hinduism	2,005	3,771
Islam	16,112	18,798
Judaism	26	34
Australian aboriginal traditional religions	0	3
Other non-Christian affiliation	418	801
TOTAL NON-CHRISTIAN	30,138	24,566
No religion	7,570	10,720
Religious affiliation not stated	9,035	7,347
TOTAL	64,958	73,739

Data source: Australian Bureau of Statistics

declined in absolute terms (3,399–3,083), yet the number had increased for Catholics (though the change to the proportional presence of Catholics was marginal, remaining at about 17.5 % of the LGA population) (Table 86.3). The Uniting Church increased its presence (1,353–1,525) as did charismatic groups such as the Pentecostals (504–604) and Baptists (829–1,032). To have had increases in the number of Presbyterian and Reformed adherents (1,204–1,696), and Uniting Church, given the national decline (see Table 86.1), was quite impressive. Within Auburn, some of the growth in Christian denominations has been associated with immigrant groups, such as Pasifika people from countries like Samoa and Tonga.

The major non-Christian groups in the Auburn area are Muslims (25.5 %), Buddhists (9.1 %) and Hindus (5.1 %). There were no adherents of an Australian Aboriginal Traditional Religion in 2006, and less than 10 in 2011. The 18,798 Muslims in Auburn have a substantial material presence. One of the most impressive examples of the presence of Muslims includes the Gallipoli Mosque which was established by an association called the NSW Auburn Turkish Islamic Cultural Centre (formed in 1976 under a slightly different name). The Centre's stated aim in the mid-1980s was "to answer religious and educational needs of all Turks in and

Fig. 86.6 Looking northeast along North Parade to Auburn Gallipoli Mosque, May 2007. Gallipoli Mosque is the place of worship for Muslims in the Auburn area, many of whom are Turkish- and Afghan-Australians. It overlooks some of western Sydney's major industrial areas, and is surrounded by suburban housing (Photo by Andrew Kinsela, used with permission)

around Auburn; to try to solve marriage, divorce and related problems through religious ways and means" (quoted in Ethnic Affairs Commission of NSW 1985: 108). It is noteworthy that such a large mosque (by Australian standards) and ambitious project was initiated to serve an ostensibly local catchment. This demonstrates the large size of the Muslim population in the Auburn and surrounding area. While the background and demographic basis of Auburn Mosque is dominantly Turkish-Australian, there is some evidence of a diversifying of Jamaah, with Muslims of many backgrounds attending the Friday Jumah prayers. Auburn Mosque occupies a 4,000 m^2 (43,000 sq ft) site, has two minarets of 39 m (128 ft) height, and a dome of 23 m (75 ft) which sits atop of eight smaller domes (Fig. 86.6). The construction cost of the mosque had originally been estimated at $4.5 M, but had grown to at least six or perhaps as much as eight million by the time it had been completed (see Dunn 1999: 359). The association holds annual Open Days where members of the public are invited to visit the mosque, to receive tours, talks and a lunch. Indeed, any group or school can arrange for guided tours of the mosque, all of which are intended to improve non-Muslims' understanding of Islam in Australia, in an attempt to break-down misperceptions and stereotypes. The local association and affiliated not-for-profit groups (such as the Affinity Intercultural Foundation) estimated that as many as 41,000 people have been on organized tours of the Gallipoli Mosque as of 2008 (Dunn and Kamp 2009: 60).

In December 2005 racist violence erupted in the coastal suburb of Cronulla in the LGA of Sutherland Shire (Noble 2009). The Sutherland Shire is one of those areas where Christian affiliation remains very high (75 %) in 2011. The non-Christian population is minor (2.7 %) and has hardly grown from the previous census in 2006 (2.3 %). In December 2005 white youths from Sutherland and other suburbs assembled in Cronulla and led a pogrom against any non-whites and non-Christians they located, especially searching out Lebanese-Australians (Noble 2009). In reaction to these attacks Lebanese-Australian youth, and others, lead revenge attacks that same evening in beach-side suburbs like Coogee and Maroubra. Later the next evening, a gun was fired outside a primary school in south Auburn (St Joseph the Worker Primary School), where parishioners were singing Christmas carols. It was reported that parents and children at the church were threatened and abused by men of "Middle Eastern appearance" (in Noble 2009). And the next evening, a Uniting Church hall in Auburn was firebombed, and burnt to the ground, and an Anglican church in Auburn (St Thomas' Anglican Church) had its windows smashed. The most religiously diverse LGA in Sydney had become a flashpoint for incidents that had their origin over 20 km (12.2 mi) away on the coast.

Muslim worshippers in Auburn walked from their mosque after their Jumah prayers to the Harold Wood Uniting Church (the site of the arson attack) to hold a solidarity meeting. Five hundred people assembled to hear from Muslim, Christian and Jewish speakers, condemning violence and celebrating inter-faith dialogue and understanding. The speakers included the President of the Uniting Church in Australia, the President of Australian Federation of Islamic Councils, and a representative of the Executive Council of Australian Jewry (Dunn 2009). The religious leaders condemned the violence, and some pointed out that Jesus Christ was of "Middle-Eastern appearance." These interfaith initiatives, as well as police surveillance of potential trouble-makers, saw an end to the inter-faith violence in the Auburn area. Indeed, the mosque tours and open days are ongoing successful interfaith initiatives.

There are organized frameworks for inter-faith dialogue within Sydney. This includes the Commission for Ecumenism and Interfaith Dialogue which is run from the Catholic Diocese of Parramatta. Parramatta LGA is contiguous to the Auburn LGA, and they together form (along with Holroyd) the central western region of Sydney. Figure 86.7 is a picture of members of the female Auburn Tigers football (Australian Football League) team (Lalor 2011). They are local Muslim women who have taken up the iconic Australian team sport of AFL, adapting the uniforms to their religious needs, and thus continuing and re-inventing the sporting traditions of central western Sydney. Parramatta has a rolling series of cultural events that celebrate non-Anglo and non-Christian cultures. These include the Parramasala festival of South Asian Arts held annually in early November. Two weeks later, there is the Loy Krathong Thai Festival which is held along the Parramatta River and its public foreshores.

There are also instances of public dissonance, some of which are apparent within the landscape. The Islamic group "MyPeace" ran billboards near major Sydney roads, in late May of 2011. The billboards included the statement that Jesus, like Mohammed, was a prophet of Islam. A Christian group, called Seek and Save

Fig. 86.7 The Auburn Tigers Womens AFL side, first season, 2011. Most of the team is of Lebanese background, but some are Fijian, Bosnian, Turkish or Afghan. Adaptations include having a female coach and requesting that the Tigers' male side have separate training nights and game days (Photo by James Croucher, used with permission)

Ministry, provided responding billboards stating "Dear Aussie Muslims, glad you want to talk about Jesus. Love to chat more." Both billboards bore an encouraging level of civility as well as an implicit interest in dialogue and in locating points of agreement. The Anglican Bishop of South Sydney, Robert Forsyth, was positive about the freedom of speech, connection, and awareness building that the billboards could represent. "Some people will be troubled to see one of their deeply held beliefs criticized or denied on a billboard and that is I guess confronting. But that's the nature of the world we live in and if we Christians are confronted we need to get over it and get on living in a multi-faith society" (quoted in Yosufzai 2011). Central western Sydney suburbs like Auburn, in a small way, can be seen as a harbinger of Australia's future, where there will be places where the majority status of Christians will fade, where a series of non-Christian faiths (Islam, Buddhism, Hinduism) will rival most Christian denominations. In these microcosms of post-secular Australia, the changing religious map will see the increased proportional presence of the non-religious.

86.6 Conclusions

The headline story of the changes in religious affiliations in Australia over the past 40 years is that the proportional presence of those with Christian beliefs was reduced by at least a third (88–61 %). Three-quarters of the Christian loss was picked up by the non-religious while one-quarter was taken up by other religions. For Australia as a whole the increase in non-Christian religions has been moderate. Non-Christian

religions have only grown by a little over three-quarters of 1 % every 5 year census period since 1971. If the trends of
the past 40 years are repeated over the 40 years into the future, to 2050, one in three Australians will be a Christian, one in eight will be non-Christian and one in two will be non-religious.

The major watershed events in the changing Australian religious maps were the human settlement of the continent more than 40,000 years before present. The next key event was the British invasion in late 1700s and early 1800s, and the establishment of the Christian churches. In the 1970s, the White Australia Policy began to seriously crumble, and this saw immigration from Asia, and later Africa, bringing non-Christian faiths. Political conflict in regions like South East Asia, the Lebanon, the Balkans, Iraq, Afghanistan and Sri Lanka have generated supplies of non-Christian immigrants to Australia. In some circumstances the Australian Government had involvements in those conflicts (Indo-China, Iraq, Afghanistan), in others there had been earlier waves of immigrants from those places (Lebanon, Balkans), and for still others the Government had helped organize the immigration programs (Lebanon, Indo-China, Albania). But non-Christians, and also Christian immigrants have come from non trouble spots too, such as through the Australia-Turkey Immigration agreement to supply family migrants for the expanding Australian economy. Sydney is the focus of immigration, consistently receiving the lion's share of permanent settlers. This is because it is Australia's biggest city, with the most developed cultural infrastructures (places of worship, cultural centers and associations), and it is a prominent global city (Olympics 2000, etc.). The immigration of non-Christians into Sydney has helped stem the decline of the non-faithful. The atheists and other non-religious are heavily concentrated in the gentrified and cosmopolitan parts of Sydney.

The average picture of Australia as a whole is not the same at the local level of individual communities. Some localities, where there is the congregation of migrant communities (Cui and Piracha 2012), will have much higher concentrations of non-Christian religions or non-religion or both. Some of them will look like Auburn. Other areas that are migrant hubs may see trends in faith and faithlessness that are not anticipated here. Some areas will be like Woollahra, affluent places with a substantial presence of the non-religious. There may be other areas where both non-Christians and the non-religious will be well above the Australian average.

The unresolved tensions regarding the religious direction and official characterization of Australia will only become more pressing with the passage of time, given the trends reviewed above. Is the land of Australia, for so long managed by Indigenous Australians according to ancient cosmologies, a secular, Christian or a multifaith country? There are strong legacies of indigenous cosmology and of Christian-centrism. Yet the nation has also been officially constructed as secularist, and at other times as a multifaith country. The trends above affirm both constituencies as increasingly present. The mixed post-secularist religious map is stark within Australia. The variety and contradiction of post-secularism is apparent across the country, as well as within cities like Sydney.

References

Australian Council on Population and Ethnic Affairs. (1982). *Multiculturalism for all Australians: Our developing neighbourhood*. Canberra: Australian Government Publishing Service.
Australian Survey Social Attitudes. (2003). *Australian survey social attitudes*. Retrieved 20 September 2009, from Australian National University: http://aussa.anu.edu.au/issp.php
Bouma, G., Cahill, D., Dellal, H., & Zwartz, A. (2011). *Freedom of religion and belief in 21st century Australia*. Sydney: Research report prepared for the Australian Human Rights Commission.
Breward, I. (1993). *A history of the Australian churches*. Sydney: Allen and Unwin.
Casanova, J. (2006). Rethinking secularization: A global comparative perspective. *Hedgehog Review, 8*(1/2), 7–22.
Connell, J. (2005). Hillsong: A megachurch in the Sydney suburbs. *Australian Geographer, 36*(3), 315–332.
Cox, H. (1965). *The secular city*. New York: Macmillan.
Cui, T., & Piracha, A. (2012). *Culture matters: An analysis of ethnic segregation and congregation in Sydney, Australia using centrographic method* (pp. 22–41). 2012 Australia and New Zealand Association of Planning Schools Conference, Bendigo.
Davie, G. (2006). Is Europe an exceptional case? *The Hedgehog Review, 8*(1//2), 23–34.
Dunn, K. M. (1999). *Mosques and Islamic centres in Sydney: Representations of Islam and multi-culturalism*. Ph.D. dissertation, Department of Geography, University of Newcastle, Newcastle.
Dunn, K. M. (2004). Islam in Australia: Contesting the discourse of absence. *Australian Geographer, 35*(3), 333–353.
Dunn, K. M. (2009). Public attitudes towards hijab-wearing in Australia. In T. Dreher & C. Ho (Eds.), *Beyond the hijab debates: New conversations on gender, race and religion* (pp. 31–51). Newcastle: Cambridge Scholars Press.
Dunn, K. M., & Kamp, A. (2009). The hopeful and exclusionary politics of Islam in Australia: Looking for alternative geographies of "Western Islam". In R. Phillips (Ed.), *Muslim spaces of hope: Geographies of possibility in Britain and the West* (pp. 41–66). London: Zed Books.
Ethnic Affairs Commission of NSW. (1985). *Eighteen years after ... The Turkish settlement experience in NSW* (Interim report). Sydney: Ethnic Affairs Commission of NSW.
Fozdar, F. (2011). The "choirboy" and the "mad monk:" Christianity, Islam, Australia's political landscape and prospects for multiculturalism. *Journal of Intercultural Studies, 32*(6), 621–636.
Frame, T. (2006). *Church and state: Australia's imaginary wall*. Sydney: UNSW Press.
Frame, T. (2009). *Losing my religion: Unbelief in Australia*. Sydney: UNSW Press.
Grassby, A. (1973). *A multi-cultural society for the future*. Canberra: Department of Immigration. Australian Government Publishing Service.
International Social Survey Programme. (1998). *International social service programme; Religion II module*. Retrieved 18 September 2009, www.issp.org/
Isaacs, J. (1982). *Australian dreaming: 40,000 years of aboriginal history*. Sydney: Lansdowne Press.
Kamp, A. (2010). Formative geographies of belonging in white Australia: Constructing the national self and others in parliamentary debate, 1901. *Geographical Research, 48*(4), 411–426.
Lalor, P. (2011, May 28). Muslim women find a new goal with AFL. *The Australian*.
Macknight, C. C. (1976). *The voyage to Marege: Macassan trepangers in Northern Australia*. Melbourne: Melbourne University Press.
Maddox, M. (2005). *God under Howard: The rise of the religious right in Australian politics*. Crows Nest/Sydney: Allen & Unwin.
Mason, J. K. (2006). *Law and religion in Australia*. Canberra: National Forum on Australia's Christian Heritage.
Nelson, J., & Dunn, K. M. (2013). Racism and anti-racism. In A. Jakubowicz & C. Ho (Eds.), *For those who've come across the seas Australian multicultural theory, policy and practice* (pp. 259–276). North Melbourne: Australian Scholarly Publishing.

Nelson, J., Possamai-Inesedy, A., & Dunn, K. M. (2013). Reinforcing substantive religious inequality: A critical analysis of submissions to the Review of Freedom of Religion and Belief in Australia inquiry. *Australian Journal of Social Issues, 47*(3), 297–317.

Ngui, S. (2008). *Freedom to worship: Frameworks for the realisation of religious minority rights.* Sydney: Department of Biological, Earth and Environmental Sciences, University of New South Wales.

Noble, G. (Ed.). (2009). *Lines in the sand: The Cronulla riots, multiculturalism and national belonging*. Sydney: Institute of Criminology Press.

Noorduyn, J. (1984). Makassarese. In R. W. Weekes (Ed.), *Muslim peoples: A world ethnographic survey* (pp. 470–472). Westport: Greenwood Press.

Office of Multicultural Affairs. (1989). *National agenda for a multicultural Australia … Sharing our future*. Canberra: Australian Government Publishing Service.

Possamai, A. (2005). *In search of new age spiritualities*. London: Ashgate.

Price, C. A. (1974). *The great white walls are built: Restrictive immigration to North America and Australasia 1836–1888*. Canberra: Australian Institute of International Affairs.

Stevenson, D., Dunn, K. M., Possami, A., & Piracha, A. (2010). Sydney: An examination of a "post-secular" world city. *Australian Geographer, 41*(3), 323–350.

Yosufzai, R. (2011). *MyPeace, your peace*, MuslimVillage.com, 27 June, http://muslimvillage.com/2011/06/27/11182/mypeace-your-peace/. Accessed 26 Nov 2012. MuslimVillage Incorporated, Lakemba.

Chapter 87
Russian Rodnoverie: Revisiting Eastern and Western Paganisms

Kaarina Aitamurto

87.1 Introduction

The first contemporary Slavic Pagan[1] groups in Russia emerged in the last decades of the Soviet era, but only at the beginning of the 1990s were they able to practice their religion openly. With the new opportunities, the religion experienced an explosive growth and nowadays there are hundreds of myriad groups practicing some form of pre-Christian Slavic spirituality. From the outset, contemporary Paganism in Russia has been closely linked to nationalist concerns and movements. It is noteworthy that the term, which the adherents most commonly use for their religion, Rodnoverie,[2] derives from the words "*rodnaya vera*" or "native faith." Given that a similar emphasis on ethnic heritage and nationalistic politics characterizes contemporary Paganism in the Central and Eastern Europe in general, it has become common to juxtapose the "nationalistic" Eastern Paganism with "liberal" Western Paganisms. However, it also has to be noted that both types can be found in virtually all countries and the difference is not always so unambiguous.

This chapter participates in the discussions about the differences between Western and Eastern Paganisms by analyzing a Rodnoverie community the Krina (a spring of water) from St. Petersburg as it resembles very clearly some important

[1] The term "contemporary Paganism" refers to religions that claim to revive the pre-Christian spirituality. In some studies, these are called Neo-Paganism(s), but because most of the adherents find this word insulting, it has become customary not to use the prefix "neo."

[2] Especially in Central and Eastern Europe, some religions reviving the pre-Christian spirituality avoid the term "paganism" because of its pejorative connotations and also because they wish to disassociate from such forms of "Paganisms" as Wicca, which they consider modern and invented religions. In consequence, many of these movements call themselves as "native faiths," for example, Rodzimowierstwo in Poland, Ridna Vira in Ukraine, Rodná víra in Czech Republic, etc.

K. Aitamurto (✉)
Aleksanteri Institute, University of Helsinki, Helsinki 00014, Finland
e-mail: kaarina.aitamurto@helsinki.fi

S.D. Brunn (ed.), *The Changing World Religion Map*,
DOI 10.1007/978-94-017-9376-6_87

Fig. 87.1 A ritual in a boat in St. Petersburg in 2011 after the international Veche that gathers Slavic Pagans from different countries. In 2011, the Veche was somewhat smaller than earlier with delegates only from Russia, Ukraine, Poland and Belorussia (Photo by Kaarina Aitamurto)

aspects concerning the topic (Fig. 87.1).[3] On the one hand, in many ways this community resembles the so-called Western, liberal form of Paganism. On the other hand, it has contacts with nationalistic groups as many of its members identify as nationalists. Therefore, the community also demonstrates how these two forms can be intertwined. Even though Russian Paganism reflects the nationalism of the outside society, it also features similar individual features of religiosity as "Western" Paganisms. Because of this individualism, Paganisms and Pagan groups are by nature heterogeneous and in this respect Rodnoverie is a very typical pagan religion. In this sense, it also reveals the way in which contemporary Paganism follows and reflects one of the major trends in contemporary religiosity that the scholars of the sociology of religion have noticed, viz., the individualization of religion.

87.2 The Emergence of Contemporary Paganism

The first contemporary Pagan religion, Wicca, was created by a retired civil servant Gerald Gardner in England in the 1950s. Though Gardner claimed Wicca to be an indigenous European pre-Christian tradition, it was for a large part based on various

[3]This chapter is based on ethnographic fieldwork carried in St. Petersburg 2005–2011 that included interviews and participant observation.

later Western Esoteric traditions of the eighteenth and nineteenth centuries. In the following decade contemporary Paganism witnessed a radical change as the ideas of reviving pre-Christian spirituality reached Northern America. While the first Gardnerian Wiccans were predominantly conservative and right-wing in their social outlook (Hutton 1999: 360–361), in America Paganism was strongly influenced by the counter-culture of the 1960s. At the same time, new forms of Witchcraft and Paganisms emerged next to the traditional Wicca. Many of these incorporated the new liberal ideas of the times into Pagan religiosity, especially influential of these related to women's liberation. One of the most radical examples of such groups was Zsuzsanna Budapest's feminist "Dianic Wicca," which defined witchcraft as a symbol and tool in feminist revolution (Budapest 1999). This new generation began to criticize Gardner's Wicca for regenerating sexist stereotypes. For example, according to Gardner the High Priestesses must be young and beautiful in order to properly feature the Goddess, but such demands were not made for High Priest, who, in Gardner's coven, was Gardner himself. The traditional Wicca was also based on the idea of the polarity of sexes as the generative force of life and fertility. In the 1960s and 1970s, such an essentialist understanding of gender was challenged by many Pagans and the ambiguity and flexibility of gender brought the issue into discussions about the religion (Starhawk 1989: 8–9, 40, 216–217). Within Western contemporary Paganism, sexual minorities have indeed been relatively well represented. They have even formed their own groups, such as Radical Faeries, which was founded in Arizona in 1979 (Adler 2006: 357–371).

Also ecological thinking shaped contemporary Paganisms in a more socially oriented and liberal direction. While earlier Paganism was predominantly seen as an esoteric religion, in the 1980s it was all the more often understood as a "nature religion." As a result, Pagans have been quite visible in many environmental actions and movements (Harvey 1997: 126–131) and for many contemporary Pagans, ecological concerns are one of the crucial reasons why they have chosen Paganism as their religion (Carpenter 1996: 395–397). Feminist and environmental activism have been combined in one of the most famous Pagan communities, "Reclaiming" from San Francisco, and in the publications of its originator, Starhawk (Miriam Simos).

As the generation of Pagans who were young in the 1960s and 1970s have aged, the radicalism of the movement has somewhat lessened. While the adherents have established careers and families, they have also become concerned about gaining approval for their religion in the mainstream society (Berger 1999: 88–99). However, the counter-cultural spirit still lives in Western Paganism and the adherents seem to be more liberal in their social and moral outlooks than the average population (Berger et al. 2003).[4]

[4]The influence of the counter-culture of the 1960s was stronger in America than in Europe and there are some controversies between conservative and liberal Pagans as well as representatives of Traditional Wicca and Eclectic Witchcraft. However, Pagans, also in Europe, have been noted to hold more liberal views than average population. See, for example, Lassander's (2009) study of the values of British, Irish and Finnish Pagans.

In the countries of the previous socialist bloc, the latest wave of contemporary Paganisms is a somewhat younger phenomenon[5] and emerged in a social context that was very different from Western Paganism. In the Central and Eastern European countries new religious movements were seldom allowed to function in public and, therefore, most of the contemporary Pagan movements formed only after the extrication of Communist rule. In Europe, drawing to the local ethnic or national tradition undoubtedly appears as a more natural form of "Paganism" than, for example, in northern America. However, the social and political upheavals of the collapse of Communism also bolstered the nationalistic flavor of these movements. In many countries the liberation from the Soviet influence increased interest in nationalist culture and various Paganisms happily took the role of the wardens and representatives of the "true" and "authentic" national heritage.

In Russia, the roots of contemporary Paganism lie in the nationalistic *dissidenstvo* of the 1970s and 1980s.[6] Nevertheless, some ideas of Paganism can also be found within the non-political intelligentsia which was interested in alternative spirituality and esotericism. Symptomatically, many Rodnoverie leaders began their spiritual search in groups of cosmists or Eastern spirituality. Though the first Pagan communities were established already in the 1980s, these were very small and invisible. Therefore, as in Central and Eastern Europe, only after the fall of the restrictions and censorship of the Communist regime were Pagans able to come public and to openly form communities and organizations.

The prevalence of nationalism and conservatism in Russian Paganism can indeed be partially explained by the fact that its emergence coincided with the serious economic crisis and the dissolution of the earlier norms and values. In his international survey of values Inglehart notices a clear rise in the conservative values in societies experiencing economic and social stability and in his analysis, post-Soviet Russia stands as an exemplary case of such development (Inglehart and Baker 2000). Furthermore, as the national past has to be considered under both painful re-negotiations and a re-evaluation in post-Soviet Russia, national pride suffers a serious blow. As a counter reaction to all these developments, Russian society witnessed a "nationalist turn" in the middle of the 1990s and the "ideological vacuum" left by the collapse of the Soviet Union seemed to be filled by nationalist ideology and rhetoric.[7]

At the beginning of the 1990s especially, the burgeoning Pagan community in Russia was highly political and reflected the uncertainty of the times. It was typical that even the smallest communities founded their own parties with programs

[5] It should be noted that some Pagan groups emerged also at the beginning of the twentieth century, for example, Zadruga in Poland, Dievturība in Latvia or Taarauseliud in Estonia. Though these groups were closed down during the Communist regime, some Pagan groups managed to function. One of the most notable of these is Lithuania "Ramuva" that was established as an "ethno-cultural organization" even in 1967.

[6] In the literature, the activity of such forefathers as Valery Yemelyanov (one of the founders of the organization Pamyat) has been discussed widely. See, for example, Shnirelman (1998).

[7] Of the Rodnoverie leaders I have interviewed, many mentioned the "ideological vacuum" first when I asked about the reasons for the rise of contemporary Paganism.

competing in their (usually conservative) radicalism. Conspiracy theories and grand esoteric theories were also quite popular. In the new millennium the Rodnoverie movement has developed in a more religious and less political direction. On the one hand, the anti-extremist laws have somewhat halted the public promotion of the most racist and xenophobic programs, for example, those that encouraged violence. On the other hand, as the movement has established itself by creating a religious identity, including ritual practices and theological notions, have developed further and become more pressing concerns. Nevertheless, even though the most visible Rodnoverie organizations disassociate themselves from open incitement of racist or xenophobic violence, there are countless small, informal groups which are engaged in racist politics and even violence (Shnirelman 2013; Pilkington and Popov 2009; Aitamurto 2011). Furthermore, nationalism has remained as one of the central leitmotifs of the movement, albeit this is often expressed as a modest wish to preserve and continue one's own cultural tradition.

87.3 "East" as a Challenge for the Study of Paganisms

Due to its nationalist and racist overtones, the rapid revival of contemporary Paganism in Post-Socialist Europe was received with mixed feelings both within Western Pagans and Western scholars studying the topic. It has been noted that in the West, racist and nationalist Pagan groups have existed already for decades (Gardell 2003). However, the majority of the Western Pagans usually do not accept these groups as a part of their community. Furthermore, the scholars of Paganism have treated racist Pagan groups as a marginal phenomenon that should not be included among modern Pagan religions, because a pluralist attitude has been seen as one of the defining elements of these groups (York 2003: 164; Berger et al. 2003: 15–16). Such definitions have been challenged by the development of Paganism in Europe, as racist Paganism has grown into a phenomenon too large to be ignored. However, instead of trying to accommodate the different forms of Paganism into some common analytical framework, scholars tend to differentiate the (nationalist) Eastern and (liberal) Western Paganisms and conduct separate discussions about them. To state it somewhat polemically, Eastern Paganisms are usually discussed as a social problem while in analyzing Western Paganism scholars more often pay attention to their positive effects, such as the empowerment of women or spreading ecological awareness.

Even though scholars studying Paganism in Central and Eastern Europe acknowledge the specific features of the area, some caveats against lumping all Paganisms in Eastern Europe in the rubric of racism and intolerance have been made. For example, Scott Simpson argues that much of the research literature unreflexively sustains the idea of Western (that is, Anglo-American and Scandinavian) Paganisms as liberal, left-leaning spirituality and the "Eastern Paganism" as a nationalistic, right-wing and conservative phenomenon. As Simpson points out, this division and especially such geographical labels as "Eastern" and "Western" do not necessarily do justice to the complex phenomenon of modern Paganisms. Furthermore, simplistic pictures often neglect such areas as France, Germany, Belgium, or the Netherlands which have strong "Western type" Pagan scenes, but also some racist Pagan-oriented groupings. Needless to say, the geographical labels of "East" and "West" also tend to omit the racist forms in, for example, the United States. In conclusion, Simpson suggests that instead of the words "Eastern" and "Western," the two main forms of contemporary Paganism are different forms of Paganism that could be categorized with some neutral terms, such as a groups "A" and "B." Furthermore, he recommends paying more attention to the ways in which these types intertwine (Simpson 2000; Aitamurto and Simpson 2013).

In turning the focus on such combinations it is easy to notice that defining Pagan movements with such categories as "nationalist," "conservative," "leftist" and "rightist" is often more or less arbitrary. A good example on the complexity of such terms is Mattias Gardell's (2003) discussion about the political left and right within racist American adherents of Germanic or Scandinavian Paganism, the Odinism. As Gardell notes, the political goals of these Pagans often contain both traditionally leftist and rightist ideas. For example, many American Odinists are at the same time both anti-communists and anti-capitalists. Next to such themes as race, family, or order, they might also address issues that have traditionally been associated with the political left, such as labor union, environmental issues or social welfare. In consequence, Gardell proposes to analyze the politics of the Odinists with a three-dimensional model which includes axises between left and right; centralism and decentralism[8]; and monoculturalism and multiculturalism (Gardell 2003: 334–338).

Gardell's suggestion is very relevant concerning Paganism in Russia as well. There are, for example, groups that subscribe to Communism (some of them even to Stalinism), but are extremely conservative in their social values. The idea of tribal socialism (*rodovoi sotsializm*) is supported by many Pagans who otherwise seem to follow very rightist politics. In a similar vein as left and right seem to be too broad categories for a nuanced analysis, dividing such concepts as "nationalism" and "conservatism" or "political" and "religious" into subcategories would enable the analysis to better address the variance in contemporary Paganisms and to explain the phenomena that first seem contradictory.

87.4 Rodnoverie

The scholars studying Rodnoverie seem to agree that there are more men than women in the movement and that a vast majority of Rodnovers are young people and live in cities. Despite the fact that contemporary Paganism is thus a very urban

[8] Gardell also discovered some leading figures of the community were earlier engaged in anarchist activity. This fact reflects the libertarian tradition in the United States that has often been adopted by extremist right wingers as well.

phenomenon, nature holds a central position in the religion. It is not uncommon that Rodnovers present ideas about forming rural settlements that would follow a Pagan worldview.[9] Not surprisingly, the movement somewhat overlaps with another nature oriented new religious movement, the *Anastasietsi*, which is known of its ecovillages (Andreeva and Pranskevičiūtė 2010). The Rodnoverie ritual calendar follows traditional agricultural festivals, the summer solstice *Kupalo* being the main event of the year in terms of the number of participants. The festivals are usually conducted in nature and especially in summertime, they last for a couple of days while the participants camp on site in their tents. The biggest festivals attract hundreds of people. Nowadays it is customary that the participants dress in traditional folkloric costumes in the rituals. The ritual is usually conducted around a fire or in front of an idol and it includes circle dance, exaltation to gods, games and a ritual drink that goes around the circle (Fig. 87.2). In addition to festivals, Rodnovers also organize conferences, seminars and workshops. Such scholarly form of activity reveals that many Rodnovers, as Pagans in general, are often literary people and many of them are either students or have university degrees.

As in virtually all contemporary Paganisms, non-dogmatism and anti-authoritarianism are characteristic features of today's Rodnoverie. In consequence, the communities and individual Rodnovers are free to independently develop their religiosity and there are countless different ways in which the "old" religion is practiced. Because of this heterogeneity and unorganized nature, it is extremely difficult to obtain reliable data on the movement, including the number of the adherents. Most of the communities are not officially registered and the smallest ones compose of only a three to four people. The scene is also very dynamic. The communities are often short-lived and their members change rapidly. On the basis of some surveys and Internet sites featuring Pagan communities and Pagans, it seems safe to say that there are at least 20–30,000 Pagans in Russia (see Aitamurto 2009). This number is probably underestimated, but it is still relatively much lower than, for example, in the United States with one million Pagans (Patheos 2014).

Despite the diversity, some uniformity has emerged within Rodnoverie. For example, the ritual practices are actively borrowed from other communities. These ideas can be appropriated from literature or photographs and videos posted on the Internet, but Rodnoverie groups also often jointly organize festivals or conferences. This kind of cooperation may develop into organized structures as well. Already in the 1990s, the first umbrella organizations were established and in the twenty-first century, the three most notable of these, the Circle of Pagan Tradition (CPT), the Union of Slavic Communities (USC) and the Veles's Circle formed a new network, *Slavyanskoe rodnoverie*.

---

[9] Some such villages have been created. One of the most earliest formed in the initiative of Aleksei Dobrovolsky (Dobroslav). A community that focuses on the reverence of goddess Makosh has already existed many years in the village of Popovka; it provides excursions for tourists who wish to attend traditional folk festivals. An ecovillage, which belongs to the organization Shag Volka, also exists near Moscow.

Fig. 87.2 Re-erecting the idol (chur) in the destroyed shrine of the community Krina in Janino in July 2011. Pagan shrines are occasionally vandalized by people who oppose Paganism, but in this case the site was destroyed by a filming group that shot a film of WWII (Photo by Kaarina Aitamurto)

What is surprising in this alliance is that at the turn of the millennium, the CPT and the USC were engaged in a bitter dispute. The root of this disagreement lied in CPT's manifest "Bittsa Appeal" that condemned "national-chauvinism." The rapprochement of these two organizations was made possible by the changes in both of them; on the one hand, the USC is much less politically oriented than at the beginning of the millennium and its nationalism less aggressive. On the other hand, the CPT has obviously felt pressures to prove its patriotism while still disapproving of the most flagrant forms of nationalism. In addition, new dividing lines within the movement have become more pressing. One of the first actions of the new alliance was a document that reproves such authors as Aleksandr Khinevich, Valery

Chudinov and Aleksandr Khinevich, who the document claims to present scientifically unsubstantiated claims about ancient Paganism and thus declared as representatives of "pseudo-Paganism" (Shizhensky 2012: 46–47). The named authors have indeed presented wild historical claims that are often ridiculed in the scholarly studies of contemporary Paganism. Apparently, for the biggest Rodnoverie organization the most pressing concern is to disassociate themselves from such authors or groups that are seen to give a sectarian or obscurant image to Paganism.[10]

87.5 The Krina

The community Krina has been formed already at the beginning of the 1990s by a Physicist Andrei Rezunkov, who is also known by his Slavic or Pagan name Blagumil. Like so many other Rodnovers, who got interested in Pagan spirituality in the 1980s,[11] Blagumil begun his spiritual search in the alternative Eastern spirituality. At the turn of the decade such publications as the *Book of Veles* became more widely known and the first Pagan leaders and groups appeared in media. As Blagumil describes, it was a time when many people suddenly realized that Russians, too, have a spiritual tradition of their own to study and follow.

At the beginning, the Krina was a very informal group of friends who occasionally travelled to countryside to explore different kinds of spiritual practices (Fig. 87.3). Revealingly, it took some time until they, for example, noticed that given that rituals were "festivals," it was appropriate to dress into a festive outfit as well. Gradually, more people wished to join the activity and the group formed into a "community" with a name, welcoming not only their own acquaintances but also other people who expressed their interest to join in.

The Krina's rituals are led by wizard Blagumil and a witch (*vedun'ya*) Inna, but there are also other strong women among the core members of the group, such as Blagumil's wife. Krina is indeed somewhat exceptional Rodnoverie community, because its events often gather more women than men. (In Eastern Paganism, there are usually men more involved than women, unlike in the Western Paganism, excluding Odinism and Asatru). However, the members of the community constantly change. In consequence, while for a long time the majority of the participants in the rituals were middle aged women, there have also been periods when young men have composed the biggest group. It should also be noticed that the word "member" may be somewhat misleading here, because the Krina mostly

[10] A similar development took place in Wicca as well. As the religion has established itself, Gardner's claim, according to which the tradition in which he was initiated, represents a direct continuation of pre-Christian religious tradition that is all the more often refuted. Instead, the majority of today's Wiccans argue that contemporary Paganism does not need a poorly substantiated historical lineage in order to be a meaningful and legitimate religion.

[11] Well-known Rodnoverie leaders who initially practiced some Eastern spirituality include Aleksei Trekhlebov, Irina Volkova (Krada Veles) and Ilya Cherkasov (Veleslav). One of the most influential communities today, Veleslav's Rodolyubie was originally called Satya Veda.

Fig. 87.3 Ritual setting in Karelia, before Kupala festival in 2009. On the left is the flag of the community and an altar; on the right, the goddess Lada and a wheel that is lighted in the ritual and rolled down to the lake; in the background is a gate through which the participants enter into the ritual space (Photo by Kaarina Aitamurto)

functions as a loose network of people who participate in the events that the community organizes. The community does not have any memberships and even though Blagumil may conduct a ritual to give a "Slavic name," this is not usually considered as an initiatory ritual entrancing a neophyte to a particular group.[12] Most of the people involved in the activity have academic education or are students and it has a very middle class atmosphere in many respects.

Creating and developing ritual practices is one of the main interests of the Krina (Figs. 87.4 and 87.5). In the 1990s, the community cooperated with a small experimental theatrical group and organized with it, for example, a big Summer solstice festival. The community publishes an annual "calendar," which is actually a small book that contains not only the dates of each year (that follow both the solar and the lunar calendar), but also small articles on Slavic gods, tradition, spirituality etc. (see, for example, Rezunkov 2004; Rezunkov 2005).

The Krina have also had a website, but it was not updated very actively. However when the new social networks became popular, the community created a group in the Russian service similar to the Western Facebook, the vKontakte group. On that site, information and discussions about the activity of the group can be posted.

[12] Recently, the community has created a form for those wishing to apply for membership. In this way it seeks to develop a more defined organizational structure, but it is too early to say whether the majority of participants will apply for membership.

Fig. 87.4 Dedi Prazdnik in Janino, November 2006. In this festival, similar to Halloween in America, the ancestors are remembered (Photo by Kaarina Aitamurto)

Fig. 87.5 Phooroni Jarili in Karelia in July 2006. In the festival, the god of sun and vegetation, Yarilo, is buried as the summer begins to decline and days become shorter (Photo by Kaarina Aitamurto)

In addition, it features photographs of the festivals and videos. In May 2011, the group Krina in vKontakte.ru had 374 members (regarding Rodnoverie and the Internet, see Gaidukov and Maslyakov 2012).

Throughout its history, the Krina has sought to identify itself as a spiritual community and to distance itself from politics. In the 1990s, they were approached by one of the most notable, politically oriented Pagan communities in St. Petersburg which proposed that they would celebrate the main annual festivals together. Blagumil did not think that they share a similar view on the religion or to rituals and politely refused the offer.

This is not to say that nationalist features could not be found in the Krina or in the worldview of its members. Studying and preserving Russian tradition is the main activity of the group. Many members of the community are interested in the study of other cultures as well and may even appropriate elements from these. However, they usually believe that people's own native tradition is the natural choice for them. It is safe to say that the majority of the members of the community identify as patriots and in rituals, salutes like "Glory to Russia!" are often heard. Even though most of the rituals are organized by the Krina independently, it has participated in some joint events with more nationalistically oriented Rodnoverie groups, such as Shag Volka or Soyuz Venedov. Occasionally, such cooperation may cause awkward situations if, for example, radical politics is a very dominant feature of a given event.[13]

Open xenophobic and racist expressions are not socially acceptable in the Krina, even though the discussions within the community often disclose the xenophobic prejudices of some of its members. Within the community, talking politics is forbidden and when, for example, some younger members begun to discuss such topics, the older members remind them of the rule.

The prohibition to talk about politics reflects the community's wish to distance itself from "political groups," but is also due to that fact that the members have a wide array of political convictions. In a social-network vKontakte.ru, one can found in the profiles of the members of the Krina the most varying descriptions of "political views," including, among others, "conservative monarchists," "socialists," "communists," "nationalists," "anarchists." Another example of variance in political leanings is a small disagreement I witnessed in a Kupala festival between two members of the community who begun to talk about money and spirituality. While the other one considered money to be the main obstacle in people's spiritual development, the other believed that spiritual development is linked to financial success. It is characteristic to the Krina that this difference in viewpoints has not appeared earlier, even though both of these people have both attended many other festivals as well.

The variance in political views reveals that politics is not a common denominator for the Krina. At the same time, the way in which "politics" is defined provides an

[13] For example, in a conference, one of the organizers suddenly stood up to present a petition called "Letter 500," a plea to ban Judaism as an extremist religion, right before Blagumil's presentation while he was already standing in front of the audience. In such instances, I have noticed that members of the Krina seem somewhat embarrassed.

interesting outlook on the values of the community. It has been noted that the divisions between "political" and "religious" are often used in very evaluative and normative ways (McCutcheon 2005). Furthermore, in such evaluative rhetoric the concept of "politics" predominantly refers to the viewpoints and arguments that seem alien or controversial. For example, opposing abortion is alternatively considered as a political convictions or moral decisions.

Also within the Krina, "politics" is often used as a derogatory term. There are themes that are not seen political in the Krina even though from the scholarly perspective these can be considered such. This is the case especially with nationalist themes, such as the disapproval of the negative influence of foreign companies or supporting the position of Russia in the global international politics. Another category that is excluded from the domain of "politics" is ecological concerns. Remarks, for example, about climate change that from analytical viewpoint could be judged as divulging green politics are not considered political, and thus prohibited, within the Krina. Thus the understanding about the category of "political" discloses which topics raise more controversies within the Krina and which ones are accepted as shared values and, therefore, do not need negotiations. Part of these values, such as ecological conviction, are related to Pagan movements while others, such as the appreciation of patriotism, reflect more the values of the outside Russian society.

Nevertheless, respect for pluralism and independent thinking are considered important values in the Krina. Once in a campfire, wizard Blagumil explained that in the eastern spirituality there are two principal ways to transmit the teaching. One of these is to literally teach the neophytes, to explain them what is the truth, to "hit the student with a stick," as he said. The other one is to gently guide people in a direction where they themselves attain some insight. In the Krina, Blagumil said, they apply the latter approach. In the activities of the group, however, accepting individual preferences and maintaining some common frame occasionally cause difficulties.

The main frames of the community rituals are planned beforehand, but there is also a meeting where people are free to present their ideas. In addition, the participants are encouraged to bring to rituals things they think suits the festival or may cause enjoyment to the participants, including various musical instruments or even toys. Occasionally, however, new participants may suggest activities that run counter to what the core members of the community consider the spirit of a given festival or the religion itself to be. In preserving some basic principles and at the same time making people feel that their suggestions are welcomed reveals the psychological skills that the core group of the community has attained during the years.

However, the entrance of new people into the community has changed the nature of the Krina and its activities. For example, in 2010, a group of young students who belong to the straight edge movement (a term used by people, who refrain from alcohol, tobacco and drugs), begun to actively participate in the rituals. By their influence, alcohol is hardly ever offered in the festivals of the community anymore.[14]

[14] Even earlier alcohol was used in a very civilized and modest way. However, there was often some wine in the festivals or a small glass of cognac offered before the meditation, which was seen as helping with the service.

Within Rodnoverie movement, the straight edge movement has indeed recently gained much ground and in that context it often features nationalist ideology. However, it also reveals some changes that have taken place in the nationalism of those within Rodnoverie. These young people argue that drinking is one of the biggest social problems in Russian society and, therefore, wish to give an alternative example. The emphasis on nationalism as a responsibility does not, of course, exclude xenophobic and aggressive nationalism, but it does represent nationalism that is more oriented toward internal criticism than towards the "other."

The straight edge movement also provides an interesting viewpoint for the analysis of "conservatism" in contemporary East European Paganisms. Though the yearning back to stricter morality certainly seems to fit in this category, in Russian society, abstinence is actually a trend that is subscribed to by many liberal and alternative youth subcultures that wish to distance themselves from the mass, which they claim to be passive and uncritical because of the overt usage of alcohol.

The term "conservatism" is indeed, perhaps, one of the categories that in the study of Paganism most needs to be unpacked and further analyzed. For example, though homosexuality is generally accepted within Western Paganisms, it would be difficult to imagine that any Russian Rodnoverie community would accept an openly homosexual member. At the same time, many Rodnovers are engaged in various youth subcultures, the biggest one of these being Metal music scene. In recent decades, Pagan Metal music has established itself as a specific genre that musically follows Black metal or folk metal. In such big cities as Moscow and St. Petersburg, some clubs organize "Pagan Metal nights" that feature smaller and less known bands with such more established ones as Severnye Vrata, Arkona and Butterfly Temple.

Also in the case of nationalism, the "codes" of the Krina are situational and often quite personalized. Although some members of the community have implicitly expressed their dislike for an immigrant from the "South," an Armenian friend of the son of Blagumil has occasionally attended the festivals. In similar vein, alternative subcultures, even such stigmatized ones as, for example, Satanism, may become a part of Rodnoverie communities if people, who have connections to these are accepted as members. Naturally, a similar evolution takes place in all religious groups, but what is exceptional in contemporary Paganism is that without any centralized authority or generally accepted dogmas, such development is much easier and faster. Thus Paganism is by its nature not a movement with one common history and destiny, but more like an idea that finds many different manifestations of which some will thrive as others wither.

The activity of the Krina also changes because they adopt influences from other groups. It is actually possible to detect some details, like their using swords and wolf skins, as well as some ritual words, such as "*byly, est, budem*" (we were, we are, we will be) as innovations borrowed from certain other communities from St. Petersburg. A couple of years ago the members of the Krina began to put the Russian flag up in their rituals next to their own community symbols. One of the reasons for this might be a wish to display their patriotism more openly and thereby in advance to answer to some accusations. In this case the development seems similar to the one that took place in the Circle of Pagan Tradition, with which the leaders of the Krina have close relationships even though they have not joined this umbrella organization.

A part of such delicate balancing between being accused of internationalism, liberalism and individualism and on the other hand, creating alternative forms of Paganism to the overtly politicized nationalistic paganism can be found in several contemporary Rodnoverie groups. Symptomatically, in recent years some Rodnovers have established contacts with the Pagan Federation, which represents the more liberal form of Paganism, while the activity of the international organization *Veche*, gather together the followers of Slavic Paganism from different Slavic countries which is known of its nationalist posture has somewhat halted.

Even though "individualism" is used as a derogatory concept in Rodnvoerie discussion, it can be argued that it is precisely individualism that is common to both Western and Eastern Paganisms. This individualism manifests itself in the freedom to create one's own religious views and subsequent variety in "Paganisms." Several scholars have noted the difficulties of finding any definitions that would apply to all forms of Paganism. For example, while most of the Pagans are polytheists, there are also Pagans who acknowledge only one Goddess or Gods, henotheists (an idea, according to which the different gods are only manifestations of one God), animists and even atheist Pagans. However, it seems that contemporary Pagans recognize each other in a way that representatives of different subcultures often do.

Another feature that combines contemporary Pagans is certain counter-cultural posture. To adhere to ancient gods in the contemporary world usually raises puzzlement and amusement. Thus, it is a choice that sets the individual somewhat apart from the "mainstream." At the same time, it is not a coincidence that contemporary Paganism has emerged only in a late modern society, where people are, as sociologists of religion have noted, all the more often constructing their own individual religion instead of cosmological explanations or moral doctrinal frameworks. They are looking for a spirituality that could answer their personal needs and help them in their personal projects of personal growth. For example, Heelas and Woodhead (2005) argue that the contemporary world is witnessing a "subjectivization of religion" which features similar tendencies as other domains of life. As educational ideals have become "pupil-centred," medicine "patient-centred," there is a growing trend in modern religiosity to shift the focus from transcendental authorities and universal models on to personal dogma and personal experience. While traditional religiosity sets internal, predetermined roles for people to live "life-as" these models suggest, the subjectivization of religion refers to listening to and following subjective emotions, intuition and reason (Heelas and Woodhead 2005: 3–10). In the study of Western Paganism, several scholars have analyzed the topic in the context of the changes in religiosity in late modern societies. They have noted that as a non-hierarchical and anti-dogmatic religion, contemporary Paganism suits well the individualization of religion (Reid and Rabinovitch 2004).

87.6 Conclusions

In Central and Eastern Europe, Paganisms (in general) is more nationalistically oriented than Paganisms in, for example, the United Kingdom and the United States. It is important to make this distinction in the study of the topic and as we study the ways in which contemporary Paganisms reflects their social context. However, following Simpson's (2000) suggestion we may also notice the myriad ways in which the type "A" and "B" intertwine. Furthermore, in order to understand why such people with so different social, religious and moral views still consider themselves as representatives of the same religiosity, it is important to notice the common themes of the varying forms of contemporary Paganism.

In analyzing Krina, it is easy to notice that the movement includes both "eastern" and "western" or conservative and liberal forms of paganism. The freedom of individual consciousness is an important principle of the movement; it disapproves of aggressive xenophobia and it is more focused on the spiritual growth of individuals than on a political goal. At the same time, it often implies that one's ethnicity determines his/her spirituality. Also occasionally there are even some hints that members consider the Russian tradition as superior to others. The members of Krina do not consider the division of nation states as artificial constructs and promoting the interests of Russia is considered an important value.

The social pressure of displaying nationalistic values has influenced the Krina, as members wish to be accepted to the Rodnoverie community. At the same time, the member genuinely consider patriotism to be one of the most important values and in this, they reflect the dominant values in Russian society. It has been also suggested that if the social security and stability would increase in Russia, nationalism as an ideology or a frame of orientation would lose its significance. This loss would be reflected in contemporary Russian Paganism as well. The case of the Krina demonstrates that even in Russia there exist many different kinds of forms of Paganisms.

References

Adler, M. (2006). *Drawing down the moon: Witches, druids, goddess-worshippers and other pagans in America*. New York: Penguin.

Aitamurto, K. (2009). Russian Rodnoverie: Negotiating individual traditionalism. In B. Rigal-Cellard (Ed.), *Religions et mondialisation* (pp. 373–390). Pessac: Presses Universitaires de Bordeaux.

Aitamurto, K. (2011). Modern pagan warriors: Violence and justice in Rodnoverie. In J. R. Lewis (Ed.), *Violence and new religions movement* (pp. 231–248). New York: Oxford University Press.

Aitamurto, K., & Simpson, S. (2013). Introduction: Modern pagan and native faith movements in Central and Eastern Europe. In S. Simpson & K. Aitamurto (Eds.), *Modern pagan and native faith movements in Central and Eastern Europe* (pp. 1–9). Durham: Acumen.

Andreeva, J., & Pranskevičiūtė, R. (2010). The conception of family homesteads in the Anastasia movement: The cases of Russia and Lithuania. *Humanitâro zinâtņu vēstnesis, 18*, 94–108.

Berger, H. (1999). *A community of witches: Contemporary neo-paganism and witchcraft in the United States*. Columbia: University of South Carolina Press.
Berger, H., Leach, E. A., & Shaffer, L. (2003). *Voices from the pagan census: A national survey of witches and neo-pagans in the United States*. Columbia: University of South Carolina Press.
Budapest, Z. (1999). *The holy book of Women's mysteries*. Oakland: The Wingbow Press.
Carpenter, D. D. (1996). Practitioners of paganism and Wiccan spirituality in contemporary society: A review of literature. In J. R. Lewis (Ed.), *Magical religion and modern witchcraft* (pp. 373–406). Albany: State University of New York Press.
Gaidukov, A. V., & Maslyakov, D. A. (2012). Rodnoverie v setyah Kontakta. In V. V. Barabanov & A. B. Nikolaev (Eds.), *Gertsenovskie chteniya 2011. Aktual'nye problemi social'nykh naukh* (pp. 311–318). St. Petersburg: OOO ElekSis.
Gardell, M. (2003). *Gods of the blood: The pagan revival and white separatism*. Durham/London: Duke University Press.
Harvey, G. (1997). *Listening people, speaking earth: Contemporary paganism*. London: Hurst & Company.
Heelas, P., & Woodhead, L. (2005). *The spiritual revolution: Why religion is giving way to spirituality*. Malden: Blackwell Publishing.
Hutton, R. (1999). *The triumph of the moon. A history of modern pagan witchcraft*. Oxford: Oxford University Press.
Inglehart, R., & Baker, W. E. (2000). Modernization, cultural change and the persistence of traditional values. *American Sociological Review, 65*(1), 19–51.
Lassander, M. (2009). Modern paganism as a legitimating framework for post-materialist values. *The Pomegranate, 11*(1), 74–96.
McCutcheon, R. T. (2005). *Religion and the domestication of dissent: Or how to live in a less than perfect nation*. London: Equinox Publishing.
Patheos. (2014). How many pagans are there? Accessed at: http://www.patheos.com/Library/Answers-to-Frequently-Asked-Religion-Questions/How-many-Pagans-are-there.html. 14 Aug 2014.
Pilkington, H., & Popov, A. (2009). Understanding neo-paganism in Russia: Religion? ideology? philosophy? fantasy? In E. Ramanauskaite, G. McKay, M. Goddard, & N. Foxlee (Eds.), *Subcultures and new religious movements in Russia and East-Central Europe* (pp. 253–304). Frankfurt am Main: Peter Lang.
Reid, S. L., & Rabinovitch, S. T. (2004). Witches, wiccans, and neo-Pagans: A review of current academic treatments of neo-paganism. In J. R. Lewis (Ed.), *The Oxford handbook of new religious movements* (pp. 514–533). Oxford: Oxford University Press.
Rezunkov, A. (2004). Chto takoe prazdnik? In A. G. Rezunkov (Ed.), *Kolovorot. Slavyanskiy solnechno-lunnyy kaledar'-mesyatseslov* (pp. 14–17). Moscow: Ladoga.
Rezunkov, A. (2005). Fenomen prazdnika kak neot'emlaya chast' Traditsionnogo narodnogo kalendarya. In A. E. Nagovitsyn (Ed.), *Vestnik Traditsionnoy kul'tury* (Vypusk No 2, pp. 19–24). Moscow: Vorob'ev.
Shizhensky, R. V. (2012). *Filosofiya dobroi sily: Zhizn' i tvorchestvo Dobroslava (A. A. Dobrovolskogo)*. Penza: Sotsiofera.
Shnirelman, V. (1998). *Russian neo-pagan myths and antisemitism* (Analysis of current trends in antisemitism. Acta no. 13). Jerusalem: Hebrew University of Jerusalem.
Shnirelman, V. A. (2013). Russian neopaganism: From ethnic religion to racial violence. In K. Aitamurto & S. Simpson (Eds.), *Modern pagan and native faith movements in central and eastern Europe* (pp. 62–76). Durham: Acumen.
Simpson, S. (2000). *Native faith: Polish neo-paganism at the brink of the 21st century*. Krakow: Nomos.
Starhawk. (1989). *Spiral dance: A rebirth of the ancient religion of the great goddess*. New York: Harper-Collins Publishers.
York, M. (2003). *Pagan theology: Paganism as a world religion*. New York: New York University Press.

Chapter 88
Evangelical and Pentecostal Churches in Montreal and Paris: Between Local Territories and Global Networks

Frédéric Dejean

88.1 Introduction

Among the abundant research on globalization and its manifestations, more and more studies are specifically interested in religious phenomena (Obadia 2010). This rise can be relocated in a wider dynamic characterized by what some observers call the "desecularization of the world" (Berger 1999) or the "revenge of God" (Kepel 1994); and, according to the authors of a recent bestseller, "God is back" (Micklethwait and Wooldridge 2010)!

This "return of God" is particularly visible in the case of Evangelical Churches, especially those belonging to the Pentecostal and Charismatic movement. (In this chapter, the term and spelling of Church refers to these movements.) This declination of Evangelical Protestantism, which insists on baptism in the Holy Ghost as manifested by the "Holy Gifts" (glossolalia, prophecy and deliverance), on its uses by the believer and on the manifestations of the "signs and wonders," "can be counted as one of the great success stories of the current era of cultural globalization" (Robbins 2004: 117). Social science has made of these Churches an epitome, which highlights the main characteristics of cultural globalization. Debates have therefore focused on a central tension summarized by homogenization/heterogenization (Robertson 1995). For some, the Pentecostal and Charismatic currents were characterized by a unique ability to duplicate themselves identically in all sociocultural settings. For its part, the contrary position insists on the malleable character of Pentecostal Churches – both in their theological contents and in their ritual practices –, which allows them to take root in local cultures to the point of adopting some practices in their own functions.

F. Dejean (✉)
Institut de recherche sur l'intégration professionnelle des immigrants,
Collège de Maisonneuve, 6220, rue Sherbrooke Est, Montréal (Québec)
H1N 1C1, Canada
e-mail: frederic.dejean@yahoo.fr

S.D. Brunn (ed.), *The Changing World Religion Map*,
DOI 10.1007/978-94-017-9376-6_88

Such a debate presents geographical aspects, for example the desire to locate spatially the proposed analyses (Featherstone et al. 1995). Globalization consecrates "the intensification of worldwide social relations which link distant localities in such way that local happenings are shaped by events occurring many miles away and vice versa" (Giddens 1990: 64). At the same time, individuals' experience of this world system is locally rooted. The neologism "glocalization" (Beyer 2007; Robertson 1995) allows us to account for recomposition effects which operate locally under the influence of larger dynamics. As Peter Beyer says, "globalization is always also glocalization, the global expressed in the local and the local as the particularization of the global" (Beyer 2007: 98).

The following pages address geographical aspects of Evangelical and Pentecostal Churches and insist on local and global dynamics, as these two scales are just two faces of the same coin. Just as globalization manifests locally, Evangelical and Pentecostal Churches contribute to global dynamics through specific places. We show in this chapter that globalization of Evangelical Churches does not lead so much to deterritorialized forms of religion, but on the contrary redefines how places and territories function. The chapter starts with a short presentation of Evangelical and Pentecostal realities worldwide; we focus on the transition from a religious universe with well-identified foci to the creation of a complex reticular system facilitated by international migrations and the emergence of transnational communities. In the second part, we present the new ways in which the local and the global are articulated. Finally, the third part focuses on reterritorialization, and we show how we can observe the Churches' local reconfiguration efforts.

88.2 Evangelical and Pentecostal Christians on the Global Stage: "The World as a Parish"[1]

88.2.1 A Worldwide Presence

If the Evangelical presence worldwide goes back to ancient times, the twentieth century is deeply marked by an acceleration of diffusion and by increasing multilateral exchanges. Pentecostalism appeared in the States around the turn of the twentieth century and spread in less than a century to all continents, from Nunavut in Canada to South Africa. The Assemblies of God, formed as early as 1914 in the wake of the Pentecostal Awakening, are today the sixth largest Christian denomination in the world (Barrett et al. 2001) and account for close to 60 million people. The pace of this dissemination is clear in certain countries and territories. In Nigeria, before the Awakening of the 1970s, there were only 300,000 Pentecostals in a country of close to 75 million inhabitants. Between 2000 and 2005, more than 40 million

[1] This expression is from John Wesley, founder of Methodism, and was recently borrowed by sociologist David Martin in his book Pentecostalism, The world their parish.

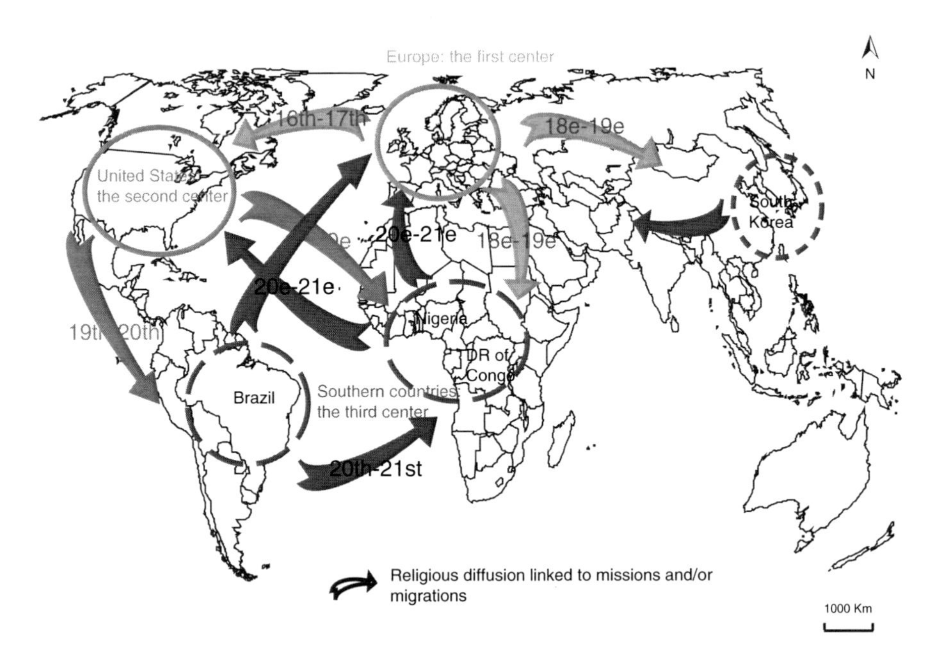

Fig. 88.1 The geohistorical diffusion of Evangelical Churches (Map by F. DeJean, 2012)

people identified as "born again" in a population of about 120 million (Marshall-Fratani 2007). According to a *Pew Forum* survey from 2006, "renewalists – including Charismatics and Pentecostals – account for approximately three-in-ten Nigerians. Roughly six-in-ten Protestants are either Pentecostal or Charismatic, and three-in-ten Catholics surveyed can be classified as Charismatic" (Pew Forum 2006: 86). It is difficult to know the exact number of Pentecostal or Neo-Pentecostal Christians, as statistics are based on heterogeneous data, and researchers don't always agree on which groups should be included. According to English sociologist David Martin (Martin 2002), there were between 250 and 300 million Pentecostals worldwide at the start of the millennium, whereas the authors of the *World Christian Encyclopedia* offer 523 million for the same period.

Figure 88.1 indicates the global spread of Protestant culture in general, and of Evangelical and Pentecostal Churches in particular. The spaces designated as "the three large geohistorical Evangelical centers" show a shift in their center of gravity. The first stage (sixteenth century) of Protestantism is European, whereas the second is based in the United States. This second stage is marked by several Awakenings, the most noteworthy being the great Pentecostal Awakening at the turn of the twentieth century. The third stage is more polycentric. Whereas the first two stages were marked by unique geographical centers, events changed during the third phase, which started in the 1950s and are still unfolding today. The main consequence of these historical stages is the complexity in many countries of the Protestant landscape, and the existence of different trends and denominations which correspond to historical layers. For instance, Fig. 88.2 highlights the situation in South Africa

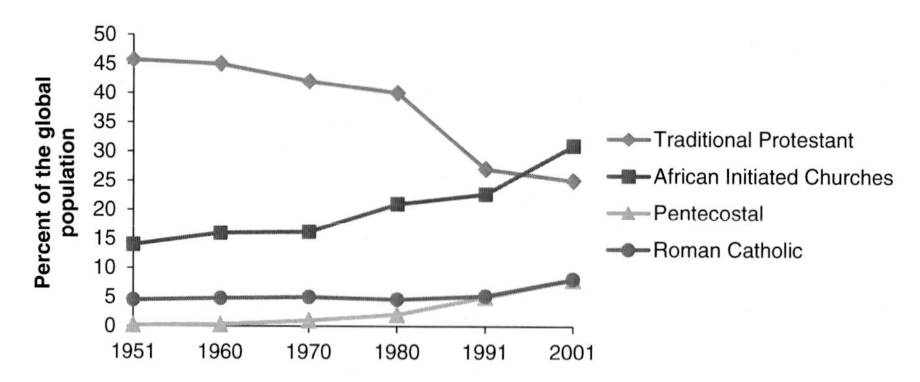

Fig. 88.2 Evolution of the Protestant landscape in South Africa (Data source: Pew Forum 2006)

where "Traditional Protestant" denominations have to compete with African Initiated churches, that is, Churches started by Africans and not by European or North-American missionaries (Meyer 2004), Pentecostal and neo-Pentecostal organizations which have been developing for the 1950s on the African Continent (Anderson 2004). In this competition, "Traditional Protestants" are clearly the losers, whereas AIC and Pentecostal are rising.

The third period is characterized by a demultiplication of reference centers, due in part to the intense use of new information and communication technologies, but also by an increase in the strength of Southern hemisphere movements and by the complexity of exchanges on a global scale. Whereas these exchanges have historically been unilateral North-south movements, there are now inverted forms, and some writers speak of "reverse missions" or "missions in reverse" (Adogame 2003; Kim 2011). These "reverse missions" gave birth to several religious success stories, as for instance the *Embassy of God*, a megachurch founded in Ukraine by Sunday Adelaja, a pastor from Nigeria (Adogame 2008; Asamoah-Gyadu 2005).

88.2.2 An Original Diffusion: The Transnationalization Process and Disaporic Networks

The Evangelical and Pentecostal presence on all continents leads us to investigate how this spread occurred, what impact this might have on beliefs and practices, and how they were able to implant themselves locally. For religious content to circulate and implant itself in a lasting way, it must possess two characteristics: (a) a "portable practice," that is, "rites that can be easily learned, require relatively little knowledge (…) and can be performed without commitment to an elaborate ideological or institutional apparatus" (Csordas 2007: 261), and (b) a "transposable message," that is, "the basis of appeal contained in religious tenets, premises or promises, can find footing across a diversity of linguistic and cultural settings" (Csordas 2007:

261). Most Churches share these two traits. In the following lines, we will look not so much at the "why" these Churches succeed, but rather at "how" religious content circulates. We emphasize the role of international migrations and the formation of transnational communities.

The spread of Evangelism and Pentecostalism is based on their numerous churches, denominations and religious or parareligious organizations (for example, non-government organizations). The novelty in the twentieth century, and this is true especially for the Charismatic and Pentecostal movements, is the multiplication of tendencies and organizations and the emergence of independent Churches often with a very local outreach. There are 740 Pentecostal denominations, 6530 non-Pentecostal mainline denominations with large organized internal Charismatic movements, and 18,810 neo-charismatic denominations and networks across the world (Barrett et al. 2001). These numbers show that Evangelical and Pentecostal organizations evolve along a logic of internal fragmentation. The corollary of this logic is a flexibility in the spread of the Christian message, which is not handled by a sole institution.

The image of a wave that sweeps everything in its wake is often used to describe the progress of Evangelical and Pentecostal Churches. It would probably be more accurate to use the image of the rhizome popularized by Gilles Deleuze and Félix Guattari (1980). The rhizome "has no beginning and no end, but only a middle, which is always growing and overflowing" (Deleuze and Guattari 1980: 31). We are witnessing today the emergence of new poles of emission, notably in Southern countries such as Brazil, Nigeria or Congo. From these poles, Churches which were originally local manage to spread to all continents, thus creating transnational Churches (Ebaugh and Chafetz 2002; Schiller et al. 1992). The use of this term allows us to insist on the space or the place of reference from which the spread originates and toward which it converges (Kearney 1995). If these Churches open local assemblies everywhere in the world, the place of origin remains a space of reference, which can take on a sacred dimension and give rise to pilgrimages. For example, devotees of the *Luz del Mundo*, a Church founded in Mexico in the 1930s, are invited to go to Guadalajara (State of Jalisco) every August, where the Church was born (Nutini 2000).

Several transnational Churches were founded in Nigeria (Fourchard et al. 2005), for example, the *Mountain of Fire and Miracles Ministries* (Hackett 2003; Moyet 2005), the *Deeper Christian Life Ministry* (DCLM), or the *Redeemed Christian Church of God* (Adeboye 2005, 2007; Ukah 2008). This last one reveals an original dynamic in the spread of the Christian message. Created in 1952, it went through very strong growth in the 1990s. In 2002, it had 5,500 parishes and about 700,000 faithful around the country. Globally, this Church is present in over 50 countries, 30 of which are in Africa (Adeboye 2007).

The existence of these transnational Churches and the ethno-cultural diversity within the communities in the Northern countries are inseparable from contemporary migratory dynamics. If some Churches dispatch missionaries to implant local assemblies, most use migrants as messengers to implant themselves in new territories (Fancello 2006). This logic of diffusion through migration is not limited to

South-north exchanges, as many immigrants who are implanted in a certain country and are members of a Church move to a new city and start a local assembly. In a city like Montreal, the ethno-cultural diversity within the Evangelical and Pentecostal movements reflects recent immigration: the city possesses about 70 Hispanic churches, approximately 50 African churches and 150 Haitian churches. This last number reminds us that Montreal is home to the second largest Haitian diaspora in the world. The first Haitian Churches opened their doors in Montreal at the start of the 1970s, at a time when Haiti was becoming one of the main sources of immigration.[2]

88.3 The New Articulations of the Local and Global

88.3.1 Internet and the Networking of Local Assemblies

> As the pastor begins his sermon, the person in charge of the audio-visual system walks up to the pulpit and whispers a few words in his ear. The pastor nods and steps down the stage. The lights go out, and the face of the founding pastor appears on the two giant screens. It is time for his weekly speech, live from Nigeria. (Fieldwork notebook, July 2011)

This scene took place during a service in a Nigerian transnational Church, *The Christ Embassy*, implanted in Montreal. It is emblematic of the founder's place within transnational Churches and of how the Internet helps maintain authority and a spirit of community despite geographic distance. The Sunday service is not just for the local community, but serves as a link to a place of reference from where the founding pastor speaks. This establishes the connection between assemblies disseminated throughout the world. Such a use of the Internet in religious activities is not rare: in Saint-Denis, a commune in the Parisian suburbs, a local assembly of the *Deeper Christian Life Ministry* (Ojo 1988) has installed in its hall an audio-visual system which allows devotees to take Biblical studies classes imparted by the founding pastor, who lives in Lagos (Dejean 2010). This allows the followers to temporarily share a common space-time, despite the distance. One of the novelties made possible by new technologies resides precisely in the simultaneous character of the collective experience (Inda and Rosaldo 2002). For a limited time, members of a same group, though spread all over the world, can live the same experience. We thus understand the importance of "live" and "real time" transmission (Virilio 2000). It is precisely because at a given time, in a given location – a room somewhere in Lagos located at the junction of geographical space and cyberspace – the pastor's word is followed from multiple locations, thereby creating a true online community (Wilson and Peterson 2002). If we are witnessing the emergence of a transnational community, it is because at the same time, members of the Church who are spread around the planet know that they are in communion. This is what is meant by the

[2] In 2011, about 10 % of immigrants to Quebec were born in Haiti (Institut de la statistique du Québec: www.stat.gouv.qc.ca/donstat/societe/demographie/migrt_poplt_imigr/603.htm).

term "transnational Church." The Church's birthplace, its neuralgic center, remains a place of reference, whose importance is maintained by visiting pastors and by the type of experience we have just described.

88.3.2 When the Local Highlights the Global

If researchers often point to the international dimensions of Evangelical Churches, we must also highlight the role these Churches themselves play in promoting this international character. The international character of a Church or a Christian organization can be both a reality and a discursive and rhetorical resource which underscores the Church's success. Not only has the Church achieved international status, but it is able to capitalize on this for its expansion. If the affirmation of Churches' international character can easily be justified in the Biblical text – Christ calls on his disciples to make his Church into a universal movement[3] – it also helps strengthen the importance and success of the Church in a highly competitive religious market. The affirmation of this international character can be made in several ways:

First, the most obvious way is to integrate the word "international" in the Church's name, even if the Church is local. The global then takes on a goal to be achieved (for example. the *Mission Charismatique Internationale* in Montreal, which changed its name around 2010, or the *Mission Internationale El Shaddai*). In certain cases, the name of the Church plays a decentering role from the country of origin toward the global (for example, the *Église Évangélique de la Grande Moisson de l'Éternel en France et dans le Monde entier au nom de Jésus*, in the Parisian area).

Second, on a more symbolic level, the international character of the Church can be highlighted by its emblem. For example, the emblem of the *Assemblée Chrétienne Parole Vivante de Montréal* has a "toutes les nations" slogan and colourful characters holding hands in a show of shared humanity which transcends ethnic and national borders.

Third, the international character of the Church's mission is also affirmed in the prayer room by the presence of flags on the wall. This is a very common practice, even in very different settings, for example, in American megachurches (Fath 2008) or in some places of worship belonging to transnational Churches. Each flag represents a country where the Church is now implanted (Van Dijk 1997). In some assemblies, especially those who have worshippers from several countries, the flags show the diversity of the faithfuls' geographical origins, thus highlighting how the Church draws its unitive power from the diversity of its origins.

[3] "And Jesus coming up spoke to them, saying, All power has been given me in heaven and upon Earth. Go [therefore] and make disciples of all the nations, baptising them to the name of the Father, and of the Son, and of the Holy Spirit; teaching them to observe all things whatsoever I have enjoined you" (Matthew 28: 18–20).

88.3.3 Traveling Pastors

> I'm sorry I haven't returned your call sooner, but I have just returned from a long visit in Central America, where I was asked to preach in several conventions." (Phone conversation with a Montreal-based pastor, October 2008)

We can see how the diffusion process of Churches and Evangelical and Pentecostal organizations is based on people's and religious goods' mobility (international migration and Internet). And this logic of mobility does not stop once the diffusion is complete. On the contrary, Evangelical and Pentecostal culture is characterized by perpetual exchange and circulation. Let's take the example of pastors. Though they are usually attached to a local assembly, their spaces of reference are rarely confined to the city where the place of worship is located. On the contrary, they are part of transnational networks which give them opportunities to travel and access religious content from abroad. The following short excerpt of a conversation bears witness to the strong mobility of pastors, who find themselves travelling in very different cultural settings. The other important element in this short excerpt is the mobilization by pastors of a "spatial capital" supporting their pastoral work. This spatial capital – built on the model of Pierre Bourdieu's "symbolic capital" – refers to the "resources accumulated by an actor which allow him to use to his advantage, depending on his strategy, the spatial dimension of society" (Lévy 2003: 126). These displacements allow pastors to accumulate religious resources (religious goods and contacts with other pastors), but also to derive a form of prestige. These travels are signs of "distinction" (Bourdieu 1979) which allow the pastors to position themselves in other religious actors' field. For a pastor, the opportunity to travel, to participate in "conventions," in "evangelization campaigns" or even in "crusades" are all ways of affirming their status on a religious scene which is not merely metropolitan or national, but rather international.

Some of these pastors don't have a local assembly and are always on the road (one the most popular of this type of Evangelist is the German Reinhardt Bönnke,[4] who draws hundreds of thousands of people for enormous "crusades of fire" on the African continent). Others head one or two Churches, and are still invited to preach abroad. We have shown (Dejean 2010) how a type of commerce is put in place, whereby pastors invite each other in an exchange of services. According to posters (Fig. 88.3) placed in public areas in Paris, pastors invited to the Parisian area are mainly from Sub-Saharan Africa, from Latin America and from North America (Coulmont 2010). Such an example shows that French Evangelism cannot be understood only at the level of the national territory, but must be placed within a complex network of pastoral exchange and mobility. Figure 88.4 shows the countries of origin of pastors invited to preach in the Paris metropolis in 2010; this map is based

[4]Born in 1940 of a Pentecostal pastor father, Reinhard Bonnke started his missionary work in Lesotho between 1967 and 1974. At the time, he received a call for "All of Africa," with the assurance that "Africa must be saved." His first healing crusade took place in Botswana in 1975: the gathering grew from a handful of faithful to a crowd of 10,000 people. Bonnke's success is due in part to his numerous testimonials of miracles during his "Crusades of Fire" (Burgess 2002).

Fig. 88.3 A poster in a street of Saint-Denis in the suburbs of Paris (Photo by F. DeJean, 2009)

on data collected on more than 100 posters placed in public spaces in Paris (Coulmont 2013). In the case of Churches linked to immigration to Western Europe, pastors do not hesitate to call upon their own family networks to invite other pastors to their local assemblies (Schiller et al. 2006).

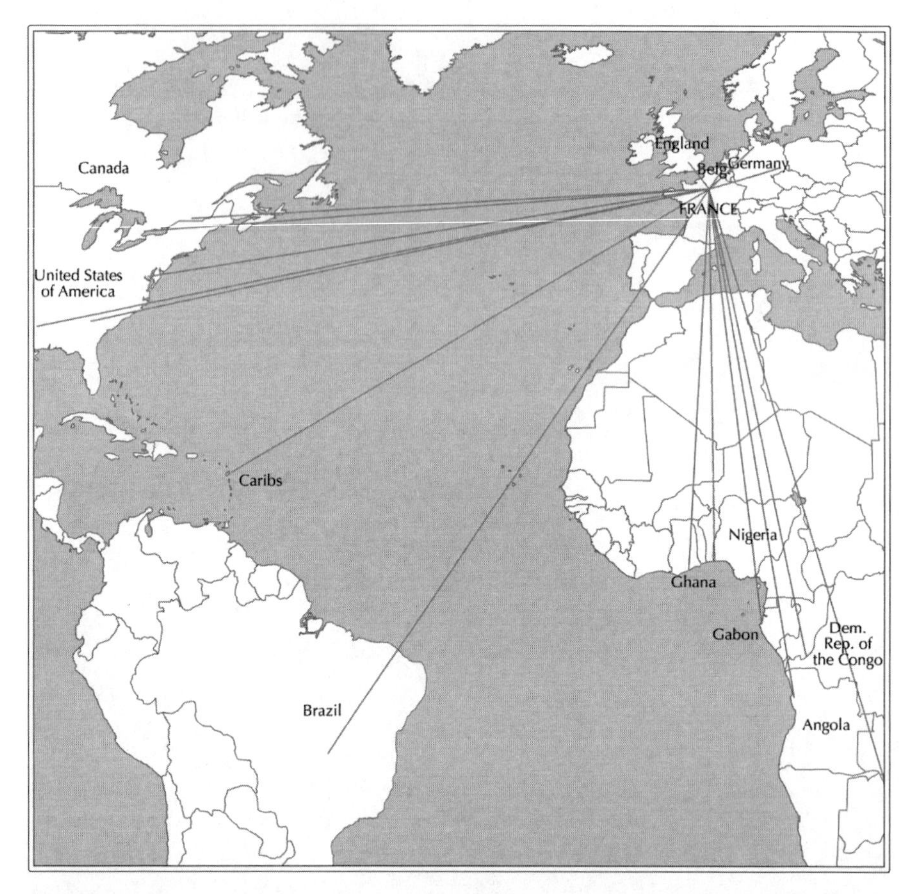

Fig. 88.4 Countries of origin of pastors invited to preach in the Parisian metropolis in 2010 (Map by Dick Gilbreath, University of Kentucky Gyula Pauer Center for Cartography and GIS, commissioned by the editor; after Baptiste Coulmont, 2013, used with permission)

88.4 Back to Space

The example of these travelling pastors indicates that global circulation affects both religious contents and the actors who produce them and circulate them on the religious market. One form of fascination for the possibilities offered by the Internet has led some researchers to talk of deterritorialization of religious phenomena for the benefit of transnational virtual communities. Such a hypothesis reinforces the idea that the new information and communication technologies do not lead to the "end of history," but rather to an "end of geography" (Virilio 2000: 9). In this third section, we question this alleged "end of geography" and show that globalization of Evangelical and Pentecostal Churches is inseparable from original ways of reinvesting local spaces, in particular urban spaces.

88.4.1 Local Forms of Religious Life in Urban Spaces

The Evangelical and Pentecostal presence is especially strong in urban spaces. In Southern countries, there is a correlation between the urbanization process and the rise of Pentecostalism (Adogame 2010; Coleman 2009). Philip Jenkins shows that the success of these Churches in Southern countries "can be seen as a by-product of modernization and urbanization. As predominantly rural societies have become more urban over the last 30 or 40 years, millions of migrants have been attracted to ever larger urban complexes" (Jenkins 2002: 85). And he adds: "These emerging churches (…) do best among young and displaced migrants in mushrooming megacities" (Jenkins 2002: 107). There is a strong demand for various services (health, education, food…) that governments are unable to provide and which Churches can offer in an efficient manner. In Northern countries, Churches created by immigrants find fertile ground in urban spaces where large majorities of immigrants are located. In a setting of migration, Churches offer a form of "social capital" which is useful for immigrants' integration process and constitute transition spaces between the country of origin and the host society (Ley 2008; Mossière 2006).

In the end, this affinity between Evangelical and Pentecostal movements and urban spaces must be grasped at several levels:

- Cities – both "City of God, city of Satan" (Linthicum 1991) – present new challenges for Churches, especially since urbanization and secularization are often correlated (Cox 1966).
- Churches must deal with urban spaces in which the legitimacy of religious signs is questioned, thus leading to a form of "secular iconoclasm" (Howe 2009).
- Urban spaces offer material and cultural resources which Churches use for their evangelical work, a phenomenon which bears witness to their strong adaptation capacity in the face of constantly changing urban realities.

The megachurch[5] is perhaps the most famous spatial form taken on by Churches in urban settings (Ellingson 2010; Fath 2008; Twitchell 2004). Whereas these megachurches first appeared in American suburbs in the "Sun belt" and on the Pacific Coast of the U.S., they are now implanted everywhere on American soil and have even spread to other countries. Thus, the world's largest megachurch is the *Yoido Full Gospel Church* in Seoul (Young-Gi 2003). Megachurches are also cropping up in Canada, Brazil, Nigeria, Ghana, France and England. Beyond the diversity of settings, they share a certain number of common traits: size (minimum 2,000 seated places in the place of worship), location (away from downtown, close to major highways in order to facilitate access from the entire city), diversity of proposed activities (religious, social and cultural) or the ability to integrate secular culture to Church activities. The buildings are "carefully maintained and slightly antiseptic professional buildings of suburban corporate American with the conveniences of a shopping mall such as an atrium dining area and food court" (Sargeant 2000: 61).

[5] A megachurch is an assembly of at least 2,000 people. The term "gigachurch" is sometimes used in reference to megachurches which draw more than 10,000 followers.

Megachurches are almost always microcosms in which the faithful have access to religious and profane activities, so there is no discontinuity between outside space and Church space. This logic is taken to its limit in the case of "campuses" built by some of the largest Churches. In Nigeria, the *Redeemed Christian Church of God* has built a *Redemption Camp* on 3.86 sq mi (10 sq km). It includes an auditorium 395 m×500 m (12,96 ft×1,640 ft) (Ukah 2008).

88.4.2 Toward a Re-enchantment of Urban Spaces: "Spiritual Mapping" and "Territorial Spirits"

If Evangelical and Pentecostal Churches have global aims, their work is carried out locally. For some currents, space and territory are at the heart of the action. This is the case for those who practice "spiritual mapping," a particular type of cartography which emerged at the very end of the 1980s and the beginning of the 1990s (Dawson 1989; Wagner 1996).

"Spiritual mapping" is based on the idea that demonic spirits don't attack only people, but also places. These demonic influences lead to the creation of "demonic strongholds" which believers – actual soldiers of God – must fight with prayer, the dominant weapon of "spiritual warfare" (Wagner 1996). These battles can be carried out in "prayer walking" sessions (Hawthorne and Kendrick 1996) during which a group of Christians walks around a building so as to "circle" the demonic forces located within.

Although spiritual mapping can be used in all spaces, cities are favorites. The demonic forces located in cities are the result of a city's history, of its past activities and of contemporary dynamics currently at work. For example, evangelization work will be confronted with the "spirit of lust" in a red-light district and to greed in a financial sector. Christians must, therefore, "besiege" these "spiritual fortresses" before they can undertake their missionary work.

"Spiritual mapping" points to the fact that urban space constitutes a major issue for certain neo-Pentecostal currents. This could be seen, for example, in the 2009 edition of *Marche pour Jésus* in the streets of Paris (Fig. 88.5). The introductory text explained that the march was not random, but bore a religious dimension: "We will leave from the Place Denfert-Rochereau, ancient place of Hell, as it used to give access to Paris' catacombs. It served as an entrance to Paris before 1870. From there was given the order of insurrection in 1944 against the Germans, and Leclerc's tanks left from this place to liberate Paris. We will then go down Raspail Boulevard. Raspail was a chemist and a member of a revolutionary secret society who famously said 'To science, the only religion of the future'." This excerpt points to the "spatial imagination" of the *Marche pour Jésus* organizers in particular, and of the way in which Evangelists do not just occupy space physically, but only symbolically and theologically.[6] Not only do they invest space and produce places, they make discourses about space which describe a specific vision of the world in which the fight between good and evil is based in space.

[6] For example, we find here a recurring element of neo-Pentecostal literature: the denunciation of the role of Freemasons in the development of city plans or in the construction of certain buildings.

Fig. 88.5 A truck used as a mobile stage for musicians at the "March for Jesus" in Paris (Photo by F. DeJean, 2009)

88.5 Conclusion

Evangelical and Pentecostal Christians maintain close ties with the globalization process. Not only do they use globalization to their ends (through international migrations which play the role of missionary enterprises), they also produce original forms of globalization of religious goods and encourage the circulation of pastors, ritual practices, theological contents and place esthetics. Evangelical and Pentecostal specificity is not so much the circulation per se as its intensity, its systematic character, the emergence of a multiplicity of emission bases – especially in Southern countries – and the forms of hybridation which are produced by these exchanges.

The study of globalization of religious goods can sometimes lead to emphasize the forms of dematerialization and deterritorialization for the benefit of networks and "cyber communities." Yet, the process of globalization of religious facts is characterized by significant impacts on territories and by new articulations between global and local scales. The term *glocalization* can therefore account in a concise way for an original dialectic between levels of analysis and functioning of geographical space. We have thus shown that globalization of Evangelical and Pentecostal Churches is always a local process and that it then spreads on a global scale. Furthermore, the global presence of

Churches bears witness to a new type of functioning for local spaces which find themselves connected through reticular logics to faraway places. Far from separating religious reality from the territory, globalization brings it back to the territory, but under different modalities, as we saw with the spread of Megachurches, which has become the confirmation of the globalized status of local assemblies, and with "spiritual mapping" and "territorial spirits." All these examples point to the fact that the creation of a Universal Church which transcends denominations and is the ultimate goal all Christians must take into consideration the particularity of territories. Furthermore, as Churches become more globalized, they must simultaneously become more local and invent new ways of implanting themselves.

References

Adeboye, O. (2005). Transnational pentecostalism in Africa: The redeemed Christian Church of God, Nigeria. In L. Fouchard (Ed.), *Entreprises religieuses transnationales en Afrique de l'Ouest* (pp. 439–465). Paris: Karthala.
Adeboye, O. (2007). Arrowhead of Nigerian Pentecostalism: The Redeemed Christian Church of God, 1952–2005. *Pneuma, 29*, 24–58.
Adogame, A. (2003). Betwixt identity and security: African new religious movements and the politics of religious networking in Europe. *Nova Religion: The Journal of Alternative and Emergent Religions, 7*(2), 24–41.
Adogame, A. (2008). Up, up Jesus! Down, down Satan! African religiosity in the former Soviet Bloc – The Embassy of the Blessed Kingdom of God for All Nations. *Exchange, 37*(3), 310–336.
Adogame, A. (2010). Pentecostal and charismatic movements in a global perspective. In B. Turner (Ed.), *The new Blackwell companion to the sociology of religion* (pp. 498–518). Malden: Wiley-Blackwell.
Anderson, A. (2004). *An introduction to Pentecostalism*. Cambridge: Cambridge University Press.
Asamoah-Gyadu, K. (2005). An African Pentecostal on mission in Eastern Europe: The Church of the "Embassy of God" in Ukraine. *Pneuma, 27*(2), 297–321.
Barrett, A. L., Kurian, G. T., & Johnson, T. M. (2001). *World Christian encyclopedia* (The world by countries). Oxford: Oxford University Press.
Berger, P. (1999). *The desecularization of the world: Resurgent religion and world politics*. Grand Rapids: William B. Erdmans Publishing Company.
Beyer, P. (2007). Globalization and glocalization. In J. A. Beckford & N. J. Demerath (Eds.), *The SAGE handbook of sociology of religion* (pp. 98–118). London: Sage.
Bourdieu, P. (1979). *La distinction. Critique sociale du jugement*. Paris: Editions de Minuit.
Burgess, S. M. (2002). *The new international dictionary of Pentecostal and charismatic movements* (Rev. and expanded ed.). Grand Rapids: Zondervan Publication House.
Coleman, S. (2009). The Protestant ethic and the spirit of urbanism. In R. Pinxten & L. Dikomitis (Eds.), *When God comes to town: Religious traditions in urban contexts* (pp. 33–44). Oxford: Berghahn Books.
Coulmont, B. (2010). *Capitals and networks: A sociology of Paris' black churches*. Paper presented at the African Churches in Europe, Mediating Imaginations: Brussels.
Coulmont, B. (2013). Tenir le haut de l'affiche: analyse structurale des prétentions au Charisme. *Revue française de sociologie, 54*(3), 507–536.
Cox, H. (1966). *The secular city: Secularization and urbanization in theological perspective*. New-York: Macmillan.
Csordas, T. J. (2007). Modalities of transnational transcendence. *Anthropological Theory, 7*(3), 259–272.
Dawson, J. (1989). *Taking our cities for God*. Lake Mary: Creation House Publishers.
Dejean, F. (2010). *Les dimensions spatiales des Eglises évangéliques et pentecôtistes dans une commune de banlieue parisienne (Saint-Denis) et dans deux arrondissements montréalais (Rosemont et Villeray)*. Université de Paris Ouest-Nanterre-La Défense, Nanterre et Montréal.

Deleuze, G., & Guattari, F. (1980). *Mille Plateaux*. Paris: Editions de Minuit.
Ebaugh, H. R., & Saltzman Chafetz, J. (2002). *Religion across borders: Transnational immigrant networks*. Oxford: Alta Mira Press.
Ellingson, S. (2010). New research on megachurches. In B. Turner (Ed.), *The new Blackwell companion to the sociology of religion* (pp. 245–266). Malden: Wiley-Blackwell.
Fancello, S. (2006). *Les aventuriers du pentecôtisme ghanéen*. Paris: Ird/Karthala.
Fath, S. (2008). *Dieu XXL: la révolution des megachurches*. Paris: Autrement.
Featherstone, M., Lash, S., & Robertson, R. (1995). *Global modernities*. London: Thousand Oaks.
Fourchard, L., Mary, A., & Otayek, R. (2005). *Entreprises religieuses transnationales en Afrique de l'Ouest*. Paris: Kartala.
Giddens, A. (1990). *The consequences of modernity*. Stanford: Stanford University Press.
Hackett, R. (2003). Discourses of demonization in Africa and beyond. *Diogenes, 50*(3), 61–75.
Hawthorne, S., & Kendrick, G. (1996). *Prayer-walking: Praying on-site with insight*. Lake Mary: Creation House.
Howe, N. (2009). Secular iconoclasm: Purifying, privatizing, and profaning public faith. *Social and Cultural Geography, 10*(6), 639–656.
Inda, J. X., & Rosaldo, R. (2002). A world in motion. In J. X. Inda & R. Rosaldo (Eds.), *The anthropology of globalization: A reader* (pp. 2–34). New-York: Blackwell.
Jenkins, P. (2002). *The next Christendom: The coming of global Christianity*. Oxford: Oxford University Press.
Kearney, M. (1995). The local and the global: The anthropology of globalization and transnationalism. *Annual Review of Anthropology, 24*, 547–565.
Kepel, G. (1994). *The revenge of God: The resurgence of Islam, Christianity and Judaism in the modern world*. University Park: Pennsylvania State University.
Kim, H. (2011). Receiving mission: Reflection on reversed phenomena in mission by migrant workers from global churches to the western society. *Transformation: An International Journal of Holistic Mission Studies, 28*(1), 62–67.
Lévy, J. (2003). Capital spatial. In J. Lévy & M. Lussault (Eds.), *Dictionnaire de géographie et de l'espace des sociétés*. Paris: Belin.
Ley, D. (2008). The immigrant church as an urban service hub. *Urban Studies, 45*(10), 2057–2074.
Linthicum, R. C. (1991). *City of God, city of Satan: A biblical theology of the urban church*. Grand Rapids: Zondervan Publishing House.
Marshall-Fratani, R. (2007). L'explosion des Pentecôtismes. *Esprit, mars-avril*, 196–207.
Martin, D. (2002). *Pentecostalism, the world their Parish*. Malden: Blackwell Publishing.
Meyer, B. (2004). Christianity in Africa: From African independent to Pentecostal-charismatic churches. *Annual Review of Anthropology, 33*, 447–474.
Micklethwait, J., & Wooldridge, A. (2010). *God is back: How the global rise of faith is changing the world*. New-York: Penguin.
Mossière, G. (2006). "Former un citoyen utile au Québec et qui reçoit de ce pays". La rôle d'une communauté religieuse montréalaise dans la trajectoire migratoire de ses membres. *Les Cahiers du Gres, 6*(1), 45–61.
Moyet, X. (2005). Le néopentecôtisme au Ghana. Les Eglises Mountain of Fire and Miracles et Christ Embassy à Accra. In L. Fourchard, A. Mary, & R. Otayek (Eds.), *Entreprises religieuses transnationales en Afrique de l'Ouest* (pp. 467–487). Paris: Karthala.
Nutini, H. G. (2000). Native evangelism in central Mexico. *Ethnology, 39*(1), 39–54.
Obadia, L. (2010). Globalization and the sociology of religion. In B. Turner (Ed.), *The new Blackwell companion to the sociology of religion* (pp. 475–497). Malden: Wiley-Blackwell.
Ojo, M. A. (1988). Deeper Christian life ministry: A case study of the charismatic movements in western Nigeria. *Journal of Religion in Africa, 18*(2), 141–162.
Pew Forum. (2006). *Spirit and power: A 10-country survey of Pentecostals*. Washington, DC: Pew Research Center.
Robbins, J. (2004). The globalisation of Pentecostal charismatic Christianity. *Annual Review of Anthropology, 33*, 117–143.

Robertson, R. (1995). Glocalization: Time-space and homogeneity-heterogeneity. In R. Robertson, M. Featherstone, & S. Lash (Eds.), *Global modernities* (pp. 25–44). Thousand Oaks: Sage.

Sargeant, K. (2000). *Seeker churches: Promoting traditional religion in a nontraditional way*. Piscataway: Rutgers University Press.

Schiller, N. G., Basch, L., & Blanc-Szanton, C. (1992). Transnationalism: A new analytic framework for understanding migration. *Annals of the New York Academy of Sciences, 645*(1), 1–24.

Schiller, N. G., Çaglar, A., & Guldbrandsen, T. C. (2006). Beyond the ethnic lens: Locality, globality, and born-again incorporation. *American Ethnologist, 33*(4), 612–633.

Twitchell, J. B. (2004). *Branded nation: The marketing of Megachurch, College, Inc., and Museumword*. New York: Simon & Schuster.

Ukah, A. (2008). *A study of the Redeemed Christian Church of God in Nigeria*. Trenton: Africa World Press.

Van Dijk, R. (1997). From camp to encompassment: Discourses of transsubjectivity in the Ghanaian Pentecostal diaspora. *Journal of Religion in Africa, 27*(2), 135–159.

Virilio, P. (2000). *The information bomb*. London/New York: Verso.

Wagner, C. P. (1996). *Confronting the powers: How the New Testament Church experienced the power of strategic-level spiritual warfare*. Ventura: Regal Books.

Wilson, S., & Peterson, L. (2002). The anthropology of online communities. *Annual Review of Anthropology, 31*, 449–467.

Young-Gi, H. (2003). Encounter with modernity: The "Mcdonaldization" and "Charismatization" of Korean mega-churches. *International Review of Mission, 92*(365), 239–255.

Chapter 89
Towards a Catholic North America?

Anne Goujon, Éric Caron Malenfant, and Vegard Skirbekk

89.1 Introduction

Changes in a country's religious heterogeneity that follow demographic change may affect nations' culture, value orientations and policies (Castells 2004; Guibernau 2007). The religious composition of a society may also have demographic effects, including behavior and family formation decisions (Goujon et al. 2007a; Lehrer 1996). North America has witnessed significant changes to its religious make-up in recent years. The Protestants in both the US and Canada have lost their absolute population majority (Figs. 89.1 and 89.2). In the US the Protestant share is estimated to have decreased from above 60 % in the 1980s to around 47 % by 2003, although it remained the largest religious group. The US Catholic proportion

The views and opinions expressed in this paper do not necessarily reflect those of Statistics Canada

A. Goujon (✉)
Wittgenstein Centre for Demography and Global Human Capital (IIASA, VID/ŐAW, WU), International Institute for Applied Systems Analysis (IIASA), Schlossplatz 1, 2361 Laxenburg, Austria

Vienna Institute of Demography/Austrian Academy of Sciences, Wohllebengasse 12-14, 1040 Vienna, Austria
e-mail: goujon@iiasa.ac.at

É. Caron Malenfant
Demography Division, Statistics Canada, 150 Tunney's Pasture Driveway, Ottawa K1A 0T6, Canada
e-mail: eric.caronmalenfant@statcan.gc.ca

V. Skirbekk
Wittgenstein Centre for Demography and Global Human Capital (IIASA, VID/ŐAW, WU), International Institute for Applied Systems Analysis (IIASA), Schlossplatz 1, 2361 Laxenburg, Austria
e-mail: skirbekk@iiasa.ac.at

S.D. Brunn (ed.), *The Changing World Religion Map*,
DOI 10.1007/978-94-017-9376-6_89

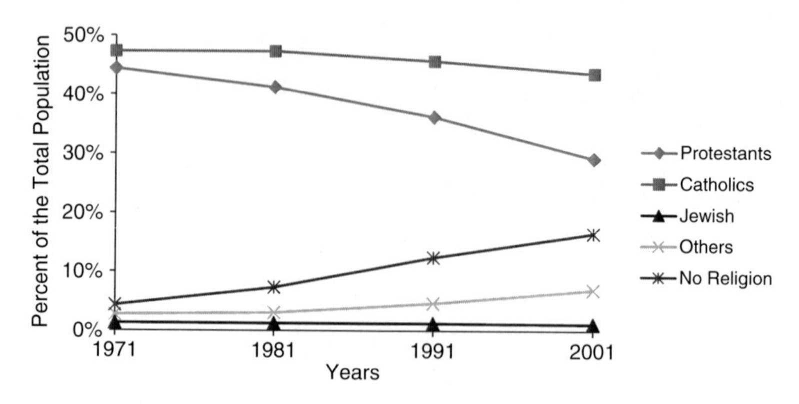

Fig. 89.1 Religious composition, Canada, 1971–2001 (Data from census, various years)

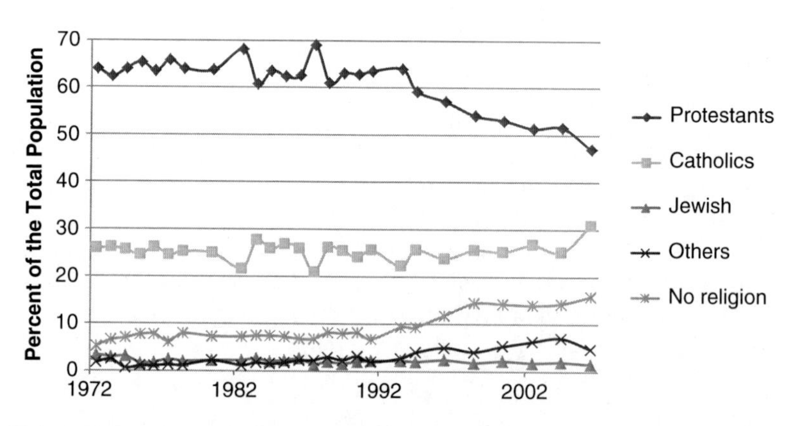

Fig. 89.2 Religious composition, United States, 1972–2006 (Data from US GSS, various years)

reached a level of 28 % in 2003, following growth among Hispanic-Catholics and decline among non-Hispanic Catholics (Skirbekk et al. 2010).

In Canada, Protestantism was the largest religion in every decennial census from 1891 to 1961. This 70-year period saw the Protestant share fall from 57 to 49 % while the Catholic proportion rose from 42 to 47 % (Statistics Canada 1993). Censuses from 1971 to 2001 have consistently shown that Catholicism is the largest religious entity in Canada, and in 2001 the Protestants were only 29 % while the Catholics were 44 % of the population (Statistics Canada 2003). In both countries, the share of the Jewish population remained fairly stable while that of the population who declared having no religion or other religious denominations increased.

In the present study we investigate the future of the religious composition in both the US and Canada – jointly. Past projections that consider religion in either Canada or the US have been carried out (Bélanger and Caron Malenfant 2005; Caron Malenfant et al. 2010; Skirbekk et al. 2010). However, these projections have had a somewhat different focus (ethno-religious change for US, ethnocultural diversity change for Canada) and a too short projection horizons (2031 for Canada and 2043 for the US) to capture longer-term effects. Earlier projections also lacked explicit consideration of the impact of potential changes in interreligious union and their possible consequences on the religious composition of the population.

We take these issues into account, use religious categories that are comparable across the two countries, and extend the time horizon to the 2060s. The longer projection period allows us to better understand how religion changes due to demographic forces, the level and composition of migration, fertility differentials, conversions and intergenerational religious transmissions, in addition to population inertia.

The joint focus on both the US and Canada allows one to better understand the commonalities and differences between these two nations which are tightly knit in terms of geography, politics, economics and culture. We also include what the case would be if the US would have Canadian parameters and vice versa to get a better understanding of the different driving forces behind the changes occurring in both countries in terms of religious composition of their population.

This study takes advantage of the richness of the Canadian long-form 2001 Census, which completion was compulsory to the population and covered 20 % of all households, as well as the Ethnic Diversity Survey of 2002 (sample size of the EDS is N=42,476), which allows us to parameterize the conversion rates. For the United States, we use pooled data from the General Social Survey from 2000 to 2006 (sample size of the GSS 2000–2006 is N=12,674).

89.2 Demographic Drivers of Religious Change

Religious mobility, fertility, migration and, to a lesser extent, mortality, are all demographic drivers of religious change. Among them, the change in the number of persons in different religious affiliations through conversion, or religious mobility, is a key one. In the US, more than one in five adults has a different religion than what they had in childhood. In Canada, similarly, about one in five persons declared a different religious denomination than that of their mother when they were aged less than 15 years old, according to the EDS. The substantial conversion rates in both countries are in line with findings from other Catholic and Protestant dominated countries which have large conversion rates (Goujon et al. 2007a; Lehrer 1996, 2004). The most common "switch" being from various religions to the "no religion" group, this phenomenon contributed over recent decades to an increase in the share of the group who report having no religion.

Religious mobility patterns are closely related to interreligious unions, and children from interreligious unions tend to adopt the mother's religion more often than the father's religion. In the US, the 1991 and 1998 GSS survey waves included a question on respondent's religion and the religion of the mother and father (N=2,289 non missing answers). These data suggest that among those with two Catholic parents, 77 % kept their parents' religion, while 89 % of those with Protestant parents kept their parents' religion. Among the unions between Catholic and Protestant, if the mother was Catholic and the father was Protestant, 43 % became Catholic and 36 % Protestant. If the mother was Protestant and the father Catholic, 59 % became Protestant and 16 % Catholic.

In Canada, 2002 EDS data on the respondent's religion and that of his/her parents when the respondent was less than 15 years old also show high retention rates when both parents share a religion, but higher for Catholics (91 %) than for Protestants (82 %). In the case of mixed unions between a Catholic and a Protestant, the mother was more influential if she was Catholic, with 52 % of the children being Catholic and 30 % Protestant. If the mother was Protestant, the percentage that becomes Catholic and Protestant was not significantly different, at about 40 % for both.

Fertility and migration patterns, the two other main demographic drivers of religious changes, have different effects on the religious demography of Canada and the US. In the US, the Catholic share increases due to fertility differentials, as the overall TFR in 2003 was estimated to be 2.1 children per woman for Protestants, while for Catholics it was 2.3 children per woman (Skirbekk et al. 2010). Catholics had significantly higher fertility than Protestants early in the twentieth century, but this waned in the second half of the twentieth century (Jones and Westoff 1979; Sander 1992). However, the large influx into the US of Hispanic Catholics with higher fertility has led to a rise in the relative fertility of Catholics in recent years and again divergence between Protestant and Catholic fertility rates. In Canada, however, these demographic forces have the opposite effect, where in 2001 fertility differentials benefit Protestants: Catholics had a TFR of 1.5 children per woman, while Protestants had a higher TFR with 1.65 children per woman (Caron Malenfant and Bélanger 2006). Catholics used to have larger families than Protestants in Canada, but the gap decreased over the course of the twentieth century (Krotki and Lapierre 1968). This relation has reversed in recent years as today, Protestants have relatively larger families.

Catholics increase their proportion through immigration to the US, as the share of Catholics among migrants is about 45 % in 2003–2006, while it is only 28 % in 2003 among the US resident population. In Canada, however, immigration reduces the Catholic proportion, since among migrants admitted between 1996 and 2001 only 20 % were Catholic (while 44 % of the total population in Canada were Catholics in 2001). Also, in both countries, the share of Protestants among the migrants (10 % in Canada and 8 % in the US) is far lower than their share in the overall population, thus contributing to decrease in the share of this denomination.

The current composition of immigration leads to an increase in the share of religious minorities (Muslims, Other religions, Hindus and Buddhists) for both countries, a decrease in the share of the Jewish population in the US and no effect on the share of the Jewish population in Canada. Muslims also gain through a higher fertility in both countries.

The share without a religious affiliation is also affected in different ways by migration in the two countries. It rises due to immigration in Canada (23 % of the immigrants admitted between 1996 and 2001 had no affiliation[1] compared to 17 % of the resident population), while this group is negatively affected by immigration in the US where the share without affiliation is less in the immigrant population than in the native population, 8 % versus 17 %. Their low fertility has a depressing effect on the population share in both countries: the TFR of the unaffiliated is 1.7 children in

[1] Immigration from China explains in large part the greater share reporting no religion within "recent" immigration to Canada.

the US (as opposed to an overall TFR of 2.1) and 1.4 children in Canada (as opposed to an overall TFR slightly above 1.5). The low fertility among those with no religion in both US and Canada is in line with many other studies showing that low religiosity corresponds with low levels of fertility (Goujon et al. 2007b; McQuillan 2004).

The evidence of a relationship between religion and mortality is inconclusive, and while some studies find religiosity to be associated with better health outcomes (for a review, see Hummer et al. 2004), on the other hand, the more religious tend to have lower education levels (Glaeser and Sacerdote 2008; Inglehart and Baker 2000) which could suggest they would have worse health outcomes (Groot and Maassen van den Brink 2007). Lerch et al. (2010) show that in the case of Switzerland, one of the few countries where religion is registered at the time of death as well as in the census, differentials in survival ratios by religion tend to be less important than other factors affecting mortality such as sex, education, and marital status.

89.3 Projection Methodology, Assumptions and Scenarios

In order to project the population of Canada and the US by religion, we estimate the initial population by age, sex and religious denomination. We also estimate age- and religion-specific fertility rates, age-, sex- and religion-specific net-migration numbers. In addition, we focus on the impact of mixed unions and religious conversion rates.[2] We employ both expected and alternate scenarios based on varying fertility, conversion and net-migration assumptions. The projections are conducted using the PDE population projection software, a tool developed at IIASA[3] and used for multistate population projections – in case of several states that interact with each other.

89.3.1 Base Population

The US estimates rely on General Social Survey (GSS) data and we choose 2003 as a starting year, due to its proximity to the 2001 Canadian census and pool GSS survey waves (2000–2006) together in order to increase sample size for the base population (N = 12,674). We compared the GSS to alternative surveys that have been conducted since the year 2000 (from Baylor, ARIS and PEW) and found that although the estimates differ, the various surveys present a broadly consistent picture, with about half the population Protestant, a quarter Catholic, and about one in eight without religion, with a scattering of smaller groups (Jewish, Hindu, Muslim, and other religious groups) (see Skirbekk et al. 2010).

---

[2] The base-year data used for the projections and analysis were the most recent at the time of submission (June 8, 2012).

[3] It can be downloaded from www.iiasa.ac.at/Research/POP/pub/software/pde/pdesetup.zip (September 13, 2012).

For Canada, the base population consists of the 2001 long-form census[4] of the population (20 % sample or about six million persons, excluding institutions), adjusted for census net under-coverage.

We use the following seven religious categories for both countries: Protestant, Catholic, Muslim, Hindu and Buddhist, Jewish, other religions, and no religion.

89.3.2 Base-Year Mortality

For mortality for the United States, we assume a single value for each age group and sex following the estimates of the National Center for Health Statistics (NCHS), available for the base year in Kung et al. (2008) and the long-range projections for life expectancy of the U.S. Census Bureau available in Hollmann et al. (2000). For Canada we follow Statistics Canada's estimates and projections by age and sex developed in their most recent population projections (Statistics Canada 2010).

89.3.3 Base-Year Net-Migration

Annual US net-migration figures come from the Population Estimates Program of the U.S. Census Bureau International Database (2010) and from their official projections, and for Canada from the Population estimates program of Statistics Canada as well as their official population projections (Statistics Canada 2010). The data for Canada refers to migration in- and out-flows whereas the US data is already aggregated to net-migration.

The religious composition of immigration was estimated as follows. For the US, due the lack of data on the religion of the immigrants, a number of assumptions were unavoidable. We retrieved the number of persons obtaining legal permanent resident status by country of birth between 2003 and 2007 (U.S. Department of Homeland Security 2009) and based their religion on the shares of population by religion in their home country using the best estimates from the World Religion Database (WRD), assuming that immigrants are randomly selected in terms of religion in their country of origin. For the US, The New Immigrant Survey (NIS) (e.g. Jasso and Rosenzweig 2006) produces fairly similar proportions of most religious groups and sensitivity analysis shows that using NIS estimates would not affect the main results of our projections (Skirbekk et al. 2010). The net numbers of migrants were further distributed by age and sex using the shares used in the 2009 national population projections, in the constant net international migration scenario which do not change over the projection period (2010–2050) (see www.census.gov/population/www/projections/2009projections.html).

---

[4]The 2001 Census is to date the most recent and comprehensive dataset on religious affiliation in Canada as the 2006 Canadian census did not include a question on religious denomination.

For Canada we retrieved from the 2001 Census the religious distribution of migrants who arrived between 1996 and 2001, by country of birth, and applied the resulting distribution to the migrants who moved to Canada between 2001 and 2006 by country of birth (according to the 2006 census). The total numbers of immigrants by religion were then distributed by age and sex, according to both the census and the Population estimates program of Statistics Canada (2001–2006). Emigrants by age, sex and religion were subtracted from the corresponding immigration figures to obtain the net international migrations religious composition assumption. In the absence of data on religious denomination of emigrants from Canada, they were assumed to be selected randomly among the religious groups.

The resulting distributions of net migrants for both countries are shown in Figs. 89.3 and 89.4.

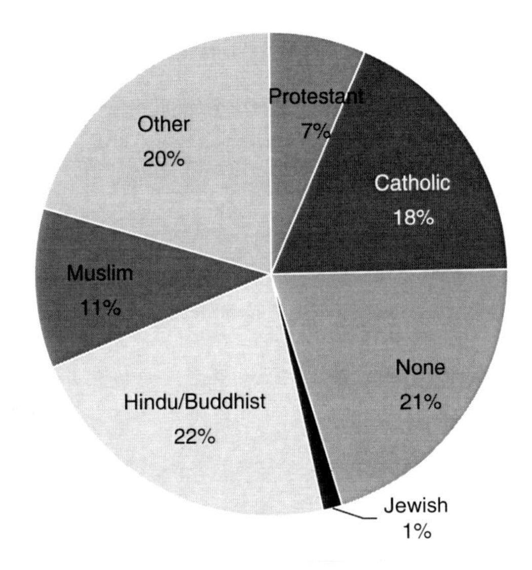

Fig. 89.3 Religious distribution of net-migration population in 2001, Canada (Source: Authors' calculations)

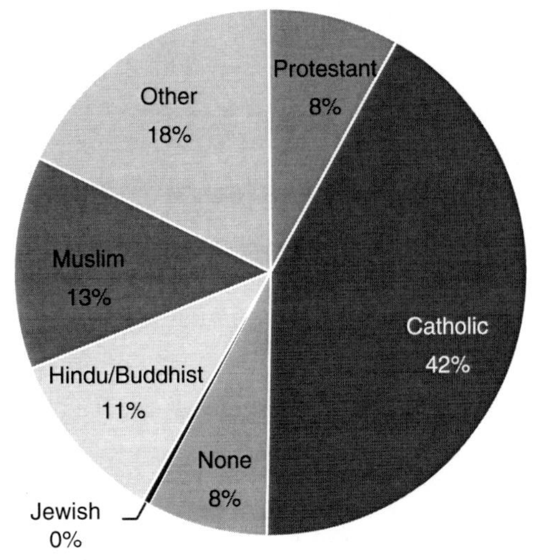

Fig. 89.4 Religious distribution of net-migration population in 2003, United States (Source: Authors' calculations)

89.3.4 Base-Year Fertility

Fertility differences by religion in the USA were estimated from GSS data on children ever born to women aged 40–59[5] for the period 2000–2006, and the differentials were then proportionally adjusted and applied to the TFR reported for 2003–2007 by the U.S. Census Bureau. For Canada, the age-specific fertility rates by religious group could be derived from census data, by applying to this database the own-children method to estimate fertility.[6] The age-specific fertility rates by religion were adjusted so that the rates for Canada overall be the same as those obtained by Vital Statistics in 2001–2007 and the same as the assumed rates (central assumption of the most recent Statistics Canada's official projections) for the following years. Fertility data for both the US and Canada is given in Table 89.1.

89.3.5 Conversion, Intermarriage and Parent-Child Religious Transmissions

To project religious mobility, we use conversion matrices computed from the Ethnic Diversity Survey for Canada and the General Social Survey for the US. The conversion matrices include some of the effect of inter-religious unions through conversion trends (some convert for marriage reasons, moreover parental inter-religious unions patterns can affect the degree and the direction of own religious switching). The Canadian data results from a comparison between the religion of the respondent and the religion of the respondent's mother when the respondent was aged less than 15 years old. In the US, the data used so far compares the religion of the respondent at two points in time: the day of the interview and during childhood.[7] Given that the religious mobility is higher among the population whose parents were in a mixed union and that the proportion of mixed unions is likely to increase in the future, we decided to establish a scenario that assumes an increasing religious mobility. To do so, we computed the probability to change religion separately for respondents whose parents were in a mixed union and for those having parents with the same religion. Those probabilities are used to create new matrices in which the weight of the patterns for persons with parents in a mixed union is double than in the baseline matrices.

The estimates for both countries concern net religious mobility (as shown in Tables 89.2 and 89.3), and were computed by sex. The data in Canada allowed for sex specific mobility rates. In the US, the sex differentials were based on the gender difference in retention across all religions, showing that men were 6 % more likely to switch out of their own religion than women (Skirbekk et al. 2010). We assume

---

[5] For Muslims, we based the differential on the 35–59 population to increase sample size.

[6] For more details on the method, please consult Grabill and Cho (1965), as well as Bélanger and Gilbert (2003) or Caron Malenfant and Bélanger (2006) for examples of application to Canadian data.

[7] The exact question in the GSS is "In what religion were you raised?"

Table 89.1 Estimated period total-fertility and differentials in Canada and the United States

Religion	Canada (2001–2005)	United States (2003–2007)
Protestant	1.63	2.12
Catholic	1.49	2.38
Jewish	1.83	1.43
Hindu-Buddhist	1.66	1.73
Muslim	2.35	2.84
Other religion	1.68	1.64
No religion	1.39	1.66
Total	1.54	2.08

Source: Authors' calculation based on GSS, U.S. Census Bureau, and the 2001 Census for Canada

Table 89.2 Estimates of religious mobility in Canada

	To						
From	Protestant (%)	Catholic (%)	Hindu-Buddhist (%)	Jewish (%)	Muslim (%)	Other (%)	No religion (%)
Protestant	76	6	0	0	0	3	15
Catholic	4	86	0	0	0	2	7
Hindu-Buddhist	6	8	59	0	0	6	22
Jewish	0	0	0	90	0	0	10
Muslim	0	0	0	0	93	0	7
Other	7	4	0	0	0	77	12
No religion	11	7	1	0	0	5	75

Data source: EDS, 2002

Note: In this table, cells with a coefficient of variation higher than 33.3 % or with sample sizes of 10 or less were suppressed. The table still includes cells with high coefficient of variations, to be interpreted with caution (cells with numbers less that 2 % and cells from Jewish to no religion and from Muslim to non religion have coefficients of variations higher that 16.6 %)

Table 89.3 Estimates of religious mobility in the US

	To						
From	Protestant (%)	Catholic (%)	Hindu-Buddhist (%)	Jewish (%)	Muslim (%)	Other (%)	No religion (%)
Protestant	81	3	0	0	0	3	12
Catholic	11	75	0	0	0	4	10
Hindu-Buddhist	12	6	55	1	2	3	20
Jewish	3	1	1	81	0	1	14
Muslim	3	0	5	0	71	7	13
Other	29	4	1	0	0 %	47	19
No religion	32	6	1	2	0	4	56

Data source: GSS, 2000–2006

that changes tend to be concentrated among the 20–24 year old, which is based on evidence from a number of countries that these are the years when religious conversions/secularization is most likely to occur (Crockett and Voas 2006; Hoge 1981; Iannaccone 1992).

89.3.6 Scenarios

We develop a set of different scenarios consistent but overall specific to the two countries where we look at the impact of different migration levels, fertility differentials and whether conversion trends continue or not. Each scenario differs from the "recent situation" scenario by only one component, in order to assess the sensitivity of results to each of them. The scenarios are presented in Table 89.4. By the

Table 89.4 Scenario matrix

		Net-migration				
Fertility	Religious mobility	Recent situation	Canadian net-migration composition	American net-migration composition	No net-migration	Double the net-migration
Recent situation	Recent situation	**H0** Recent situation	**H1** Canadian net-migration composition	**H2** American net-migration composition	**H3** No net-migration	**H4** Double the net-migration
	Canadian religious mobility	**H5** Canadian religious mobility				
	American religious mobility	**H6** American religious mobility				
	No religious mobility	**H7** No religious mobility				
	Mixed unions	**H8** Mixed unions				
Early convergence	Recent situation	**H9** Early fertility convergence				
Late convergence	Recent situation	**H10** Late fertility convergence				

Source: Authors

sheer difference in population size of Canada and the United States, it is clear that the results are strongly influenced by the United States whose population is almost ten times larger than the Canadian population (318 million vs. 34 million). That is why we included some contra-factual simulations of US and Canadian demographic variables on religious mobility and migration composition.

The scenarios envisaged are the following:

Fertility

1. Recent situation: Fertility differentials by religious denomination remain constant at levels shown in Table 89.1.[8] The overall fertility will change through changes to the different weights of the several religious groups in the total population. In Canada, the TFR will reach 1.81 by the end of the projection period and 2.10 in the US – compared to 1.54 and 2.08 in the starting period.
2. Early convergence: The differentials existing in the fertility of women by religious denomination disappear by 2013–2017 in the US and by 2011–2015 in Canada to reach 2.1 in the US and 1.7 in Canada and stay at that level until the end of the projection period.
3. Late Convergence: The fertility differentials remain constant as in the recent situation scenario and only disappear in 2033–2037 in the US and in 2031–2035 in Canada.

Religious Mobility

1. Recent situation: The transition rates between the seven religious groups shown in Tables 89.2 and 89.3 are kept constant until the end of the projection period.
2. Canadian religious mobility: The Canadian religious mobility rates shown in Table 89.2 are applied to the US.
3. American religious mobility: The US religious mobility rates shown in Table 89.3 are applied to Canada.
4. No religious mobility: We assume that from 2001 for Canada and 2003 for the US, everybody stays in the religion they are born, inherited from their mother.
5. Mixed unions: Conversion rates are increased by reweighting the religious mobility matrices to double the importance of conversion rates of children from mixed unions, based on the Canadian case. For the US, the increase in religious mobility estimated for Canada is applied to the US matrices.

Net Migration

1. Recent situation: The total migration follows the projections of the respective statistical agencies. In the US, it means that the net-migration is increasing from 874,000 in 2003 to 2,055,000 in 2050, and further to 2,300,000 in 2063, following the increase observed in the last period. In Canada, following population estimates

[8] More precisely, and as mentioned previously, for Canada the rates presented in Table 89.1 are adjusted to meet by 2011–2015 the official central assumption of Statistics Canada (1.7 children per woman at the national level) and are maintained constant after. The adjustment does not affect the differentials.

and numbers obtained from the central assumption of the official Statistics Canada's projections (Statistics Canada 2010, scenario M1), it would increase from 215,000 in 2001 to 347,000 in 2061. The composition of immigration presented in Fig. 89.3 and 89.4 is kept constant until the end of the projection.

2. Canadian net-migration composition: We apply the religious composition of Canada net-migration by country of birth (based on the 2001 census) to the US composition.
3. American net-migration composition: We apply the religious composition of the US net-migration by country of birth (based on the WRD database) to the Canadian composition.
4. No net-migration: We assume that from 2011 in Canada and 2013 in the US, net-migration would be zero.
5. Double the net-migration: We assume that from 2011 in Canada and 2013 in the US, net-migration would double the value of the recent situation migration variant.

89.4 Results

The results[9] are first analyzed at the level of North America. Our business as usual scenario – recent situation – shows that North America would not be Catholic by the middle of the century but it would be close at the end of the projection period in 2062 where the Catholics would make up to 32 % of the population, compared to 34 % for the Protestants (Fig. 89.5). However most of the changes in the relative share of the Catholics and Protestants in North America is due to the decline in the share of the Protestant population rather than in the increase of the share of the Catholic population which remains quite stable: from 30 % in 2002 to 32 % in 2062 whereas the share of Protestants decreases dramatically from 45 to 34 % during the same period of time.

The first half of the twenty-first century would see a rapid increase of some religious minorities such as the Muslims – from 0.6 % in 2002 to 5 % in 2062 – and the Hindu Buddhists – from 1.2 % in 2002 to 3 % in 2062 – due mostly to positive net-migration and to the relative higher level of fertility of Muslims in both countries and also of Hindus and Buddhists in Canada.

The share of the secular group would slightly increase until 2012 to 17 % and then decline slightly to 16.5 % in 2062, victim of a disadvantageous representation in the migration in the US and moreover of the low fertility of that group that is actually causing his importance to decline over time as was shown by Skirbekk et al. in 2010 for the United States only, in spite of important transition from the dominant religion to that group – the secularization trend.

[9] Since projections are carried out for the period 2001–2061 for Canada, and 2003–2063 for the US, we assumed that the results for North America are for the period 2002–2062, as an average of the two periods.

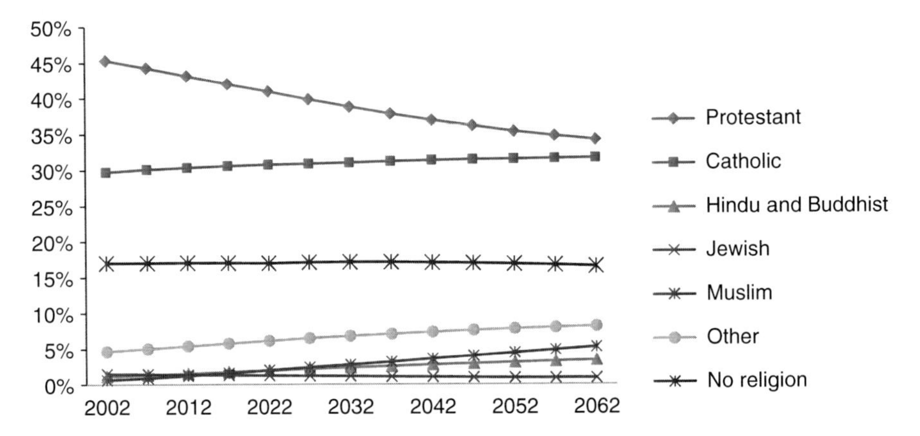

Fig. 89.5 Projected 2002–2062 North America population, by religious category, using the "Recent Situation" scenario (Source: Authors' calculations)

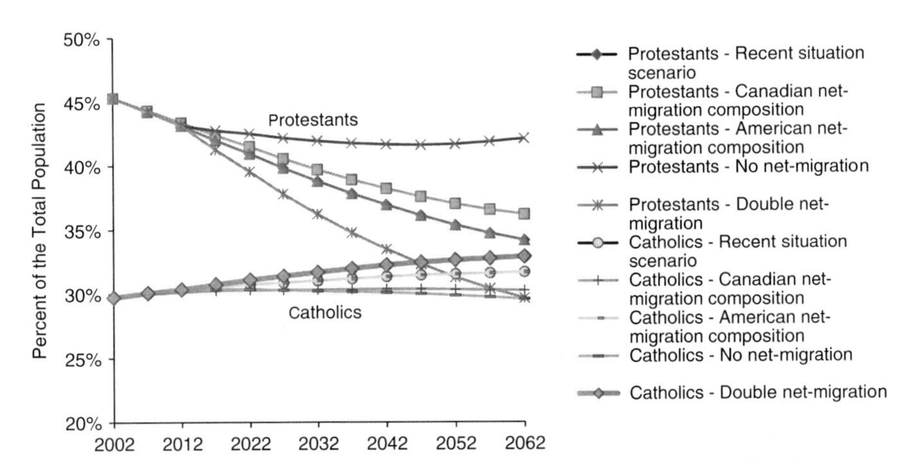

Fig. 89.6 Projected 2002–2062 Catholic and Protestant share of the population of North America with various net migration scenarios (Source: Authors' calculations)

If we compare the different forces at play in the projections, we can see that future assumptions of migration and religious mobility (Figs. 89.6, 89.7, and 89.8) have a larger bearing on the future North American religious landscape compared to fertility. However the relative importance of these forces is different if we compare the Protestants whose share is more sensitive to migration, with the Catholics whose representation would be more influenced by changes in the intensity of the religious mobility. In the case of a zero-net-migration from 2012 onwards, the share of the protestant population would stabilize at 42 % of the total population throughout the projection period (from 45 % in 2002) while this would mean a constant share of the Catholic population at 30 %. If migration doubles from 2012 onwards,

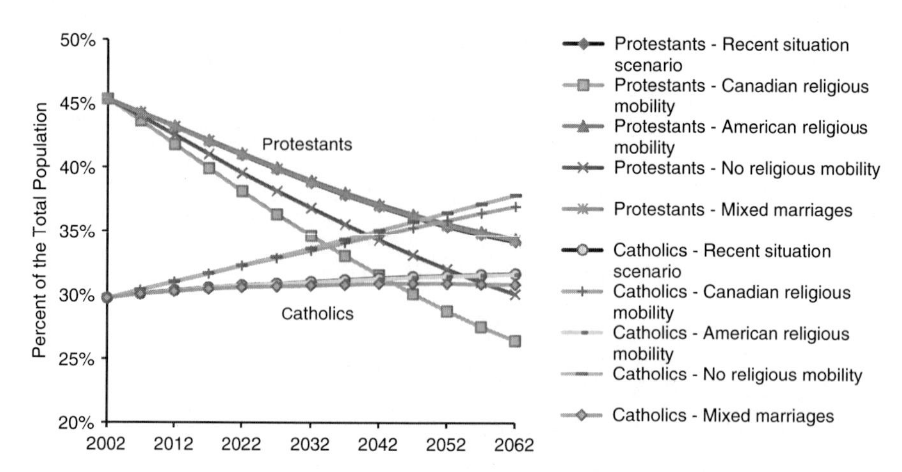

Fig. 89.7 Projected 2002–2062 Catholic and Protestant share of the population of North America with various religious mobility scenarios (Source: Authors' calculations)

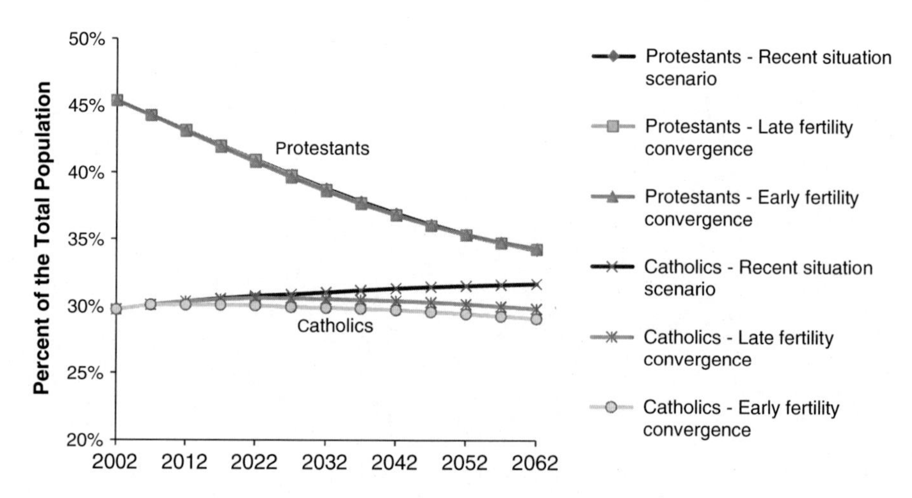

Fig. 89.8 Projected 2002–2062 Catholic and Protestant share of the population of North America with various fertility scenarios (Source: Authors' calculations)

then the proportion of Protestants would go down to 30 % and that of the Catholics would increase slightly to 33 %. It is worth noting that the absence of religious mobility from 2012 onwards would have the same effect as a doubling of migration for the share of the Protestant population, but even more it would mean for the Catholics a substantial increase to 38 % of the population where it would become the religion with the highest membership in North America. Surprisingly a late or early convergence of all fertility to the same level for women belonging to different religious denominations would not affect the share of the Protestant group in the total population – and that of the Catholics group very slightly – and this is due to the fact that those two groups have in fact very similar fertility in both countries and the realignment would not have such a substantial effect for the two dominant religious group as it would for the other denominations as we will see later.

Switching the net-migration religious composition of Canada and the US does not have an important effect on the share of the religious denominations, implying that those shares are very close, which is a comforting result that in the absence of data on the religion of the migrants, it actually makes overall sense to use the religious proportion of the country of birth of those migrants, and this is as well the case for minorities that are otherwise very sensitive to migration as can be seen from Table 89.5. On the other hand, the two countries are having different experiences in terms of religious mobility and the Protestants would lose tremendously if North America was to adopt the religious mobility of Canada and its share would be down to 26 % in 2062 (compared to 34 % in the recent situation scenario – from 45 % in 2002) and the Catholics up to 37 % (compared to 32 % in the recent situation scenario – from 30 % in 2002).

It is interesting to see that the population with no religion is not very sensitive to all of the scenarios and that its share stays more or less constant through all the years (see Table 89.5). Most interestingly and significant, it is the early fertility convergence scenario rather than any other scenario on religious mobility that would lead to the highest share of the population with no religion to 18.5 % in the total population, which means that this group is really victim of its fertility behavior since it has very low fertility levels in both countries, way below replacement level.

Another religious denomination that in all scenarios will have a reduced share in the total population is the Jewish community (see Table 89.5), which is estimated to shrink from 1.5 % today to 0.7–1.1 % in 2062. The share of the Muslim population is projected to increase from 0.6 % to between 2 and 7 % of the total population in 2062 depending on the scenario. The projections show that the Canadian immigration has a lower Muslim share than the American one and that American Muslims tend to lose more than Canadian Muslims through religious mobility. As expected, more than fertility, migration is one of the main engines of Muslim growth in Northern America which would not reach 2 % of the total population by 2062 in the absence of migration, but would reach more than 7 % in the case of a doubling of migration.

In 2062, the third religious group in North America after the Protestants and the Catholics would be a group of other religions whose share could increase to 10 % in the case of a Canadian religious mobility confirming this trend that Canadians and also U.S. citizens are moving away from traditional religions (Protestantism and Roman Catholicism) to other religious denominations. Actually the scenario of no religious mobility would have the same effect as the business as usual scenario (recent situation).

The scenario implementing an intensification of mixed unions is not influencing greatly the share of the different religious groups. At most the difference between the recent situation and the mixed unions scenario is 1.5 percentage point and this in the case of the category without religion.

Table 89.5 Share of the different religious denominations in the total population, North America, 2002 and 2062, 11 scenarios

Scenario	Protestant	Catholic	Muslim	Hindu-Buddhist	Jewish	Other	No religion
2002	45.3	29.7	0.6	1.1	1.5	4.7	17.0
2062							
Recent situation	34.2	31.7	5.2	3.4	0.9	8.1	16.5
Canadian immigration composition	36.2	30.3	4.3	2.6	1.1	7.5	18.1
American immigration composition	34.2	31.7	5.2	3.4	0.9	8.1	16.5
Canadian religious mobility	26.4	37.0	5.7	2.9	0.7	9.6	17.6
American religious mobility	34.4	31.6	5.2	3.5	0.9	8.0	16.3
No religious mobility	30.1	37.9	5.9	3.9	0.7	8.2	13.3
Late fertility convergence	34.3	29.8	4.6	3.6	1.0	8.7	18.0
No migration	42.1	29.6	1.9	1.4	1.0	5.8	18.2
Early fertility convergence	34.3	29.1	4.4	3.6	1.1	8.9	18.5
Double migration	29.7	32.9	7.2	4.5	0.8	9.5	15.5
Mixed unions	33.3	31.0	5.2	3.3	0.9	8.4	18.0

Source: Authors' calculations

Since Canada is relatively small relative to the US in the projection results for North America, this section is dedicated to specific results of the projections for Canada. According to the recent situation scenario (Fig. 89.9) both the Protestant and the Catholic denominations would be decreasing by more than 15 percentage points between 2001 and 2061, and the no religion group would increase its share to become the second group, after the Catholics, with more than 20 % of the total population in 2061. Canada would thus be very different in terms of religious composition than what it was one century ago, when the country was predominantly Protestant and when more than 90 % of Canadians were either Protestants or Catholics. This change in the religious landscape of Canada would be due mostly – but not only -to religious mobility as well as immigration in which both Catholics and Protestants are underrepresented and almost all other groups (including no religion) overrepresented. Indeed, the only scenarios in which Protestants maintain an advantage on the no religion group by 2061 are those alternating different religious mobility and immigration paths: American religious mobility, No religious mobility

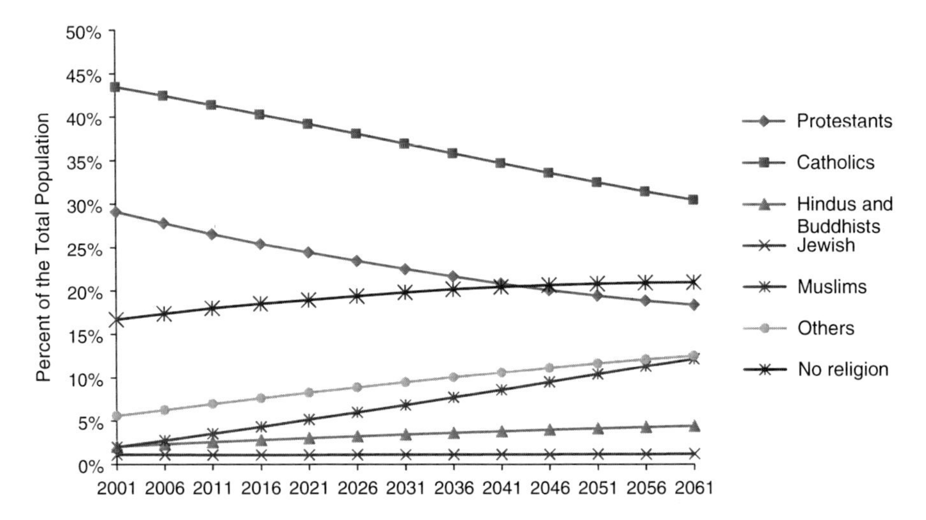

Fig. 89.9 Projected 2001–2061 Canadian population, by religious category, using the "Recent Situation" scenario (Source: Authors' calculations)

and No migration (data not shown).[10] But even according to these scenarios, the trends for the Protestant, Catholic and the no religion groups would remain in the same direction.

The US and Canada are overall on a different path of religious diversification: whereas in the US, religious denominations would be still dominated by the Catholics and Protestants, Canada would see the share of those with other religions (Muslims, Hindus and Buddhists, and others) triple according to most scenarios until the 2060s, from around 10 % to around 30 %, while the same share would increase from 6 to 15 % in the US, in almost all scenarios. In most scenarios, the share of Hindu-Buddhists would reach between 4 and 6 %, while that of Muslim would reach between 10 and 13 % in Canada by 2061. Very sensitive to international migrations, these shares could be higher or lower if immigration were to increase or decrease significantly. This is especially true for the Muslim group, whose share in the Canadian population could be 6 % without immigration but 15 % if immigration doubles.

The results obtained for Canada can be compared, for validation purposes, to those prepared by Statistics Canada (Caron Malenfant et al. 2010) at mid-projection term, in 2031, the assumptions of the recent situation scenario being overall very close to that used for the reference scenario of the national statistical agency. Even if the two data series differ in their methods and in the fact that, as opposed to that of Statistics Canada, the current projections exclude the institutional population and do not assume any increase in the non-permanent resident population, they are very

[10]Another factor is that Protestants and, to a lesser extent Catholics, are also older on average than the other groups (Statistics Canada 2003) which will lead to a high number of deaths when the baby boomers will reach old ages, in the first half of the twenty-first century.

consistent with regards to the projected religious composition of the Canadian population. Both project a Catholic population that would represent about 36 % of the Canadian population in 2031, and shares of Protestants and no religion that would be very close each other with about 20 %. They are also consistent with regards to the other religious groups, with almost identical proportions of Muslims (7 %), of Hindus and Buddhists (between 3 and 4 %), persons of Jewish denomination (1 %) and other religions (10 %).

89.5 Conclusion and Discussion

What could be the religious landscape of North America in 50 years from now? Is North America likely to become predominantly Catholic within the century as it happened to Canada between the 1961 and 1971 censuses? Our projections of the religious composition of Canada and the United States reveal that for most combinations of hypotheses regarding future changes of fertility differentials, conversion and secularization rates, and migration, Catholicism would not be the main religious group in North America by mid-twenty-first century, but would be close to it as shown in Fig. 89.10. Both Canada and the US would continue to be dominated by Christians, but the relative weight of the Protestants and Catholics is likely to change and religious minorities would grow in size. For Canada, we find in all scenarios that Catholicism is likely to keep its dominant position well into the second half of the twenty-first century. However, although it is likely to remain the largest religion, its share of the population is likely to decline over the coming decades, particularly if immigration remains high and secularization continues.

It is important to mention that we only measure self-declared denomination, not the implication of having a given religious denomination, or the substantial differences in religious content and practice of each religion in the two countries. Religious affiliation could for some be a matter of self-identification with a cultural group, which could be the case among many French Canadians or Hispanic Catholic Americans. Formal affiliation does not necessarily imply that individuals will follow religious recommendations in terms of fertility, for instance the policies of the pope or leaders within various religious. We do not know whether adherence to a certain religion is likely to reflect a conservative political view or attitude. We also do not consider the large differences in fertility within religions (e.g. Protestant fundamentalists have higher fertility than liberal Protestants, Orthodox Jewish have more children than moderate Jewish), which also will affect fertility levels and conversion trends in future decades.

The content of Catholicism differs greatly between the two nations, and while US Catholic growth is driven by Mexican migration, Mexican migration is not significant in Canada. Our range of scenarios stresses that the religious switch is driven by demographic components and policies affecting these – if there should be strong reductions in migration to the US, there would be a lower share of Catholics in North America by 2060.

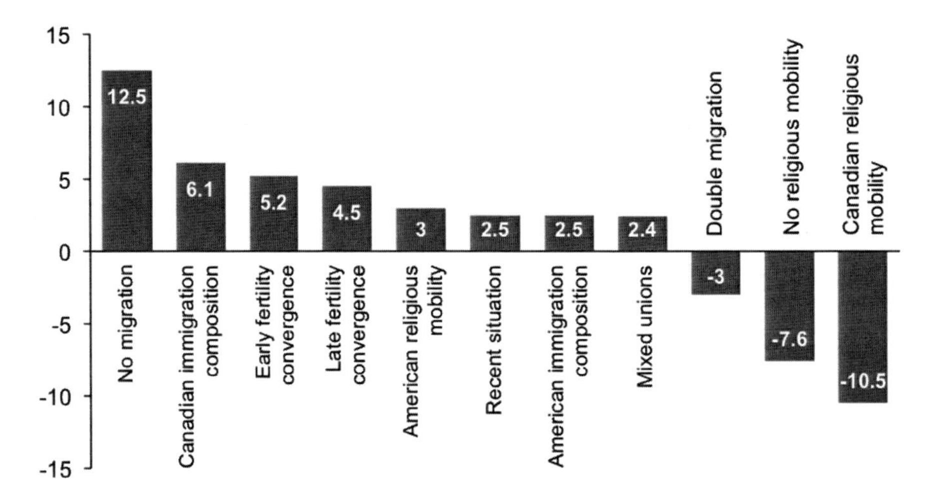

Fig. 89.10 Projected difference between Protestant and Catholic shares of the population in 2062 (in percentage points) (Source: Authors' calculations)

Unless there should be a rapid shift in conversion, migration or fertility levels, there would be a growth in minority religions (Islam, Hinduism and Buddhism, Other Religions) which results from continued immigration, a younger age structure, higher fertility and low losses through secularization and conversion to other religions. The Jewish share may decline in the US (due to low fertility, low immigration, and high secularization rates) but remain stable in Canada (due to continued immigration and relatively higher fertility). The population without religious affiliation is young, and is favored by religious mobility, but low shares of non-affiliated persons among immigrants to the US as well as low childbearing levels in both Canada and the US are likely to stall the growth in secularization in the coming decades.

Acknowledgements We would like to thank Laurent Martel and Patrice Dion for useful comments, Marcin Stonawski for his generous assistance with the data as well as participants in the VID colloquium on April 21st, 2011 (Vienna, Austria) where preliminary results were presented for wise suggestions.

References

Bélanger, A., & Caron Malenfant, É. (2005). *Population projections of visible minority groups, Canada, provinces and regions: 2001–2017* (Statistics Canada Cat. 91-541-XIE). Ottawa: Statistics Canada.

Bélanger, A., & Gilbert, S. (2003). The fertility of immigrant women and their Canadian-born daughters. *Report on the demographic situation in Canada, 2002, Statistics Canada* (Cat. 91–209). Ottawa: Statistics Canada.

Caron Malenfant, É., & Bélanger, A. (2006). The fertility of visible minority women in Canada. *Report on the demographic situation in Canada, 2003 and 2004* (Statistics Canada Cat. 91–209). Ottawa: Statistics Canada.

Caron Malenfant, É., Lebel, A., & Martel, L. (2010). *Projections of the diversity of the Canadian population* (Statistics Canada Cat. 91-551-XWE). Ottawa: Statistics Canada.
Castells, M. (2004). *The power of identity. The information age: Economy, society and culture* (Vol. II). Chichester: John Wiley and Sons Ltd.
Crockett, A., & Voas, D. (2006). Generations of decline: Religious change in twentieth-century Britain. *Journal for the Scientific Study of Religion, 45*, 567–584.
Glaeser, E. L., & Sacerdote, B. I. (2008). Education and religion. *Journal of Human Capital, 2*(2), 188–215.
Goujon, A., Skirbekk, V., Fliegenschnee, K., & Strzelecki, P. (2007a). New times, old beliefs: Projecting the future size of religions in Austria. *Vienna yearbook of population research 5*, 237-270.
Goujon, A., Skirbekk, V., & Fliegenschnee, K. (2007b). *New times, old beliefs: Investigating the future of religions in Austria and Switzerland.* Work session on demographic projections, proceedings, Bucharest, 10–12 Oct 2007. Eurostat methodologies and working papers (pp. 355–370).
Grabill, W. H., & Cho, L.-J. (1965). Methodology for the measurement of current fertility from population data on young children. *Demography, 2*, 50–73.
Groot, W., & Maassen van den Brink, H. (2007). The health effects of education. *Economics of Education Review, 26*(2), 186–200.
Guibernau, M. (2007). *The identity of nations*. Cambridge: Polity Press.
Hoge, D. R. (1981). *Converts, dropouts, and returnees*. New York: The Pilgrim Press.
Hollmann, F. W., Mulder, T. J., & Kallan, J. E. (2000). *Methodology and assumptions for the population projections of the United States: 1999 to 2100* (Population division working paper no. 38). Washington, DC: Population Projections Branch, Population Division, U.S. Census Bureau.
Hummer, R. A., Ellison, C. G., Rogers, R. G., Moulton, B. E., & Romero, R. R. (2004). Religious involvement and adult mortality in the United States: Review and perspective. *Southern Medical Journal, 97*(12), 1223–1230.
Iannaccone, L. R. (1992). Religious practice: A human capital approach. *Journal for the Scientific Study of Religion, 29*, 297–314.
Inglehart, R., & Baker, W. E. (2000). Modernization, cultural change, and the persistence of traditional values. *American Sociological Review, 65*(1), 19–51.
Jasso, G., & Rosenzweig, M. R. (2006). Characteristics of immigrants to the United States: 1820–2003. In R. Ueda (Ed.), *Companion to American immigration* (pp. 328–358). Malden: Blackwell Publishing.
Jones, E. F., & Westoff, C. F. (1979). The end of Catholic fertility. *Demography, 16*(2), 209–218.
Krotki, K., & Lapierre, E. (1968). La fécondité au Canada selon la religion, l'origine ethnique, et L'état matrimonial. *Population, 23*, 815–834.
Kung, H.-C., Hoyert, D. L., Xu, J., & Murphy, S. L. (2008). Deaths: Final data for 2005. *National vital statistics reports,* Vol. 56, No. 10. Hyattsville: National Center for Health Statistics.
Lehrer, E. (1996). Religion as a determinant of fertility. *Journal of Population Economics, 9*(2), 173–196.
Lehrer, E. (2004). Religion as a determinant of economic and demographic behavior in the United States. *Population and Development Review, 30*(4), 707–726.
Lerch, M., Oris, M., Wanner, P., Forney, Y., & Dutreuilh, C. (2010). Religious affiliation and mortality in Switzerland, 1991–2004. *Population (English edition), 65*(2), 217–250.
McQuillan, K. (2004). When does religion influence fertility? *Population and Development Review, 30*(1), 25–56.
Sander, W. (1992). Catholicism and the economics of fertility. *Population Studies, 46*(3), 477–489.
Skirbekk, V., Goujon, A., & Kaufmann, E. (2010). Secularism, fundamentalism, or Catholicism? The religious composition of the United States to 2043. *Journal for the Scientific Study of Religion, 49*(2), 293–310.

Statistics Canada. (1993). *Religion, the daily, statistics Canada's official release bulletin* (Statistics Canada Cat. 96-304E). Ottawa: Statistics Canada.

Statistics Canada. (2003). *Religions in Canada, 2001 Census: Analysis series* (Statistics Canada Cat. 96F0030). Ottawa: Statistics Canada.

Statistics Canada. (2010). *Population projections for Canada, provinces and territories, 2009 to 2036* (Statistics Canada Cat. 91–52). Ottawa: Statistics Canada.

U.S. Department of Homeland Security. (2009). *Yearbook of immigration statistics, 2006*. Washington, DC: GPO.

Chapter 90
Changing Geographies of Immigration and Religion in the U.S. South

Patricia Ehrkamp, Caroline Nagel, and Catherine Cottrell

90.1 Introduction

In August 2011, some 20 bishops, clerics, and other religious leaders joined the U.S. Federal Government and Hispanic advocacy groups in a lawsuit intended to challenge the State of Alabama's restrictive immigration law in federal court. The law, passed by the Alabama legislature in 2011, made it a criminal offense for immigrants to be in Alabama without proper documentation, and also criminalized service provision to undocumented immigrants. Religious leaders argued that their ministerial and outreach activities—which include provision of English as a Second Language (ESL) classes, Spanish-speaking Bible study groups, and the transportation of immigrant children to after-school programs—could be construed as criminal acts if they were shown to include undocumented immigrants. Church leaders argued that the law was both a throwback to the Jim Crow era, in which state power was used to terrorize and to oppress African Americans, and an assault on the constitutional freedoms of religious organizations.

The controversy over the Alabama legislation bears witness to the new landscape of immigration and religion, as well as to the deep entanglement of faith and citizenship in the U.S. South. While the last great wave of immigration to the U.S., from

P. Ehrkamp (✉)
Department of Geography, University of Kentucky, Lexington, KY 40506, USA
e-mail: p.ehrkamp@uky.edu

C. Nagel
Department of Geography, University of South Carolina, Columbia, SC 29208, USA
e-mail: cnagel@mailbox.sc.edu

C. Cottrell
Department of Geography and Earth Sciences, Aberystwyth University, Aberystwyth, SY23 3DB, UK
e-mail: cac19@aber.ac.uk

S.D. Brunn (ed.), *The Changing World Religion Map*,
DOI 10.1007/978-94-017-9376-6_90

about 1880 to 1924, largely bypassed the U.S. South, contemporary flows have transformed several Southeastern states into major immigrant gateways, reflecting important geographical shifts in post-war economic development (Marrow 2011; Winders 2005, 2012). Immigration has profoundly complicated the black-white racial divisions that dominated the region for centuries. This reconfiguration of race is taking place in many spheres of society, but is especially notable in places of worship, which, in the context of the U.S. South, have been almost entirely separated along black-white lines since the nineteenth century. Immigration certainly has not eliminated racially segregated patterns of worship, but the presence of immigrants has contributed to increasing diversification of existing communities.

Christian faith communities have responded in a variety of ways to new immigrant communities, and in doing so, have actively participated in the re-shaping of social boundaries. In a context in which the state has largely left the work of immigrant integration to local voluntary organizations, faith-based organizations play a crucial role in responding to immigrants' material needs and in facilitating their incorporation into American society. At the same time, some congregations have sought to effect deeper social transformations by integrating immigrants into congregational life. This process can become contentious as congregations deal with the practicalities of altering worship styles and well-established ways of doing things to accommodate newcomers. It also becomes contentious as congregations decide at what point secular law should curtail ministry to immigrants who lack legal documents.

As Christian faith communities grapple with their responsibilities toward immigrants in their congregations, immigration also brings a wider diversity of faith identities and faith-based claims. Southern cities have had sizeable Jewish communities since the nineteenth century; today, they are also home to growing Muslim, Hindu, and Sikh communities who are making their mark on metropolitan landscapes with purpose-built places of worship. As with the presence of immigrants in Christian congregations, the presence of non-Christian groups in the South raises a number of questions about who belongs and on what terms. As recent (mid-2012) reactions to the construction of a mosque in Murfreesboro, TN, suggest, some Christian faith communities characterized non-Christian groups as un-American, while others have urged greater 'tolerance' of non-Christian groups in the name of religious freedom. Still others have highlighted the 'common ground' that exists between faith traditions.

In this chapter, we examine the internal workings of faith communities in the U.S. South and the ways these are deeply enmeshed in every-day productions and negotiations of societal membership, citizenship rights, and immigrant integration. We begin with an empirical overview of immigration and the ways it has complicated the region's social and political landscape in the U.S. South. We then discuss the diversity of immigrant faith communities and the very different ways that established faith communities have tried to incorporate immigrants. Finally, we draw on our research on faith communities in Charlotte, NC, Greenville-Spartanburg, SC, and Atlanta, GA to show in greater detail the different ways that faith communities—both Christian and non-Christian—are producing diverse conceptions of social difference and social membership. Our general aim is to convey how ideas about citizenship are molded in faith-community contexts and the ways that these processes are shaped by particular regional histories.

90.2 Immigration and a Changing U.S. South

Many recent studies have documented the growth of the U.S. South's foreign-born population, but, as Winders (2005) aptly notes, few have critically analyzed what new diversity means for Southern communities or for immigrants themselves in terms of shifting constructions of identity and racial dynamics. In the U.S. as a whole, the immigrant population has increased rapidly since the 1965 Immigration and Nationality Act, which removed quota restrictions on immigrants from the Eastern Hemisphere. Today, around 13 % of the U.S. population is foreign-born—a similar percentage as that at the peak of the last major wave of immigration, between about 1880 and 1924. In contrast to immigration in the late nineteenth century, which was dominated by immigrants from Europe, today's immigrants come mainly from Latin America (especially Mexico) and Asia.

While the traditional immigrant-receiving states—California, New York, Florida, Texas, New Jersey, Illinois, and Massachusetts—continue to attract a disproportionate number of immigrants, their immigrant population growth rates have been leveling out. Meanwhile, the immigrant population growth rates in Southern and Midwestern states have been on the rise. Between 1990 and 2000, the foreign-born populations in Georgia and North Carolina grew by 233 % and 274 % respectively—among the highest rates in the country during this period. Between 2000 and 2010, the fastest growth rates were to be found in South Carolina, Alabama, and Tennessee. Cities like Atlanta, Georgia and Charlotte, North Carolina have become important immigrant gateway cities (Singer et al. 2008).

Many immigrants living in the U.S. South are unskilled laborers who work in agriculture, poultry processing, domestic work, landscaping, and construction (Marrow 2011; Odem 2004). But the region also attracts highly skilled immigrants. For example, the Research Triangle in North Carolina attracts researchers, doctors, and engineers from India, China, and the Middle East, while the BMW, Mercedes Benz, and Michelin production sites established in the 1990s near Atlanta, GA and Greenville-Spartanburg, SC bring in managers and executives from Germany and France. For this reason, Germans appear among the top three foreign-born populations in South Carolina in the 2010 census.

In order to lure such corporate foreign investment, and to bolster economic growth strategies, cities in the U.S. South have sought to refashion themselves through celebratory notions of "diversity." Such new diversity discourses manifest themselves in now-ubiquitous "international festivals." For over 20 years, the Latin American Coalition in Charlotte, NC, has organized an annual Latin American Festival that now attracts over 25,000 visitors to the downtown area. Atlanta's Peachtree Latino Festival occurred for the 12th time in 2012, and the Greenville-Spartanburg area hosts an annual International Festival, among others.

90.3 Immigration and Religious Diversity in the U.S. South

The growing racial and ethnic diversity that immigration has brought to the U.S. South has significantly altered the region's religious landscape, which has been dominated for centuries by Protestant Christianity. It is important to take note of the salience of Protestant Christianity in the social life of the region. The region as a whole boasts above average religiosity as compared to the rest of the U.S. and the highest percentage of residents that are considered "very religious" (Newport 2012). More than residents of other regions, people in the U.S. South are likely to characterize themselves as "evangelical" (Ownby 2005) and there is a widespread assumption that the public square is Christian, as seen with on-going efforts to institute prayer in public schools and to place the Ten Commandments in public buildings in Southern states (Wells 2004).

While dominated by Protestant Christianity, however, the region's religious landscape has also taken shape around major historical fault lines. For more than 250 years, Protestant churches were deeply implicated in the formation, maintenance, and contestation of racial hierarchies. In the pre-Civil War period, Protestant churches were key supporters of slavery and favored the South's secession from the Union in 1860. After the Civil War, blacks, who, as slaves, had often worshipped in bi-racial but segregated churches, established their own denominations, including the African Methodist Episcopal Church. White churches (mostly Baptist, Presbyterian, Episcopal, and Methodist), meanwhile, continued to support the existing racial order and were, with few exceptions, largely complicit in the creation of Jim Crow laws that denied full civil and political rights to emancipated slaves (Wilson 2005).

Most mainline and evangelical Protestant churches continue to be racially homogenous, and many of them remain very socially and politically conservative (Manis 2005). Yet this long-standing pattern of racially separate religious worship is being challenged from multiple directions; so, too is the dominance of Protestant Christianity in the U.S. South. One notable trend has been the substantial growth of Roman Catholicism. Catholic churches began to multiply in the 1950s, due in part to the migration of Northerners to burgeoning Southern metropolitan areas. The growth of the Catholic Church accelerated in the 1970s and 1980s, with the inflow of Vietnamese Catholic refugees and of thousands of (predominantly) Latino labor migrants. This pattern can be seen in Greenville, SC, where the city's two Catholic churches (the first consecrated in the 1870s and the second in 1939) were joined by five more churches between 1952 and 1989. Many of the Catholic churches in the region have added Spanish masses in the last two decades (the Archdiocese of Atlanta alone lists over 60 parishes that offer at least one mass per week in Spanish), and a number of these parishes established separate Latino or Hispanic ministries with pastors or deacons who are native Spanish speakers. Most dioceses, as well, have at least one Vietnamese Catholic congregation with its own priest.

Along with the growth of Catholicism has been the appearance of Orthodox and Protestant churches associated with particular ethnic-national groups.

Syrian Orthodox, Serbian Orthodox, Coptic Orthodox, Korean Presbyterian, and Vietnamese Assembly of God churches, for instance, can be found in most metropolitan areas in the South. Some of these new churches have their own building and facilities, but smaller immigrant churches rent space from established faith communities. For example, in Charlotte, North Carolina, the Cross of Victory Church, which was established by and serves Congolese refugees, holds its services on the campus of Trinity Fellowship Church (Breen Bolling 2004). In such cases, the faith communities remain separate, with immigrant pastors serving part-time or as volunteers. In some cases, congregations are able to "upgrade" facilities over time, as communities grow through continued immigration.

Another important phenomenon that has complicated the existing religious landscape in the South has been the creation of self-consciously multi-ethnic and multicultural Protestant churches, many of which are non-denominational. This is a nationwide trend, but is especially notable in the U.S. South in light of the region's segregationist history. Some of these churches are rooted in the Pentecostal tradition, which historically has had a mixed record on racial segregation, but which has fully embraced a message of racial reconciliation and diversity. Pentecostal missionaries have made significant inroads in Latin America since the 1980s, and some 25 % of all immigrants from South America are evangelical Protestants, with Pentecostalism being one of the fastest growing religions in that region (The Economist 2007). Attending a large neo-Pentecostal service in Greenville, SC, during our fieldwork, we were struck by the social mix of the congregation and by the prevalence of posters and wall murals declaring the need to "destroy the walls of racism that divide us."

Other multi-ethnic/multicultural churches are more closely aligned with Baptist traditions. During our fieldwork, we met with a number of Baptist-trained pastors of large congregations, some of them housed in purpose-built structures and others in converted shopping malls and warehouses. Many of these churches are devoid of the traditional accoutrement and symbols of Protestant churches, that is, crosses, steeples, and pews. The worship style is contemporary, the teaching biblically-centered and oriented around practical deployment of biblical principles in everyday life. The aim of these churches is to welcome the "un-churched" and to appeal to those put off by traditional church worship and dogma. In their contemporary, non-traditional feel, these churches attract a mix of whites, blacks, and immigrant groups, including many Latinos. Unlike most churches, as we will discuss shortly, they generally emphasize the "togetherness" of worshippers, and some of them provide simultaneous translation of services into Spanish so that Latinos can participate in the main service (Garces-Foley 2007).

More established Protestant churches in the South have responded in a variety of ways to the "competition" for immigrants. All of the mainline denominations today have created Latino, immigrant, and/or minority outreach ministries, which operate nationally, regionally, and in individual congregations. For instance, the Charlotte Presbytery (part of the Presbyterian Church U.S.A., the most liberal of the Presbyterian denominations in the U.S.) established a Hispanic ministry a few years ago that has planted missions on the urban-rural fringe where many Latino laborers work in agriculture, poultry processing, and construction. The Atlanta Tri-Presbytery, as well, maintains a "New Church Development Commission" that provides guidance and resources for establishing new churches that serve immigrants. Some large Baptist churches have also been involved in "church planting" efforts in immigrant neighborhoods. We met with the pastor of a wealthy Baptist church affiliated with the Southern Baptist Convention that has been financing mission efforts in the Spartanburg area specifically aimed at Latinos. Their aim is to purchase a church structure to house a new congregation and to provide financial support until the church is able to maintain itself as an independent congregation. This reflects a broader strategy among Baptist churches to "grow by caring" (Greer 1993), and speaks, as well, to the Southern Baptist Convention's public repudiation in the 1990s of racism.

In addition to planting churches and missions, many established Protestant churches have opened their doors to immigrant congregations, usually of the same denomination but led by a separate, immigrant pastor. We spoke, for instance, with pastors whose churches host Burmese, Japanese, Latino, and Chinese congregations. In some cases, the "ethnic congregations" remain almost entirely separate from the main congregation whose facilities they use. Indeed, many ethnic congregation leaders intend to build their own churches when it becomes financially feasible to do so. In other instances, white pastors are keen to integrate immigrants more fully into the larger congregation. Some express interest in creating a single, multicultural congregation, but they generally feel that it is more realistic simply to include immigrants in fellowship activities and to encourage the children of immigrants to attend youth group activities, while preserving separate non-English-language services. We discuss the implications of these different approaches in greater detail below.

90.4 Non-Christian Diversity

Christianity, then, has become increasingly diversified in the South (and elsewhere in the U.S.) as a result of immigration and other trends in religious practice. As Christianity diversifies in the South, it also faces the growth of non-Christian faiths. For instance, there are at least 64 Muslim organizations and 24 Buddhist centers or temples in Georgia alone (www.pluralism.org). The growth of these communities has been made visible with the construction of places of worship, many of them drawing on architectural styles from immigrants' places of origin. Following the residential patterns of immigrants, and the sprawling and multi-centric structure of cities in the U.S. South, many of these temples, mosques, and gurdwaras are located in suburban areas (Fig. 90.1).

The Atlanta suburb of Lilburn, GA exemplifies this suburban pattern. Lilburn, with a population of not quite 12,000, boasts three Hindu and Vedic temples, as well as two mosques, one serving a Shi'a Muslim community, the other serving Sunni Muslims. Among these places of worship is the currently largest Hindu temple in

Fig. 90.1 Masjid Omar mosque, Lilburn, GA (Photo by P. Ehrkamp)

the U.S., the Bochasanwasi Shri Akshar Purushottam Swaminarayan Sanstha (BAPS) Shri Swaminarayan Mandir, an impressive structure that extends across 29 acres (11.7 ha) (Fig. 90.2).

Hindu temples generally serve Indian American communities, though it is important to note that 'Hindu' encompasses many different regional faith traditions, and temples will often accommodate several, quite distinctive, devotional practices. Likewise, some Buddhist temples welcome non-immigrant devotees, but others are immigrant-oriented (mostly Chinese, Vietnamese, and Cambodian), and temples sometimes serve devotees from a particular national and/or ethnic group. Muslim communities in the U.S. South typically encompass many different nationalities (including native-born converts), though some are dominated by a single group (Leonard 2003).

90.5 Immigration, Religion, and Societal Membership

The transformations taking place within denominations, within individual congregations, in the Southern religious landscape overall are interwoven with wider social dynamics relating to identity, integration, and citizenship. For over a century, U.S. scholars have been scrutinizing the relationship between religion and the integration of immigrants into "mainstream" American society. Early sociological accounts viewed religion as an element of traditional, "old country"

Fig. 90.2 Hindu BAPS Mandir Lilburn, GA (Photo by P. Ehrkamp)

culture that buttressed group identity, but that inevitably changed as immigrants established themselves in the new society (Gordon 1964; Herberg 1955; Thomas and Znaniecki 1920; Young 1932). Such analyses suggested that religion in American society served more a social function than a spiritual one. Having waned somewhat in the 1970s and 1980s, interest in immigrant religion surged in the 1990s, as post-1965 immigrants—and their religious customs and places of worship—became a more visible feature of American cities (Ecklund 2005; Levitt 2003, 2007; Menjivar 1999; Stepick 2005; Stepick et al. 2009). Contemporary scholars, in particular, highlight the diasporic relationships—both material and spiritual—that are sustained through immigrants' places of worship and their religious practices (Levitt 2007; Menjivar 1999). But scholars are also interested to know how religious communities mediate processes of incorporation and are themselves subject to these processes. Warner (1994) for instance, has argued that immigrant faith groups adapt and take on the characteristics of mainline Protestant institutions, including active lay membership, regularly scheduled services, a congregational structure, and myriad community services and activities (see also Yang and Ebaugh 2001). This shift is especially notable in the case of non-Christian faith communities, which, in the U.S. context, are increasingly serving multiple community functions.

In recognizing the institutional shifts that take place within faith communities, scholars are also interested in the civic functions that they serve. A number of recent studies have explored the extent to which faith communities, which vary in terms of spiritual orientations, leadership styles, hierarchical structure, and so on, foster civic-mindedness and civic participation. This scholarship recognizes that while faith communities serve an important role cementing community ties and in fostering particular identities, they can also provide spaces of interaction and encounter between immigrants and non-immigrants and can foster in immigrants the values and outlooks that encourage political integration and active citizenship (Foley and Hoge 2007; Kniss and Numrich 2007; Ley 2008). Our own research stems directly from these inquiries into the role of immigrant places of worship as sites of interaction and citizenship formation. Yet rather than equating citizenship with a preconceived set of civic virtues, our research has sought to uncover the ways in which faith communities produce particular ideas about citizenship and debate among themselves what it means to be a member of the society-at-large (Ehrkamp and Nagel 2012). From our perspective, citizenship—and relatedly, integration or "assimilation"—are relatively fluid concepts that are actively formulated and debated in various spheres and spaces of every-day life, including those associated with faith communities. There are, in this sense, multiple understandings that circulate in society and different ways that spiritual beliefs inform wider conceptions of societal belonging (Ehrkamp and Nagel 2014).

90.6 Faith Communities and Negotiations of Citizenship in the South

The shifting religious landscape of the South has set into motion numerous negotiations over the terms of membership in society and the extent to which certain beliefs, practices, and identities are compatible with the mainstream. These negotiations are brought into sharp relief in interactions between Christian and non-Christians, especially in cases where non-Christians have tried to build places of worship in predominantly white, Christian neighborhoods. While some of the non-Christian faith community leaders with whom we have spoken experienced a warm welcome they have received by their neighbors, others spoke of the hostility that has greeted their presence. One member of a South Carolina Islamic center told us that an arsonist had burned down their mosque a few years previously; the arsonist, as it turned out, was a member of the church next to the mosque. When the Muslim community subsequently tried to purchase land from two brothers to build a mosque, one of the brothers refused to sell, but finally relented under pressure from the other. Such events, it seems, are not rare. A mosque in Marietta, GA, a suburb of Atlanta, for instance, was burned twice by arsonists in 2010. And in many instances, cities and towns have fought to keep out religious difference through zoning and building codes. For example, the city of Lilburn, GA, lost a law suit brought by the Federal government for discrimination against a Muslim group. The Federal lawsuit stated that the city's refusal to rezone property for a planned mosque expansion violated the Religious Land Use and Institutionalized Persons Act (RLUIPA) of 2000 by treating Muslims differently than other religious groups. The RLUIPA prohibits land use regulations that treat religious institutions differently from nonreligious institutions, and prohibits discrimination against any assembly or institution based on religion or religious denomination.

In a context where, as we alluded to earlier, the public square is seen to be Protestant, non-Christian groups sometimes engage in public outreach efforts designed to highlight the contributions they make to the wider community. Some Hindu community leaders participate in "international" festivals held to showcase "cultural diversity," viewed as a highly marketable attribute by many municipal leaders in the South. Dance performances, Indian food stalls, and other cultural performances are ways that Hindu temple leaders and members promote what they often describe as a mutually enriching multicultural society. Similarly, the BAPS Shri Swaminarayan Mandir in Lilburn, GA, offers guided tours, audio tours, and has a visitor center. It also invites visitors to celebrate Diwali and the Hindu New Year holidays and expects sufficient numbers of visitors to offer off-site parking with a shuttle service to the Mandir in order to avoid traffic jams. Yet it is not clear that this path to multicultural membership is open to all non-Christian groups. Our contact in the South Carolina Muslim community stated that members of the Islamic center had been reluctant to engage in any kind of outreach activity given the hostility they had faced. Instead, they had decided to focus their efforts on building a community school and developing more solidarity among Muslims—solidarity which, he believed, would give them a stronger basis for interacting with the wider community.

Christian congregations offer up a very different set of dynamics, and scenarios vary widely between churches. Some churches we have visited, for instance, are located in neighborhoods that are transitioning from being predominantly working-class or middle-class white to predominantly working-class Latino and African American. As the neighborhoods have changed, the pastors have tried to change the style of worship to make services more appealing to newcomers. The aim in such cases is to have everyone worship together, in the same room, at the same time—"sharing the pews," as one of our informants put it. This kind of model runs into a number of difficulties and obstacles because of language barriers, cultural differences and expectations in terms of the style of worship; the selection of music, in particular, can be a major bone of contention. Existing, long-time members of congregations are at times not willing to accept newcomers that have nothing in common with them but their shared faith.

We spoke with one pastor in Charlotte who was trying to reinvent an aging, middle-class, white church as a multi-ethnic church that would better serve the immediate neighborhood. Her efforts, however, had led to the exodus of several remaining white members. Her church, and a similar church we visited in Greenville, have small membership rolls, but have become heavily engaged in serving their non-white neighbors, creating after-school and summer programs for mostly minority youths. Members of the Greenville church whom we interviewed talked about the changes in their own way of thinking that this had entailed. They described the way in which their encounters with Latinos and African Americans had drastically altered their views of racialized minorities and had exposed them to "other cultures'" which had made them more sympathetic to the hardships faced by these groups. Members with whom we spoke were generally supportive of the pastor's efforts to serve undocumented immigrants, which had brought him into confrontation with the police. While describing their admiration for the strong family values shown by their Latino neighbors, they also commented on some of the tensions that had emerged over cultural differences and stereotypes. They remarked, for instance, that Latinos do not show the same regard for cleanliness as Anglos do. Part of their work, then, was to help "teach" their Latino neighbors to adhere to American values and ways of doing things, clearly taking an assimilationist approach.

The Greenville church maintains separate Anglo and Hispanic worship services, and unlike the pastor in Charlotte, the Greenville pastor did not seem inclined to radically change the worship service to bring groups together. His argument was that differences need to be bridged, but they do not need to be erased. This viewpoint has been repeated in many of our conversations with pastors, who emphasize the "right" of Latinos and other immigrant groups to worship in their own style and in their own language. Their responsibility as pastors, as they see it, is to provide space for different worship styles and to reach as many people as possible with the Gospel—a responsibility confirmed by numerous biblical directives to spread the word among all peoples and nations, and languages. (Multi-lingualism has been an explicit policy of the Catholic Church, which instructs churches to provide culturally appropriate forms of worship for parishioners; the Catholic Diocese of Charleston, in addition, now requires parish priests to learn Spanish).

Many of the immigrant pastors (as well as immigrant congregants) we interviewed embraced a similar view. Those who are affiliated with a larger, predominantly Anglo congregation reiterated the cultural importance for immigrants to worship in their own language. Yet interviews with immigrant pastors also revealed some tensions between immigrant and non-immigrant communities over language and other issues of difference. Some suggested that they had been relegated to a second-class status within the church. One group of Latino ministers in an Anglo-dominated Charlotte megachurch expressed their frustration with the unwillingness of the Anglo congregation to recognize the plight of undocumented immigrants and their seeming hostility toward the church's occasional bi-lingual services. Others made humorous, but nonetheless, pointed remarks about the racist attitudes of white congregants. One Chinese Baptist pastor who headed a Chinese immigrant congregation suggested that he felt marginalized within the larger church, somewhat like a poorer cousin and not necessarily taken seriously by white, longer-term members of the church. He felt tolerated but not accepted. Not surprisingly, this pastor was hopeful that he might be able to establish his own, immigrant-only church eventually. Indeed, even some Anglo pastors commented on their congregants' desire to go on missionary trips in Africa and Latin America, but their reluctance to reach out to ethnic and racial minorities in their immediate neighborhood.

Language separation mostly applies to worship services and adult Bible study classes, however. Most of the churches we visited offered inclusive programs for children and youths, among them Bible study classes on weekends, homework tutoring, sports leagues, creative arts programs, music programs, and dance. Churches' programs for children and youth varied depending on the size and membership of individual churches. A number of churches have Vacation Bible Schools during the summers that immigrant and non-immigrant children partake in, and some took youths on mission trips to deprived urban areas during spring break. All these activities, including Bible study classes, mixed immigrant and non-immigrant children because pastors and congregants felt that there were no language barriers to overcome among children and adolescents. But our interviewees also reported that it was occasionally difficult to include everyone in these youth programs, in part because immigrant youths were much more likely to have jobs outside of school and had less spare time than white, middle-class youths who were free to spend their time joining church activities.

Many of these difficulties in creating multi-cultural churches were brought to the fore in the 2012 Multi-Ethnic Summit, held in Spartanburg, SC, at a non-denominational megachurch. Sponsored by the Billy Graham Foundation, the purpose of this summit, first held in 2001, is to aid mainly evangelical congregations in becoming more inclusive. The summit thus provided workshops on such topics as "Immigration: Today's Moral Dilemma," "Transforming Churches Toward Cultural Diversity," and "Intercultural Encounter and Cultural Partnership across Cultural Lines." One important theme was how to address cultural diversity without demanding assimilation or simply being color blind. Contrary to denominational churches' practice of establishing separate ethnic or immigrant ministries, one presenter suggested that churches "should strive to be multicultural and be one group," because "we are all part of the body of Christ and should encounter and interact with each other" (field notes, April 20, 2012)—a Christian understanding of multiculturalism that approximates some scholarly notions of multiculturalism that cannot or should not ignore cultural differences in order for people to participate equally in society (Carens 2000; Modood 2007). Yet there were objections from some audience members who voiced their opinions that some people's practices were "sin," suggesting that they were not willing to tolerate *all* cultural differences.

The question of how to deal with undocumented immigrants also generated disputes. Some of the presenters argued strongly in biblical terms against the construction of certain groups as "illegal," highlighting both the injustices of U.S. immigration law and the primacy of Christian compassion over legality. But, again, our research participants did not universally embrace such a view. As we have seen in our interviews in churches, the issue of undocumented immigration is a very thorny one that many pastors would rather ignore, but are unable to do so in light of state anti-immigrant legislation. Most of those we have asked say that they do not ask about anyone's legal status, but nor do they go out of their way to critique U.S. border politics. Contrary to the approach taken by some of the presenters at the multi-ethnic summit, many of our interviewees have tended to cite Bible passages about following the law although they do not view it as their job to enforce it. Given the thorniness of the issue, it is not surprising that one pastor we interviewed expressed relief that the Burmese refugees who have joined his congregation are legal; this, he suggested, had made it far easier for the congregation to support the community.

90.7 Conclusion

Immigration has brought about considerable ethnic, racial, and religious diversity to the cities and towns of the U.S. South in recent decades. This increasing diversity is reflected in urban and suburban landscapes as new places of worship emerge. But it is also reflected in the changing compositions of existing congregations and parishes that often undergo intense negotiations as to how best to address this new diversity, and meet immigrants' social and spiritual needs without offending or losing existing members of communities of faith. Many of our interviewees view the diversity of newcomers as enriching experience. But biblical imperatives to minister to newcomers regardless of legal status, language, and national origin also meet with objections based on understandings of secular law and of the public square as Christian. In this context, as our work shows, shared faith may help bridge some differences, but does not necessarily overcome tensions around legal status and cultural practices. What kinds of differences become acceptable, and what type of multiculturalism such acceptance of difference enables will shape outlooks for society in the U.S. South that reach far beyond the walls of any single place of worship, and beyond any single faith.

Acknowledgments This research was in part supported by the National Science Foundation under Grants No. BCS-1021907 and BCS-1021666. Any opinions, findings, and conclusions or recommendations expressed in this material are those of the authors and do not necessarily reflect the views of the National Science Foundation. We thank Derek Ruez for his research assistance.

References

Breen Bolling, C. (2004). Congolese celebrate in Charlotte – community gathers to share food, art, song, memories of home. *The Charlotte Observer,* p. 1L. Charlotte.

Carens, J. (2000). *Culture, citizenship, and community: A contextual exploration of justice as even-handedness*. Oxford: Oxford University Press.

Ecklund, E. H. (2005). 'Us' and 'them': The role of religion in mediating and challenging the 'model minority' and other civic boundaries. *Ethnic and Racial Studies, 28*, 132–150.

Ehrkamp, P., & Nagel, C. (2012). Immigration, places of worship and the politics of citizenship in the US South. *Transactions of the Institute of British Geographers, 37*, 624–638.

Ehrkamp, P., & Nagel, C. (2014). 'Under the radar': Undocumented immigrants, Christian faith communities, and the precarious spaces of welcome in the U.S. South. *Annals of the Association of American Geographers, 104*, 319–328.

Foley, M. W., & Hoge, D. R. (2007). *Religion and the new immigrants: How faith communities form our newest citizens*. Oxford: Oxford University Press.

Garces-Foley, K. (2007). New opportunities and new values: The emergence of the multicultural church. *Annals of the American Academy of Political and Social Science, 612*, 209–224.

Gordon, M. M. (1964). *Assimilation in American life: The role of race, religion, and national origins*. New York: Oxford University Press.

Greer, B. A. (1993). Strategies for evangelism and growth in three denominations (1965–1990). In D. A. Roozen & C. K. Hadaway (Eds.), *Church and denominational growth* (pp. 87–111). Nashville: Abingdon.

Herberg, W. (1955). *Protestant-Catholic-Jew*. Garden City: Doubleday.
Kniss, F., & Numrich, P. D. (2007). *Sacred assemblies and civic engagement: How religion matters for America's new immigrants*. New Brunswick/London: Rutgers University Press.
Leonard, K. (2003). American Muslim politics. *Ethnicities, 3*, 147–181.
Levitt, P. (2003). You know Abraham was really the first immigrant: Religion and transnational migration. *International Migration Review, 37*, 847–873.
Levitt, P. (2007). *God needs no passport: Immigrants and the changing American religious landscape*. New York/London: The New Press.
Ley, D. (2008). The immigrant church as an urban service hub. *Urban Studies, 45*, 2057–2074.
Manis, A. M. (2005). The civil religions of the South. In C. R. Wilson & M. Silk (Eds.), *Religion and public life in the South: In the evangelical mode* (pp. 165–194). Walnut Creek: AltaMira Press.
Marrow, H. B. (2011). *New destination dreaming: Immigration, race, and legal status in the rural American South*. Stanford: Stanford University Press.
Menjivar, C. (1999). Religious institutions and transnationalism: A case study of Catholic and evangelical Salvadoran immigrants. *International Journal of Politics Culture and Society, 12*, 589–612.
Modood, T. (2007). *Multiculturalism: A civic idea*. Cambridge/Malden: Polity Press.
Newport, F. (2012). Mississippi is most religious U.S. state. In *Gallup Poll* (Vol. 2012). Princeton: Gallup.
Odem, M. E. (2004). Our Lady of Guadalupe in the New South: Latino immigrants and the politics of integration in the Catholic church. *Journal of American Ethnic History, 24*, 26–57.
Ownby, T. (2005). Evangelical but differentiated: Religion by the numbers. In C. R. Wilson & M. Silk (Eds.), *Religion and public life in the South: In the evangelical mode* (pp. 31–62). Walnut Creek: AltaMira Press.
Singer, A., Hardwick, S. W., & Brettell, C. (2008). *Twenty-first century suburban gateways: Immigrant incorporation in suburban America*. Washington, DC: Brookings Institution Press.
Stepick, A. (2005). God is apparently not dead: The obvious, the emergent, and the still unknown in immigration and religion. In K. I. Leonard, A. Stepick, M. A. Vasques, & J. Holdaway (Eds.), *Immigrant faiths: Transforming religious life in America* (pp. 11–38). Lanham/Oxford: AltaMira Press.
Stepick, A., Ray, T., & Mahler, S. J. (2009). *Churches and charity in the immigrant city: Immigrants and civic engagement in Miami*. New Brunswick: Rutgers University Press.
The Economist. (2007). Lighting on new faiths or none. *The Economist*. http://www.economist.com/node/9116934. Accessed 10 June 2012.
Thomas, W. I., & Znaniecki, F. (1920). *The Polish peasant in Europe and America: Monograph of an immigrant group. Organization and disorganization in America* (Vol. V). Boston: Richard G. Badger.
Warner, R. S. (1994). The place of the congregation in the American religious configuration. In J. P. Wind & J. W. Lewis (Eds.), *American congregations* (pp. 54–99). Chicago: University of Chicago Press.
Wells, T. (2004). Growing religious diversity in South Carolina: Implications for the Palmetto State. *Pluralism Project at Harvard University*. http://www.pluralism.org/affiliates/student/wells/SC_Religious_Diversity.pdf. Accessed 10 June 2012.
Wilson, C. R. (2005). Introduction: Preachin', prayin', and singin' on the public square. In C. R. Wilson & M. Silk (Eds.), *Religion and public life in the South: In the evangelical mode* (pp. 9–26). Walnut Creek: AltaMira Press.
Winders, J. (2005). Changing politics of race and region: Latino migration to the US South. *Progress in Human Geography, 29*, 683–699.
Winders, J. (2012). Seeing immigrants: Institutional visibility and immigrant incorporation in New Immigrant Destinations. *Annals of the American Academy of Political and Social Science, 641*, 58–78.
Yang, F., & Ebaugh, H. R. (2001). Transformations in new immigrant religions and their global implications. *American Sociological Review, 66*, 269–288.
Young, P. V. (1932). *The pilgrims of Russian town*. Chicago: University of Chicago Press.

Chapter 91
New Ecclesiologies and New Ecclesio-geographical Challenges: The Emergence of Post-ecclesiological Modernity

Grigorios D. Papathomas

91.1 Introduction

The sixteenth century opens a new period in History and the Theology of the Church and Christianity. This period which could literally be characterized as "post-ecclesiological" for reasons which will be discussed below. The beginning of this period could, indicatively, be dated back to the time of the Reformation (1517), though, of course, many precursory signs had already appeared much earlier, especially in the ecclesiology formed at the time of the Crusades (1095–1204).

The five following centuries (sixteenth to twentieth centuries) provide us with enough historical evidence and theological facts to define this as *new* and *innovative* period, compared to the completely different ecclesiological practice which preceded it, but also an *unprecedented* age, hitherto unknown, which sealed the end of the Ecclesiology, as lived and developed by the Church during the previous fifteen centuries.

After this observed ecclesiological deviation and its introduction, which was *de facto* and not *because of some ecclesiological evolution* towards a "post-ecclesiological" age, it was natural for various new ecclesiologies to appear/emerge, such as *confessional* ecclesiologies (Protestants), *ritualistic*[1] ecclesiologies (Catholics), and *ethno-phyletic* ecclesiologies (Orthodox Christians), or better yet, to respect the order of their historical appearance and also ritualistic, confessional and ethno-phyletic ecclesiologies. These are essentially *hetero-collective* ecclesiologies, which were constituted according to militant and surrogate principles and which have been dominating since then not only to characterize all of ecclesial life, but also

[1] By the term ritualism, I mean the different rites (the ancient liturgical traditions) which continued to coexist in the bosom of the Roman Catholic Church and on which are founded religious groups or ecclesial entities, in parallel, overlapping and universal.

G.D. Papathomas (✉)
Faculty of Theology, University of Athens, Athens, Greece
e-mail: grigorios.papathomas@wanadoo.fr

S.D. Brunn (ed.), *The Changing World Religion Map*,
DOI 10.1007/978-94-017-9376-6_91

to dictate the statutory texts shaping the existence and functioning of all Churches of that age and also today.

Today, we are in an historico-theological position to distance ourselves from the facts of the historical and ecclesiological past and can re-examine the causes which provoked these ecclesiological deviations. We propose to directly discuss here *in a purely dialectic and critical mindset and without any polemic temptation*, the three ecclesiologies which are so different in their origin and their perspective and yet have a common denominator, are alike, contiguous and coexist, albeit without any communion or identification between them. This common denominator goes by the name *co-territoriality*, a serious ecclesiological problem recorded during the whole second millennium, the same millennium which was also confronted with numerous unsolvable *issues* of an exclusively *Ecclesiological* nature. This is in contrast to the first millennium the *Christological issues* were basically resolved. In other words, when a Christological problem appeared, the Church during the first millennium intervened conciliarly and resolved it, something which, as will become clear, does not occur in the second millennium. These three ecclesiologies are the following:

1. The Ecclesiology of the Crusades (thirteenth century)
2. The Ecclesiology of the Reformation (sixteenth century)
3. The Ecclesiology of Ethno-Phyletism (nineteenth century)

We now examine this interrelated, newly appearing and heterocentric ecclesiological trilogy in more detail.

91.2 The Ecclesiology of the Crusades (Thirteenth Century)

As an ecclesiological fact, the reciprocal *rupture of communion of 1054* only concerned the two Patriarchates of the Church, that is, the Patriarchate of Rome and the Patriarchate of Constantinople. However, this rupture extended itself *de facto* to the other Patriarchates of the East as the Crusades quickly characterized it as a *Schism*. It was proven later that this term referred to a unique fact which from an ecclesiological and canonical point of view could legitimize the establishment of new *homonymous Churches* on territories of already existing Patriarchates and Churches of the East given that the *rupture of communion* by itself could not legitimise such a thing.

Indeed, the political movement of the Crusades gave a new twist to the *rupture of communion of 1054* and proclaimed it to be a *schism,* that is, the canonical and ecclesiological fact which considers an ecclesial body as being *detached* from the whole and, consequently, *inexistent* in a given location; but it created and pushed the order of ecclesiological things in a new direction. Thus two categories of Churches were created alongside the two pre-existing Patriarchates of the East. *Homonymous Latin Patriarchates* were established in the East (the Latin Patriarchate of Jerusalem which was established at end of the first crusade, 1099 and later also the Latin

Patriarchate of Antioche (1100), and the *non-Autocephalous*[2] Catholic Church of Cyprus [1191], etc.). This fact occurred by itself – if we accept that we have a *rupture of communion* and not a *schism* – officially engenders *the ecclesiological problem of co-territoriality* (1099).

However, this unprecedented emergence of co-territoriality does not stop here. Alongside all these *Latin ecclesial entities* were also established *Latin ritualistic Patriarchates* and *Eastern Catholic Churches* (Maronite Patriarchate, Melchite Patriarchate, Syrian Catholic Patriarchate, etc.), under the *hyperoria* ("across the boundaries") and the hierarchical *isosceles* (equivalent) jurisdiction of the Patriarchate and the Pope of Rome *on one and the same territory*.

The *jurisdiction was hyperoria* which was always in the case of the *rupture of communion* since new Latin and ritualistic Patriarchates were being created in the canonical territories of the Eastern Church. But it was also *isosceles*, because, although the Patriarchates were all equal among themselves, they were all *subordinate* to and *commonly dependent* on the Patriarchate of Rome. This ecclesiological aberration was also unheard of and has been maintained to this day (cf. the existence of two different types of Church in the same territory (*conviventia*), but also is of two totally independent Codes of Canon Law not communicating with each other). It is during this very time that a new conception of the Primacy of the Patriarch and the Pope of Rome appeared, one quite different than the ecclesiological experience of the first millennium. We can consider the Patriarch and the Pope of Rome as both are in fact "Primus inter *inferiores*" (*mono-jurisdictional primacy*) while, in the ecclesiology and praxis of the Church of the first millennium, the First Patriarch (the President) of the *ecclesial communion of the five Patriarchs* (*conciliar Pentarchy*) established during the 4th Ecumenical Council of Chalcedon (451) was "Primus inter *pares*" (*communional and synodal primacy*). This discussion is beyond the scope of the present text. In other words, one structure of a pyramid type came to replace the structure of the type constellation.

Since the thirteenth century the Ecclesiology of the Catholic Church introduced for the first time in history an ecclesiological form, viz., the establishment of a Church at a location of *dual co-territoriality*. On one hand it was co-territoriality with Patriarchates, with which it is not – or even may be – in *rupture of communion*, and on the other hand, co-territoriality with other self-established Churches of different *ritus*. The latter, however, are in complete communion or, as it is acceptable to say, *united* with Rome, though they all coexist as ecclesial bodies and entities in one and the same territory. This is how we end up, since from the end of the middle ages having Catholic Churches of different *ritus* on the same land. This state is what we could more precisely call *internal co-territoriality* (*ad intra*). But we also end up with a Latin Roman Catholic Patriarchate together with other ritualistic Roman Catholic Patriarchates at a place where a Patriarchate already exists (recall for example the case of Jerusalem). This is *external co-territoriality* (*ad extra*).

[2] See "The time of Xenocracy in Cyprus (1191–1960)." (2000). Historico-canonical note, Hydor ek Petras [Crete], Vol. 12–16, pp. 205–209 (in French).

This *dual co-territoriality* results from the political situation created by the Crusades and imposed and perpetuated itself until the Reformation. In other words, from the thirteenth to the sixteenth century we have, on one side, ecclesiastic *mono-territoriality* and *mono-jurisdiction* in Western Europe on the land of the Patriarchate of Rome. On the other side we have encouragement by the latter of *ecclesiastic co-territoriality* followed by the exertion of *hyperoria (multi)jurisdiction* on the territories of other Churches of the East which, from that point onwards both *internal* and *external* co-territoriality are established (and coexisted). In these new ecclesiological idioms, one could perceive the beginnings of the development of *global ecclesiology*, starting primarily after the Reformation.

However, despite the political pressure of the time, the stance of theology, which lives with the vision of re-establishing *ecclesial communion* and resolving the ecclesiological problem, remains strong in the Western Christian world. The two Councils, that is. the Council of Lyon (1274) and of Ferrara-Florence (1438–1439), gathered together bishops (who called each other *brother* during these Councils) who were in *rupture of communion* and not in a situation of *schism* (otherwise there would be no point in summoning such Councils). There is also the continuing settlement of monks from the East on Mount Athos from the beginning of the fourteenth century which clearly shows that the desire for an ecclesiological solution to the *rupture of communion* was still alive, despite all the *politically dictated, though still solvable, co-territorial behaviour*.

91.3 The Ecclesiology of the Reform (Sixteenth Century)

It was the Reform which caused the emergence of the ecclesiological problem of co-territoriality on the territory of the Patriarchate and Church of Rome. Indeed, in the sixteenth century, this ecclesiological aberration of co-territoriality was for the first time conveyed to Central and Western Europe, fragmenting both internally and territorially the Patriarchate of Rome, just as the other Churches of the East which had previously been internally fragmented. Here, it is worth remembering[3] how

[3] See our article entitled "The oppositional relationship between the locally established Church and the ecclesiastical 'Diaspora' (Ecclesiological unity faced against 'co-territoriality' and 'multi-jurisdiction')," in Synaxis, vol. 90 (4–6/2004), pp. 28–44, and in Archim. Grigorios D. Papathomas (2006), Ecclesiologico-Canonical Questions (Essays on the Orthodox Canon Law), Thessaloniki-Katerini, "Epektasis" Publications (series: Nomocanonical Library, n° 19), Chap. III, pp. 107–144 (in Greek). Also, "La relation oppositionnelle entre Église établie localement et "Diaspora" ecclésiale (L'unité ecclésiologique face à la co-territorialité et la multi-juridiction)," in L'Année canonique [Paris], t. 46 (2004), pp. 77–99, in Contacts, t. 57, n° 210 (4–6/2005), pp. 96–132, in Ast. Argyriou (Textes réunis par), (2005). Chemins de la Christologie orthodoxe, Paris, Desclée (coll. Jésus et Jésus-Christ, n° 91), Chap. 20, pp. 349–379, in Ast. Argyriou (Textes réunis par), (2005). Chemins de la Christologie orthodoxe, Paris, Desclée (coll. Jésus et Jésus-Christ, n° 91), (2005), 20, pp. 349–379, and in Archim. Grigorios D. Papathomas. (2005)., Essays on Orthodox Canon Law, Florence, Università degli Studi di Firenze Facoltà di Scienze Politiche "Cesare Alfieri" (coll. "Seminario di Storia delle istituzioni religiose e relazioni tra Stato e Chiesa-Reprint Series," n° 38), Chap. 2, pp. 25–50 (in French).

co-territoriality emerged *confessionally* and how it contributed to the aggravation of this ecclesiological problem.

The ecclesiological experience of the first millennium occurred in a given location, the unique *canonical* criterion permitted the establishment and existence of a "local" or "locally" established' Church that was *exclusive territoriality* and *ecclesiological mono-jurisdiction*. The Reformation was then not so much because of its *spatial* separation from the Church of the West, from whence it came, but rather because of its different *mode of existence*. It introduced a new criterion needed for the establishment of a Church, a criterion ecclesiologically and canonically inconceivable previously. Indeed, the newly formed ecclesial communities of different confessions, whose existence at that time was entirely autonomous, needed an ecclesiological hypostasis, which could neither be based on the ecclesiological experience of the Church, such as it was until then, nor on the institutional structure of the *local Church-diocese*. The reason for this was simple: these communities started existing and *coexisting* on a territorial region where a Church was already present, that is, a Church already endowed with ecclesiological territorial identity (Church at a location – *Ecclesia in loco*: Church *that is at* Rome).

It was crucial, however, to find a way on one hand for these Communities to be *Church*, which is in fact why the Reform took place. On the other hand it was important to have *some element to differentiate them* from the pre-existing Church, with which they did not want any identification whatsoever. Martin Luther did not have any intention to create a new Church, but it was impossible to do differently. The use of any local designation would not only cause confusion, but would also require the adoption of equivalent institutional structures such as a bishop, diocese and territorial name. That was what happened in the Crusades when a *schism* (*sic*) had already been declared *a priori*, which legitimized the exact reproduction of the pre-existing structures and designations of the Patriarchates and the Churches of the East.

However, the Reform neither outwardly proclaimed a *schism* with the Church of the East, from whence it "came," nor engaged in an ecclesiological procedure of *rupturing communion* or any analogous process. *It was interested in obtaining an ecclesiological hypostasis but, as a Reform, it definitely wanted to differentiate itself.* In Lutheranism and Calvinism, that is, in traditional Protestantism where dogma is emphasized above all, a dependence of the Church *exclusively* on the *Confession of Faith* (*Confessio Fidei* [cf. Confession of Augsburg – 1530]) is observed. So the Reformation chose, fatally but necessarily, the *adjectival designation* coming from the *confession* of each Protestant leader, avoiding at first the use of a *local designation*. And the need for *confessionalismus* in Ecclesiology was established as well as the *confessionalisation* of the Church, first inside Protestant area, and then outside it. In short, the schism of *ecclesiological unity* in the West caused the emergence of *confessionalismus* and resulted in the newly formed Churches being designated by their *confession* rather than their *territory*; not after the name of a location, but using a *confessional designation* and an *adjectival designation* (for example, *Lutheran* Church, *Calvinist* Church, *Methodist* Church, *Evangelist* Church, etc.).

In summary, the Reform unintentionally enlarged and systematised *co-territoriality* as a form of ecclesiological existence, but then its self-fragmentation into further confessional Churches revealed within Ecclesiology the same corruptive symptom. With astonishing similarity the same characteristic ecclesiological symptom of *dual co-territoriality* appeared here as well, that is, *external co-territoriality* due to the coexistence of each confessional Protestant Church with the Catholic Church from which it came and *internal co-territoriality* since several Protestant Churches *coexisted* on the same territory and in the same city (*conviventia*) without achieving the fullness of communion attained by an ecclesial body in one location as envisioned by the *Pauline Ecclesiology* of the New Testament, viz., the *exclusive basis* (*sola scriptura* and *fundamentum fidei*) of Protestant Ecclesiology. Therefore, there was also not even more *mono-confessionalism* within the *Protestant Family* (*Confession*). In the beginning, however, there was only one and unique confession, but confessionalistic self-fragmentation and non-formal proliferation. And so, despite the vigorous proclamation on behalf of the Protestants that Pauline Ecclesiology is the only New Testamentary truth, the confessional Ecclesiology of co-territoriality is, nevertheless, found within it, not only annihilating every Pauline and New Testamentary vision of the establishment of a Church at a given location, but also relativizing the constantly repeated position of the *sola scriptura*.

91.4 The Ecclesiology of Ethno-phyletism (Nineteenth Century)

For Orthodox Christians things were even more complex and much can be said about the issue. However, we will limit ourselves to two aspects: (a) the existence of *internal co-territoriality* in Orthodox Ecclesiology, to which an extra negative ecclesiological characteristic is added, viz., the *multi-jurisdiction level*, and (b) the non-existence of *external co-territoriality*. We begin with the latter since, in practice, the choice of this ecclesiological position appeared first historically.

First, despite contradictory views between Orthodox Christians on the Orthodox Church today, the year 1054 is not characterised as a *schism*, but rather as a *rupture of communion*. The Orthodox Church never declared it as such throughout the entire second millennium. Thus we have an ecclesiastical event as a schism which is not justified by historical and canonical sources. So "interruption of the communion" is not an accomplished schism. Apart from the fact that "all lasting schisms lead to heresy" (a phrase attributed to John Chrysostom (c. 347–407), famous preacher and Patriarch of Constantinople) and consequently to the complete detachment from the ecclesial body, the Orthodox Church should declare a schism. It would have had to take the same ecclesiological actions as the Church of Rome after the Crusades, and to establish an "Orthodox Patriarchate of Rome," something which, staying completely consistent with itself, it has not done for the last millennium and unwaveringly continues to refuse to do. In addition, for the same reason, it would not have accepted that the three common Councils of the Second Millennium were held, or at least it would not have taken part in them (Lyon 1274, Ferrara-Florence 1438–1439, Brest-Litovsk 1596). (Actually, the third Council of Brest-Litovsk 1596 was summoned during the same century as the beginning of the Reformation). However, the Council of Trento (1545–1563), gave the definitive *coup* to the politics of church union that was promoted at that time. Since the seventeenth century ecclesiological disruption within the body of the Catholic Church in conjunction with the religious wars in the East engendered other priorities and things took a different turn, something which clearly showed up in the Second Vatican Council 1962–1964.

Therefore, it is an ecclesiological error when Orthodox Christians use the term "schism" to refer to the events of 1054. It is about a borrowed terminology and a characterization from a homeopathic reaction. This is another characteristic of the "Babylonian captivity of Orthodox Theology."[4] Thus, the refusal of the Orthodox Church to declare the "rupture of communion of 1054" as a "schism," and also, by extension, the refusal to establish an "Orthodox Patriarchate of Rome," reveals that it lives in hope of *re-establishing communion* and for this reason only, does not practice *external co-territoriality*. We ought to recognize then, regarding this issue, that not only *Pauline Ecclesiology*, but also *conciliar* and *patristic Ecclesiology* "of a single Church at a given location" are clearly preserved in the Orthodox Church and its Ecclesiology.

However, the same view does not apply in the case of *internal co-territoriality*. We ought to state that even on this issue the Orthodox Christians have surpassed the Catholics and Protestants' ecclesiological deviation, since, apart from co-territoriality, they also exert and practice *co-jurisdiction* as well as *multi-jurisdiction* (*multilateralist* and *hyperoria*). (We pretend to be in communion, without there being actual communion since, as we shall observe below, extreme care and vigilance are taken to privilege *ethno-phyletic assets* and not an *ecclesiological communion*). This point precisely shows that contemporary orthodox ecclesiology is an ecclesiology with stratifications and symmetrical deviations, revealed not only in orthodox ecclesiological practice across the world today, but also in the statutory practice of the Orthodox National Churches as we shall see below. Two examples of statutory dispositions with non-ecclesiological content are sufficient to highlight the enormity of the existing ecclesiological problem. It would be useful to recall one article from the Statutory Charters of a Hellenophone and Slavophone Church, that is, the Statutory Charter of the Church of Cyprus and the Statutory Charter of the Church of Russia, in order to put them in the perspective of our ecclesiological research.

- Members of the Orthodox Church of Cyprus are:
  - all Cypriot Orthodox Christians, who have become members of the Church through baptism, and who are *permanent residents* of Cyprus (the juridical principle of *jus soli*) as well as

[4] See Florovsky, G. (1939). "Patristics and modern theology." In, H. S. Alivisatos (Ed.) Procès-Verbaux du Premier Congrès de Théologie Orthodoxe à Athènes-1936 (pp. 239–240). Athens: A. S Pyros.

- *all those of Cypriot origin* (the juridical principle of *jus sanguinis*), who have become members of the Church through baptism, and *are currently residing abroad* (Article 2, Statutory Charter of the Church of Cyprus- 1980).

- The jurisdiction of the Russian Orthodox Church extends to people of *orthodox confession* residing in the USSR [1988]; residing on the *canonical territory* of the Russian Orthodox Church [2000], as well as

  - *people*[5] *who reside abroad and who voluntarily accept its jurisdiction* (Article I, § 3, Statutory Charter of the Church of Russia- 1988 and 2000).

Both articles are representative of Statutory Charters with three main and common non-ecclesiological properties:

First, the jurisdiction of these Churches extends itself, deliberately and principally, to people, just as in the ecclesiology of the Reform, and not exclusively to territories. In other words and without further analysis, the exertion of ecclesiological jurisdiction on people simply means that this single statutory fact gives these Churches the right to intervene, by definition, into the canonical bounds of other locally established Churches. While we all know that autocephaly, according to Pauline Ecclesiology, is granted to a given location, to a territory with explicit boundaries and on purely geographic criteria, today usually geo-state and not a nation. So the notion of autocephaly is essentially that found in the New Testament Ecclesiology in contrast to the Old Testament insofar as the latter identifies the chosen people with the nation. Consequently, the jurisdiction of a locally established Autocephalous Church is exerted on a specific territory and never on an entire Nation, much less on scattered people. "People," therefore, are defined not on "canonical territory" which a Church invokes only in self-defence against "intruders" who, conforming to their Statutory Charter, plan to establish an exterior (hyperoria) co-territoriality on its "canonical territory." This is done to prevent external ecclesiastical interventions on its own ecclesial territory on the part of some other jurisdiction (or some other "confession") acting according to the same principles, since this Church itself statutorily practices such ecclesiastic interventionism on the canonical territory of other Churches.

Second, the Churches in question statutorily declare that they are unwilling, for any reason, to limit the exertion of their jurisdiction to territories situated within their canonical boundaries as they should ecclesiologically since not only are they both locally established Churches but also because of the principle of Autocephaly, which determines their ecclesiological and institutional existence, demands it. However, they insist on expanding beyond their canonical boundaries, since their Statutory Charters gives them this right.[6] In ecclesiological practice, this is called

[5] This presumably implies the faithful.

[6] In the same mindset, the Patriarchate of Russia has easily kept its recent promises, given everywhere (Western Europe, Estonia, Russian "hyperoria" Church, etc.) to provide a "large (sic) ecclesiastic autonomy." A recent event explains this mindset. Four documents were published, concerning the restoration of unity between the Patriarchate of Russia and the Russian "hyperoria"

institutional interference and, most of all, institutional and statutory confirmation of co-territoriality. In other words, this practice is an institutional ecclesiastic attempt to reinforce co-territoriality within ecclesiology.

Third, and most importantly, these Churches, when referring to territories outside their boundaries, knowingly and purposely make no clear distinction between territories plainly of the "Diaspora" and principal "canonical territories" of other locally established Churches. By extension, this particular statutory reference to people obliterates the elementary canonical distinction of "canonical territories" and "territories of the Diaspora," thus creating not only the definition of internal co-territoriality (this time founded on a statutory basis with the results of a multilateral hyperoria multi-jurisdiction) but also on another anti-ecclesiological phenomenon and this characteristic: the notion and practice of global ethno-ecclesial jurisdiction. This newly formed idiom, just like in the case of the Catholic Church of the Middle Ages, begins to define a global Ecclesiology which is limited to a national(ist) level this time. It also results in the formation of numerous global orthodox national Ecclesiologies.

Thus, despite inherent contradictions the Statutory Charters of the Churches of Cyprus and Russia introduce a dual ecclesiological-canonical system for the exertion of their ecclesiastic jurisdiction, a system which is built ecclesiologically speaking, on an inherent contradiction:

Internally, within the boundaries of the body of the locally established Church, they ecclesiologically exhibit "canonical territory," that is, territoriality and mono-jurisdiction.

But, externally, beyond the boundaries of the body of the locally established Church, they statutorily claim "hyperoria jurisdiction" that is, co-territoriality and multi-jurisdiction.

This fact in itself, by definition, constitutes a corruption and an alteration of the Ecclesiology of the Church and results, if I may to use the expression, in an ecclesiological hotchpotch. On this point, the Ecclesiology of the Church of the New Testament, of the Canons and the Fathers, bears no relation, none at all, to the Statutory Charters and vice versa. In this way, we affirm the famous adage which underlines the eonistical priorities of the Christians: "Siamo primo Veneziani e poi Christiani" (translation: Principally, we are Venetians and then Christians).

"The fullness of time has come" (Gal 4, 4), and we must realise that the statutory ecclesiology of National Orthodox Churches today is deeply problematic. The deficiency of the Statutory Charters is not so visible inside a country, although the recent theory about ethno-cultural "canonical territory" – which reminds us of the

Church. From these published documents, it appears that the current leaders of the Russian "hyperoria" Church have abandoned all previous grievances against the Patriarchate of Moscow. In exchange for recognising the Patriarchate of Moscow's jurisdiction, the Russian "hyperoria" Church has, "with respect to economy," obtained a status of "auto-administration." allowing it to exist as a specific ecclesial structure in different parts of the world where it is established, in parallel with the diocesan structures of the Patriarchate of Moscow which already exist on these same territories (SOP, n° 300 (7–8/2005), pp. 21–22).

international juridical principle of the *jus soli* – does expose a few problems. However, this deficiency is more tangible outside the country, in the territories which we refer to, though we should not, as part of the "diaspora." The problem also lies in the fact that these Statutory Charters contain elements which are not only ethno-phyletic, but also of confessional, juridical and most of all, non-canonical and non-ecclesiological. They remind us more of a section from a more general ethnocratic manifesto than they reflect the Ecclesiology and Theology of the Church. These official statutory texts of the twentieth century once again attest to the "Babylonian captivity of Orthodox Theology" of the Church. This time they are related to state nationalism and the dominant national ideology, and to its metamorphosis into an ethno-theology which consequently engendered ethno-ecclesiology as the dominant characteristic of the post-ecclesiological age for Orthodox Christians. Of course, this age is not characterized by the term itself, but by the reality the term reflects, viz., a reality, which on a more profound level can be found in the priority given to ecclesiastic ethno-culturalism (ethnoculturalismus).

As actors of "multilateralism" (multilateralismus), for reasons which today are known, clear and obvious, Orthodox Christians today blame the Crusades of Western Christians, but they are unable to recognize that their ecclesiological stance, statutorily and institutionally, which follows in the footsteps of the Crusades and their Ecclesiology. An ecclesiological, not ethno-phyletic, look at the cases of co-territoriality, for example, in Estonia, Moldavia or the Former Yugoslavian Republic of Macedonia (FYROM) suffices to point out the ecclesiological-canonical confusion which rules over orthodox geo-ecclesiastical circles today.

In summary let us examine a related issue associated with the mentality that the ethno-phyletic content and ethno-cultural perspective of such statutory Charters spreads.

Essentially, the Church has always been Eucharistic and, as far as geographical areas are concerned, territorial in the expression of its identity and its presence in history. Pauline ecclesiology, as well as the whole patristic ecclesiology which followed, has never designated a "local" or "locally established" Church in any other way than through a geographical name as the terms themselves indicate. The defining criterion of an ecclesial community, an ecclesial body or an ecclesiastic circumscription, has always been the location and never a racial, cultural, national or confessional category. A Church's identity is described, and has always been described, by a local designation, that is. a local or locally established church (for example, Church which is at Corinth (1 Cor 1, 2; 2 Cor 1, 1), Church of Galatia (Gal 1, 2), Patriarchate of Jerusalem, Patriarchate of Rome, Church of Russia, etc.). But a Church preceded by a qualitative adjective (for example, Corinthian Church, Galatian Church, Jerusalemite Church, Roman Church, Russian Church, etc.) has never previously existed as it exists today. And this is because, in the first case, we always refer to the one and only Church established at different locations (for example, the Church being at Corinth, at Galatia, at Rome, in Russia, etc.), whereas in the second case it appears not to refer to the same Church, since it is necessary to describe it using an adjective (ethno-phyletic or confessional category) in order to define it and to differentiate it from some other Church: Serbian,

Greek or Russian Church – just as we say Evangelic, Catholic, Anglican or Lutheran Church. We have seen that the Lutheran Church, having lost its local "canonical" support for reasons which were confessional and related to the expression of its identity, resorted to other forms of self-definition. Similarly, within the territory of the "Orthodox Diaspora," while we cannot in any way say "Church of Serbia of France," which would be ecclesiologically unacceptable, specifically because it would cause total confusion between the Churches. We can instead, for purely ethno-phyletic reasons relating to the expression of its identity, easily say, as we do, not only orally but also in institutional and statutory texts, "Serbian Church of France"[7] or "Russian Church of Estonia."

91.5 The One Church and the Many Churches

The conclusion of this brief ecclesiological analysis of the usage of adjectival designations is that we have one, and only one, Church in Corinth, only one Church in Galatia, and only one Church in Jerusalem. However, these are not three different Churches, but one Church, the one and the same Church of the Body of Christ, which is found in Corinth, Galatia and Jerusalem. In this sense there are no, and cannot be any, "sister Churches" as separate ecclesial bodies, but one unique Church in different locations. In this ecclesiological context, the word "sister" is completely unwarranted, because it creates two bodies where only one can exist. This designation does not exist in the Ecclesiology of the first millennium. The use of this term presupposes and, most of all, implies unsaid confessional or cultural projections in the one indivisible Body of the Church. In precisely the same way, we do not have a Russian Church, a Bulgarian Church, a Jerusalemite Church; these would be three Churches and not one. But we have one Church, one and the same Church of the Body of Christ, found in Russia, Bulgaria, Jerusalem. This explains why each ethno-ecclesiastic Statutory Charter is heading, through its position and its premises, towards a deviant perspective and not towards the communion of locally established Churches, as was the case beforehand with the Canons of the Church which were universally common and the same for everyone.

Comparing the principles which govern the three aforementioned Ecclesiologies, it is remarkable to note the external elements they have in common. With the Catholics, for example, the adjectival designation of the locally established Church stems from the ritus, i.e., the designation of the respective Church as "Maronite," or "Melchite," "Greek Catholic," "Uniate," etc. With the Protestants, similarly, the adjectival designation of the locally established Church stems from the confession,

[7] Extract from our article, op. cit., in Synaxis, vol. 90 (4–6/2004), pp. 32–33, in Archim. Grigorios D. Papathomas, Ecclesiologico-Canonical Questions (Essays on the Orthodox Canon Law), Chap. III, pp. 115–116, in L'Année canonique [Paris], t. 46 (2004), pp. 81–82, in Contacts, t. 57, n° 210 (4–6/2005), pp. 102–103, and in Archim. Grigorios D. Papathomas, Essays on Orthodox Canon Law, Chap. II, pp. 29–30.

that is, the designation of the respective Church as "Lutheran," "Calvinist," etc. By exact analogy, the same happens in the National Orthodox Church, where the messianism of the Nation, another form of a confession of faith, consciously or subconsciously prevails, while, at the same time, a perverse relation and dependence of the Church on the Nation and the dominant national ideology is observed. And so, derived from this dependence on the State-Nation, the adjectival designation follows naturally, that is, Serbian, Romanian, Russian, for each Church respectively.

This new and unheard of phenomenon of ecclesiastical adjectival designation can be explained with little difficulty as, subconsciously, since the ecclesiological center of gravity moved from being territorial to ethno-phyletic, or, in the corresponding case in the West, ritualistic or confessional. We have replaced the local designation with an adjectival designation, corresponding to the deviant ecclesiological experience, and, if adjectival categories are used, driven by precisely the same need for self-designation motivating the use of confessional adjectival categories. However, as far as Ecclesiology is concerned, there is no such thing as a ritualistic or confessional Church or, in the corresponding case, a national ethno-phyletic Church.

Even though these terms may appear to be equivalent (isomorphic), for example. the Church of Romania or the Romanian Church, and though the difference in terminology may seem quite superficial, we maintain, according to what we have seen above, that there is a real and significant difference between using the name of a place and using an adjectival epithet because these reflect two different conceptions of the Church, revealing either ecclesiological or deviant and heterocentric subconscious intentions. However, as far as the actual content is concerned, the chasm separating them is very vast, just as is the chasm between the "ecclesiological" and the "non-ecclesiological".

91.6 Comparative Approach of the Triple Question

These three divergent ecclesiologies, developed during the last eight centuries of the second millennium (thirteenth to twentieth centuries), have essentially led the Church into the post-ecclesiological age. This is the age in which we live, in which we try to give superficial solutions, either through Councils like the Second Vatican Council and the proposition to increase Ecumenism, or through increasing efforts to federalise Protestant Churches, or even by the fruitless attempt to summon a Pan-Orthodox Council, which has been in preparation, to no avail, for almost half a century. It is certain that the solution will neither be ritualistic or ecumenistical, nor confessional or federative (fusion within the confusion), and certainly not ethno-phyletic or multi-jurisdictional, but will definitely have to be ecclesiological and canonical, which may appear distant, if not utopic, in today's age of post-ecclesiality which has been characterised as the age of modern Christianity, a Christianity which remains woefully multilateralist and non-ecclesiological.

In this comparative approach to the issue, we could add the fact that the emergence of the Reform imposed a de facto situation of co-territoriality, creating, where a Church (Patriarchate) of the West already existed and after the passing of the Religious Wars and, much later, with the emergence of Ecumenism, the evident and uncontested asset of co-territoriality of modern ecclesiology. Since then co-territoriality becomes the exclusive de facto ecclesiological situation for everyone and a perennial ecclesiological fact, unanimously accepted, and, finally, a constitutive element of territorial expression of every locally established Christian Church and Confession. Also all the ecclesial locally established communities gave the impression that they prefer to be and to live in statu confessionis more than in statu Ecclesiae. Thus today co-territoriality constitutes the basic common characteristic of all the Ecclesiologies of Christian Churches:

- For the Catholic Church, let us recall one example. In Jerusalem there are five Catholic Patriarchates, all coexisting, governed by two unilateral Codes of Canon Law.[8] The emergence of Uniatism is also part of the same ecclesiological problem, as well as Rome's efforts to sustain co-territoriality, born by the practice of Uniatism.
- Protestant Churches multiply themselves informally on the same land and across the world trying to solve the problem through federalisations.
- For the Orthodox locally established Churches, let us also recall one example: in Paris there are six coexistent orthodox bishops with equivalent or synonymous – sometimes even homonymous – overlapping ecclesiastic jurisdictions (despite this being explicitly forbidden by the Ecclesiology of the 1st Ecumenical Council

[8] The pathology of the Ecclesiology of the Catholic Church is evident due to the existence of two Codes of Canon Law, the Latin Code and the Eastern Code, which both allow ritualistic and cultural (personal) co-territoriality as an ecclesiological given for the establishment of a Church or an Ecclesial Community, irrespective to the pre-existence of another Church, not only of another confession (hetero-confessionalistic), but even of the same confession (homo-confessionalistic) or of the same rite (homo-ritualistic). In our opinion, the coexistence of two Codes, independent from each other (cf. priest marriages, forbidden by one but allowed by the other, according to a purely geo-cultural criterion), fully reflects the mentality of the post-ecclesiological age. It was inconceivable for every Church Council, ecumenical or local, to formulate two categories of dogma or two categories of canons, tailored to two different categories of people, according to cultural, ritualistic or confessional criteria, as happened during the Second Vatican Council. The same preaching of Christ addresses also the Primitive Church either to the Jews either to the Pagan. In this sense, Vatican I, which published a Code, was more progressivistic than Vatican II, which published two Codes – indeed, two divergent Codes. This is not a matter of inculturation, but of the discriminatory behaviour vis-à-vis faithful and peoples. However, it is true that the Second Vatican Council undertook numerous attempts and positive efforts to escape from the disastrous situation which the post-ecclesiological age imposed and relentlessly continues to impose. The adoption of two Codes, unilateral and independent from each other, shows that there is still a lot of work left for the Catholic Church to resolve the ecclesiological problem of co-territoriality, firstly in its own bosom, then beyond it, by an ecumenical cooperation with the other Churches.

of Nicaea [325][9] and the 4th Ecumenical Council of Chalcedon [451][10]), and all the aforementioned statutory facts of co-territoriality.

- To these few representative examples could be added the Ecclesiology of the World Council of Churches (WCC), with its conscious deliberate pluralistic coexistence as dominant ecclesiological criterion, and, let us not forget, the communion of Anglican Churches, the Armenian Churches and the self-called "Orthodox Catholic Church of France" (ECOF).
- Also, the 17 different Old Calendarist Churches in Greece exhibit, to an astonishing degree, the same characteristic symptom of dual co-territoriality (external with respect to the Orthodox Church of Greece, but also internal with respect to the relations these 17 homonymous and self-proclaimed "Genuine (sic) Orthodox Churches of Greece" have between each other), and, let us not forget the "Russian Hyperoria Church" with the exercise of a world ecclesiastical jurisdiction and with a behaviour, by definition, of co-territoriality.

Consequently, the problem for the Churches face is not primarily ritualistic, confessional or ethno-phyletic, but above all an ecclesiological problem and a problem of ontological communion of the Churches in Christ.

91.7 The Three Ecclesiologies

Never before during the 2,000 year history of Christianity has there been such a broad and far-reaching violation of the Church's Ecclesiology as the one experienced during the "post-ecclesiological" age of the last eight centuries (thirteenth to twentieth centuries). The blame lies with all of us, Catholics, Protestants and Orthodox Christians. The organization of the Churches according to a code, a confession or a national status has ignored, and continues to ignore, repeatedly and deliberately, the ecclesiological canonical tradition stemming from the vital ecclesiastical praxis of the Church of Christ, as inherited from the New Testament, the Ecumenical and Local Councils and the Fathers. Instead it draws its inspiration, though it ought not to, from the realities and conditions of the eonistic "post-ecclesiological" age, without there being the possibility or even the slightest will to find our way back from "how far [we] have fallen" (Rev 2, 5).

As can be concluded from the previous analysis, if it really proves to be true, that is the Crusades effectively created, de facto, a new ecclesiastic situation that influenced – not to say imposed on, Ecclesiology and its evolution. Then Reformation brought forward the problem of ecclesiological co-territoriality, a problem which had already been present since the time of the Crusades (1st

[9]Canon 8/Ist: […] For in one church there shall not be two bishops.

[10]Canon 12/IVth: "It has come to our knowledge that certain persons, contrary to the laws of the Church, having had recourse to secular powers, have by means of imperial prescripts divided one Province into two, so that there are consequently two metropolitans in one province; therefore the Holy Synod has decreed that for the future no such thing shall be attempted by a bishop, since he who shall undertake it shall be degraded from his rank."

Crusade – 1099). The main characteristic of this new ecclesiological situation was the establishment of co-territorial Churches instead of territorial Churches. Therein lies the ecclesiological problem of co-territoriality. In other words, Churches not being in full communion, but rather coexisting with other Churches. Churches with a ritualistic, confessional or ethno-phyletic and, most of all, non-ecclesiological basis and hypostasis (ritualistic, confessional and ethno-phyletic conviventia). A ritualistic, confessional or ethno-phyletic hypostasis which defines and dictates the Codes of Canon Law, the official texts of Protestant Confessions, the Statutory Charters of Orthodox National Churches but also their underlying ecclesiology. These constitute the image and the characteristics of the currently prospering and flourishing "post-ecclesiological" age.

This study discusses how in Modern Times, Orthodox Ecclesiology has strongly been influenced by fully developed protestant Ecclesiology, and less so by Catholic ecclesiology, due to the latter's uni-dimensional ecclesiastical structure on a global scale, engendered by the rupture of communion of 1054 and the ulterior ecclesiological development centred on a single Patriarchate-Church across the world. Perhaps this also explains the easy coexistence of Protestants and Orthodox Christians in the World Council of Churches (WCC), the crowning achievement of the post-ecclesiological age (Table 91.1).

This is the **ecclesiological puzzle** illustrating the meaning, the characteristics, but also the perspectives of the "post-ecclesiological" age. Out of these three Ecclesiologies:

- The Catholic Church has never condemned ritualistic Ecclesiology (thirteenth century) as a deviation from the Ecclesiology of the Church. On the contrary,

Table 91.1 The ecclesiology during the post-ecclesiological age

Catholic Church	**Poly-ritualism; co-territoriality**
External	Establishment of Churches on the territories of other Churches (intra-ecclesial conviventia)
Internal	Churches of ritus form, acceptance of the co-territorial Uniatism and mutual territorial overlap at a single location (intracatholic-ritualistic conviventia)
Protestant Churches	**Multi-confessionalism; co-territoriality**
External	Establishment of Churches on the territories of other Churches starting from the day of their confessional birth (intra-ecclesial conviventia)
Internal	Churches formed by the informal multiplication of Communities and their mutual territorial overlap at a single location (intraprotestant-confessional conviventia)
Orthodox locally established Churches	**Multi-jurisdiction; co-territoriality**
External	Ø
Internal	Churches and ecclesiastical jurisdictions of ethno-phyletic and cultural multi-jurisdictional form and their mutual territorial overlap at a single location (intraorthodox-ethnophyletic conviventia)

Source: Grigorios D. Papathomas

Ecclesiological ritualism continues to inspire the different ritualistic Catholic Churches and determine their beginnings.

- Protestants also never condemned confessional Ecclesiology (sixteenth century) as deviating from Pauline Ecclesiology. On the contrary, Ecclesiological Confessionalism even continues to inspire Protestant Churches and determine their beginnings, after moving definitely from the Biblical Pauline Ecclesiology. So, although theologically unjustified, the very absence of any condemnation diminishes their responsibility.
- Orthodox Christians, however, when ethno-phyletic Ecclesiology started flourishing and prospering (nineteenth century), immediately summoned the Pan-Orthodox Council of Constantinople and condemned Ecclesiological Ethno-Phyletism as heresy (1872). Heresy!… Out of all Christians, only Orthodox Christians had the theological courage to take action conciliarly and condemn such a deviating form of Ecclesiology as heresy, revealing the magnitude of the ecclesiological awareness pervading them at least at that time. After that Council, however, almost all National Orthodox Churches had nothing to show for themselves, statutorily or canonically, other than ethno-phyletic Ecclesiology, that is, statutorily speaking, the heresy they condemned conciliarly. So today, everyone behaves ethno-phyletically, acts ethno-phyletically, and organizes their "ethno-ecclesial diaspora" (sic), while continuing to organise themselves ethno-phyletically to this day (twenty-first century).

This is why Orthodox Christians, in contrast to the Catholics and the Protestants, will be held inexcusably responsible for having adopted such an anti-ecclesiastic behaviour, despite the ad hoc conciliar decisions and recommendations which contribute to the fragmentation of the Church body wherever it is invited and established over the world.

This clearly and strongly attests to the fact that the age we are living through is unmistakably post-ecclesiological, in the time when we know very well that Ecclesiology concerns the mode of existence of the Church. If this is really so, at a time where everyone (Catholics, Protestants and Orthodox Christians) speaks of Eucharistic Ecclesiology, the following question can be asked: in the time of improper Ecclesiology how far is the Eucharist possible? For the Fathers of the Church, if faith was improper, the Eucharist was impossible! But what of the case of Ecclesiology?

Finally, the three Ecclesiologies we have explored share the same pathology, regardless of differences in their theology or confession or even Church so that when speaking of the pathology of a Church's ecclesiology, the same principles are generally valid for the ecclesiology of other Churches too. This includes all their consequences, taking, nonetheless, the specifics and proportions of each Church into account. Thus, there are three "sister" ecclesiologies (by analogy to "sister Churches"), sharing similar and analogous characteristics… three Ecclesiologies which are not in communion, simply because they are disjointed. Three "sister" Ecclesiologies which are completely unrelated to the Ecclesiology of the Church… The New Testament will have to be… rewritten, to theologically justify contemporary Ecclesiologies and their practice... The reestablished in Christ people of God in the New Testament is against every exclusiveness and foreign vis-à-vis any isolationism (particularismus). Because of a subjective personal or collective choice in charge of the rest of the all, the tendency is to isolate a part of the all (particularismus) and through this isolation, to separate, and finally, to divide a body and, by extension, the unity of a body.

The cultural demands of peoples today in our multicultural society are more powerful than the ontological answers that Churches provide. Churches will have to choose whether to conserve the Pauline Ecclesiology of the New Testament which has guided them for fifteen centuries or to give in to the confessional, ritualistic, cultural or nationalist demands of the post-ecclesiological age, which have become the unquestionably established ecclesiology of the present. Certainly, and by the look of things of the future in the latter case, the Church of Christ will be trailing behind the tragically eonistic course of the peoples; the fault will lie with the Churches rather than leading the eschatological way already traced out by the Resurrection (Rev 22, 20).

The votes of France and of Holland during the European referendum (29/5/2005 and 31/5/2005 respectively) were whether to accept a common European constitution. By rejecting this referendum, it was demonstrated that these two countries freed themselves from nationalism and rigid "etatism [statism]." Both have played a leading role in the European ideal and construction, which genuinely fought the nationalist past in Europe. In short they could escape their past. So how could countries still under Europe's influence ever succeed? Not only did these countries not free themselves, but, to this day, they also, by some ecclesiastic institutional means or other, claim that it is the idea of the State-Nation, in other words, the nationalism of the State, or better yet, the phyletic nationalism which determines the ecclesiology of the Church and the canonical resolution of every ecclesiological issue. In this case, the voice of the Canons of the Church and her Ecclesiology can scarcely be perceived in the face of the powerful echo of the current Orthodox ethno-ecclesial Statory Charters. So this voice can nary be heard in the turmoil caused by the corrupted ecclesiological echo in this post-ecclesiological age.

91.8 Summary: New Ecclesiologies and New Ecclesio-geographical Challenges – The Emergence of Post-ecclesiological Modernity

The disunity of the Churches makes it impossible for them fully and effectively to bear Christian witness in the public sphere. In the course of the second Christian millennium, the three major Christian traditions – Roman Catholic, Protestant and Orthodox – have come to distance themselves from the territorial principle of ecclesiology according to which the Church must be one "in each place." From the time of the Crusades (1095–1204), the Roman Catholic Church began to establish Latin Patriarchates parallel to the pre-existing Oriental Local ones and create the ecclesiological problem of *co-territoriality* (1099). Gradually, and especially since the introduction of "Uniatism" (1596), Catholic ecclesiology came to allow churches of different ritual traditions to exist within a single territory. This anti-ecclesiological and anti-canonical *conviventia* creates a new epoch for the Church, an epoch which is obviously post-ecclesial. Therefore, Protestantism, emphasizing the "confession of the faith" which created the ecclesiological problem of *confessionalism* (1517) as the foundation of the Church came to admit the *co-existence* (*co-territoriality-conviventia*) in a single place of churches of different confessions. As for Orthodoxy, it did not consider the interruption of communion with the Western Church (1054) as a full *schism*, and did not, therefore, attempt to create anything resembling a parallel "Orthodox Patriarchate of Rome." But since the nineteenth century, the emigration of Orthodox Christians to regions outside the traditional territory of their respective churches, together with the growth of Ethno-Phyletism (1872), led to the creation of multiple Orthodox bishoprics (*co-territoriality-conviventia*), based exclusively on ethnic criteria (*multi-jurisdiction*), in full communion with each other. National Orthodox Churches sometimes go so far as to claim a kind of extra-territoriality to enable them to minister to their compatriots abroad.

This research makes a contribution to the ecclesio-canonical problem of *co-territoriality* through the three major Christian Ecclesiologies of the second Christian millennium:

1. The Ecclesiology of the Crusades (thirteenth century)
2. The Ecclesiology of the Reform (sixteenth century), and
3. The Ecclesiology of Ethno-Phyletism (nineteenth century)

While Catholics have never distanced themselves from their "Ritualistic" ecclesiology, nor Protestants from their "Confessionalism," the Orthodox did formally and synodally condemn "Ethno-Phyletism" in 1872. For that very reason, the survival of ethno-phyletist tendencies in Orthodoxy church practice is all the more reprehensible. The only way forward for all three confessional families is to return to the sound principle of Pauline ecclesiology in the quest for unity in each place. In this post-modern world of (religious) individualism and (ecclesiastical-ecumenical) relativism, only a witness of true unity and far away from one post-ecclesial geographical *conviventia* can viably make the churches' voices heard in the universal public sphere.

Chapter 92
Hinduism Meets the Global Order: The "Easternization" of the West

Åke Sander and Clemens Cavallin

92.1 Introduction

Over the last several years the topic of "religious diasporas" and their impact on the West has received considerable scholarly attention, particularly in relation to such issues as globalization, migration and increasing ethnic and religious diversity (Esman 2009; Knott and McLoughlin 2010; Safran 2007).[1] And while much of this attention has been focused on Muslims and so-called Islamic fundamentalism, "Hindus" and their various cultural and religious traditions have received a good deal of consideration as well, especially with respect to the question of the so-called "resurgence of religion."[2]

With regard to Muslim traditions, if the rise of Islamophobia, right-leaning political movements and other related phenomena are any indication, it would appear that the reaction to their presence in the West has been largely negative (Allen 2010; Sander 2010, 2011). Yet even those who warn against the impending "Islamization of the West" tend to characterize this "threat" more in terms of its external challenge to Western cultural, political and social ideals than its capacity to deeply penetrate the Western psyche and fundamentally transform "indigenous" attitudes, beliefs, values, standards and customs (for example,

[1] A Google search (1 February 2012) listed over 50 million hits for the term "diaspora." See also www.diasporas.ac.uk for confirmation of the strong interest in diaspora studies.

[2] The global revival of religion has been lately chronicled in a number of important books (for example, Berger 1999; Shah et al. 2011; Micklethwait and Wooldridge 2009; Kaufmann 2010; Stark and Finke 2000; Thomas 2005; Juergensmeyer 2000, 2008; Casanova 1994); with specific reference to Hinduism, see, for example, Jacobsen and Kumar 2004; Bauman 2008; Narayanan 2006; Oonk 2007; Vertovec 2000).

Å. Sander (✉) • C. Cavallin
Department of Literature, History of Ideas and Religion, University of Gothenburg, Göteborg, Sweden
e-mail: aake.sander@religion.gu.se; clemens.cavallin@religion.gu.se

S.D. Brunn (ed.), *The Changing World Religion Map*,
DOI 10.1007/978-94-017-9376-6_92

Bawer 2006; Caldwell 2009; Fallaci 2002; Sander 2010; Ye'or 2005). In reality, the transformational impact of Islam on Western *consciousness* has been and will likely remain rather small.

In considering Hindu traditions, on the other hand, the opposite seems to have been the case. Especially since the 1960s, the general Western reception of Indic philosophies and religions has been quite positive and is best portrayed not as a *phobia* (an aversion or dislike), but rather as a *philia* (a fondness or fascination). Colin Campbell (2007) notes in this connection that the ideas, beliefs and practices of Indian religious traditions have today so extensively penetrated various sectors of Western culture and consciousness that it is not unreasonable to characterize this half-century romance, this fundamental shift in worldview towards the "Eastern ideal type," as the "Easternization" of the West. He then singles out two factors as having been most responsible for the emergence this phenomenon: (1) technological and economic advances that have facilitated mass movements and aided the transference of people, artifacts and ideas (that is, processes of globalization); and (2) the shortcomings, failures and inadequacies of the secularist materialism and scientism of the West. This is to say that the project of modernization, secularization and post-modernization[3] has, among other things, resulted in the alienation of many Westerners from both traditional Western religions and secular ideologies, thus creating a hunger to recapture the spiritual and existential meaning that has been lost along the way—a need so compelling that it seems almost genetically based (Berger 1999; Frankl 1987).

The breakdown of the West's "old traditions," in other words, has opened up an ideological space for a range of alternatives, among which the Eastern traditions have figured quite prominently, especially those of a Hindu or Buddhist orientation, the fundamentals of which strongly resonate with the growing Western interest in health, well-being and self-realization. These types of alternatives have in many respects proved to be "just what the doctor ordered" for ideologically exhausted postmodern Westerners, with their penchant for individual expression, questioning authority, freedom of choice and all-consuming consumerism (Kurth 1999).

92.2 Globalization

Over the last several decades the term globalization has become a "fashion word" within the humanities and social sciences and its meaning has been much discussed (Eisenstadt 1999, 2000, 2003; Eriksen 2007; Michie 2003; Ritzer 2007; Scholte 2005). Indeed so many explications of the term have been proposed, written about, analyzed, redefined and deconstructed that it has come to mean everything and

[3]Whether you chose to use the prefix "late," "ultra," "liquid" or "post" in connection with "modernity" depends whether you side with Gidddens (1999) and view the development of these processes as continuous, or with Bauman (1998) and view them as discontinuous (cf. Christiano 2007). Here we side with Bauman.

nothing at the same time, leaving some to conclude that it has basically lost its terminological effectiveness. The position adopted here is that despite these well-founded reservations, the term still retains a certain heuristic value when referring to those processes of economic, technological, political and social innovation that have caused the world to increasingly function as one interconnected interdependent interactive community (think Internet), as well as to the profound reshaping of human consciousness that these changes have entailed.[4] And along with this radical transformation has come a type of "condition" that is unique to our postmodern times; it entails a heightened and somewhat disillusioned awareness of the vast plurality of religious and other worldviews that has caused us to relativize, personalize, subjectivize and privatize them all (Sander and Andersson 2009).

Returning once again to the concept of globalization, apart from its more descriptive and empirical applications, it has become the controversial center of an intense and highly charged public debate. For those in favor of globalization, it represents the royal road to an international civil society that will usher in an era of global peace, democracy and prosperity, which includes the upliftment of the developing world. For those that are opposed, however, it represents the darkening road to the hegemony of capitalist-driven culture, unsustainable growth and consumption, the domination of the masses by a wealthy elite and the control of natural resources to the disadvantage of all save "the 1 %" (Bauman 1998, 2006; Buckman 2004; Fukuyama 1992; Nederveen 2004; Hacker and Pierson 2010). Regardless of where one stands on this issue, the fact remains that largely due to developments in information technology (and especially the Internet, social networking and the like), globalization has taken a quantum leap forward over the last 30 years, with no apparent end in sight. This, of course, is not meant to imply that globalization is a strictly recent phenomenon. On the contrary, *the phenomenon* is commonly understood to be far older than *the concept,* in a sense, as old as humanity itself (*homo viator*).

With regard to globalization's present manifestation, however, the result in almost every corner of the world has been the rapid activation of an array of economic, political and social changes that have challenged prevailing notions of national sovereignty and tested local cultural, ethnic and religious traditions. According to some, the speed and scope of these worldwide transformations have been dramatic enough to warrant descriptions such as "cultural earthquake" or "cultural revolution." We are also inclined to view this event as potentially representing the first truly global revolution in human history.

---

[4]Although it is beyond the scope of this particular chapter, we will here alert the reader to an important series of discussions concerning whether the processes of modernity and globalization should be conceptualized in terms of singular or multiple forms (Berger and Huntington 2002; Eisenstadt 1999, 2000, 2003). The basic idea, argued by Eisenstadt, is that there are some common features that all "modern" societies share, helping to distinguish them from "non-modern" societies, but that these features attain multiple forms in different regions of the world, hence the term multiple modernities. In other words, Eisenstadt considers the processes of modernization to be multilinear rather than "traditionally" unilinear (cf. Davie 2007: 89ff). And as far as our viewpoint is concerned, we believe that this is fundamentally the correct way to go.

As earthquakes tend to disrupt physical and artifactual landscapes, revolutions tend to disrupt traditions, beliefs and other important aspects of peoples' accustomed lives, with many experiencing such disruptions as threats to their very identity, security and survival. And as long-established norm and value systems come under increasing psychological, sociological and political pressure, feelings of stress and alienation naturally arise among the many individuals that have been affected. Under such circumstances, with their personal and social life-worlds so thoroughly shaken, people tend to respond in different ways. Some (largely urban, well-educated, well-resourced and well-connected persons) tend to see the occurrence as a welcome opportunity, others (perhaps the average person on the street) merely attempt to cope and live as best they can, while yet others react in militant opposition (generally under the banner of religion or nationalism or a combination of the two). The common assumption is that for certain persons in this type of situation traditional religion becomes attractive as a social and political mobilizer (Berger 1999; Eisenstadt 1999; Karner and Aldridge 2004; Sander and Andersson 2009). Samuel Huntington (1996: 75–76) has summarized the complex interplay between (post)modernization and religion as follows:

> In the early phases of change, Westernization thus promotes modernization. In the later phases, modernization promotes de-Westernization and the resurgence of indigenous culture in two ways. At the societal level, modernization enhances the economic, military and political power of the society as a whole and encourages the people of that society to have confidence in their culture and to become culturally assertive. At the individual level, modernization generates feelings of alienation and anomie as traditional bonds and social relations are broken and this leads to crises of identity to which religion provides an answer.

92.3 Religion and Globalization

Despite the fact that the relationship between religion and globalization has been complex and difficult to theorize about (Beyer 1994, 2006; Beyer and Beaman 2007), it can be roughly categorized in terms of three enabling interactions: (1) globalization has enabled the spread of religions, which in their very act of spreading have themselves enabled the process of globalization; (2) globalization has enabled religions to recruit, mobilize and induce participants to act on behalf of various social and political agendas; and (3) religions have enabled individuals to cope with and react to the stress that comes with highly transformational cultural change.

Over the last several decades, academic discourse among sociologists and political scientists has been largely focused on religion's exploitation of globalizing processes and technologies for political purposes, with a specific emphasis on Muslims and Islamic fundamentalism. However, it can be equally argued on the basis of their reciprocal relationship that the impetus to spread religious traditions beyond their original borders has itself contributed to the development of these processes and technologies, particularly in globalization's early phases (Esposito et al. 2008; Juergensmeyer 2003; Oonk 2007; Robertson 1992; Beyer 2006). The historical connection between the two has largely manifested itself in terms of conquest, trade, mission, migration, diasporas and the transnational spread of religious ideas, practices and beliefs, all of which have been encouraged by religious leaders in their desire to spread (or missionize) the faith.

For centuries the direction of these processes primarily moved from West to East in the form of the Christianization, Protestantization and Westernization of the non-Western world. More recently, however, and especially over the last half-century, that direction can be said to have largely reversed, and at an accelerating pace. Within Christianity, for example, there are today *as many* missionaries traveling north from the Christian south; and according to John Esposito (2001), the current directionality of exchange between the Western and Islamic worlds is best portrayed as a "multi-lane super highway with two-way traffic." Likewise, when it comes to exchanges between India and the West, it is undeniable that since the 1960s (with a decided movement forward from the period of British rule to then), the influence of various Indic traditions on Western culture, society, philosophy and spirituality, and even on the Western psyche itself, has been surprisingly profound. Indeed so unexpected is today's penetrating outcome that it would have been beyond the wildest imaginings of even the late nineteenth and early twentieth century missionaries who traveled from Britain, Scotland and other parts of Europe to "civilize" India's "heathen" populations. The impact of Indic traditions on the West is, of course, the primary subject matter of this article, and will be returned to in greater detail below. Before this, however, we will take a closer look not at the partnership, but rather at the friction between religion and globalization and the resistance of the former to the latter, with a special focus on our highly transformational postmodern times.

Globalization in our times has been fueled by the development of previously inconceivable mediums of transportation, communication and informational exchange and is intrinsically linked to a particular postmodernist ideology that affirms the rightness and desirability of values such as pluralism, multiculturalism and individualism, while also professing a strongly held belief in the relativity of all cultures, ethical systems and ideas. By considering only the last of these features, it is not difficult to see why traditional religions might feel threatened by such a world-view and utilize their power of mobilization in an effort to resist both it and the crisis of meaning (or "postmodern condition") that it entails: the one brought about by the destabilization of previously relied upon truths, assumptions, norms, values, manners and customs that can no longer be unproblematically applied (Karner and Aldridge 2004). According to thinkers such as Berger (1999), for individuals that value continuity, stability and predictability, the postmodern condition can be a very uncomfortable state of affairs, and one that enhances the appeal of religious movements promising certainty and the continuance of traditional ways of life.

For purposes of analysis, the various reactions of individuals and/or social groups to globalization (and post-modernization) have been placed on two scales that purport to measure:

1. Degrees of tolerance for ambiguity (Wilkinson 2006; Norton 1975; Furnham and Ribchester 1995); and
2. Degrees of open-mindedness (Rokeach 1960; Sander and Andersson 2009).

While most people are obviously located at different points along these scales (distributed according to something like a gauss curve), the following description focuses only on the extreme ends.

Beginning at the one extreme, we find those who are highly intolerant of ambiguity and seriously troubled by pluralism: Lifton's (1993) "fundamentalist selves," (the alter ego of his "protean selves") who tend to interpret information marked by new, complex and vague meanings as a real or potential threat and as a significant source of cognitive, emotional and normative discomfort (Adorno 1950). These are the persons that tend to defend local particularistic identities, embrace intolerant forms of fundamentalism (or cultural protectionism) and uncritically accept conspiracy-theory propaganda (Hjärpe 2003: 106; Jones 2002). If and when they become involved in religious and/or political mobilization, they tend to gravitate towards movements that advocate various forms of nationalism, ethnic particularism, religious fundamentalism, racism, sexism, homophobia, social exclusion and clash-of-civilizations thinking. In general, they also tend to be of the closed-minded type.

At the other extreme, on the other hand, we find Lifton's "protean selves", those who are highly tolerant of ambiguity, and who thus look upon postmodern values and thinking, as well as the various forms of relativism that have brought the phenomenon about, with unqualified enthusiasm and positivity. These are the "truly" postmodern, cosmopolitan prototypes, with an intellectual and emotional openness to difference that is practically reflected in their daily lives (Vertovec 2010). For such persons, variations of culture, nationality, ethnicity, race and other essentialized categories are not of primary concern when identifying and dealing with others or when developing their own sense of self and identity. Their relationship with others, in other words, is based on an awareness of the hybridity and interconnectedness of the present world, which is perceived (sometimes somewhat naively) as enriching rather than threatening. Such persons, if religiously inclined at all, tend to gravitate towards broad "spiritual revolution" movements such as the so-called "New Age" (Heelas and Woodhead 2005). They also obviously tend to be of the open-minded type.

While this description provides a largely psychological explanation for the reasons that people fall at one or the other end of these scales, one can obviously theorize about this in other ways as well, such as according to notions of social movement and social mobilization.[5] It is also understood that certain individuals assume extreme positions on the basis of a conscious, well-reasoned, rational process of decision-making, grounded upon political and ethical considerations. Such types, however, appear to be in the minority, especially when it comes to the closed-minded/fundamentalist end of the scale. With regard to attempts at explanation, we are aware that they place us in the

[5] With regard to social movement theory, see, for example, Hannigan 1991; McLennan 1995; Hirst 1994; for other theories of social mobilization, see Castells 1998: 19–81; Davie 1994: 10–28; and Ebaugh and Chafetz 2002: 165–191.

middle of the more general "agent-structure problem," one of the "most fundamental" in the social sciences today (Giddens 1999).[6]

This notwithstanding, it is interesting to note that both "ambiguity intolerant" anti-liberal fundamentalism and "ambiguity tolerant" liberal postmodern cosmopolitanism are presently on the rise and in competition throughout world. As to the first of these trends we have such manifestations as Protestant fundamentalism in the U.S., Hindutva in India, Islamism in Arabia, Islamophobia in Europe and the tendency towards ghettoization and particularism among diaspora populations in the West. As to the second, there appears to be a correlation between transnational, transcultural experiences on the one hand and the development of cosmopolitan attitudes on the other (Mau et al. 2008). This, of course, is not to say that an increase in the personal experience of other cultures and religions (along with their carriers) necessarily results in an increase of cosmopolitan attitudes and values; indeed this only seems to be the case when certain specific conditions have been satisfied.[7]

In Shall the Religious Inherit the Earth? (2010), Eric Kaufmann's exhaustive demographic study, the author suggests that it is time to deposit all Huntington-type clash-of-civilizations theories in the graveyard of failed conceptions and accept the fact that the real clash in the twenty-first century is between the two antithetical types mentioned above: postmodern cosmopolitan liberals on the one hand and moral conservative fundamentalists on the other.[8] He also notes, however, that we are entering a period of "unprecedented demographic upheaval" that will likely have quite unsettling consequences for those that long for the victory of the former over the latter (2010: 269). Kaufmann's basic statistical finding is simple, straightforward and relatively unassailable: the more religious that people tend to be, regardless of their tradition, their level of education, their economic status and so forth, the more children they tend to have. This is indicated, among other things, by the fact that both religious countries and religious diasporas tend to have higher and more rapid levels of population growth than either of their secular counterparts. According to Kaufmann, because immigrants tend to be more religious than the postmodern peoples of their host countries, who tend to be more materialistic, self-absorbed, freedom loving, career-oriented and averse to the maintenance of large families, the effects of immigration and procreative multiplication will eventually reverse the Western secularization process. Drawing upon a profusion of empirical data, Kaufmann foresees the gradual diminishment of liberal secularism and Western modernity, concluding that the religious will inherit the earth (Kaufmann 2010: 269).

[6] For discussions on this problem, see, for example: Archer 1996, 2000, 2003; Berger and Luckmann 1967: 40ff; Parekh 2000; Rothstein 2003: kap. 2; Swidler 2001a, b.

[7] See, for example, Amir 1969, 1976; Triandros and Vassiliou 1967; and Jaspars and Hewstone 1982.

[8] The position reflected in this paraphrase of a statement by Stark and Finke (2000), who added that we should whisper "Requiescat in pace" over Clash-of-Civilization's grave, has been forcefully argued by Erik Kaufmann (2010) as well. Samuel Huntington and his followers never seemed to realize that "the dynamic global image of our times is not the clash of civilizations, but the mark being of civilizations and people" (Eck 2001: 4).

92.4 Religious Studies and the New Diasporas

Largely due to the manifold movements of peoples across national, territorial and cultural boundaries over the last 40 years,[9] religious pluralism has become part of the fabric of most contemporary societies. Despite this fact, much of the research in sociological, anthropological and political science that purports to examine immigrant communities and populations have not taken "the religious dimension" into proper account (Eck 2001; Stark and Finke 2000: 1; Ebaugh 2003; McLaughlin 2010). Within Religious Studies, for example, the primary focus has been on the classical country-of-origin traditions of a given immigrant community's religion as well as on that religion's traditional sacred texts. The same can be said of Diaspora Studies (McLaughlin 2010), which have largely focused on the migrant-related problems of minority diaspora groups, viz., those that are concerned about such issues as social, cultural and religious prejudice, labor- and housing-market discrimination and imbalances in educational and other societal opportunities (Sander and Larsson 2007).

The previous predominance of the now largely abandoned "secularization thesis" provides much of the explanation for this lack of scholarly interest since it was widely assumed that processes of modernization would cause religion to become irrelevant and eventually disappear (Berger 1968: 3; Wallace 1966: 265; Wilson 1966; Bruce 2002; Sander and Andersson 2009). Another possible reason concerns the strongly held Western assumption that immigrants coming from the "underdeveloped" to the developed world would gradually come to recognize the superiority of modern Western culture and thus eagerly embrace the lifestyles and values of their host societies. However, as is becoming increasingly clear, this simply has not occurred. Indeed the distinctive features of immigrated religious traditions have tended to become more rather than less visible over the last several decades, not only in terms of institutional structures such as temples and mosques, but also in term of costumes, symbols, ceremonies, traditional practices and unique dietary, funerary and gender-related requirements (Juergensmeyer 2003; Casanova 1994; Eck 2001). The intense debate that is currently under way in numerous European countries over the so-called "head-scarf problem" serves as a case in point (see Bowen 2007; McGoldrick 2006; Roy 2007; Rosenberger and Sauer 2012; Scott 2007). Beyond this, an increasing number of researchers report that immigrants tend to become not less but more religious after settling into their host countries, considering religion to be an extremely important part of their individual and communal identities and lives (see Kaufmann 2010).

Today the secularization thesis' obvious misreading of the modern situation, the wrongheadedness of other secularist assumptions and the global impact of religion on various cultural, social and political developments have collectively resulted in

[9]The total number of foreign migrants worldwide stands today at an estimated 231 million, up from 76 million in 1960 and 174 million in 2000 (http://www.oecd.org/els/mig/World-Migration-in-Figures.pdf; http://esa.un.org/unmigration/TIMSA2013/documents/MIgrantStocks_Documentation.pdf).

a renewed and growing interest in the academic study of religion. And here it must be said that contemporary researchers are no longer approaching religion and religiosity as mere dependent variables, but are rather viewing these phenomena as highly significant factors in the lives of individuals and communities, especially those of the diasporic kind (McLaughlin 2010; Sander and Andersson 2009). In many ways, however, the scholarly study of religion and religiosity is a newly developing discipline that continues to undergo conceptual, theoretical and methodological transformation in the search for its own identity. Numerous specialists in various fields, for example, consider many of the terms used in this chapter to be unclear, ambiguous or even questionable, with words such as "globalization," "culture," "ethnicity," "identity," "diaspora" and even "religion" and "religiousness" remaining the focus of heated debate. This point becomes clear when perusing the large number of recent scholarly publications on religion produced by researchers from such disciplines as psychology, sociology, political science, geography, religious studies and international relations.[10] Although we are fully aware of these definitional problems, delving into them here would cause us to greatly overstep this chapter's space limitations.

92.5 The South Asian Diaspora

Not too long ago the world was conceived as being segregated by fairly strict religious and geographical boundaries, with Christianity located in Europe and the Americas, Islam in the Middle East, Hinduism in India and so forth. And even though these religious divisions have never been without their exceptions and ambiguities, in the globalized world of the twenty-first century their existence is far more rare: most religions have members in virtually every society and few can avoid the awareness that "almost everyone is everywhere" (Juergensmeyer 2003; Beyer 2006; Beyer and Beaman 2007; Eck 2001). According to Jeurgensmeyer (2003), the globalization of religion has transpired in accordance with three basic patterns of dispersal and interaction:

- *The diaspora pattern,* exemplified by the distribution of migrants in self-contained communities outside their "homeland."
- *The transnational pattern,* exemplified by the missionary activities of minority religious traditions oriented towards the conversion of whole regions.
- *The pattern of religious pluralism,* exemplified by the emergence of new forms of religion as a result of cultural and religious interaction and hybridization. (New Age, New Religious and Hindu Guru movements exemplify this pattern of "syncretism" and "hybridization" between old and new Eastern and Western traditions.)

[10] See the special bibliography in K. Knott and S. McLoughlin (Eds.) 2010.

This perspective brings us back to the major theme of this chapter in the sense that it has been the relatively large migration of Indic peoples to the West that has been primarily responsible for the physical globalization of Hinduism. In the case of the United States, this movement became especially pronounced following the passage of the 1965 Immigration Act, which resulted in a substantial increase in immigration from India. Indeed it was during this period that a variety of Indian-based spiritual movements, groups and gurus travelled to America to spread their teachings and establish their own missions among the counter-cultural youth of the 1960s. Some of these movements, after having established themselves in the West, returned to India with Western followers in an attempt to strengthen their institutions at home (a topic that we will briefly return to below).

Although for various theoretical, empirical and conceptual reasons it is difficult to reliably ascertain the number of Indian peoples living outside South Asia, the relatively recent transplantation of a number of Indic diasporas had led to a situation in which almost all important forms of Hinduism and Buddhism, along with their temples, monasteries and other institutional facilities, can be found in most major cities of the West.[11] Based upon standard sources such as the World Religious Database, the Pew Forum, the World Bank, the United Nations and Gallup, the estimated number of Indic persons living outside South Asia stands at around 20 million. Among these, approximately three million live in the United States, around 1.8 million live in Europe (with 1.3 million living in the U.K.) and around 1.3 million live in Canada, these being the three Western regions with the largest Indian populations.

The importance of these migrants and the various religions and spiritual movements they represent is indicated by the fact that since the late 1970s there has been a blossoming of both academic and public interest in all things "Indian:" Indian spirituality, Indian teachers, Indian philosophy, Indian medicine, Indian diasporas, Indian movies, music, fashion, cooking and art (Campbell 2007; Knott and McLoughlin 2010; McLaughlin 2010; Oonk 2007). Moreover, since the late 1980s there has been a dramatic increase in the number of India-related literatures produced, distributed and read, especially those relating to Indian philosophy and religion (both scholarly and otherwise). This particular phenomenon is one that has been neglected in almost all diaspora and migration studies (Ebaugh 2003; McLaughlin 2010). It is also one that reflects the alienation experienced by many Westerners relative to their previously relied upon religions, understandings, ideologies and values, and that substantiates their growing interest in various aspects of Asian philosophy and religion, often presented in hybridized forms by a variety of health, therapy, wellbeing and self-realization movements (for example, Carrette

[11] For further discussion on the reliability of Indian demographic figures, see Brown 2000; Sander and Larsson 2007; Sheffer 2003: 99ff; for further discussion on Hinduism and Buddhism in the West, see Eck 2001; Juergensmeyer 2003; Vertovec 2000.

and King 2005; Heelas 2008; Lyon 2000).[12] Perhaps the most recent example of this concerns the enormous success of "mindfulness" (*sati*), a form of Theravada Buddhist meditation that has penetrated a large segment of this "industry" in the West. Originally based upon Jon Kabat-Zinn's 1970s pain-patient studies, the movement began to flourish after the 1990 publication of his *Full Catastrophe Living*, which introduced the training method MBSR (*Mindfulness Based Stress Reduction*).

Karner and Aldridge (2004) have noted in this connection that processes of globalization have enhanced the role of religion in the search for identity, belonging and (we would add) meaning. The specific contribution of migrated traditions in this regard is to some degree explained by the fact that globalization establishes a connection between the local and the global, as has been pointed out by Peter Beyer (2006: 24), with reference to Roland Robertson (1992): "globalization is at the same time the universalization of the particular and the particularization of the universal."

The process is said to operate over time in a sequential, circular manner, beginning when a particular(ized) local religion deterritorializes, spreads and globalizes to other parts of the world by way of political, economic, technological and religious forces and strategies. In the course of this process, the tradition universalizes by shedding many of the cultural, ethnic and social forms that are specific to its origins. This universalized form once again particularizes or *"glocalizes"* (Robertson 1995; Beyer 2006) in order to become "functionally significant" (Gurwitsch 1964: 148f) in its various new local situations, which, of course, have their own specific cultural, ethnic and social forms.

Referred to as the "hybridization of a tradition" (Brubaker 2005; Knott and McLoughlin 2010; Nederveen 2009), it is a process whereby persons located in the receiving context select those elements of the incoming tradition that they consider to be most valuable and useful, and then reshape them such that they become applicable to their own environments. In other words, the particularization of the universal repeats but also transforms it, thus relativizing and localizing the original. Theoretically, this newly particularized, hybridized and localized religious form can undergo yet other sequences of globalization, universalization, particularization, hybridization and localization, either in new geographical contexts or after having migrated back to its original location. As has been noted by many scholars and researchers in this field (for example, Beyer 2006: 75f), the globalization of Hindu traditions serves as a prime example of the practical workings of this process.

[12] There has been much discussion about the great impact of "holistic," "psychologized," "health seeking," "therapeutic" and "spiritual health" ideas (Heelas 2003) that have been inspired by Eastern religions and culture, but formulated through a decidedly Western prism (that is, the general view of the West on health and well-being and the particular view of the West's health sector on the same). See, for example: Flanagan and Jupp 2007; Hanegraaff 1998; Hermansen 2000; Löwendahl 2002; Hedges and Beckford 2000; DeMarinis 2003; Kickbusch and McQueen 2007; Ugeux 2007.

Before moving on to the topic of Easternization, however, it should be mentioned that when the receiving context is religiously plural, the process becomes more complex. If, for example, the globalization of a particular religion is primarily driven by the spread of diaspora communities and its expression remains largely within the framework of those communities, change may be less dramatic since minorities in pluralistic societies tend to cling to their tradition's original localized cultural, ethnic and social forms.[13] If, on the other hand, the religion of a minority community to some degree "crosses over" and becomes accepted by members of the majority society, the processes of universalization, de-traditionalization and re-traditionalization will be far more pronounced.

Another type of complexity can be found in the phenomenon of "internal religious pluralism," in which the immigrated tradition is represented by multiple forms of religious, cultural, ethnic and social expression. This development can be seen in the case of Islam in Europe, where mostly younger and second generation Muslims strive to transcend the plurality of regionally-based Islamic forms by prioritizing "religion" over "custom" so as to derive either "a pure and universal" or a "melded European" brand of Islam (Cesari 2010; Maréchal 2003; Sander and Larsson 2007). Both attempts, of course, represent different forms of hybridization (Sander and Larsson 2007; Sander and Andersson 2009).

92.6 Easternization of the West and the Westernization of the East

The very notions "East" and "West" constitute one of the most intriguing aspects of Colin Campbell's thesis on the apparent *Easternization* of the West since the 1960s. Interestingly, and in fact, the Eastern influence has been an integral part of Western religious and intellectual life from the very beginnings of Western civilization, dating all the way back to the ancient Greek Pythagoreans followed by the early Christian gnostics and up to the New Age movements of today, all broadly corresponding with what is generally described as Western Esotericism (Faivre 1994). According to Campbell's East-west dichotomy, these marginalized heterodox Western traditions incorporate the typically Eastern element of metaphysical monism, which stands in direct contrast to Western materialistic dualism, whether in its religious (Christian) or its secular (Enlightenment) dress.

In the nineteenth century, the most significant fusion of Western esotericism and Eastern religious traditions was found in the Theosophical Society, established in New York in 1875 by the Russian spiritualist Helena Blavatsky among others (Washington 1993). Although the Society's point of departure was clearly the incorporation of Eastern into Western traditions, its focus shifted more towards the East after moving its headquarters to Adyar, India in 1882. During the 1890s, however,

[13] See, for example, Mol's (1979) conception of religion as the "sacralizer of identity" and "harnesser of change."

there was a split between the Society's U.S.-based Western branch and the one located in India. Among the leading figures that remained in India, Henry Steel Olcott (1832–1907) successfully strove to revive Buddhism in Sri Lanka and Annie Besant (1847–1933) became involved in the struggle for India's independence, up to the point of being elected president of the Indian National Congress in 1917. The theosophical project can in many ways be described as a precursor of the modern New Age phenomenon, which also favors Western esoteric, heterodox traditions and reinterpreted Eastern religious beliefs and practices over traditional orthodox Christianity (Hammer 2001).

At the same time that Western esotericism was moving East, Swami Vivekananda (1863–1902) took a major step towards the West through his appearance at the 1893 World's Parliament of Religions in Chicago. Perhaps the most known representative of Hinduism in the West, Vivekananda intrigued his conference audience by promoting a modern form of *advaita vedantic* non-dualism that stressed the monistic oneness of all things and presented Hinduism as a tolerant, ecumenical and universalist tradition that accepted the truth of all religions and had no interest in the making of converts. This particular interpretation of Hinduism, it turns out, was to some degree the result of a significant process of local transformation and reinterpretation, largely brought about by the British colonial presence in West Bengal and the social, cultural, intellectual and technological modernization that followed in its wake. Vivekananda was well versed in Western intellectual traditions and began his career as a reform-oriented skeptic. However, after an eventful meeting with Ramakrishna, a significant Hindu mystic who taught the unity of all religions, he experienced a religious conversion and became this guru's most prominent disciple.

In this sense, the successful exportation of Hinduism in 1893 was at least partially a consequence of Bengal's century-long exposure to various processes of globalization and Westernization. And since there are a number of Indic religious traditions and schools of philosophical thought, one can ask how it was that *advaita vedantic* non-dualism, in particular, became the philosophical foundation of a modern, universalist (and in many ways de-ritualized and non-iconic) form of Hinduism? The question takes on even more significance when considering non-dualism's longstanding philosophical and ontological differences with the *bhakti* movements that were dominating much of the Hindu religious scene at the time, movements that represented variously qualified forms of non-dualism as well as what some have described as "personalist dualism" (Sardella and Gupta 2015).

Campbell (2007) argues that the Western concept of "godhead" entails a radical separation (or dualism) between God and God's material creation, indicating a conception of transcendence that asserts the existence of an epistemological gulf between human beings and the divine, thus directing humanity's focus to the phenomenal world. Through the process of secularization, he continues, the religious version of this dualism has been weakened, if not altogether destroyed, and a secular form of materialist dualism had taken its place in the contemporary context (Campbell 2007: 62f). The Eastern ideal type, on the other hand, is viewed as being based upon the conception of an impersonal divine force that permeates the universe, conceived more as an emanation than as a separate creation. Thus, according to Campbell's understanding, the Eastern ideal type contains neither the conception of a personal supreme god nor the conception of revelations: "… no Eastern divinity resembles this figure in being a truly transcendent all-powerful, and otherworldly creator figure" (Campbell 2007: 64).

The question is whether Campbell's understanding is truly representative or whether it merely provides a narrow, one-sided portrayal of India's philosophical and religious traditions? The history of Indo-philosophic discourse answers this question with a resounding affirmation of the latter alternative. Dating back to at least the eleventh century with Ramanuja and the thirteenth century with Madhva, various forms of qualified non-dualism leading by degrees to outright dualism have been developing right along side, and in opposition to, Shankara's unqualified non-dualism and its various historical offshoots, up to and including Vivekananda's *advaita vedantic* non-dualism. These alternate schools arose out of a 1,000 years of philosophical reflection and are today representative of the growing power of *bhakti* devotionalism in the second millennium of the Common Era. And with their claim that "metaphysical personhood is the foundation of Vedantic thought," they also pose a direct challenge to Campbell's assumption that the Eastern ideal type exclusively conceives the divinity as "an impersonal, formless and all-pervading substance known as *Brahman*" (Sardella and Gupta 2015).

It was one such form of *bhakti* dualism (known as Gaudiya Vaishnavism) that made its way to New York City in 1966 via A. C. Bhaktivedanta Swami Prabhupada, the founder of the International Society for Krishna Consciousness (ISKCON), the global organization popularly known as Hare Krishna. Here it is interesting to note that the eventual migration of this teaching throughout the Western world and beyond, including its return to India in the form of numerous committed Western adherents, was facilitated by a fusion of certain egalitarian and universalist notions embedded in the medieval writings of its originators and other similar (but perhaps more developed) notions promoted by Westerners in India during the time of the British Raj (Sardella 2010). Thus, as with Vivekananda, processes of Westernization (for example, Western values and intellectual traditions), modernization (for example, the printing press and advanced principles of management, organization and distribution) and globalization (for example, advanced modes of communication and transportation) played a role in enabling Gaudiya Vaishnavism to step not only beyond territorial borders, but also beyond religious borders (for example, the caste system, Brahmanism, gender barriers, etc.) so as to make its successful journey West (Sardella 2010). In founding ISKCON, Bhaktivedanta made Gaudiya Vaishnavism available to all persons regardless of race, color, gender or creed, offered brahminical initiation to both female and male disciples, and inaugurated a priesthood based on merit rather than birth, thus following the lead of his own guru Bhaktisiddhanta Saraswati, who had already enacted most of these reforms in India before him (Sardella 2010).

In more recent times, apart from its Western followers and character, ISKCON has become one of the primary institutions catering to diaspora Hindus in various countries throughout the world, and especially in the West. Moreover, Gaudiya

Vaishnavism's international presence has grown beyond ISKCON to include a number of fairly successful institutions as well as numerous websites, blogs, wikis, etc. (another sign of globalization's impact) (Rochford 2007: 194ff). The Swaminarayan sect provides yet another example of an Indic-oriented institution that has developed a highly successful relationship with Hindu (although mainly Gujarati) diasporas throughout the world, something that is evident from the organization's gigantic temple projects in Delhi and London (Williams 1984).

In light of the above discussion, it appears that Campbell's definition of the Eastern Ideal Type is limited in scope and does not properly account for the variety of religious, theological and philosophical ideas that comprise what has come to be known as Hinduism; nor does it capture the various nuances of religious change that have occurred and continue to occur in the meeting between East and West. Globalization entails both the Westernization of the East and the Easternization of the West, and thus encompasses important *"east in the west"* and *"west in the east"* traditions. Here the primary point is that although the construction of Eastern and Western ideal types may have a certain heuristic value in terms of providing a framework that simplifies the process of creating models of cultural transformation, there is a critical need to go beyond this framework to a perspective that highlights and attempts to identify the various nuances, currents, differences and mutual influences that are involved in the intermingling of East and West, and the hybridizations that this interactivity produces.

In this sense, the history of Hinduism is a history of cultural transformation, meeting and exchange. It took shape through the challenge of the *shramana* movements of the first millennium BCE, and the concomitant birth and development of Buddhism, which forced Vedic traditions to reformulate themselves as can be seen in the tension between the householder and the renunciant, between worldly success and spiritual liberation. Later, the impact of Islam and the revival of *bhakti* movements decisively changed the political and religious landscape, which was also true of the British colonial presence and has remained true of today's global framework. In tandem with the influence on the West of Vivekananda's Eastern notions of universalism, spiritual oneness, the guru etc., Western nationalistic notions certainly contributed to the aims, if not the spirit, of the aggressive anti-colonial Hindu nationalism that took hold in nineteenth century India. Eventually, this process gave birth to the *hindutva* form of political Hinduism, illustrating how processes of globalization, Westernization and modernization can be used to mobilize an entire people on behalf of a political cause.

Through the development of global infrastructures and mediums of communication, transportation, migration and transmission, globalization causes cultures to react in either a particularistic or a universalistic manner. While requiring homogenization, it nonetheless makes differences clearly felt. In terms of the Hindu diaspora, the central question revolves around the preservation and transmission of cultural, ethnic and religious identity; and in this, one can observe attempts aimed at preserving unique regional distinctions as well as those aimed at developing more ecumenical forms of Hinduism. For the most part, the presence of Hindu religious traditions in the West remains almost entirely dependent upon Hindu diasporas, with the exception of institutions that have managed to attract a number of Western adherents, followers and casual participants (for example, ISKCON).

The Easternization discourse brings attention to other Indic-related phenomena as well. One example would be the Western success of various forms of yoga, primarily as channels to health, psychological well-being and fitness rather than enlightenment and salvation. Especially in the United States, the triumph of a consumerist wellness culture is a fairly obvious fact of societal life. To use Campbell's ideal type dichotomy (while admitting its limitations), it is Eastern in its emphasis on the divine impersonal energies within, but Western in its materialistic intentions and aims. In other words, while the term *yoga* has acquired a certain nimbus through its connection to ancient Indian wisdom and enlightenment, in the West it largely represents a variety of techniques promising various forms of materialistic advancement and success, from health to fitness to psychological equanimity to prosperity etc. (Doniger 2014: 116.125).

92.7 Conclusion

Especially over the last half-century, the ever-expanding wave of global travel and communication, via both the physical and the virtual highways of the world, has created a situation in which forms of religion, non-religion, anti-religion (and everything in between) converge, mingle and interact, at times exchanging, at times transforming, at times agreeing and at times conflicting along the way: from vampire-cults to New Atheists to Fundamentalists to the Amish. And in terms of the topic at hand, what was originally Eastern has now become Western and what was typically Western has now been exported to the South East and beyond. It is this profound reality of vast and deep global interconnectedness that all forms of religious, social, political, scientific, intellectual and cultural thinking must now come to grips with.

And in terms of religion's "coming to grips," two primary strategies (or approaches) have been discussed above: (1) the strategy of affirming one's local, particularistic religious identity and (2) the strategy of reformulating one's religious tradition so that it strikes a more universal and inclusive tone. The former of these strategies connects a given religious tradition to a specific geographical and cultural form. Here conversion to a given religious tradition is automatically conversion to that religion's culture, including its language, mores, values, customs, etiquette and cuisine. With the latter universalistic option, on the other hand, religion is re-designed and designed again in ways that make it applicable to a variety of geographical and cultural contexts. Understanding this involves an intense study of the tradition in an attempt to identify its non-negotiable (or core) as well as its accidental (or peripheral) elements (Ramadan 1999).

In the tension between protectionism (or particularization) on the one hand and universalization (or homogenization) on the other, traditionalist and fundamentalist strains of religions such as Hinduism, Islam and Christianity are facing increasing pressure as globalization continues to take its seemingly inexorable course. In the specific case of Hinduism in India, there has been a strong emphasis on the fusion of culture, national identity and religious practice, leading to a restriction on Christian conversion, which is viewed as a threat to this fusion. The other side of the Hindu coin consists of movements such as the ones mentioned above, which propagate more universal, glocalized and hybrid strands of Eastern religion and philosophy to Western audiences, often with only the request that they immerse themselves in practice (without the demand for conversion).

Of course, the tension between protectionism and universalization is not only a Hindu phenomenon, but is being increasingly felt throughout the Islamic as well as the Christian world. In many ways it parallels the present global economic circumstance: either we move toward a tighter integration of markets or we risk weakening the global system with a new wave of fierce protectionism motivated by dysfunctions in the system. The answer is not easy to come by as the entire globalization project has been recently thrown into doubt by the mountains of debt accrued by economies throughout the world and the potential collapse of the *euro* along with many of the Southern European countries that adopted it. Regardless of which scenario will dominate the future, one thing seems fairly clear: the locus of global economic power is shifting towards the East, with China and India apparently leading the way. Given this circumstance, it would not be too surprising if we were to witness a similar development in the realms of culture and religion as well. If Peter Berger has taught us anything, he has taught to "wait and see."

References

Adorno, T. W. (Ed.). (1950). *The authoritarian personality*. New York: Harper.

Allen, C. (2010). *Islamophobia*. Farnham: Ashgate.

Amir, Y. (1969). Contact hypothesis in ethnic relations. *Psychological Bulletin, 71*, 319–342. Also In E. Weiner. (1998). *The handbook of interethnic coexistence* (pp. 162–181). New York: Continuum.

Amir, Y. (1976). The role of intergroup contact in change of prejudice and ethnic relations. In P. A. Katz (Ed.), *Towards the elimination of racism* (pp. 73–123). New York: Pergamon.

Archer, M. (1996). *Culture and agency: The place of culture in social theory*. Cambridge: Cambridge University Press.

Archer, M. (2000). *Being human: The problem of agency*. Cambridge: Cambridge University Press.

Archer, M. (2003). *Structure, agency, and the internal conversation*. Cambridge: Cambridge University Press.

Bauman, Z. (1998). *Globalization: The human consequences*. London: Polity Press.

Bauman, Z. (2006). *Liquid fear*. Cambridge: Polity Press.

Bauman, M. (2008). *Global Hindu diaspora*. A bibliography of books and main articles.

Bawer, B. (2006). *While Europe slept: How radical Islam is destroying the West from within*. New York: Anchor Books.

Berger, P. (1968, April 25). A bleak outlook is seen for religion. *New York Times*. p. 3.

Berger, P. (1999). The desecularization of the world: A global overview. In P. L. Berger (Ed.), *The desecularization of the world: Resurgent religion and world politics* (pp. 1–18). Washington, DC: Eerdmans, Ethics and Public Policy Center.

Berger, P., & Huntington, S. (Eds.). (2002). *Many globalizations: Cultural diversity in the contemporary world*. New York: Oxford University Press.

Berger, P., & Luckmann, T. (1967). *The social construction of reality*. London: Penguin.
Beyer, P. (1994). *Religion and globalization*. London: Sage.
Beyer, P. (2006). *Religions in global society*. London: Routledge.
Beyer, P., & Beaman, L. G. (Eds.). (2007). *Religion, globalization, and culture*. Leiden: Brill.
Bowen, J. R. (2007). *Why the French don't like headscarves: Islam, the state, and public space*. Princeton: Princeton University Press.
Brown, M. (2000). Quantifying the Muslim population in Europe: Conceptual and data issues. *International Journal Social Research Methodology, 3*(2), 87–101.
Brubaker, R. (2005). The 'diaspora' of diaspora. *Ethnic and Racial Studies, 28*(1), 1–19.
Bruce, S. (2002). *God is dead: Secularization in the West*. Oxford: Blackwell.
Buckman, G. (2004). *Globalization: Tame it or scrap it? Mapping the alternatives of the anti-globalization movement*. London: Zed Books.
Caldwell, C. (2009). *Reflections on the revolution in Europe: Immigration, Islam and the West. Can Europe be the same with different people in it?* London: Allen Lan.
Campbell, C. (2007). *The easternization of the West: A thematic account of cultural change in the modern era*. Boulder: Paradigm Publishers.
Carrette, J. R., & King, R. (2005). *Selling spirituality: The silent takeover of religion*. London: Routledge.
Casanova, J. (1994). *Public religions in the modern world*. Chicago: University of Chicago Press.
Castells, M. (1998). *Nätverkssamhällets framväxt. Del 1. Informationsåldern. Ekonomi, samhälle och kultur*. Göteborg: Daidalos.
Cesari, J. (Ed.). (2010). *Muslims in the West after 9/11: Religion, politics and law*. New York: Routledge.
Christiano, K. (2007). Assessing modernities: From 'pre-' to 'post-' to 'ultra-'. In J. A. Beckford & N. J. Demerath (Red.), *The SAGE handbook of the sociology of religion* (pp. 39–56). London: Sage.
Davie, G. (1994). *Religion in Britain since 1945: Believing without belonging*. Oxford: Blackwell.
Davie, G. (2007). *The sociology of religion*. London: Sage.
DeMarinis, V. (2003). *Pastoral care, existential health, and existential epidemiology: A Swedish postmodern case study*. Stockholm: Verbum.
Doniger, W. (2014). *On Hinduism*. New York: Oxford University Press.
Ebaugh, H. (2003). Religion and the new immigrants. In M. Dillon (Ed.), *Handbook of the sociology of religion*. Cambridge: Cambridge University Press.
Ebaugh, H. R. F., & Chafetz, J. S. (Eds.). (2002). *Religion across borders: Transnational immigrant networks*. Lanham: AltaMira.
Eck, D. L. (2001). *A new religious America: How a "Christian country" has now become the world's most religiously diverse nation*. San Francisco: HarperSanFrancisco.
Eisenstadt, S. N. (1999). *Fundamentalism, sectarianism, and revolution: The Jacobin dimension of modernity*. Cambridge: University Press.
Eisenstadt, S. N. (2000). Multiple modernities. *Daedalus, 129*(1), 1–30.
Eisenstadt, S. N. (2003). *Comparative civilizations and multiple modernities. Part 1*. Boston: Brill.
Eriksen, T. H. (2007). *Globalization: The key concepts*. Oxford: Berg.
Esman, M. J. (2009). *Diasporas in the contemporary world*. Cambridge: Polity Press.
Esposito, J. L., & Voll, J. O. (2001). *Makers of contemporary Islam*. Oxford: Oxford University Press.
Esposito, J. L., Fasching, D., & Lewis, T. V. (Eds.). (2008). Introduction. In *Religion and globalization:World religions in historical perspective* (pp. 1–35). New York/Oxford: Oxford University Press.
Fallaci, O. (2002). *The rage and the pride*. New York: Rizzoli.
Faivre, A. (1994). *Access to western esotericism*. Albany: State University of New York Press.
Flanagan, K., & Jupp, P. C. (Red.). (2007). *A sociology of spirituality*. Aldershot: Ashgate.
Frankl, V. E. (1987). *Man's search for meaning: An introduction to logotherapy* (Rev. ed.). London: Hodder and Stoughton.
Fukuyama, F. (1992). *The end of history and the last man*. London: Hamish Hamilton.
Furnham, A., & Ribchester, T. (1995). Tolerance of ambiguity: A review of the concept, its measurement and applications. *Current Psychology, 14*(3), 179.

Giddens, A. (1999). *Modernitet och självidentitet*. Göteborg: Daidalos.
Gurwitsch, A. (1964). *The field of consciousness*. Pittsburgh: Duquesne University Press.
Hacker, J. S., & Pierson, P. (2010). *Winner-take-it-all-politics: How Washington made the rich richer—And turned its back on the middle class*. New York: Simon & Schuster.
Hammer, O. (2001). *Claiming knowledge: Strategies of epistemology from theosophy to the New Age*. Leiden: Brill.
Hanegraaff, W. J. (1998[1996]). *New age religion and Western culture: Esotericism in the mirror of secular thought*. Albany: State University of New York Press.
Hannigan, J. (1991). The social movement theory and the sociology of religion: Towards new synthesis. *Sociological Analysis, 52*(4), 311–331.
Hedges, E., & Beckford, J. (2000). Holism, healing and the new age. In S. Sutcliffe & M. Bowman (Eds.), *Beyond new age: Exploring alternative spirituality* (pp. 169–187). Edinburgh: Edinburgh University Press.
Heelas, P. (2003). *Bringing the sacred to life: The crisis of traditional religion and the rise of wellbeing spirituality*. Oxford: Blackwell.
Heelas, P. (2008). *Spiritualities of life: New Age romanticism and consumptive capitalism*. Malden: Blackwell.
Heelas, P., & Woodhead, L. (2005). *The spiritual revolution: Why religion is giving way to spirituality*. Malden: Blackwell.
Hermansen, M. (2000). Hybrid identity formations in Muslin America: The case of American Sufi movements. *The Muslim World, 90*, 158–197.
Hinnells, J. R. (2000). *Zoroastrian and Parsi studies: Collected works of John R. Hinnells*. Aldershot: Ashgate.
Hirst, P. (1994). *Associative democracy: New forms of economic and social governance*. Cambridge: Polity Press.
Hjärpe, J. (2003). *Tusen och en natt & den elfte september. Tankar om Islam*. Stockholm: Prisma.
Huntington, S. (1996). *The clash of civilizations and the remaking of world order*. New York: Touchstone.
Jacobsen, K. (Ed.). (2004). *South Asian religions on display: Religious processions in South Asia and the diaspora*. London: Routledge.
Jacobsen, K. A., & Kumar, P. P. (Eds.). (2004). *South Asians in the diaspora: Histories and religious traditions*. Leiden: Brill.
Jaspars, J., & Hewstone, M. (1982). Cross-cultural interaction, social attribution and inter-group relations. In S. Bochner (Ed.), *Cultures in contact: Studies in cross-cultural interaction* (pp. 127–156). Oxford: Pergamon Press.
Jones, J. W. (2002). *Terror and transformation: The ambiguity of religion in psychoanalytic perspectives*. Hove: Brunner-Routledge.
Juergensmeyer, M. (2000). *Terror in the mind of God: The global rise of religious violence*. Berkeley: University of California Press.
Juergensmeyer, M. (2003). *Global religions: An introduction*. Oxford: Oxford University Press.
Juergensmeyer, M. (2008). *Global rebellion: Religious challenges to the secular state, from Christian militias to Al Qaeda* (Rev. ed.). Berkeley: University of California Press.
Karner, C., & Aldridge, A. (2004, Fall/Winter). Theorizing religion in a globalizing world. *International Journal of Politics, Culture and Society, 18*(1/2), 5–32.
Kaufmann, E. (2010). *Shall the religious inherit the earth? Demography and politics in the twenty-first century*. London: Profile Books.
Kickbusch, I., & McQueen, D. V. (2007). *Health and modernity: The role of theory in health promotion*. New York: Springer.
Knott, K., & McLoughlin, S. (Eds.). (2010). *Diasporas: Concepts, intersections, identities*. London: Zed Books.
Kurth, J. (1999). Religion and globalization. *The Templeton Lecture on Religion and World Affairs, 7*(7). www.people.umass.edu/~beemer/pdffiles/FPRI.pdf
Lifton, R. (1993). *The protean self: Human resilience in the age of fragmentation*. New York: Basic Books.

Löwendahl, L. (2002). *Med kroppen som instrument: en studie av New age med fokus på hälsa, kroppslighet och genus*. Ph.D. dissertation, University of Lund, Stockholm.
Lyon, D. (2000). *Jesus in Disneyland: Religion in postmodern times*. Cambridge: Blackwell Publishers.
Maréchal, B. (Ed.). (2003). *Muslims in the enlarged Europe: Religion and society*. Leiden: Brill.
Mau, S., Mewes, J., & Zimmerman, A. (2008). Cosmopolitan attitudes through transnational social practices. *Global Networks, 8*, 1–24.
McGoldrick, D. (2006). *Human rights and religion: The Islamic headscarf debate in Europe*. Oxford: Hart.
McLennan, G. (1995). *Pluralism*. London: Open University Press.
McLaughlin, S. (2010). Religion and diaspora. In J. R. HInnells (Ed.), *The Routledge companion to the study of religion* (2nd ed., pp. 558–580). London: Routledge.
Michie, J. (Ed.). (2003). *The handbook of globalisation*. Northampton: Edward Elgar.
Micklethwait, J., & Wooldridge, A. (2009). *God is back: How the global revival of faith is changing the world*. New York: Penguin Press.
Mol, H. (1979). Theory and data on the religious behavior of migrants. *Social Compass, 26*(1), 31–39.
Narayanan, V. (2006). Hindu communities abroad. In M. Juergensmeyer (Ed.), *The Oxford handbook of global religions* (pp. 57–65). New York: Oxford University Press.
Nederveen, P. J. (2004). *Globalization or empire?* New York: Routledge.
Nederveen, P. J. (2009). *Globalization and culture: Global mélange* (2nd ed.). Lanham: Rowman & Littlefield.
Norton, R. W. (1975). Measurement of ambiguity tolerance. *Journal of Personality Assessment, 39*(6), 607–619.
Oonk, G. (Ed.). (2007). Global Indian diasporas: Exploring trajectories of migration and theory. In G. Oonk (Ed.), *Global Indian diasporas: Exploring trajectories of migration and theory* (pp. 9–27). Amsterdam: Amsterdam University Press.
Parekh, B. (2000). *Rethinking multiculturalism: Cultural diversity and political theory*. Cambridge, MA: Harvard University Press.
Ramadan, T. (1999). *To be a European Muslim: A study of Islamic sources in the European context*. Leicester: Islamic Foundation.
Ritzer, G. (Ed.). (2007). *The Blackwell companion to globalization*. Oxford: Blackwell.
Robertson, R. (1992). *Globalization: Social theory and global culture*. London: Sage.
Robertson, R. (1995). Glocalization: Time-space and homogeneity-heterogeneity. In M. Featherstone (Ed.), *Global modernities* (pp. 25–44). London: Sage.
Rokeach, M. (1960). *The open and closed mind: Investigations into the nature of belief systems and personality systems*. New York: Basic Books.
Rochford, E. B. (2007). *Hare Krishna transformed*. New York: New York University Press.
Rosenberger, S., & Sauer, B. (Eds.). (2012). *Politics, religion and gender: Framing and regulating the veil*. London: Routledge.
Rothstein, B. (2003). *Sociala Fällor och Tillitens Problem*. Kristianstad: SNS.
Roy, O. (2007). *Secularism confronts Islam*. New York: Columbia University Press.
Safran, W. (2007). Concepts, theories and challenges of diaspora: A panoptic approach. *Societá Italianan per lo studio della Storia Contemporanea*. www.sissco.it/index.php?id=1311
Sander, Å. (2010). Welcome to Eurabia. *Are Islam and Muslims taking over Europe? Some reflections on the claims of the Eurabia literature*. Paper delivered at the International Association for the History of Religion (IAHR) conference, Toronto.
Sander, Å. (2011). Eurabia i ett nötskal. In S. Olsson & S. Sorgenfri. *In Perspektiv på islam: en vänbok till Christer Hedin* (1. uppl.). Stockholm: Dialogos.
Sander, Å., & Andersson, D. (2009). Religion och religiositet i en pluralistisk och föränderlig värld – några teoretiska, metodologiska och begreppsliga kartor. In D. Andersson & Å. Sander (Eds.), *Det Mångreligiösa Sverige – ett landskap i förändring* (pp. 35–147). Lund: Studentlitteratur. Andra rev.

Sander, Å., & Larsson, G. (2007). *Islam and Muslims in Sweden: Integration of fragmentation? A Contextual study*. Berlin: LIT Verlag.

Sardella, F. (2010). *Bhaktisiddhanta Sarasvati: The context and significance of a modern Hindu personalist*. Göteborg: Department of Literature, History of Ideas, and Religion, University of Gothenburg.

Sardella, F., & Gupta, R. (2015). Modern reception and text migration of the Bhagavata Purana. In R. Gupta & K. Valpey (Eds.), *The Srimad-Bhagavatam: A companion volume*. New York: Columbia University Press.

Scholte, J. A. (2005). *Globalization: A critical introduction* (2nd ed.). New York: Palgrave Macmillan.

Scott, J. W. (2007). *The politics of the veil*. Princeton: Princeton University Press.

Shah, T., Toft, M., & Philpott, D. (2011). *God's century: Resurgent religion and global politics*. New York: W.W. Norton.

Sheffer, G. (2003). *Diaspora politics: At home abroad*. New York: Cambridge University Press.

Stark, R., & Finke, R. (2000). *Acts of faith: Explaining the human side of religion*. Berkeley: University of California Press.

Swidler, A. (2001a). *Cultural repertoires and cultural logics: Can they be reconciled?* Anaheim: Annual Meeting of the American Sociological Association.

Swidler, A. (2001b). *Talk of love: How culture matters*. Chicago: University of Chicago Press.

Thomas, S. M. (2005). *The global resurgence of religion and the transformation of international relations: The struggle for the soul of the twenty-first century*. New York: Palgrave Macmillan.

Triandros, H. C., & Vassiliou, V. (1967). Frequency of contact and stereotyping. *Journal of Personality and Social Psychology, 7*, 316–328.

Ugeux, B. (2007). The new quest for healing: When therapy and spirituality intermingle. *International Review of Mission, 96*(380–381), 22–40.

Vertovec, S. (2000). *The Hindu diaspora: Comparative patterns*. London: Routledge.

Vertovec, S. (2010). Cosmopolitanism. In K. Knott & S. McLoughlin (Eds.), *Diasporas: Concepts, intersections, identities* (pp. 63–68). London: Zed Books.

Wallace, A. (1966). *Religion: An anthropological view*. New York: Random House.

Washington, P. (1993). *Madame Blavatsky's baboon: Theosophy and the emergence of the western guru*. London: Secker & Warburg.

Wilkinson, D. (2006). *The ambiguity advantage: What great leaders are great at*. London: Palgrave Macmillan.

Williams, R. B. (1984). *A new face of Hinduism: The Swaminarayan religion*. Cambridge: Cambridge University Press.

Wilson, B. (1966). *Religion in secular society: A sociological comment*. London: Watts.

Ye'or, B. (2005). *Eurabia: The Euro-Arab axis*. Madison: Fairleigh Dickinson University Press.

Part VIII
Globalization, Diversity and New Faces in the Global South

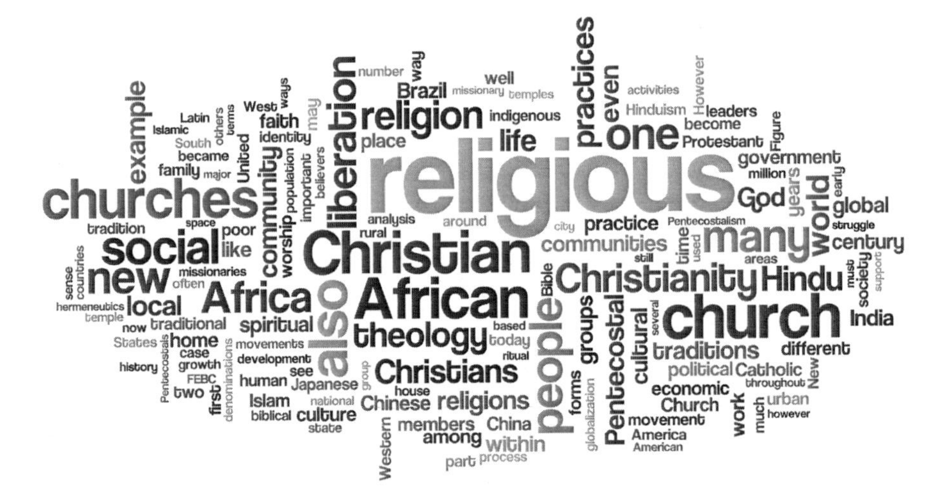

Chapter 93
The World's Fastest Growing Religion: Comparing Christian and Muslim Expansion in the Modern Era

Philip Jenkins

93.1 Introduction

In 1973, Jean Raspail published the book *The Camp of the Saints*, a futuristic fantasy that has subsequently become a cult classic for the far Right on both sides of the Atlantic (Raspail 1975). Raspail imagines how the Third World's black and brown people invade and overwhelm the white North, which has been rendered defenseless by the rise of gutless Western liberalism. Obviously, we can criticize the book from many points of view, but a modern reader will be astonished that a book published so relatively recently would fail to present its apocalyptic thesis in the religious terms that we now find so familiar. In recent years, the book has been quoted ever more widely as Europe's growing Muslim populations have become more visible, but that Islamic element is nowhere in its pages. Although the faceless Asian masses of *Camp of the Saints* are fighting to erase Europe's failed God, they have no alternative of their own, and are fighting neither for Allah or Krishna. They are waging race war, not jihad.

In historical terms, it seems, the narrative of a global religious upsurge emerged only yesterday, but it has subsequently gained great power. The narrative of a West being overwhelmed by the religions of what we would now call the Global South (a term coined in 1980) has become familiar among both Christians and Muslims. The political lines, however, break down in curious ways. Many Christians look to the rising masses of Africa and Asia to reaffirm and strengthen their own traditional views, especially in matters of morality and sexuality. Such conservatives happily await the arrival of a camp of authentic saints, black and brown Christians who

P. Jenkins (✉)
Institute for the Study of Religion, Baylor University, Waco, TX 76798, USA
e-mail: philip_jenkins@baylor.edu

S.D. Brunn (ed.), *The Changing World Religion Map*,
DOI 10.1007/978-94-017-9376-6_93

cannot come soon enough. Some Muslims too, by no means only extremists, hope that rising Muslim numbers will contribute to the spread of Islam throughout Europe. In their respective ways, each narrative assumes a failed West that will be redeemed – rather than conquered – by a purified South. The conquest, such as it is, is to be spiritual (Jenkins 2011).

Such an approach makes certain demands on the narrative. The religion's growth must be explosive, unprecedented and unexpected, and ideally, it should emphasize human decision rather than accident – it should stress conversions more than demography. But in either instance, the story must be unique. "Our" growth must be the greatest of its kind. And that narrative works until "we" meet the other religion, those other saints in their own distinctive camp, who also boast uniqueness.

Among Wikipedia's many advantages over traditional encyclopedias, it is much more flexible and innovative in its approach to topics covered, and some of the headings can be genuinely surprising. I particularly note the entry on "Claims to be the Fastest-Growing Religion," which collects many such claims, which variously award the prize to Islam, Christianity, Buddhism, the Falun Gong, Scientology, and other competitors (Claims to be the Fastest-Growing Religion 2012). Obviously, even compiling such a list suggests the tongue-in-cheek implication that most such claims, at least, are bogus or misinformed. But considering such a discussion does raise a significant point about religions in general, namely, that accurate quantitative information about any faith is hard to come by, and any plausible comparisons are exceedingly difficult. Wisely, scholars who know the problems in assessing the figures for any one faith are nervous about daring to make comparisons, about illuminating *obscurum per obscurius*.

While respecting such concerns, the lack of comparison also has its own drawbacks, in allowing members of different religions to tell their stories in isolation, without a sense of wider context. This raises the danger of describing events in one tradition as if they were distinctive or even unique, whereas in reality they had close contemporary parallels in other faiths. Assume for the sake of argument that we writing from a Christian point of view, about trends in Christian history, and we note what appears to be spectacular growth and expansion. From a comparative perspective, we have to ask a simple question: compared to what? The answers can be instructive and even sobering, and not in such a simplistic matter as winning the notional prize for fastest growing religion.

What happens, then, if we compare the story of Christian growth in the modern era with that of Islam? How does such a comparative approach affect our sense of the relative successes or tribulations of the two faiths? Above all, are there ways in which the experience of one religion can cast light on that of the other, or raise questions about the directions of research?

I should say at the outset that I will be making considerable use of the valuable resources supplied by the World Christian Database (WCD). As a broad guide to the overall religious picture, the Database is reliable, although I have difficulty in accepting the exact validity of these numbers, especially in certain regions. (I discuss these issues at greater length in the Appendix at the end of this chapter.)

93.2 Christianity Moves South

The story of Christian expansion over the past century or so is now well known. Briefly, Christian numbers have remained fairly constant in the traditional heartlands of Europe and North America, while booming spectacularly elsewhere (McLeod 2006; Sanneh 2007; Noll 2009). Growth has been particularly marked in Africa, above all, but also in Asia and Latin America (Table 93.1).

The most important trend we notice from these figures is the precipitous *relative* decline of North America and Europe as Christian heartlands. This does not mean that Christian numbers in these regions have shrunk, quite the contrary. Rather, these religious blocs have been overwhelmed by the relative growth of Christian numbers elsewhere, above all in the Global South: that is, the continents of Africa, Asia and Latin America. According to the evidence presented by the WCD, between 1900 and 2010, the number of Christians in Europe grew by 29 %, a substantial figure. In Africa, though, the absolute number of recorded believers grew in the same period by an incredible 4,930 %. The comparable growth in Latin America was 877 %. The growth for particular denominations was even more startling. During the twentieth century, Africa's Catholic population grew from 1.9 million to 130 million – an increase of 6,708 % (Allen 2009; Jenkins 2011).

The number of African believers soared, from just 10 million in 1900 to 500 million by 2015 or so, and (if projections are correct) to an astonishing billion by 2050. Put another way, the number of African Christians in 2050 will be almost twice as large as the total figure for all Christians alive anywhere in the globe back in 1900 (Isichei 1995; Hastings 1996; Jenkins 2006).

Twentieth century Christianity was decidedly a Euro-American faith. Combining Christian numbers in Europe and North America, these continents accounted for 82 % of all believers in 1900, and even by 1970, that figure had fallen only to 57 %. Since that point however, change has been very marked. Today, Euro-American Christians make up 38 % of the worldwide total, and that figure could reach a mere 27 % by 2050 (Jenkins 2011).

Actually, even those figures gravely understate the scale of the change, because the Christians listed as "European" or "North American" today include large

Table 93.1 The changing distribution of Christian believers (in millions)

Region	1900	1970	2010	2050
Africa	10	143	493	1,031
Asia	22	96	352	601
North America	79	211	286	333
South America	62	270	544	655
Europe	381	492	588	530
Oceania	5	18	28	38
TOTAL	558	1,230	2,291	3,188

Data source: World Christian Database, www.worldchristiandatabase.org/wcd/

communities from the Global South. By 2050, for instance, perhaps a quarter of the people of the U.S. will have roots in Latin America, and 50–60 million Americans will claim a Mexican heritage. Another 8 % of Americans will have Asian ancestry, and usually those communities – Korean, Chinese, Vietnamese – have strong Christian elements. In Europe also, those enduring Christian populations will include sizable immigrant communities – African, Asian and Afro-Caribbean. In 2050, therefore, even our "Euro-American" Christians will include Congolese believers living in Paris, or Koreans in Los Angeles.

If we envisage the Christianity of the mid-twenty-first century, then at least in numerical terms, we have to think of a faith located much nearer the Equator. Again according to WCD statistics, by far the largest share of the world's Christian population in 2050 will be African, with 32 % of the global total. South Americans will make up 21 % of the whole, a number that grows if we include people of Latino origin in North America. In short, well over half of all Christians alive in 2050 will be either African or Latin American. When we recall the distribution of Christians as recently as 1970, that is an incredible global change to occur in basically just two or three generations. It is not surprising that some Christian observers speak of a New Pentecost, an outpouring of the Spirit on all nations.

93.3 The Growth of Islam

But how do such figures look when we compare the development of Islam in the same era? The answer is quite surprising. Both religions have acquired vastly more adherents in the past century, but in some ways, Muslims have significantly outpaced Christians. When considered as a share of global population, Christian numbers have proved strikingly stable over the past century. In the year 1900, about one-third of the world's people were Christians, and that proportion remains more or less unchanged today. Moreover, if we project our estimate forward to the year 2050, that proportion should still be about one-third.

But if we consider Muslim numbers in the same terms, as a share of global the world's people, then that religion has enjoyed a far more impressive surge. In 1900, the 200–220 million Muslims then living comprised some 12–13 % of humanity, compared to 22.5 % today, and a projected figure of 27.3 % by 2050 (WCD). Christians in 1900 outnumbered Muslims by 2.8–1. Today the figure is 1.5–1, and by 2050 it should be 1.3–1.

Put another way, there are four times as many Christians alive as there were in 1900; but over the same period, Muslims have grown at least seven-fold.

93.4 Differential Demographics

So how can Christian numbers be exploding, but still be left so far behind Muslims in the rate of expansion? Much of the answer lies in differential demographics, namely, that some parts of the world are growing much faster than others. Briefly,

European numbers have been growing very slowly indeed in comparison with those of Africa, Asia and Latin America, and that is very good news indeed for a faith based chiefly in Asia and Africa, as Islam was historically.

Leaving aside religious affiliations, let us look just at raw numbers. Back in 1900, Europeans made up around a quarter of the world's population. By 2050, that number will probably be closer to 8 %. Africa in contrast had around 130 million people in 1900. That number passed the billion mark by 2005, and by 2050 the number could reach anywhere from 2 to 2.25 billion.

Just to take one example, in the lands that would become Kenya, the population in 1900 was a mere one million, but that figure has now swollen to around 40 million, in little over a century. By 2050, Kenya could have 80 million people or more. In 1900, there were three Europeans for every African. By 2050, there should be three Africans for every European. If a faith attracted the loyalty of a certain number of Africans (say) in 1900, and merely kept that market share, then that faith would be very much stronger in numerical terms. And in part, that is the story of Islam over the past century.

93.4.1 Islamic Demographics

We can see the expansion in any part of the Islamic world (In what follows, I have relied on Hefner 2010 and Robinson 2010). In 1900, Muslims made up perhaps nine million of Egypt's ten million people. Today, the proportion of Muslims is certainly higher, but the overall population has swelled from 10 million to perhaps 79 million. Iran has experienced comparable growth, from 10 million in 1900 to 66–70 million today. As Muslims comprise the overwhelming mass of that nation, that represents a huge gain in Islamic numbers.

Indonesia presents a similar case (Noer 1973). In 1900, the Dutch East Indies had a population of around 42 million, rising to 70 million by 1940 and 150 million by 1980. (From 1949, the country became known as Indonesia.) Today the total population is estimated at 240 million. While we do not have detailed religious statistics for earlier years, the best contemporary estimate suggests that around 80 % of Indonesians are Muslim. If we assume that that figure also held good in earlier eras, then the number of Muslims in this region would have grown, roughly, from 34 million in 1900 to 56 million in 1940, 120 million by 1980, and 190 million today. I do not defend the precise accuracy of this figure, nor the assumption that rates of Islamic loyalty would have remained precisely steady over the past century. Obviously they did not, and what I am offering is a working model. However, the basic point remains. Since 1900, Muslim numbers have increased dramatically in the Dutch East Indies/Indonesia, from around 34 million to 190 million. The number of Muslims just in Indonesia today is not far short for what the *global* total was back in 1900.

The Indian sub-continent offers another example of expansion (Hardy 1972). In 1900, Muslims were a major component of the population of British India, concentrated in what would be the later states of India, Pakistan and Bangladesh.

This area contained at least 65 million Muslims, perhaps as many as 80 million. Today, however, the same region has between 450 and 480 million Muslims, a rate of increase quite comparable to that in Indonesia.

In large measure, then, Muslim growth occurred because Muslims were concentrated in those regions that maintained very high fertility rates throughout the twentieth century. A rising tide lifts all religion.

93.4.2 Conversions

Now, that is by no means the whole story. For both Christians and Muslims, demographic growth was vastly reinforced by evangelism and conversions, a story in which Christians won significantly greater successes.

Let us for instance look at the lands that became Nigeria. In 1900, these territories had around 16 million inhabitants, who included 4.2 million Muslims and around 180,000 Christians. Muslims, in other words, represented 26 % of the population, compared to about 1 % for Christians. By 1970, however, Muslims outnumbered Christians very slightly – by 41 % of the population, as opposed to 40 %. As a share of population, the number of Christians had grown 40-fold during this 70 years period. As differential demographic rates played no apparent role in this transformation, what we are seeing is a solid rate of conversions to Islam, compared to an overwhelming mass conversion to Christianity, concentrated among certain peoples, especially the Igbo. Today, both faiths command the loyalty of about 45 % of Nigerians.

Stressing demographic change must not distract us from the very real success of Christian missions in spreading at least the seeds of faith, which in many areas would produce harvests far beyond the wildest expectations of the sowers. It is a token of their success that Christian efforts attracted such fervent resistance in many areas, but also emulation. In the Muslim world, this meant the revival of Dawa, the "Call" of faith, which in the early twentieth century was associated with such names as Rashid Rida (1865–1935) in Egypt, and the Sufi thinker Maulana Muhammad Ilyas (1885–1944) in British India (Metcalf 1982). In the 1920s, Maulana Ilyas began the *Tablighi Jamaat*, which in Christian terms would become a worldwide missionary movement still very active to the present day. Driving such efforts was genuine fear of Christian successes and admiration of their techniques – however limited the successes that such missions might have had in converting Muslims specifically.

However we tell the story then – whatever emphasis we place on conversions as opposed to birth rates – there is no doubt that the southward shift of population is critical to understanding global religious change. Put another way, let us imagine what the picture of Christian expansion would look like if we took Europe out of the story. (I mean just the land-mass of Europe, rather than people of European stock throughout the world). The number of non-European Christians in 1900 would be about 180 million, compared to 1.7 billion today, or 9.5 times as many. That represents a growth rate for the century larger than that of Islam.

If we are asked about the fastest growing religion in modern history, then, we can answer unequivocally: Christianity outside Europe, and Islam.

93.5 New Areas of Growth

In Christian history, the main story of the past century has been the huge shift to areas of growth far removed from the traditional heartlands of the faith. Something similar is also true of Islam, which of course originated in the Arab Middle East, but subsequently expanded mightily. By the fourteenth century, Islam was firmly established in East and South Asia, and was growing in North Africa. In modern times, though, these later additions to Dar al-Islam have dwarfed the Arab world. Of the world's eight most populous countries, only one – Egypt – is Arabic by ethnicity and language (Table 93.2).

When beginning this study, I originally thought I would find a comparable movement within Islam, away from the Arab world towards traditional outliers, in the further reaches of Africa and Asia. In fact I did not, on anything like the Christian model. Although numbers grew in Indonesia and the Indian subcontinent, the proportions of Muslims living in those regions remained fairly stable over time. Islam, then, offers nothing as dramatic as the very rapid movement away from its traditional centers of population, from Europe to Africa. While a movement has occurred, it has been more gradual, and the decisive shift was already in progress centuries ago. The Indian sub-continent comprised about 30 % of all Muslims in 1900, and they represent a very similar proportion today.

The real change, such as it was, came in Africa (Table 93.3). Nigeria well illustrates this movement. As we have seen, the Muslim share of that population grew from 26 % in 1900 to 45 % today, but that growth in market share gives no idea of

Table 93.2 The world's largest Muslim nations, 2010

Nation	Number of Muslims (in millions)
Indonesia	184
Pakistan	178
India	166
Bangladesh	146
Turkey	74
Iran	74
Egypt	74
Nigeria	72

Source: World Christian Database, www.worldchristiandatabase.org/wcd/

Table 93.3 The changing distribution of Muslims worldwide (in millions)

Region	1900	1970	2010	2050
Africa	34.5	147	422	816
Asia	156	415	1,083	1,637
North America	–	0.8	5.6	11
South America	0.05	0.4	1.6	2.8
Europe	9	18	40	46
Oceania	–	0.07	0.5	1.4
TOTAL	200	581	1,553	2,515

Source: World Christian Database, www.worldchristiandatabase.org/wcd/

the change in absolute numbers. In 1900, the lands that would comprise Nigeria had around 16 million, but that population is now 158 million, and that figure could by 2095 be approaching 290 million. The effect on Islamic strength is apparent. Nigeria had four million Muslims in 1900, as against 72 million today. As a share of the whole Muslim world, that proportion grew from 2 % in 1900 to almost 5 % today. From being a thinly populated outlier on the edges of the Muslim world, Nigeria is increasingly a key player in that realm – just as it is in the emerging Christian world. Demographics and conversions work together.

The lack of reliable estimates for the number of Muslims in Africa at various periods makes it difficult to generalize this trend. The WCD suggests that in 1900, the whole continent of Africa had around 35 million Muslims, or 17 % of the whole. Today that figure has increased to 27 %, and it may be as much as one-third by 2050. Other sources, however, suggest that Africa's Muslim population was already a good deal higher in 1900 than the WCD postulates, making any such statements questionable. We can indeed speak of relative growth in Africa, its exact scale is hard to determine.

93.6 Foundations of Faith

Although both faiths shifted their geographical centers of gravity, they grew in different fashions. In most cases, Christian growth in Africa and Asia developed on fresh ground. Of course older Christian traditions existed – the Ethiopian church, and the Thomas Christian traditions of India – but most converts came from animist backgrounds (in Africa particularly) or from roots in other faiths such as Hinduism (Jenkins 2008). They were thus new Christians, in a society without older Christian frameworks.

The Muslim story is of course different, because of the centuries old strength of Islam in Africa, in South and East Asia. Islam in these areas was in no sense a freshly imported religion with foreign connotations, but was a known and respected local reality. Issues of inculturation thus played out in a totally different setting. By the time Muslim numbers swelled in the twentieth century, Africans (for example) had several centuries of experience seeking to accommodate Islam to local realities. Nigerian Muslims proudly recall such distinguished predecessors as the great reforming revolutionary Uthman Dan Fodio, who created the mighty Sokoto Caliphate of the early nineteenth century (Ballard 1977).

This local tradition was best expressed in the Sufi orders which had played such a central part in the original conversion of these territories centuries before, and which continued to shape Islamic thought and life up to the present day. In North and West Africa for instance, we find the potent faith centered on the marabouts, whose tombs and shrines are so basic a part of the landscape. Uthman Dan Fodio was a teacher of the *Qadiriyya* school of Sufism.

93.7 Religions and Empire

Because of these local roots, Islam and Christianity also occupied very different relationships to the colonial empires that represented a fundamental reality in most of the Islamic world of the early twentieth century. At least in its origins, Christianity was strongly associated with empire, certainly in Africa, and the persistent danger was that it would be regarded as the white man's religion, to endure only as long as those empires did. Islam played a quite different role. Not, certainly, that Islam was of necessity an anti-colonial ideology. Across Asia and Africa, the colonial empires depended absolutely on the loyalty of Muslims, particularly soldiers and civil servants, who found little problem in faithfully serving infidel rulers.

But Islam also had a strong anti-imperial strand that was rooted in Sufi sects and popular Islamic movements. From the late eighteenth century onwards, much of European military history and lore was formed in conflict with Muslim populations – by the British in India, the French in North Africa, the Russians in the Caucasus, the Dutch in the East Indies, the Spanish in Morocco, the Italians in Libya.

From the 1830s onwards, expanding European empires everywhere encountered militant nationalist movements, which were commonly motivated by Islam and the rhetoric of jihad. What the British remember as the Indian Mutiny of 1857 was to thousands of its participants a jihad against the British infidel. In the 1860s, Muslim rebels in southwest China sustained an independent sultanate in rebellion against the Chinese imperial regime, and other Muslim revolts flared elsewhere in China's weakly controlled far West. In 1873, the East Indian sultanate of Aceh began a 30-year revolution that the Dutch never wholly suppressed. At the height of its success, the Muslim secessionist regime in Aceh was quite as independent of imperial control as was the Mahdist state in Sudan. Among the greatest of the Muslim opponents of empire was Imam Shamil, who fought doggedly against Russian expansion in the Caucasus from 1834 to 1859. His contemporary Abd al-Qadir occupies the same heroic position in the mythology of modern Algeria. In the 1920s, Omar al-Mokhtar led mujahedin resistance against the Italian occupation of Libya, becoming a hero for Libyans and others. Moroccans fondly recall Abd al-Krim, whose forces slaughtered thousands of Spaniards, and whose guerrilla tactics inspired Ho Chi Minh. Across the Muslim world, anti-imperial activism was inextricably linked with specifically Muslim heroes, who explicitly drew their inspiration from the cause of religion.

In some ways, these various movements resemble the forces of contemporary Islamist radicalism, not least in the centrality of the concept of jihad. Nineteenth century Iranian regimes proclaimed the holy quality of their struggles against infidel Russians and British. Then as now, revolutionary movements received the support of Muslim scholars and clergy, and Sufi orders were critical to the organization of radical activities.

These precedents affected the later attitude of Muslim nations towards empire, to the West, and to the Christians whom they often associated with Western imperialism. Muslims, above all, never felt any need to assert their local and patriotic credentials. Although they certainly spawned their own nationalist movements (notably the Chilembwe revolt in Nyasaland in 1915), Christians were slower to identify their cause with the authentic voice of Africa, or of individual nations. Only when they had developed their own corpus of saints, martyrs and spiritual heroes could they make such an assertion. Christians too certainly outpaced Muslims in developing flourishing indigenous cultures, marked above all by hymns and vernacular literature.

93.8 Prophets and Saints

Having said this, we do find some parallels between the modern expansion of both faiths. In both cases, Muslim and Christian, individual spiritual leaders proliferated during the imperial years, and portrayed themselves according to their respective traditions. Christians produced prophets; Muslims became Sufi sheikhs.

The early twentieth century is a critical period in the modern history of Christianity, especially the years around the First World War and the great epidemics that followed that conflict. In Africa, that was the point at which prophets and evangelists took the faith wholeheartedly into their own hands, translating it into local cultures and worship styles, and creating churches thoroughly rooted in African soil. In doing so, they began the mass conversion of the continent. This was for instance the great era of the local "Ethiopian" churches, spawned in response to the victory of the Ethiopian state over invading Italian forces in 1896.

Although we can point to hundreds of Christian evangelists and prophets, a few heroic names stand out, such as Liberia's William Wadé Harris, and Simon Kimbangu in the Congo (Martin 1975; Walker 1983). At the time, colonial authorities deeply distrusted the new churches. Chiefly, they feared possible sedition, but they were also wary of any syncretistic mixing of Christianity with animist beliefs. Simon Kimbangu spent 30 years in a Belgian colonial prison; French authorities kicked Harris out of the Ivory Coast. But those churches and their offshoots flourished, and have spread widely across modern Europe.

Among Muslims too, at very much the same time, colonized African communities took that faith into their own hands, packaged it in familiar forms, and made it immensely popular (Glover 2007; Hanretta 2009; Seesemann 2011). The best-known Muslim equivalent of the Christian prophets was the saintly Cheikh Ahmadou Bamba (1853–1927), from Senegal. At the end of the nineteenth century, he founded a pious Sufi order called the Mourides, the *Muridiyya*, founded in mystical devotion to God. Cheikh Bamba taught a practical message of hard work, charity and pacifism, founded on the principle "Pray as if you will die tomorrow, and work as if you will live forever." His movement drew heavily on African roots, with its cultivation of local saints and shrines. Like African Christians, the Mourides stand or fall on their promise of healing in mind and body.

French colonial authorities viewed Cheikh Bamba about as sympathetically as the Belgians saw Simon Kimbangu, and the Cheikh likewise spent long years in custody, whether in exile or under house arrest. But his movement too was vindicated by history. The Mouride Way today claims several million members – about the size of the Kimbanguist church – and like them, Mourides are spread around the globe. Mourides and their order are particularly strong in southern Europe and in American cities.

93.9 Globalization and Mainstreaming

If the historical experiences are so different, it is remarkable that the modern histories of Islam and Christianity have so many points of resemblance – at least in the Global South. Despite these disparate origins, the fact of Christianity's imperial origins and Western connections has clearly not damaged it at all, or stemmed its mass appeal. Both faiths remain very strong throughout the global South, in part because of the extreme weakness of states who could contemplate the many social functions currently supplied by religious groupings.

Also, both faiths have been marked by certain parallel features, including a persistent struggle between native and globalized forms of faith. In the case of Islam, the key new fact has been the ascendancy of new and more stringent religious forms emanating especially from the Arabian peninsula, and reflecting the strict attitudes of Saudi Wahhabism, but also the Indian Deobandi tradition. (Again, I draw heavily on the essays in Hefner 2010.) From 1979, the oil states invested heavily in spreading their particular forms of Islam around the world, in part to compete with the local and more syncretistic forms associated with the Sufis. New mosques proliferated across Africa and Asia, but generally following Arab models rather than the rich local context. In some cases, militant Islamists directly fought and terrorized Sufis, whom they regarded as near-infidels. In a few instances, Sufis fought back effectively.

Although such violent intra-faith confrontations have not marked the Christian world, conflicts between global and local are well known. The closest parallel would be the spread of U.S.-derived models of Pentecostalism and megachurches, which largely overwhelmed the once thriving native-oriented churches, the Ethiopians, Zionists and other AIC's. As in the Muslim case, the globalized forms enjoyed certain advantages, including vast financial sources; command of the most popular forms of technology, media and communication; and a critical aura of modernity. As with Muslims, Christian churches are drawn towards transnational norms.

93.10 Conclusion

Both Muslim and Christian narratives are flawed in suggesting the uniqueness of their particular experience, still less any claims to "miraculous" quality or divine guidance. But any consideration of modern historical expansion points to striking parallels between Christianity and Islam, and especially the mixture of demography and evangelism in driving growth. These two stories, the tales of the two camps of saints, are clearly two faces of a single coin.

Appendix: Using the World Christian Database

Although enormously useful, WCD data suffer from two main problems. The WCD exaggerate the number of Christian believers to be found in traditionally Christian countries that historically supported state churches, which particularly affects our sense of the level of Christian belief in Europe. For example, the WCD gives the number of British Christians as 50 million, which basically comprises every resident of the country not openly identified with some other religion. It suggests nothing about actual Christian practice or commitment, or even the number of people who might admit to some kind of Christian identification, however lukewarm, in a survey. However, much Europe's importance in the Christian world seems to have fallen over the past century, the WCD statistics actually understate that decline.

Equally problematic are the WCD estimates for countries where Christianity is strictly regulated or regarded with widespread suspicion by government or rival religious communities. Nobody doubts that countries like India and China have sizable Christian populations over and above portrayed by the official statistics of those nations. But how large are these shadow populations? Many observers would be suspicious of the very large Christian populations implied for China (115 million) and India (58 million). In India, for instance, official government data suggest a Christian population of around 23 million, which is universally known to be a grave underestimate. Millions of Christians, especially among the poor, are nervous about openly admitting their faith in the face of potential persecution from fundamentalist Hindu groups. But a consensus of informed estimates puts India's real Christian population at around 40 million, rather than 58 million. Chinese data are even more open to speculation, and the WCD number, if 115 million, stands at the summit of likely estimates. I personally would place the probable number of Chinese Christians well below that, perhaps at a half of the WCD figure. I may well be wrong, but in this matter, neither I nor the scholars of the WCD really have any firm data on which to rely.

References

Allen, J. L., Jr. (2009). *The future church*. New York: Doubleday.
Ballard, M. (1977). *Uthman dan Fodio*. London: Longman.
Claims to be the Fastest-Growing Religion. (2012). *Claims to be the Fastest-growing religion*. Retrieved October 23, 2012, from http://en.wikipedia.org/wiki/Claims_to_be_the_fastest-growing_religion
Glover, J. (2007). *Sufism and jihad in modern Senegal*. Rochester: University of Rochester Press.
Hanretta, S. (2009). *Islam and social change in French West Africa*. Cambridge: Cambridge University Press.
Hardy, P. (1972). *The Muslims of British India*. Cambridge: Cambridge University Press.
Hastings, A. (1996). *The Church in Africa, 1450–1950*. Oxford: Clarendon.
Hefner, R. W. (Ed.). (2010). *The new Cambridge history of Islam* (Vol. 6). Cambridge: Cambridge University Press.
Isichei, E. (1995). *A history of Christianity in Africa*. Grand Rapids: Eerdmans.
Jenkins, P. (2006). *The new faces of Christianity*. New York: Oxford University Press.
Jenkins, P. (2008). *The lost history of Christianity*. San Francisco: HarperOne.
Jenkins, P. (2011). *The next Christendom* (3rd ed.). New York: Oxford University Press.
Martin, M.-L. (1975). *Kimbangu*. Grand Rapids: Eerdmans.
McLeod, H. (Ed.). (2006). *Cambridge history of Christianity: World Christianities 1914–2000*. Cambridge: Cambridge University Press.
Metcalf, B. D. (1982). *Islamic revival in British India*. Princeton: Princeton University Press.
Noer, D. (1973). *The modernist Muslim movement in Indonesia 1900–1942*. Singapore: Oxford University Press.
Noll, M. A. (2009). *The new shape of world Christianity*. Downers Grove: InterVarsity Press.
Raspail, J. (1975). *The camp of the saints*. New York: Scribner.
Robinson, F. (Ed.). (2010). *The new Cambridge history of Islam* (Vol. 5). Cambridge: Cambridge University Press.
Sanneh, L. O. (2007). *Disciples of all nations*. New York: Oxford University Press.
Seesemann, R. (2011). *The divine flood*. New York: Oxford University Press.
Walker, S. S. (1983). *The religious revolution in the Ivory Coast*. Chapel Hill: University of North Carolina Press.
World Christian Database. http://www.worldchristiandatabase.org/wcd/

Chapter 94
The Emerging Geography of Global Christianity: New Places, Faces and Perceptions

Robert Strauss

94.1 Introduction

In the early 2000s, Philip Jenkins, historian from Pennsylvania State University, awakened his Western readers to the "Next Christendom," a global Christianity emerging from the Southern Hemisphere. Born in Port Talbot, Wales in 1952, Jenkins' early academic career was in criminology. After 20 years of academic work in State College, Pennsylvania, he changed the focus of his research to global Christianity and its associated religious movements, which culminated in the book, *The Next Christendom: The Coming of Global Christianity* (2002).

Surely and steadily, Buenos Aires, Addis Ababa, and Seoul had joined Rome, Geneva, and New York as focal points of the Christian faith (Mbiti 1995: 154). These cities illustrate that an undeniable geographic shift had occurred. Latinos, Africans, and Asians were emerging as new leaders in the global mission enterprise (Fig. 94.1). Passion, community transformation, and honor were taking their place along side of doctrine, church polity, and righteousness. The analysis by Jenkins (2002/2011) builds upon the foundation of Stephen Neill (1900–1984), bishop in the Anglican Church, who traced the geographic expansion of Christianity from its origins in the Middle East to the Colonies of the New World (1991). Neill was born in Edinburgh, Scotland, but served as an Anglican missionary in Tamil Nadu, India. Regarded as an expert on the historical expansion of Christianity in India, he taught at a theological seminary in the Tamil vernacular, one of the more difficult languages in India.

The next Christendom from new places has introduced new leaders, roles, and theological perspectives. While this chapter provides an overview of the shift that has occurred, this book itself explores the depths of new issues that are emerging from the Majority World, not only related to Christianity, but also other world

R. Strauss(✉)
President and CEO, Worldview Resource Group, Colorado Springs, CO 80962, USA
e-mail: rstrauss@wrg3.org

S.D. Brunn (ed.), *The Changing World Religion Map*,
DOI 10.1007/978-94-017-9376-6_94

Fig. 94.1 Mission leaders from the global north and south (Photo by Robert Strauss)

religions and spiritual experiences. The old paradigm of nations sending and receiving missionaries has been turned on its head. Today, everyone is going everywhere with the good news that not only addresses the redemption of individuals, but also a revitalization of local communities. For example, in the eighteenth and nineteenth centuries, Northern Europe sent Protestant missionaries to India and China. Today, churches are flourishing in India and China, which in turn are sending missionaries abroad. The message of today's cross-cultural workers is holistic, focusing on individuals, families, communities, and nations.

While geography at a macro level is less significant today because of ease of travel, technology, and virtual connections, geo-culture at the micro level is more important than ever. Rather than exporting Western forms to the rest of the world, cross-cultural workers are critically contextualizing[1] (Hiebert 1986) the function and meaning of redemption among host societies using locally familiar forms.[2] This

[1] Paul Hiebert (1986) defines critical contextualization as a process whereby local customs are both phenomenologically respected and also critically evaluated by local religious leaders and cross-cultural workers based on coherence with Sacred Scripture. For example, the Hindu Vedas present one Ultimate Reality with an infinite number of possible manifestations. Therefore, religious activity conditioned by social structure and past associations is respected. The result is innumerable gods developed in countless forms. In contrast, the Bible of the Christian faith presents only one way to God (Acts 4:12).

[2] Ralph Linton (1936) introduced the concepts of form, function and meaning. Form refers to the style, shape, manner or procedure. Function speaks to purpose while meaning is the significance and importance. To illustrate, a handshake may be a ritual in athletics that means, "You did a good job!"

fact is elaborated below under The New Perspectives. In other words, as everything changes, some things stay the same. Furthermore, geopolitics is now key to the further expansion of Christianity. A Christianity that is perceived as Western or from the North may not be embraced as readily or deeply as Christianity from the South. To understand how geopolitics is impacting the expansion of Christianity, one must consider the shift of Christendom to the global South.

94.2 The Shift

94.2.1 A Review of Past Analyses by Philip Jenkins

The first edition of *The Next Christendom* by Jenkins was a response to the historical geographic shift in Christianity to the global South. Jenkins was not the first or only religious studies scholar to notice the global shift (2002: 4–5). Others were calling attention to it in published academic works. Walbert Buhlmann (1976) had already written, *The Coming of the Third Church*, which analyzed the emerging church from the "Third World." Among others, British ecclesiastical historian Edward Norman had lectured and written in the early 1980s about Christianity in the Southern Hemisphere. Missiologists were also already aware of the emerging mission movements from the Majority World (Walls 1996; Engel and Dyrness 2000). In point of fact, and not surprisingly, those involved in the mission community were keenly aware of the emerging church the global South. As far back as the 1850s, Henry Venn (1796–1873), an Anglican clergyman of the Church Missionary Society, spoke of the coming "euthanasia of the mission." By this peculiar phraseology, Venn was speaking about a strategy of "exiting," whereby mission agencies would depart and native pastors would govern native indigenous churches [see also *The Planting and Development of Missionary Churches* by John Nevius (2003/1886)]. These strategies were laying the foundation for the Next Christendom.

However, it was not clear a decade ago, nor is it clear today, how many Christian believers in the West are aware of the shift. Nevertheless, a Euro-American religion is now global. In 1900, one may have imagined a typical Christian as an Italian, German, or American, but over a century later one would more likely see a Brazilian, Ugandan, or Filipino. Hilaire Belloc, the noted Catholic historian and collaborator with G.K. Chesterton, wrote, "Europe is the Faith and the Faith is Europe" (1920: 191). Not so anymore and certainly not going forward.

For example, parishioners of the First Presbyterian Church in Colorado Springs meet on Sunday mornings for worship, looking at elaborate stained glass windows before them and listening to a repertoire of sacred hymns from a grand pipe organ. Founded in 1872 by a Presbyterian missionary named Sheldon Jackson, the church today is pastored by Dr. Jim Singleton, a scholar of church history.

At the same time, 20,000 km (12,200 miles) southwest in the jungles of Papua New Guinea, Maile, a church elder among the Bisorio tribal people, teaches his people from a Bible translated into the Bisorio vernacular. The riverbank setting at

Table 94.1 The changing distribution of Christian believers

	Number of Christians (millions)			
Area	1900	1970	2010	2050
Africa	10	143	493	1,031
Asia	22	96	352	601
North America	79	211	286	333
Latin America	62	270	544	655
Europe	381	492	588	530
Oceania	5	18	28	38
TOTAL	558	1,230	2,291	3,188

Data source: http://www.worldchristiandatabase.org/wcd

the headwaters of the Sepik River is several days' travel from the capital city of Port Moresby. In 1982, two missionaries from the United States founded the church, whose structure is made of wooden polls with a thatched roof of leaves. Maile and other Bisorio church leaders travel tireless throughout inhospitable terrain in the surrounding mountainous region to spread the Good News of redemption. This local story from the Sepik River area of Papua New Guinea is repeated regionally and nationally all across the global South.

Based on research provided in the World Christian Database (from the Center for the Study of Global Christianity, Gordon-Conwell Theological Seminary), Table 94.1 generalizes the numbers of Christians distributed by continent. A staff of scholars led by Todd Johnson maintains the World Christian Database, which is updated quarterly and represents 238 countries and 13,000 ethnolinguistic peoples. The research data (highlighted areas) show where and when the shift has and is taking place.

Around 1900, over 82 % of Christian believers were from the North (460 of 558 million). By 2010, that distribution declined to less than 39 % (874 of 2,291 million). In other words, today the majority of Christian believers are geographically from the South. By the middle of the twenty-first century, only about 25 % (863 of 3,188) of Christian believers will be from the North.

94.2.2 *A Corresponding Geographic Shift in the Locus of the Global Mission Enterprise*

Accordingly, the locus of the global mission enterprise likewise has shifted. The emerging mission movements from the Majority World have taken their place on the stage of the global mission enterprise. For example, the church in China has launched the "Back to Jerusalem" movement (Chinese: 传回耶路撒冷运动), a vision for local churches throughout China to send 100,000 Chinese missionaries back along the Silk Road to Jerusalem (Hathaway et al. 2003). The campaign is not new, but was conceived in the 1920s by Chinese students at the Northwest Bible Institute, established by James Hudson Taylor II (1894–1978), the grandson of Hudson Taylor (1832–1905) who was the founder of China Inland Mission.

Since independence in 1947, the national mission movement from India has rapidly developed. Today the Indian Missions Associations coordinates over 250 national Indian missionary agencies, through which local churches have sent over 40,000 Indian missionaries. Dr. K. Rajendran, the longtime General Secretary of Indian Missions Associations, later became the Chairman of the Board of Directors of the World Evangelical Alliance (Rajendran 2005).

Based in Hyderabad, India, the Indian Missions Associations launched a training arm in 1992, the India Institute of Missiology. Today the Institute accredits curriculum for 60 national Indian missionary training centers. Dr. C. Barnabas, interestingly with a Ph.D. in chemistry, has directed the Institute since its inception. In each academic term, 2,000 Indian nationals are in training, many earning BA, MA, and Ph.D. degrees in the academic discipline of missiology.

Tradition tells us that Thomas the Apostle carried the Gospel to India in the first century AD. In South India these converts to Christianity were known as Nasrani Mappila.[3] It is interesting to think that the Gospel of Christ began spreading throughout India long before it reached geographic Europe. Of India's 1.2 billion people, approximately 2.5 % or 30 million are professed Christians.

In Latin America, the Cooperación Misionera de Ibero-America is a broad network of churches, mission agencies, and missionary training centers throughout all of Ibero-America. Today, approximately 10,000 Latino missionaries are associated with the network. In Brazil, Chile, Argentina, Paraguay, Guatemala, and other countries, there are biblical seminaries, missionary training centers, and locally based mission agencies that are passionate about advancing the Kingdom of God.

These are but a few examples. Similar stories can be told about South Korea, Nigeria, the Ukraine, and many other nations where local churches are sending cross-cultural workers elsewhere in the expansion of Christianity.

Second only to the United States, South Korea sends out more missionaries than any other global country. From the continent of Africa, Nigeria is the leading nation participating in the global mission enterprise with approximately 5,200 national Protestant Christian missionaries. Based in Jos, Nigeria and led by Executive Secretary Reverend Timothy Olonade, the Nigeria Evangelical Missions Associations envisions mobilizing, training and sending out 50,000 Nigerian Christian missionaries to northern Nigeria, North Africa, and the Arab Peninsula (see www.nematoday.org).

---

[3] Nasrani is a local term for Christians, while Mappila is a title of respect.

94.2.3 Factors That Precipitated the Shift

The historical phenomenon of the geographic shift may be reliably analyzed in various ways. An "exit strategy" approach to church planting has resulted in a "passing of the baton" to local native leadership (Steffen 1997). Also, as cross-cultural workers increased in their cross-cultural competencies, the process of critical contextualization produced locally familiar forms and functions imbued with ancient meaning from the Sacred Scripture (Hiebert 1986). However, no single activity in global missions has strengthened the local indigenous native churches more than the translation of the Scripture into the local vernacular. No scholar has argued this more forcefully than Lamin Sanneh, especially in his book, *Whose Religion is Christianity: The Gospel Beyond the West* (2003). Sanneh, a son of royalty, was raised in Gambia, West Africa. Today he is a professor of Missions and World Christianity at Yale University (Fig. 94.2).

More broadly and most notably, the history of Christian expansion beginning in the seventeenth century is tied to colonization (Neill 1991). For example, the King of Denmark sponsored the Danish-Halle Mission, beginning in 1706. Influenced by his court chaplain, Dr. Lutkens, King Frederick IV sent two missionaries from the Lutheran clergy and graduates of Halle University in Wittenberg, Germany to work in the Danish colonies near Tranquebar. Frederick was concerned that the Roman Catholic Church was sending missionaries to the new colonies, but the Protestants had not. One of the two, Bartholomew Ziegenbalg, translated the New Testament into Tamil.

Finally, as nations secured their independence in the nineteenth and twentieth centuries, consistently local, regional, and national expressions of Christianity emerged in new places with new people and perspectives (Tucker 2004) (Fig. 94.3).

Philip Yancey (2001) offers a fresh explanation for the global shift that transcends human explanation. It occurs to him that the observable pattern is a historical phenomenon of God moving geographically where He is wanted. The Bible is clear that God's eternal program for humankind is that the whole earth be filled with

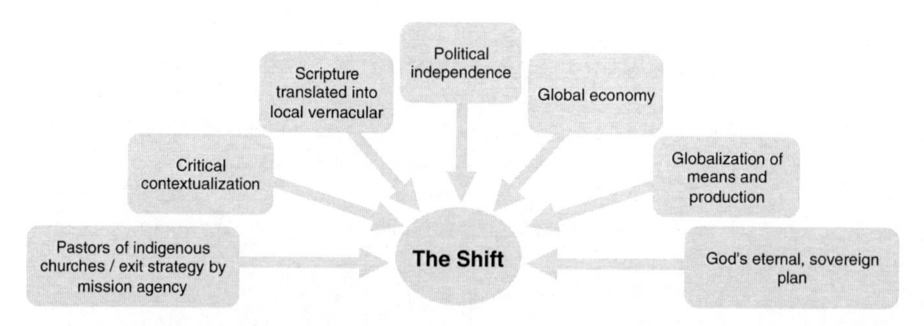

Fig. 94.2 Factors precipitating the global shift (Source: Robert Strauss)

Fig. 94.3 Diego and Ailin, new faces from the South (Photo by Robert Strauss)

people who know and love Him. Note these references throughout the Old and New Testaments that demonstrate the global scope of God's intention (read the surrounding context of these references for the fuller narrative):

- Genesis 12:3 "in thee shall all families of the earth be blessed" (a statement to Abram)
- Joshua 4:1–7, 24 "that all the people of the earth might know"
- 1 Samuel 17:46 (young David slaying Goliath) "that all the earth may know"
- Psalms 33:8 "all the inhabitants of the world"
- Isaiah 45:22 "all the ends of the earth"
- Haggai 2:7 "all nations"
- Mark 11:17 "Is it not written, My house shall be called of all nations the house of prayer?"
- Luke 2:10 "bring you good tidings of great joy, which shall be to all people"
- Revelation 5:9 "hast redeemed us to God...out of every kindred, tongue, people, and nation"

94.3 The New People

94.3.1 Collectivism vs. Individualism

The Christian believers from the South tend to be more group-oriented rather than individualistic. This fact is grounded in empirical research among human subjects. Geert Hofstede (1980) and Trompenaars and Hampden-Turner (1997), among other social researchers, describe the United States as the most individualistic culture in the world. As such, Christian believers in the West tacitly assume that faith is a personal matter. Whether an individual is able to articulate it or not, there are tacit assumptions related to my faith, my beliefs, my church, and my pastor. The next Christendom is not making the same assumptions. Followers of Christ in the emerging mission movements from the Majority World are not thinking in terms of "I," "me," and "my."

94.3.2 Holism vs. Dichotomism

Distinct from the United States, Canada, and Northern Europe, most people in the Majority World are holistic in their view of life. The tribal animist in Papua New Guinea does not consciously think of his incantations to the spirit world, whether to personal spirit beings or impersonal spirit forces, as an activity separate and distinct from his physical work in the garden. Both are part of normal daily life. In South India, a woman from a Hindu background may sleep with her face pointed toward the East, the direction from which energy flows. For her, religion is not just a theological creed, but also a way of life. Similar to the man in Papua New Guinea, she assumes that impersonal spirit forces are all around, both negative and positive, impacting harm and happiness.

In the emerging global South, Christian believers tend not to dichotomize life. They are not separating their religious faith from day-to-day life experiences. From the reason crops in the field grow to the cause of death, they would assume some sort of interface between this world and another.

94.3.3 Relationships vs. Tasks

Empirical research as well as our own personal experiences have shown us that Western cultures are task oriented with a primary focus on the bottom line. Despite repeated training in participative management, servant leadership, and Theory X and Y,[4] we still emphasize the task to be done as per proven business processes in a specified timeframe according to a line-item budget. These business components are carefully articulated in written low context documents – the request for proposal, statement of work, technical and cost proposals, non-disclosure agreement, contract, and after-action review.

The next Christendom is much more oriented to personal relationships that have been built over time in varied contexts. Written contracts are not necessary, but if drafted they are merely starting points at best because relationships and mutual activity evolve. Partnerships between the North and South will be forged informally through long-term friendships, time, shared experiences, and building trust. This is high context, where status is not achieved through performance, but acquired through affiliations.

94.4 The New Perspectives

94.4.1 What Are the New Roles?

Not only are men and women from the Majority World assuming status and roles as leaders in the global mission enterprise, but also local roles have emerged as critically important. For example, if a Western mission agency or a non-governmental organization wants to launch an initiative in the south of India, today it would be folly on the part of the Western entity and inconceivable from the perspective of local South India Christian leaders that such an endeavor would be attempted apart from local needs assessment, endorsement, planning, implementation, and sustainment.

In Bangalore, India, the capital of Karnataka, Dr. Jayakumar Ramachandran is a key Christian leader. A Telugu-speaker born into a business family, whose father was a strict Hindu and mother a Christian, Jayakumar has founded a mission agency, a missionary training center, extension centers in the northeast of India, medical clinics, and orphanages. Residents in the Banaswadi Layout of Bangalore know him as a patron in relationships of clientelism. He has been the Chairman of the Board of Directors for the India Institute of Missiology. With a Masters of divinity from Dallas Theological Seminary in Dallas, Texas and a Ph.D. in missiology earned in India, Jayakumar has been an advocate for standards of academic rigor in national missionary training centers throughout India.

A large U.S.-based missionary agency launched a church-planting endeavor in the south of India in 2003, but chose not to seek sponsorship from church leaders. After 8 years of fervent activity, virtually no access has been gained into local ministries.

[4] In 1960 Douglas McGregor of Harvard University introduced Theory X and Theory Y in the book, the Human Side of enterprise. Approaches to management and motivation, Theory X assumes an authoritative manager must force and direct work. Theory Y assumes people are self-motivated and thus a participative style of work management is preferred.

In contrast, another agency headquartered in Colorado initiated a partnership in 2004 with the India Institute of Missiology to provide expertise in a story-based worldview approach to cross-cultural ministry. Access to local leaders originated with the World Evangelical Alliance[5] where Dr. K. Rajendran provided inimitable transfer of trust. The leaders of the agency were introduced to Dr. C. Barnabas and Dr. Jayakumar Ramachandran. Authentic relationships became the foundation for partnerships, a flourishing local ministry, and sustained impact.

In Ibero-America, Dr. Omar Gava from Villa Carlos Paz, Argentina, is the Coordinator of Missionary Training for Cooperación Misionera de Ibero-America International. His Doctorate of Ministry with an emphasis on andragogy was earned at the Facultade Teologica Sur America in Londrina, Brazil. With over 40 years of practical ministry experience, Gava coordinates all missionary training curriculum design and delivery. He founded Recursos Estrategicos Globales (Global Strategic Resources) in Argentina, a civic organization that provides missiological resources to all the missionary training centers in South America, Central America, Mexico, and the Iberian Peninsula.

94.4.2 What Are the New Relationships?

In the next Christendom, we see a return to the model of the early church, which was comprised of multiple ethnicities. In the New Testament there is no record of a Gentile-only church. Despite racial resistance particularly from Jewish leaders in the first century, who understood that the Jews were the only people of God, the narrative of the biblical book of the Acts of the Apostles tells a story of inclusivity. Note the selected passages from Acts shown in Table 94.2.

At the birth of the early church, Jews from 15 nations had gathered in Jerusalem, as specified by Old Testament requirement for the Feast of Pentecost. From the beginning of Acts (1:8), Christians believed God had instructed them to spread the Gospel of Grace to all nations. As the narrative of the book of Acts unfolds, men, women, children, plus all ethnicities and social classes were recipients of the Good News. In contrast to Roman culture where even seats in the Coliseum were designated according to social class, seating in the Christian church was not. The hope of renewing this inclusivity is in the next Christendom.

---

[5] Launched in London, England in 1846, the World Evangelical Alliance (WEA) is the world's largest association of evangelical Christians serving a constituency of 600 million people. Today it is comprised of 1128 national evangelical alliances organized into seven regions of the world. The purpose of WEA is multifold: (a) to live and proclaim Kingdom values, (b) to provide identity and voice to global followers of Christ, (c) to seek holiness, justice and renewal at every level of society (family, community and nation), and (d) to glorify God.

Table 94.2 The story of inclusivity in the Acts of the Apostles

Chapter:Versus	Description of scripture text demonstrating inclusivity
2:5–11	Jews from 15 nations (note Parthia, Mede, and Elam in modern Iraq)
2:16–21	Sons, daughters, young men, and old men
5:14; 8:12	Men and women
6:1	Hellenized Jews (note a change in social structure to accommodate)
6:7	Priests from Judaism
8:4–5	Philip the evangelist sent to Samaritans (mixed race of Jews and Gentiles)
8:36–38	Ethiopian from North Africa
10:1–11:18	Italians
10:6	A tanner (in India – *chamar*, lower caste)
13:1	Simeon called Niger (name means "black" – leader of church in Antioch)
13:1	Lucius of Cyrene (from North Africa – leader of church in Antioch)
16:14, 27	Lydia (merchant) and a jailer (low social class)
17:4, 12, 34	Prominent women
18:8	Synagogue rulers
21:5	Wives and children included in the prayer send-off of the Apostle Paul

Source: Robert Strauss

94.4.3 What Is the New Theology?

Throughout Latin America, Christian believers tend to express their faith through demonstrable activity ("signs and wonders" from Romans 15:19) and with genuine passion. They read about the miracles in the Bible, but do not discard them as for another dispensation. They regard them as applicable to their very life situations today. These assumptions about theology contrast with American fundamentalism where, starting in the 1920s, a clear distinction was been made between fundamentalists and holiness groups (Carpenter 1997; Hofstadter 1963; Marsden 1980, 1987, 1991).

In Africa, south of the Sahel, the spread of the Christian Gospel into indigenous regions has been rapid, resulting in groups of believers who have integrated their spiritualist customs with the new teachings from Scripture. Existing worldview assumptions about reality – classification, definition of self, definition of other, relationships, cause and effect (Kearney 1984: 106) – have not been displaced by a rival story from the Bible. Selected bits and pieces of the biblical narrative have been meshed together with existing core assumptions. Although alarming to some missiologists, this syncretism is a reality among African Christian believers throughout the continent (Ranger 2008: 155).

Analyzed from an intercultural point of view, culture is the learned, shared patterns of perception and behavior. Figure 94.4 displays a model of culture represented by four concentric circles. Outwardly, one is able to observe human behavior, which is patterned, shared, and learned. People know how to greet one another. While the function remains somewhat the same throughout the world, the form differs. In Milwaukee, Wisconsin the form may be a handshake. In Buenos Aires, Argentina, the form will be a kiss on the cheek. Sociocultural institutions promote and prohibit behavior.

Beneath these observable aspects of culture are perceptions in the form of shared values and core worldview assumptions. At the value level, people have strong feelings about what ought or ought not to be. At the worldview level, people have assumptions about what is real, what is not real, who self is, who "other" is, relationships, cause and effect, and how time and space function.

Regarding Africa and many other regions of the world, people have quite readily assimilated the outward demonstrations of the Christian faith, whether these are Christian terms or rituals. Missionaries have been able to establish Christian institutions that serve host societies. However, it has been much more difficult to impact people at the levels of values, identity, and core worldview assumptions. Hence, change at these deeper levels has not occurred quickly. A villager in Senegal may go to the Christian pastor for help through prayer to God and at the same time solicit the local shaman to placate the offended ancestors through traditional incantations (see Fig. 94.4).

In India, the Christian Gospel has spread rapidly through cities and rural regions by mass conversion movements among homogeneous social networks (Pickett 1933; McGavran 1970). Entire villages and people groups converted to Christianity, especially among lower caste peoples. Two factors facilitated receptivity to the Christian message: (a) the ubiquitous social stratification of Hinduism and (b) the inclusivity of the Christian Gospel.

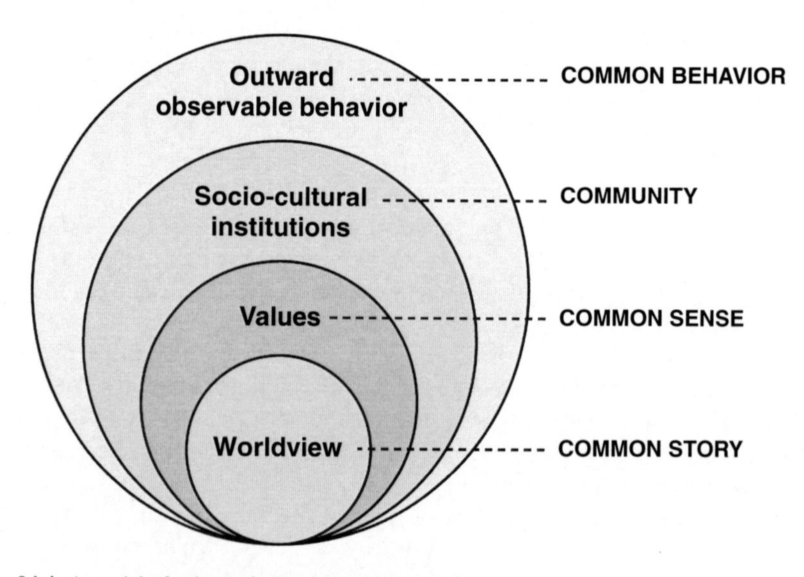

Fig. 94.4 A model of culture (Source: Robert Strauss)

94.5 Conclusion

Phillip Jenkins awakened us to the next Christendom, a global Christianity where influence and compassion emanate from new places, people, and perspectives. Where do we go from here? Numerous questions have been posed, both explicitly and implicitly, by the historic shift. How will the Global North answer these questions? On what basis will the North embrace inclusivity? How will egalitarian approaches of the West merge with the hierarchical approach of the East? To what degree will the Global South endorse gender equality and under what circumstances? As partnerships develop and resources are shared, what strategies will emerge to manage reciprocity and dependency? In what ways may the Global South curb the increasing secularization of the Global North? What is the Next Christendom's stance on human rights, homosexuality, and freedom?

American anthropologist Clifford Geertz writes, "I don't think things are moving toward an omega point; I think they are moving toward more diversity" (Page 2010: 127). And yet, *e pluribus unum*. Are we ready?

References

Belloc, H. (2009/1920). *Europe and the faith*. Charlotte: TAN Books and Publishers.

Buhlmann, W. (1976). *The coming of the third church*. Maryknoll: Orbis.

Carpenter, J. (1997). *Revive us again: The reawakening of American fundamentalism*. New York: Oxford University Press.

Engel, J., & Dyrness, W. (2000). *Changing the mind of missions*. Downers Grove: InterVarsity Press.

Hathaway, P., Yun, B., Yongze, P., & Wang, E. (2003). *Back to Jerusalem: Three Chinese house church leaders share their vision to complete the great commission*. Atlanta: Authentic Publishing.

Hiebert, P. G. (1986). *Anthropological insights for missionaries* (17th ed.). Grand Rapids: Baker Academic Publishing.

Hofstadter, R. (1963). *Anti-intellectualism in American life*. New York: Knopf.

Hofstede, G. (1980). *Culture's consequences*. Newbury Park: Sage Publications.

Jenkins, P. (2011). *The next Christendom: The coming of global Christianity* (3rd ed.). New York: Oxford University Press.

Kearney, M. (1984). *World view*. Novato: Chandler & Sharp Publishers.

Linton, R. (1936). *The study of man*. New York: Appleton-Century Company.

Marsden, G. (1980). *Fundamentalism and American culture: The shaping of twentieth-century evangelicalism 1870–1925*. New York: Oxford University Press.

Marsden, G. (1987). *Reforming fundamentalism*. Grand Rapids: Eerdmans.

Marsden, G. (1991). *Understanding fundamentalism and evangelicalism*. Grand Rapids: Eerdmans.

Mbiti, J. (1995). In K. Bediako (Ed.), *Christianity in Africa*. Edinburgh: Edinburgh University Press.

McGavran, D. (1970). *Understanding church growth*. Grand Rapids: Eerdmans Publishing.

Neill, S. (1991). *A history of Christian missions*. London: Penguin.

Nevius, J. (2003/1886). *The planting and development of missionary churches*. Hancock: Monadnock Press.

Page, S. (2010). *Diversity and complexity*. Princeton University Press.

Pickett, W. (1933). *Christian mass movements in India*. New York: Abingdon.

Rajendran, K. (2005). The future global new Christian leader. In *Christian leadership*. Retrieved from www.southasianconnection.com/articles/39/1/How-do-Global-New-Christian-Leaders-look-like/Page1.html

Ranger, T. (2008). *Evangelical Christianity and democracy in Africa*. New York: Oxford University Press.

Sanneh, L. (2003). *Whose religion is Christianity?* Grand Rapids: Eerdmans.

Steffen, T. (1997). *Passing the baton: Church planting that empowers*. La Habra: COMD.

Trompenaars, F., & Hampden-Turner, J. (1997). *Riding the waves of culture*. New York: McGraw-Hill.

Tucker, R. (2004). *From Jerusalem to Irian Jaya: A biographical history of Christian missions*. Grand Rapids: Zondervan.

Walls, A. (1996). *The missionary movement in Christian history*. Maryknoll: Orbis Books.

Yancy, P. (2001). *God at large*. Retrieved from www.christianitytoday.com/ct/2001/february5/40.136.html?start=2

Chapter 95
Deterritorialization in Havana: Is There an Alternative Based on Santeria?

Yasser Farrés, Alberto Matarán, and Yulier Avello

95.1 Introduction

Several studies have been insisting on the global processes of urban homogenization and loss of identities (for example, Sassen 1991; Auge 1992; Muñoz 2008). Different perspectives have conceptualized this phenomenon, but there still exists a noteworthy rejection for accepting this reality. This is not surprising. Around the 1950s, theoretical debates underlined the homogenizing nature of the practices generalized by Functionalism, but those tendencies continued anyway. Nevertheless, an interest in the multidimensional nature of those processes is increasing; considerations on the social, environmental, and economic dimensions existing beyond the aesthetic have become fundamentals for proposing of solutions.

However, the current perspectives cannot explain the totality of hierarchies articulating the global reproduction of those processes. In that sense, the following essay describes a relationship between those phenomena and the "disenchantment of the world," which is given excess credit by the modern science and became a civilization discourse based on the supposed universal legitimacy of an abstract person that, actually, response to the Western modern canons. That is the reason for

Y. Farrés (✉)
Department of Philosophy, University of Zaragoza, Campus San Francisco, Pedro Cerbuna, 12, 50009 Zaragoza, Spain
e-mail: yasserfarres@gmail.com

A. Matarán
Department of Urban and Spatial Planning, University of Granada, Campus Fuente Nueva, Severo Ochoa s/n, 18071 Granada, Spain
e-mail: mataran@ugr.es

Y. Avello
COPEXTEL S.A., Ministry of Informatics and Communications, Havana, Cuba
e-mail: arq.avello@gmail.com

S.D. Brunn (ed.), *The Changing World Religion Map*,
DOI 10.1007/978-94-017-9376-6_95

presenting the possibility for alternatives based on the spiritualities of some cultures of resistances. In order to achieve this aim, the paper focuses the role of "Santeria" or "Regla de Osha-Ifá" for the cultural identity at the municipality of Regla in Havana, Cuba. Empirical research presents a system of relationship between some "use-perception-transformation processes" related to certain rituals and some components of the human environment. The chapter also contains a brief introduction to the Cuban *transculturation* process and the Afro-Cuban religions while the notions of "deterritorialization of metropolis" and "territorial coloniality" support a theoretical/conceptual framework.

95.2 The Deterritorialization in Architecture, Urbanism, and Planning as a Product of the Global "Territorial Coloniality"

The concept of "deterritorialization of metropolis," developed by Magnaghi (2011), is particularly remarkable among the existing approaches on the global processes of urban homogenization and loss of identities. He defines that phenomenon as the generalization, against autochthonous territorial values and traditional cultures, of a model of megalopolis characterized with the "metropolis form."[1] According to him, technology supports a process called "liberation of city" with respect of the territory and culture. That is, what is a favorable situation for losing cultural identities and traditions which generates unequal uses of space regarding the limited technological access that social groups could have. Magnaghi argues that this process is inseparable from the "liberation of territory" (use of it as a simple support for economic activities and functions which are increasingly independent and uprooted of the place and its specific environmental, cultural or identity qualities, along with the presumption of creating a second artificial nature). Both processes share two characteristics: "decontextualization" (term referring to the destruction of the landscape identity caused by the rupture of relations between new forms of settlement and places), and "degradation" (not only environmental, but also social and economic).

Instead of discussing the deterritorialization of the metropolis concept and to avoid futile discussions about what a metropolis is, this chapter uses these concepts: "deterritorialization in Architecture, Urbanism, and Planning" and also "deterritorialization of built environment." Both seem o be a more appropriate multiscalar approach and also follow the argument of *the global reproduction of self-referred typologies with no references to specific local cultures and territories*

[1] "...an urban structure with a strongly dissipative and entropic nature; with no physical boundaries or limits to growth; strongly destabilizing and hierarchical; homogenizer of the territory it occupies; ecocatastrophic; devaluing of the individual qualities of places, deprived of aesthetic quality, and reductionist in to models of life" (Magnaghi 1989).

(Farrés 2010; Farrés and Matarán 2012). This argument does not represent a basic argument of Magnaghi's concept.

There are several additional conceptualizations, some which are economy-based and others insist on cultural aspects. A common position here is to present homogenizing changes as a sui generis product of the global capitalist logics while at the same time referring globalization and neoliberalism as main factors for this phenomenon's expansion. In some way, this consensus defines a certain *hypothesis of the capitalist exclusiveness for territorial, urban and architectural homogenization* (Farrés 2010; Farrés and Matarán 2012) which affirms the global capitalism as the origin of those processes. The thinking is also present both in Magnaghi, who follows the conception of capitalism as a deterritorialiser machine defended by Deleuze and Guattari (1987) and in 100 additional concepts that Taylor and Lang (2004) grouped in a long list where global, international, world, transnational, and other adjectives seems to recognize global capitalism as the definitive reason for those transformations.

Nevertheless, the processes of homogenization and loss of identities also exist in socialist countries like Cuba (Farrés and Matarán 2012). This point might suggest that real socialism is merely a state capitalism, but it would be a sterile dualist discussion. In contrast, if theoretical studies define those processes as products of modernization, innovative arguments will appear from critical theories; for example, including the criticism of the "modern/colonial capitalist/patriarchal world-system" (Grosfoguel 2005, 2006a, b), a point of view that goes beyond economic reductionism and culturalism (Grosfoguel 2007). In that sense, there is an ongoing discussion about the "territorial coloniality," which explains that *the deterritorialized praxis obeys the hegemony of the epistemic model displayed by Western in the modern/colonial world system* (Farrés 2010; Farrés and Matarán 2012). This holistic perspective permits a better understanding of the hierarchies existing in the production of human spaces as cultural phenomena.

"Territorial coloniality" is also a useful analysis category because the supposed necessity of modernization is still a thread for practices, but there is no sufficient attention to the linked existence of modernity and coloniality. That term, particularly expressing the general concept of "coloniality" as defined by Santiago Castro-Gómez (2007), refers to *the conjunct of power patterns in territorial praxis that achieves a hegemonic conception of territory over other "inferiorized" concepts* (Farrés 2010). Resulting of western universalisms (see Grosfoguel 2006b), those patterns articulate a triangular structure among the "coloniality of territorial power," the "coloniality of territorial being" and the "coloniality of territorial knowledge." The coloniality of territorial power is the intersubjective space where a group of people obtains the enunciation power to define what is territorially correct. The coloniality of territorial being is the hegemony of the "urban being" over any other form of human existences or "non-urban being." The coloniality of territorial knowledge is the hegemony of some specific knowledge deciding how to design and inhabit the territories, cities, and architectures.

95.3 The Territorial Coloniality as a Product of the "Disenchantment of the World:" Routes for the Reterritorialization Associated with Religion

It is possible to say that the territorial coloniality executed by western universalisms is a product of considering the environment not as a *subject* but an *object* (Farrés 2010). Such a conception is associated to the "disenchantment of the world," which is given excess credit by the modern science. Recent territorial conflicts between modernization processes and non-western cultures in Global South countries illustrate that problem. For instance, the contradictions between the *Aymaras* ancestral traditions and the developmental policies presented by Evo Morales in Bolivia, even though their new national Constitutions included concepts from Andean cosmologies (see Rossell 2009). These governmental territorial policies are based on the Latin notion of *territorium* (terra-torium, that is, "land belonging to someone") while the territorial existence for those autochthonous civilizations turns around concepts like *Pachamama* (Mother Earth) or *Sumak Kawsay* (Good Life), which consider people belong to nature, not the reverse.

A comparison between the current mainstream cities and those existing almost two centuries ago can illustrate the impact of the "disenchantment of the world" in Architecture, Urbanism, and Planning. Indeed, the coexistence of typological varieties is verifiable more in the ancient cases than in today. For example, North European Christian cities models were different than contemporaneous Islamic, Eastern, or pre-Columbian cities. Each was so genuinely territorialized in social, cultural, environmental, and technical factors that biunivocal relations existed between each typologies of settlements and the corresponding society's worldview. On the contrary, traditional relations disappear nowadays while the reproduction of global models has no restrictions due to technological transferences, material exports and oil energy uses. Professionals do not easily accept these circumstances.

If the studies consider homogenization and loss of identity as two products of the global dominant modernizer culture, it would be easy to understand that the possibilities for an alternative exist into the *cultures of resistances*. There are conditions for new territorial policies based on *subaltern epistemologies* or *border thinking* produced from the borders of the universalized epistemic model (Mignolo 1999, 2000; Grosfoguel 2009). But it is necessary to comprehend those communities' experiences. Therefore, an archaeology and ecology of knowledge (Foucault 1969; Sousa Santos 2007) is convenient, particularly, to appreciate the potential role of *religion*[2] to facilitate a multiplicity of relations between people and the environment.

It is necessary to analyze built environments' history as a part of humankind's evolution to understand that territories, cities, urban spaces, buildings, and decorative elements are more than expressions of culture or containers for human activities.

---

[2]The use of "religion" will avoid misunderstandings, but the term should demand a discussion because it is a modern atheist classification applied to non-scientific knowledge. Speaking about "cosmologies" or "philosophies" seems to be more useful and productive.

This might seem obvious, but a kind of dualism-signed modern architecture when designing in response to an abstract universal person (Lecorbusier's Modulor) stripped of race, sexuality, and religion, that is, what culture and tradition mean. In that sense, it is important to comprehend meanings and values projected into the built environment resulting from specific *use-perception-transformation processes*, which, curiously, have become a way for resisting homogenization in recent times.

Both the universal and local histories remain imprinted in the structures of human environments offering possibilities for societal self-acknowledgment as well as reasons for recognizing patrimonial values in those components of the human material culture. However, the institutionally recognized heritage usually refers spaces belonging to dominant social groups. In fact, there is a Historiography of Architecture and Urbanism which talk about "relevant architectures"[3] while other "minor architectures" are disregarded. This situation is changing, but a long path has to be covered and new approaches are necessary, especially if the historical studies aim to get operative sense and turn into fundamentals for real changes. In order to make that happen, a starting point for such an analysis of built environment is to accept that both the responses to materials needs and spiritual needs are imperative in studying the origin of the human environments. This position can become the basis for new territorial, urban, and architectural projects, and represents a crucial hypothesis to confront the dominant epistemology that generalizes deterritorialized human environments in the current modern/colonial world-system.

95.4 Santeria and Territorial Identity at the Eastern Area of Havana

One of those "subalternized" cosmologies is *Regla de Osha-Ifá* or *Santeria*, an Afro-Cuban religion resulting from the confrontation between African slaves' religion and the dominant Catholicism during colonialism in Cuba. According to this worldview, a person must exist in balance and harmony with a previous life and the environment, as well as with his/her own representations of it (Avello 2006). This conception turns into different kinds of relations (filial, economical, socio-cultural, and with divinities or mystical forces).

For *santeros* and *santeras* (men and women practicing Santeria), it is essential to worship *Orishas* by mean of adorations, providing food, and doing specific ritual on specific dates as a part of the liturgy. To provide food for Orishas ("dar de comer a los Santos") is one of the most important rites for expectant responses (open path to prosperity, help in case of problems, etc.). Many members of this religious community participate in this festive rite, as well as friends or invited people even

[3] In order to avoid misunderstanding, this essay will use the term architecture as a generic word to refer to spaces created by humans independently of it scales (interior spaces, architectural spaces, urban spaces, territorial spaces, etc.).

though they may be not religious. All this requires specific spaces where the natural component used to be present because of mystic meanings.

The words "Orishas" and "Santos" (Saints) are interchangeable in this religion in Cuba. During colonialism, Spanish tried to impose Catholicism to African slaves and prohibited them from a freedom of worship. Therefore, Africans used images of Catholic saints when venerating their own deities. This process, known in Cuban Cultural Studies as *transculturation,* generated the so-called Afro-Cuban syncretism.[4] In a similar way, African slaves had no own religious spaces, but awarded meaning to the existing components of the built environment and nature.

Even after the slaves' freedom (1880–1886), afro descendent people could not openly express their religion, which continued to be marginalized until recent times due both to the dominant Catholicism until 1959 and the official materialism of the Marxist Cuban Revolution. Nevertheless, the Afro-Cuban religion community was growing during this entire time. Recently, different changes have created conditions for the acceptance of religion in the Cuban society. It might be an improved condition for a cosmology that produced a system of human values, social relations, relations between humans and environment, material and immaterial heritages as well as noteworthy contributions to the identity of settlements. It is the case of the current municipalities of Guanabacoa and Regla, we have two relevant afro descendants' territories with a common history in the eastern zone of Havana. Regla, which native name was Guaicanamar, "in front to the sea," belonged to the existing *cacicazgo*[5] of Guanabacoa (Fig. 95.1).

The origin of the current municipality of Guanabacoa goes back to 1554 when Spanish conquerors established a settlement for rehousing indigenous inhabitants from the surrounding areas. It began to die because of the new slave conditions and other reasons. African slaves replaced this group while an important colonial town that provided of its corresponding Catholic institutions; so much, that it received the title of "Villa de la Asunción de Guanabacoa" from Philip V of Spain (see Vidal 1887; Castellanos 1948; Rodríguez 1992). In the case of municipality of Regla, it is known that in 1573 it existed a passage settlement. However, the town's foundational date is March 3, 1687 when Don Pedro Recio de Oquendo donated part of his lands to raise a chapel for venerate *Our Lady of Regla* (Cárdenas and Aguiar 2010).

Due to its port condition, Regla became an unload pointing for slave commerce, a not attractive activity to be included in the port of its nearby *Villa de San Cristóbal de La Habana*, the main city of the western side of Cuba. At the same time, because of its geographical location, Regla became a crucial connection between Havana and Guanabacoa. Those historical and geographical circumstances made Guanabacoa and Regla typical cases where Afro-Cuban religion remained oppressed by Catholicism, a condition that continued when afro descendants began to work at the port after the

[4] Other syncretic creeds are: (1) Regla Congo or Mayombe, as well known as "Palo Monte"; (2) Regla Arará; and (3) Abakuá Secret Society. The "abakuá" or "ñáñigos" (men practicing this last religion) found their work on the worship to their ancestors and what they did, while consist of an association for mutual help similar to masonry (see Fernández 2001).

[5] Cacicazgo is a territory governed by a cacique, the leader of each indigenous community in pre-Columbian Caribbean.

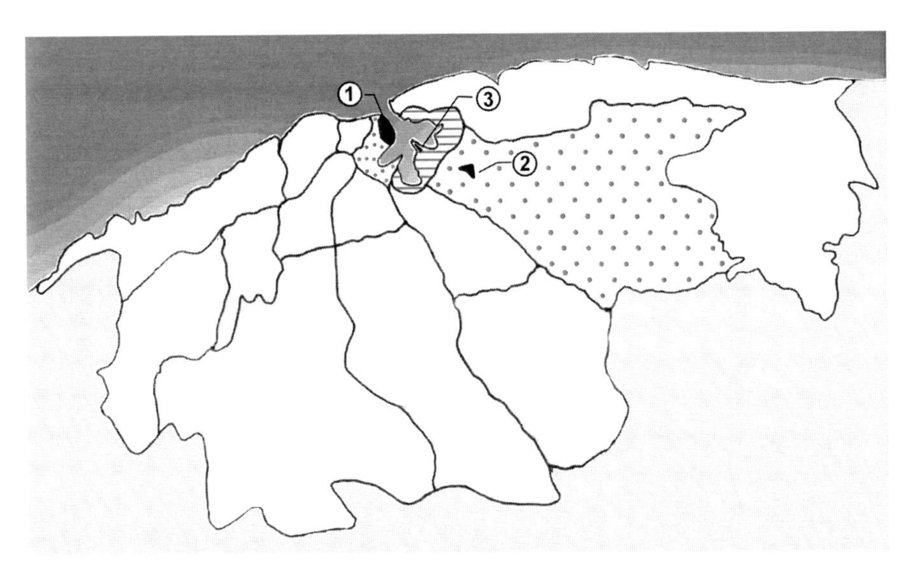

Fig. 95.1 The current municipalities of (*1*) La Habana Vieja, (*2*) Regla, and (*3*) Guanabacoa and their foundational sites (Map by Yasser Farrés)

end of slavery. Similar cases exist in other Cuban port areas, but this text will analyze the case of Regla due to its connotation for the Santeria practicing community in Havana.

95.4.1 Components of the Environment with Significances for Santeria in Regla

Osha-Ifá rituals have central relationships with the components of nature like wind, rainbow, stars, rain, as well as with human body and even places or sites where the use of space has religious connotations (Valdés 2002). That conception suggests this as a primitive religion; however, it has constantly changed because of the new human environment conditions, which were different from the original settings in African villages. This circumstance forced *santeros* and *santeras* to establish parallel places when enslaved in Cuba. In the case of Regla, some natural components of the environment as well as some human-created components (Fig. 95.2) have become so emblematic that people have strongly appropriated of it. Indeed, there is a systemic relationship between the uses and meanings of that system of spaces.

Many authors study the natural components of environment with significance for Santeria because veneration to nature is a shared characteristic in all African cultures. Slave found analogies among Cuban plants, animals, or geographical features and those existing at their ancient lands; however, due to the urban placement of practicing people is not easy to be close to those elements. That is why some spaces in Regla became important for the rest of that community in the metropolitan area of Havana.

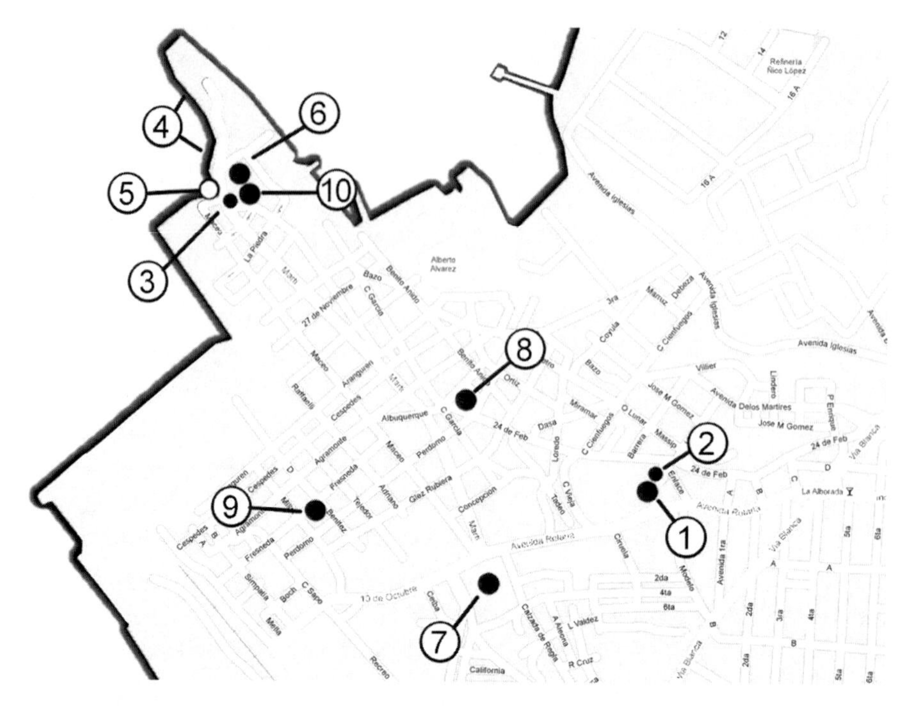

Fig. 95.2 Localization of the components of environment with significance for Santeria at the municipality of Regla: (*1*) Lenin Hill, (*2*) Queen Palm at Lenin Hill, (*3*) Ceiba de Regla, (*4*) the coast, (*5*) Reglas' Pier, (*6*) Sanctuary of Our Lady of Regla, (*7*) Cemetery, (*8*) Market, (*9*) Atandá well, (*10*) Panchita Cardenas' house-temple or Ilé-ocha (Map by Yasser Farrés)

Fig. 95.3 View of the coast next to the promenade from Havana, indicating (*1*) The Ceiba of Regla; (*2*) the catholic Sanctuary of Our Lady of Regla; and (*3*) the Reglas' Pier (Photo © Libertad Rodríguez Otero, used with permission)

The coast next to the promenade is one of the most important natural components of the environment for Santeria in Regla (Fig. 95.3). It has become an accessible place for rituals and meditations where devote followers of *Yemayá*, queen of the sea, offer fruits and animal sacrifices to this Orisha. It is also a place for venerating

Olocum. In the colonial past, there were processions in the streets where *cabildos*[6] were going towards the coast to strip people of negatives "charges" by using plants and coconuts, which they threw into the sea once charged of "bad things." Likewise, a maritime procession used to expose the Virgin of Regla around the harbor in the form of a circular trajectory.

The so-called "Lenin Hill" is also important. Its meaning is that of a wilderness (el monte). Because of it importance for surviving, African ancestors accepted wilderness as a place of magic-religious events: deities and spirits inhabit it. "There are the orishas Elleguá, Oggún, Ochosi, Oko, Aye, Shangó, Allagguna, and Eggún –the spirits, Eleku, Ikus, Ibbayer (…). It is full of spirits!" (Cabrera 1993: 17). Humans must ask for permission to use these resources. In the case of Regla, the "Lenin Hill" heights permit one to see the whole territory while it also became a place for meditation, unique at that municipality.

Afro-Cuban Santeria practice is also applied to symbolic analogies for trees. For example, *ceiba* (or *seiba*) is considered "mother of all trees," probably because of its robust trunk and proportions. It exists in isolation, not in groves and always in clear sites. This quality has turned it into a sacred place; in fact, there is a widespread belief that there is no possibility for electric shock when standing under a ceiba: "…many old peasants honestly swear that no lightning has ever damaged a ceiba" (Roldán Oriarte 1940: 488). According to mythology, deities inhabit ceiba's crown while invocated ancestor spirits are sitting on the branches. Roots are also mystic because *Oro Iña* emerged from it and lives there, and because it goes into the bowels of the mother earth.

According to an oral tradition, a *babalawo* (an Osha-Ifá priest) sowed a ceiba in front of the Sanctuary of the Virgin of Regla. That place gained religious and cultural importance while generated a space for dancing, praying, singing, and feeding to Orishas. "La Ceiba de Regla" is known in the whole city with sings that identity if in the municipality. This place represents the resistance of Afro-Cuban culture in front of Catholicism as well as the syncretism.

The queen palm is another remarkable tree (Fig. 95.4). It has been a source for both animal food and construction material for natives, "cimarrones" (fugitive slaves) and farmers. Its configuration attracts lightning. It is not a common tree, but Changó's place of refuge (Cabrera 1993: 216–217). The queen palm also has a specific meaning for *ñáñigos* or *abakuás* (Cabrera 1993: 267). The particularly well known *Queen Palm of "Lenin Hill"* is at a high site, from where Changó can protect warriors and hunters. Santeros and santeras go to this place to offer food under the tree in a ritual of song and prayer. Other queen palms exist in that place, but this is the most important.

Similar to natural components is the built environment which has meaning for Santeria, not exactly because of a logical "metropolization" and modernization of that religion but forced by slavery and subsequent marginal conditions in urban

[6] It is an association of Africans from the same ethnicity and their descendants that was permitted by colonialists. It was considered it as a space for festive meetings, although it was a religious-mutualist association (see Bolivar 1990).

Fig. 95.4 The Queen Palm of the "Lenin Hill" (Photo © Yulier Avello Pereiro)

areas. In Regla the streets were open stages for several religious activities organized by *cabildos* in the past. In recent days, processions and other rites take place in streets. For example, practicing people place offerings at the intersections for *Elegguá*, an orisha that clears the way for people, as well as for *Oggún* and *Oshosi*. Similar to the coast, people throw negatives "charges" into the street.

Belonging to the architectural scale are two remarkable components of the built environment: the Sanctuary of Our Lady of Regla and Regla's pier. The sanctuary it is the most important religious component of the human environment for both Catholicism and Santeria. Raised in honor to the Catholic patron of sailors, which Santeria practicing relates to Yemayá, we discover that Cuban and foreigner devotes are used to visiting this place. On the other hand, despite its ruinous state, the pier is still an icon in the municipality. Inaugurated in 1912, it was the main entrance for visitors and settlers. It had different use them, but became an important stop along the route of the processions mentioned before.

Specifically, the route of the ancient cabildo's procession included a cemetery that is located on Calixto García Street and González Rubiera Avenue. Known as

Fig. 95.5 Cemetery of Regla (Photo © Yulier Avello Pereiro)

"the old cemetery," this is the place of the "negros de nación" mortal remains.[7] People used to dance, brake coconuts, and throw fruits and even some copper coins at that place of honor. Later, about 6:00 pm, the cabildo goes to the New Cemetery to repeat that rite, before going to santeras' houses (Hernández 1980).

Santeros and santeras believe that many orishas inhabit the cemetery: *Oyá* is at the eastern entrance and *Yewa* is around tombs while *Obá* is at the graves. That is why similar rites used to take place when a santero or santera dies. However, that tradition is disappearing in this place, probably, "… because there is no interest or willpower for carrying a sarcophagus on shoulders across the city through cemetery for doing the respective rite. Only some ñáñigos, but not all, can be found doing those ceremonies inside cemetery while dancing and singing in honor to the deceased"[8] (Fig. 95.5).

The market of Regla is one of the buildings that has meaning for Santeria even though it is not a religious space (Fig. 95.6). Some informants said that an initiated person in Osha-Ifá (*Iyabó*) must visit this place with his/her *babalocha* (godfather) or *iyalocha* (godmother), who will introduce him/her to the four street intersections of the block where the market is. After a ritual, those practicing people must enter the market and visit stallholders to steal insignificant things (grains of beans, rice, etc.); a symbolic action that should advise them what not to do it during his/her life. Other informants have never seen that rite in this particular market, however it is recognized that it happens in "Mercado de Cuatro Caminos," in the municipality of Cerro, a bigger and more important example is among the practicing people from the entire Havana City (Avello 2006). In any case, "Mercado de Regla" is a territorially important store for buying plants, tools, and equipment for practicing Santeria.

[7] "Blacks of nations" is a term referring to slaves from ancient African nations or territories.

[8] Interview to a Babalawo at Regla done by Avello (2006).

Fig. 95.6 The market in Regla (Photo © Yulier Avello Pereiro)

Another relevant human-created component is Atanda's well (Fig. 95.7). Oral traditions affirm that an African built the well where the *Shangó Tedum cabildo* was located, probably when Atandá was heading that association and blessed it as a part of a ceremony. Most religious people know about this 9 m (29.5 ft)-diameter and 40 m (131 ft)- deep hole, placed on the top of a hill. Today, some houses surround the well because of urban evolution. Its water is limited and contaminated, although occasionally used for religious activities.

The house-temple or 'ilé-ocha' is probably the most relevant human-created space for Santeria, which is why it is one of the components of built environment that must be specifically emphasized. People in Regla have historically recognized the Panchita Cárdenas' house-temple (Fig. 95.8), which is next to the Sanctuary. Its altar is open to the street.

Once free, Africans adapted the existing housing typologies for use as a temple or other occasional social activities (for example, to accommodate the cabildos). They had no possibilities to create specific buildings, therefore, houses had multifunctional spaces. Santeros and Santeras adapted their humble properties for Orishas veneration, rituals of initiation, and other specific ceremonies. The model of a house-temple was simple; several rooms and a courtyard were used both for living and for practicing religion. However, in most cases, there are no sufficient area; therefore, functions share spaces although zones remain differentially characterized. A four-corner house (rectangular or square) is recommendable, because corners are *Elegguá*'s territory. The spatial structure is simple but a complex system of significance is associated to its components, as presented in Table 95.1.

Fig. 95.7 The water from Atanda's well is occasionally used for religious activities (Photo © Yulier Avello Pereiro)

Fig. 95.8 Panchita Cárdenas' house-temple (Photo © Libertad Rodríguez Otero, used with permission)

Table 95.1 Significance of some components of a house-temple

Component	Significance associated to religious use
Front door	It is the first important component: everything (good or bad) goes in or goes out through the door. The orisha Olorun inhabits the exterior side. He represents the sun that gives back colors when sprayed with water. The warriors (Elegguá, Oggún, Ochosi and Osain) are protecting the house from the interior side
Living room	It is the most social area, where "brothers and sisters in religion" do a variety of ceremonies and meetings
Altar room	Statues of Catholics Saints and Virgins are on altars, surrounded by flowers, candles, and even light bulbs. The guardian angel of each practicing person is hierarchical. Correspondingly, Orishas remain organized into a rigorous hierarchy inside of the glass cabinets or "canastillero"; resting with their stones and other attributes in bowls made of "guira" (a kind of fruit), clay or porcelain. It is necessary to put de lid on those "soperas" and cover them with mantles and the corresponding orisha's tools (machetes, flags, arrows…) are in the surrounding
Igbodú	Other representations of Orishas exist in this room, used for animal sacrifices and other secrets or private ceremonies, for example, initiation to Santeria
Iyawó room	A person initiated in Santeria (Iyawó) stays for a week at the house-temple. A specially organized room will close during a previous period to become space for individual meditation and secret ceremonies
Kitchen	Many religious activities in Santeria include feast. It is necessary to cook for people, and after that, for Orishas. Each Orisha demands his own kind of food, separately elaborated. Kitchen must be dimensioned for that exigency
Courtyard	A symbolical burial of dead practicing people takes place at the house-temple's courtyard. Sometimes, the burial of sacrificed animal is the way of feeding an orisha
Drains	This component for dumping wastes is the mean for invocating the orisha Eggún

Source: Authors

95.4.2 A Territorial Identity Based on a System of Relations Established at a Multiscalar System of Spaces

Natural and human-created components of environment define a complex system of religious spaces in Regla, which results from a process of public apprehension towards space. Similar to what happens inside the ilé-ocha (where a system of meanings and uses relates to interior spaces), another system of meaning and uses links the spaces compounding environment. Both scales are connected. The ritual of "feeding deities" articulates that relationship by mean of a group of activities to be done which begins in the wilderness, includes both the four street intersections near the house of the practicing person and the space in front of the house and finishes it with activities inside the front door.

Both ceremonies and offerings to deities start in the wilderness. Similar to the followings stages, an animal is sacrificed (maybe a chicken or hen) for feeding the Orisha. Once in the four street intersections near to the practicing person's house,[9]

[9] It is important to underline that the four corner of any space (room, house, blocks, square … etc.) are important because cardinal points are tenets of all religious activities.

they use prayers, songs and animal sacrifices for invoking Elegguá, Oggún and Oshosi (the warrior orishas, protectors of people). A similar ritual occurs in the space in front of the house. The front door becomes the finishing stage for the ritual of feeding deities. Prayers, songs, and other activities leave "bad things" out while entering the "aché" (benediction of deities or cosmic energy).

Even though those activities may occur in any Cuban city, village or human settlement, it is possible to say that Santeria practicing people have been turning Regla into a recognizable cultural identity in Havana City. Those use-perception-transformation processes defines a territorial identity that citizens have recognized as having a cultural value.

95.5 Conclusion

This essay has explained that the current deterritorialization in Architecture, Urbanism and Spatial Planning not only results from a long-term history related to the "disenchantment of the world," but also represents the hegemony of the globally reproduced western modernizer project. That process of homogenization and loss of identities is a consequence of designing in response to an abstract universal person stripped of race, sexuality or religion, as well as considering the territory as a support for economical activities and other functions.

Nevertheless, despite global "territorial coloniality," this study illustrated the possibility of finding territorial practices from subaltern positions that can establish other kinds of relationship between humans and environment. According to the information presented on the case of Havana, Santeria seems to be one of those cases that deserve a highest recognition, because of the genuine relations that santeros and santeras use to establish with their environment (both natural and built). Due to the strong understanding between practicing people and their environment, Santeria seems to be a valid route for changing the hegemonic tendency of deterritorialization, homogenization and losing of identities. It represents a starting point for decolonizing the territorial existence. It means producing a variety of 'territorial being' moreover the universalized western canon.

The idea of applying the philosophy of the "decolonial turn" (Maldonado-Torres 2007) to the architectural and urban studies is a completely new approach; and particularly, the notion of "territorial coloniality" as a category for analysis. In fact, there are no sufficient references for turning this analysis into a decolonial professional practice in Architecture, Urbanism or Spatial Planning. However, in order to introduce proactive positions for changing territorial coloniality, a recommendation is that urban designers and architects should seriously consider the necessities of the all "colonial subject" (santeros and santeras in this case, but indigenous people in another context).

In the case of Santeria, it is possible to assume the santeros and santeras have spatial necessities as starting points for architectural, urban and territorial projects. In that sense, architects and urban designers should try to understand their spatial conceptions. The paper presented by Rodrigues (2009) represents a valid reference for going into this matter. She elaborated an analysis on the spatial conception in Candomblé that could be useful because of the deeply relationships existing between that Afro-Brazilian and the Afro-Cuban Santeria.

That recognition of Santeria for decolonizing the human territorial existences in Cuba must also include an acceptance of the Afro-Cuban aesthetic, which Cuban architects used to disregard when considering it as naïve or *kitsch*. However, this is another point for further analyses in order to comprehend the changing map of this religion.

Acknowledgment The authors extend their sincerest thanks to the anthropologist Ernesto Valdés Jane, Director of "Proyecto Orunmila" (www.proyecto-orunmila.org), who advised the previous fieldwork done by Yulier Avello in the municipality of Regla, which is one of the origins of the present paper. Similarly thanks to architect Libertad Rodriguez Otero, from the "Oficina del Historiador de la Ciudad de La Habana" for her photographs.

References

Auge, M. (1992). *Non-lieux. Introduction á une anthropologie de la surmodenité*. Paris: Edition de Seuil.

Avello, Y. (2006). *Identidad, ambiente construido y religión en Regla* (Thesis in option to grade in architecture, directed by Eliana Cárdenas & Yasser Farrés). Havana: Faculty of Architecture-ISPJAE.

Bolivar, N. (1990). *Los orishas en Cuba*. Havana: Ediciones Unión.

Cabrera, L. (1993). *El monte*. Havana: Editorial Letras Cubanas.

Cárdenas, E., & Aguiar, A. (2010). Iglesia de Nuestra Señora de Regla. *Palabra Nueva. Revista de la Arquidiósesis de La Habana*, 197. http://www.palabranueva.net/contens/1006/000102-4.htm

Castellanos, G. (1948). *Relicario histórico. Frutos coloniales de la vieja Guanabacoa*. Havana: Librería Selecta.

Castro-Gómez, S. (2007). Decolonizar la Universidad. La hybris del punto cero y el diálogo de saberes. In S. Castro-Gómez & R. Grosfoguel (Comp.), *El giro decolonial: reflexiones para una diversidad epistémica más allá del capitalismo global* (pp. 79–91). Bogotá: Siglo del Hombre Editores.

Deleuze, G., & Guattari, F. (1987). *A thousand plateaus. Capitalism and schizophrenia*. Minneapolis: University of Minnesota Press.

Farrés, Y. (2010). *Descolonizar el territorio. Consideraciones epistémicas para el caso de La Habana* (Thesis in option to the Master of Advances Studies in Urbanism, Planning and Environment, directed by Alberto Mataran). Spain: University of Granada.

Farrés, Y., & Matarán, A. (2012). Colonialidad territorial: para analizar a Foucault en el marco de la desterritorialización de la metrópoli. Notas desde La habana. *Tabula Rasa, 16*, 139–159.

Fernández, J. M. (2001). *La Habana crisol de culturas y credos*. Havana: Editorial Ciencias Sociales.

Foucault, M. (1969/2002). *The archaeology of knowledge*. London/New York: Routledge.

Grosfoguel, R. (2005). The implications of subaltern epistemologies for global capitalism: Transmodernity, border thinking and global coloniality. In R. Applebaum & W. I. Robinson (Eds.), *Critical globalization studies* (pp. 283–293). New York/London: Routledge.

Grosfoguel, R. (2006a). From postcolonial studies to decolonial studies: Decolonizing postcolonial studies: A preface. *Review (Fernand Braudel Center), 29*(2), 141–142.

Grosfoguel, R. (2006b). World-systems analysis in the context of transmodernity, border thinking, and global coloniality. *Review (Fernand Braudel Center), 29*(2), 167–187.
Grosfoguel, R. (2007). The epistemic decolonial turn. Beyond political-economy paradigms. *Cultural Studies, 21*(2–3), 211–223.
Grosfoguel, R. (2009). A decolonial approach to political economy: Decolonial approach to political economy: Transmodernity, border thinking and global coloniality. In J. Suárez-Krabbe (Ed.), *Kult 6 – Special issue epistemologies of transformation: The Latin American decolonial option and its ramifications* (pp. 10–38). Denmark: Department of Culture and Identity at Roskilde University.
Hernández, G. (1980). *Mito y realidad del cabildo de Regla*. La Habana: Liceo Artístico de Regla.
Magnaghi, A. (1989). Da metropolis a ecopolis: Elementi per un progetto per la città ecologica. In M. Manzoni (Coord.), *Etica e metropolis*. Milano: Guerini.
Magnaghi, A. (2011). *El proyecto local. Hacia una conciencia del lugar*. Barcelona: Edicions UPC.
Maldonado-Torres, N. (2007). On the coloniality of being. *Cultural Studies*, 21(2), 240–270. http://dx.doi.org/10.1080/09502380601162548
Mignolo, W. (1999). *The darker side of the renaissance: Literacy, territoriality, and colonization*. Ann Arbor: University of Michigan Press.
Mignolo, W. (2000). *Local histories/Global designs: Coloniality, subaltern knowledges, and border thinking*. Princeton: Princeton University Press.
Muñoz, F. (2008). *Urbanalización: Paisajes comunes, lugares globales*. Barcelona: GG.
Rodrigues, Lia P. (2009). Space and the ritualization of Axé in Candomblé. *Kult 6 – Special issue: Epistemologies of transformation: The Latin American decolonial option and its ramifications* (pp. 85–99). Denmark: Department of Culture and Identity at Roskilde University.
Rodríguez, M. (1992). Crecimiento urbano de Guanabacoa. *Arquitectura y Urbanismo, 13*(3), 29–36.
Roldán Oriarte, E. (1940). *Cuba en la mano*. Havana: Imprenta Ucar, García y Cía.
Rossell, P. (2009). El proyecto de Evo Morales más allá de 2010. *Nueva Sociedad*, 21. http://www.nuso.org/upload/articulos/3607_1.pdf
Sassen, S. (1991). *The global city, New York, London, Tokyo*. Princeton: Princeton University Press.
Sousa Santos, B. (2007). Beyond abyssal thinking: From global lines to ecologies of knowledges. *Review, 30*(1), 45–89.
Taylor, P. J., & Lang, R. E. (2004). The shock of the new: 100 concepts describing recent urban change. *Environment and Planning, 36*(6), 951–958.
Valdés Janet, E. (2002). *Tratado de las comidas – ofrendas, ceremonias y ritos – a las posiciones*. Havana: Proyecto Orunmila.
Vidal, F. (1887). *Historia de la Villa de Guanabacoa*. Havana: Imprenta La Universal.

Chapter 96
Calling a Trickster Deity a "Bad" Name in Order to Hang It? Deconstructing Indigenous African Epistemologies Within Global Religious Maps of the Universe

Afe Adogame

96.1 Introduction: Historicizing the Study of Religions of Africa

The historiography of African religions and spiritualities provide a significant template for comprehending and deconstructing indigenous epistemologies within global academic studies of religion. African religions are used here as a generic term to embrace the indigenous religions, African Christianities and African Islam. The historical trajectory of the study of religions in Africa has evolved through several phases, each involving different purposes and points of view. Jan Platvoet et al. (1996: 105) categorizes these overlapping epochs paradigmatically, as *Africa as Object*, when its religions were studied virtually exclusively by scholars, and other observers from outside Africa; and as *Africa as Subject*, when the religions of Africa had begun to be studied also, and increasingly mainly, by Africa scholars.

Descriptions and theories of Africa's religious history have been essential elements of the cultural contacts since the very first encounters and remain so up to the present (Ludwig and Adogame 2004: 2). Within these historical phases, the colonial and missionary machineries invented and produced ways of knowing and meaning-making that anchored and facilitated processes of subjugation, exploitation, and expropriation. Particular alien forms of reasoning were entrenched while also laying claims to a "civilizing mission." The "European" knowledge that was introduced into Africa came on collision course with indigenous knowledge systems in a spate of ideological contestation culminating in a *bricolage* of knowledges. The knowledge funneled through the colonial process took centre-stage, assuming a dominant epistemology that marginalized and almost silenced alternative

A. Adogame (✉)
School of Divinity, University of Edinburgh, Edinburgh EH1 2LX, Scotland, UK
e-mail: a.adogame@ed.ac.uk

S.D. Brunn (ed.), *The Changing World Religion Map*,
DOI 10.1007/978-94-017-9376-6_96

worldviews and conceptualizations of the universe. Such a hegemonic way of knowing and meaning-making was even presumed to be capable of turning indigenous epistemologies on their head.

Legacies of the European Enlightenment filtered thought patterns that legitimized tropes of otherness and binaries of difference espoused as tradition versus modernity, primitive versus civilized, superiority versus inferiority complex into the very fabric of the dominant knowledge. It was characteristic of the forms of reasoning that it privileged and superimposed on other cultures. This dominant knowledge was liberating, transforming but also entrapping. The contestation that ensued in the production of religious knowledge produced a chasm of epistemological richness and bankruptcy at the same time. Indigenous religious epistemologies hardly witnessed their obituary in the face of the knowledge-encounter that ensued.

The growth and transformation of indigenous religions, Islam or Christianity can be better grasped when considered within the locus of mutual religious interaction, competition and influence. Islam and Christianity saw the introduction of new religious ideas and practices into indigenous religions. The encounter transformed indigenous religious thought and practice, but did not supplant it; indigenous religions preserved some beliefs and ritual practices, but also adjusted to the new sociocultural milieu. As a result of social and cultural change, some indigenous beliefs and rituals were either dropped or modified due to the impingement of European and Arab cultures, Christianity and Islam. The change also led to the revivification and revitalization of other aspects of the indigenous religious cultures. In many cases, Islam and Christianity became domesticated on the African soil. Thus, the contact produced new religions, with some appropriating indigenous symbols and giving them a new twist (Adogame 2007: 536).

This scenario produced multiple discourses and theories of knowledge within the academic study of religion in Africa, and through this knowledge-production is continually negotiated in ways that result in the reification of some meaning-making systems, the invention of others, and in a kind of "hybridized" epistemologies. The contestation that ensued in the production of religious knowledge produced a chasm of epistemological richness and bankruptcy at the same time. Indigenous religious epistemologies hardly witnessed their obituary in the face of the knowledge-encounter that ensued. This (re)production and contestation of ways of knowing and meaning-making has dire implications for unpacking and decolonizing indigenous religious epistemologies.

As Platvoet et al. (1996) demonstrate, travelogues, missionary and the colonial historiography, of late eighteenth and early nineteenth century, pioneered the study of and writing about the religions of Africa. These early genres were essentially non-scholarly collections of random observations, superficial opinions, and inaccurate information often impregnated with cultural bias and prejudices. Such accounts by Victorian travelers, such as Sir Samuel Baker, Richard Burton, James Hunt; and Christian missionaries including Thomas Bowen, David Livingstone had its target audience. They were designed to appeal to the popular Western mind, and so were written for this specific public. While some accounts denied Africans any modicum of religion, others made African religions appear as a morass of bizarre beliefs and practices. Most scholars in this phase were "children of their age;" ultimately regarded Africans as culturally "degraded," thus reinforcing popular prejudice. Nevertheless, their accounts were useful to the extent to which they served as an "information bank" upon which several scholars later depended.

The traits of prejudice that dominated these accounts continue to haunt some learned minds and the academia today. The basic terminologies such as "primitive," "nonliterate," "premodern" which still find some space in contemporary scholarship are hardly value-free. They still characterize Africa as the sharp opposite of the West, or the "heart of darkness" (Conrad 1995), thus reinforcing a negative perspective. It is therefore expedient that (African) scholars re-interrogate the concepts and terminologies we employ in describing African religions and spiritualities.

The earliest phase was supplanted by arm-chair ethnographers and evolutionary anthropologists who propounded theories on the origin and evolution of human culture following evolutionary paradigms. One feature of this phase was classical approaches championed by nineteenth century theories such as by Edward B. Tylor and James G. Frazer. This era produced a barrage of opprobrious labels including animism, fetishism, idolatry, primitivism, totemism, superstition, heathenism, and magic to designate the indigenous religions of Africa. These incongruous terms stamped indigenous religions of Africa with appearance of sameness and primitiveness, and a stigma of inferiority, especially in comparison with Islam and Christianity.

With the decline of evolutionary theory and the advent of social anthropology, systematic fieldwork studies of African societies took root in the late nineteenth century. Anthropological approaches developed in different directions such as the fieldwork approach of Malinowski and the social-functionalist theory of Radcliffe-Brown. Evans-Pritchard developed a new approach characterized by a shift from function to meaning. A new crop of British scholars including John Middleton and Victor Turner emerged. French anthropologists focused on African cosmological systems and implicit philosophies demonstrating that African religious systems are not simply reflections of socioeconomic relations, but form coherent and autonomous spheres of thought and action. A notable example is the work of Marcel Griaule (1965) among the Dogon of Mali. These systematic field studies slanted fieldwork studies according to author's nationality, and imposed a "colonialist" structure upon the interpretation of African social and religious systems. Different colonial policies, background and experiences significantly affected the study of and research on religion in post-colonial African contexts.

The 1950s and 1960s onwards marked the era of integrated and consolidated research on the religions of Africa, the transition from *"Africa as object"* to *"Africa as subject."* The word *religion* is a late-comer to the scholarly discourse about Africa. It was only in the late colonial period of the 1950s that scholars began to use the terms *religion* and *philosophy* to characterize African religions in a positive way. Attention shifted to more recent limited forms of cultural and religious change through specially designed fieldwork projects, utilizing oral traditions, political history, and contemporary socio-religious analysis. More fruitful results came from the historical dimensions of African religion (Ranger and Kimambo 1972). A small number of philosophically and theologically oriented comparative studies by both

European and African scholars developed. Their interpretations greatly influenced European understanding of African religions. They interpreted African religious concepts according to Western philosophical categories (Placide Tempels, Alexis Kagame). Scholars such as Robin Horton (1984, 1971) reject the use of Western theological and ontological categories because they do not always correspond to African concepts and religious experience. More descriptive surveys, theological and ethnographic surveys of African religions by Geoffrey Parrinder (1954) and African scholars such as Joseph Danquah (1968), John Mbiti (1969), and Bolaji Idowu (1973) became popular, although they suffered from another set of problems. They pretended to be exhaustive and to cover too many societies and types of religious phenomena. They focused too much on "beliefs" and neglected the sociocultural and ritual fabric within which they are embedded. Thus, by producing a pyramidal structure, they reduced African religions to a set of "doctrines" analogous in structure to Western faiths.

Another generation of more critical and constructive perspectives such as by Okot p'Bitek (1971), Ali Mazrui (1986) and Wole Soyinka (1992) emerged. They criticized the former of hellenizing and christianizing African religions, "dressing up" African deities in Christian robes. Both coteries of African scholars objected to Eurocentrism, but assumed a culturally nationalistic posture in defining Africa religions. New thematic, comparative approaches; feminist and global perspectives have contributed to the study of African religions as a significant aspect of global scholarship (Olupona 1991, 2000, 2004; Kalu 2007). In fact, African scholars now play a central role on the international scene, sometimes providing the practitioners' perspective. Since the academic scenario has changed drastically, why then do we need to deconstruct indigenous religious knowledge? Is this a case of reinventing the wheel? I shall come back to address this concern in the following paragraphs. For the moment, suffice it to mention that European scholars did not only dominate this endeavor, viz., the academic study of religions in Africa, they impinged their methodologies and brought their worldviews and epistemologies to bear. In fact, the academic study of religion in Africa has its roots outside the continent, just as the very category of religion itself has a European history.

96.2 Toward a Deconstruction of Indigenous African Epistemologies

It is expedient that scholars of African religions, societies and cultures constantly re-interrogate the dynamic phenomenon that forms the basis of our research. We must be reflexive about the theories we utilize, the very concepts we embrace and the conclusions we come up with. Religions, societies and cultures are hardly static and unchanging; they are dynamic and are constantly in flux. Religion is not a museum piece or tourist-trod monument, but a vibrant force in the lived experiences of many people around the world; many religious traditions such as indigenous African religions are presently experiencing a renaissance. Old approaches to understanding and explaining religions are being increasingly challenged. Theories and interpretations from classical works on religions of Africa may be turning moribund and need urgent revisiting on the basis that these beliefs and rituals have been changing and transforming all the time.

96.2.1 Misconceptions

The very concept "religion," like "Africa" is a Western invention, an academic construct involving both misconceptions and changing perceptions, that hardly does justice to the complexity of African spiritualities (Mudimbe 1988; Wiredu 2006). While we continue to use the concept to embrace African spiritualities, we should be aware of its limitations and tendency to obscure its dynamism. To explain African spiritualities and religious life in western categories can be informing, illuminating and offering useful insights; just as it could be misleading and obscuring. African modes of thought, ritual patterns and symbolism that are integral to their religious worlds are sometimes puzzling to western ethnography. The perception of Esu as the Devil/Satan in contemporary Yoruba Christian parlance and in their everyday belief and ritual practice is a misnomer which arose from a mistranslation and perpetuation of a wrong interpretation in Christian missionary lexicon. Esu Elegbara is one of the most popular *orisa*, a trickster deity within Yoruba cosmological myths and philosophy. Esu is the keeper of *ase* (vital force) with which Olodumare (the Supreme Being) created the universe; a neutral force who controls both the benevolent and the malevolent supernatural powers; the guardian of Orunmila's oracular utterances. Esu is devoid of any emotions, supports only those who perform prescribed ritual sacrifices and acts in conformity with the moral laws of the universe. Without his intervention, the Yoruba people believe, no ritual sacrifice will be efficacious. As a neutral force, he straddles all realms and acts as an essential factor in any attempt to resolve the conflicts between contrasting, but coterminous forces in the world. This distinctive identity and ambivalent functions makes Esu stand out in the Yoruba pantheon of divinities and explains why it attracts devotees throughout Yorubaland and in the Yoruba religious diaspora.

In the 1850s, Ajayi Crowther, the first African Bishop, published a Yoruba grammar and commenced translation of the Bible into Yoruba language, erroneously picking Satan/Devil as the Christian equivalent of Esu. This misconception and manipulation was reinforced in the latter compilation and production of the first dictionary of the Yoruba language in the late 1930s. This Euro-Christian demonization stamped Esu with a precarious identity. This, however, formed part of a larger project of demonizing African religions and cultures. Several other examples can be drawn to demonstrate this tendency, but the misperception of Esu suffices here owing to space. The fact that African Christianities and also scholars have adopted this misconception lends credence to the urgency to revisit adopted terminologies that are often used uncritically.

96.2.2 Modernization and Globalization

The experience of colonization and sustained interaction with the West has produced and continue to perpetuate an imagined culture in transition from tradition to modernity. The chasm created between tradition and modernity is now being turned on its head as such binaries of opposition are no longer very convincing. It is more useful in some sense to talk of the modernity of African cultures, traditions and religions in an era of social-cultural flux. A proper grasp of indigenous spiritualities and religious epistemologies is central to the conceptual issues that modernization then and globalization now raises. A historical excursion of the distortion must needs to be balanced with an explanation of why redressing the distortion and suppression is an imperative.

Indigenous peoples, religions and spiritualities are not only local but global in terms of geography, their membership demography, the universality of their ideologies, their value systems, and their growing concern for humanity. As indigenous religions are the majority of the world's religions, a redress will help us better to understand the (in)human condition that is more and more characterized by global insecurity, war, wanton destruction, climate change, eco-vandalism, natural disasters and wide ranging crises. If the assertion that the world is rapidly becoming a global village is anything to go by, then the world perhaps has more reason and responsibility than ever to treat indigenous peoples and their religions with respect and reciprocity.

Indigenous peoples today stand at the crossroads of globalization. In many ways, they challenge the fundamental assumptions of globalization. The impact of globalization on indigenous peoples and how indigenous peoples are part of the globalization process and have shaped it has included their religious beliefs and ritual systems. As the world is now more interdependent, religious effects are now more noticeable. Thus, a proper grasp of the texture, shape and complexity of the religious traditions and cultures improves our understanding of indigenous peoples in conditions of globality. It also points to the significance of religion in contextualizing indigenous societies in the context of ongoing globalization processes. One important aspect of the interconnectedness between globalization and the religious cultures of sub-Saharan Africa lies in the fact that African religions both influence globalization and respond to the challenges and opportunities which globalization presents. Indigenous African religions are integral to the processes of globalization; they assess and reflect on the impact of globalization and global change on their modus operandi, their religious praxis and cosmologies; but at the same time avail themselves of the opportunity for self-repositioning within the global religious scene. The interface of religious cultures of sub-Saharan Africa with globalization needs to be located against the backdrop of the interlocking relationship and mutual enhancement of at least the triple religious heritage – indigenous religions and spiritualities, African Islam and African Christianities (Adogame 2009).

Indigenous religions have significantly influenced world art, sculpture, painting and other cultural artifacts, which populate the world's famous museums, gallery, libraries, and art exhibitions. The commodification of indigenous art and religious objects is on the increase. European language vocabulary has been further enriched through loaned words such as "shaman," "tabu," "voodoo" from indigenous religions. Although usually dislocated from its "religious" context, horticultural, culinary and medical knowledge has significant input from the indigenous African peoples and their epistemology. The global dimension of indigenous religions in Africa is manifesting itself in varied forms, in some cases transcending the continent into the African diaspora. Migration, tourism and the appropriation of new media technologies has facilitated the introduction of indigenous religions into new cultural contexts. Thus, the status of indigenous religions and new, contemporary religious forms, such as African Christianities and Islam, needs to occupy a significant place in current discourses on African religions and societies. The relevance and urgency of the issue to central questions of local and global public policy cannot be underestimated.

96.2.3 The Concept of Religion

The perception of religion as a phenomenon completely separate from culture is not a suitable reflection of the embedded nature of "religion" in African cultures (Adogame 2007: 534). In contrast to western thought patterns, there is not a rigid dichotomy between sacred and mundane domains as these spaces are fluid and interconnected. "The concept of religion is a sufficiently artificial or synthetic construct that its very creation is itself an implicit theorization of cultural realities" (Arnal 2000: 22). In most African societies, religion is variously conceptualized as a spiritual, epistemological and philosophical phenomenon. Beyond the typical focus on religion as a coterie of belief and ritual patterns, the treatment of religion as an epistemological phenomenon further helps to shed new light on studies of African cultures and societies. We need to pay attention to the significance of cosmological ideas as expressions of moral values in relation to the material conditions of life and the total social order. Religion viewed in this way allows for a deeper understanding of the complex interaction between Africans and non-Africans such as Europeans, an encounter largely based on frequently incompatible worldviews. As a category of analysis for the study of societies, religion is, therefore, quintessential to our understanding of African cultures in a global context.

Processes of transformation and change are not merely driven by socio-political agendas, but are informed distinctly by the knowledge closest to the people. The socio-cultural landscape, including indigenous knowledge and technologies of diverse communities and groups, correlates directly with the dynamics of change and the transformation of individuals' as well as groups' cultural meaning systems and senses of belonging. The spiritual, physical and animal worlds, local geographies are central in this consideration too. Therefore, any analysis or interpretation of existing or changing cultural patterns and societal institutions, terrestrial and extra-terrestrial worlds, how they are conditioned and who conditions them, cannot claim validity without full recognition of the important role indigenous knowledge systems play – a fact largely discarded by colonial knowledge hegemony.

Recent works such as *European Traditions in the Study of Religion in Africa* (Ludwig and Adogame 2004) emphasizes that current discussions about Africa must no longer be influenced by Eurocentrism and paternalism; rather, the goal is a dialogue in which all participants [African and European] function, albeit at the same level. This dialogue should promote democracy, human rights and civil society (2004: 1). The book provided a cross-section of European, and to a lesser extent African, views on religions in Africa, indicating the lacunae that need to be filled in subsequent endeavour. We raised this challenge elsewhere querying: Has the academic study of religion in Africa overcome a failure of nerve and summoned sufficient courage to chart an independent intellectual destiny? Have African scholars of religion been bold enough to shake off the "chains of mental slavery" to proceed to develop "African traditions" in the study of religion in Africa? (Adogame et al. 2012: 1). This recent book has perhaps brought the conversation to "full circle" by exploring African traditions in the study of religion in Africa.

As the philosopher, Mogobe Ramose, remarks matter-of-factly:

> It is still necessary to assert and uphold the right of Africans to define the meaning of experience and truth in their own right. In order to achieve this, one of the requirements is Africans should take the opportunity to speak for and about themselves and in that way construct an authentic and truly African discourse about Africa. (Ramose 2003: 1)

Ramose was probably over-ambitious in canvassing for the construction of an "authentic" and "truly" African discourse about Africa. While this task is obviously far to seek, many will undoubtedly query the authenticity and truth-claims of an African discourse as a mirage. Nonetheless, he was spot on in avowing that some of the spokespersons for Africa (European scholars) seem to have dwarfed African scholarly voices to a somewhat perpetual silence, even about themselves. In the same vein, David Chidester aptly probes whether or not African scholars of religion can chart their own destinies in developing African traditions in the study of religion in Africa:

> If nineteenth-century comparative religion was fashioned at the intersection of academic discourse and imperial force, has the study of religion subsequently undergone a process of intellectual decolonization? Has it become self-critical of its own interests? (Chidester 2004: 86)

It must be borne in mind that European scholars have contributed significantly to the understanding of African religions, but also paved the path in its obscurity and public misunderstanding. Any suggestion of Africans telling their "own story," (re) constructing new or Africa-centred knowledge should not sound as if the doomsday is over, and the panacea is here. Certainly, while we stress the need for African scholars to take vantage positions in telling "their own story" and in (de)constructing indigenous religious epistemologies, there is an inherent danger in this endeavour if caution is not adhered to. I do not suggest that all Africans necessarily speak in the same language, have the same voice or reason in the same way. In actual fact, such a venture flies in the face of reason and logic as Africa is characterized by "different folks" playing "different strokes."

The nineteenth century western (European) perception of cultural uniformity and the sociopolitical notion of race ignored the separate linguistic, cultural, and ethnic identities that characterize African societies. Such perception of Africa as a monolithic entity is still rife. Africa is one continent with several worlds, one characterized by complex cultural, religious, linguistic varieties, and diverse historical experiences (Martin and West 1999). It is home to innumerable ethnic and social groupings. All these ethnic groups have cultures, each different from the other, but which together represent the mosaic of cultural diversity of Africa (Adogame 2007: 533). Beneath the silver lining of cultural diversities and ethno-historical peculiarities lay a stream of affinities, commonalities in terms of values, worldviews, and ritual patterns.

96.2.4 Identity and Diversity

Africans embrace a wide sense of religio-cultural identity hence I emphasize here both the unity and diversity of the religions of sub-Saharan Africa. I present the diversity of indigenous religious traditions in terms of broad unifying themes, recognizing that African religious life is both intensely diverse at local levels, but profoundly similar when seen in regional, comparative perspective. A number of characteristics emerge through the bewildering variety of forms and expressions. Thus, even if there is not only one African religious expression, there is an African "genius for religion," a shared creativity of Africans in expressing their individual and collective experiences through religion. There is also a proclivity to find new expressions for the old and new feelings, new answers for old as well as new questions, new configurations of meaning as a highly creative process (Lawson 1985).

The challenge for Africans to take a vantage position in contesting dominant epistemologies and in knowledge (re)production is a pertinent one indeed. It will certainly allow more Africans to speak for themselves (cf. Kalu 2007). But this "speaking for oneself" could lead to romanticizing and essentializing African spiritualities and religious cultures. As we have shown above, some pioneering African scholars were trapped in homogenizing tendencies. Even Okot p'Bitek (1971) who lashed out on European and African scholars was also a child of his time. His atheistic background shaped his thoughts and ideas about African religion profoundly.

96.2.5 Rituals and Beliefs

Several Africans, including intellectuals have been educated to repudiate and deny the reality, symbolism and efficacity of indigenous rituals. Many would even deny the reality of a spiritual cosmos existing side by side with a mundane one. Some have probably lost the essence and relevance of religious rituals for reinvigorating, sustaining social cohesion, social life and well-being. But what does this mean? How viable is it to sustain resilient indigenous beliefs and praxis in the face of a falsity and the conscious demonization that pervades? We can perhaps grasp resilient beliefs and rituals better when we comprehend the religious worldviews that underlie them, their historical and mythical sources, and what sustains them. Religious practices in several respects symbolize social reality. The symbolism emerges out of the fundamental nature of religious beliefs. The prevalence of symbolism in ritual in Africa derives from the conception of relations between humans and spiritual entities. As Appiah remarked,

> If the emphasis in western theory on the distinctively symbolic character of traditional religious thought and practice is misleading, it is worth taking a moment to consider why it should have been so pervasive. And the answer lies, I think, in the character or religion in the industrial cultures in which this theorizing about religion takes place. (1992: 114)

Appiah further identifies [European] Christianity as a religion that defines itself by doctrine. The history of the church is largely the history of doctrines. In contrast, to define indigenous religions, and even African Christianities, simply from its belief system and in terms of doctrine or [written] theology will be putting it upside down. Religion or spirituality is usually not thought out in the agora of theology, but lived out in the marketplace of Africa. The ritual attitude and emphasis of indigenous practitioners and new African Christianities (African Instituted Churches and Pentecostal/Charismatic churches) partly explains why they are erroneously dismissed by some western scholars on grounds of biblical inerrancy and for having a this-worldly orientation. The rationale for such a mistaken identity, one-sided interpretation is that African churches are not seen to engage major doctrinal discourses. There are, however, exceptional cases such as the African churches recent responses to apartheid, slavery, gay priesthood, same-sex relationship discourses. In other instances, the doctrinal discourses are not of any magnitude comparable to the church councils and controversies that characterize early church history. The absence of such could actually give wrong signals as if African Christians are non-reflexive about their doctrines and praxis. This is hardly the case. African Christians interrogate their beliefs and practices in reliving their faith, as they confront their day-to-day existential problems. The doctrine that is central to Christianity, does not precisely mean beliefs, rather it means the verbal formulae that express belief. Their theology is not in the books but on their heads, in their thoughts, utterances and day-to-day actions and life modes.

Africans generally celebrate life, their religion, they dance it, sing it, and act it. Prayer and song texts are hardly written down in books and recited or sung. Prayers are mostly rendered extempore rather than from a prayer book or a formal compilation of prayer genres. Songs/choruses are often sung from their heads spontaneously rather than from hymn/song books. African religions are living, expressive religions. Through rituals, people act their religion. This distinctive pattern of renewing and articulating faith throws a crucial methodological challenge to scholars. To understand and interpret their complex ritual worlds, we would require methodologies that are cut out to unearth and conceptualize their day-to-day ritual dimension and how this is informed by beliefs, rather than look out for any stereotyped theology of a sort.

Indigenous knowledge refers to the knowledge of ritual performances constitutive of their meaning-making, belief systems, as well as the substantive dimension of their livelihood constructions. Just as the dichotomy between sacred and secular is sometimes blurring, it is problematic to compartmentalize indigenous epistemologies. It embraces beliefs, practices, technologies, values and ways of knowing and sharing, in terms of which, communities have survived for centuries. It is informed by and relates to all domains of life and the environment, including the creative and artistic aspects of music, dance and oral tradition. It also includes philosophy, ethics and worldview – concepts of life, death, cosmos, environment, spiritual world, spirituality, divination, transfer of religious knowledge, rites of passage, aetiologies of sickness and disease, and traditional healing systems.

As knowledge is a reflection of the society of which it is a part, indigenous religious knowledge continues to be valued by African peoples. Knowledge is cumulative, thus African societies often adapt their knowledge systems www.exampleessays.com/essay_search/knowledge_systems.html to suit changing circumstances. Indigenous societies share common affinities in their religious worldviews, belief in spiritual entities, the use of concepts to represent them, in ritual attitudes towards their manipulation and control. Cosmologies or oral narratives transmit their worldview values and describe the web of human activities within the spiritual cosmos. Myths represent a vital source for understanding religious cosmologies, creation of the universe, origin of man, societal norms and ethos. Indigenous religions are concerned with underlying life-forces, vital forces, or mysterious powers. The belief in transcendental reality, a supreme being, divinities, spirits, ancestors, magic, sorcery and witchcraft are central; although the names, functions, rankings in hierarchy and emphasis on each have contextual variation.

The religious world is characterized by a multiplicity of divinities, spirits and ancestors; and beliefs and practices concerning them are a dominant element. Most Africans, whether converted to Christianity or Islam or not, still share belief in their deities and ancestors in an ontology of invisible beings. These beliefs in incorporeal spiritual forces mean that most Africans cannot fully accept those scientific theories in the West that are inconsistent with it. The beliefs and rituals associated with spiritual forces constitute a distinctively indigenous pattern of religious thought and action. They are transmitted through oral traditions, myths, legends, art, paintings, symbols, sculpture, poetry, proverbs, songs and dances transgenerationally. These sources should dictate the methodologies appropriate for studying African spiritualities and religious cultures.

Rituals are geared towards ensuring and sustaining cosmic harmony at individual, collective levels. The pursuit of health, fertility, and a balance between humans and with nature constitute some of their basic concerns. The well-being of individuals, the community or social groupings is attained through a process of "explanation, prediction and control." As Robin Horton puts it "religious beliefs are theoretical systems intended for the explanation, prediction and control of space-time events" (1971: 94). Horton's point was that the fundamental character of African religious systems is that the practices arise from the belief in the powers of invisible agents. His view, then, is that religious beliefs of traditional peoples constitute explanatory theories and that traditional religious actions are reasonable attempts to pursue goals in the light of these beliefs – attempts, in other words, at prediction and control of the world (Appiah 1992: 120).

Horton, on the other hand, seems less adequate in his comparison between African traditional religion and western science. Although he notes important differences between the social contexts of theory formation and development in precolonial Africa and post-Renaissance Europe, his claim for difference summarized as a cognitive world of traditional cultures "closed" and that of modern cultures "open" is suspect. As he writes, "It is that in traditional cultures there is no developed" awareness of alternatives to the established body of theoretical tenets; whereas in scientifically oriented cultures, such an awareness is highly developed (Wilson 1970: 153). Evidence abounds within African traditional modes of thought, of styles of reasoning (Appiah 1992: 126). Horton's stress on the 'closed' nature of traditional modes of thought looks less adequate in the face of Africa's complex history of cultural changes and of Hallen's *babalawo*, or in the presence of the extraordinary metaphysical synthesis of the Dogon elder, Ogotemmeli drawn from Griaule [1965] cosmology (p. 127). The Yoruba *babalawo* – the diviner and healer – whom Hallen cites is critically appreciative of the tradition he believes in.

For the average westerner, to call something "religious" is to connote a great deal that is lacking in traditional religion, such as a written theology, and not to connote much that is present, such as religious praxis. Their ritual structures draw largely upon a philosophy of relationships. An individual's passage through life is monitored, marked and celebrated from pre-birth, parturition, childhood, transition to adulthood, adulthood, marriage, old age, death and the living-dead. The rituals associated with these life-stages are significant within indigenous cultural matrix. Ritual action is very central to the lives of African peoples in enhancing the relationships to the powers of life. Religious practices usually took the form of ritual in which objects are empowered to facilitate the process.

In describing religious worlds of African religions and how people live and act within these frameworks, it is important to show special places (sacred places) that provide a ritual environment for the performing of religious acts. Such worlds consist of special roles that define the purpose of the actors in the religious drama. They consist of special powers, presences or beings with which the actors form prescribed relationships within the dramatic settings. Once we are familiar with religious places, roles, and powers, then we are prepared to describe and analyze religious activities – how they live and act in a religious world. Then, we can also describe the many religious symbols that are present in, and inform, the actions that characterize the stages on life's way – symbolic acts and objects. We have to also interrogate the transformation processes in which old places, roles, powers are taking on new meanings; while new places, roles, and powers are acknowledged. Scholars attention should focus on how religious places, roles, powers and actions are expressions of a coherent system of thought that inform the conduct of the lives of the people who participate in them (Lawson 1985).

96.3 Conclusion

In conclusion, indigenous religious epistemology is a complementary and informative knowledge repository for a coterie of disciplines particularly in the humanities and social sciences. African scholars of religion and society, with renewed vigour and determination, must begin to revisit their roles in critiquing, (re)producing and mainstreaming old/new knowledges on/about religions and spiritualities in Africa. By deconstructing indigenous epistemologies, we perhaps begin to chart a path to redressing the marginalization of indigenous religious knowledges and to launching it from obscurity to mainstream academic inquiry.

References

Adogame, A. (2007). Religion in sub-Saharan Africa. In P. Beyer & L. Beaman (Eds.), *Religion, globalization and culture* (pp. 533–554). Leiden: Brill.

Adogame, A. (2009). Practitioners of indigenous religions in Africa and the African diaspora. In G. Harvey (Ed.), *Religions in focus: New approaches to tradition and contemporary practices* (pp. 75–100). London/Oakville: Equinox.

Adogame, A., Chitando, E., & Bateye, B. (Eds.). (2012). Introduction. African traditions in the study of religion in Africa: Emerging trends, indigenous spirituality and the interface with other world religions. In their *Essays in honour of Jacob K. Olupona* (pp. 1–13). Surrey/Burlington: Ashgate.

Appiah, A. K. (1992). *In my father's house: Africa in the philosophy of culture*. New York/Oxford: Oxford University Press.

Arnal, W. E. (2000). Definition. In W. Braun & R. McCutheon (Eds.), *Guide to the study of religion* (pp. 21–34). New York: Cassell.

Chidester, D. (2004). 'Classify and conquer': Friedrich Max Müller, indigenous religious traditions and imperial comparative religion. In J. K. Olupona (Ed.), *Beyond primitivism: Indigenous religious traditions and modernity* (pp. 71–88). New York: Routledge.

Conrad, J. (1995). *Heart of darkness and other stories*. Hertfordshire: Wordsworth Editions Limited. Reprint.

Danquah, J. (1968). *The Akan doctrine of God: A fragment of Gold Coast ethics and religion*. London: Frank Cass.

Griaule, M. (1965). *Conversations with Ogotemmeli: An introduction to Dogon religious ideas*. Oxford: Oxford University Press.

Horton, R. (1971). *Patterns of thought in Africa and the West: Essays on magic, religion and science*. Cambridge: Cambridge University Press.

Horton, R. (1984). Judaeo-Christian spectacles: Boon or bane to the study of African religions. *Cahiers d'Études Africaines, 96*(24/4), 391–436.

Idowu, B. (1973). *African traditional religion: A definition*. London: SCM Press.

Kalu, O. (Ed.). (2007). *African Christianity: An African story*. Trenton: Africa World Press.

Lawson, T. E. (1985). *Religions of Africa*. San Francisco: Harper and Row.

Ludwig, F., & Adogame, A. (Eds.). (2004). *European traditions in the study of religion in Africa*. Wiesbaden: Harrassowitz Verlag.

Martin, W. G., & West, M. O. (Eds.). (1999). *Out of one, many Africas: Reconstructing the study and meaning of Africa*. Urbana/Chicago: University of Illinois Press.

Mazrui, A. (1986). *The African triple heritage*. London: BBC Publications.

Mbiti, J. S. (1969). *African religions and philosophy*. London: Heinemann.

Mudimbe, V. (1988). *The invention of Africa: Gnosis, philosophy and the order of knowledge*. Bloomington: Indiana University Press.
Olupona, J. K. (Ed.). (1991). *African traditional religions in contemporary society*. New York: Paragon House.
Olupona, J. K. (Ed.). (2000). *African spirituality, forms, meanings and expressions*. New York: The Crossroad Publishing Company.
Olupona, J. K. (Ed.). (2004). *Beyond primitivism: Indigenous religious traditions and modernity*. New York: Routledge.
Parrinder, G. (1954). *African traditional religion*. London: Sheldon Press.
p'Bitek, O. (1971). *African traditional religion in western scholarship*. Nairobi: East African Literature Bureau.
Platvoet, J., Cox, J., & Olupona, J. (Eds.). (1996). *The study of religions in Africa: Past, present and prospects*. Cambridge: Roots & Branches.
Ramose, M. B. (2003). The struggle for reason in Africa. In P. H. Coetzee & A. P. J. Roux (Eds.), *The African philosophy reader* (2nd ed., pp. 1–8). London: Routledge.
Ranger, T., & Kimambo, I. (1972). *The historical study of African religion, with special reference to east and central Africa*. London: Heinemann.
Soyinka, W. (1992). *Orisha liberates the mind. Wole Soyinka in conversation with Ulli Beier on Yoruba religion*. Bayreuth: Iwalewa.
Wilson, B. (Ed.). (1970). *Rationality*. Oxford: Basil Blackwell.
Wiredu, K. (2006). Toward decolonizing African philosophy and religion. In E. P. Antonio (Ed.), *Inculturation and postcolonial discourse in African theology* (pp. 291–331). New York: Peter Lang.

Chapter 97
Christianity in Africa: Pentecostalism and Sociocultural Change in the Context of Neoliberal Globalization

Samuel Zalanga

97.1 Introduction

In the study of the role of religion in society, social scientists often encounter two extreme standpoints. On one side of the spectrum are those who are categorically dismissive of religion, viewing it as an illusion and a cultural instrument in the hands of a ruling class that uses it to control and oppress the poor and socially marginalized social groups (Feuerbach 2008). The other side of the spectrum is occupied by scholars who view their role as apologists for the positive and constructive role that religion plays in society by providing people with a deep sense of meaning, belonging and stability in life (Durkheim 2001). Between the two extreme positions is the approach often identified with Max Weber, one that suspends judgment on the exact role of religion a priori, because the influence and consequences of religion on society remains an empirical question depending on the social structure, social context, and the way social classes and status groups operate and interact with each other (Weber 1946: 267–444; Weber 2001). This chapter takes the Weberian approach in its examination of the role of Pentecostal Christianity in Africa today.

Given that Africa is a huge continent that is internally diverse, with additional layers of diversity within each country, this chapter will provide some relatively broad generalizations of trends, knowing fully that the general observations are manifested in nuanced ways across the continent. The main goal of the chapter is to examine the emergence of Pentecostalism and Charismatic Christianity in Africa, the manners and ways the believers live out their faith, and the consequences of that type of faith on their lives, culture, social institutions, the economy, state, and society.

S. Zalanga (✉)
Department of Anthropology, Sociology and Reconciliation Studies,
Bethel University, St. Paul, MN 55112, USA
e-mail: szalanga@bethel.edu

S.D. Brunn (ed.), *The Changing World Religion Map*,
DOI 10.1007/978-94-017-9376-6_97

Pentecostal or charismatic Christianity is not monolithic in Africa; it is an amorphous group that is composed of a variety of practices. Moreover, Christianity has a long history in Africa, one that is centuries old, especially the North African region (for example, Morocco, Algeria, Tunisia, Libya and Egypt), which at one point used to be part of the Roman Empire (Johnson 1976). Some of the church fathers were from Roman North Africa (for example, Saint Augustine, Origen, Cyprian, Clement of Alexandria). The Athanasian Creed was named after the Bishop of Alexandria whose name was Athanasius. Saint Augustine of Hippo was also from Hippo, which is in present day Algeria, North Africa (Johnson 1976). The Coptic Church in Egypt and the Ethiopian Orthodox church all have long historical links to early Christianity that predate the Protestant Reformation and European colonialism and missionary work in Africa (Robinson 1982). Christianity spread to many parts of Sub-Saharan Africa between the late nineteenth century and early to middle twentieth century, which was the era of missionary adventures in Africa.

The era of missionary adventures was followed by colonial rule. In many African societies the missionaries worked hand in hand with the colonial government providing necessary support for each other's missions (Shorter 1970). Given the racist worldview of many Europeans in the Victorian era, Christians and missionaries, as products of their time, reflected the dominant cultural perspectives of Europeans towards Africans and other non-Western peoples (Davidson 1984). Many missionaries supported racist colonial policies because doing so allowed them to pursue their primary goal of saving souls. This, consequently, led them to cover-up, or ignore, many colonial atrocities, such as those in the Belgian Congo (Bate 2004).

After most African countries became independent in the 1960s, more missionaries went to Africa and many new and different church denominations emerged. Some of the denominations were splinter groups from missionary churches, while others were African independent churches that emerged as a protest against Western missionary efforts (Daneel 1987). There are currently four types of churches into which the African Christian landscape can be divided. The first group is the legacy of Western missionary established churches, which is called mainline denominations (for example, Catholic, Anglican and Methodist churches). The second group is composed of "Ethiopian-type" churches, which are churches that broke away from Western churches and worked hand in hand with colonial rulers. As a general rule, the "Ethiopian-type" churches are theologically sympathetic to the struggles of the oppressed and their desire for social liberation. "They are non-prophetic movements and do not place a great deal of emphasis on the Holy Spirit and all the extra-ordinary activities assigned to the Holy Spirit in other new denominations" (Moyo 2007: 315). Examples of churches in this category are Ethiopian African Methodist Episcopal Church, the Order of Ethiopia, and the Presbyterian Church of Africa. The third category of churches is the "spirit-filled" churches, also called Zionist churches, because they include Zion in their names, given that they descended from Christian Catholic Apostolic Church that originated in Zion, Chicago, Illinois (United States) in 1896. By 1904, the church had sent missionaries to South Africa. Examples of churches in this category are Aladura in Nigeria, which are known for prayers, and the Harris church, in Cote d' Ivoire (Morrson et al. 1989: 76;

Ranger 1986: 3). These churches "are prophetic in character and place a great deal of emphasis on the work of the Holy Spirit, who manifests himself in speaking in tongues, healing, prophecy, dreams, and visions and who helps to identify witches and cast out evil spirits" (Moyo 2007: 315).

While the leadership of the churches attempted to indigenize Christianity in African culture, they essentially approached religion from a predominantly functionalist or practical paradigm. This means that Christianity offers its believers not only salvation, but also protection against magic, evil spirit, curses, and witchcraft. While this concern is true of every orthodox Christian group, some denominations underscore this benefit more than others. The churches in this category discourage participation in traditional African cultural activities or celebrations. In effect, salvation for this group of believers is not just spiritual; it must also have some practical benefits. Believers in this category recognize the role and function of the priest in traditional African religion, which is co-opted into the new belief system, rendering the traditional priest redundant (Kiernan 1995: 23–25).

The fourth category of churches, called Pentecostal/Charismatic, in combination with the third category, will be the main concern of this chapter. These churches operate within a broad framework generally known as evangelical protestant denominations. The Pentecostal and Charismatic churches either imitate, or are inspired by, similar evangelical churches, which are based in the United States (especially the Bible Belt), and European countries. The Assemblies of God in the United States is a very good example of the denomination in reference here.

Generally found in Nigeria, Ghana, Kenya, Zimbabwe, Zambia and South Africa, these churches are primarily urban-based, though they have rural branches. The influential leaders of the church are in urban centers and their members are predominantly young people with high aspirations, who appear overly confident, are highly educated by African standards, but encounter many frustrations in an African society characterized by either stagnant economies or economies that are growing, but with highly unequal distribution of wealth and opportunities. The church teachings also play an important role in the lives of the members by providing a theodicy of success and failure.[1] By theologizing about evil spirits, ancestral curses, and witchcraft, the congregants are provided explanation for why some succeed and why others fail, which is an essential coping mechanism in a society where corrupt and sinful people prosper, while the God-fearing people suffer (Brouwer et al. 1996: 151–178).

There are several reasons why many Pentecostal churches have burgeoned in the past three decades in Sub-Saharan Africa. First, for many Christians, orthodox churches with their colonial legacy did not reflect the cultural or social aspirations and values of the people. In contrast, the level of despair and hopelessness in many parts of Africa compel people to search for spiritual power more than would

[1] Theodicy is "a concept used by Max Weber to explain systems of belief that help to explain human suffering, inequalities, sickness, and other negative aspects of human life and society." See M. O. Emerson, W. A. Mirola & S. C. Monahan (Eds.). (2011). Religion matters: What sociology teaches us about religion in our world. Boston: Allyn & Bacon, p. 103.

otherwise be the case. Gifford highlights the problems by quoting the frustration of Ghana's Pentecostal preacher, Mensa Otabil who began a conference session by lamenting:

> 'Look around the continent, the situation seems almost hopeless. It seems one big continent of war, strife, hunger, malnutrition, pain, famine, killing, ignorance. If you look at CNN it seems all the bad news comes from Africa.... When you look at yourself as an African, it is easy to think that God has cursed you.' Consequently, he claims, Africa's biggest problem now is its inferiority complex: 'We are a people who feel inferior and wallow in our own inferiority.' This is frequently the context in which Africa's Pentecostalism is developing. (Gifford 2001: 77–78)

Consequently, these African believers created spirit-filled churches that would allow them to sing and dance during worship. Second, it has been argued that mainline missionary churches did not make an effort to understand traditional African spirituality and its functions in the life of the African within his or her cultural context. Most elements and aspects of traditional African spirituality were similarly condemned as heathen, and often attempts were made to forcefully condemn or forbid them. Yet, the new faith, coming to the continent, failed to fully replace traditional African spirituality, and the roles that it performs. Third, many Africans felt that missionary churches preached about the equality of human kind on biblical grounds, but because they modeled discrimination against Africans on the basis of race, their mode of living contradicted what they preached, as they discriminated against Africans (Davidson 1989). Consequently, Africans felt the need to start their own churches where they could escape discrimination. In South Africa, for instance, African independent churches were an attempt to escape apartheid and its systematic oppression of Blacks (Davidson 1989).

An argument that has also been posed is that when the Bible was translated into African languages, many people who were literate read the Bible and interpreted it on their own. In the process, they saw many aspects of Old Testament culture to be more closely aligned with African culture rather than modern European culture (for example, respect for parents and by implication ancestors, polygamy etc.). Similarly, the nature of leadership in many orthodox churches was such that black people were denied leadership positions of power and responsibility. In addition, women were excluded from fulfilling important roles in the church and in the denominational structures. Thus, new denominations were created by Africans in which they were free to participate and exercise their leadership gifts and creativity, and where women were allowed a more active role within the church. Examples of Christian churches and religious movements in this rubric are: Alice Lichina in Kenya, the Nyabingi in Uganda and Kenya, and the movement for healing and possession in Mozambique (Mikell 1997: 26). Given how Pentecostal and Charismatic Christianity is very vibrant in Sub-Saharan Africa, it is worthwhile to briefly examine the sociological literature on what makes people convert to a religion, and once converted, what factors are most likely to help them sustain their commitment. Furthermore, what is the theological content and moorings of African Pentecostal and Charismatic Christianity? These are the issues examined in the next section.

97.2 Conversion to a Religion and African Pentecostal Theology

In this section I would like to provide an overview of the literature that addresses why people convert to a religion, before summarizing the key theological teachings and emphases of African Pentecostal churches. In terms of how people become religious, Susan Kwilecki, Professor of Religious Studies at Radford University, Virginia, summarized arguments, which she labeled as the "interactional axiom," through which individuals cultivate faith commitment. Her analysis, and resulting theory, addressing why and how people become religious is also useful for understanding what social-psychological forces drive individuals to commit to Pentecostal and Charismatic Christianity in Africa. According to Kwilecki:

> An individual first learns he or she is surrounded by supernatural forces as a member of a group (family, religious community, subculture, culture). Collective traditions supply the raw materials from which personal religions are constructed. (Kwilecki 1999: 38)

In explaining what helps believers to sustain or strengthen their commitment to a faith, Kwilecki argued that whether that happens or not depends on "the nature and number of psychic functions religion performs" (Kwilecki 1999: 50). This means that the more a particular religion is able to perform many functions in a person's life, the more that religion will be centrally important in the person's life.

Another factor that helps believers sustain or strengthen their commitment to a faith is the usage of time-honored spiritual techniques (Kwilecki 1999: 51). No religious commitment can be sustained without engaging in religious rituals. Participation in rituals is meant to increase the strength of a person's religious beliefs, as well as the meaning system of a religion to the believer. Kwilecki's analysis emphasizes that religion is socially embedded in society and that people become religious and sustain their faith through social processes. This insight should constantly be taken into cognizance as one engages in an analysis of Pentecostal Christianity in Africa.

There are three major and highly influential motifs, which have been identified by Gifford, that have had a major impact on the theology of African Pentecostal and Charismatic Christianity. The first dominant theology is the *prosperity gospel*. This motif is primarily applied to fund-raising in the United States. It teaches that the God of Abraham, Isaac and Jacob is a rich God, and if Christians want to be rich, they must sow seeds of faith, in the form of financial contributions. This teaching is often supported by using select scriptures from the Bible. Furthermore, this strategy of raising money has made it possible for many preachers in the United States and Africa to fund very expensive church ministries, for example, Oral Roberts in Oklahoma, United States (Harrell 1985: 424). In many respects, the faith gospel claims that if you can say something and believe it as true, you can claim it. In the United States this kind of theology became popular during the economically expansive decades of the 1950s and 1960s, when people were feeling confident in their prosperity. But in Africa, this theology works even during times of suffering, because people see their giving as a mystical strategy for rising out of poverty, and other types of suffering. Interestingly, many Americans critique non-Western prosperity gospel teaching without realizing that its roots are planted in, and inspired by, the teachings of Pentecostal ministers in the United States. Gifford describes the prosperity-gospel thinking in the following manner:

> According to faith gospel, God has met all the needs of human beings in the suffering and death of Christ, and every Christian should now share the victory of Christ over sin, sickness and poverty. A believer has a right to the blessings of health and wealth won by Christ, and he or she can obtain these blessings merely by a positive confession of faith. As regularly articulated, several well-known names have helped create it: most notably, E.W. Kenyon, A.A. Allen, Oral Roberts, T.L. Osborn, Kenneth Hagin, Kenneth Copeland, John Avanzini. Each of these has made his own contribution. (Gifford 2001: 62)

Gifford agreed with Coleman that the prosperity gospel, rooted in American neo-Pentecostal denominations, fits well with elements of traditional African religion as observed by Peel (1993) in the faith communities of the Yoruba people of Nigeria. Traditional Yoruba religion is functional and practical in the sense that the purpose is to bring about prosperity broadly conceptualized, including financial blessing, children, and peace (Olupona 1993: 240–273). What this means is that there is a local cultural dimension to the prosperity gospel in traditional African societies, but there is equally an influence of the American faith gospel. Indeed, an aspect of the teaching of the prosperity gospel literature, which has been disseminated all over Africa, particularly in the 1980s and 1990s, was that in the "end times," God would takeaway wealth and riches from unrighteous people and give it to the righteous (that is, Pentecostal Christians). Some African leaders who subscribe to this kind of Pentecostal theology speak publicly in its favor.

The Pentecostal prosperity gospel has been disseminated throughout Africa via church spiritual conventions and revival meetings. Chiluba who was the president of Zambia in the 1990s, and a Pentecostal believer, declared his country a Christian nation in 1991. In support of the prosperity gospel, he asserted that when Zambians give to God, they will not only receive personal blessings, but even the nation of Zambia will be blessed and, as a result, the country would not need future access to foreign loan.[2] The prosperity gospel teachings clearly have the influence and imprint of the theology of Kenneth Hagin and Kenneth Copeland (Brouwer et al. 1996).

The message and ideas of the prosperity gospel paradigm was well received by African Pentecostal pastors, because it helped generate finances they needed to build the expensive sanctuaries, with flamboyant seating and musical equipments. There was elective affinity between the needs and desires of African Pentecostal pastors and the content of the prosperity message espoused by American Pentecostal ministers. Furthermore, the prosperity message resonated well with members of the congregation, because the message promised a strategy for getting rich amidst poverty.

The second theological area that is emphasized in African Pentecostal churches is *deliverance ministry*. This was very popular in the 1990s and remains so to this day. Gifford described the theological essence of deliverance teaching in this way:

> The Christian may have no idea of the cause of the hindrance, and it may be through no fault of his own that he is under the sway of a particular demon. It often takes a special man of

[2]These publications are in Zambia.

> God to diagnose and then bind and cast out this demon. Thus, in the mind of many of its exponents, this deliverance is a third stage, beyond being born again, beyond speaking in tongues. (Gifford 2001: 65)

Ministers who engage in this ministry believe that although Lazarus was saved in heaven, the real life he deserved on earth was the prosperous life of Abraham. This means that deliverance can change an otherwise pathetic life situation into a prosperous one here on earth, before going to heaven. In effect, Lazarus suffered in this world because he did not receive deliverance. There are, however, different dimensions and interpretations of deliverance. For instance, in an influential book published by Aaron Vuha of Evangelical Presbyterian Church of Ghana, entitled: *The Package: Salvation, Healing and Deliverance* (1993: 36), the author provided a detailed account of how demons operate in the world. He noted that demons were originally angels before they became spirits. They operate when people are sleeping and they dwell in rivers, mountains, rocks, trees, and in humans. He also provided examples of their consequences: "phobias, complexes, allergies, chronic diseases, repeated hospitalization, repeated miscarriages, non-achievement in life, emotional excesses, and strikingly odd behavior" (Gifford 2001: 66). Vuha's book identified the inability to engage in a sustained marriage as a manifestation of a demonic spirit in one's life. He explained how demons enter human beings, how to identify them, and how to cast them out. In addition to the dramatic claims of possession and exorcism, according to "deliverance ministry," one is constantly under the attack by demons, so there is a need to be constantly vigilant. The book also contains testimonies of people who have been delivered from demons.

Although demons and deliverance are an integral part of African traditional religion, when Pastor Matthew Addae-Mensah introduced this ministry in Ghana after an encounter with a Nigerian minister involved in spiritual deliverance from demons ministry, many Ghanaian pastors and Christian believers were skeptical about the authenticity of such a ministry. It was not until an American Pentecostal minister, Derek Prince, came to Ghana and affirmed the teaching, that deliverance ministry gained support. It was then that Addae-Mensah's ministry was granted legitimacy. Yet, it was not Derek Prince alone who was promoting deliverance ministry in the United States. Other U.S. ministers involved in such ministry are Marilyn Hickey and Roberts Liardon (Gifford 2001: 69). An insight from Addae-Mensah's story is that African Christians have a propensity to refuse belief in something preached to them by a fellow African, but will accept the legitimacy of the same teaching when expressed and articulated by an American. As Max Assimeng, Professor of Sociology at Legon affirmed: "*Things are truer if un-African, so we quote Americans. It is traditional, but projected in modern dress. The more foreign, the more serious, true, powerful it is*" (cited in Gifford 2001: 69).

There seems, however, to be some disagreement among African ministers engaged in spiritual deliverance ministry. For instance, Pastor Eastwood Anaba of Broken Yoke Foundation in Bolgatanga, Northern Ghana, critiqued aspects of the theology that are attributed to demons by some ministers (for example, diseases). In particular, he was critical of a book written by a deliverance minister in Nigeria, namely, Emmanuel Eni whose book "*Deliverance from the Power of Darkness*,"

(1987) made many esoteric claims about the way demons operate in the spiritual world (Meyer 1995; Ellis and ter Haar 1998). Pastor Eastood Anaba's main criticism was that pastors involved in deliverance ministry, like Emmanuel Eni, do not distinguish when, and which, negative events are attributed to demons, but rather generalize that all negative events are the influence of demonic forces (Gifford 2001: 70; Anaba 1985, 1993, 1994).

The third major theological teaching that is emphasized in African Pentecostalism is *Christian Zionism*. Gifford described the essence of this theology, which is rooted in the American experience, and therefore bears an American imprint, in the following manner:

> Part and parcel of pre-millennialist dispensationalism is the idea that God has never abandoned Israel: God works through two agents on earth, the church and Israel. Thus so many biblical references to Israel refer to precisely that – the modern state of Israel established in 1948. Since God will accomplish his end-time purposes through Israel, and Israel is a prerequisite of Christ's return, Israel must be defended by every means possible. This leads to unquestioning support, on supposed biblical grounds, for everything the modern Israeli government wants or attempts. (Gifford 2001: 74)

It is indeed true that Christian Zionism has been a major issue in the experience of evangelical Christianity in the United States, and given the influence and size of American missionaries in Africa, it is not surprising that ideas of Christian Zionism would be replicated in Africa (Walls 1991). According to an organization in the United States, called *Christians United for Israel* (CUFI), led by Pastor John Hagee of Saint Antonio, Texas, there is a biblical mandate in support of the tenets of Christian Zionism. According to CUFI:

> The biblical mandates for supporting Israel began with Genesis 12:3. I will bless those who bless you and I will curse those who curse you. Secondly, David said in Psalms 122:6, 'Pray for the peace of Jerusalem. They shall prosper that love you.' Because of the fact that in history, if Jerusalem is at war, the world is at war. If there's peace in Jerusalem there's peace in the world. When Israel became a state, in 1948, I remember well sitting at the table in our home and we heard that announcement come over the radio. And my father said, 'This is the most important biblical day in the 20th Century. For all the prophets of the Old Testament have now been vindicated and Israel has been born.' (Moyers 2007)

Gifford stated that there are certain factors that can help in better understanding the emergence of Christian Zionism in America. First, the United States has used Israel as a nation to indirectly represent America's national interests in the Middle East. This notwithstanding, some scholars argue that unconditional support by U.S. for Israel sometimes works against U.S. national interests (Buchanan 2008). Second, many Christians who insist on unconditional support for Israel, want a Christianity that projects strength in global affairs. Christian Zionists were impressed by Israel's phenomenal success in the Six Days War when Israel defeated her Arab enemies, at a time when the United States was withdrawing from Vietnam, with no impressive performance (Gifford 2001: 74–75). Furthermore, many of the leading members of America's evangelical right, which included Pat Robertson (Pentecostal Minister), Jerry Falwell, and W. A. Criswell of the Southern Baptist convention, expressed staunch support for Israel and their sentiments permeated many Pentecostal churches in the United States, which in turn influenced Pentecostals in Africa

(Brouwer et al. 1996: 145–149). Although some African Pentecostal ministers, such as Ezekiel Guti of Zimbabwe Assemblies of God Africa (ZAOGA), received some theological training in the United States (Maxwell 2006: 88–89), the main mechanism through which American Pentecostals have influenced the training of African Pentecostal ministers has been through sponsorship. The majority of sponsorship has been carried out through funding theological training colleges and institutes in Africa, which is less expensive than training individuals in the United States (Brouwer et al. 1996: 179–208). In Zimbabwe, for instance, the three biggest Pentecostal denominations: Family of God (FOG), Zimbabwe Assemblies of God in Africa (ZAOGA), and Apostolic Faith Ministries (AFM) all have a seminary funded by foreign sponsors. FOG has *Evangel Bible College* (Assemblies of God), ZAOGA has *Africa Multinational for Christ College* (AMCC), and AFM has *Living Waters Seminary*. These three institutions are responsible for training many of the ministers leading the Pentecostal movement in Zimbabwe.

There is also another way in which Christian Zionism plays out in African politics and Christianity. The movement coincided with the period of Colonel Muammar Gaddafi, of Libya, and Ayatollah Khomeini's rise in political influence in Africa, and at a time when the United States was trying to restrict Arab Muslim expansion into Africa, especially radical Islam, after the Iranian revolution of 1979. Consequently, in a country like Nigeria where Pentecostalism has a rich and fertile social soil and there is a major religious divide between the predominantly Christian South and predominantly Muslim North, Pentecostal churches in the country became a medium for expressing the antagonistic attitude that many U.S. Pentecostals and evangelicals have held toward Muslims and Arabs. Liberia, under Samuel Doe, is a good example of how Christian Zionism worked in Africa to promote antagonistic attitude towards Muslims and Arabs (Gifford 2001: 75–77; Brouwer et al. 1996: 145–150). Even in a country like Zambia that is over 70 % Christian, the Pentecostals spoke out publicly against Muslims and Arabs indicating that there had been external pressure or influence on Zambian Christians, particularly since Islam has not been a threat in the nation of Zambia (Gifford 2001: 74–79).

97.3 Theological and Social Principles Underpinning Pentecostal and Charismatic Christianity in Africa: An African Perspective

According to Dr. Rev. Tokunboh Adeyemo,[3] there are four theological and social principles that underpin the ministry and practices of Pentecostal churches in Africa (Miller and Yamamori 2007). The first principle is that they are "*scratching where*

[3] Dr. Reverend Adeyemo worked with the Association of Evangelicals in Africa and died in 2010. Prior to his death, he was the retired General Secretary of the Association of Evangelicals in Africa based in Nairobi, Kenya. He was also the former Executive Director of the Center for Biblical Transformation in Kenya. Pentecostals in Africa as in Latin America are considered part of the evangelical religious movement.

the people are itching." They are also "*contemporary and contextual*." By this, he meant that Pentecostal and Charismatic churches, along with their leaders, are in the trenches with their parishioners. The church leaders identify with the parishioners and focus on their existential struggles and challenges. Thus, Pentecostal and Charismatic Christianity is directly relevant to the people it serves.

Second, African Pentecostals and Charismatic believers, according to the Late Dr. Adeyemo, take the Bible at "*face value*." They read the Bible and believe that all the scientific events and miracles that have been recorded in the Bible are equally applicable to our world today. They believe that God's power is the same yesterday, today, and tomorrow. The signs and wonders in the Bible are not just meant for the past, but are equally applicable for today's world as well. He noted that African Pentecostals and Charismatic believers engage in heresy by interpreting the Bible literally. But this kind of heresy is different from the type in the West, which grew out of the scholarly tradition of higher criticism of scripture that resulted in many people denying the deity of Christ and the miracles in the Bible.

Third, African Pentecostal Christianity is uniquely committed in its belief and faith in prayer. Adherents believe that prayer can bring about miracles, healing, and answers to problems in life. They engage in all-night prayer by the thousands, from 9 p.m. on Fridays to 6 a.m. on Saturdays. They pray about every aspect of their lives without exception and expect healing or intervention by God. Prayer for African Pentecostals is a source of strength, power, and renewal. Fourth, the Pentecostal experience gives these Christians a sense of identity and self-esteem, which is very empowering. They are ordinary people who are socially marginalized, but in the church, they gain a new identity, after they are spirit-filled and empowered, to believe that they are now special people, princes and princesses, in the kingdom of God. The new identity that is created is embedded in new socially constructed physical and moral spaces.

97.4 Pentecostalism and the Social Construction of the Geography of Sin and Moral Space

When one situates Pentecostalism within the context of African cultures, and its status vis-à-vis the idea of progress, the movement treats the village and the rural as not merely a neutral geographical space, but rather the epitome of backwardness and fertile ground for evil forces to flourish – that represent backwardness and demonic oppression. Given the nature of urbanization in Africa, it is to clarify whether urbanization in Africa creates a total break in the mindsets of urbanites compared to rural residents (Kleniewski 2006: 159–166). Consequently, the large number of African people who live in rural areas and who are engaged in the village social network of relationships and mutual cultural obligations are perceived as either more sinful, more susceptible to sin, and in particular, easy prey for becoming hosts to evil and demonic spirits. This idea stands in contrast to life in the city, which represents progress and enlightenment, especially certain physical spaces of the city associated with modernity. Life in the village, with its corresponding social relations, is defined as an arena of evil forces that can undermine one's journey into God's prosperity. Pierre-Joseph Laurent describes the backward characterization of village life in Burkina Faso in the following manner:

> Change, development, or even individual control over modes of accumulation necessitates keeping the process of kinship redistribution at bay, here rendered imaginable by conversion to the Assemblies of God. Conversion leads to effective protection against the constraining forces of village communalism, protection acquired by belief in the 'power' of God. Jesus is a 'large rock' behind which adepts with individualist ambitions may hide themselves.

In effect, the individual is being encouraged to move well beyond the constricted network of social relationships in the village or tribal community, and become the perfect material for *homo-economicus*, which will lead to a reconstitution and redefinition of personal identity, based on the new social network in the church, which is more cosmopolitan and modern. The challenge that Pentecostal churches face is whether the kind of teaching they engage in can, deliver on the promises made to the members of their congregation without structural changes in society. It is interesting to note that by enabling church members to create new identities beyond the family and tribal community, Pentecostalism is succeeding where the post-colonial state has either failed or performed poorly (Maxwell 2006). Consequently, in doing that, the church has had the tendency to denigrate traditional African culture, which is the issue examined in the next section.

97.5 Pentecostalism and the Demonization of Traditional African Culture

From an objective anthropological and sociological point of view, Pentecostals in Africa appear to be replicating past colonial and missionaries' efforts and results. In the nineteenth century, missionaries and colonial rulers characterized African culture as frozen and in the hands of heathens, the kind of biblical language used to describe people living in the Promised Land, before the arrival of Israelites in the Old Testament. Here is a situation where Africans, because of the theological lens they use to view their culture, perceive it in an inferior manner without a nuanced critical analysis informed by anthropological insights. In terms of the time orientation favored by Pentecostals, they are forward looking and care less about history, except the history of Christianity, upon which they would do well to carefully reflect on. This deep philosophical and anthropological hatred of traditional African culture raises an important concern for some Africans. In the past, Europeans and missionaries did the same thing in the nineteenth and early twentieth century to Africans, and they were attacked for being Eurocentric, prejudiced, and racist. Today fellow Africans are doing exactly the same to other Africans, but under the banner of Pentecostal spiritual renewal. One does not have to take a wholesale and objectively hateful posture towards African culture before he or she can make a theologically/socially constructive case for desirable social change on the African continent.

An important issue that this position of African Pentecostals raises, and about which many Pentecostals have no time to deeply reflect, is this question: does one need to be a modern European before he or she can become a true Christian, or is there a way that one can become a Christian while still being an African? (Daneel 1987). One good example of a church that believes one can be a Christian while still being an African is *the Church of Jesus Christ according to the Prophet Simon Kimbangu* in the Democratic Republic of Congo (Mazrui 1986: 152–156). Interestingly, many of the mainline denominations that grew out of colonial missionary churches have long abandoned their wholesale prejudiced disregard of traditional African culture. They are now making greater effort to make Christianity meaningfully rooted in African culture. Mainline denominations appear to be working hard to synthesize Christianity with traditional African culture insofar as the elements of the cultural heritage do not fundamentally negate the main tenets of Christianity. In making an effort to integrate the two, these denominations are rectifying their previous mistakes, having learned from their errors (Daneel 1987: 46).

Pentecostalism can lead to polarization, and an explicit rejection of traditional and historical cultural reality within an African community. For example, when the government of Ghana attempted to revive traditional pre-colonial African heritage through annual celebrations, the leaders of Pentecostal churches filed a law suit with Ghana Human Rights Commission, against the government, asserting that they were being forced to participate in evil practices that dilute their faith.[4] In effect, the Pentecostal Christians believed the efforts of the Ghanaian government represented a violation of their human rights, especially their freedom to worship. What this means is that Africans, in the name of Pentecostal Christianity, wanted to be granted extra human rights in order to condemn and disrespect their African heritage, and the rights of others. For Pentecostals, a precondition for Africa's development is that the African past has to be domesticated and controlled through the process of spiritual deliverance. But, if being human is impossible to fully realize without a culture, does that not mean that Pentecostals have a very shallow view and understanding of culture?

The Western allies of African Pentecostals often express, with pride and enthusiasm their past heritage, starting from the Ancient Greeks, through the early church that emerged in the context of Hellenistic culture, made possible by Alexander the Great's conquest (Johnson 1976). Looking at the history of Christianity in the Roman Empire, there are many practices and theological beliefs in contemporary Christianity that originated in, or were derived and adapted from, Platonism and Aristotelian thought via the work of church fathers like Saint Augustine and Saint Thomas Aquinas (Allen 1985). There are also practices that were drawn from Roman pagan religions as the church tried to strategically adapt itself to its social-cultural and political realities in the various areas. If ideas and practices from ancient Greek and Roman pagan cultures can be adapted to Christianity, what makes it impossible to do the same with African pagan culture? Why is it that the Christian religious rituals that were used to sanctify pagan ideas and practices derived from

[4]Personal Interview, Summer, 2012 in Accra. The research was funded by Bethel University, Saint Paul, Minnesota.

ancient Greek and Roman religious tradition cannot equally be applied to African culture? One interesting observation that comes out of the preceding analysis is that Pentecostalism is an agent of selective modernization of society. This is the issue examined in the next section.

97.6 Pentecostalism as an Agent and Process of Selective Modernization

Another lens through which one can examine Pentecostalism in Africa is to critically evaluate its role as an *agent of modernization*. Inkeles and Smith (1974), in their book *Becoming Modern*, identified four important independent variables that are institutions, which are necessary to promote modernization, and by which the mindset of people can be transformed in order to acquire the social-psychological characteristics of modernity. The institutional variables are: school and education, exposure to mass media, urbanization, and factory work experience. It must be noted that many Pentecostal churches have established parochial elementary and secondary schools, as well as universities.[5] In Nigeria, Bishop David Oyedepo's denomination, Living Faith (or Winners' Chapel), has established Covenant University located in Canaan Land, Ota, in Ogun State of Nigeria. The university is ranked higher than some of the universities established by the Federal Government of Nigeria based on its quality of education, facilities and faculty, but it is very expensive and unaffordable by people from poor socioeconomic backgrounds. In addition to the university, Winners' Chapel also has Faith Academy (Secondary School), and Kingdom Heritage School (Primary School). Across Africa, there are many universities established by Pentecostal denominations. Generally, students that attend these universities are Christians who are comfortable with the Pentecostal religious tradition. Participation in some religious rituals is mandatory in these religiously inspired universities, making it difficult for non-believers to attend.

Many Pentecostal churches have also heavily invested in mass media, whether those outlets are denomination-related newspapers published by a denomination's printing press or radio, and television broadcast owned or sponsored by the church. Headquarters of Pentecostal churches or their main sanctuary for worship are often located in major cities, such as Lagos, Accra, Harare, Nairobi, Kampala, Lusaka, Johannesburg, and attract people from different social backgrounds. Indeed, many of the cities have people from different African countries that come to worship, thereby creating a cosmopolitan atmosphere. Many Pentecostals in Africa, when compared to the general population, are more educated and work in organizations that operate like factories, based on modern rational principles of organization.

But the most impressive way Pentecostal churches represent themselves as gatekeepers, and the embodiment of modernity, is through the humongous churches

[5] Some examples of Pentecostal Universities in Africa are: Redeemers University (Nigeria), Pentecost University (Ghana), Covenant University (Nigeria), and Uganda Pentecostal University.

they build with the capacity to seat thousands of people.[6] They also equip such churches with the most sophisticated audio-visual technology, so that everyone present can be a part of the worship service and sermon. By African standards, this is not a small, but an impressive achievement. In some cases, the Sunday worship services are aired live on radio and television.[7] They also organize sophisticated campaigns to convert pagans, Muslims, and uncommitted Christians. The design and organizations of such campaigns are impressive and effective in converting people. They surely take rational planning and organization seriously which impresses people, particularly given the poor organization in many programs of postcolonial African states. Pentecostal churches across Africa have a huge gospel music industry which produces CDs and DVDs. The gospel music is meant to edify, to preach the gospel, and also encourage people to transform their lives.

If Pentecostals want to transform society, their preferred mechanism for doing so is to transform an individual's personality, attitudes and worldview. The transformation of an individual's personality is carried out through the teaching of Christian ethics with specific reference to monogamous marriage, sexual morality, judicious and spiritual use of money, and dressing appropriately to glorify God, by not exposing one's body in the case of women, which is somewhat patriarchal and patronizing. Pentecostals understand the risk associated with temptations embedded in a successful modern life, but they trust that the proper internalization of Christian ethics and training are a bulwark against such temptations. They spend much effort protecting and safeguarding themselves against the influences of tradition and some aspects of modernity, which are viewed as negative influences that threaten the individual Christian. It is a self that needs to be shielded because it is fragile and can easily be undermined by the evils of tradition or the seductiveness of modern life.

Some Pentecostal churches do not just preach, but they support their members by providing them with micro-loans and training them on how to succeed in managing their own small businesses.[8] The goal of this practice has been to help needy members of the congregation become financially independent. This means that there are

[6] Examples are Bishop Oyedepo's Winners' Chapel; T.B. Joshua's Synagogue Church of All Nations International, in Nigeria; Ray McCauley's Rhema Bible Church (Johannesburg); Ebenezer Miracle Worship Center, Kumasi; and Royal House Chapel, Accra, Ghana.

[7] Examples of television broadcasts in Africa are: TB Joshua's Emmanuel TV (Nigeria), Pastor Chris Oyakhilome' programs that are streamed online and broadcasted on television (pastorchrisonline.org). There are also the following channels in Uganda that air the programs of local and foreign preachers: Top TV, Lighthouse TV and Channel 44. In some countries like Ghana, preachers purchase air time from government television, thereby getting their programs broadcasted. These are just a few examples. Most Pentecostal churches have their programs online or circulate their programs through devices such as cd-video. Many of these programs are broadcast from Africa but watched live by Africans in the Diaspora or online at one's convenience.

[8] For examples of how Pentecostal Churches are engaged in social activism to transform the lives of people, check: D. E. Miller & T. Yamamori. (2007). Global Pentecostalism: The new faces of Christian social engagement. Berkeley: University of California Press. The DVD that accompanies the book has video interviews with several Pentecostal ministers from various African countries about their level and degree of social engagements to support the poor and socially marginalized, while trying to save souls as well.

Pentecostal denominations that are involved in constructive community development projects. Here are three illustrations of such Pentecostal churches.

97.7 Empirical Examples of Pentecostals' Involvement in Community Development Projects

The first example of an African Pentecostal and Charismatic church leader involved in constructive and meaningful community development is represented by the work of Pastor George Karambuka, who is a Pentecostal church leader and health-care worker (Wasamu 2013). His church ministry in Nairobi, Kenya, which has been in existence for 15 years, involves mobilizing uncircumcised adult men in Kenya to be circumcised, because "research indicates that circumcision helps prevent the transmission of HIV/AIDS from women to men." HIV infection rates are higher in Southern and Eastern Africa compared to other regions in the continent, because among other reasons, there are many uncircumcised men in Eastern and Southern Africa. Pastor George Karambuka fundamentally supports orthodox Christian teaching on "premarital abstinence and marital fidelity," but still recognizes the value and need for broader precautions from a public health point of view. His ministry uses multiple strategies by focusing on the following three goals (a) "help individuals see their need for HIV testing, treatment, and counseling;"(b) "integrate the gospel into all levels of care;" (c) "train Christian leaders to care holistically and through the church" (Wasamu 2013).

Edward C. Green, a leading scholar in HIV/AIDs research and the author of *Broken Promises: How the AIDS Establishment Has Betrayed the Developing World* (Green 2011), agrees with Pastor George Karambuka in so far as the circumcision promoted is voluntary. Green affirmed that there is research-backed data supporting the claim that the circumcision of adult males reduces the infection rate of HIV/AIDS. Pastor Karambuka is the officer in charge of the "Loco Health Center" in Nairobi, Kenya and because of the success of his work and dynamic leadership, leaders of many other Christian churches and denominations (Pentecostal and mainline) are referring their members to his clinic. Yet, some Bible scholars, such as Martin Wesonga, Dean of Academic Affairs at the Bishop Hannington Institute of Theology and Development in Mombassa, Kenya, have maintained that Pastor Karambuka's ministry is theologically misguided. What is needed, according Martin Wesonga, is the circumcision of the heart and not the body. Furthermore, other scholars argue that "the church should focus on what the Bible teaches and not offer alternatives." Surely, Pastor George Karambuka and his ministry represent a constructive, meaningful, and progressive approach to healthcare under the serious context of HIV/AIDS pandemic in East and Southern Africa (Wasamu 2013).

The second example is also from Nairobi, Kenya. In terms of engagement with society, Rev. Oscar Muriu of Nairobi Chapel has asserted that denominations should go beyond theological training. By and large most denominations do not have the resources or the luxury of sending every person who wants to be a pastor to seminary for theological training. Consequently, most of the potential church workers are trained through mentoring and apprenticeship. After a short period of training and apprenticeship, trainees are sorted, based on gifts and aptitudes, and assigned to a role within the church that best suits them. The churches train people to become missionaries, church planters, or workers in para-church organizations, even though a few are given the privilege of receiving formal theological training in college. Rev. Muriu argued that the weakness of a community is often the window of opportunity, through which the church can gain credibility by attending to the needs of the community. Pentecostal churches have learned from experience that sometimes what the church intends to do for a community may not necessarily be what the community considers to be its most important challenge. There are four cardinal principles which inform the Pentecostal ministry: planting churches, missions, teaching from the Word, and social ministry. Social ministry means trying to attend to the social problems that ordinary people encounter in a society.

The third example of constructive and meaningful community development involvement is exemplified by the approach adopted by Kampala Pentecostal Church, in Uganda, which is led by Rev. Chris Komagum. The church initially had more than 7,000 members, but when leaders of the church introduced the cell-based model of community church, some members did not like that and they left the church. The goal of the Kampala Pentecostal Church was that no cell-group would be made up of more than ten people so that there could be deeper discussion and accountability among members, as well as closer inter-personal acquaintance. This would ensure that care, love, and understanding would be shown to neighbors. When the church under the inspiration of someone from the United States adopted the idea of *holistic ministry*, many people found the concept and strategy attractive, and church membership increased to more than 8,000. Indeed, the membership of the church rises to 10,000 people during summer vacations when college students are home for summer holidays. The different administrative levels of organization within the Kampala church have provided members the opportunity for leadership training, and growth in society.

For Pentecostals in Africa, being born again is synonymous with being prosperous. They are chosen people by God, and this is supposed to be empirically proven and evidenced through external expression, such as: the kind of clothes they wear, the kind of houses they live in, the kind of hotel they use for their revival programs, and the kind of cars they drive. Their pastors and members have no restraint in wearing fashionable expensive clothing if it does not expose one's body, in the case of women. Outward appearance is considered to showcase the blessings of God in one's life and is also viewed as a beacon for attracting others to the faith. As Christians, the Pentecostal believers are expected to abandon extended family networks and traditional obligations, which are considered to be carriers of evil spirits. The Pentecostal believer becomes a member of a covenant community – the church. In fulfilling these requirements, the church becomes a socially constructed moral space where the individual's personality is reconstructed, his or her ethical values are transformed, and the accomplishments of progress of a modern society are celebrated while also critiqued. Given the cosmopolitan nature of Pentecostal messages, and based on social processes in which they are embedded, believers construct their social identity at a level beyond the nation state.

A major limitation of Pentecostalism in general, but particularly its African brand, is its naiveté about assuming that even when power and wealth are achieved or realized, they do not have the capacity to transform one's consciousness. While it is possible for one to acquire power and wealth and still remain faithful to one's spiritual values, study of the evolution of the Assemblies of God in the U.S. has raised an alternative view that warrants caution for those who take a simplistic position on this issue. The authors highlighted the frequency with which religious movements start as revival movements, or sects challenging the status quo, and innovate new ways of experiencing and relating to God that lead to the revitalization of society through the mechanism of personal transformation. But, building on Thomas O'Dea's theory (1961) of the routinization and institutionalization of charismata, Poloma and Green provided empirical evidence to support the fact that the Assemblies of God, a major Pentecostal denomination in the United States with great influence in Africa, is now more concerned about stability and pattern maintenance. They arrived at this conclusion because many Assemblies of God evangelical churches are now suspicious of any organized revival opportunity that is gathering momentum within the denomination. Yet, without continuous opportunities for revival, according to the authors, religious movements begin to lose capacity for renewal, personal transformation, and the inspiration for benevolent acts. They maintained that as the members of the denomination became part of the middle and upper classes, they lost their original faith as represented by the mission and ethos of classical Pentecostalism.

97.8 African Pentecostalism in the Era of Globalization

Pentecostal Christians in Africa are often compared to other Christian denominations, which are more highly globalized in their mindset and outlook. Many of these denominations in Africa like to include "international" or "world" in their name. One way that globalization has impacted contemporary Pentecostal Christians in Africa is through the easier means of traveling abroad (within or beyond Africa), studying in the United States or Europe, and listening to or watching a sermon of any renowned Pentecostal preacher online. One can listen or watch a Pentecostal preacher in the United States online.[9] Church members can also listen to or watch the sermon of their local pastor online or on DVDs or CDs. And they can do this at any time. During the 1980s and 1990s computers were not common in Africa, but today, in spite of the extensive digital divide, many people can download a preacher's

[9] Examples of Pastors that could be watched on television or online are: TB Joshua, Benny Hinn, T.D. Jakes, Joyce Meyer, Creflo Dollar, Kenneth Copeland, Kenneth Hagin, Pat Robertson et al. With the exception of TB Joshua who is a Nigeria, all the others are Americans.

message and listen to it or watch it anywhere, and at any time it is convenient, on an Ipad or smart phone, through the use of satellite modems and other technologies.

Globalization has also created conditions that have compelled millions of people to migrate. Many Pentecostal immigrants from Africa come to the United States, while maintaining contact with their families and churches back in their countries of origin. Levitt (1988) has brilliantly documented how immigrants in Boston, originally from Dominican Republic, return home to introduce new ideas of spirituality in their churches, while their presence in the U.S. enriches the religious landscape, as concluded in their study of the Assemblies of God denomination in the United States. From Poloma and Green's study, it is evident that benevolent acts, on the part of the members of the Assemblies of God, are primarily occurring at the micro-level in the majority of churches. The theology and biblical understanding of most churches in the Assemblies of God denomination, as well as their members, are more evangelical than Pentecostal. This constrains them from confronting structural injustices and sin that are perpetuated by sinful institutional structures remotely manipulated by powerful interests. Poloma and Green characterize the situation as follows:

> The Assemblies of God (AG)'s uncritical acceptance of a conservative political stance, at least in the United States, is not consistent with the nature of the potentially radical Pentecostal experience. The Azusa Street revival, the event that catapulted the Pentecostal gospel, according to some historical accounts, empowered blacks and women long before the Civil Rights and Feminist Movement of the 1960s. However, this breaking down of dividing walls was short-lived as organized Pentecostalism mirrored the same problems of racism and sexism found in the dominant culture.

In this respect, it is the immigrant Assemblies of God churches in the United States that are more willing to confront and deal with issues of social justice broadly conceptualized. Obviously, this is one reason why the Assemblies of God congregations are relatively socially homogenous; these immigrant congregations have a broader and deeper commitment to social justice than the typical evangelical-type Assembly of God congregation. Not surprisingly, the leadership of the church does not include a single woman or minority person, in spite of the increasing number of Latino, African, and Asian members within the religious movement. Indeed, the immigrant Pentecostals constitute the major source of growth for the religious movement in the United States in the twenty-first century.

Similarly, there are Pentecostal denominations in Nigeria, Ghana, Zimbabwe, and elsewhere that are not only sending missionaries to plant and pastor churches in other African countries, but they also have church branches in the United Kingdom, Western Europe, and the United States (Gornik 2011). Examples of churches Gornik studied in New York are: Ethiopian Church, Redeemed Christian Church of God, Presbyterian Church of Ghana, and the Church of the Lord (Aladura). Although these churches operate in the United States, their national headquarters are back in African cities like Harare, Lagos, Accra, and Nairobi. Consequently, there have been exchanges of influences and ideas between the two sides of the world. The pastors in charge of the church branches, in the United States, the United Kingdom, or Europe return to their home countries in Africa on regular basis for consultation at the world headquarters. The African leaders of the denomination also visit the United States, the United Kingdom, and European countries regularly for revival meetings and to evaluate the progress of the congregations. In some cases, the African denominations that open branches in other African countries have had problems adapting their administrative structure to the new environments and social contexts (van Dijk 2001: 216–234). Some of the churches have faced serious challenges in Africa, because they have been accused of collecting offering in one country and sending part of it to the headquarters of the church in another country. Here is an example documented by Cedric Mayrargue:

> It must, however, be noted that foreign origin can be the source of problems within these movements. The conflicts which have often broken out in the Pentecostal Church in Benin originate, aside from matters of personal rivalry, from accusations made against the dominant influence of the central Ghanaian Pentecostal Church. This movement has been reproached for sending part of the funds collected by its Beninese branch directly to headquarters in Accra. The authoritarian manner with which it appoints and moves ministers in Benin has also been criticized by some people who have left the church. (Mayrargue 2001: 289)

Globalization has also exposed Africans to the phenomenon of satellite television. Some of the best channels in African countries are the same ones used by United States televangelists to preach. It is not uncommon to see more than seven different Pentecostal preachers, on a Sunday, on a television channel that is accessible to the public and watched in Africa. Often the service is simulcast (simultaneous broadcast), and as the sermon is preached in the United States, the preacher primarily attends to the audience in front of him or her, even though the sermon is being watched in faraway Africa. The content of such sermons, such as Joyce Meyer preaching confidently in a continent where patriarchy is still very strong, is an indirect but significant cultural innovation in Africa's religious landscape, which is quite different in its constitution from the American one.

African Pentecostals, with their global orientation, challenge ethnic and national conceptions of identity as being parochial. This is manifested in the way they project their members' spirituality outward, beyond their nation by situating it within an imagined religious community of global Pentecostals. Given that Nigeria has the largest national population in Africa, many of whom are Christians, it is not surprising that Nigerian Pentecostal Churches have grown and expanded to a level where they have created transnational networks through their departments of foreign missions. Some of these churches have trained other African and Western nationals and sent them back to their country of origin to establish churches.[10] Denominations such as TB Joshua's Synagogue Church of All Nations International, The Redeem Church of Christ, and Bishop Oyedepo's Winners' Chapel (all with headquarters in Nigeria), have held crusades in various African countries, and some have established branches of their churches in those countries.

[10] Examples here are TB Joshua's The Synagogue Church of All Nations International, The Redeem Church of Christ, which has nearly one hundred branches in Manchester, United Kingdom. Bishop Oyedepo's Church has branches in different parts of the world, not just Africa, but in Europe and the United States.

Yet, Matthew Ojo's historical research entitled, *The End-Time Army: Charismatic Movements in Modern Nigeria* (2006), indicates something that could be referred to as "*Nigerian Manifest Destiny*."[11]

> God has had occasions (sic) recently to reiterate by words of prophecy and through revelations what his plans are for Nigeria and through Nigeria: God intends that Nigeria should be the beacon of the Gospel in Africa. Looking through the West African belt, Nigeria is surrounded by poverty-stricken and grossly under developed nations. It is an act of Divine Providence that Nigeria stands out different as the richest nation in terms of human and material resources in this belt. This is for no other purpose than to enable the CHURCH champion God's ultimate will for Africa. It then behooves on the CHURCH to identify the track along which God is moving, align itself to God so as to bring NIGERIA, AT THIS PRECARIOUS STAGE OF HER DEVELOPMENT, INTO GOD'S REDEMPTION PLAN. (cited in Matthew Ojo 2006: 119)

In principle, there is nothing wrong with Nigerian Pentecostals sending missionaries to other African countries. In practice, however, the language used by many churches indicates that they believe God has specially chosen Nigeria to perform this role in Africa, and given how the country is comparatively and relatively blessed in both material, natural, and human resources, this reinforces the idea of *Manifest Destiny*. According to such churches, it was the duty of Nigerian Pentecostals to evangelize the whole of Africa, and through evangelism, Africa will be Christianized and saved from demonic and spiritual forces that have put the continent in bondage, hindering the progress of the continent. This mission project took off during the 1970s oil-boom years in Nigeria, which coincided with a politically active and interventionist Nigerian foreign policy in the Southern African region that was still under White minority rule. Nigeria supported liberation movements against White minority regimes in Luanda, Pretoria, and Salisbury (Ojo 2006: 118). Apparently, the *Manifest Destiny* syndrome is not just a monopoly of Victorian England or eighteenth and nineteenth century America (Gossett 1997: 310–338).

At its core, the Manifest Destiny syndrome has a subtle sense of cultural and spiritual hubris that is cultivated among people who look around and believe that they have something unique, and that which makes them better than others. Consequently, these people see themselves as more Godly than others, and imagine that they have a unique role to accomplish. Unfortunately, in the process of performing the mission role, the cultural invasion and hubris that it entails is easily ignored.[12] In their critical commentary of Africa, Nigerian Pentecostals are not only critical of other non-Pentecostal Christians, but perceive Muslims as representing the forces of darkness.

---

[11] The author is applying the logic of Manifest Destiny in U.S. history as discussed in the book entitled: Race: The History of an Idea in America by Thomas F. Gossett. The American concept of Manifest Destiny is based on Anglo-Saxon racial superiority. What I am borrowing from the concept is its essentialist idea of group superiority and therefore destiny, which is at the core of the concept of Manifest Destiny. Do Nigerian Pentecostals and Charismatics have a concept of Manifest Destiny at one point in history based on some essentialist assumptions and special status of Nigeria among African nations? Yes, there is evidence to support this observation as can be seen in the quote cited.

[12] For a good example of the point being made here, see Michael M. Phillips, "Unanswered prayers: In Swaziland, U.S. preacher sees his dream vanish," The Wall Street Journal, December 19, 2005.

African Pentecostals are overly concerned vis-à-vis their United States counterparts about creating discourses of identity that transcend their locality and nations. For them it is not just a matter of faith, but prestige. Research indicates that while religious groups are committed to promoting world peace and social inclusion, it is far easier to recruit and sustain members if sentiments are used in teaching and preaching, that is, sentiments that refer to a specific denomination or congregation as unique and special, as compared to promoting an abstract imagined community of Pentecostals. The concrete feelings and sympathies of belonging to some specific religious community are more effective in mobilizing people to contribute cash or in-kind donations than a vague abstract idea of an imagined community. This is especially true when denominations are competing for the same market share of believers and supporters. In this respect, there is a certain degree of tension between the need of a religious group to belong and find meaning in its local context, and the demands of its denomination as a global entity committed to global peace and the social inclusion of all. Roberts (2004) described such tension as follows:

> Church officials may choose to embrace the emerging global culture, acknowledge and celebrate global economic interdependence, and foster greater tolerance and open-mindedness toward those who are "different." The alternative course would be to intensify the solidarity between the religion and the national or ethnic group and to stress the differences between "us" and "them." The latter course is likely to buoy the members. (Roberts 2004: 390–391)

Globalization is made easier in any area of culture and society when there is a common discourse or language that binds people together. Globalization has enhanced this through introducing technology that enables portable transfer of cassette tape-recorders, gospel music CDs, and sermons on DVDs. In this process, a common language and discourse is developed, which can serve to bind many people together, even though they come from different races, countries, and nationalities. However, just as Pentecostalism introduces new discourses and language that bring people together, it also promotes a new concept of citizenship and rights among Pentecostals. Given that in a Diaspora Pentecostal church in Europe, United Kingdom or the United States, one can be an illegal immigrant and still get an official church wedding, these rights are not based on the statutory requirements of citizenship. Related to this is also the fact that the Pentecostal concept of citizenship in Africa is often based on the Christian concept of righteousness.

One of the key goals of African Pentecostals is to create a Christian nation, because in their vision of Christianity and eschatology, they view this path as the only meaningful goal. These believers perceive other religious groups, and even other Christian traditions, as inferior (Brouwer et al. 1996: 151–178). Consequently, the only "right" option available to a progressive society is to convert believers of other faith traditions to Pentecostal Christianity. This type of approach to citizenship and rights can easily lead to hubris and a desire to culturally colonize others, even though the intention of the conversion and colonization project is presented as innocuous and benevolent. The approach presumes a hierarchy of cultures, based on the criteria of ranking exclusively determined by Pentecostals based on their type of religious values. They want to colonize the public sphere with their faith values even when they live in a plural or liberal democratic society. They also use military language in how they conceptualize spreading their faith to create a Pentecostal theocracy. As an example, here is how one Pentecostal pastor, in Nigeria, articulated his vision of Pentecostal involvement in the Nigerian and global public sphere:

> Clearly, we will have to contend and conflict with wicked spirits in heavenly places and rulers of the darkness of this world and wrest their control over entire cities, regions, nations and continents to enable us to do our job of preaching the gospel to all nations of the world. The Holy Spirit has recently been calling the Church (the body of Christ) to prepare for unprecedented warfare through many Christian leaders. The 1990s will definitely witness the most intense (sic) spiritual warfare the church has ever been involved in over its entire 2,000 year history. There is no de-militarized zone. You will have to fight. Be of good cheer our Lord has already given us the victory. (As quoted in Ruth Marshall-Fratani 2001: 101)

Given that Pentecostals desire the creation of a Christian nation, they can often be at logger heads with the vision or assumptions that undergird a secular society and a liberal democratic nation state, as is exemplified by how President Chiluba of Zambia declared his country in 1991 to be a Christian nation, because he was a Christian and a Pentecostal. He believed that in declaring Zambia as a Christian nation, he was honoring God. In their rhetoric and in practice, African Pentecostal leaders find it difficult to maintain the boundaries between what Max Weber would call the legal-rational foundation of the modern liberal-democratic state and the covenant community they desire to create based on the shared solidarity of their faith in a globalized world. The concept of a global civil religion and citizenship is outside their vision for a community of believers that are righteous and brought together by faith in Jesus Christ. In the contemporary globalized world, the dominant ideology that undergirds the globalization project is neoliberal capitalism. The ethics of neoliberal capitalism penetrates the culture and institutions of African societies, thereby changing the public sphere and everyday life-world of the people, that is, the private sphere. How does Pentecostal Christianity fit into this unfolding social reality? This is the issue addressed in the next section.

97.9 African Pentecostalism and the Neoliberal-Global Empire

The current world system is under the hegemony of neoliberal globalization. Neoliberalism, in terms of public policy, can be characterized as the policy instruments of the "Washington Consensus," which is a product of an agreement among the World Bank, the IMF, and the U.S. Treasury concerning how to reform and run the global economy, which evolved after the collapse of Soviet communism. There are ten policy instruments in this reform. The policy instruments are: (a) fiscal discipline; (b) reduction of public expenditure that is devoted to social services, such as health, education, and infrastructure; (c) cutting taxes, expanding the sources of taxes, and the collection of taxes; (d) creating a market for competitive currency exchange rates; (e) securing property rights; (f) deregulating the economy; (g) liberalization of trade; (h) privatization of hitherto government managed and administered public services; (i) eliminating barriers to direct foreign investment; and (j) liberalizing the financial markets (Rodrik 1996: 9–41).

There is no doubt that the free market has opened opportunities for many people, but it is also a fact that the changes in the new international political economy have negatively impacted the human welfare and standard of living of many people in different parts of the world, both in developed and developing countries (Stiglitz 2006; Shipler 2005; Ehrenreich 2011). In the new global economy, what matters most is one's standing in the marketplace. One's standing in the marketplace is defined by: (a) whether the person has useful skills or human capital that is valued in the marketplace; (b) whether the person has discretionary surplus income that can attract an investor's or a business person's attention; and (c) whether the person produces a commodity or service that is central to the operation of the economy. Lacking any one of these aspects, in the new global economy, a human being may, in theory, have human rights, but in practice, it is as if he or she does not exist. Under this social ethic, human dignity is conceptualized in terms of a person's capacity to play an effective role in sustaining the market exchange system, which is perceived as the fundamental building block of a society (Hayek 1944; Mises 1949; Scruton 2006: 208–231). In effect, the new free market ideology and regime has put in place conditions that have either excluded the poor from effective participation in society or have made it exceptionally difficult for them to do so.

Given the tenets of African Pentecostal theology, that is, the prosperity gospel, spiritual deliverance, and Christian Zionism, what kind of moral and ethical critique do African Pentecostals have to advance concerning the current neoliberal hegemonic system of our time? Often Pentecostals and evangelicals will say that a human being is someone created in the image of God, and is loved, through grace, by God. In addition, the destiny of a successful Pentecostal Christian believer is to be rich, because of the victory of Christ on the cross, after his death and resurrection. But the empirical data suggests that inequality is widening within Africa (Todaro and Smith 2012: 229–235). Indeed, the gap of inequality has widened within developed countries (Brynjolfsson and McAfee 2011: 28–52), between and among developing countries, and between some developing countries and developed countries (Stiglitz 2006).

In the past 30 years, African Pentecostals have not had any concerted systematic theological discourse to directly confront the kind of moral and ethical critique that secular scholars, and theologians in other Christian traditions have put forward to interrogate and critique neoliberal globalization.[13] They focus on the individual and the role of demonic and evil spirits, as they are embodied in culture and tradition, and as obstacles to the progress of individuals. They have no incisive understanding of the epistemological presuppositions of neoliberal globalization and its implications

[13] For examples in this respect, see: John Mihevc. (1995). The market tells them so: The World Bank and economic fundamentalism in Africa. London: ZED Books; Third World Network; Charles McDaniel. (2007). God & Money: The moral challenge of capitalism. New York: Rowman & Littlefield Publishers.

for the dignity of human kind. A support for this line of reasoning has been provided by Paul Gifford, in his book entitled: *African Christianity: Its Public Role* (Gifford 1988). On the public role of the varieties of African Christianity, especially, as it has affected economics and politics, Gifford asserts:

> Most varieties of Christianity in Africa, judging from their preaching or literature, are not partners in Mihevc's discourse. African Christians ask the same questions as Mihevc: Why are Africans suffering? Where does evil come from? What might redemption or salvation mean in these circumstances? But their answers only very infrequently consider African political structures or Western influences like the international banking industry. Their principal answers are expressed in terms of lack of faith or blockages caused by demons. That this is understandable cannot be denied, and it makes 'a certain sense.' Most Africans manifest a 'primal' mentality, and we have shown that churches meeting these primal needs are understandably flourishing. Mihevc may be right in suggesting that African theology needs a strong base in the Enlightenment; but he is surely incorrect in implying that such a theology is already flowering. (Gifford 1988: 332)

Often the criteria African Pentecostals use for measuring the success of a good Christian is similar to the ones neoliberals use in evaluating success in the marketplace or economy, that is, empirical results indicating accomplishment in material accumulation. Both neoliberals and Pentecostal Christianity are competing in their effort to transform the world. They both have eschatological vision of human history, with Pentecostals thinking about end times and paradise being the ultimate end of the eschatological journey for Christians. Neoliberals view the ultimate end of the secular eschatological journey as the age of high mass consumption, as outlined by W.W. Rostow's *Five Stages of Economic Growth* (1960). In helping humanity to reach the ultimate stage in these two eschatological visions and journeys, both Pentecostals and neoliberals compete in transforming the individual and society.

Indeed, deducing from the behavior of ministers in Africa and their parishioners, one main goal of Pentecostals is to obtain greater access and success in the current structure of the global capitalist economy so that when they succeed, they can use it as a testimony of God working in their lives. They do not raise fundamental ethical questions or concerns about the structure of the neoliberal hegemonic system and its consequences for humanity. Yet the forces behind the neoliberal paradigm do not have special care for the values and presuppositions of Pentecostalism or Christianity in general. Pentecostals think within the system and not about the system.

As an example, *Forbes* magazine provided the list of the five richest Nigerian ministers, and not surprisingly, it turned out that they are all Pentecostal ministers (Nsehe 2011). In the same article, Nsehe documented what the Pentecostal ministers were worth, as part of their spiritual business empire, in a country where inequality and poverty are high. Here is the net worth of the richest five Pentecostal ministers in Nigeria: Bishop Oyedepo ($150 million); Chris Oyakhilome ($30–$50 million); T. B. Joshua ($10–$15 million); Matthew Ashimolowo ($6–$10 million); Chris Okotie ($3–$10 million). Some of these ministers lead denominations that have branches in the United Kingdom, the United States, and other African countries.[14]

[14]The richest clergy in Nigeria spend their money ordering private jets (for example, Bishop Oyedepo of Winners Chapel, has four private jets), or invest in a printing press for publishing their books, which is a major source of revenue. They also purchase real estate property in the United

Pentecostals in the United States, by and large inspire African Pentecostal and Charismatic groups. Given that they are a subgroup within the broader evangelical movement in the United States, some insights from scholars and opinion leaders of the religious right, addressing social justice, can help us gain a general idea of the thinking of influential evangelical Christians, especially in the Bible Belt. One critical emphasis among evangelicals of the right is the need for free markets or for allowing the market to be the main determinant for the allocation of scarce resources. In effect, the market process of exchange becomes the only morally and ethically acceptable way for deciding who gets what and how much? Whatever is the outcome of the commutative exchange process is considered legitimate and should not be called into question on some "cosmic" ethical grounds, irrespective of the unequal nature of the outcome. Beisner (1988) made this case strongly, as follows:

> Just as personal justice is individual conformity with the standards of rightness, so social justice is societal conformity with the standards of rightness. Understanding this should prevent our falling into the mistaken idea that social justice has something to do with a particular distribution of goods, privileges, or powers in society. Real social justice, on the contrary, attends only to the question whether goods, privileges, and powers are distributed in conformity with the standards of rightness. Whatever factual distribution results from conformity with those standards is just regardless how far it strays from conditional equality – the real idea behind many uses of the term social justice today.

It should be noted that some leaders of renowned African Pentecostal denominations received their training in the Bible Belt of the United States, and also received extensive financial support for their ministries from supporters in that region.[15] The case in point here is Ezekiel Guti, the leader of Zimbabwe Assemblies of God Africa (ZAOGA). He received basic theological training in the United States before returning to Zimbabwe. He was later aided by supporters from the United States who helped construct a theological training school for his denomination.[16] Yet, when Guti received financial assistance from the United States, he did not treat it as organizational

---

States and the United Kingdom. There is no clear cut distinction between what the church owns and what the founders own. Generally, what the church owns is what the leader owns. Many ordinary Nigerians see nothing wrong with this practice because they believe it is God's reward for the minister's service who is a "servant" of God.

[15] Ezekiel Guti and his denomination received foreign funding for the construction of churches, Bible schools, and expansion of his ministry to other African countries. He received funding from the following organizations: Christ for the Nations Institute, Dallas; Native Churches Crusade in Dallas, and Forward in Faith Mission International, which is also in Dallas (Maxwell 2006: 91–92).

[16] Ezekiel Guti started attending Berean College in Dallas, Texas, but could not continue his education because of the lack of financial resources. Later, a person named, Sande, helped him get admission into Christ for the Nations Institute, in Dallas, Texas. According David Maxwell, "At Christ for the Nations Institute, Guti took a diploma in Charismatic Orientated Biblical Studies." The course was vocational rather than academic. Evangelism, Church Growth, Charismatic Gifts and Worship were the major priorities... Scriptures were learn[ed] rather than interpreted and examinations took the form of simple multiple-choice questions. Much of the learning was done through listening to audio-tapes, although students were expected to read Gordon Lindsay's tracts and booklets including some of his more idiosyncratic interpretations of Old Testament prophets, who he believed foretold the invention of the television set and the motor car. Credits were gained through attendance rather than examination, and these were measured with a time clock and swipe card. There were penalties for tardiness (Maxwell 2006: 89).

money, but rather used much of it for patronage in order to boost his personal "strongman" aura. Guti's ministry in Zimbabwe is just one example. The following quote suggests and illustrates the question of wealth and material accumulation among Pentecostals in Zimbabwe as a whole:

> Although the likes of Celebration Ministries International (CMI)'s Tom Deuschle, Tudor Bismark's New Life Covenant Church (NLCC), Christ Ministries (CM)'s Godwin Chitsinde, Ezekiel Guti's Zimbabwe Assemblies of God Africa (ZAOGA) and United Family International Church (UFIC)'s Prophet Emmanuel Makandiwa were not readily available for comment, some of these "prophets and prophetesses" have made quite a tardy sum of money for their institutions and themselves. Amid increasing charges that the church is now a convenient place or way to make a quick buck through the perversion of Christian dogma, CMI, NLCC, UFIC and ZAOGA are among the names or churches that have taken a leading role in raising top-dollar through various initiatives. These include newsletters and other audio-visual material or paraphernalia, regular conferences, transport business and even educational facilities. According to a South African publication, Makandiwa preaches about 'possessing the promised' and justifies his car-upgrade(s) on the philosophy that whenever 'the devil steals from you, he replaces what was stolen two-fold'. This was after he had crashed his Mercedes S320 limo and quickly replaced it with a bigger one, the paper said ("Richest Pastors," 2013).

Notwithstanding the truth of select scandals concerning church leaders, there have been positive contributions, which both Pentecostals and evangelicals have made through the opportunities created by globalization, in Africa. These religious organizations play an important role in bringing attention to, and helping solve or alleviate, some global problems, particularly in areas of Africa like Sudan, an area which has been going through conflict and is in great need of humanitarian support. These groups have also been involved in other social issues, of which even skeptics have had to admit that they have made a positive difference (Fowler et al. 2010: 119–167).

97.10 Conclusion: Lessons and Reflections on the Role of Pentecostalism in African Societies

The analysis of Pentecostalism in Africa in this chapter raised numerous critical questions that cannot all be exhaustively digested. But by way of conclusion, one can draw some important lessons that leaders can constructively reflect upon and extend these lessons to higher and broader levels. The first lesson one can draw from the analysis above is in the form of a major critique of Pentecostalism in terms of how it fits within the contemporary existential reality of living out the Christian faith in the modern world, as well as the question of what constitutes a successful Christian life. In this respect, Hollenweger makes this observation:

> They see the world as a cosmic and moral duality. There is no room for the natural. Everything is either divine or demonic. They reject historical critical research in favor of their own experience. We are successful, they say, so what is the need to ask the truth question? This is of course a syncretism that is no longer theologically acceptable. They

> replace the biblical principle of truth for the North American yardstick of success, and argue that if you are successful your success must be good. Against this, many critical Pentecostals argue that Jesus did not ever make a fuss about demons. (Hollenweger 1999: 180)

The preceding quote suggests the unwillingness of the great majority of mainstream African Pentecostals to engage in rigorous research that tries to honestly and courageously investigate what is happening within the tradition. Many Pentecostal churches are more interested in self-affirmation, using their material success as empirical evidence that God is working in their lives.

Indeed, it can be vehemently asserted that when one examines the life of the Apostle Paul, who wrote most of the New Testament, it was characterized more by suffering for Christ than the accumulation of material wealth as an indicator of God's presence or blessing in his life. In addition, demonology was not a focal point of the Apostle's ministry. Consequently, the fact that material prosperity and demonology have become focal points in the theology of Pentecostalism in Africa indicates a distortion of Scripture and an aberration from authentic teachings of the New Testament.

The second lesson one can draw from the study of mainstream Pentecostal Christianity in Africa is that it is a brand of Christianity that is comfortable with the existing structures of neoliberal globalization. African Pentecostals are more likely to engage in charitable acts toward others, either by helping the poor in kind or teaching them skills that can enhance individual human capital, so that they can better compete in the "dog-eat-dog" world of neoliberal globalization, where competition for success and scarce resource is ruthless. While these types of social engagements, which are by-products of their faith are commendable, they do not raise fundamental questions about the substantive issue of social justice, which is a separate concern from charity, good as charity is.

It is pertinent to note that in evangelical Christianity in general, many denominations conflate charity with social justice. Many evangelical churches and Christians are afraid of any discussion of social justice, because they assume it is tantamount to promoting socialism or communism. This thinking, however, has undermined the capacity of many churches and denominations to create a social space for serious and rigorous discussion concerning issues of social justice because of the assumption that charity is synonymous to social justice.

Pentecostal theology as it exists in Africa today thus lacks the kind of sophisticated structural critique of neoliberal globalization that comes from the Catholic tradition, as articulated in the following quote in Pope John Paul II's 1988 encyclical letter entitled "solicitude rei socialis" (that is, the Social Concerns of the Church):

> One must denounce the existence of economic, financial, and social mechanisms which, although they are manipulated by people, often function almost automatically, thus accentuating the situations of wealth for some and poverty for the rest. These mechanisms, which are maneuvered directly or indirectly by the more developed countries, by their very functioning, favor the interest of the people manipulating them. But in the end they suffocate or condition the economies of the less developed countries. (cited in Todaro and Smith 2003: 125)

Pentecostals are inextricably entangled with the prosperity gospel such that the power of the Holy Spirit falls short in their sacred spiritual devotion and reflection to give them revelation or insight to see beyond the benefits of neoliberal globalization to the winners in the world. They fail to see the profound injustice that many human beings are forced to live with, in the name of progress.

The third lesson from the analysis of Pentecostal Christianity in Africa is that Pentecostalism in the continent expresses what Saint Augustine calls, "*libido dominandi.*" As Richard John Neuhaus explained it: "*libido dominandi* is the lust for power, advantage, and glory" (Tonkowich 2008). Pentecostals use language that suggests it is their own way or no other way. They use military language that suggests the conquest and domination of the world in the name of Jesus and in the power of the Holy Spirit. Their main criterion of citizenship in the modern nation state is Christian, or more precisely, Pentecostal righteousness. Anyone lacking such righteousness is considered to be either a second-class citizen or just good for a mission project, as part of the grand strategy for creating a "New Jerusalem" in their own image. Given that by all means the Pentecostal ministers get rich fast, and some of them are better known for their material accumulation and economic entrepreneurship than anything else, many brands of African Pentecostalism represent the pursuit of lust to conquer and dominate (*libido dominandi*), dressed in Christian spiritual garb.

Fourth, Pentecostalism cannot be understood in Africa, or any part of the world (if the preceding analysis is anything to go by), without understanding the intersection of the global and the local. Also, it cannot be understood without appreciating the horizontal learning and linkages that take place within the Global South, the African continent in this respect. African Pentecostalism is significantly influenced by Western and American Pentecostalism. One cannot account for the receptivity of foreign ideas of Pentecostalism in Africa without understanding the preexisting culture, social problems, social structure, and aspirations of Africans and Africa, which have helped to create an elective affinity for the reception of Pentecostal messages from abroad. Pentecostal leaders, along with the blessings and patronage they confer, conform to the role of the traditional priest and strongman in traditional African politics, state, and society. Miracles are in high demand in a continent where the social structure is increasingly becoming rigidly stratified along class lines, and where the political system is less responsive to the needs and yearnings of ordinary poor people. The patron-client relationship is a feature of many African societies, and Pentecostalism in many respects replicates that in a new form and in a new social context. Based on current discourse on globalization in the social sciences, what seems to be happening in the case of Pentecostalism in Africa is what Pieterse (2013) called *cultural hybridization*. He describes it in the following manner, which is pertinent in understanding the phenomenon of Pentecostalism in Africa.

> Cultural hybridization is something like music remixing. It takes elements from separate cultures and mixes them to create a new culture, producing a third entity that is not the same as its antecedents. This differs from assimilation (homogenization) in that the elements of the original cultures remain distinct and there is space for choice in determining how the new whole will be shaped. Hybridization can be subversive, as it gives people a way to

> redefine their identity without altering all of the fundamentals. Whereas some may interpret culture as unchanging and static, defining it according to traditional practices, hybridization allows change to occur while tradition is maintained. It often occurs along cultural margins, spaces at the boundaries of cultural practices, as different ways of life cross over and mix. (Pieterse 2013: 29)

It is obvious that African Pentecostalism borrows from the West, especially, the Bible Belt of the United States, but then it adapts what it borrows and enculturates it in African context creating something not totally foreign but also not totally African. It is in this respect that we see the entrepreneurial spirit and creativity of Pentecostal ministers. As to whether this entrepreneurship and creativity is used for creating a more just society, that is a different and an empirical question. An example of how such spiritual entrepreneurship can go out of control is illustrated by a Pastor of Living Faith Church (Winners' Chapel, in New York City), who established a branch of the church almost as a franchise. Yet when disagreement over control of the church and its resources started between him and the headquarters of the church in Nigeria, Bishop Oyedepo, the founder of the Church, sent a delegation from Nigeria to New York City, which illegally took over control of the church's bank account when the founder of the church in New York City, Pastor Onalye, was the only signatory. The issue later became a criminal case that is being investigated by the police in New York.

The fifth lesson that could be drawn from the analysis of Pentecostalism in Africa is that inequality and social marginalization have widened in many countries and among many social groups within the period that Pentecostal Christianity exploded in Africa. The rural-urban divide in Africa has expanded under neoliberal globalization because the main criterion for deciding where and how to invest scarce resource is the rate of return on investment, rather than human needs. Similarly, many urban slums have burgeoned in the past 30 years as Pentecostalism has expanded in Africa. This means that it is possible for a religion preaching the love of God for all humanity to expand while simultaneously, social inequality and social marginalization are increasing. Tony Ubani had this to say about Nigeria with respect to the expansion of religion accompanied by human suffering:

> .... But in Nigeria, all the warehouses have been bought over by churches. What an irony! We pray all the time in Nigeria and yet Nigeria is sliding backwards. South Africa is marching forward. Our leaders sit at the front rows in churches and give fat offerings and tithes. Unlike in the days of old from the Bible, no pastor or Man of God can look a corrupt leader in the face and tell him the truth. (Ubani 2013)

Along the same line, mega-churches have also increased in the United States in the past 30 years, and evangelical Christians have increased their participation in partisan politics while social inequality and social insecurity have seriously increased for vulnerable social groups in the country within the same period (Brynjolfsson and McAfee 2011: 28–52). This obviously shows that there is a need for Pentecostals in Africa and Evangelicals in the United States to examine themselves and their approach for socially transforming the world. Is the mission of Christianity just to save souls or to transform the whole of humanity in a holistic way? What sacrifices do Christians need to make to transform society, and are they willing to do that? Is the mission of social transformation aimed at imperial cultural conquest and domination of the world by Christians, as Christian reconstructionism and dominion theologians envision the world? (Martin 1996: 353–355)

The sixth major lesson from the analysis of Pentecostalism in Africa is that the continent continues to suffer from its characterization not only as a dark continent in the negative sense espoused by Europeans (Blaut 1993; Eze 1997), but also by Pentecostal ministers in Africa. European scholars in the nineteenth and early twentieth centuries characterized Africa as a moral space (not simply a geographical area), representing evil and backwardness because of the continent's evolutionary lag, lack of historical consciousness, and the lack of capacity to think rationally, analytically, and theoretically (Bell 2002: 37–58). Pentecostals see demonic and territorial forces controlling the affairs of the continent, thereby making it the most backward and crisis-ridden continent in the world. In the cosmological worldview of Pentecostals, there is a global geography of the distribution of sin, because sin is more highly concentrated in Africa, which is evidenced presumably by the higher existence of poverty, witchcraft, political instability, and underdevelopment. The Pentecostal churches see themselves as outposts of spiritual redemption that were originally located in the Western world, especially the Bible Belt of the United States, which then radiates to other parts of the world. In this theological discourse, rural areas in Africa are presumably the worse places to live in because of the dangers of demons and witchcraft that find such spatial areas more convenient to dwell.

The seventh important lesson one can draw from the analysis of Pentecostalism in Africa is that even though it is difficult for many Christians in Africa and the United States to accept the empirical reality that people may believe in God but have different understandings of who God is, the reality is that it is not the mere belief in God that matters, but rather how those beliefs affect one's worldview and social engagements. The specific nature of the beliefs is a critical issue, because people hear or read different things about who they are and what their role is in the universe (Froese and Bader 2010: 36). Pentecostalism represents only one way of thinking and worshiping God in Africa, and even then, there are some nuances of understanding who God is within the broad scope of Pentecostalism in Africa.

The preceding analysis of Pentecostalism also leads one to the conclusion that there is reason to raise critical concerns about the flamboyant celebration of global Christianity and the claim that the center of gravity of Christianity has shifted from the Western world to the Global South. Often such claims are supported by relatively easy to compile quantitative measures indicating the rapid expansion, if not triumph, of Christianity in the non-Western world. Yet, the findings in this chapter suggests that ethnographic studies of African Pentecostalism and Charismatic Christianity are needed because they introduce us to certain qualitative measures of Christianity, which are more difficult to gather but very incisive in evaluating the cultural depth and institutional and cultural consequences of Christianity on the day to day lives of people in Africa (Maxwell 2006).

Given the discussion and analysis in this chapter, there is strong evidence that the rapid quantitative expansion of Christianity in Africa is not necessarily accompanied by commensurate qualitative growth and changes as one would normally expect if things have been working very well. Related to this, there is a need for further studies on the social and theological realities of Pentecostalism in rural Africa compared to similar dynamics in large cities.

Finally, the ninth conclusion is the failure of Pentecostals to learn from the crisis of modernity vis-à-vis faith as it has unraveled in the history of Western civilization. By and large, Pentecostalism in Africa embraces modernity. It recognizes that there are temptations in the modern world, but such temptations are nothing in comparison to the Pentecostal power that ensures their victory over the temptations of modernity. Victory means enjoying all the benefits of modernity while neutralizing and avoiding all its negative consequences. However, the naiveté African Pentecostals is in failing to appreciate the fact that the church, if it is to remain a true church, no matter its size and expansion, is supposed to be a "*covenant community*." On the other hand, if and when modernity succeeds, as it is doing under the auspices of neoliberal globalization and its sponsors (The World Bank, the International Monetary Fund, Wall Street, and the U.S. Treasury), the end result will be expansion in the creation of a "*contract society*" with corresponding social relationships, which are counter to the vision of a covenant community. This represents a shift from what Ferdinand Tonnies (1963) described as community to society (i.e., *Gemeinschaft* to *Gesellschaft*). As a cautionary warning to African Pentecostals who take modernity for granted, the quote below from Max Weber, based on his analysis of the process of rationalization, is worth reflecting on. The process of rationalization is championed best by neoliberal globalization, which today represents the most advanced expression of modernity. Weber asserted:

> The market community as such is the most impersonal relationship of practical life into which humans can enter with the one another. This is not due to that potentiality of struggle among the disinterested parties which is inherent in the market relationship. Any human relationship, even the most intimate, and even though it be marked by the most unqualified personal devotion, is in some sense relative and may involve a struggle with the partner, for instance, over the salvation of his soul. The reason for the impersonality of the market is its matter-of-factness, its orientation to the commodity and only to that. Where the market is allowed to follow its autonomous tendencies, its participants do not look toward the person of each other but only toward the commodity; there are no obligations of brotherliness or reverence, and none of those spontaneous human relations that are sustained by personal unions (1978: 636).

Weber's analysis here is a warning that market rationality or the process of rationalization cannot be stopped; indeed, it has no limit or end. Consequently, it can spill over into the church and faith, which can be indirectly commoditized. Moreover, the canon of rationality, once adopted established, and applied to how the churches conduct their affairs, can initiate a process of relative disenchantment, resulting in churches operating like business corporations.

In addition to Weber, there are also relevant insights from Georg Simmel's "*The Metropolis and Mental Life*," (1903) where he analyzed the social-psychological consequences of living in a metropolitan environment, which is the center of the capitalist money economy that American evangelicals and African Pentecostals take for granted. Nancy Kleniewski, a renowned urban scholar in the United States, summarized the main thrust of Simmel's argument in "*The Metropolis and Mental Life*" as follows:

> The starting point of Simmel's analysis was the observation that people living in a city must interact frequently with strangers. He thought that these frequent interactions overstimulated

> the nervous systems of urban dwellers, causing them to withdraw mentally as a kind of self-preservation technique. Simmel argued that urban interactions thus tended to be colder, more calculating, more based on rationality and objectification of others than relationships in smaller communities. (Kleniewski 2006: 28)

In spite of his analysis of the socio-psychological impact of metropolitan life over people, which comes across as negative in tone, Simmel still prefers urban over rural life due to the lack of privacy in rural communities.

The preceding observation has serious implications for the kind of theological education and training that African ministers, especially in the Pentecostal tradition, receive. Pentecostals should think not just about the Bible, miracles, and prosperity but also about the relationship between the effort to authentically live out their faith in today's world of increasing rationalization and secularization. In a rationalized world, there is more emphasis on cost-benefit analysis, and means over ends. By taking these realities for granted, Pentecostal Christianity renders itself naïve, simplistic, and oblivious to some of the major challenges that human kind faces in the twenty-first century. There is surely room for discourses within Pentecostalism, when properly understood, that can critique the contemporary social order characterized by injustice, while preparing people for eternity, which is a priority goal that they pursue seriously. Unfortunately, this is not currently the case in most African Pentecostal faith communities. One looks forward to a new generation of Pentecostal leaders who would not only preach about demonology (whatever it is), prosperity, and miracles, but also rigorously engage with the existential challenges and questions that Africa faces, and do so in faith, humility, and love, based on God's common grace to all humanity.

References

Allen, D. (1985). *Philosophy for understanding theology*. Atlanta: John Knox Press.

Anaba, E. (1985). *Elevated beyond human law: Through touch of the spirit*. Accra: Publisher Not Provided.

Anaba, E. (1993). *God's end-time militia: Winning the war within and without*. Accra: Publisher Not Provided.

Anaba, E. (1994). *Releasing God's glory from earthen vessels*. Accra: Publisher Not Provided.

Bate, P. (2004). *Congo: White king, red rubber, black death* (Audio Visual Material, Documentary). The documentary could be watched freely at the following website: http://topdocumentaryfilms.com/congo-white-king-red-rubber-black-death/. It provides concrete historical evidence and documents supporting the Belgian atrocities in the Congo. The historical evidence includes film footages of the dehumanizing treatment of the Congo people under Belgian colonial rule and archival documents.

Beisner, E. C. (1988). Prosperity and poverty: The compassionate use of resources in a world of scarcity. In M. Olasky (Ed.), *Turning point in worldview series no. 5* (p. 47). Westchester: Crossway Books.

Bell, R. H. (2002). *Understanding African philosophy: A cross-cultural approach to classical and contemporary issues*. New York: Routledge.

Blaut, J. M. (1993). *The colonizer's model of the world: Geographical diffusionism and Eurocentric history*. New York: Guilford Press.

Brouwer, S., Gifford, P., & Rose, S. D. (1996). *Exporting the American gospel: Global Christian fundamentalism*. New York: Routledge.
Brynjolfsson, E., & McAfee, A. (2011). *Race against the machine: How the digital revolution is accelerating innovation, driving productivity, and irreversibly transforming employment and the economy*. Lexington: Digital Frontier Press.
Buchanan, P. (2008, July 15). A phony crisis and a real one. *AntiWar.Com*. http://www.antiwar.com/pat/?articleid=13134. Accessed 12 Jan 2013.
Daneel, M. L. (1987). *The quest for belonging: Introduction to a study of African independent churches*. Gweru: Mambo.
Davidson, B. (1984). *Africa: The story of a continent: Program 5* (The Bible and the Gun), VHS Material. New York/Manchester: Homevision Videos.
Davidson, B. (1989). *Modern Africa: A social and political history*. London: Longman.
Durkheim, E. (2001). *The elementary forms of religious life*. New York: Oxford University Press.
Ehrenreich, B. (2011). *Nickel and dimed: On (not) getting by in America*. New York: Picador Publishing.
Ellis, S., & ter Haar, G. (1998). Religion and politics in Sub-Saharan Africa. *Journal of Modern African Studies, 36*, 175–201.
Emerson, M. O., Mirola, W. A., & Monahan, S. C. (2011). *Religion matters: What sociology teaches us about religion in our world*. Boston: Allyn & Bacon.
Eni, E. (1987). *Delivered from the power of darkness*. Ibadan: No Publisher Provided.
Eze, E. C. (1997). *Race and the enlightenment: A reader*. Hoboken: Wiley-Blackwell.
Feuerbach, L. (2008). *The essence of Christianity*. Mineola: Dover Philosophical Classics.
Fowler, R., Hertzke, A. D., & Olso, L. R. (2010). *Religion and politics in America: Faith, culture and strategic choices* (4th ed.). Boulder: Westview Press.
Froese, P., & Bader, C. (2010). *America's four Gods: What we say about God and what that says about us*. New York: Oxford University Press.
Gifford, P. (1988). *African Christianity: Its public role*. Bloomington: Indiana University Press.
Gifford, P. (2001). The complex provenance of some elements of African Pentecostal theology. In A. Corten & R. Marshall-Fratani (Eds.), *Between Babel and Pentecost: Transnational Pentecostalism in Africa and Latin America* (pp. 62–79). Bloomington: Indiana University Press.
Gornik, M. R. (2011). *Word made global: Stories of African Christianity in New York City*. Grand Rapids: William B. Eerdmans Publishing Company.
Gossett, T. F. (1997). *Race: The history of idea in America*. New York: Oxford University Press.
Green, E. C. (2011). *Broken promises: How the AIDS establishment has betrayed the developing world*. Walnut Creek: Left Coast Press.
Harrell, D. E. (1985). *Oral Roberts: An American life*. Bloomington: University of Indiana Press.
Hayek, F. A. (1944). *The road to serfdom*. Chicago: University of Chicago Press.
Hollenweger, W. J. (1999). Crucial issues for Pentecostals. In A. H. Anderson & W. J. Hollenweger (Eds.), *Pentecostals after a century: Global perspectives on a movement in transition* (pp. 176–196). Sheffield: Sheffield Academic Press.
Inkeles, A., & Smith, D. H. (1974). *Becoming modern: Individual change in six developing countries*. Cambridge: Harvard University Press.
Johnson, P. (1976). *A history of Christianity*. New York: Simon & Schuster.
Kiernan, J. (1995). African traditional religions in South Africa. In M. Prozesky & J. de Gruchy (Eds.), *Living faiths in South Africa* (pp. 15–27). Cape Town: David Philip.
Kleniewski, N. (2006). *Cites, change & conflict: A political economy of urban life* (3rd ed.). Belmont: Thomson Higher Education.
Kwilecki, S. (1999). *Becoming religious: Understanding devotion to the unseen*. London: Associated University Press.
Levitt, P. (1988). Local level global religion: The case of the U.S.-Dominican migration. *Journal of the Scientific Study of Religion, 37*, 74–89.
Marshall-Fratani, R. (2001). Mediating the global and local in Nigerian Pentecostalism. In A. Corten & R. Marshall-Fratani (Eds.), *Between Babel and Pentecost: Transnational Pentecostalism in Africa and Latin America* (pp. 278–315). Bloomington: Indiana University Press.

Martin, W. (1996). *With God on our side: The rise of the religious right in America*. New York: Broadway Books.
Maxwell, D. (2006). *African gifts of the spirit: Pentecostalism & the rise of a Zimbabwean transnational religious movement*. Athens: Ohio University Press.
Mayrargue, C. (2001). The expansion of Pentecostalism in Benin: Individual rationales and transnational dynamics. In A. Corten & R. Marshall-Fratani (Eds.), *Between Babel and Pentecost: Transnational Pentecostalism in Africa and Latin America* (pp. 274–292). Bloomington: Indiana University Press.
Mazrui, A. A. (1986). *The Africans: Triple heritage*. London: BBC Publications.
McDaniel, C. (2007). *God & money: The moral challenge of capitalism*. New York: Rowman & Littlefield Publishers.
Meyer, B. (1995). Delivered from the "Powers of Darkness:" Confessions of satanic riches in Christian Ghana. *Africa, 65*, 236–255.
Mihevc, J. (1995). *The market tells them so: The World Bank and economic fundamentalism in Africa*. London: ZED Books, Third World Network.
Mikell, G. (1997). Introduction. In G. Mikel (Ed.), *African feminism: The politics of survival in Sub-Saharan Africa* (pp. 1–50). Philadelphia: University of Pennsylvania Press.
Miller, D. E., & Yamamori, T. (2007). *Global Pentecostalism: The new face of Christian social engagement*. Berkeley: University of California Press.
Mises, L. v. (1949). *Human action: A treatise on economics* (3rd ed.). Chicago: Henry Regnery Company.
Morrson, D. G., Mitchell, R. C., & Paden, J. N. (1989). *Understanding black Africa*. New York: Irvington Publishers.
Moyers, B. (2007, October 7). Christians united for Israel. Transcript. *Public Broadcasting Service, Television Broadcast*. http://www.pbs.org/moyers/journal/10052007/transcript5.htm. Accessed 14 Mar 2013.
Moyo, A. (2007). Religion in Africa. In A. A. Gordon & D. L. Gordon (Eds.), *Understanding contemporary Africa* (4th ed., pp. 317–350). Boulder: Lynne Rienner.
Nsehe, M. (2011, June 7). The five richest pastors in Nigeria. *Forbes Magazine*. http://www.forbes.com/sites/mfonobongnsehe/2011/06/07/the-five-richest-pastors-in-nigeria/. Accessed 13 Jan 2013.
O'Dea, T. (1961, October). Five dilemmas in the Institutionalization of religion. *Journal for the Scientific Study of Religion,* 1, 30–39.
Ojo, M. (2006). *The end-time army: Charismatic movements in modern Nigeria*. Trenton: Africa World Press, Inc.
Olupona, J. K. (1993). The study of Yoruba religious tradition in historical perspective. *Numen, 40*, 240–273.
Peel, J. D. (1993). An Africanist revisits magic and the millennium. In E. Barker, J. A. Beckford, & K. Dobblelaere (Eds.), *Secularization, rationalism and sectarianism: Essays in honor of Bryan R. Wilson* (pp. 81–100). New York: Oxford University Press.
Pieterse, J. N. (2013). Globalizing cultures. In K. E. I. Smith (Ed.), *Sociology of globalization* (pp. 29–37). Boulder: Westview Press.
Ranger, T. O. (1986). Religious movements and politics in Sub-Saharan Africa. *African Studies Review, 29*, 1–70.
"Richest Pastors". (2013). Richest Pastors and Prophets in Zimbabwe. *My Zimbabwe*. http://www.myzimbabwe.co.zw/zim-gospel/1749-richest-pastors-and-prophets-in-zimbabwe-and-africa.html. Accessed on 13 Jan 2013.
Roberts, K. A. (2004). *Religion in sociological perspective* (4th ed.). Belmont: Thomson.
Robinson, J. M. (1982). *The Nag Hammadi library*. New York: Harper & Row.
Rodrik, D. (1996). Understanding economic policy reform. *Journal of Economic Literature, 34*, 9–41.
Rostow, W. W. (1960). *The stages of economic growth: A non-communist manifesto*. London: Cambridge University Press.
Scruton, R. (2006). Hayek and conservatism. In E. Feser (Ed.), *The Cambridge companion to Hayek* (pp. 208–231). New York: Cambridge University Press.

Shipler, D. (2005). *The working poor: Invisible in America*. New York: Vintage Publishers.
Shorter, A. (1970). *Cross & flag in Africa: The "White Fathers" during the colonial scramble (1892–1914)*. Maryknoll: Orbis Books.
Simmel, G. (1903). The metropolis and mental life. In K. Wolff (Ed.), *The sociology of Georg Simmel* (pp. 409–424). New York: Free Press.
Stiglitz, J. E. (2006). *Making globalization work*. New York: W. W. Norton.
Todaro, M. P., & Smith, S. C. (2003). *Economic development* (8th ed.). New York: Addison-Wesley.
Todaro, M. P., & Smith, S. C. (2012). *Economic development* (11th ed.). New York: Addison-Wesley.
Tonkowich, J. (2008). *Libido Dominandi: St. Augustine and the Lust for Domination*. Washington, DC: The Institute on Religion & Democracy. http://www.boundless.org/2005/articles/a0001674.cfm. Accessed on 19 Jan 2013.
Tonnies, F. (1963). *Community and society*. New York: Harper & Row. [Originally published in 1887 as Gemeinschaft und Gesellschaft.]
Ubani, T. (2013, January 30). Only old men go to Church. *Vanguard Newspaper*, Nigeria. http://www.vanguardngr.com/2013/01/only-old-men-go-to-church/?utm_source=feedburner&utm_medium=email&utm_campaign=Feed%3A+vanguardngr%2FdIeb+%28Vanguard+News+Feed%29&utm_content=Yahoo!+Mail. Accessed 31 Jan 2013.
Van Dijk, R. (2001). Time and transcultural technologies of the self in the Ghanaian Pentecostal diaspora. In A. Cotren & R. Marshall-Fratani (Eds.), *Between Babel and Pentecost: Transnational Pentecostalism in Africa and Latin America* (pp. 216–234). Bloomington: Indiana University Press.
Vuha, A. K. (1993). *The package: Salvation, haling and deliverance*. Accra: No Publisher Provided.
Walls, A. F. (1991). World Christianity, the missionary movement, and the ugly American. In W. C. Roof (Ed.), *World order and religion* (pp. 147–172). Albany: State University of New York Press.
Wasamu, M. (2013). African pastors lead crusade for circumcision: Religious leaders join effort to circumcise 20 million adult men in flight against HIV. *Christianity Today.Com*. http://www.christianitytoday.com/ct/2012/december/circumcision-crusade.html. Accessed 13 Jan 2013.
Weber, M. (1946). *From Max Weber: Essays in sociology* (Trans. and introduction by H. H. Garth & C. Wright Mills). New York: Oxford University Press.
Weber, M. (1978). *Economy and society* (G. Roth & C. Wittich, Eds.). Berkeley: University of California Press.
Weber, M. (2001). *The Protestant ethic and the spirit of capitalism*. New York: Routledge.

Chapter 98
Negotiating Everyday Islam After Socialism: A Study of the Kazakhs of Bayan-Ulgii, Mongolia

Namara Brede, Holly R. Barcus, and Cynthia Werner

98.1 Conceptualizing Multiple Scales of Islam

As the beginning of a new century dawned in Central Asia, so too did new manifestations of ideas and practices related to nationalism, religion, homelands and identity. Across the region, the formation of new nation-states has produced new questions about post-Soviet identities. One dimension of change is the renewal and reconfiguration of religious identities and practices. Long suppressed by Soviet policies, these identities and practices are now taking on new forms, spanning both the heterogeneity of local and traditional beliefs and practices as well as a broader, global or pan-Islamic movement (McBrien 2009; Montgomery 2007). Caught at the cross-roads of change, people living in once remote places, such as Bayan-Ulgii *aimag* in western Mongolia, are now engaged in a global negotiation of Islamic practice and identity. Keeping in mind the uniqueness and locality of Bayan-Ulgii as well as its connections with and similarities to other Turkic Central Asian communities, this chapter seeks to fill the gaps in scholarly knowledge about Islam among Mongolian Kazakhs and relate this information to ongoing conversations about the role of religion in the regional and cultural geographies of Central Asia.

Specifically, we consider increased levels of adherence to universal practices associated with Islam such as almsgiving and *namaz* and the endurance of local Islamic practices such as shrine visits, which are increasingly challenged in the

N. Brede (✉) • H.R. Barcus
Department of Geography, Macalester College, St. Paul, MN 55105, USA
e-mail: nbrede@gmail.com; barcus@macalester.edu

C. Werner
Department of Anthropology, Texas A&M University, College Station, TX 77843, USA
e-mail: werner@tamu.edu

S.D. Brunn (ed.), *The Changing World Religion Map*,
DOI 10.1007/978-94-017-9376-6_98

transnational religious dialogs, yet remain as key forms of identity and practice at local scales (Louw 2006). We argue specifically that religious practices in Central Asia, as illustrated by this case study, are polyvocal and continually shifting, and much more complex than that suggested by simplistic dichotomies used by some analysts: "orthodox" vs. "local," "official" vs. "folk," and "high" vs. "low" Islam. Our case study of religious practices in western Mongolia provides further illustration of how the meanings and moralities associated with Islamic practices and rituals throughout Central Asia are being contested in post-socialist spaces (Abashin 2006; Rasanayagam 2011).

In this chapter we utilize primary ethnographic data collected during the summers of 2008 and 2009 to explore the ways in which Kazakhs in Bayan-Ulgii *aimag* negotiate Islamic practice and identity in an increasingly globalized context. We integrate an interdisciplinary and critical approach to religion, using the geographic concept of scale in order to describe the intersection of global, national, and local discourses and how these shifting influences are incorporated into religious identities and experiences in daily life. We consider how the Kazakhs in Bayan-Ulgii, similar to Muslims in other regions of the world, understand Islam and "Muslimness" in unique ways as these understandings are constantly deconstructed and reconstructed through discursive power and social dynamics. With this in mind, Islam in Bayan-Ulgii is shown to be a polyvocal and dynamic symbol with many meanings and also a cornerstone of local, national, and transnational identities for Mongolian Kazakhs (Privratsky 2001).

The remainder of this chapter is organized into six additional sections. Section 98.2 provides a comparative historical overview of Islam in Central Asia that is followed by a discussion of our theoretical underpinnings in Sect. 98.3. Sections 98.4 and 98.5 provide a description of our study area, data and methods. In Section 6 we offer findings, concluding with Sect. 98.7.

98.2 Comparative Historical Overview of Islam in Central Asia

Central Asia is an enormous region with a complex history and its limits are difficult to define; it is often viewed as encompassing only the post-Soviet republics of Kazakhstan, Kyrgyzstan, Tajikistan, Turkmenistan, and Uzbekistan (Levi 2007: 15). Mongolia differs from these states culturally, historically, and politically, and was never formally part of the Soviet Union. Despite these differences, western Mongolia is included in broader definitions of Central Asia because its inhabitants share cultural, national and religious traditions, connections, and historical experiences with other Turkic Central Asians (for example, Brunn et al. 2012; Dani and Masson 1992). While locally experienced and practiced, Islam in Central Asia shares some commonalities across historic time periods and geographic regions. We offer here a brief overview of Islam in the region in order to provide historic and geographic context for our case study.

98.2.1 Islam in the Pre-socialist Period

Although Islam is the predominant religion in the contemporary period, a number of different religions, including Zorastrianism, Buddhism, Tengrism, Manichaeism, and Shamanism, have played important roles at various times in the history of Central Asia (Foltz 2000). Islam was first introduced to Central Asia in the eighth century, with the expansion of the Abbasid Caliphate and the conversion of ruling elites in cities like Bukhara, Khorezm and Samarkand. In the following centuries, these cities, already great hubs of trade, art, science and medicine, also became renowned as centers for Islamic practice, architecture, and thought (Levi 2007: 20). The religious authority of the clerical *Ulama* and ascetic Sufi orders was highly influential in social and political life even into the early Russian period, maintaining social order through legal and religious codes of *shariat* and *adat*. Pre-modern Islamic discourse was fluid, multivocal, and integrated into the emergent dynamics of local societies (Abdullaev 2002: 246). Communities like Bukhara encompassed a range of Islamic understandings that nevertheless came together around a perceived common religious identity, "living according to both normative and heterodox values" (Privratsky 2001: 74). The spread of Islam to nomadic peoples including the Kazakhs and the Kyrgyz was more gradual, and conversions were generally led by Sufi sheikhs and dervishes (Heyat 2004: 277). Although nomadic and settled people of Central Asia adopted Islamic practices, they kept elements of pre-Islamic faiths, such as Zoroastrianism and Shamanism; this syncretism continues to play a role in the daily life of Central Asians up to the present (Levi 2007: 18).

Pre-socialist Islamic practices, especially among nomadic groups occupying the Central Eurasian steppe, were poorly documented, yet a few generalizations can be made based on writings by Islamic scholars and Western travelers. In settled communities and cities, many religious activities centered on mosques and included completion of the five pillars, pilgrimages to local saint shrines like Turkistan, and formal Qur'anic study (Khalid 2007: 183). During this period, the public performance of Islam was a largely male activity; women tended to remain in the domestic sphere and wore heavy cotton *paranji* robes and horsehair veils called *chachvon* (Northrop 2007: 95). For nomadic groups like the Kazakhs and the Kyrgyz whose religious activities could not be organized around a stationary mosque, devotional exercises focused on life-cycle rituals like circumcision, marriage, burial and holiday feasting (Adams 2007: 200).

98.2.2 Religion in the Socialist Period

In the wake of the Bolshevik revolution in 1917, Russian imperial territories were restructured into Soviet Republics, a process that provoked political and social upheavals that changed everyday life and everyday piety in Central Asia. Inspired by Marxist attitudes towards religion, Stalinist-era purges disrupted local religious practices and institutions through the killing or silencing of political and clerical elites and the religious systems in which they participated (Khalid 2007: 99). At the same time, the Bolsheviks set up ethnic-based territories through a process that involved the rhetorical and structural imposition of essentialist differences between groups whose linguistic and cultural identities had been historically fluid (Levi 2007: 30). Although Mongolia has been independent since its separation from China in 1921, its communist policies between 1924 and 1990 also led to social restructuring and the repression of Buddhism and Islam. The repressions were particularly strong under Choibalsan, a leader who is frequently compared to Stalin (Diener 2007).

The socialist assault on "religion," "tradition" and "backwardness" began in 1926, and was paralleled by similar policies in Mongolia under Choibalsan (Diener 2007; Khalid 2007). Mosques were destroyed, the courts and *madrasas* (Islamic schools) of the *Ulama* were closed, and religious experts were killed, imprisoned, deported to labor camps, or forced into hiding (Khalid 2007: 71). Many religious practices that did continue were curtailed, restricted to groups of "believers" who had to register with the state. Such groups were forbidden to do charity work, organize community activities or do anything other than operate places of worship. These policies not only restricted religion, but also imposed specific imperialist discourses. As Khalid writes, "the assumptions about religion that underlay the law—that it was [exclusively] a corporate enterprise undertaken by believers coming together in tangible organizations—derived from Christianity, but were now extended in Soviet practice to all religions" (2007: 73). All religions were similarly condemned by the state, which actively propagated an ideology of scientific atheism (Kendzior 2006). Given that social benefits accrued to those who affiliated with the Communist Party, public displays of religion were risky. During our interviews, we were repeatedly told that people used to be afraid to practice their religion in visible ways. In practice, however, people throughout Central Asia found creative ways to continue local religious practices, such as disguising circumcision parties as birthday parties (Kehl-Bodrogi 2006: 249). Similarly, several of our informants described how elderly relatives would secretly pass on religious knowledge to the younger generation at home. Despite such efforts to maintain religious knowledge, many of the Mongolian Kazakhs we interviewed noted with regret that their knowledge of Islam is limited and imperfect due to the repression of Islam under socialism. As one informant explained, "the only Muslim practices that we Kazakhs did during socialism were shrine visitation, burial, *Kurban Ait*, and circumcision."

98.2.3 Religion in the Post-socialist Period

After the 1991 demise of the Soviet Union, Central Asian populations experienced another wave of radical and highly varied economic, social, and religious changes. Throughout the region, Central Asian Muslims have been rediscovering their national and religious identities (Abashin 2006; Fathi 2006; Heyat 2004;

Kehl-Bodrogi 2006; Roberts 2007: 339). Throughout the former socialist bloc, one of the most noticeable changes has been the explosion of foreign influences—including Islamic missionaries who provide spiritual advice, Islamic organizations who offer educational opportunities for young people, and wealthy Muslims who provide financing for the construction of new mosques (Ghodsee 2009; Liu 2011: 121). In response, state policies—from the relatively noninvasive approach of Kazakhstan to the intense repression and isolation of Turkmenistan—have been particularly important in determining the way that people "return" to Islam. On the one hand, in the name of national security, states like Uzbekistan have continued to suppress what leaders perceive to be "extremist" Islam in ways that have probably caused more terror than the groups they claim to be controlling, whose aims are usually peaceful and community-minded (Adams 2007; McGlinchey 2011). On the other hand, in states with relatively greater religious freedoms (notably Kazakhstan, Mongolia, and Kyrgyzstan), there has been a tremendous resurgence in Islamic practice in both traditional/local and new/transnational forms (Privratsky 2001). In addition to differences at the state level, individual practices have continued to vary depending on gender, age, resources, and social networks (Abashin 2006; Fathi 2006; Heyat 2004; Roberts 2007).

98.3 Conceptual Framework: Considering the Role of Scale

Fragner (2001) argues that interconnected local, national, and transnational categories are useful in describing patterns of sociopolitical identity formation. Using the analogy of a Russian *matrioshka*-doll, he describes how identities are acted out at the local level, incorporating transnational, regional and local influences within themselves in an "ultimately changing pattern" (Fragner 2001: 342). His discussion of the role of scale in Central Asian regionalism and nationalism is an idea that can be equally well applied to religious identity. Abdullaev's geographic study of Central Asian Islam, though political in focus, advances Fragner's analysis and applies it to religion. Utilizing the idea of a "cultural nexus," he describes ways in which multiple scales of influence are acted out on the stage of intra-communal power dynamics (Abdullaev 2002). In this essay we use Fragner's concept of scale as a framework for understanding the diverse and evolving religious lives in Bayan-Ulgii, Mongolia, and the interconnectedness of multiple scales of religious practice and religious influences in this region of the world.

At the broadest scale, religious transnationalism has both historic roots and contemporary manifestations. The idea of the *Umma*, or universal community of believers, has been present in Islamic thought since the time of the Prophet Muhammad, and is a common theme in Muslim scholarship (Geertz 1971). Universalistic discourses and movements, though by definition transnational in scope, have specific contexts, histories, and aims and operate differently in different communities (Khalid 2007; Fragner 2001). Because of the paucity of connections to other Islamic countries and schools of thought during the socialist period, the resurgence of transnational ties to Islam and Islamic identity is a relatively new phenomenon in post-socialist Central Asia. Nevertheless, transnational discourses are an increasingly powerful influence on local religious experience as foreign organizations and internationally educated imams and religious professionals have become primary religious authorities in the region (Roberts 2007).

In the contemporary period, religion, nationalism, and state power are interconnected in Central Asian countries. The national scale of religious life can be seen in the ways that governments approach the subject of religion and the ways that citizens employ religion in discourses of national identity. The socialist division and isolation of Central Asian republics led many people to consider "Muslimness" a subset of national culture rather than a potentially transnational identity, inextricably fusing national and religious identities (Khalid 2007: 83). Though overt state displays of power are not present in Mongolia, the Kazakh inhabitants of Bayan-Ulgii aimaq are nevertheless connected to Islam on a national scale through their linguistic and cultural ties to Kazakhstan, as well as personal ties to Mongolian Kazakhs who migrated to Kazakhstan since 1991 (Diener 2007; Finke 1999). It is important, therefore, to understand the roles of nation, state and Islam in considering the position of Mongolian Kazakhs as a religious and ethnic minority in Mongolia.

Although national and state power dynamics and discourses are clearly important in shaping religious experiences, local religious experiences are also shaped by tensions between transnational forms of Islam and local forms of Islam. While national and transnational influences also come into play in local Islam, "unofficial" locally normative Islamic forms can vary greatly from place to place, having long histories intertwined with pre-Islamic traditions, pre-socialist religious discourses, and efforts to retain religious heritage in the face of socialist repression (Privratsky 2001). The new generation of Islamic authorities is becoming increasingly critical of what they consider to be heterodox practices and rituals, such as shrine worship and ancestor worship (Fathi 2006).

Questions of authenticity and syncretism remain at the forefront of these highly charged discussions of locally accepted practices and ideas. Whereas traditional Islamic authorities like *moldas* commonly recognized practices such as visiting the shrines of saints or family ancestors, tying ribbons to sacred trees and lighting candles for the dead, many modern advocates of transnational Islamic ideas consider such activities to be heretical pre-Islamic survivals (Abramson and Karimov 2007: 325). Western scholars often take a middle ground between these positions and assert that, while some traditions have pre-Islamic origins, they are nevertheless integral to local conceptions of Islam and "Muslimness" (Khalid 2007; Rasanayagam 2011)

The reality of local post-socialist religious change is far more complicated than a collective "return" from atheism or a gradual transformation from "Islam without belief" to a more unitary Sunni faith (Roberts 2007). Broader discussions of Islam in Central Asia incorporate both "scripturalist universalism" and shrine offering rituals passed on by grandparents (Roberts 2007). Moreover, different people and communities appropriate the polyvocal symbols of Islam into their beliefs, practices and identities in unique ways. For example, weddings that might initiate with sacred ceremonies overseen by imams often end in parties with open bars.

While some people fast during Ramadan and go on *Hajj* to Mecca, others toast the Prophet with a shot of vodka at *Kurban Ait* (Montgomery 2007). In Kazakhstan and elsewhere, it is common to see women in miniskirts walking side by side with friends wearing *hijab* headscarves, and for both to consider themselves to be Muslims (Abazov 2007: 73).

Throughout the region, religious expression is contextualized historically and socio-politically and mediated through the lenses of states, imams, traditional authorities, family members, and other sources of influence (Roberts 2007: 352; Abdullaev 2002). Central Asians, by and large, view Islam as an ascriptive element of their ethno-national and local identities, a taken-for-granted category that is nevertheless being rethought in light of new teachings imported from countries like Saudi Arabia and Turkey (Montgomery 2007).

98.4 The Study Area: Bayan-Ulgii *Aimag*, Mongolia

The predominantly Kazakh-speaking Mongolian *aimag* (province) of Bayan-Ulgii forms part of Mongolia's Western Region, alongside neighboring Khovd and Uvs *aimags* (Fig. 98.1). This high, arid, landlocked area of the Altai Mountains shares borders with Russia and China, and is separated from Kazakhstan to the west by a 60 km (36 mi) strip of Chinese-Russian border. Many inhabitants of the region, both Kazakh and Mongolian, are engaged with semi-nomadic livestock production as a means of subsistence and a primary source of income (Finke 1999: 104; Diener 2007). In 2000, Bayan-Ulgii was home to 80,776 Kazakhs, who represented around 88.7 % of the *aimag* population (MNSO 2001).

At 4.3 % of the country's total population, the region's Kazakhs are the largest ethnic minority in Mongolia, a country dominated by the Khalkha Mongol ethnic group. Ulgii, a city of around thirty thousand inhabitants that straddles the Khovd River, is the provincial administrative center and a hub of social, economic, and religious organization for Mongolia's Kazakhs. Like other Mongolian *aimags*, Bayan-Ulgii's peripheral areas are divided into smaller territorial units called *soums*, each of which is organized around a small central town providing commercial and community services and a boarding school for the children of herders. In recent years, new mosques have been constructed in these *soum* centers (Finke 1999: 107). Despite their history of mobility, Mongolian Kazakhs are situated in "a peripheral unit in a peripheral region of a peripheral state;" their lives, after the fall of socialism, were plagued with problems of transportation, limited connectivity, and economic underdevelopment (Diener 2007: 164). The shift from communist to democratic political systems and the transition to a capitalist rather than a command economy influenced every aspect of daily life. Though the Kazakhs in Bayan-Ulgii benefited from increased religious and personal freedom and mobility, poverty rates increased from 1 % in 1989 to 33 % in 1998, and basic infrastructure, social services, and healthcare fell victim to budgetary crises (Portisch 2006: 3). By the mid to late

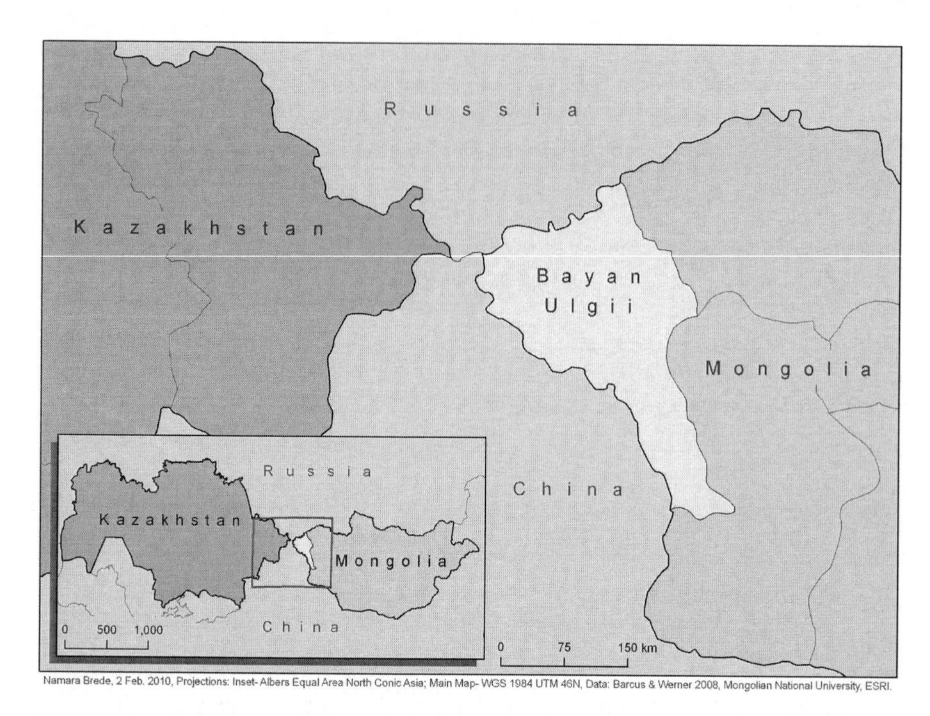

Fig. 98.1 Bayan-Ulgii *aimag*, Mongolia (Map by N. Brede)

1990s there were some improvements as consumer goods from China and other countries became much more readily available. In addition, trade and tourism increased as the Mongolian government worked out new political and economic agreements with trade partners in Asia, Europe and America (Barcus and Werner 2010; Diener 2007; Werner and Barcus 2009).

While these new transnational connections, development, and social changes have brought significant changes to this *aimag*, the overall population of Bayan-Ulgii has remained small and the Kazakhs have maintained a strong ethnic identity. Due to the region's relative isolation, the World Association of Kazakhs declared in 1992 that the Kazakhs of Bayan-Ulgii had retained their "traditional culture and language" more effectively than any other Kazakh community inside or outside of the territory of Kazakhstan (Diener 2007: 159).

Influenced by post-socialist nationalist sentiments as well as geopolitical considerations, the newly independent state of Kazakhstan instituted policies supporting diasporic Kazakhs (*oralman*) to "return" to their "homeland." With these ideological and economic incentives, over 60,000 Mongolian Kazakhs migrated to Kazakhstan in the early 1990s (Diener 2007: 160). Our research on Kazakh migration from Bayan-Ulgii has shown that many factors are implicated in this process, making it more than a simple out-migration from an undesirable area to a desirable one (Barcus and Werner 2007: 9). While some migrants have benefited from the move to Kazakhstan, others have experienced a variety of challenges, including discrimination from Kazakhstani Kazakhs, linguistic limitations with Russian, distance from family, difficulty adjusting to a different climate, and homesickness. As a result, approximately one-third of these early migrants have returned to Mongolia (Diener 2007; Barcus and Werner 2007), Despite the return migrations, Bayan-Ulgii has experienced a notable "brain drain," as only 30 % of people educated in Ulaanbaatar or abroad return home, and many families see the availability of secondary and post-secondary education as one of the primary incentives to migrate (Barcus and Werner 2007: 10; Diener 2007: 168).

In light of these changes, and the region's relatively benign political stance towards religion, Bayan-Ulgii provides an interesting case study of post-socialist religious, social and cultural development. Islamic education is the one exception to this rule of educational out-migration; most Mongolian Kazakhs educated in foreign *madrasas* return to teach in Mongolia (Diener 2007). These foreign-educated Islamic experts are instrumental in bringing transnationally normative viewpoints and discourses to Mongolian Kazakh Islam. While a few studies make passing reference to religion and religious practices in Bayan-Ulgii (see for example Barcus and Werner 2007; Portisch 2006), only Finke (1999) provides a somewhat substantive account, although this was not the primary focus of his study. This essay, therefore, provides a unique look at how everyday Islamic practices have been reinterpreted and reconfigured in western Mongolia following the demise of socialism and the influx of external Islamic influences.

98.5 Data Collection and Methodology

This investigation is part of a larger field study addressing migration decisions of Mongolian Kazakhs in a post-transition context. This broader study (see Barcus and Werner 2007, 2010; Werner and Barcus 2009 for more details) is based on a mixed-methods approach that combines a questionnaire based on a stratified snowball sample with ethnographic approaches, such as semi-structured interviews and participant observation. One component of the study considers how religious practices, such as shrine visits, reinforce place attachment, and therefore contribute to immobility. We use semi-structured interviews and participant observation to examine religious practices in western Mongolia. During June and July of 2008, we conducted 28 "life history" interviews with Kazakhs living in western Mongolia. Interview questions covered a range of topics, including education, work history, health, family, and religious practices. Our sample included men and women from different social backgrounds, ranging in age from 18 to 75. During June and July of 2009, we conducted an additional 21 semi-structured interviews with lay informants and 10 semi-structured interviews with imams and religious professionals. The majority of these interviews were conducted in the Kazakh language, though a small number of interviews were conducted in English (a language that many young people are now studying). An ethnographic approach to data collection was employed as ethnography offers logistical and epistemological advantages over more extensive techniques, "allowing interviewees to construct their own accounts of their experiences by describing and explaining their lives in their own words" (Valentine 2005: 111). The time spent living with families in western Mongolia and working with our local research assistants has also informed our understandings of local religious practices. Such interactional knowledge is useful when asking questions about small-scale social processes and their connection to multi-scalar networks.

98.6 Pious Practices, Places, and Prohibitions

98.6.1 Newly Constructed Mosques as Sites for Changing Religious Practices in the Post-socialist Period

In general, Islamic religious practices in Bayan-Ulgii have changed significantly since the end of socialism and the society as a whole is becoming more diverse in terms of religiosity. As new beliefs and practices are introduced through increased contact with the broader Muslim world, some Mongolian Kazakhs, including some of the newly trained religious authorities, are challenging "traditional" practices, while other Kazakhs are simply blending new and old beliefs and practices.

The construction of new mosques in the town of Ulgii and the *soum* centers is certainly one of the most visible signs of an Islamic revival in western Mongolia. Finke (1999) observed this phenomenon during his research in the late 1990s, and our ethnographic research suggests that religious activity and mosque construction has continued through the late 2000s (Fig. 98.2). Interviews with imams in Ulgii and surrounding *soums* revealed that there are now over 30 mosques in Bayan-Ulgii, financed by governmental and private donations from Turkey, Saudi Arabia, France, and other countries. In light of this development, and the importance that our respondents placed on mosques, this section explores six religious activities centered on these important locations.

As centers of piety, mosques are versatile, providing many services and acting as hubs for religious education and new religious practices like *namaz* (the daily recitation of prayers) *zhuma namaz* (Friday prayers), and almsgiving. In addition they host activities like circumcision rituals, weddings, and *Kurban Ait* celebrations that historically were performed at home, in secret, or in other places. This co-option of religious authority and religious space by new institutions represents a change in public and private religious expression in Bayan-Ulgii.

Fig. 98.2 A newly constructed rural mosque (Photo by H. Barcus, 2009)

98.6.2 Namaz: A Daily Obligation

Across all informants, the most frequently mentioned standard for overall piety was the five-times daily *namaz* prayer. By the late socialist period, only a small percentage of the population recited prayers five times a day. This is changing in the post-socialist period. According to one of our interlocutors, Shokan, "if you pray *namaz*, you are a very religious person." Finke also regards *namaz* as a benchmark, claiming, "the number of people praying regularly is very small ... the same is true of fasting and giving alms" (1999: 136). Though more people pray now than during Finke's fieldwork, the idea that *namaz* is a predictor for other pious activities—particularly those performed in mosques—appears well founded: all seven respondents who claimed to pray *namaz* also fasted during Ramadan and gave alms. Similarly, McBrien (2009) notes a strong correlation in Kyrgyzstan between women who recite prayers five times a day and women who wear *hijab*.

During our interviews, we asked people to talk about how their own religious practices compared to other family members. Although it is difficult to make demographic generalizations based on a relatively small sample size, our research suggests that those who pray daily tend to be either elderly or young, and are predominantly male. Of the seven people interviewed for this study, Saparzhan and Amantai were older rural men (66 and 75 years of age) who had prayed since 2003 and 1999 respectively. Both were the only members of their households other than their grandchildren to do so, and both felt an obligation to pray "for their families." In contrast, Shokan, Khanmurat and Temir were young men between 18 and 21 who lived or studied in urban areas and began praying independently of their parents in late high school. Finally, Alia and Altingul were the only two women in the study who recited prayers daily. Alia was an elderly woman who claimed that she had prayed throughout the socialist period, while Altingul was a 40 year-old *madrasa* teacher who started to recite prayers in 1994.

During interviews with imams, we were told that the majority of individuals who show up to recite daily prayers at the mosque are younger and older men. The imams, however, acknowledged that the domestic nature of most prayer (except for *zhuma namaz*) makes it difficult to reach any definitive conclusions about the demographics of the most pious Muslims. During *zhuma namaz* at the Ulgii Central Mosque in June 2009, around 200 people were in attendance, and most were men in their teens and twenties. While several older men were also present, less than a dozen men between the ages of 30 and 50—with the exception of the imam and his assistant—were present. As in most Sunni mosques, women were separated from men during prayer and given a separate, smaller space to pray, emphasizing the gendered nature of the mosque as a religious space (Roberts 2007; Khalid 2007). Women who attend prayers at mosques are expected to don a headscarf inside the mosque; some, but not all, opt to wear *hijab* at other times as well (Fig. 98.3).

Despite the demographically limited nature of *namaz*, respect for the practice is widespread. Even secular Kazakhs consider religion important to their national

Fig. 98.3 Young women and girls studying the Qur'an at a local mosque. Many women and girls only wear the headscarf (*hijab*) for certain occasions, such as these lessons (Photo by N. Brede, 2009)

identities. One of our interlocutors, Muratbai, does not pray himself, but he has praise for those he considers to be "very religious" (Montgomery 2007). A short vignette from an Ulgii park adjacent to the Central Mosque illustrates this point. At dusk, the park was filled with people of all ages sitting with friends, talking, and drinking beer. The moment the call to prayer was heard, however, all activity ceased. None of the occupants of the park made signs of praying *namaz*, but the respectful silence served as a reminder of the increasingly public nature of Mongolian Kazakh Islam. Both newly practicing and more traditional Kazakhs seem to welcome this shift; however, it is primarily an urban phenomenon. Rural areas, with fewer mosques, fewer young people, and more traditional practices, experience different changes (Finke 1999).

98.6.3 Almsgiving: A Popular New Form of Piety

Like *namaz*, almsgiving, or *zakat*, is a religious activity that seems to have been less common during socialist years in Bayan-Ulgii. Unlike *namaz*, however, the third pillar of Islam enjoys widespread popularity today. All but two of our informants had given alms at some point, and most did so at least once a year. The month of Ramadan was considered to be the most appropriate time to give alms. Over half of our informants also mentioned giving alms at other times of the year, usually as a way to ensure good luck for a special occasion, such as a journey. Before each of our trips to rural field sites, for example, our Kazakh research assistants would visit the Ulgii Central Mosque to give alms and receive a blessing for the trip from the imam.

Almsgiving takes several forms, demonstrating the local variety of even new and internationally normative pious practices. Whereas most informants gave small amounts of money (usually a few thousand Mongolian *togrog*, or between 1 and 5 USD), some rural respondents gave sheep instead. This was particularly common among herders, whose primary assets are livestock, rather than cash (Diener 2007; Finke 1999). Livestock alms were often given by families or groups of families rather than individuals in order to offset the high value (around $US 40) of an animal. Imams often preferred such donations as well, since animals provided extra food to give to the poor and facilitated animal sacrifices during *Kurban Ait* festivities at the mosque (Alibek).

Despite the popularity of almsgiving, disagreements over imams' use of alms illuminate issues of authority and trust between some Mongolian Kazakhs and Islamic professionals. One elderly rural informant, Muratkhan, explained how he preferred to donate a sheep to the local mosque during Ramadan, but his middle-aged sons argued that he should give directly to the poor instead because they did not trust the imam. An interview with the imam who ran that particular mosque presented the other side of the story, and partially confirmed the suspicions of Muratkhan's sons: while a portion of the alms were used for charitable purposes (as intended by the concept of *zakat*), most of the resources were used for building upkeep and salaries. Despite these local tensions, most of the people we interviewed were happy to support almsgiving, and felt their donations fulfilled an important obligation and brought them blessings and good luck.

98.6.4 The Holy Month: Practices During Ramadan

The majority of the individuals we interviewed described how they became more pious during the Islamic holy month of Ramadan by fasting, praying, and almsgiving. They also described how they go without sleep on the 27th night. All but three of our 31 informants participated in some of these religious activities relating to Ramadan, and over half fasted from dawn to dusk each day. Of those who fasted during the holy month, most also recited special *tarawih* prayers at a mosque. For several, this was the only time they engaged in organized prayer. According to the imams we interviewed, *tarawih* prayers are the most popular prayers at mosques throughout Bayan-Ulgii, with attendance higher than even *zhuma namaz*. Along with *tarawih* prayers, informants of all ages in both rural and urban settings said that they stayed awake all night on the 27th night of Ramadan. According to one source, this was a popular practice for Central Asian Muslims even during socialism (Adams 2007).

Ramadan practices have a long socialist and pre-socialist history, and gendered social tendencies developed over this time can be seen in modern forms of piety. Based on our research, men seem more likely to recite *namaz*, especially in visible places such as mosques, while women seem more likely to fast during Ramadan. The feasts and celebrations used to break fast, and the prayers associated with Ramadan, were performed primarily at home—the traditional locus of feminine piety (Northrop 2007). Though most of our interlocutors started to pray after the transition, the case of Kulpynai, a Khovd University student, illustrates the continued importance of socialist-era Ramadan practices for gendered, ethnic and intergenerational identity. Growing up, Kulpynai was impressed with her mother and grandmother's maintenance of "Kazakh traditions" by fasting during Ramadan. In fifth grade, years before she went to the mosque to learn from the imam, her mother taught her to fast and also instructed her in other "Kazakh Islamic values." Although her familial and clerical education complemented each other, she said that she was somewhat embarrassed to admit that she learned more from her mother than from the imam.

In one sense, Kulpynai's religious experience is reminiscent of the Tajik Communist official who enlisted his female family members to fast and pray for him by proxy in order to be a "good Tajik" (Khalid 2007). In both cases, Ramadan practices were performed and learned in a feminine, domestic space, and evoked both religious and ethnic identity. In Kulpynai's case, however, socialist-era secrecy has given way to what W.E.B. Du Bois might call a "double consciousness … of two warring ideals in one [body]," a sense of self-denigration also expressed by other women who learned religious practices and discourses at home (Du Bois 1903: 45). Even as she continued to employ traditional, domestic, and feminine approaches to piety, Kulpynai acknowledged these approaches as inferior to the transnational discourses of imams. In spite of such internal and external contradictions, however, she enjoyed fasting during Ramadan "because it is good to fast with other Muslims"—a case of practice enforcing *communitas* with other Mongolian Kazakh Muslims regardless of religious orientation.

98.6.5 Kurban Ait: Everyone's Favorite Festival

With regard to *communitas*, ethnic identification, and the integration of local and large-scale religious discourses, few Islamic practices are locally revered more than *Kurban Ait*, celebrated on the tenth day of the Islamic month of *Dhu al-Hijjah* (Adams 2007; Montgomery 2007). Though celebrations of *Kurban Ait* (also known as *Eid al-Adha*) vary in length between Muslim countries, in Central Asia and Mongolia the feast lasts at least 3 days and involves ritual animal sacrifice, *toi* parties, and family gatherings (Adams 2007). More than any of the previously mentioned religious activities, *Kurban Ait* was widely practiced in Bayan-Ulgii during the socialist period. Finke's observations suggest that *Kurban Ait* was already popular in the early 1990s, alongside non-Islamic festivals like Central Asian New Year celebration *Navruz* (linked to Zoroastrianism) and the Mongolian national festival *Naadam* (1999: 137).

The popularity of *Kurban Ait* has increased since the 1990s. When we asked people when they started to adopt various religious practices, we noticed a trend in which *Kurban Ait* usually predated other practices, such as reciting *namaz* and fasting during Ramadan, by several years, and participation in the religious feast was not correlated with other forms of piety. The rural herder Saparzhan did not begin praying *namaz* or visiting ancestral shrines until his grandson told him to do so in 2003, yet he recalls sacrificing a sheep on *Kurban Ait* every year during socialism. Eighteen year-old Shokan, for instance, also recalls how his parents had sacrificed a sheep every year during *Kurban Ait* even before he was born, however he only began studying Islam and praying *namaz* at the behest of his older siblings in 2007. During socialism these sacrificial sheep were used to prepare food at home. Although *Kurban Ait* had an established presence during the socialist years, it has been integrated into new practices of mosques and *madrasas*. According to Aybek, the president of the Mongolian Islamic Association (MIA), every mosque that can afford to do so sacrifices between one and a dozen sheep each year for *Kurban Ait* and gives the meat to the poor. Most of these sheep are donated by members of the congregation who provide the sheep as part of the Muslim obligation to give alms, if able to do so. International and local donors have been known to provide funds to purchase sheep for these festivities. As such, a celebration that was secretly practiced during socialism has become much more visible and public in the post-socialist years. By adopting and promoting this practice, religious authorities including the MIA have further increased the popularity of an already popular Islamic practice in Bayan-Ulgii.

This expansion of a private celebration to a public sphere controlled by new religious authorities has implications in terms of content as well as form. Throughout many parts of Central Asia, families continue to celebrate *Kurban Ait* at home, yet new prayers, forms, and ideas are added to the old celebration while certain acts, like drinking vodka during the *Kurban Ait* feast, are falling by the wayside among many families (Roberts 2007). These new elements may represent what Deweese calls the "re-Islamization" of practices by "figures intent upon a more rigorous observance of Islamic rites and a clearer evocation of Islamic doctrine…to reassert their authority in an environment from which they had been largely excluded in socialist times" (Deweese 1995: 3). *Kurban Ait*, like all of Bayan-Ulgii's Islamic practices, is undergoing discursive and structural changes as a result of changing sources of authority. Nevertheless, the practice is unique in its endurance.

98.6.6 Circumcision: A Powerful Marker of Identity

Like other Central Asian Muslim groups, Mongolian Kazakhs consider circumcision an important marker of ethnic and religious identity, and the significance of the practice during both socialist and post-socialist eras has been well-documented (Finke 1999). Privratsky discusses attitudes toward the custom at length in his exploration of "collective memory" in Kazakhstan, noting that most families circumcised their sons, and those that did not were scorned as betrayers of ethnic unity (2001: 63). Finke confirms the importance of circumcision for Mongolian Kazakhs, asserting that, "the only Moslem [sic] custom rather strictly observed in socialist times was circumcision" (1999: 136). Our interviewees verified the continued importance of circumcision and also indicated that this is another example of a religious practice that has moved from the domestic sphere to the public sphere, and is now closely affiliated with mosques, religious authorities, and transnational religio-cultural influences.

Our interviews suggest that circumcision was an almost universal practice among Kazakhs of Mongolia. Only one individual, an Ulgii man in his mid-twenties, claimed to be uncircumcised. Most informants (or their male relatives) were circumcised as children, though the age at circumcision varied greatly. Several young men, both urban and rural, claimed to have been circumcised as babies; one Ulgii woman said she circumcised her sons at age 6, and one college-aged informant described how his brother was circumcised as a teenager in 1990. Our interviews suggest that this variability is best explained by the fact that the procedure was difficult to perform during socialism. For example, one rural informant, who worked as a mining engineer, said he had to secretly circumcise his sons at home in the middle of the night because he would be fired if his supervisors found out. Ironically, he noted that that the supervisors were also Kazakh, and also circumcised their sons in secret.

Compared to the socialist period, modern circumcision is easier and more available—particularly in Ulgii, where doctors perform the procedure at clinics and hospitals. In rural areas, however, a lack of physicians and specialists brings the practice under the purview of mosques. Several rural residents, including the rural imams, explained how a specialist from a Turkish religious organization comes to mosques in each *soum* and circumcises the region's young boys on a semi-annual basis. Apparently, the specialist has been coming to Mongolia since the early 1990s and performs up to four-dozen circumcisions a day as a free service. After circumcision, families organize elaborate circumcision feasts (*sundet toi*), entertaining guests and celebrating this important life-stage ritual associated with Islam.

The presence of a Turkish specialist heralds an unprecedented standardization and trans-nationalization of circumcision practices, a departure from a past in which the procedure was performed at home and in secret as a sign of ethno-religious identity and resistance (Privratsky 2001). By making circumcision not just a public but also a non-Kazakh act, this state of affairs brings a new level of international connectivity to a symbol inscribed on the bodies of its carriers (Foucault 1978). For a rural population that remains poor, pastoral, and geographically isolated, the trans-nationalization of this powerful yet private symbol reinforces the newly connected nature of the religious and ethnic identities that it connotes.

98.6.7 Marriage and Life Stage Rituals

Similar to circumcision, marriage is a religiously significant life stage marker that creates and delineates ethnic identity as well as gendered experiences. In Mongolia, where Kazakhs comprise less than 5 % of the total population, the selection of a marriage partner continues to reaffirm the importance of ethnicity and religion (Finke 1999: 138). People regularly told us that they would only marry a Kazakh (and therefore a Muslim). During our fieldwork, we only encountered one reference to inter-ethnic marriage, and in this case, a Mongolian man was forced to convert to Islam in order to marry a Kazakh woman. These strong ethnic boundaries are not merely a question of parental control. During an informal conversation, one young Kazakh woman mentioned that she was dating a boy she met at college, and had not yet told her parents. When asked if he was Kazakh or Mongolian, she was clearly shocked that we had asked the question: "of course he's a Kazakh, I would not want to date someone who is not Muslim."

This overlap of ethnic and religious endogamy among Mongolian Kazakhs sheds light on narratives about marriages that were formed during socialism. Several older informants described how they were married in secret by a *molda*, suggesting that this religious ceremony held more meaning for them than the civil registration. Just like other religiously significant activities formerly performed in secret, marriage has increasingly become the purview of imams, who perform marriage ceremonies prior to the *uilenu toi* marriage feast. Even today, however, the religious part of the marriage is a fairly private event compared to the large wedding reception (Fig. 98.4). However, invitations to wedding feasts today evince a much more open approach to the religiosity of marriage and are often replete with Islamic symbolism, such as Arabic script, stars, crescents, and other Islamic imagery.

Fig. 98.4 A newly married couple celebrating with relatives in a yurt during the *betashar* (face-opening) ceremony, a small family celebration that usually takes place before the large wedding reception (*uilenu toi*) (Photo by C. Werner, 2009)

98.6.8 Shrines and Sacred Sites as "Traditional" Centers of Piety

Though the practices described in the previous part of this chapter are diverse, they are united in their increasing orientation toward mosques as local nodes of piety, connectivity, knowledge, and authority (Foucault 1978). Devotional practices at *zirat* shrines, however, are an exception to this rule, since devotees travel to locations scattered around the landscape to engage in such activities. The importance of shrines, including saintly shrines and natural sites, are well documented in the literature on Central Asia (Abramson and Karimov 2007; Fathi 2006; Kehl-Bodrogi 2006; Louw 2006; Privratsky 2001). In Western Mongolia, where there is an absence of historic mosques and shrines associated with the burials of significant Muslim saints, such as in Bukhara and Turkestan, shrine practices are popular and enduring at a domestic level. Acts of devotion at shrines remain mostly a family affair, conducted at the burial sites of family members and other local sacred places. Almost all respondents visit ancestral shrines, and even people whose family burial shrines are far away still expend the time and effort to visit them on occasion.

Only one of the people we interviewed expressed opposition to the practice of shrine visitation, using rhetoric similar to that of some conservative international groups to claim that Muslims should not build burial shrines or decorate burial places in any way. Such ideologically motivated condemnations of shrine practice emphasize the contested and dynamic nature of shrine visitation. Many of the tendencies of trans-nationalization, "re-Islamization," and increasing post-socialist popularity that affect other rituals based in mosques also influence *zirat* praxis (DeWeese 1995). In light of the popular, enduring, and contested quality of these rituals, informants were asked a number of questions on the subject, and responses confirmed the importance of *zirats* to both traditional and modern piety.

98.6.9 Location and Significance of Shrines

Shrines are unique among places of religious practice in Bayan-Ulgii; they are personalized, and the location of such sites can be highly significant for devotees. Writing about southern Kazakstan, a region known for multiple sacred sites frequently visited by religious pilgrims, Privratsky discusses this subject in depth, considering shrines as foci of religious territoriality that mark the landscape (and by extension its inhabitants) as both Muslims and Kazakhs and simultaneously offer a sense of historical connection to religious tradition through family lines (2001). Despite the absence of renowned saintly shrines, the shrines in Bayan-Ulgii serve many of these same functions. As Altingul explained, burial and prayer at gravesides are a few explicit forms of religious expression that remained during socialism—a bridge between past and present Muslim identity in a single spiritually meaningful place.

Zirat shrines in Bayan-Ulgii are clustered around historic family pastures of the province's residents, reinforcing the recentness of urbanization in a place that remains largely nomadic and rural (Diener 2007). Visits to shrines simultaneously connect Kazakhs with their national, local, familial and religious heritage, in addition to providing spiritual fulfillment. Visits to rural shrines also provide opportunities for urban residents, as well as migrants to Kazakhstan, to stay connected with family members who still live in the countryside. Kazakhs sometimes express pride in the location of their family shrines. Admitting that one shrine visitation was the only religious practice he performed, one man in Ulgii told a story about his ancestor who was buried at a significant Kazakh battle site. The continuity of shrine visitation during the socialist period added to this sense of connection to, and pride in, the past; several respondents visited shrines during socialism, and even young people who had learned the prayers for shrine visitation at mosques expressed pride in older family members who had visited shrines during socialism.

A visit to the largest cemetery in Ulgii with two imams from the Ulgii Central Mosque confirmed the ongoing popularity, widespread official sanction, and historical continuity of shrine-based practices. The cemetery, whose oldest graves correspond with the early socialist-era development of Ulgii in the 1930s, sprawls across a vast area that encompasses nearly a square mile (Fig. 98.5). The individualized and elaborate nature of burial shrines is one of the cemetery's most striking

Fig. 98.5 The Ulgii cemetery seen from a distance (Photo by C. Werner, 2009)

Fig. 98.6 An ornate burial shrine beside simpler plots (Photo by N. Brede, 2009)

features, and shrines range from simple raised beds to huge hexagonal constructions embellished with ornate Islamic imagery (Fig. 98.6). Although elaborate shrines are more common in sections of the cemetery built after 1990, even shrines from the 1940s display explicit Islamic images like stars, crescents and Arabic script. According to our imam guide, burial sites were among the few locations where such symbols could be used in communist Mongolia.

New Islamic symbols are also common around other rural sacred sites, though unlike *zirats* they are not authoritatively recognized. One example of such a local devotional site, a sacred spring traditionally considered to have healing powers, was encountered while staying with a Kazakh family in their pasture in Khovd *aimag*. Though not as elaborate as other such sites, or *mazars*, in other countries, the spring was littered with small Islamic devotional items like those available in the Ulgii Bazaar, and strips of cloth were tied around clumps of grass (DeWeese 1995; Montgomery 2007). A nearby rock believed to cure warts if rubbed against the foot was similarly festooned. The prevalence of items common to other Central Asian sacred sites like cloth strips, in addition to explicit Islamic symbols, suggests a connection between the healing site and other popular conceptions of Islam.

98.6.10 Pious Acts at Shrines

In addition to the location and personal significance of shrines, we asked people to discuss the practices they engaged in while visiting shrines. Initial observations of healing springs, rocks, and other sacred places seemed to suggest a pattern of popular site-specific piety including practices like tying ribbons on trees and leaving food for spirits, following Ambramson & Karimov's work in Uzbekistan (2007). These practices, however, were less widespread than we expected and more common at natural sites. In comparison, the most popular shrine activity among informants was "reading *Ayat*"—the recitation of *Ayat al-Kursi*, a Qur'anic verse widely considered valuable for both the individual who recites it and for the spirits of the dead (Alonso and Kalanov 2008).

While the absence of some common Central Asian religious practices at shrines is attributable in part to the influence of new authorities who prefer Qur'anic recitation to other forms of piety, reading *Ayat* appears to pre-date the teachings of such authorities. Several older informants spoke of reciting *Ayat* during the socialist period or hiring *moldas* to recite the prayer at occasions of death, burial, and remembrance. Knowledge of the Arabic prayer seems widespread; informants of all ages and levels of religious training recited *Ayat* in Arabic and those who could not expressed embarrassment about their inability (Fig. 98.7). As with other practices that survived socialism, new religious authorities have played a role in promoting this broader familiarity with *Ayat*. The imams we interviewed noted that the short classes they offered on praying *Ayat* were among their most popular. Further, the Ulgii Central Mosque pays a man to live year-round at the entrance to the city's cemetery in order to assist shrine visitors who do not know the correct rituals and prayers.

Despite the relative uniformity of *Ayat* as a form of practice, the meanings that the recitation held for different informants demonstrated its polyvocality as a religious symbol. Of the 15 people who knew how to recite the prayer, 12 said they did so for the benefit of the spirits of dead ancestors, yet the quality of this benefit was disputed. Four informants said they prayed *Ayat* to help their ancestors or relatives get into paradise, but another two (both educated at the independent Hussein mosque, run by an imam who vehemently opposes shrine worship as "idolatry") stated that they did *not* read *Ayat* for that purpose and that it was impossible to help a dead person get into heaven. Most imams, however, agreed that people could pray for their ancestors to get into paradise as long as they did not ask their ancestors to give them anything in return.

Nevertheless, five informants, two of whom had learned to pray *Ayat* in mosques, said that they prayed to receive blessings, good luck and good health

Fig. 98.7 Most signs in western Mongolia are in multiple languages. This sign for a local mosque is in Arabic, Kazakh and English (but not Mongolian) (Photo by N. Brede, 2009)

from their ancestors. The prevalence of this idea despite the efforts of imams to discourage it and present a different conception of shrine devotion indicates the continued importance of trans-generational teaching and learning. In the pre-socialist period, local Sufi authorities promoted a wider variety of shrine practices than are currently exhibited in Bayan-Ulgii, and praying for favors from the dead was considered acceptable (Abramson and Karimov 2007). Alia, an Ulgii woman from a pre-socialist clerical family, exemplified the modern continuation of such ideas, having maintained them along with practices like *namaz*, Ramadan fasting, and *Kurban Ait*. Alia taught her younger family members to pray in Arabic on the anniversaries of the death of family members, since "It helps give good wishes to dead people, and in return we Kazakhs can ask the spirits of the dead for help."

98.6.11 Occasions of Shrine Devotion

Just as with other forms of piety in Bayan-Ulgii, the frequency of shrine visits and *Ayat* are quite variable. Some people visit family *zirats* regularly for multiple reasons. One of the most common occasions was prior to travel and migration, a consistent theme in the life of the increasingly transnational Mongolian Kazakh community. Over half of respondents said that they went to family shrines when their relatives visited from abroad, and that they would do the same before traveling or migrating. This tendency emphasizes two important aspects of local piety. First, shrine visitation can serve to reinforce family networks, bringing multiple generations together around a common activity in a place that evokes narratives of family history (Privratsky 2001; Turner and Turner 1978). Secondly, through shrine visits, migration and transnational movement are framed and contextualized in a religious way. According to one young woman named Mairagul, "it is good to remember our family and our religion when we are about to move away."

Holidays like *Kurban Ait* and the month of Ramadan also provided opportunities for families to gather and travel to shrines. Informants frequently mentioned the importance of visiting family *zirats* on Muslim holidays, and several considered the practice necessary for the completion of such celebrations. In this sense, shrine visitation can be seen as another dimension of festival practices in addition to activities at home and at mosques initiated through both familial and transnational religious networks. As with other holiday activities popular during socialism, new religious authorities officially sanction *zirat* practices at these times. Several of the imams we interviewed actively encourage shrine visitation during Ramadan and *Kurban Ait*.

A final occasion for shrine visits—burial—is addressed in more detail below. It should be noted, however, that visiting the graves of deceased family members 1 week, 40 days, and 1 year after death, as described by Finke, remains common in Bayan-Ulgii (1999: 137). If such visits were not possible, many informants made a special effort to pray *Ayat* at home on these dates. Several others spoke of lighting candles for the dead inside their homes, a practice common across much of Central Asia that some new religious authorities and institutions condemn (Montgomery 2007). Such dates can even affect the timing of major trips. For example, one rural informant's brother, who was unable to come from his home in Kazakhstan in time for a burial, scheduled his next visit to coincide with the 1-year anniversary of the death.

98.6.12 Burial and Funerary Practice

Considering the importance and meaning that Kazakhs place on burial shrines, it is not surprising that burial itself is an important part of popular Mongolian Kazakh religious life. Islamic burial and shrine construction was one of the most visible religious practices during the socialist period, accompanied by festivities and mourning practices that Finke describes as "very ritualized...[and always] connected with the slaughter of animals and the entertainment of guests" (1999: 137). From our interviews, we discovered that burial and mourning remain elaborate and important, however such practices have increasingly fallen under the purview of imams and mosques, and have correspondingly undergone some changes. Additionally, burial practices are contested; informants disagreed about the location and timing of burial and the decoration of burial sites. These issues are notable because they relate to discourses of religious and national belonging, homeland, and tradition.

Although the majority of people we interviewed practice shrine veneration and mourning rituals, they are not in total agreement regarding the specifics of burial practices. On the one hand, people generally agreed that the body of the deceased should be buried in the ground, it should be "pure" (*taza*) before burial, and that either a *molda* or an imam should conduct the burial. On the other hand, they have differing opinions about the time that should be taken between death and burial—a question informed by interpretations of the Islamic mandate for burial within 24 h (Roberts 2007). Though beholden to the opinions of religious experts in this matter, most informants did not consider observance of the 24-h limit to be critical. Serik, for example, whose *molda* grandfather had performed burials for decades, said that burial times were flexible and could be postponed for as long as 2 or 3 days in order for family members to arrive and attend the ceremony. On the other hand, the conservative Saudi-educated imam Khanibek vociferously condemned the laxity of Serik's approach and argued that no exceptions should be made to the 24-h rule; dead people, in his opinion, should immediately be buried in undecorated places. Interestingly, most of the other imams tended to disagree with Khanibek's hard-line opinions. Even Aybek, the president of the MIA, said exceptions could be made, mentioning a prominent local imam who died in Turkey and whose body took several days to return to Ulgii for burial.

The case of the imam whose body was transported from Turkey to Ulgii highlights the relevance of burial shrine location to multi-scalar discourses of identity and religiosity. Shrines, as noted earlier, serve as foci for the construction and reaffirmation of local and Kazakh identities and the strengthening of ties between family members and ancestral places. The decision of where to bury a dead family member thus has far-reaching implications for patterns of devotion, social networks, history, identity, and sense of place (Privratsky 2001). Though imams like Khanibek may gain some headway, the ethnographic data suggest that laypeople and authorities wish to continue traditional burial patterns. Most informants agreed with the statement, "it is important to be buried in your birthplace," and many noted that women tend to be buried with their husband's family in the traditional patrilocal fashion (Finke 1999; Diener 2007).

Religious conviction and national ideology are not the only factors influencing people's opinions about burial practices; for many, convenience was the primary criterion for decisions about burial sites and timing. When a group of six Ulgii respondents were asked how they would pick a place to bury a family member who died in Kazakhstan, all mentioned that the cost of transporting the body would be beyond the means of most families, and three said that moving a body was unnecessary since "Kazakhstan is the homeland." The other three informants (all of whom were over 45 years of age) said they would bring the body of a relative home for burial alongside other family members if they were financially able. Burial location seemed particularly important for older informants, and one older woman (Alia) who had recently migrated to Kazakhstan said that she wanted her body to be buried "in my homeland [Mongolia], with my family." With the growth of discourses that portray Kazakhstan, rather than Bayan-Ulgii, as the "homeland" of Mongolian Kazakhs, the orientation of future generations with respect to burial practices and locations is unclear; in the words of Gulzhan, a middle-aged informant "old people want to be buried in their birthplace, but young people don't care; they don't like to think about death."

98.7 Concluding Thoughts

In this chapter we provide an overview of changing religious practices in Bayan-Ulgii *aimag*, Mongolia using a conceptual framework that integrates transnational, national and local perspectives on religious practice and identity. A few general trends are noteworthy. First, almost all informants took for granted the unitary nature of Muslim and Kazakh identity, a national-scale conceptualization of religious ascription that had great currency at the community and individual levels (Asad 2009; Privratsky 2001). In part because of the idea that ethnicity and religion were linked (an important part of life and identity for this minority community, and a concept actively promoted during socialism), Mongolian Kazakhs tended to be amenable to increased involvement with new, "official" religious discourses, activities and ideas. This tendency, echoed in the rapid (re)adoption of public religious practice by much of the population, militates against the idea that "transnational" and "local" Islam represent *conflicting* interests (Gellner 1992; Montgomery 2007). New foreign-educated religious elites, however, have unquestionably become the most important influence in the social fabric of Mongolian Kazakh religion in the post-socialist period. While many Mongolian Kazakhs, particularly in rural areas, maintain practices frowned upon by such elites, even "traditional" Kazakhs have integrated new nationally and transnationally oriented practices and ideas into private, public, and family religious life.

With respect to religious practice itself, private, public, mosque-based, and shrine-based activities were the primary practical modes by which informants interpreted the idea of piety. While some groups, such as young men, exhibited a preference for what Geertz might call "scripturalist" Islam, (upon which previous scholars of Islam in Bayan-Ulgii have exclusively focused) most people were primarily interested in community or interactive activities like *Kurban Ait* and family shrine visitation (Geertz 1971). Judging from the testimony of lay and imam informants, practices like *namaz* have grown rapidly in popularity in most areas, but for most individuals the most important religious practice in mosques is almsgiving. Nevertheless, even the "traditional" or local practices that were passed down through socialism, like shrine veneration and circumcision, have gone through processes of change and alignment with the agendas and influences of transnational religious authorities similar to those seen in other countries (Alonso and Kalanov 2008; Khalid 2007).

Moreover, individuals influenced by authoritative discourses who refrain from drinking, pray *namaz* daily, and complete the five pillars are widely recognized as "good Muslims" and "religious people," indicating that their vision of orthopraxy carries a great deal of popular currency. Even middle-aged people who live out a vision of "Muslimness" rooted in the "low-key" moral ideals of *taza jol* send their children to mosques in order to provide them with knowledge of their "national heritage"—ironically imparted by imams educated in Turkey and other foreign countries (Privratsky 2001). This tendency indicates a shift in Islamic education, as in pious practice, away from the family sphere and toward mosques—a change that has radically altered the roles of older people and women as sources of religious and cultural knowledge (Finke 1999).

A multitude of questions about Mongolian Kazakh religious life remain unanswered and would benefit from future work. Future projects might include greater participant observation of popular religious events like *Kurban Ait*, and perhaps investigations of domestic or affective piety such as those conducted by Privratsky (2001). With respect to religious authority, which seems only likely to increase in importance, future work could productively focus on the specific ideological and organizational tendencies of the MIA and its interaction with independent imams in Ulgii. Lastly, there are related areas of inquiry, tangentially related to these emerging changes in religious practice that might indicate more pervasive cultural changes. Such questions might consider how Mongolian Kazakhs' new global awareness and engagement is influencing the use of traditional instruments and music in everyday life, or whether the naming of children will continue to reflect traditions in Mongolia or be linked more closely with Kazakhstani Kazakh traditions, or more broadly, new Islamic practices.

Acknowledgments This research was funded by the National Science Foundation BCS-0752411 and BCS-0752471.

References

Abashin, S. (2006). The logic of Islamic practice: A religious conflict in Central Asia. *Central Asian Survey, 25*(3), 267–286.

Abazov, R. (2007). *Culture and customs of the Central Asian republics*. Westport: Greenwood Press.

Abdullaev, E. (2002). The Central Asian nexus: Islam and politics. In B. Rumer (Ed.), *Central Asia: A gathering storm?* (pp. 245–298). Armonk/London: M. E. Sharpe.

Abramson, D. M., & Karimov, E. E. (2007). Sacred sites, profane ideologies: Religious pilgrimage and the Uzbek state. In J. Sahadeo & R. G. Zanca (Eds.), *Everyday life in Central Asia: Past and present* (pp. 319–338). Bloomington: Indiana University Press.

Adams, L. (2007). Public and private celebrations: Uzbekistan's national holidays. In J. Sahadeo & R. G. Zanca (Eds.), *Everyday life in Central Asia: Past and present* (pp. 198–212). Bloomington: Indiana University Press.

Alonso, A., & Kalanov, K. (2008). Sacred places and "folk" Islam in Central Asia. *Research Unit on International Security and Cooperation (UNISCI) Discussion Papers, 17*, 173–185.
Asad, T. (2009, October 23). *Thinking about religious belief and politics.* Speech delivered at Macalester College.
Barcus, H., & Werner, C. (2007). Trans-national identities: Mongolian Kazakhs in the 21st century. *Geographische Rundschau International Edition, 3*(3), 4–10.
Barcus, H. R., & Werner, C. (2010). The Kazakhs of western Mongolia: Transnational migration from 1990–2008. *Asian Ethnicitym, 11*(2), 209–228.
Brunn, S. D., Toops, S. W., & Gilbreath, R. (Eds.). (2012). *Atlas of Central Eurasian Affairs*. New York: Routledge.
Dani, A. H., & Masson, V. M. (1992). *History of civilizations of Central Asia* (6 vols). Paris: UNESCO Publishing.
Deweese, D. (1995). Shrine and pilgrimage in Inner Asian Islam: Historical foundations and responses to Soviet policy. *The National Council for Soviet and East European Research Title VII Program* (pp. 1–6). Bloomington: Indiana University Press.
Diener, A. C. (2007). Negotiating territorial belonging: A transnational social field perspective on Mongolia's Kazakhs. *Geopolitics, 12*, 459–487.
Du Bois, W. E. B. (1903). *The souls of black folk*. New York: Avenel.
Fathi, H. (2006). Gender, Islam, and social change in Uzbekistan. *Central Asian Survey, 25*(3), 303–317.
Finke, P. (1999). Kazakhs of western Mongolia. In I. Svanberg (Ed.), *Contemporary Kazaks: Cultural and social perspectives* (pp. 103–139). New York: St. Martin's Press.
Foltz, R. (2000). *Religions of the Silk Road: Overland trade and cultural exchange from antiquity to the fifteenth century*. London: Palgrave Macmillan.
Foucault, M. (1978). *The history of sexuality: An introduction* (R. Hurley, Trans.). New York: Pantheon Books.
Fragner, B. G. (2001). The concept of regionalism in historical research on Central Asia and Iran (a macro-historical interpretation). In D. Deweese (Ed.), *Studies on Central Asian history in honor of Yuri Bregel* (pp. 341–354). Bloomington: Indiana University Institute for Inner Asian Studies.
Geertz, C. (1971). *Islam observed: Religious development in Morocco and Indonesia*. Chicago: University of Chicago Press.
Gellner, E. (1992). *Postmodernism: Reason and religion*. London: Routledge.
Ghodsee, K. R. (2009). *Muslim lives in Eastern Europe: Gender, ethnicity, and the transformation of Islam in postsocialist Bulgaria*. Princeton: Princeton University Press.
Heyat, F. (2004). Re-Islamisation in Kyrgyzstan: Gender, new poverty and the new moral dimension. *Central Asian Survey, 23*(3–4), 275–287.
Kehl-Bodrogi, K. (2006). Who owns the shrine? Competing meanings and authorities at a pilgrimage site in Khorezm. *Central Asian Survey, 25*(3), 235–250.
Kendzior, S. (2006). Redefining religion: Uzbek atheist propaganda in Gorbachev-era Uzbekistan. *Nationalities Papers: The Journal of Nationalism and Ethnicity, 34*(5), 533–548.
Khalid, A. (2007). *Islam after communism: Religion and politics in Central Asia*. Berkeley: University of California Press.
Levi, S. (2007). Turks and Tajiks in Central Asian history. In J. Sahadeo & R. G. Zanca (Eds.), *Everyday life in Central Asia: Past and present*. Bloomington: Indiana University Press.
Liu, M. (2011). Central Asia in the post-cold war period. *Annual Review of Anthropology, 40*, 115–131.
Louw, M. (2006). Pursuing 'Muslimness': Shrines as sites for moralities in the making in post-Soviet Bukhara. *Central Asian Survey, 25*(3), 319–339.
McBrien, J. (2009). Mukadas's struggle: Veils and modernity in Kyrgyzstan. *Journal of the Royal Anthropological Institute, 15*(S1), 127–144.
McGlinchey, E. (2011). *Chaos, violence, dynasty: Politics and Islam in Central Asia*. Pittsburgh: University of Pittsburgh Press.

MNSO (Mongolia National Statistics Office). (2001). *2000 Population and housing census: The main results*. Ulaanbaatar: Mongolia National Statistics Office.

Montgomery, D. W. (2007). Namaz, wishing trees, and vodka: The diversity of everyday religious life in Central Asia. In J. Sahadeo & R. G. Zanca (Eds.), *Everyday life in Central Asia: Past and present* (pp. 355–370). Bloomington: Indiana University Press.

Northrop, D. (2007). The limits of liberation: Gender, revolution, and the veil in everyday life in Soviet Uzbekistan. In J. Sahadeo & R. G. Zanca (Eds.), *Everyday life in Central Asia: Past and present* (pp. 89–102). Bloomington: Indiana University Press.

Portisch, A. O. (2006). *Kazakh domestic crafts production in Western Mongolia: Everyday relevance and meaning*. Paper presented at CESS 7th annual conference, Ann Arbor, MI.

Privratsky, B. G. (2001). *Muslim Turkistan: Kazak religion and collective memory*. Richmond: Curzon Press.

Rasanayagam, J. (2011). *Islam in post-Soviet Uzbekistan: The morality of experience*. Cambridge: Cambridge University Press.

Roberts, S. R. (2007). Everyday negotiations of Islam in Central Asia: Practicing religion in the Uyghur neighborhood of Zarya Vostoka in Almaty, Kazakhstan. In J. Sahadeo & R. G. Zanca (Eds.), *Everyday life in Central Asia: Past and present* (pp. 339–354). Bloomington: Indiana University Press.

Turner, E., & Turner, V. (1978). *Image and pilgrimage in Christian culture*. Oxford: Basil Blackwell.

Valentine, G. (2005). Tell me about…: using interviews as a research methodology. In R. Flowerdew & D. Martin (Eds.), *Methods in human geography: A guide for students doing a research project* (2nd ed., pp. 110–127). Harlow: Pearson Education Limited.

Werner, C., & Barcus, H. R. (2009). Mobility and immobility in a transnational context: Changing views of migration among the Kazakh diaspora in Mongolia. *Migration Letters, 6*(1), 49–62.

Chapter 99
How the West Was 'One' (Hinduism and the Aquarian West)

Martin J. Haigh

99.1 Introduction

In the recent geography of religions, one of the most dramatic developments has been the expansion of the Dharmic traditions in the Western and post-Communist worlds. This movement has seen traditions previously restricted to Asian immigrant minorities become every day features on Main Street, North America, and in Everytown, Europe. The religious discipline of Yoga is taught in a myriad health spas, meditation is a much vaunted modern therapy, Buddhist tracts are found on the Mind, Body and Soul counters of most good bookstores, and the dancing Hare Krishnas are part of the landscape of most major cities. None of this was imagined by the futurists of the 1950s and the extent and character of this cultural migration and change is, even today, not widely comprehended.

> Yoga is today a thoroughly globalised phenomenon. A profusion of yoga classes can be found in virtually every city in the western World and (increasingly) throughout the Middle East, Asia, South and Central America and Australasia.... (Singleton and Byrne 2008: 1)

Of course, it is widely understood that this situation is a legacy of the 1960s 'Hippy Counter Culture,' which lives on under the 'New Age' umbrella with its traditions of self-help, holistic universalism, hobby spirituality, de-stress, spa, therapy and business development (Drury 2004). This 'New Age' or 'Aquarian' movement emerged in the Sixties and Seventies in response to disillusionment with the shallow materialism of Western culture and everyday life.

This chapter argues that the emergence of the Dharmic traditions in the West is inextricably linked to the evolution of Aquarian, 'New Age,' thinking. It employs Jan Nattier's three-category "Import/Export Baggage" model of how "*Buddhism came to Main Street*" USA (Nattier 1997). Focusing on Hinduism,

---

M.J. Haigh (✉)
Department of Social Sciences, Oxford Brookes University, Oxford OX3 0BP, UK
e-mail: mhaigh@brookes.ac.uk

S.D. Brunn (ed.), *The Changing World Religion Map*,
DOI 10.1007/978-94-017-9376-6_99

rather than the larger tradition of Western Buddhism, it observes four ethnographic types of practitioner: New Agers, Western Converts, Disenfranchised Diaspora, and Hindu Immigrants. It reflects on the ways that Hindu thinking transformed as it crossed the major East-West barrier one-way (Bauman and Saunders 2009) and, briefly, how it transformed again as it returned to India from the West, the so-called 'Pizza Effect.'

Out of respect for the 1960s Zeitgeist, this chapter uses Timothy Leary's: "Tune in, turn on, drop out" mantra as its framing motif (Leary 1965). The "Tuning In" section deals with the New Age movement and the rise of western Hindu movements after Vivekananda. "Turning On" tackles the late 1960s and the leading role of the Beatles pop group in making Indian religion fashionable in the West. It also provides case studies of Nattier's three categories. Those Imported by Westerners are represented by the new Vedanta/Tantric religion of Adidam. Traditions Exported by gurus from India are represented, first, by the early-arriving Self Realisation Fellowship of Swami Yogananda and, then, by the Transcendental Meditation Movement of Maharishi Mahesh Yogi and by ISKCON, the Hare Krishna Movement, both of which are strongly linked with the Beatles. The Baggage category of movements, those that arrived in the West with Hindu immigrant communities, is represented by the HTSSC, Hindu Temple Society of Southern California. Finally, the Dropping Out section considers those Hindu religious influences in the West that have escaped their original contexts to become embedded in Western Society, most notably Yoga and Meditation. The chapter concludes with a reflection on the phenomenon of cultural hybridization and its relationship, not to the globalization implied by the phrase 'Pizza Effect,' but to its benign fellow traveller – moral cosmopolitanism (Hill 2000).

99.2 The Dawning of the Age of Aquarius

The story of the movement of Hindu ideas from India into the Western can be told in many ways and attached to various charismatic individuals and historical accidents (cf. Forsthoefel and Humes 2005). However, there is always a deep cause – the scene was already set, the ways prepared. As with all accidents, these were events that were just waiting to happen. In this case, the Western culture had self-created fertile grounds. The Sixties backdrop was the threat of nuclear holocaust, the new revelations of impending environmental calamity (which finally demonstrated that our people had reached the limits of the Earth, a new notion at the time), rejection of the futility of war, especially the war in Vietnam, and rejection of the mores and regimentation of what we now know as the Corporate World with its trademark dissembling insincerity and totalitarian insistence on conformity in thought and dress that Adi Da dismisses as "mummery" (Da 2007: 194). Additional factors were secularisation, the decline and TV-trivialisation of organized religion and the emergence of the Boomers, a generation of Post-Second World War babies. David Lynch's "Final Battle" section of his "Industrial Symphony" presents a generational creation myth – an ambiguous flurry of air-raid sirens, searchlights, bomber planes, and a shower of babies, which is followed, while "The World Spins," by the re-appearance of an angel (Lynch 1990). Wilber (2002) lampoons the worldview of the Baby Boomers as an unhappy combination of deep unquestioning narcissism, a hedonistic obsession with 'therapy' and welfare, melded with liberal egalitarianism, a budding undirected spirituality, and a desire to transcend, what he so nicely terms, the everyday "meatscape."

Doubtless, a key factor in the successful growth of Hindu ideas from this period was a growing sense of the shallowness of this Western material "meatscape" culture. Probably, this was fuelled by the early symptoms of that acceleration and compression of time and that cult of fame and 'success' that characterizes our present era. However, unlike today, where shallowness itself can be seen as a goal, those times nurtured a strong feeling that, actually given the stress involved, the glittering prizes offered by conventional society were not worth the having. Modern society was no more than a scurrying, angst-ridden, meaningless, charnel house of mindless materialism. There simply had to be a life with more depth and more meaning.

The movie "Koyaanisqatsi" offers an iconic representation (Reggio 1982). This consists of wordless slow motion and time-lapse photography of cities (and other landscapes) set against a nagging score by Philip Glass. In the Hopi language, "Koyaanisqatsi" means "crazy life," life out of balance, and this film sets out a detached, almost drug-haze-like, critique of modern life as turmoil, disintegration, destruction, pointlessness in vapid motion. The dawn of the New Age saw an upsurge in interest in the ideas of other cultures, real and imagined: Celtic, Atlantean, Native American, Asian and African. All attracted interest as spiritual seekers sought a simpler, more meaningful, way of life and/or to recover an imagined, peaceful, lost Golden Age.

99.3 Tuning In: The New Age

Sixties psychedelia guru, Timothy Leary writes:

> This period of robotisation is called the Kali Yuga … [when an individual becomes] hooked to a series of external actions, he is a dead robot, and it's time for him to die and be reborn … [So,] Drop out – detach yourself from the social drama, which is as dehydrated and ersatz as TV. Turn on – find a sacrament that returns yourself to the Temple of God, your own body. Go out of your mind, get high. Tune in – be reborn. Drop back in to express it. Start a new sequence of behaviour to express it. But the sequence must continue. You cannot stand still. Death, Life, Structure… (Leary 1965: 3)

So begins the Quest for a new, mode of existence, a life in balance, a life that is no longer mechanical but spiritual. You are the center of your Universe, so treat yourself well, integrate Mind, Body, Spirit, Self, and relax into your 'Spa World.' You are a pilgrim soul. Your mission is to find yourself, your deeper spiritual self, your Holy Grail, whatever that may be. You are the Hero in your own Odyssey. You make the Rules or maybe no-rules. There is something deeper, more spiritual, truly meaningful out there, but you may need help to find it. Go look for that help wherever it may be found. It may be a teacher, it may a technique, it may be a place or a chemical substance. This is an eclectic path, but there are many other travellers. Ignore the robots of organized religion and social convention – be yourself.

> Float like a God, not shuffle like a robot … throw out everything that is not 'tuned in' to your highest ambition. Make your body a Temple, your Home a shrine. You are a God, live like one! Grow with the flow. (Leary 1982: 89)

The World Religions are your spiritual DIY [Do It Yourself] store (Redden 2005). No one has a monopoly on the Truth, so seek Truth across the spectrum of spiritual traditions: mystical Christian, Buddhist, Hindu, Islamic (Sufi), Jain, Jewish, Wiccan, Native American, Celtic, Sikh, Pagan, Shamanic/Animist and so on.

If you find you have 'Turned On', then 'Drop In' (at least for a while) to one or other of these traditions. If not, then build your own package. Take what you need for your own personal development (Spangler 1996; York 2001). Death, Life, Structure: "prayer is the compass" but Turning On "requires diligent yoga" (Leary 1965: 4). Notice the Hindu subtexts throughout Leary's thought but also its Universalism. Recognize that the

> New Age is strongly committed to eroding the barrier between religious traditions, concentrating on the trans-cultural dimensions of human awareness rather than on beliefs that divide human beings… [New Agers seek] spiritual and philosophical perspectives that will transform humanity and the world… Their essential quest is…. for a holistic worldview that offers both insight and hope. (Drury 2004: 11)

Hence, the New Age metanarrative: the personal quest, the voyage of personal awakening, emergence and enlightenment. Its method is experimentation, which also implies a willingness to take risks, to make mistakes and to be seen a fool (cf. Sharma 1993). Its goal is reconnection, not merely with Nature but with Truth, an ultimate reality, which is not material but spiritual. But, it also requires a way of breaking with the "Judaeo-Christian-Marxist-puritan-literary existentialist … suggestion that the conformist cop-out is reality. Dropping out is the hardest yoga" (Leary 1965: 6). Kerouac (1958) remarked: "I had indeed learned how to cast off the evils of the world and the city, just as long as I had a decent pack on my back." In the Sixties, especially after the Beatles pop-group's adventures in Rishikesh, the physical manifestation of this "casting off" became the Hippy Trail, which ran overland from London to Kathmandu (and then down to Goa, when the weather grew cold). This was a means of escape and along it ran currents of thought, spiritual, musical and narcotic (Campbell 2008, 2010).

However, long before the mid-1960s liberalization of immigration policies that allowed India's Gurus and others to enter the United States, Hinduism had established a foothold through Swami Vivekananda. Vivekananda emerged in the West at the Parliament of World Religions in Chicago 1893, where his message of Universalism and tolerance with a Hindu aspect was warmly received (Vivekananda 1893). He went on to lecture widely across the U.S., built a following, and Vedanta Societies became established in several states, especially in southern California. The Los Angeles branch of Vivekananda's Rama Krishna Order, a tiny tranquil enclave in Hollywood, founded by Swami Prabhavananda in 1930 was a hub of

Fig. 99.1 Vedanta Society of California Temple, where Adi Da gained enlightenment – permanently realizing "The Bright" (Photo by Martin Haigh)

intellectual activity during that decade (Drury 2004; Jackson 1994). The Vedanta societies, like the Theosophists of the late Nineteenth century, did much to bring Hindu thought to the U.S., including the most influential translation of the Yoga Sutras of Patanjali (Prabhavananda and Isherwood 1953). Drury (2004) sets Swami Vivekananda's Vedanta between Mme Blatavsky's Theosophists and Gurdjieff's "Fourth Way" in his chapter on New Age antecedents.

Today, the Vedanta Society is staffed almost entirely by western converts, while its temple speaks volumes about the transformation of Hinduism in the West. Certainly, it has a vaguely Indian exterior (Fig. 99.1), but its interior is set out in pews facing a large curtain. When this covers the icons of Swami Vivekananda and his guru, Sri Ramakrishna, the chamber could easily be taken for a Protestant chapel.

Also noteworthy is the Society's relationship to Swami Vivekananda's teachings. These urge followers to worship the God that is in every human, which is why the Ramakrishna Order in India is deeply involved in social work. By contrast, the Vedanta Societies in California and the U.K. dedicate themselves to spiritual growth and detachment. Indeed, the U.K. centre at Bourne End near High Wycombe, observes "Franciscan" vows of silence. Its residents are, again, largely Western converts. The symbolism of its temple room, again set out with Protestant chapel-like rows of seats, is also interesting. It is a long white tunnel with small images of Swami Vivekananda and Sri Ramakrishna at its far end.

99.4 Some Theory

Probably, it is easier to make sense of what happened next with the eye of theory. First, the ethnographic, there are four main types of people involved in the Hindu communities of the West (cf. Nattier 1998). There are immigrant communities, groups that have moved wholesale to the West and carried their own, usually closed to outsider, traditions with them. There are the individual migrants, student sojourners, economic migrants, and children of migrants, who find themselves adrift in an alien culture, separated from others by class, caste, or upbringing, who are seeking a spiritual home. There are the New Agers, a multiethnic but mainly Western contingent of people who are engaged in their own agendas, but who wander into, through and out of different traditions, pausing occasionally to adopt, adapt and/or Turn On to aspects, usually a technique, that can be used for their own ends. Finally, there are Western converts, who have Dropped Out of mainstream Western Religion, often via the New Age, in order to pursue a new tradition seriously and, typically, enthusiastically.

Second, there are the three main modes of the geographical transmission of aspects from the Dharmic religions into the West (Nattier 1997). Jan Nattier writes that:

> Religions – not just Buddhism – travel in three major ways: as import, as export, and as "baggage." (They may also be imposed by conquest, which, happily, is not a factor in this case). Those transmitted according to the "import" model are, so to speak, demand driven: the consumer (i.e. the potential convert) actively seeks out the faith. "Export" religions are disseminated through missionary activity, while "baggage" religions are transmitted whenever individuals or families bring their beliefs along when they move to a new place. It is these divergent styles of transmission, not matters of doctrine, practice, or national origin that have shaped the most crucial differences within American Buddhism. (Nattier 1997: 73)

Western Hinduism is the same. So, Nattier's model recognises: Imports – traditions brought into the West by Westerners themselves, Exports – traditions brought into the West by Hindu missionaries and Baggage – traditions arriving with immigrants as part of their home culture.

Of course, the Nattier model, while helpful, is far from perfect. As Kemp (2007: 107) notes for Buddhism: the "Unique character of the Western or "World" version is created by the interplay of sending, fetching and baggage mechanisms at work in both categories and these fuel the development of the tradition." These traditions are transformed and reshaped in transit.

The extent of such changes is greatest in traditions that have arrived by the Export route where the missionaries have adopted Western forms in order to make their message and tradition more acceptable to Western converts. The example already mentioned is Vivekananda's Vedanta Society's Western temples, which are set out after the fashion of Christian chapels.

An opposite trajectory is apparent in Import traditions. These grow from wholly alien roots in their Western home culture and this pervades everything they do, but self-conscious of this, these movements seek authenticity and work hard to adopt as much as possible of their source tradition's forms, mores and structures.

Of course, the baggage traditions are normally closed to those not born into the appropriate community and caste. Unlike Buddhism, only a tiny minority of Hindu traditions are open to converts. However, even these closed traditions find themselves reshaped by compromise with their new social habitat.

So, whatever their mode of transmission, all three types end up creating some kind of cultural hybrid, which, in due course, may assume a life of its own, spreading out internationally and even re-entering its original cultural space in new and modified forms. The popular name given to this process is the 'Pizza Effect' after the globalized fast food that emerged from Italian American origins and is now re-introduced to Italy (and bemused Italians worldwide) often via American franchised restaurants (Mittelman 1999). To date, Baggage and Import traditions have proved less influential than Export traditions. So, this chapter deals with these two categories first.

99.5 Baggage: The Hindu Temple Society of Southern California (HTSSC)

Just off Mulholland Drive in Hidden Hills, Malibu, the Hindu Temple Society of Southern California (incorporated 1977) has constructed a classic South Indian temple. Like its newer cousin, the Sri Balaji Temple in Tividale, West Midlands, U.K., this structure is constructed in classical Chola style (cf. Tamil Nadu). This, "one of the largest and most authentic Hindu temples in the Western Hemisphere" (HTSSC 2014: 1), is created in compliance to the ancient codes of the Silpa Shastras (Fig. 99.2).

Its main deity is Sri Venkateswara (a form of Sri Visnu) and it promotes a traditional devotional (Bhakti) theism. However, as in Tividale, its rites and rituals serve a pan-Hindu community and the temple acts as a center for a group larger than that of its core sect. The U.K.'s Sri Balaji Temple contains subsidiary shrines for several popular deities and devotes one of seven hillocks to each of the major world religions. In Malibu, the largely open air Vaisnava temple is accompanied by a separate, lower, in-door, area devoted to Lord Siva and His pantheon. Of course, the congregation is overwhelmingly South Indian and, as with most Hindu baggage traditions, this community is, essentially, closed to the West. However, while you have to be born into membership, in general, its congregation is friendly to western outsiders and pleased to explain its qualities to visitors (Narayan 2004).

However, even this baggage tradition is not unchanged. It is both influenced by and adapted to the Westernisation of its congregation and there begins the issue. India's Hindu temples are not, in general, congregational. *Congregationalization*, is the tendency for temples to serve as community centers, which is not their function in India. However, in the West, temples often form a focal point for diaspora families trying to preserve their culture through mutual reinforcement and the education of Western-born offspring (Raj 2004: 6). Bauman and Saunders (2009) categorize several further

Fig. 99.2 Malibu Hindu Temple (Photo by Martin Haigh)

changes. These include: *Ecumenization*, which is the generalization of Hindu rituals, a reducing of the regional and community variability that is found in India toward something that is more generally acceptable to the sum of an immigrant community; *Protestantization*, which means adopting some of the forms, patterns and timetables of the Western religious mainstream, meetings in evenings, weekends, and so forth; and finally, *Westernization* that evolves from the need of an immigrant community to explain itself to the cultural majority and from a certain freedom to rethink what their tradition means or, at least, to fit it to local lifestyle patterns.

99.6 Import: Adidam

Adidam is the creation of Adi Da Samraj, a Western guru, who embodies both the ideas of Timothy Leary (1965) and the ancient Indian tradition of the Avadhut, an antinomian purveyor of "crazy wisdom," whose archetype is the sage Sri Dattatreya (Rigopoulos 1998; Haigh 2007). Adi Da's Hindu philosophy has many affinities with Swami Vivekananda's Advaita Vedanta teachings but it is strongly influenced by Tantric ideas from Kundalini/Siddha Yoga (Jones and Ryan 2007). Adidam's tiny community of mainly Western converts includes celebrity supporters, most notably, the philosopher Ken Wilber, which gives it an impact that far outweighs its numerical strength (Sleeth 2006).

Adi Da, initially Franklin Jones (1939–2008), was born in suburban Long Island, NY. While at college, Da dropped into Sixties–style experimentation with hallucinogenic drugs and was a legal subject in hospital trials of such drugs for the US Military Veterans Administration. He called these experiences 'self-validating.' Ever the spiritual seeker, Adi Da connected, first with Swami Rudrananda, aka Albert Rudolph, a disciple of Swami Muktananda, who taught an eclectic version of Tantric Kundalini Yoga in New York. After visiting India in 1969, Swami Muktananda granted Adi Da the right to initiate others into the Siddha Yoga tradition (Jones 1972).

Among many key moments in Adi Da's hagiography are his attainment of a permanent state of enlightenment called "the Bright," through Union with the Goddess, and his vision of "The Dawn Horse." which took place in the small Hollywood Temple of the Vedanta Society of Southern California (see Fig. 99.1). Here, while observing a Spiritual Master demonstrating the Yogic power of manifesting objects, Da found a mystical shape appearing in front of him, a perfectly formed, living and breathing miniature horse. Later, seeing a TV documentary on "Eohippus" (meaning 'dawn horse' in Greek), the small ancestor of today's horse, he named this vision the 'Dawn Horse.' He writes:

> I was at once the adept who performed the miracle of manifesting the horse, and I was the one who was party to the observation of it and its result. And I did not have any feeling of being different from the horse itself. I was making the horse, I was observing the horse, and I was being the horse. (Da 1985: 69)

Later, He explained that the vision was a sign of Adi Da's own Person, taking form in and as the conditional universe. A devotee website summarises His position with a quotation from Sri Dattatreya, speaking through the Tantric *Tripura Rahasya* 20, verses 128–133. "The perfect among the sages is identical with Me. There is absolutely no difference between us."

Adi Da went on to found a bookshop and spiritual centre in Los Angeles before, later, to escape media attention, helped by a donation from celebrity supporter Raymond Burr, retiring to Naitauba Island, Fiji. Here, in 1986, Da achieved His "Divine Emergence," His personal recognition of His own God-hood (Devotees 1995). After many more years of teaching, Da gave up his body in November 2008.

Adidam's aim was to help devotees find their personal paths of escape from Narcissistic self-hood toward the goal of self-realization of the "The Bright," and their union with the Divine Consciousness through devotion to Himself as the embodiment of its True Heart. In the Dawn Horse Testament, a key text, Adi Da writes:

> This Testament is My Intention to Awaken the Consciousness of every being to Its Ultimate Real (and Necessarily Divine) Condition…. The "Centerpole" of this Testament is the Heart Itself, the Consciousness That Is Transcendental, Inherently Spiritual, and Necessarily Divine … To Understand this Testament is to be Released from the egoic vision and its point of view. Let it be so. Feel and speak this Message… Signal your Heart that it is time to Awaken, As You Are. (Da 1985: 1)

Devotees speak of His transmission of "*shaktipat*," a spiritual power capable of altering states of consciousness (Devotees 1995). Da's teaching technique, called "Crazy Wisdom" was designed to shock devotees out of their complacency.

However, it was Adi Da's teaching methods and lifestyle that triggered His fall from grace in the material world. His fall is best explained in terms of the following funny-sad "*Sex Slave*" "*Sues Guru*" clipping from the *San Francisco Chronicle* of 1985.

> Mark Miller was a college tennis star obsessed with finding "absolute truth" when he picked up a book called The Knee of Listening in 1976. Six months later, Miller was a devoted follower of Franklin Jones, a fat New York-born guru who held court at an old hot springs near Clear Lake. His girlfriend, a former cheerleader and Playboy centerfold, had become one of the guru's nine wives. "We were naive, trusting kids from Southern California," Miller told The Chronicle. "We believed that Playboy was an art form and that Franklin Jones was God." (Butler 1985: 1)

For many months, Adi Da remained a focus of scandals similar to those that surrounded "Osho," Bhagwan Rajneesh, and eventually, the media attention drove Adi Da away from California.

Ken Wilber commented:

> [Da] makes a lot of mistakes. These are immediately reinterpreted as great teaching events, which is silly. And then he gets mad and goes into sort of a divine pout .... he has holed up in Fiji, become very isolated, which I think could be disastrous, for him and for the community. The entire situation has become very problematic. (Wilber 2006: 1)

Of course, "problematic" was the euphemism that sociologists of that time were using for Jonestown and highlights the measure of concern. Certainly, disillusioned cult members portray Adi Da as an abusive, drug and sex-fuelled, ego-maniac (Conway 2007). By contrast, His writings suggest that He saw Himself as the epitome of those four Vedic "Great Sayings," the Mahavakyas, which argue as follows: I am my soul, my soul is Brahman, the divine consciousness, therefore I am Brahman ("Aham Brahmasmi" in Sanskrit). The fourth concludes by saying "That thou art also," but Adi Da's version is "Aham Da Asmi." In this book's introduction, he writes of becoming free from the Ego and of being the Bright, the Light of the Divine Universal that pervades the universe with Love and of being alone (Ruchira Avatar Adi Da Samraj 1998).

Da's legacy to New Age philosophy emerges from his, typically Hindu, version of the evolution of human consciousness, a seven stage model that rises through three bodily stages of recognition of the universal energy, physical, emotional and mental, then to a fourth stage of bodily surrender via Love-Communion, a stage of mysticism, followed by ego-death and a seventh of final transfiguration into the radiance of God (Da Free John 1981). This model has been cited by Ken Wilber in support of his new geography, the Integral Philosophy map of the levels of human consciousness (Wilber 2006). In his last years, Adi Da added greatly to his canon by writing a startlingly straight-forward and perceptive Advaita Vedanta way towards Global Citizenship. His "Not Two Is Peace," which makes better sense in Sanskrit, may become Adi Da's final mahavakya (Da 2007).

99.7 Export: The Self Realization Fellowship

Adidam remains among the more colorful and disturbing of the New Age West's Hindu Import traditions. There are so many others but their combined influence is small compared to that of the three following "Export" traditions (Partridge 2004). These traditions were all brought to the West by *bona fide* Hindu spiritual missionaries in the spirit of Swami Vivekananda and the first among these is credited with keeping the Hindu flame alive in America during those years of restrictive immigration quotas (1922–1965) that separate Swami Vivekananda from the boom years of the psychedelic Sixties (Jackson 1985).

Paramahamsa Yogananda, often heralded as Vivekananda's successor, was the first Hindu teacher of yoga to make the U.S. his home (1920—1952). A former university lecturer in philosophy, he lived a model life, blessedly free from scandal. He arrived in the United States in 1920 as India's delegate to an International Congress of Religious Liberals (Jackson 1994). Finding a warm reception, he stayed on lecturing to large audiences (Thomas 1930). In 1920–1921, he established the Self Realisation Fellowship in Los Angeles to teach meditation and yoga, albeit not simply the Patanjali-based Yoga that is now dominant, but a different Vedic style called Kriya Yoga. Kriya Yoga traces its roots back to Sri Yogananda's teacher, the "Himalayan Divine Yogi," Mahamuni Babaji, who is considered an incarnation of Sri Krishna and through him to roots in the *Yoga Upadesh* of Yogi Parasara and Sri Dattatreya's *Yogarahasyam* (Giri Babaji 1994). Like the Vedanta Societies, Swami Yogananda became very influential, especially in the western U.S. His *Autobiography of a Yogi* (Yogananda 1946) became a key reading for spiritual seekers and was named (by Philip Zaleski, editor of the annual "*The Best Spiritual Writing*" series of HarperSanFrancisco) as one of the "*100 Most Important Spiritual Books of the 20th Century*" (1999), an honor shared with Gandhi's autobiography (Brusat and Brusat 1999).

Paramahansa Yogananda's Kriya Yoga philosophy, like that of Adi Da, may contain aspects of Vivekananda's Universalism and of Tantra, but his Self Realisation Fellowship became as much New Age as Hindu. One of its charming features is that it has an atypically sunny and positive view of the future. Where most religions, including traditional Hinduism, see the world sliding downwards into its end-times, Sri Yogananda sees the future as a rising curve to perfection and our New Age, the Dwapara Yuga, as lifting us above the Kali Yuga's Iron Age of Quarrel (Yogananda 1999a, b; Richard 2007). Sri Yogananda's followers, to this day, are mainly Western converts; the list has included celebrities such as Elvis Presley and the Beach Boys. Even more than the Vedanta Society, the Self Realisation Fellowship's Californian Centres mix Eastern and Western influences. Despite one being located in a Windmill (Fig. 99.3), Self Realisation Fellowship temples are set out with pews and little ornamentation, much like Protestant Chapels, while their Altar panel centers upon Sri Krishna and Jesus Christ side by side. Figure 99.4 illustrates the famous

Fig. 99.3 Self Realisation Fellowship California: Lake Shrine's Windmill Temple (Photo by Martin Haigh)

and influential Lake Shrine, at Pacific Palisades on the ocean end of Sunset Boulevard, near Los Angeles, which includes an impressive memorial to the Mahatma Gandhi (Fig. 99.4).

From the 1960s the Self Realisation Fellowship became overshadowed by two new Hindu movements in the West, both again linked to pop-music stars, the Beatles. These were the Transcendental Meditation movement of Maharishi Mahesh Yogi and the Hare Krishna Movement of ISKCON's A.C. Bhaktivedanta Swami Srila Prabhupada. Both are Export traditions, but the first follows more closely the recognizable philosophical lineage and New Age leanings of the Self Realisation Fellowship and Vedanta Societies.

99.8 Export: Transcendental Meditation: The Maharishi Business Model

The Maharishi Mahesh Yogi, better known as the 'Beatles Guru' from their celebrated visit to Rishikesh during 1968, but whom they met in Wales in 1967, was a Physics graduate from Allahabad University (1942), who became a disciple, then

Fig. 99.4 Self Realisation Fellowship California: Lake Shrine's Gandhi Memorial (Photo by Martin Haigh)

assistant, of Swami Brahmananda Saraswati, then Shankaracharya (leader) of the Jyotir Math, one of India's five major centers of Advaita Vedanta ("All is One" non-dualism). His core philosophy was hard line Advaita Vedanta but, in the West, his movement is also an archetype of the New Age. The Maharishi's followers have been and remain mainly Western converts, most notably in recent years, the film director David Lynch.

The New Age, of course, regards the world's religions as a kind of storehouse of techniques and the Maharishi became successful in the West precisely by selling a technique, Transcendental Meditation (TM). Thus, TM is presented, not as religion, but as a scientific method to link the individual with "*sat-chit-ananda*" (the truth consciousness bliss that is the core of the universe). Followers from the universities that the TM movement has established have taken a lead in scientific research to establish and validate the medical benefits of the method, work that continues to the present day. For example, among a large number of medical studies, controlled trials show that the practice of TM reduces examination stress in college students

(Travis et al. 2009) and that the technique has clinically significant effects on the reduction of both systolic and diastolic blood pressure (Anderson et al. 2008). There are few negative reports; although Otis (1984) suggests some new subjects may experience heightened anxiety or a habituation akin to drug addiction.

The Maharishi's arguments, however, contain the Advaita Vedanta presupposition that the world is a psychic construct and his TM techniques tap into this, thus enabling yogic experts to channel bliss into the world. His technique of 'Yogic flying' (hopping) was supposed to be a 'scientific' route to conflict-free politics and a cure for all social and managerial ills in government. His conclusion was that all social problems could be solved if enough people were linked to the Universal Consciousness that maintains everything and its evolution.

The basic argument of his political movement was that if you change minds, you change consciousness and you change the world. The Natural Law Party was founded in 1992 and contested elections in 74 countries before fading away after 2001. At its peak in the U.K., the Party attracted 0.3 % of the vote while, in the 2000 presidential elections in the U.S., it attracted 84,000 votes.

By contrast, the TM universities, like the Maharishi University of Management in Fairfield, Iowa, remain small but successful; this university is often ranked among the best colleges in the United States. A distinctive and much criticized feature of TM is that it sells its products and short courses at high prices, although not its college education, where fees are modest. TM's main European Centre is the Maharishi University of Management, TM's World Headquarters, Vlodrop, Netherlands, where the Maharishi passed away in February, 2008. This is well known for selling anti-stress and management courses to business people, much like many other classic New Age therapy ventures.

The movement is criticized both for suggesting that achieving spiritual peace without spiritual discipline is easy and for suggesting that spiritual peace is a commodity to be bought and sold. Other Yogis and strict Hindus consider that charging fees for instruction in TM is unethical and they deplore the selling of "commercial mantras." However, his followers claim that this is an adaptation for western culture, which does not value anything that is given freely and values things more if they are expensive. Of course, the Maharishi was of Vaishya (trader) caste, so, by tradition, he was better qualified to sell than to teach meditation or give away mantras. However, the TM path describes a classic New Age trajectory. As David Spangler notes: "In Thirty Years, the New Age has gone from a prophesy to profits, from a vision to Visa, from a consciousness to a commodity, which, actually, is not an unusual development arc in Western Culture" (Spangler 1996: 223).

Meanwhile, meditation has become a feral Hindu tradition. In the 1970s, more than five million people practiced TM and it was the dominant form of meditation. Today, meditation is much more popular, propounded by a very large array of Dharmic traditions, notably different Buddhist sects, but also by the (mainly western converts) of the post-Hindu Brahma Kumari's "World Spiritual University," and by a huge array of New Age practitioners. In general, meditation is widely recognized as beneficial against epilepsy, stress, anxiety and certain menstrual problems (Arias et al. 2006). However, as the number of meditation practitioners has expanded, "TM" as trade-marked, has faded away.

99.9 Export: ISKCON, The Hare Krishna Movement

Until the mid-1960s, the Hindu contributions to Western society were all somewhat cerebral. Their imported 'Wisdom from the East' was all the product of the kind of Advaita Vedanta universalism that recognized the non-difference between embodied people, the unity of everything, and that everything was One with a formless supreme spirit, Nirguna Brahman. The bulk of Hindu tradition is, of course, rather different. It recognises a God or many forms of God and/or God in many forms, Saguna Brahman, but it also recognizes the social dharmas of tradition: caste, gotra and community. This closed kind of religion is as characteristic of the Baggage Tradition Hinduisms as it is of Hinduism in its homelands. However, parts of this theistic tradition have also reached the West.

The most important emerge from sects that regard the chief measure of a person to be neither birthright nor knowledge (Jnana Yoga) but devotion (Bhakti). This belief, again, frees these traditions to travel outside their source communities, across the boundaries of caste, nation and even race, to spread a more popular and populist form of Hinduism based on the worship of a particular deity or pantheon. Since these traditions also contain the Hindu concept of karma, another idea that has gone feral in the West, and reincarnation, this also allows for the kind of universalism that derives from the doctrine that embodied consciousness can neither be created not destroyed. Instead, it migrates from body to body from life to life.

The Law of Karma means "what goes around comes around" and it contains the possibility that the atman, the spiritual self or soul, is recycled after death and may reappear in any of 8,400,000 species of life. This implies that the same spiritual soul is present in every living thing, not just every human. The Puranic scriptures all include examples of devotees who are reincarnated as various kinds of animals and, by extension, they describe various forms of animals that achieve liberation through devotion. So, while those who were not born into the Brahmin caste and an appropriate community in India may be considered to be 'suffering' as a legacy from their previous lives, there is nothing to prevent them achieving success in their present life. The tradition is not bound by community, culture, race or even species.

Anyway, from 1965 onwards, a new ingredient was added to the Export mix, Bhakti Vedanta. This recognizes a personal deity, a deity with a personality and forms. Its contribution is a religion based in the realms of devotion and devotional service and that conceives religion as transcendental love affair. It also offered the New Age seekers a new, simple technique and this was the foundation for its success.

99.9.1 Export Globalized: ISKCON (International Society for Krishna Consciousness)

ISKCON, better known as the Hare Krishna Movement, has classic New Age origins. Its founder and Acharya, Srila A.C. Bhaktivedanta Swami Prabhupada, was a Vaishnava renunciant and preacher from Bengal who belonged to a tradition that had long harbored a wish to take its message out of India and to the whole world. In 1965, Srila Prabhupada created an opportunity to bring his brand of Bhaktivedanta to the streets of, first, New York City and later San Francisco and Los Angeles. Here, by chanting mantras in the park, he constructed a following among the hippies and Leary dropouts of the early American counterculture (Satsvarupa dasa Goswami 1983).

Once again, the secret of success was a technique. In this case, the technique was the chanting of the Hare Krishna Maha Mantra both as an individual form of meditation (Japa), but more influentially, as a collective experience called Kirtan. From its early days, the movement also provided both food and a home for its mainly young followers. More importantly, its technique provided a substitute for the expensive and dangerous chemical fixes of Timothy Leary. This subtext is evident in some of ISKCON's early advertising slogans, which include: "Chant and Be Happy" and "Stay High Forever" (Satsvarupa Goswami 1983). Working within the proto-New Age counterculture and with rootless but idealistic youngsters, the movement quickly 'caught the wave' of those times and the imagination of many. It expanded across North America and, once again with material assistance from the Beatles pop group, into the U.K. (Prime 2010; Fig. 99.5) and beyond. One unique outcome was, for the one and only time, to have a Sanskrit Mantra as the most

Fig. 99.5 Kirtan at Bhaktivedanta Manor (UK), Janmashtami Celebration 2011 (Photo by Martin Haigh)

popular Western "pop-song" of its day, in this case – 1970, although a modern version resurfaced in the American charts as recently as 2010.

ISKCON emerges from the teachings of a Bengali Hindu revivalist and populist theologian, Shri Krishna Caitanya Mahaprabhu, who is considered the fifteenth century 'Golden Yuga-Avatar' of Sri Krishna. Sri Caitanya (1486–1533) popularized the chanting of the Mahamantra (from the Kali-Samtaranopanisad, the 103rd Upanisad of the Krsna-Yajur-Veda), which was given by Lord Brahma to ward off the evil influences of the current "Age of Iron." "Hare Krishna, Hare Krishna, Krishna, Krishna, Hare, Hare/Hare Rama, Hare Rama, Rama, Rama, Hare, Hare" has become the most famous of all mantras in the West.

Sri Caitanya's philosophy is a version of Theistic Vedanta called Acintya Bhedabheda, which means the inconceivable oneness and difference between the Supreme Self and individual embodied self, whose proper role is as the eternal devoted servant to the Supreme. Sri Caitanya insisted on the unity of Godhead, within Sri Krishna, and that God can be realised by sublime devotion and loving service. In this tradition, devotees create their own path by holy living and self-improvement and they advance through devotion without distinction due to caste, gender or race. Sri Krishna takes many forms and accepts many incarnations but His original transcendental form is the cowherd youth of Vrindavan: Govinda – Gopala. The Srimad *Bhagavata Purana 10*: describes Sri Krishna's romantic pastimes with Srimati Radharani and her cowherd (Gopi) girlfriends in the forests of Vrindavan on the banks of the Yamuna River. The climactic event of the tradition is Lord Krishna's "Rasa Lila," a moonlit "Dance of Divine Love," where the Lord dances personally with each of His (Gopi) devotees (Schweig 2005). Each devotee's goal is to become one of Sri Krishna's eternal associates and their aim is to construct a personal relationship with Sri Krishna of which the highest kind is conjugal paramour.

ISKCON, like the other Export movements, is a cultural hybrid. In its early years, it worked with youngsters that had absolutely no idea or experience of Indian culture and had to be introduced gently to the Movement's sets of rules by the Guru (Prime 2010). The original ISKCON way of life was something based closely on the Indian monastic model of its sixteenth century founders but, naturally, it was not easy for Western teenagers to adopt the lifestyle of Hindu renunciant monks, which is why the average devotee in ISKCON lasted only three years before 'blooping-out.' There were also problems associated with the Movement's imported views on gender relations (Bryant and Ekstrand 2004). Subsequently, gradually, the tradition mellowed in the West becoming at once more tolerant, less monastic, and much more focused on meeting the spiritual needs of its congregations. Today, westerners are a minority at ISCKON temple events, which are dominated by the families and, especially, the young people of the Hindu diaspora.

However, from the start, the ambitions of ISKCON were very different from that of other Export Movements. Srila Prabhupad's aim was not simply to spread Krishna devotion around the world, but to revive his religion back in its Indian heartland (cf. Nattier 1997). His shock troops for this were his "Dancing White Elephants," Western devotees brought back to spread the word, first in Kolkata, but later across India. Ultimately, ISKCON found a new following among India's

Fig. 99.6 ISKCON Temple, New Delhi (Photo by Martin Haigh)

Westernized 'urban alienate.' Today, there are major new temples in Bangalore, New Delhi (Fig. 99.6), and many other major cities, while ISKCON is re-centered in Sri Caitanya's homeland of Mayapur, West Bengal.

99.10 Dropping Out: Feral Ideas and the Pizza Effect

Today, India is being transformed by the explosive growth of its middle classes and its cities. Together, these processes have helped created a vast urban alienate of educated migrants and their children, who are separated from their traditional communities and who construct their lives in modern India. These people are the main recruiting ground for re-exported traditions of which the prime example is ISKCON.

ISKCON in India provides an excellent example of 'The Pizza Effect.' Just as American Pizza has been re-imported to Italy, similarly, ISKCON has brought its Westernized Hinduism back to India. Here, it has built a major presence among the westernised middle classes of its major cities. Today, in Sri Krishna's Holy Land, Vrindavan, UP, India, giant billboards line the National Highway to advertise new luxury homes in terms of their distance from ISKCON's Vrindavan Temple (Fig. 99.7).

Fig. 99.7 Billboard Hare Krishna Residency, Vrindavan, adjoining ISKCON Temple (Photo © Abhinava Goswami, used with permission)

Certainly, despite its unending commitment to authenticity, this Western Hindu import has Western aspects that the original did not. This includes its openness to all races, its (relatively) equal treatment of women, and a very western style of marketing, which, in New Delhi as in Los Angeles, includes dioramas with sound effects, lights and, in New Delhi, for reasons that are not too clear, metal robots. The following, funny-happy, news clipping provides an extreme example from the UK's *Telegraph* newspaper (Anon 2009).

> Hindu monk Swami Bhakti Gaurava Narasingha has set up the world's first surfing monastery. Swami Narasingha, 64, leader of the "Surfing Swamis", born Jack Hebner in Florida rides on a surfboard inscribed with the Hindu holy word "Om" and runs a surfing ashram, a prayer community, where the water sport is an extension of the prayers. Set up in 1994, the ashram has about 150 members, mostly from Karnataka. The idea of the Mantra Surf Club arose 4 years ago while teaching bodysurfing. "They all got excited about it and I called a friend of mine & said the next time you come to India, bring me a couple of surfboards."

ISKCON in the West has also changed. Once ashram-based, it is now mainly congregational and many large temple houses, like the one in Montreal, are almost empty. Scandals in the 1980s and doctrinal divisions have reduced the numbers of North American devotees, although the movement continues to spread, especially in eastern Europe and the former Soviet Union. Meanwhile, in both the U.S. and U.K., the movement has revived by finding new congregations among the Indian Diaspora. Narayan (2004: 6–7) notes:

> Although the prime movers of ISKCON are white converts, Indian Hindus usually show great respect for the strength of ISKCON members' devotion, their knowledge of Sanskrit, their strict vegetarianism and their elaborate detailed rituals.
>
> Since the 1970s, when Hindu immigrants began to settle down in America in significant numbers, many informants pointed out that the Hare Krishna temple was the only temple they found in the region and it soon became the first temple homes of many new Hindu settlers. On Sunday afternoon they could show up for the worship of Krishna, complete with devotional chanting and a lecture on the Gita, usually delivered by a well-trained Caucasian devotee wearing the faded orange robes of a Hindu monk. They could also enjoy a Sunday vegetarian meal and observe the great festival days that made them nostalgic about their homes in India. Therefore, the ISKCON temple was a transitional space where they could settle and feel at home before organizing or even building a new temple. It must be pointed out, though, that the Hindu immigrants have continued to participate in the life of the ISKCON temples, transforming these temples into multi-ethnic Hindu communities.

ISKCON Temples have also found a growing niche serving the Westernized children of these diaspora communities, who increasingly dominate congregations and events. Today, while some academics talk about the Hinduization of the movement, others regard ISKCON as a prototype for a new more cosmopolitan world Hinduism. It is worth noting, in this connection, that the largest ISKCON Temple in the world was recently opened by the diaspora Indian community of Fiji (www.iskconfiji.com/).

99.10.1 Yoga Goes Feral

While it could be argued that ISKCON could itself be regarded as a feral tradition, its only offshoot that might be regarded as a true New Age technique is that of communal chanting or 'Kirtan.' Kirtan has become adopted, although not yet very widely, both as an aspect of Yoga and as a self-help therapy (Bruder 2004; Gass and Brehony 1999; Black 2008).

Meditation is also naturalized. It is now part of much New Age practice.

However, the most successful feral escape is the physical practice of yoga itself. Yoga appeared in the West in early nineteenth century and gained popularity with the vegetarian and health fads of 1930s. It became trendy in the 1960s when several Yogis began teaching in the West. Key players (again), included: The Beatles, who were first exposed to Yoga while filming "Help!" (1965), when Swami Vishnu-Devananda gave them his "*Illustrated Book of Yoga*." Singleton and Byrne (2008: 1) write:

> Though the yoga diaspora began well over a century ago, it is only since the mid-1990s that it has taken on the global proportions that make it such a visible – and profitable – enterprise today.... Precise practitioner statistics are hard to come by and unreliable, but it is estimated that in 2004 there were 2.5 million practitioners of yoga in Britain alone, a truly exponential increase from previous years.

In the U.S. there are thought to be between 12 and 20 million Yoga practitioners, 77 % are women, 45 % have been active for more than 2 years and 59 % are younger than 35 years of age. Today, more than 85 % of all fitness centres offer Yoga and there are more than 77,000 yoga teachers (Yoga in America Market Survey 2008).

Belatedly, some rival traditions have awoken to the fact that Yoga is not just about physical well-being and peace of mind, the standard New Age sales package. It is an ancient religious tradition. So, there is a backlash, which is much resisted, especially by the overwhelmingly female communities of practitioners for whom Yoga is, primarily, a means for relaxation and fitness like Pilates and arts from other spiritual traditions. In rural Britain, a ban on yoga classes at two Baptist Churches in Taunton, UK, for being "unchristian," has been challenged as a breach of the U.K.'s Equality Act 2006 by the Hindu Council U.K. (Blake 2007). In Malaysia, a National Fatwa Council decree that made yoga forbidden to Muslims (because it fosters a blasphemous union with God) came under fire from women's groups such as Tai Chi is "The Sisters of Islam," who protested that it was just an exercise like Tai Chi (Al Arabiya 2008). Tai Chi is another example of a tradition that has travelled to the West in simplified and adapted forms (Ryan 2008). Meanwhile from far away Antarctica, Johnson (2004: 162) complains: "God has a tough gig …The few Catholics, Protestants and Mormons are surrounded by drunks, scientists, drunk scientists and heathens who hold yoga classes in the Chapel."

99.11 Conclusion

So, was the West really "One" by the invasion of Hindu ideas (cf. Thomas 1930)? It is said that 98 % of Americans believe in God and that 20 % are New Agers (Barnia 1996). Hindu traditions remain a significant component of the New Age mix and in the churn of New Religious Movements, not least those from the Age of Gurus that followed the 1960s (Forsthoefel and Humes 2005). Today's culture may be more conservative, more corporate and less receptive to new spiritual ideas, but its Aquarian sub-currents remain strong.

However, a more accurate answer to the question might be – "No," it was not the West but Hinduism that started to become "One" as it evolves into a true world religion. Even in India, Hinduism is changing. Scholars talk of the emergence of a homogenized, 'all-India' Hinduism but, in India, regional, linguistic, caste and sectarian particularities remain strong. Outside India, in the West, especially the U.S., Hindu ecumenization has been stronger because the relative dispersal and disparate character of the small populations of immigrants, as well as their cosmopolitan character, has encouraged some blending of traditional mores. Here, modern Baggage Traditions, such as the BAPS Swaminarayan Sanstha, and new Export traditions like ISKCON hold disproportionate sway, and the process of change is further accelerated by the trend for the western Hindu temples of many persuasions to act as community centers and to provide the institutional framework for communities (Bauman and Saunders 2009).

The case of ISKCON, of course, takes Jan Nattier's model to a new level. ISKCON, which created a new multiethnic Hindu community in the West during the 1970s, quickly re-exported itself to India and, as the Surfing Swamis of Mulki demonstrate, created a new multicultural theme in Global Hinduism, a true Pizza-effect scenario. This has found especial resonance with the new legions of culturally dispossessed Hindus. These are, first, the expatriate and first generation Indians of the Western diaspora and, second, India's own middle class urban –alienate, who seek a modern version of their indigenous religion.

Beyond this, the most dramatic incursions of Hinduism into western society have come from the immigration of Hindu spiritual disciplines. The Hindu invasion of western culture has been driven by techniques: meditation, kirtan and especially yoga, along with some of the spiritual ideas that support these methods. The medium of transmission remains adoption by New Age groups and the mode of propagation is through its eclectic colonies in the spas and fitness clubs.

Of course, while the fact is not much discussed, the 'New Age' is heavily engendered. Its chief supporters, worldwide, are women and it has strong roots in a feminine spirituality. One speculates that its prevalence goes largely un-noticed in male-dominated societies because its advances are made invisibly, through contacts made in the 'women's locker room' or its cultural equivalent.

However, what is certain is that Hinduism is contributing to a much larger global cultural movement. A David Spangler (1996: 153) writes:

> The transformation into Aquarius is marked not by an event, but by a change in consciousness and perspective, it is a redefinition of ourselves, our world and the sacred in a way that brings all three into a greater wholeness and co-creative relationship.

Hinduism has become both a World Religion and part of a hybrid cultural movement, a kind of benign globalization, perhaps eventually, a truly universal religion of the kind envisioned by Swami Vivekananda. This process could be termed Cosmopolitanisation. Jason D. Hill (2000: 7) notes:

> Moral cosmopolitans are out to detribalize the world. Hybridization is a moral goal because it destabilizes zones of purity and privilege . . . Moral cosmopolitanism is concerned with the not-yet self, the self that ought to exist. (Hill 2000: 8)

This goal is one that seems close to both the New Age aspiration and the Moksha Dharma disciplines of self-realisation promoted by exported Hinduisms.

Acknowledgement Thanks go to Ms. Jennifer Parsons-Kerr and the children for helping me explore the temples of Southern California.

References

Al Arabiya. (2008, November 23). *Yoga hugely popular in Muslim Malaysia*. Cairo/Kuala Lumpur: Al Arabiya News Channel (AlArabiya.net, AFP). Retrieved January 2010, from www.alarabiya.net/articles/2008/11/23/60645.html

Anderson, J. W., Liu, C., & Kryscio, R. J. (2008). Blood pressure response to transcendental meditation: A meta-analysis. *American Journal of Hypertension, 21*(3), 311–316.

Anon. (2009). Monastery for surfers. *Telegraph* (UK) 9:23 AM, 09 Nov 2009. Retrieved January 2011, from www.telegraph.co.uk/news/worldnews/asia/india/6530585/Monk-sets-up-monastery-for-surfers.html

Arias, A. J., Steinberg, K., Banga, A., & Trestman, R. L. (2006). Systematic review of the efficacy of meditation techniques as treatments for medical illness. *Journal of Alternative and Complementary Medicine, 12*(8), 817–832.

Barnia, G. (1996). *The index of leading spiritual indicators*. Dallas: Word Publishing.

Bauman, C., & Saunders, J. B. (2009). Out of India: Immigrant Hindus and South Asian Hinduism in the USA. *Religion Compass, 3*(1), 116–135.

Black, S. (2008). *"Chant and be happy": Music, beauty, and celebration in a Utah Hare Krishna community*. Tallahassee: Florida State University, College of Music, Unpublished Masters thesis. Retrieved January 2012, from http://etd.lib.fsu.edu/theses_1/available/etd-11102008-163528/unrestricted/BlackSThesis.pdf

Blake, D. (2007, September 4). Hindu council attacks 'Illegal' church ban on yoga. *Christian Today*. Retrieved January 2010, from www.christiantoday.com/article/hindu.council.attacks.illegal.church.ban.on.yoga/12812.htm

Bruder, K. (2004). *Following the voice into silence: Chant-in-Kirtan as an interactive modality for self-help*. Paper presented at International Communication Association, annual meeting, New Orleans, LA. Retrieved January 2011, from www.allacademic.com/meta/p113340_index.html

Brussat, F., & Brussat, M. A. (1999). *100 best spiritual books of the twentieth century*. Traverse City: Spirituality & Health Magazine Review. Retrieved January 2010, from www.spirituality-health.com/newsh/items/review-feature/item_6545.html

Bryant, E. F., & Ekstrand, M. L. (2004). *The Hare Krishna movement: The postcharismatic fate of a religious transplant*. New York: Columbia University Press.

Butler, K. (1985, April 3). "Sex slave" sues guru: Pacific Isle orgies charged. *San Francisco Chronicle*, p. 1.

Campbell, D. (2008). *The paradise trail*. London: Headline Review.

Campbell, D. (2010, March 23). The hippy trail. *The Observer*, p. 19. Retrieved March 2010, from www.guardian.co.uk/world/2010/mar/28/goa-hippy-trail

Conway, T. (2007). *Adi Da and his voracious, abusive personality cult*. Retrieved May 2011, from www.enlightened-spirituality.org/Da_and_his_cult.html

Da, F. J. (1985). *The dawn horse testament*. Middletown: Dawn Horse

Da, Adi, The World Friend. (2007). *Not two is peace: The ordinary people's way of global cooperative order*. Middletown: Is Peace 723.

Devotees. (1995). The divine life and work of Adi Da (the Da Avatara). In Adi Da (1982) *The hymn of the True Heart Master* (pp. 18–93). Middleton: Dawn Horse Press.

Drury, N. (2004). *The new age: Searching for the spiritual self*. London: Thames & Hudson.

Forsthoefel, T. A., & Humes, C. A. (Eds.). (2005). *Gurus in America*. Albany: SUNY Press.

Gass, R., & Brehony, K. (1999). *Chanting: Discovering spirit in sound*. New York: Broadway Books.

Giri Babaji, Swami Satyeswarananda. (1994). *Kriya Sutras of Baabji*. San Diego: The Sanskrit Classics, Commentaries Series.

Haigh, M. (2007). Sri Dattatreya's 24 gurus: Learning from the world of Hindu tradition. *Canadian Journal of Environmental Education, 12*(1), 127–142. Retrieved October 2011, from http://cjee.lakeheadu.ca/index.php/cjee/article/viewFile/627/518 (Also: http://eric.ed.gov/PDFS/EJ842786.pdf).

Hill, J. D. (2000). *Becoming a cosmopolitan: What it means to be a human being in the new millennium*. Lanham: Rowman & Littlefield.

HTSSC. (2014). *Malibu Hindu temple*. Calabasas: Hindu Temple Society of Southern California. Retrieved September, 2014, from http://malibuhindutemple.org/pictures_from_previous_events.aspx

Jackson, C. (1994). *Vedanta for the west: The Ramakrishna movement in the United States* (2nd ed.). Bloomington: Indiana University Press.

John, D. F. (1981). *The bodily sacrifice of attention* (pp. 29–30). Middletown: Dawn Horse Press.

Johnson, N. (2004). *Big dead place: Inside the strange and menacing worlds of Antarctica*. Los Angeles: Feral House. Retrieved January 2011, from http://feralhouse.com/press/bigdeadplace/

Jones, F. (1972). *The knee of listening*. Lakemont: CSA Press. Retrieved (with expanded 2004 edition), May 2011, from www.beezone.com/AdiDa/KneeofListening/book/tableofcontents.html

Jones, C. A., & Ryan, J. D. (2007). Adi Da Samraj, teacher of Crazy Wisdom. In C. A. Jones & J. D. Ryan (Eds.), *Encyclopedia of Hinduism* (pp. 6–7). New York: Facts on File Inc.

Kemp, H. (2007). How the Dharma landed: Interpreting the arrival of Buddhism in New Zealand. *Journal of Global Buddhism, 8*, 107–131.

Kerouac, J. (1958). *The Dharma bums*. New York: Penguin (1986).

Leary, T. (1965). *Turn on, tune in, drop out* (Chapters 11–22 of "*Politics of ecstasy*"). Berkeley: Ronin (1999).

Leary, T. (1982). *Your brain is God* (excerpted from: "*Changing Your Mind among Others*"). Berkeley: Ronin (2001).

Lynch, D. (1990). *Industrial symphony no. 1: The dream of the brokenhearted*. Los Angeles: Propaganda Films. Retrieved September 2014, from: http://www.youtube.com/watch?v=1q3v9zT2SG0

Mittelman, J. H. (1999). Resisting globalisation: Environmental politics in East Asia. In K. Olds, P. Dicken, P. F. Kelley, L. King, & H. Wai-chung Yeung (Eds.), *Globalisation and the Asia-Pacific: Contested Territories* (pp. 70–85). London: Routledge.

Narayan, A. (2004). *Who is a Hindu? The search for a Hindu identity in America*. Paper presented at American Sociological Association, annual meeting, San Francisco, CA. Retrieved January 2010, from www.allacademic.com/meta/p109697_index.html

Nattier, J. (1997). Buddhism comes to main street. *Wilson Quarterly, 1997*(2), 72–81. Retrieved January 2011, from http://buddhism.lib.ntu.edu.tw/FULLTEXT/JR-ADM/nattier.htm

Nattier, J. (1998). Who is a Buddhist? Charting the landscape of Buddhist America. In C. S. Prebish & K. K. Tanaka (Eds.), *The Faces of Buddhism in America* (pp. 183–195). Los Angeles: University of California Press.

Otis, L. S. (1984). Adverse affects of transcendental meditation. In D. H. Shapiro & R. N. Walsh (Eds.), *Meditation, classic and contemporary perspectives* (pp. 201–208). New York: Aldine.

Partridge, C. (Ed.). (2004). *New religions: A guide*. Oxford: Oxford University Press.

Prabhavananda, S., & Isherwood, C. (1953). *How to know God: The yoga aphorisms of Patanjali*. Los Angeles: Vedanta Press (Vedanta Society of Southern California).

Prime, R. (2010). *When the sun shines: The dawn of Hare Krishna in Britain*. Los Angeles: Bhaktivedanta Book Trust.

Raj, A. (2004, August 14). *Indian diaspora in North America: The role of ethnic networks and organizations*. Annual meeting, American Sociological Association, San Francisco, CA, Paper. Retrieved, January 2010, from www.allacademic.com/meta/p109175_index.html

Redden, G. (2005). The new age: Towards a market model. *Journal of Contemporary Religion, 20*(2), 231–246.

Reggio, G. (1982). *Koyaanisqatsi: Life out of balance*. San Francisco: Institute for Regional Education and American Zoetrope. Extracts retrieved, March 2011, from www.youtube.com/watch?v=fY4L5npPdao

Richard, P. (2007). *Dwapara Yuga and Yogananda: Blueprint for a new age*. Dallas: The Noble New.

Rigopoulos, A. (1998). *Dattatreya: The immortal guru, Yogin and Avatara*. Albany: SUNY Press.

Ruchira Avatar Adi Da Samraj. (1998). *Aham Da Asmi (Beloved, I am Da)*. Middleton: Dawn Horse Press.

Ryan, A. (2008). Globalisation and the 'internal alchemy' in Chinese martial arts: The transmission of Taijiquan to Britain. *East Asia Science, Technology & Society: An International Journal, 2*(4), 525–543.

Satsvarupa dasa Goswami. (1983). *Prabhupadat, your ever well-wisher*. Los Angeles: Bhaktivedanta Book Trust (2001).

Schweig, G. M. (2005). *Dance of divine love: The Râsa Līlâ of Krishna from the Bhâgavata Purâṇa, India's Classic Sacred Love Story*. Princeton: Princeton University Press.

Sharma, A. (1993). *Experiential dimension of Advaita Vedanta*. New Delhi: Motilal Banarsidass.

Singleton, M., & Byrne, J. (Eds.). (2008). *Yoga in the modern world: Contemporary perspectives* (pp. 1–14). Abingdon: Routledge Hindu Studies.
Sleeth, D. B. (2006). *The internal ego: A comparative study of the ego in the work of Freud, Jung, and Adi Da*. San Francisco: Saybrook University, Unpublished Ph.D. dissertation. Retrieved May 2011, from http://dbsleeth.com/uploads/Dissertation.pdf
Spangler, D. (1996). *A pilgrim in Aquarius*. Forres: Findhorn Press.
Thomas, W. (1930). *Hinduism invades America*. New York: Beacon Press. Retrieved May 2011, from www.archive.org/stream/hinduisminvadesa013865mbp#page/n4/mode/1up
Travis, F., Haaga, D. A. F., Hagelin, J., Tanner, M., Nidich, S., Gaylord-King, C., Grosswald, S., Rainforth, M., & Schneider, R. H. (2009). Effects of transcendental meditation practice on brain functioning and stress reactivity in college students. *International Journal of Psychophysiology, 71*, 170–176. Retrieved August 2010, from www.transcendental-meditation.co.uk/articles/Effects%20of%20TM%20on%20brain%20functioning%20and%20stress%20in%20students.pdf
Vivekananda, S. (1893). Addresses at the Parliament of Religions, 1893. In: *The complete works of Swami Vivekananda* (Vol. 1, pp. 1–25). Kolkata: Advaita Ashrama (1989 Edition). Retrieved January 2011, from www.youtube.com/watch?v=lxUzKoIt5aM
Weil, R. (2009). *Yoga*. Medicenenet.com, USA. Retrieved January 2010, from www.medicinenet.com/yoga/article.htm
Wilber, K. (2002). *Boomeritis*. Boston: Shambhala Publications.
Wilber, K. (2006). *Integral spirituality*. Boston: Integral Books.
Yoga in America. (2008). *Yoga in America market survey*. Retrieved January 2010, from www.yogajournal.com/advertise/press_releases/10
Yogananda, S. (1999a). The end of the world. In International Publications Council (Ed.), *The world in transition* (pp. 3–21). Los Angeles: Self Realization Fellowship.
Yogananda, S. (1999b). The need for universal religious principles. In International Publications Council (Ed.), *The world in transition* (pp. 101–112). Los Angeles: Self Realization Fellowship.
Yogananda, S. (1946). *Autobiography of a Yogi*. Los Angeles: Self Realisation Fellowship (1998 Edition). Retrieved December 2010, from www.ananda.org/inspiration/books/ay/
York, M. (2001). New Age commodification and appropriation of spirituality. *Journal of Contemporary Religion, 16*(3), 361–372.

Chapter 100
Hinduism and Globalization

Rana P.B. Singh and Mikael Aktor

100.1 Introduction: History of Hinduism

India has its own history of contrasts: ecological, religious, linguistic, historical, political and eco-psychological. India, the homeland of the three major religions, viz. Hinduism, Sikhism and Buddhism has a long history of civilization, going back to c. 2600–1900 BCE tracing the urban culture along the Indus River in the west India. Later the Harappa culture spread toward farther east and south as revealed by archaeological findings at many sites, including Mehargarh (west), Novsaro, Kot Diji, Amri, and Surkotada, Lothal (south). By c. 1800 BCE the Vedic culture was introduced, which later became the base of evolving Indian culture. By c. 1200 BCE many hymns of the *Veda*s were composed, the first and oldest called the *Rig Veda*. The remaining three *Vedas* are *Yajura*, *Sama* and *Atharva Vedas*. In the later phases, c. 1000 BCE, the Vedic culture expanded and spread over the Ganga Valley in north India. With the start of settled life, scholastic traditions also became very strong leading to writings of the *Upanishads* that deal with the philosophy of human-nature interaction and the relationship between humanity and divinity. Along these lines varieties of thoughts and ways developed in their own regional settings, which altogether generally are referred to as Hinduism. Michaels states that

> There is neither one founder of the religion nor one church nor one religious leader. Nor is there one holy book or one doctrine, one religious symbol or one holy centre. As a result, no one binding religious authority could emerge... Hinduism is not a homogeneous religion

R.P.B. Singh (✉)
Department of Geography, Banaras Hindu University, Varanasi, UP 221005, India
e-mail: ranapbs@gmail.com

M. Aktor
Institute of History, Study of Religions, University of Southern Denmark,
Campusvej 55, 5230 Odense, Denmark
e-mail: aktor@sdu.dk

S.D. Brunn (ed.), *The Changing World Religion Map*,
DOI 10.1007/978-94-017-9376-6_100

at all, but is rather a potpourri of religions, doctrines and attitudes toward life, rites and cults, moral and social norms. (Michaels 2004: 3)

It was during later periods, following the Buddha (sixth or fifth century BCE) who tried to resolve the superstitions and ritualistic frame of Hindu traditions, when Buddhism was introduced. It was also when the great epics of the *Ramayana* and the *Mahabharata* were composed. The great threat and also the enculturation from the West started with the invasion of Alexander the Great (c. 327–325 BCE), when the Mauryan dynasty (c. 324–185 BCE) was at its zenith and when Ashoka patronized the Buddhism. Since the beginning of the Current Era and several centuries many text and treatises were composed referring to the life-issues, values, society and administration. All the nature symbols and divine spirit were given anthropomorphic form which led to the foundation for the growth of various forms of divinities. The most popular were Vishnu (in the forms of Krishna and Rama), Devi (goddess), and Shiva. Also there were several indigenous and folk deities, locally worshiped independently or sometimes together with the above three main divinities, as represented in Minakshi temple of Madurai (Fig. 100.1). During this phase of growth there also evolved the hierarchical system of four functional groups, called *Varna* ("color-based group"): Brahmins ("priests"), Kshatriya ("warriors"), Vaishya ("producers"), and Shudra ("servants"), which with the passage of time became segmented into hierarchy-based functional sub-groups of social classes, called *jati* (caste) (Fig. 100.2).

Fig. 100.1 A section of one of the temple towers of the Minakshi-Sundareshvara temple in Madurai, Tamil Nadu. South Indian temple towers are famous for their multi-colored elaborate depictions of the Hindu pantheon. The polytheism of Hinduism reflects a multi-faceted world view (Photo by Mikael Aktor, April 2011)

Vedic Hinduism: Sociogony, Social Organization				
Cosmic Body (Purusha)	Human Body (*Sharir*)	Social Body (*Varna*)	Functional Body (*Kriya / Karma*)	Salient Qualities (*Gunas*)
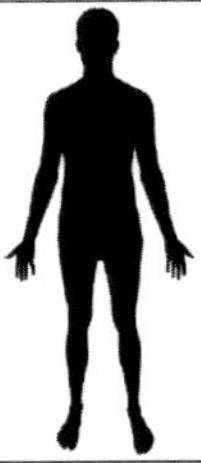 Heaven	Head	Brahmins	Priests: deliberative	Thought: perception, speech
Mid-space	Heart	Kshatriyas	Warriors: defensive	Strength: energy, courage
Earth	Thighs	Vaishyas	Producers: commercial	Support: sexuality, appetite
Underworlds	Feet	Shudras	Servants: servitorial	Service: protection, help

Fig. 100.2 Vedic Hinduism: Sociogony, Social Organization (after Singh, 2009, p. 71)

Fig. 100.3 A Banyan tree (*Ficus benghalensis*) in Delhi Zoo; the tree with its thick net of aerial roots is often regarded as the national tree of India. In Hinduism it is a sacred tree often functioning as a temple in its own right with small figurines of deities placed on the ground (Photo by Mikael Aktor, May 2011)

From the four *Varnas* developed 36 major castes and several regional sub-divisions, totaling 3,742 castes (*jatis*) and sub castes (*up-jatis*) and an additional 4,635 communities (Singh 1999: 58). In this process there also developed various sects and sub-groups of religious sects and traditions, numbering around 200. Hinduism is like a Banyan tree with many branches and roots that continuously spring up their own roots when it is fully grown (Fig. 100.3). The branches of Hinduism such as Vaishnavism, Shaivism, Shaktism, Jainism, Buddhism, etc. have developed their own deep roots, and many of them became independent in passage of time. Though there are many branches, roots, leaves and flowers, the Banyan tree is only one! Hinduism, like the banyan tree is a single unified entity with diversified beliefs and customs, of course it has both its glories and pitfalls.

In a way Hinduism may be referred to geographically as a defined group of distinct but related "regions" working together and resulting in a multiplicity or spatial mosaicness (von Stietencron 1989: 21). This promoted the most common practice of liberality in matters of religion having freedom for each individual to practice any of the paths what he wants to follow, and also to combine different religious approaches and ways of performances in his personal quest for the divine to satisfy oneself. This led to religious plurality and multiple layering and liberality, which is an important message in the era of globalization.

Hindus constitute a billion plus inhabitants of the planet, roughly one-fifth of the world's population today and when the modern Indian innovativeness in technology mindedness combines with the universal, spiritual, vibrating, dancing, life force energy of the tradition, they are in a position to change the shape of the world for the betterment of generations to come. Thus Hinduism ("Hindu traditions" together in one frame) is an ongoing learning, transforming, and transferring process forever rooted in the past, continuity and practice in present, and with hope and vision moving into the future.

The four stage *Varna* frame, when extended into the four-level structure of stages and actions in Hindu life, can be formulated on the basis of the Puranic description of lifecycle, viz. *dharma* (moral duty) and *karma* (phenomenal action) interaction. Of course, they act side-by-side, they are no way identical, but rather operate in a complex system. At different levels the lifecycle in Hinduism follow the system of the counterpart (Fig. 100.4), representing the integrated frame of daily ongoing actions related to religio-ritual practices and duties. Considering an average life-span of 100 years, the periods of life along with the subsequent actions are divided into four stages. Although they are not identical, they have similarities.

The *Grihyasutra* prescribes five great sacrifices which one is required to perform everyday as daily routine forming part of his/her *lifecycle*. The human being is at the base having as a main sacrifice and responsibility to serve humanity symbolized with water. The two side axes above the base refer to the worship of knowledge and spirits; and two top axes symbolizing gods and manes. These five sacrifices are further correlated with five elements which constitute the total life: earth, water, fire, sky/ether, and air, as described in Hindu mythologies. According to the *Bhagavadgita* (7.4) the structure of the human and cosmic relationship can be divided into eight-fold divisions which include the above five elements in addition to mind (*mana*), understanding (*buddhi*), and self-sense (*ahamkara*). All these sets are controlled by the ninth part, that is, eternal realization (*chaitanya*) (Fig. 100.5). The man and cosmic relationship in a life-span is theologically reflected with five essential sacrifices (*yajna*) symbolically represented with five elements which together constitute the organic life. These five elements and sacrifices are influenced by three mental states and overall governed by eternal realization.

The on-going ritual practice in a space-time-continuum can be expressed with the structure of *ritual-mandala* which expresses the levels and intensity of believers

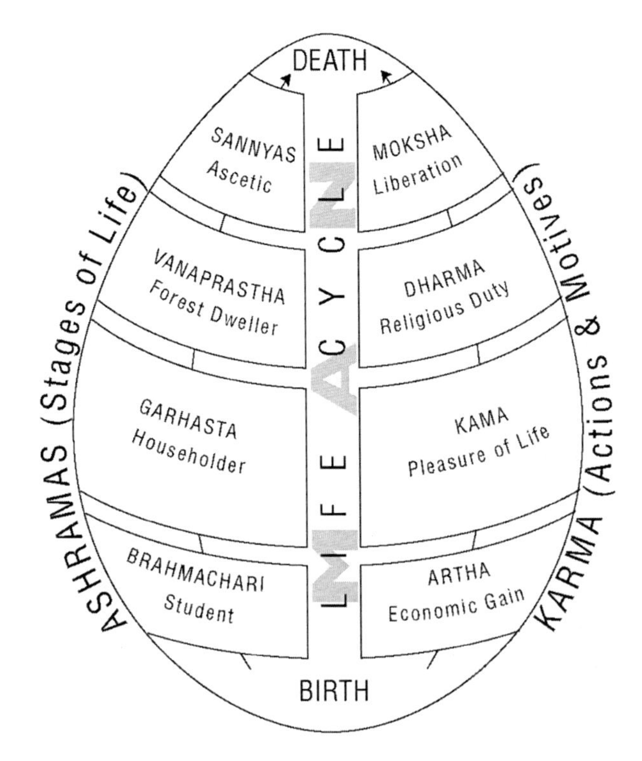

Fig. 100.4 Lifecycle in Hinduism (after Singh, 2009, p. 79)

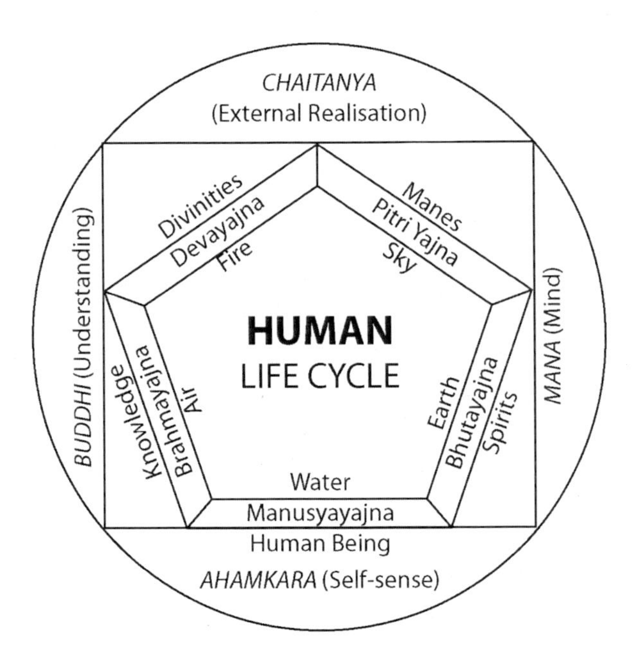

Fig. 100.5 Human and cosmic relation in Hinduism (after Singh, 2009, p. 81)

from low to high, and from secular to sacred. With the fourfold taxonomy of ritual practices, there appears a system of "whole" that represents the rituals in practice, geometrically described with four patterns of ascending, descending, pyramidal, and concentric rings (see Fig. 100.5). In each case "the symbolic significance of rituals organized in these patterns are parallels by number, ritual purity or other dimensions of worshippers and their gods" (Singh 1984: 108). In the process of space-time variation in daily life of Hindu society, there appears a complex system of ritual practices which distinctively form four geometric shapes. The counter-dependent and interdependent processes work together, and the variations can easily be observed in intensity of Scale, and Status that further reflect the segmented reality of the changing dimensions of "self" and his/her "world" in time-bound frame. These interrelationships finally merge to form a total system of *ritual-mandala* (Fig. 100.6).

Of course, there does not exist strict orthodoxy in Hinduism; however, there are several principles that share a commonality among the various sects. Virtually all Hindus believe in:

- The three-in-one god known as "*Brahman*," which is composed of: Brahma (the creator), Vishnu (the Preserver), and Shiva (the Destroyer).
- The Caste System (4 *Varnas*, 36 *Jatis*, c. 2400 *sub-Jatis*).
- *Karma*, the law that good begets good, and bad begets bad. Every action, thought, or decision one makes has consequences, good or bad, that will return to each person in the present life or in one yet to come.
- Reincarnation is also known as "transmigration of souls" or "*samsara*." This is a journey on the "circle of life," where each person experiences as series of physical

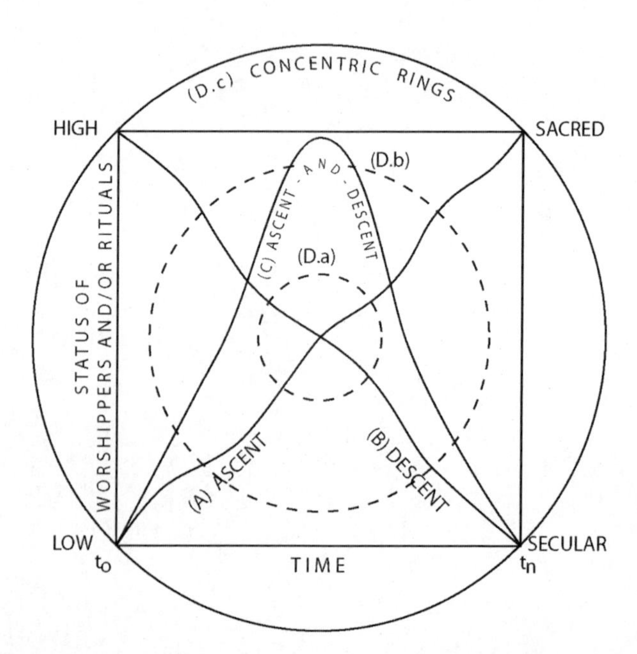

Fig. 100.6 Ritual Mandala and time geometry in terms of secular-sacred hierarchy (after Singh, 2009, p. 247)

births, deaths, and rebirths. With good *karma*, a person can be reborn into a higher caste or even to godhood. Bad *karma* can relegate one to a lower caste or even to life as an animal in their next life.
- *Moksha* (*Nirvana*). This is the goal of the Hindu. Nirvana is the release of the soul from the seemingly endless cycle of rebirths.

There also exist and continue to exist threefold spaces of Hinduism, of course they have areas of overlapping and also sometimes contestations for identity and superiority:

(a) "Village Hinduism" prevailing in rural India (inhabiting 68 % of the population) made up of grassroots, "*Little tradition*" Hindu ritual practices including shamanistic traditions of ecstatic experience but with some observance of pan-India mainline Hindu practices;
(b) Literate, or scripture-based "Sanskrit, Vedic Hinduism" is a "*Great tradition*" variety represented by Brahmin priests, *pandits*, itinerant ascetics or monastic practitioners who propagate the ancient mythology and dominate the network of Brahmanical system; and
(c) "Renaissance Hinduism" or Neo-Hinduism ("reformative") is popular among the urban alienate, a portion of the new urban middle class, (often followers) of Ramakrishna (1836–1886), Vivekananda (1863–1902), Satya Sai Baba (1926–2011) and many others, and active in the missionary programmes in India and abroad (cf. Larson 1995: 20–21; Chapple and Tucker 2000: xxxix).

Hindu adherents, standing fourth (ca. 1.08 billion as of 2007) in the numerical hierarchy at global level, after Christianity (2.1 billion), Islam (1.5 billion), and Secular/Nonreligious/Agnostic/Atheist (1.1 billion), maintain closer ties and links with other religious groups like Buddhism (376 million), Jainism (4.2 million) and Sikhism (23 million), the religions that originated and grew up in Indian soil long back in history, a process that resulted in a mutual cohesiveness. As shown in Table 100.1, the Hindu percentage of population varies by country.

In India the Hindu community shares about 80 % of total population, but that percentage is slowly decreasing. Contrary to this pattern is the Muslim population, which is increasing in numbers and its percentage (Table 100.2).

Over the course of time such people organized themselves for socio-cultural activities by establishing their religious-cultural centers for community services that also extend support to Hindu community organizations in India. Such centers and temples act not only as community cultural centres, but also raise funds for charity works in India and remain central to philanthropic giving.

A scenario of diasporic Hindus acknowledges that

> The articulation of identity by the Hindu migrants to Great Britain and North America, for example, has typically focused more on religion than on language or ethnicity. Religion represents a practical and socially acceptable source of identity in these places and maintains for the individual a strong sense of connection to the certainties of a familiar worldview and ethos. For Hindus, moreover, religious identity often provides the foundation for a larger diasporic community at local scales than would language or ethnicity. (Stump 2008: 374)

Table 100.1 Hindu percentage of the population, by country

Country	% Hindu
Nepal	81.0
India	79.8
Mauritius	50.0
Fiji	33.0
Guyana	33.0
Suriname	27.4
Bhutan	25.0
Trinidad and Tobago	22.5
United Arab Emirates	21.3
Bangladesh	12.4
Sri Lanka	12.1
Kuwait	12.0
Qatar	7.2
Malaysia	7.0
Réunion	6.7
Bahrain	6.2
Oman	5.7
Singapore	5.1
Belize	2.3
Indonesia	2.0
Seychelles	2.0
Pakistan	1.8
United Kingdom	1.7
Canada	1.0
Kenya	1.0
New Zealand	1.0

Data source: Wikipedia (2013)

Obviously, the articulation of a common Hindu identity often brings together individuals whose traditional expressions of Hinduism have differed, the result being processes of simplification, homogenization, or innovations in religious practice as believers emphasize the commonalities in their understandings of tradition.

Still about one-half of the Hindu population in India is not really educated or literate in a strict sense and the majority exhibits little concern for the present century or economy. Yet they are still listening to mythologies and practicing rituals without a critical and rational testing of their relevance today, and they still hope that an universal order be maintained in this era of *Kali-Yuga* by coming of savior, Kalki.

Indian culture has never been isolated from the outside world. At no time did India or Indian kingdoms close itself off from interaction with the larger world like what happened in periods of East Asian history. Seals produced during the Harappan civilization (2500–2000 BCE) have been found in the countries of the Persian Gulf

Table 100.2 Population and decadal growth of Hindus and Muslims in India, 1961–2011

Religious community	Percentage of total population						People 2001, million	People 2011, million	Decadal growth rate, %			
	1961	1971	1981	1991	2001	2011			1971–1981	1981–1991	1991–2001	2001–2011
Hindus	84.4	83.5	83.1	82.4	80.5	79.8	827.6	965.5	24.2	25.1	20.3	16.7
Muslims	9.9	10.4	10.9	11.7	13.4	14.6	138.2	177.3	30.8	34.5	36.5	29.3

Data sources: Census of India, various reports

attesting to a trade with the Middle East at this early stage. The post-Alexander Seleucid Bactria spread its influence into northwest India after its independence 250 BCE and became a melting pot of Greek and Indian culture and religion. Early Buddhist art in North India was dominated by the Gandhara style that was one of the aesthetic results of this interaction. The trade of spices, textiles and incense connected India with Rome and with the larger Europe for many centuries. During the same centuries Buddhism spread from India towards East and North and became a major vehicle for intellectual exchange with larger Asia through a well-established network of monasteries and universities. Hindu and Buddhist culture also spread South East over the seas and established itself through Indo-China and Indonesia. From about 1000 CE India became increasingly influenced by Islamic culture and thus was connected to the Middle East and North Africa. Vasco da Gama's arrival to Calicut (Kozhikode) at the Malabar Coast in 1498 boosted the trade with Europe but at the same time paved the way for European colonization during the next centuries. Christian missionaries were generally on board on the ships together with military, trade and administrative personnel, and through conversion new Christian communities were added to the original Saint Thomas Christians or Nasranis in Kerala. With the British colonization of India from 1798 and onward the mutual influence between India and Europe of religious as well of as secular ideas increased. Hindu reformers like Ram Mohan Roy were open to inspiration from Christian and Western social ideas while Vivekananda later experienced great success in exporting Vedanta to the West. After the Second World War the gradual increases in the world economy made it possible for large groups of young spiritually seeking Westerners to travel to India and for Indian religious leaders to settle in Europe and America where they established Hindu religious institutions and networks that today are permanent parts of the religious landscape in the West.

100.2 Globalization and Hinduism

The process of globalization has also promoted cultural interactions and the acceptance of cultural differences and the establishment of universal norms of behavior in which religions have also intersected in a significant and complex ways, including movement and propagation of religious ethics, and promotion of tolerance, transformation and acceptance of other ideas at the global scale. What is unique about this wide history of cultural and religious globalization is that Hinduism remained such a strong tradition in India. Whereas the polytheistic religions of the Mediterranean, Arabian and North African cultural spheres vanished under the Christian Churches and the Islamic Ummahs, the Hindu traditions of India were capable of resisting missionary efforts. We should also remember that unlike Christianity, Islam and Buddhism, Hindu traditions do not owe their existence to one founder from whom a more or less uniform set of doctrines was established. These two factors, its polytheistic mythology and its openness of doctrine, have together contributed to the great diversity of the Hindu traditions. This diversity expresses itself in the remarkably rich and large narrative culture of India. With the disappearance of Greek, Roman, Germanic and other indigenous Indo-European mythologies, the European stock of religious narratives was severely reduced. In spite of the revival by artists and literati during the European Renaissance, it never became a living culture again. This is in strong contrast to India. Whereas Mediterranean mythologies died during the first half of the first millennium CE, Hindu traditions have had 2,000 years more to proliferate, enter into new syntheses and become even more multiple. Even those Hindu reformers like Ram Mohan Roy who insisted on the singleness of God, coupled by a critique of so-called Hindu "idolatry," never succeeded in winning the minds and hearts of the masses.

Hindu traditions are woven into a common history and have been systematized by priestly and other elites in such a way as to preserve a recognizable unity in a seemingly irreconcilable diversity. This process has been described as an interplay between local so-called "small traditions" and a pan-Indian "great tradition" whereby local groups or religions have aspired to recognition through adaptation of characteristics of the great tradition, or in terms of Indian ethnography, as a process of "*sanskritisation*" (Srinivas 1952). The process attests to the rich and diverse demography that throughout Indian history has been one of the major factors of Indian social dynamics.

Even today with modern media of communication Hindu traditions and the Hindu self-awareness remain strong. The rich and great narrative tradition mentioned already is undoubtedly an important factor. This tradition is a major means of socialization not only in homes where parents tell their children the stories of Rama or Ganesha, but even also in school books, TV series, Bollywood production, songs, cartoons and many other forms of popular culture.

Based on the essence of Hindu ethics, Gandhi's method of "truth force" (*satyagraha*) led to India's independence from England in 1947 and also inspired Martin Luther King Jr.'s non-violent methods in the American civil rights movement in the 1950s and 1960s. Moreover, in the nineteenth century, the American writers Emerson and Thoreau had been affected by, and in turn encouraged, the study of Hinduism. Since the 1960s the growth of *yoga* and meditation movements to promote physical and mental health in the United States and Europe indicates that Hindu methods can be beneficial even when unaccompanied by faith in the theoretical worldview of Hinduism.

Historically, there have been deep and complex tensions between Hindu reactionaries and extreme Islamists that continue even today in different contexts. These stem largely from the clash between the Muslim monotheistic and Hindu polytheistic outlooks. Nevertheless, Hinduism reveals a strong acceptance of other faiths. Swami Vivekananda (1893) illustrated this in his lectures in the World Parliament of Religions at Chicago, which encouraged acceptance and understanding of Hinduism and its teachings. Since mid twentieth century with the pushing forces of globalization, the ethics and basic Hindu philosophy of Human-Nature interrelationship has been in process of wider acceptance.

The proliferation of Hindu temples now spread over the North American religious landscape appears as part of a new process of globalization for Hinduism in an era of transnational religions. Many Hindus today are urban middle class people with religious values similar to those of their professional counterparts in America and Europe. Just as modern professionals continue to build new churches, synagogues, and now mosques, Hindus are erecting temples to their gods wherever their work and their lives take them. Despite the perceived exoticism of Hindu worship, the daily lifestyle of these avid temple patrons differs little from their suburban neighbors (cf. Waghorne 2004). The construction of huge temples and replication of grand temples of the south in the north Indian big cities is another way of mass acceptance and using Hindu sensibility for religion and religious tourism.

Nevertheless it would be wrong to say that Hindu traditions are unaffected by modern globalization processes. According to recent globalization studies there seems to be two opposite effects of globalization on religion. One is the tendency witnessed in certain more conservative religious environments to close themselves off from others and to emphasize the borders to the surrounding society the other is the tendency toward openness of other more liberal religious groups (Beyer 1994: 86ff). Both of these main tendencies can be seen in the religious landscapes of India. The conservative tendency is regarded as a reaction to the pressures of globalization in the form of increased fragmentation and competition from other ideologies. Globalization essentially means that global differences are shared by everyone everywhere. The world has become a single place. This increase of complexity (socially, conceptually, ideologically, economically, etc.) triggers defensive and nostalgic reactions for the more conservative religious groups. For the more liberal groups, on the other hand, globalization is regarded as an opportunity for strengthening the profile of the group by engaging in global issues of peace, environment, climate debates and similar areas that tend to be left unsolved by politicians and that are in need of long term solutions.

Thus, it would not be wrong to see the growth of Hindu Nationalism during the last three decades in the light of globalization. Notwithstanding the fact that most of the Hindu communalist movements have their roots in the pre-1947 struggle for independence, the vitality and growth that we have witnessed during these decades present significant evidence of conservative reactions to globalization. At the other end of the spectrum we find many religious movements that have opened themselves to non-Indians and have articulated their message in terms of global values of peace, non-violence, spirituality and human development. Many of these Hindu movements are also established in the West with large groups of Western followers. In the West they have for some time been influential in actually changing the religious landscape and the religious priorities of established Western religion. Thus, in an attempt to adapt to the development, the Christian Churches in many Western societies have had to look towards such Hindu movements in their own countries in order to understand the changes going on in Western popular religion. It is by no means unusual for Westerners to practice *yoga*, believe in *karma* and at the same time have their children baptized in the local church. Hindu traditions like *yoga* and Hindu ideas like *karma* and reincarnation have for long been import products along with Ayurvedic therapy and Buddhist inspired Mindfulness. There can be no doubt, however, that over time the influences created by long time globalization processes between Western and Indian religion and thinking have worked both ways.

To be sure Hinduism is a remarkably diverse religious and cultural phenomenon, with many local and regional manifestations. However, within this universe of beliefs several important themes emerge that are to be considered as key teachings for the environmental issues at global level in the twenty-first century:

- The earth can be seen as a manifestation of the goddess, and must be treated with respect, in a way supporting the Gaia theory. Many Hindu rituals recognize that human beings benefit from the earth and offer gratitude and protection in response.
- The five elements (*pancha-mahabhutas*), space, air, fire, water, and earth are the foundation of an interconnected web of life and represent primal energy (see Fig. 100.5).
- *Dharma*, often translated as "moral duty," can be reinterpreted to include our responsibility to care for the earth and service to humanity.
- Our treatment of nature directly affects our *karma*. Moral behavior creates good karma, and our behavior toward the environment has karmic consequences.
- Simple living is a model for the development of sustainable economies and promoting global brotherhood.
- Mahatma Gandhi exemplified many of these teachings, and his example continues to inspire contemporary social, religious, and environmental leaders in their efforts to protect the planet.

The greatest loss recorded in Hindu traditions during the colonial period was the loss of the old ethic of eco-justice, which refers to the sanctity of life and cosmic interconnectedness (ecological cosmology) that extends to the sense of a global family or a universal brotherhood *(vasudhaiva kutumbakam).* This ethic helped society to maintain an order between *dharma* (moral code of conduct) and *karma* (right action). In the course of acculturation, the ideology of materialism, consumerism and individualism, which was always proscribed in traditional Hindu thought, has been accepted by contemporary society. At the other extreme, and perhaps in consequence, the movement of the revival of ancient cultural values is being turned to fundamentalism by some groups. The old principle of *satyameva jayate* ("Only truth triumphs") is now replaced by *arthameva jayate* ("Only wealth triumphs") (Singh 1999: 59). The foreign cultural domination of India during the last 700 years and the influence of imported culture during recent globalization processes have played a major role in this form of transformation (Dwivedi 1990: 210). Hinduism in India today faces several crises, among the notable ones are:

- Decline of spiritual sense (*dharma*) of Hinduism: lacking an understanding of nature worship, and associated meanings, messages and contexts of rituals;
- Using public show and celebrations as expression of the religion;
- Predominating dogmatic orthodoxy to Hinduism, and emotionally blackmailing of the innocent-poor masses;
- The prevalence of superstitions and thereby marginalizing reasoning and rationality in lack of insights from science;

- Rising Middle class and their lust for a Western irresponsible capitalist culture;
- A lack of social consciousness, self-realization and service to sufferers, and social services, and
- The threatening pace of religious fanaticism and fundamentalism.

As the impact of globalization and the new technology like television and Internet are taking more pace the Hindu society continues to urbanize and begins to influence forms of worship, resulting to less segmentation in Hinduism. In a way, and maybe for the first time in history, Hinduism is turning into a unified religion through acceptance of pan-India level festivals and popularization of several regional festivals like *Karwa Chauth* and *Chhatha*, in a more standard form in big cities like New Delhi and Mumbai. Also the caste-affinity and related conservative rule are loosening, and newer sects like Swaminarayan coming as a grand show pieces through their architectural grandeurs (cf. Luce 2011: 312–313). Quite naturally technology helped in speeding up the nationalization of Hindu practices and the acquaintance the mythological stories of Hindu gods. The wide acceptance and popularization of pilgrimage-tourism is also supported by Hindu consciousness of their identity, globalization and propagation by the media (cf. Singh and Haigh 2015, f.c.). Nevertheless, in the passage of time Hinduism today has been unable to check the negative consequences like the break-up of joint families, disrespect of elders, excessive consumerism and materialism, individualism, abandonment of moral values (*dharma*) and lust for money. The sense of self-retrospection is slowly being replaced by blaming colonialism and westernization; traditional values of the ancient past are uncritically accepted as worthy frame for the future; ritual scenario of Hindu traditions are used as tools for political support; and secular democracy is narrated through religious vision, and so on.

100.3 Conclusion

Perhaps, when the forces of globalization have spent themselves, the arboreal shoots will take root again and spring into other stable tree trunks (in the Banyan) through which the same life force will continue to flow. In India, religion is currently used as a tool for 'secular democracy' with the support of 'secularization of religious ideas.' The Indologist Gerald Larson (1995: x) opines that the post independence 'secular state of India' is to a 'significant degree a forward caste Neo Hindu state, or, in other words, that the "secular state" in the Indian context has a number of religious aspects and may even represent in some respects a religious entity.' The philosopher S. Radhakrishnan (1959: vii–viii), who was the second president of India, argued that the Indian idea of secularism 'does not mean irreligion or atheism or even material comforts. It proclaims that it lays stress on universality of spiritual values which may be attained in a variety of ways. … This is the meaning of a secular conception of the state though it is not generally understood.' In maintaining secular democracy, the Hindu sense of tolerance, a living tradition, 'has contributed vitally. … More important, is the attitude of "live and let live" toward all manifestations of religious diversity' (Smith 1963: 149). During the last decades there have been recorded several incidences of Hindu-Muslim riots and religious contestations. However, it is hoped that in time Hinduism, with its inherent virtues of tolerance, ethical values, and concept of *dharma* linked to the four ends of life, will resuscitate itself and rise from own ashes like the phoenix and that one of the world's oldest religions will live on (Bhela 2010–11: 100). Frawley (2008) remarked that

> The Hindu tradition embraces both spiritual and scientific knowledge, both religion and culture, not dividing them up such as has occurred in the West. Such an integral Hindu vision is quite in harmony with the dawning planetary age. But it is not as yet articulated in a practical manner. Nor do Hindus understand the threats to their culture that globalization brings.

He has rightly mentioned that

> The Hindu response to planetary concerns, particularly of an environmental and ecological nature, has been rather weak. India has already been devastated in terms of its ecology and this trend is increasing at an alarming rate. But there is little organized Hindu effort to counter this, the great Hindu tradition of Bhumi Puja (worshipping the earth) and love for Mother Earth notwithstanding. (Frawley 2008)

Nevertheless the Western criticism that Hinduism suffers from moral relativism, expressed in millions of gods, lots of scriptures, many *gurus*, total freedom, etc. is still true in different ways and varying degrees. However, the majority of Hindus never accepts such hard realities and believes that Sanskrit is a divine language and the ancient texts are the voices of gods (cf. Malhotra 2011: 211–215).

Hinduism has already been comfortable historically and culturally in accepting, absorbing and getting transformed whatever taboos and traditions came in contact – resulting into constantly expanding its belief systems. Hindus are taking solace in the fact that Hinduism always has been at ease or comfortable with plurality; so nothing is surprisingly new in the modern plurality created by globalization. Hindu traditions are a complex web of multiplicity – multiple Gods, multiple practices, multiple ways, multiple means and also simultaneous multiple ontological structures of monotheisms, monisms, polytheisms, and panentheisms – altogether that converges into multiple wholes of mosaicness (cf. Biernacki 2010: 1). Hinduism around the world absorbs facets of "modernity (coping with science and technology) and post-modernity (the erosion of traditional values mostly due to globalization and cross-cultural influences). However, due to its open-minded theology, and the Hindu penchant for absorbing and reinterpreting new innovations, contemporary researches into cosmology and landscape ecology do not threaten the tenets of the belief systems as is the case for the prophetic monotheisms that rely upon the Genesis narrative" (cf. Chapple 2010). Cybernetic and automation technology have been embraced as a new worship tool, with rituals available online and with live webcam and television broadcasts from India's most holy sites that have helped in shrinking the worlds of Hindus otherwise scattered all over the globe.

References

Beyer, P. (1994). *Religion and globalization*. London: Sage.

Bhela, A. (2010–11). Globalization, Hinduism, and cultural change in India. *Asia Journal of Global Studies, 4*(2), 93–102.

Biernacki, L. (2010, July 28). *Future of Hinduism: A rich and strange metamorphosis: Glocal Hinduism*. Patheos resources. Retrieved February 27, 2012, www.patheos.com/Resources/Additional-Resources/A-Rich-and-Strange-Metamorphosis.html

Chapple, C. K. (2010, June 29). *Future of Hinduism: New and old in conversation: Hinduism's balancing act*. Patheos resources. Retrieved February 27, 2012, www.patheos.com/Resources/Additional-Resources/New-and-Old-in-Conversation-Hinduisms-Balancing-Act.html

Chapple, C. K., & Tucker, M. E. (Eds.). (2000). *Hinduism and ecology: The intersection of earth, sky and water*. Cambridge: Harvard University Press.

Dwivedi, O. P. (1990). Satyagraha for conservation: Awakening the spirit of Hinduism. In J. R. Engel & J. G. Engel (Eds.), *Ethics of environment and development* (pp. 201–212). Tucson: University of Arizona Press.

Frawley, D. (2008). *Hindu response to globalization*. Retrieved January 26, 2012, Web: http://findarticles.com/p/articles/mi_hb155/is_1_17/ai_n28885242/

Larson, G. J. (1995). *India's agony over religion*. Albany: State University of New York.

Luce, E. (2011). *In spite of the gods: The strange rise of modern India*. London: Abacus.

Malhotra, R. (2011). *Being different: An Indian challenge to western universalism*. New Delhi: HarperCollins.

Michaels, A. (2004). *Hinduism: Past and present* (trans: Harshav, B.). Princeton: Princeton University Press.

Radhakrishnan, S. (1959). *Eastern religions and western thought*. London: Oxford University Press.

Singh, R. P. B. (1984). Toward phenomenological geography of Indian village: A dialogue of space-time experiences. In R. L. Singh & R. P. B. Singh (Eds.), *Environmental appraisal and rural habitat transformation* (pp. 103–115). Varanasi: National Geographical Society of India, Pub. 32.

Singh, R. P. B. (1999). Rethinking development in India: Perspective, crisis and prospects. In D. Simon & A. Närman (Eds.), *Development as theory and practice* (pp. 55–75). London: Addison Wesley Longman.

Singh, R. P. B. (2009). *Geographical thoughts in India: Snapshots and vision for the 21st century*. Newcastle upon Tyne: Cambridge Scholars Publishing.

Singh, R. P. B., & Haigh, M. (2015). Hindu pilgrimage and a contemporary scenario. In S. D. Brunn (Ed.), *The changing world religion map (f.c.)*. New York: Springer.

Smith, D. E. (1963). *India as a secular state*. Princeton: Princeton University Press.

Srinivas, M. N. (1952). *Religion and society amongst the Coorgs of South India*. Oxford: Clarendon.

Stump, R. W. (2008). *The geography of religion: Faith, place and space*. Lanham: Rowman & Littlefield Publishers.

Vivekananda, S. (1893). *Four lectures in the world parliament of religions*: Response to welcome (11 Sept.), Why we disagree (15 Sept.), Paper on Hinduism (19 Sept.), and Religion not the crying need of India (20 Sept.). Chicago: World Parliament of Religions. www.vivekananda.net/BooksBySwami/CompleteWorks/. Accessed 20 Feb 2012.

von Stietencron, H. (1989). Hinduism: On the proper use of a deceptive term. In G. D. Sontheimer & H. Kulke (Eds.), *Hinduism reconsidered* (South Asian studies no. 24, pp. 11–27). New Delhi: Manohar Publications.

Waghorne, J. P. (2004). *The diaspora of the gods: Modern Hindu temples in an urban middle-class world*. New York: Oxford University Press.

Chapter 101
The Diasporic Hindu Home Temple

Carolyn V. Prorok

Now it is time that gods emerge from things by which we dwell....

Ranier Maria Rilke 1925 (The first lines of a poem cited by D.F. Krell in Martin Heidegger 1977)

101.1 Introduction

A common perception of Hindu sacred space and places of worship often revolves around magnificent specimens of monumental architecture. Voluptuously sculpted Khajurao, whimsical bas relief of elephants on parade at Mahabalipuram, and the sublime towers of Dakshineswar are among Hinduism's[1] greatest

[1] The term Hindu is largely a British colonial invention that, in some ways, is meant to cover most if not all indigenous faith traditions from the Indian subcontinent (Doniger 2009). Strangely to us today, colonial Brits included the localized practice of Islam in this category. Sikhs and Jains have been somewhat successful in distinguishing themselves from this broad category. Many other faith communities simply accept the designation while others struggle to reject this label. Dalits, who represent countless village traditions, are among the more activist today (Kumar 2004). Still, an extraordinary philosophical literature and a rich ritual tradition have developed over several millennia that are more or less linked through a number of common sensibilities. In this sense, the term holds enough relevance for our purposes here. In any case, we are stuck with this term for now. Throughout this essay, the term Hindu will largely encompass the classical Brahminical tradition with which many in the world are familiar as well as the village traditions upon which it is often juxtaposed. Brahminical Hinduism, aka Sanatan Dharma, encompasses the classical Hindu pantheon of great gods and goddesses and the complex ritual heritage that attend them. The great temples are likely to be Brahminical. Village Hinduism varies widely in local practice, includes the sacrifice of animals, usually to the local goddess that protects the village, and it is strongly linked to local natural elements (Wiser 1971; Dubois 1906; Schwartz 1967; Mines 2010). These two broad classifications are not mutually exclusive as they often intersect, intertwine, and interpolate each other in actual practice. The terms Great Tradition and Little Tradition were coined by Western anthropologists decades ago to distinguish them and may be outdated in usage today. Thus, the terms Brahminical and Village traditions will be used here, though reference may be made to usage of the earlier terms.

C.V. Prorok (✉)
Independent Scholar, Box 111, Slippery Rock, PA 16057, USA
e-mail: carolyn.prorok@sru.edu

S.D. Brunn (ed.), *The Changing World Religion Map*,
DOI 10.1007/978-94-017-9376-6_101

architectural achievements.[2] Temples may also be somewhat modest in scope and décor. Such temples are found throughout the subcontinent and abroad. Less well-known, yet so numerous as to exist in the millions in India, are shrines to specific deities in the villages, under trees, along roadways, and capturing open spaces in urban zones from street corners and road dividers to sharing the wall of compounds and parks/gardens. They are often small (1–3 m^2), but they are colorful and attract many passersby for a brief encounter with God. Despite countless places for faithful worship in India and abroad, most Hindu families do not visit them often or regularly. Rather, their own homes are centers for worship and ritual observance.

Practically invisible to non-Hindus and non-family members are home altars that are essential for the spiritual and material well-being of households.[3] Practicing Hindus maintain an altar or a shrine in available space inside the home Khare (1976). In addition, another altar may be built outside along an interior wall or in a garden. The home is the *a priori* sacred space and place in Brahminical Hindu life, particularly in its supporting role for the ritual reproduction of the Hindu family. Predicated on the establishment of this critical sacred space and place, an architecturally distinctive temple culture in diaspora emerges considering relevant regional styles in India, local cultural aesthetics and available materials, and globalized representations of Hinduism. A synergy between the micro (individual home) and the macro (global Hindu diaspora) sacralization of space and place is key to our understanding of the continued vibrancy of Hindu communities in lands often unsupporting or even antagonistic to their religious life.

The following essay is not a claim that Hindus exclusively produce and reproduce their homes as sacred centers.[4] A case can be made that all *homes*, in the broadest sense, are sacred to some degree. Instead, I present particular ways in which Hindus create and recreate the sacrality of their homes as critical to their very existence as Hindus. My presentation relies largely on 30 plus years of field work in Hindu communities around the world, and more importantly, the incredible hospitality of countless families who invited me into their homes and their lives over the years.[5]

[2] Khajuraho, eleventh century, northern Madya Pradesh (southeast of Delhi); Mahabalipuram, eighth century, northeastern Tamil Nadu (south of Chennai); Dakshineswar, nineteenth century, eastern bank of the Hugli River (south central Kolkata).

[3] Lindsay Jones (2000: 141) notes that, "…for most Hindus, the focus of their ritual lives was not the famous temples but the small altar that existed in virtually every private home."

[4] Catholics, especially of an older generation, may have a shrine inside the home or yard. Evangelical Christians invite God into their home through prayer, and many Muslims set aside a place for daily prayer. Espiritistas, Santeras, and Mambas of Afro-Caribbean faiths often use their homes in the U.S. as ritual centers. Homes in traditional cultures around the world have some degree of sacred organization or functionality. Moreover, the home may be an initial place to worship for any religious community when they are a minority in a new land and with limited resources. The particular ways in which this plays out will have its own dynamic.

[5] I want to extend my deepest appreciation to the many families in India and the diaspora for taking me into their homes. My life is so enriched by all of you. Basday, Lucy, Lydia, Vaidehi, Mala, Neeraja, Kiran, Ruth, Bhadra, Mrs. Ismail, Kanchan, Daphne, Dinesh, Amoy, Nirmala, Sherry Ann, and Padma are especially dear to my heart. Importantly, Rose Salville-Iksic deserves special recognition for her critical role in French translation, the day in—day out survey of La Réunion family temples, and her cheerful support in all things geographical.

101.2 Brief Background and Framework

From the 1830s to 1917, Britain transported hundreds of thousands of people from the Indian subcontinent to its far-flung tropical empire as laborers (for example, the Caribbean, Southeast Asia, Eastern and Southern Africa, Pacific and Indian Ocean islands). In addition, migrations from the subcontinent deep into Southeast Asia (for example, Bali) have occurred since early medieval times. Finally, after WWII, widespread migration to North America, Europe, Australia, New Zealand, and the Arab Gulf States has resulted in a substantial, contemporary diaspora. Combined with earlier migrations, more than 30 million people of Indian (both colonial and contemporary India included) origin are living in dozens of countries. About 60 % of this population identifies as Hindu though this proportion changes with each community abroad. Thus, being Hindu and being of Indian origin should not be conflated. Many people of Indian origin are Christian, Muslim, or Sikh. Some are secular or identify with other minority faiths. Moreover, some Hindus actively practice multiple faith traditions. Although the role of the home in Hindu life and spirituality is more or less universal, the following essay will focus on communities that self-identify as Hindu in the former colonies that received indentured laborers from the Indian subcontinent in the nineteenth and early twentieth centuries.[6]

The humanistic approach in cultural geography will be used as a framework for understanding the experience of Hindus abroad as many continue to struggle to reproduce their way of life in an alien world.[7] Philosophers such as Edmund Husserl (1960), Martin Heidegger (1977), Gaston Bachelard (1964), and geographers such as Yi-Fu Tuan (1977, 2001), Anne Buttimer (1976, 1993), David Seamon (2000), Seamon and Mugerauer (1985), and Edward Relph (1976) have developed considerable insights into how we humans experience the world, create and recreate meaningful places, and in the end, make it possible to survive beyond our mere physical need of food, water, and shelter. Critical to human existence is a sense of *place* such that our experiences—and perceptions of those experiences—(re)produce varying feelings of *belongingness* to a neighborhood/town/village, community, environment, nation, or even our very home. Our active engagement in creating such *belongingness* is inextricably linked to our ability to *dwell*—to build our lived world from that which is meaningful, satisfying, and ultimately reflective of what it means to be authentically American, or Tibetan, or Hindu in the world. In this sense, we will explore how building the home temple is an authentically Hindu act

[6] The British were the first European colonizers to transport people from India to their colonies as a source of labor on plantations (for example, sugar, cacao, rubber) and to build railroads Laurence (1994). The French followed but to a lesser degree. The Dutch took advantage of the British system and arranged for laborers in their Caribbean holdings. The Dutch later used laborers from the island of Java in the same way. My field work has primarily occurred in India, Trinidad (of Trinidad and Tobago), Suriname, Malaysia, Zimbabwe, and La Réunion. Other fieldwork in the U.S., UK, Netherlands, Sri Lanka, Singapore, Nepal, Australia, and New Zealand is not the focus of this essay, but the ideas here do hold some relevance.

[7] This chapter could just as easily have used a feminist approach, or a post-colonial, post-modern, or political economy approach to understanding Hindu home temples in diaspora Jones (1997).

of *dwelling* that has been essential to people's survival as a distinctive community into the twenty-first century, despite the persistent onslaught of proselytization, discrimination, and exploitation under nineteenth and twentieth century (post)colonial conditions.

101.3 Binding the Home

My mother trained her baby boomer brood to take off our shoes when we came into the house. For her, it was a strategy that forestalled pulling out the mop and vacuum cleaner more than once a day. For Hindus, it has nothing to do with soiling the floors in the conventional sense. Given my childhood training, taking off my shoes at the thresholds of my Hindu friends' homes was a reflexive and unquestioned ritual. As my experiences and understanding of Hindu cultural and religious life in India developed over time, I finally realized that my barefooted friends and their house guests brought street grime into the home on the soles of their feet (they wipe down the floors regularly). So I finally asked, "Why do you leave your shoes at the threshold?" An answer that made complete sense: "shoes are made of leather" was proffered. Of course; I knew this to be the case for temples. This was my first insight to the temple and the home being equivalent as reflected in people's habitual engagement with each one. The remains of dead animals are ritually polluting[8] and their handling and usage play a significant role in social organization (for example, vegetarianism and caste).

The world outside of the threshold of homes and temples is fraught with countless polluting factors. One can protect the spiritual integrity of homes and temples, and by extension our own bodies, by ritually fortifying their boundaries and by social pressure to maintain "clean" habits. In the latter case, bathing and fasting before morning prayers and avoiding animal products (except milk which is sacred) in one's diet are just two examples. In the latter case, diasporic Hindus practice a wide variety of adapted vegetarianisms and, in the former case, families ritually mark

[8] The concept of ritual pollution is operative in most societies in a variety of ways and to different degrees. Here in the United States, insects or bugs if you will, are polluting as a social factor. People with bugs (for example, bed bugs, roaches, fleas, lice) in their living space or on their bodies are considered socially inferior—remember the reference our society still uses to indicate some marginalized people as having "cooties"? It is not my intention to minimize or rudely represent this sensibility. Instead, it is merely an indication of how we humans mark and marginalize the "other" in our daily lives. In Brahminical Hindu life, consuming most animal products, manufacturing products from dead animals, and handling animal/human remains for disposal all place a person in spiritual jeopardy, and by extension, implicates many aspects of social life. Contact with polluting substances, and or persons polluted by them, can affect one's socio-spiritual status. Avoiding and/or ameliorating one's contact with proscribed substances and/or people (thus protecting oneself and one's family) through daily practice and cleansing rituals is continual. Keep in mind that such sensibilities are difficult to erase in any society, but government policies and activist groups in India are working to eliminate discrimination based upon these age-old social attitudes. Issues surrounding ritual pollution have some elements calcified and others eliminated altogether from one diasporic community to another.

Fig. 101.1 A young woman recreates rangoli from rice flour each morning at the threshold of her family's home in Bangalore, India (Photo by Carolyn V. Prorok 2005)

Fig. 101.2 Rangoli in St. Benoît, La Réunion—an island in the Indian Ocean that is a French department (Photo by Carolyn V. Prorok 2010)

the boundary of their home space in a number of ways. One way that is common throughout India and to some degree in the diaspora, is the creation of white or colorful geometric and floral patterns that protect the threshold of the home from unwanted spiritual intrusions. Often called *rangoli*, different regions of India have their own terms for these beautiful designs (Figs. 101.1 and 101.2). Friends who live in high-rise buildings make *rangoli* right in the hallway or just inside their apartment doors.

Fig. 101.3 A simple family altar is tended daily at the corner of a Hindu home in Pulau Pinang, Malaysia. Surrounding the altar is a patch of Tulsi or sacred basil—*ocimum sanctum* (Photo by Carolyn V. Prorok 1994)

In addition, a *torana* (gateway) of mango leaves can be strung across the top of the doorway (especially on holy days), images of deities can be carved into the door or any other entry to the living space, small altars can be established at the corner of the house or yard wall (Fig. 101.3), *mala* (garlands) of fresh flowers (Fig. 101.4) or sacred seeds can be hung in any number of places, traditional symbols of Hinduism (e.g., *ohm*) can be displayed, or the family can erect symbols of a recently occurring ritual event at the home such as the *jhandi* (pennant on a bamboo pole), which is mostly found in Bihar, India and Trinidad (Fig. 101.5).

The perimeter of the house and associated open or walled area around the house may have only one protective object or it might have many of them in multiple forms. Not usually visible to even guests visiting in the front room will be an altar in the home. It is here that images and artifacts revolving around the ancestral patriline, the deity that has been worshipped for generations by the patriline, and any other deity or *swami* with which family members have formed a relationship can be found. Maintaining the altar, making daily offerings, and reciting relevant prayers every morning and evening are primarily the responsibility of the eldest woman in the home although any family member can participate. Men may or may not actively engage with the shrine on a daily basis, but they do have ritual responsibilities for specific events. This gendered division of ritual labor is rooted in the belief that women, as natural repositories of *shakti* (the creative energy that makes this material world possible (Menon 2002; Diesel 1998; Pintchman 2001; Yagi 1999)), must, through her actions, guarantee the spiritual health (and by extension the material health) of the household (Fig. 101.6).

The virtuous wife has great reserves of *shakti*. While the sexual implications of *virtue* are always foregrounded (Kelly 1991; Puri 1997), for a practicing Brahminical

Fig. 101.4 Mala is making garlands from posies in the garden at her home in Chennai, India (Photo by Carolyn V. Prorok 2005)

Hindu, virtue is implicated at a number of levels that protect and enhance one's *shakti*, which must always serve the perpetuation of the patriline. A weakened, or diminished, *shakti* weakens the husband and by default, the patriline (Sekine 1999; Brodbeck 2009; Menzies 2004; Nagar 1998, 2000; Kurian 1998; Gupta 2006; Goldman 2009; Pfaffenberger 1980; Orenstein 1970; Pearson 1993; Pinkney 2008; Reddock 1994; Shepherd 2002; Gupta 2009). Women bear an incredible burden in this regard, yet for those who successfully cleave to this role a sublime power can be wielded within the confines of family life. Women who are perceived to be unable to sincerely submit to this expectation are too easily discarded.[9]

[9]Traditional, conservative families who practice Brahminical Hinduism are most likely to enact this relationship between a woman's shakti strength and the expected health of the patriline. Many social proscriptions on women's mobility, social relations, and access to resources such as education derive in part from the desire of the in-laws to reserve her shakti for the benefit of the extended family. Eventually, her socialization into married life becomes self-regulating, and as her own male children marry, her status will improve within the family yet continue to exhibit many of the restrictions commonly accepted as a woman's role in life. Keep in mind that non-Brahminical practices may or may not make the same demands or regulate a woman's daily life in the same manner or to the same degree. Also, many individual families have liberalized their practice,

Fig. 101.5 MahaKali protects a home in southern Trinidad. The jhandi (pennants) at this home reflect multiple Kali Puja—different from the Durga puja but resulting in the same effects. Puja is the standardized Brahminical ritual that is a vehicle for worship of specific deities in the Hindu pantheon. The number of jhandi indicate the frequency of puja events at this home. Note the flowers used as offerings and for mala at the base (marigolds and hibiscus). Tulsi (*Ocimum sanctum*) is found to the left of the marigolds. She is considered a worthy focus of worship as well as a vehicle of worship (Photo by Carolyn V. Prorok 2003)

Maintaining the home shrine such that the spiritual integrity of the family is strengthened requires virtuous behavior and intentions. A virtuous wife channels her husband's intentions and expectations. She defers to her husband in all matters, yet it is ultimately the woman's responsibility to monitor the social, emotional, spiritual, and physical boundaries of the home such that it is a protected space

---

especially in the education of their daughters. Thus, while the status of women in India and its diaspora, in general, has been low, much progress is being made in all sectors of each society. Today, Trinidad & Tobago has an Indo-Trini woman as prime minister. In diaspora, the conditions vary greatly from one region to another.

Fig. 101.6 An interior home shrine in Harare, Zimbabwe. Notice the photo of a female ancestor on the right side of the table. The poster is of the great god Shiva and his consort Parvati—it can be bought on any city street in India, from diasporic shops that import it, or ordered off the internet. Satya Sai Baba's code of conduct is pasted to the right of the poster. Ritual materials are stored on a shelf below the table (Photo by Carolyn V. Prorok 1999)

within which the family is to thrive. Since Vedic times (3,000–4,000 years ago), ideal Brahminical practice required a ritually pure living space that is simultaneously hospitable, even under what would be considered intolerable and polluting invasions into the interior, privacy of the home A Hindu wife is ultimately responsible for providing this hospitality even as she must protect and ritually bind the contours of the home. Jamison (1996), a Vedic philologist, describes the extraordinary intrusions a proper Brahminical home—read wife—had to endure in order to maintain ultimate ritual purity while polluting agents abounded; even those introduced by her own family members. This sensibility continues to this day throughout India and the Hindu diaspora even though its idealized forms have shifted and have had to accommodate radically divergent cultural, social, and religious contexts. As a simple, yet clear example, consider my experience:

> A bell is ringing. It is a tinkling bell that starts and stops. I open one eye to soft light seeping through the shuttered window. The monsoon sky is dawning on Bombay. This is my first morning in India, and I am exhausted from the long journey yet anxious to greet the new day. My friend's mom picked me up at the airport in the middle of the night and now she is quietly chanting a mantra at the little shrine just outside the doorway of the room where I sleep. (Field notes, May 13, 1988)

Amma never laid down that night. She told me that she only wanted to bathe and offer her prayers. Later that summer, returning to Mumbai (Bombay) after circumambulating the subcontinent, I developed a close relationship with Amma such that she revealed to me her deep anxieties about my initial visit. When her son (a friend at grad school) called her to make arrangements for me, she could not refuse him despite her anxieties. He told her not to worry, but worry she did. You see, there was the concern that I might smoke, eat meat, or drink alcohol *in her home*. Moreover, I might bring strange men into the house. American women on the silver screen and in the popular Indian imagination did all of those things, and I was/am an American woman. I was received with extraordinary hospitality and grace, yet she needed to fortify herself for the emotional and spiritual energy to be expended on my behalf, and more importantly, on behalf of her son and husband. Amma was able to reveal these concerns to me after getting to know me personally as I was/am a vegetarian, a non-smoker, and fully able to eschew alcohol for any length of time. And, of course, her perception of what constitutes "strange men" was/is a non-starter. Whew! That's a relief. Whether or not I would have taxed the spiritual and emotional energy of the home is actually irrelevant. I was already welcomed before I had arrived. Amma kept a virtuous home, and a virtuous home is God's home.

Most women and men in the nineteenth-early twentieth century diaspora did not practice Brahminical Hinduism upon arrival and therefore were not subject to the same kind of idealized practices as described above. Yet diasporic women were and are still subject to narrowly defined cultural and social mores regarding their sexuality. Moreover, Brahminical Hinduism adapted to a wide range of challenging factors in order to become ascendant in many diasporic communities (see section below and notes), thus expectations concerning the many ways in which a woman expresses her virtuous character also had to adapt to local conditions. Maintaining a diasporic home as a Hindu home is a precarious endeavor that may require heroic effort.

101.4 Home Is Where the Gods Live

The family altar is a critical element in homes that practice at least some degree of Brahminical Hinduism in India and abroad. Village houses of lower caste Hindus in India traditionally did not have such ritualized practices as families were considered part and parcel of the village itself as *home* and thus protected by the local goddess through rituals and offerings at her shrine. This sensibility was (and still is in many villages) so strong that marriage practices are village exogamous (you must marry someone from another village).

In many diasporic communities, especially at a time when public temples have not yet developed, the homes of specific families may become the focus of regular ritual events such as monthly *Katha Puja* (full moon ritual) or important events such as *Divali* and *Shiv Ratri* (great night of *Shiva*). Once this process begins, the homes take on even greater ritual significance for the family and community. As the years go by, a portion of the house or yard is transformed into a temple that can accommodate fairly large groups of people (Fig. 101.7). Eventually, homes, already the spiritual center of the family, transform into actual temples and become places of public worship. This process also includes the family temples built by priests to serve their followers. Even when a community organizes to build a monumental,

Fig. 101.7 A public temple in a private home in New Nickerie, Suriname. This temple space is on the second floor of a home. The family slowly developed it over several decades, and they continue to live on the first floor (Photo by Carolyn V. Prorok 1997)

public temple, the practice of transforming one's home or some part of one's property into a temple has already been woven into the very fabric of social, political, economic, cultural, and religious life. Thus, the practice continues to this day. With their temples, diasporic Hindu communities can thrive.

When people from the Indian subcontinent were first transported to Britain's far flung tropical empire in the nineteenth century, the majority of laborers were from villages and they tended to be lower caste with a very small proportion being higher caste and Brahminical in religious practice Laurence (1994). This means that most laborers likely practiced animal sacrifice to goddesses, continued to have rituals that resulted in fire-walking and spirit possession, and they would have attempted to transfer what were localized, specific deities from their villages though this would be very difficult given their place specific character. Brahminical Hinduism does not tolerate many of these practices. Moreover, British (also French and Dutch) colonial policies did not encourage *any* Hindu practices and often discouraged it with discriminatory regulations. This dynamic often bolstered the role of Brahminical Hindus who organized and actively challenged discrimination. In this way, Brahminical practice came to dominate in some diasporic communities.[10]

[10] Read Vertovec (1992) and van der Veer and Vertovec (1991) to grasp the complexity and significance of this process in the Caribbean. They maintain that the Trinidadian form of Brahminical practice has evolved into its own identifiable, ethnically based system where a north Indian, Vaishnavite form of Brahminical Hinduism eventually became ascendant and Village Hindu practices profoundly marginalized today—though still evident if you know where to look. In La Réunion, Village practices such as animal sacrifice, spirit possession, and fire-walking are still important and widespread, even though much pressure against these practices by Brahminical Hindus is on the rise—so much so, that priests are being imported from India to displace local "creole" priests. On

Living in barracks on plantations was not so much a spiritual issue (beyond its discomforts) for village Hindus, but having no village shrine or goddess to protect them was problematic. Also, they would have come from different villages in India, so establishing an altar had to be collaborative. The greater problem was that their altars and other ritual sites were often disrespected, vandalized, or even destroyed by plantation owners and other local people; this made life in the colonies even more difficult.[11] Once indentureship was completed and families could establish their own separate homes, worship at the home was safer and more under their control. Thus, the very pressure of colonial life, policies, and negative social context vis-à-vis Hindu practices, set the stage for the Brahminical minority to gain leverage within the community. Given the already established role of home as temple in Brahminical practice, and the need of local communities of Hindus to protect their worship sites, home temples became a common feature of the landscape and somewhat standardized within any given colony such as Trinidad or La Réunion (Figs. 101.8 and 101.9). The cultural region of origin (for example, Bihar in northern India or Tamil Nadu in southern India), the particular practice of the Brahminical presence, the local and internal politics and economics of temple building, or even the cultural impact of the local colonial population of non-Hindus all played a role in the character of the home temples as they emerged in larger numbers throughout the twentieth century. In addition, from the late 1800s, *Swami* from India would go on tour and bring to the colonies and newly independent nations a window into the popular practices emerging in India or to refocus a community on their traditional heritage. As Brahminical Hindu material culture, philosophy, and practice became globalized in the last third of the twentieth century, it too would have its impact on the material character of home temples as well as the practices of individual families.

Producing a landscape replete with home temples is the result of many scales of interaction over the past 150 years. If you can imagine scaled layers and accretions of familial interactions:

1. The individual and family struggles to meet its material, social, emotional, and spiritual needs under difficult conditions. Some family members may struggle to hold onto traditional ways while others are "modernizing" or even adopting new sectarian practices. Others may be acculturating at a rapid rate to the local non-Hindu culture and others might be converting to Christianity or Islam. Some family members are travelling long distances (again) to other countries to get an education or find better opportunities. There are periods of time when most members of the family are not on the same page at the same time regarding

the other hand, read Mearns (1995) and Prorok (1998) to better understand how village practices from India prevail in Malaysia even as Brahminical organizations struggle to influence them and represent Hindus as a group to the government. As you go from one community to another around the world, the degree of Brahminization varies. More importantly, caste based social organization and practice varies widely from one former colonial community to another.

[11] Malaysian rubber plantations stand as an exception to this situation. Recruitment of laborers tended toward keeping people from the same, or neighboring, villages together and promising them the support of a temple on the plantation as an inducement to migration.

Fig. 101.8 A public/private temple in central Trinidad. The Vishwanath Cultural Center is a family temple that is dedicated to the philosophy and practice of the Divine Life Society. The Divine Life Society has a massive public temple on a major highway, but you will see home temples dedicated to a specific sectarian group across the island. It is important to note that most home temples are not dedicated to an organization. While this family temple is substantial, its features are common among family temples across the island (Note the sacred Neem tree to the left *Azadirachta indica*) (Photo by Carolyn V. Prorok 2003)

critical issues. Add in the regular responsibilities of putting food on the table or protecting the health of children, and you have a situation that requires a stabilizing practice. Moving beyond the home altar to building a family temple can bring the family together. It can also serve to give thanks to God for surviving a critical event. The act of making one's family shrine or temple publically viewable impacts one's social status within the community.

2. The family is embedded in a community of other Hindu families who also have similar struggles. Competition and cooperation between and among families, neighbors, communities across the nation or colony, and the internal struggle

Fig. 101.9 Family temple in Saint-André, La Réunion. Saint-André has a high concentration of Hindus, and not surprisingly, countless family temples such as this one. They are ubiquitous and tend to support traditional village practices like animal sacrifice and worship of folk deities. A torana of mala tied together form the patterned bunting over the doorway, and Madurai Veeran (literally warrior of Madurai—a warrior saint and folk deity in southern India) on his steed is painted on the roof. A sacred mango tree (*Mangifera indica*) is behind the temple. In La Réunion, most public temples display classical, Brahminical architecture and a few family temples do as well. Trinidad and La Réunion each developed their own folk style of temple building through family based home temples. Trinidad's public temples are uniquely Trinidadian. I know of only one temple with classical Brahminical (agamic) architecture in Trinidad. It was built by the globalized Hindu sect of Dattatreya led by Sri Ganapati Sachchidananda (Photo by Carolyn V. Prorok 2010)

between Brahminical and Village practices all figure in decisions regarding the home altar and/or building a home temple. It may become difficult to go to the temple down the road due to conflict in the community, or new social pressures emerge that would restrict the way one's family practices at home, or the internal politics of a Hindu organization puts new pressures on a family. One might need to build one's own yard temple in order to join or avoid the current situation. Then again, two or more families might pool their resources to build a temple that they will share and thus form an allied front.

3. Individual families and their communities must often interact as a minority within a larger political, cultural, and economic system of non-Hindus. In the Caribbean exists a substantial population of African-Creole communities or consider the majority, native Malay people in Malaysia who have their own struggles, needs, and political intentions. Each diasporic community has experienced differing degrees of integration with their larger, plural societies. In Trinidad, two prime ministers of Indian origin have successfully maintained their position

of leadership for a term of office while the Fijian and Guyanese have elected an Indian to the highest office only to see them quickly overthrown in coups d'état. Some communities will never experience real political influence. Can you imagine never having the hope of real political power? Or, you do achieve political power only to have riots or an overthrow of the government strip you of that power. In many cases, this devolves into an inability to affect policies that drive economic conditions. We must come back to putting food on the table. If the family invests in building its own temple, then maybe the Gods will work harder to help us. Maybe we can let everyone who passes by our home know who we are—know that we are *here*.

4. Think of the significant impact that the globalization of our economies, political ideologies, cultural practices, and religious representations, in this case Hindu, has on all local communities around the world. Hindus can now go online to see and buy posters, ritual materials, and multiple ways to worship. They can even participate remotely in *puja* events at particular temples in India or elsewhere. Many sectarian Hindu groups now proselytize on the internet. They are often Brahminical in structure or character, but also may incorporate a village practice that has been reworked, or they may have a universalizing philosophy that incorporates all the other religions of the world (great and small) into a more or less Hindu framework. Look for Satya Sai Baba, Sri Ganapati Sachchidananda, Arya Samaji, or Divine Life missions as just a few examples. You can google these groups and many others to see how they have established themselves around the world and particularly in places with large communities of diasporic Hindus. Another aspect of globalization is a shift towards fundamentalism which operates on a feedback loop between devotees on the Indian subcontinent and in the diaspora Stump (2000). Add to these dynamic forces, the co-opting of Hindu imagery and practices by non-Hindus to sell everything from beer to shoes, the Christianizing of yoga, or to just decorate one's clothes or body with tattoos of Hindu gods and goddesses. Christina Aguilera, an American pop singer, emerges out of a lotus flower just like the great goddess *Lakshmi* on her recent album cover. Such co-opting of Hindu sacred imagery is especially disconcerting and even painful for many Hindus Doniger (2009).

Transporting laborers from the Indian subcontinent to plantation colonies in the nineteenth century was itself part and parcel of a globalizing process Younger (2010). Some laborers imagined an opportunity to own land and have a better life so far from their natal villages; others were pushed by widespread starvation and political upheaval, while still others were duped or kidnapped into service. The very origins of one's familial journey, both geographically and experientially, intersects with contemporary conditions in one region or another to produce a particular trajectory towards defining *home* as a Hindu home. Is it back in Bihar, or in Trinidad—is it born out of trauma or anticipation—and, could it be that it is yet to be created? The synergistic interplay between and among these scales of interaction produce the degree to which any family is able to negotiate a sense of belonging, of being at *home,* even after four or five or six generations. To have all of the possible combinations of internal (to the family) and external factors be in balance simultaneously is impossible. The conflation of *home* and *temple* in daily life has been, and continues to be, a foundational strategy for survival as a Hindu family and as a community.

101.5 At Home Abroad[12]

Heidegger explains our impetus to *dwell*—to be free and at peace—to be *home*—to be where we protect from and are protected from danger. *Home* is not simply the house in which you live; it can refer simultaneously to your neighborhood, town, region, nation or the world at large. It should and can be the place where you are yourself, where you do not have to explain who you are, or why you are who you are. *Dwelling* is not a passive acceptance of the world as it is around us. Instead, it is the active, intentioned process by which we create *home* in the context of the world as it is around us. A house filled with conflict and abuse is not *home* in Heidegger's concept of dwelling. Instead, we make imperfect attempts at emotional, intellectual, social, sensorial, cultural, spiritual, and physical integration with the environment of our living in the world. In this way, we struggle to achieve a satisfying quality of life. Husserl reminds us that the homeworld is always juxtaposed against an alienworld that cannot be ignored. The alienworld may seem far away or it could be just beyond the threshold of your door. The power of homeworld in one's life is, in part, predicated on the perceived and actual powers of the alienworld over against which it is defined. Geographers have grappled with similar concepts for decades as we seek to better understand the power of *place* in human experience.

To this day, even after several generations of living as a Trinidadian or Malaysian or French or Surinami citizen, Hindus must continually (re)present themselves as a worthy part of the national social whole (Arya 1968; Bakker 2003; Prorok 1994, 2004; Puri 2004; Ghasarian 2001; Munasinghe 2001: 63; Herzig 2006; Horowitz 1963; Mearns 1995; Benoist 1998; Connelly 1983; Younger 2010; Ramanathan 1995; Rajoo 1984; Shepherd 1993). Developing a sense of belonging, of attachment to this land (for example, Trinidad or Malaysia), while maintaining one's Hindu identity is hard work under such conditions. Acculturating to the values, customs, and or belief systems of the larger society is a common strategy. Yet such acculturation can be a contradiction to the values, beliefs, and practices at home. A complete focus on the values and practice of the house in which one lives, that is, those that are in line with living with God, may place one at odds with friends at school, co-workers, or even local authorities (for example, being fined for having a public

[12] Seamon (2011) and Steinbock (1995) provide an explication of Husserl's concept of homeworld/alienworld as used in this essay. In addition, go directly to sources listed for the philosophers and geographers in this section. Also, look for work by Bourdieu (1980, 1998, 2001), Merleau-Ponty (1958), and Mugerauer (1994). I have provided relevant references in the source material.

procession of the deities on a holy day). In each diasporic community around the world, one can observe the collective choices about what to conserve, what has to be modified, and what must be eschewed. The emergence of a uniquely Trinidadian style of temple architecture illustrates this process.[13]

Relph describes varying degrees of insidedness and outsidedness that capture the conundrum of many Hindus in diaspora. Achieving what he calls "existential insidedness," or a complete and fulfilling sense of belonging to a place, may be difficult for many of us, but it is particularly difficult for Hindus in diaspora. To have this sense of belonging to one's home, as a Hindu, likely requires one to take on the full and substantial responsibility of living in accordance with the Gods, while achieving the experience of such insidedness in the country one calls home challenges the very foundation of one's identity as a Hindu. Thus, stepping across the threshold of one's house takes you physically into homeworld, and walking through the gate into the street, takes one into the alienworld, a world one must encounter in order to get an education and support the family. The degree of alienation experienced in society at large, in part, affects the ways in which individuals and family units are able to function effectively, thus implicating their well-being in the long term. This dynamic devolves back into the process of sacralizing the home through the building and rebuilding of their temples. Elaboration of the temples in size, types of material used, architectural features, decorative elements, and even the dimensions of the *murtis* (physical representations of the deities) is ongoing and a direct challenge to alienating conditions.

If one converts to the dominant religion of the country, or acculturates to such a degree that one can no longer effectively maintain the home altar and/or temple through proper behavior and rituals, then being Hindu does not survive outside of fragmentary elements that linger in daily life or on special occasions. Women are often monitored for the degree to which they have acculturated, particularly if it implicates their virtue (Kelly 1991; Puri 1997, 2004). A prime example occurs during Carnival in Trinidad when Hindu priests critically examine women's participation, the suggestive lyrics to songs that describe Indian/Hindu women, or when some Hindus become emotionally distraught with the use of Hindu imagery in sexually explicit non-Hindu celebrations.

On the other hand, being Hindu survives through full engagement with *dwelling* as a Hindu. Not surprisingly, this can have an alienating effect on the perceptions of non-Hindus in society at large. Until the larger society fully and unconditionally accepts Hindus and Hinduness as part and parcel of what it means to be a Trinidadian or Malaysian or South African, then the only way in which the community survives is by cultivating the potential for existential insidedness—belonging to a home-

[13] Read Prorok (1991) to see how a uniquely Trinidadian temple architecture emerged over a 100 year period.

world—in the limited and limiting space of one's actual home.[14] The house in which one lives potentially becomes one's only refuge of, or at the very least, a circumscribed experience of insidedness—of belonging—once the deities are installed and the breath of life (*prana*) is imparted to them through appropriate rituals and daily observance. Patrilineal ancestors remain linked to the contemporary family and future generations are guaranteed. All manner of trials and tribulations can be faced with greater confidence. Joyful celebrations are more deeply satisfying when one's home is also God's home. In the U.S. we say that *home* is where the heart is. For Hindus, *home* is where the Gods live.

References

Arya, U. (1968). *Ritual songs and folksongs of the Hindus of Surinam*. Leiden: EJ Brill.

Bachelard, G. (1964). *The poetics of space*. Boston: Beacon. Press.

Bakker, F. L. (2003). The mirror image: How Hindus adapt to the Creole Christian world of the Caribbean. *Studies in Interreligious Dialogue, 13*(2), 175–186.

Benoist, J. (1998). *Hindouismes Créoles: Mascareignes, Antilles*. Paris: éditions du C.T.H.S.

Bourdieu, P. (1980). *The logic of practice*. Stanford: Stanford University Press.

Bourdieu, P. (1998). *Practical reason*. Stanford: Stanford University Press.

Bourdieu, P. (2001). *Masculine domination*. Stanford: Stanford University Press.

Brodbeck, S. (2009). Husbands of earth: Ksatriyas, females, and female Ksatriyas in the Striparvan of the Mahabharata. In R. P. Goldman & M. Tokunaga (Eds.), *Epic undertakings* (pp. 33–64). Delhi: Motilal Banarsidass Publishers.

Buttimer, A. (1976). Grasping the dynamism of lifeworld. *Annals of the Association of American Geographers, 66*(2), 277–292.

Buttimer, A. (1993). *Geography and the human spirit*. Baltimore: John Hopkins University.

Connelly, N. D. (1983). *Temporary middlemen: Hindu and Moslem Asians in Bulawayo*. Unpublished Ph.D. dissertation, Southern Methodist University, Dallas.

Diesel, A. (1998). The empowering image of the divine mother: A South African Hindu woman worshipping the goddess. *Journal of Contemporary Religion, 13*(1), 73–90.

Doniger, W. (2009). *The Hindus: An alternative history*. New York: Penguin Press.

Dubois, A. J. A. (1906). *Hindu manners, customs, and ceremonies* (3rd ed., trans: Beauchamp, B. K.). Oxford: Clarendon Press.

Ghasarian, C. (2001). We have the best gods! The encounter between Hinduism and Christianity in La Réunion. *Journal of Asian and African Studies, 32*(3–4), 286–295.

[14] It must be recognized that Hindus, like all human communities, have family situations where addiction and abuse occurs. Thus, the home may still not hold the promise of existential insidedness, of belonging, as described by Relph, Husserl, and Heidegger. It is also worth noting that the larger society's evolution towards greater acceptance of Hindus and Hinduism is ongoing, though the process is not yet complete. In Trinidad and Tobago, the nation elected its second Hindu as prime minister in 2010 yet the religious and cultural heritage of Hindus is still not included in any meaningful way in the definition of what it means to be Trini or Trinbagoan. Malaysia guarantees religious freedom in its constitution yet Hindus have difficulty getting permits to hold public religious events, and authorities remove temples they believe stand in the way of development. Mosques are not subject to the same practice. These examples and many more from diasporic lands illustrate the difficulty in experiencing full inclusion as a Hindu.

Goldman, S. J. S. (2009). Sita's war: Gender and narrative in the Yuddhakanda of Valmiki's Ramayana. In R. P. Goldman & M. Tokunaga (Eds.), *Epic undertakings* (pp. 139–168). Delhi: Motilal Banarsidass Publishers.

Gupta, C. (2006). The icon of mother in late colonial North India. In C. Bates (Ed.), *Beyond representation: Colonial and postcolonial constructions of Indian identity* (pp. 100–122). New Delhi: Oxford University Press.

Gupta, M. D. (2009). *Family systems, political systems, and Asia's 'missing girls': The construction of son preference and its unraveling* (World Bank policy research working paper, #51484). Washington, DC: The World Bank.

Heidegger, M. (1977). *Basic writings: From being and time (1927) to the task of thinking (1964)* (Edited & collected by D. F. Krell). San Francisco: Harper and Row.

Herzig, P. (2006). *South Asians in Kenya: Gender, generation and changing identities in diaspora* (Kultur, Gesellschaft, Umwelt, Band 8). London: Transaction Publishers.

Horowitz, M. M. (1963). The worship of South Indian deities in Martinique. *Ethnology, 2*(3), 339–346.

Husserl, E. (1960). *Cartesian meditations: An introduction to phenomenology* (trans: Cairns, D.). London: Martinus Nijhoff Publishers.

Jamison, S. W. (1996). *Sacrificed wife, sacrificer's wife: Women, ritual, and hospitality in ancient India*. New York: Oxford University Press.

Jones, L. (2000). *The hermeneutics of sacred architecture: Experience, interpretation, comparison* (Vol. 1 & 2). Cambridge: Harvard University Press.

Jones, J. P., Nast, H., & Roberts, S. M. (Eds.). (1997). *Thresholds in feminist geography: Difference, methodology, representation*. New York: Rowman & Littlefield Publishers.

Kelly, J. D. (1991). *A politics of virtue: Hinduism, sexuality, and countercolonial discourse in Fiji*. Chicago: University of Chicago Press.

Khare, R. S. (1976). *The Hindu hearth and home*. New Delhi: Vikas Publishing House.

Kumar, V. (2004). Understanding Dalit diaspora. *Economic and Political Weekly, 39*(1), 114–116.

Kurian, R. (1998). Tamil women on Sri Lankan plantations: Labour control and patriarchy. In S. Jain & R. Reddock (Eds.), *Women plantation workers* (pp. 67–88). New York: Berg.

Laurence, K. O. (1994). *A question of labour: Indentured immigration into Trinidad and British Guiana, 1875–1917*. London: James Currey Publishers.

Mearns, D. J. (1995). *Shiva's other children: Religion and social identity amongst overseas Indians*. New Delhi: Sage Publications.

Menon, U. (2002). Making Sakti: Controlling (natural) impurity for female (cultural) power. *Ethos, 30*(1/2), 140–157.

Menzies, R. A. (2004). *Symbols of Suhag: Paradigms of women's empowerment in Hindu domestic ritual stories*. Unpublished Ph.D. dissertation, University of Iowa, Iowa City.

Merleau-Ponty, M. (1958). *Phenomenology of perception* (trans: Smith, C.). London: Routledge.

Mines, D. P. (2010). The Hindu gods in a South Indian village. In D. P. Mines & S. Lamb (Eds.), *Everyday life in South Asia* (2nd ed., pp. 226–237). Bloomington: University of Indiana Press.

Mugerauer, R. (1994). *Interpretations on behalf of place*. Albany: SUNY Press.

Munasinghe, V. (2001). *Callaloo or tossed salad? East Indians and the cultural politics of identity in Trinidad*. Ithaca: Cornell University Press.

Nagar, R. (1998). Communal discourses, marriage, and the politics of gendered social boundaries among South Asian immigrants in Tanzania. *Gender, Place and Culture: A Journal of Feminist Geography, 5*(2), 117–139.

Nagar, R. (2000). I'd rather be rude than rule: Gender, place and communal politics among South Asian communities in Dar es Salaam. *Women's Studies International Forum, 23*(5), 571–585.

Orenstein, H. (1970). Death and kinship in Hinduism: Structural and functional interpretations. *American Anthropologist, 72*, 1357–1377.

Pearson, A. M. (1993). *"Because it gives me peace of mind": Functions and meanings of vrats in the religious lives of Hindu women in Banaras*. Unpublished Ph.D. dissertation, McMaster University, Hamilton.

Pfaffenberger, B. L. (1980). Social communication in Dravidian ritual. *Journal of Anthropological Research, 36*(2), 196–219.

Pinkney, A. M. (2008). *The sacred share: Prasada in south Asia*. Unpublished Ph.D. dissertation, Columbia University, New York.

Pintchman, T. (2001). The goddess as fount of the universe: Shared visions and negotiated allegiances in Puranic accounts of cosmogenesis. In T. Pintchman (Ed.), *Seeking Mahadevi: Constructing the identities of the Hindu great goddess* (pp. 77–92). Albany: SUNY Press.

Prorok, C. V. (1991). Evolution of the Hindu temple in Trinidad. *Caribbean Geography, 3*(2), 73–93.

Prorok, C. V. (1994). Hindu temples in the western world: A study in social space and ethnic identity. *Geographia Religionum, 8*, 95–108.

Prorok, C. V. (1998). Dancing in the fire: The politics of Hindu identity in a Malaysian Landscape. *Journal of Cultural Geography, 17*(2), 89–114.

Prorok, C. V. (2004). Om Shanti, Om Kar: A Hindu temple in Harare, Zimbabwe. *Confluence, 33*(3), 4.

Puri, S. (1997). Race, rape, and representation: Indo-Caribbean women and cultural nationalism. *Cultural Critique, 36*, 119–163.

Puri, S. (2004). *The postcolonial Caribbean: Social equality, post-nationalism and cultural hybridity*. New York: Palgrave.

Rajoo, R. (1984). Sanskritization in the Hindu temples in west Malaysia. *Journal Pengajian India, 2*, 159–169.

Ramanathan, K. (1995). *Hindu religion in an Islamic state: The case of Malaysia*. Unpublished Ph.D. dissertation, Universiteit van Amsterdam, Amsterdam.

Reddock, R. E. (1994). *Women, labour and politics in Trinidad & Tobago*. London: Zed Books.

Relph, E. (1976). *Place and placelessness*. London: Pion.

Schwartz, B. M. (Ed.). (1967). *Caste in overseas Indian communities*. San Francisco: Chandler Publishing.

Seamon, D. (2000). Concretizing Heidegger's notion of dwelling. In E. Führ (Ed.), *Building and dwelling* [Bauen und Wohnen] (pp. 189–202). Munich/New York: Waxmann Verlag GmbH/ Waxmann.

Seamon, D. (2011). *Phenomenology and uncanny homecoming: Homeworld, alienworld, and being-at-home in Alan Ball's HBO television series, six feet under. Uncanny homecomings: Narrative, structures, existential questions, theological visions*. Retrieved March 17, 2012, from www.academia.edu/1153800/Being-at-Home_and_Husserls_Homeworld_Alienworld_in_Alan_Balls_HBO_Television_Series_Six_Feet_Under

Seamon, D., & Mugerauer, R. (Eds.). (1985). *Dwelling, place and environment: Toward a phenomenology of person and world*. Dordrecht: Martinus Nijhoff.

Sekine, Y. (1999). Rethinking the ambiguous character of Hindu women. In M. Tanaka & M. Tachikawa (Eds.), *Living with Sakti: Gender, sexuality and religion in south Asia* (pp. 221–242). Osaka: National Museum of Ethnology.

Shepherd, V. A. (1993). *Transients to settlers: The experience of Indians in Jamaica, 1845–1950*. Leeds: Peepal Tree.

Shepherd, V. A. (2002). *Maharani's misery: Narratives of a passage from India to the Caribbean*. Bridgetown: University of the West Indies Press.

Steinbock, A. J. (1995). *Home and beyond: Generative phenomenology after Husserl*. Chicago: Northwestern University Press.

Stump, R. W. (2000). *Boundaries of faith: Geographical perspectives on religious fundamentalism*. Lanham: Rowan & Littlefield Publishers.

Tuan, Y. (1977). *Space and place: The perspective of experience*. Minneapolis: University of Minnesota Press.

Tuan, Y. (2001). Cosmos vs. hearth. In P. C. Adams (Ed.), *Textures of place: Exploring humanist geographers* (pp. 319–325). Minneapolis: University of Minnesota Press.

van der Veer, P., & Vertovec, S. (1991). Brahmanism abroad: On Caribbean Hinduism as an ethnic religion. *Ethnology, 30*(2), 149–166.

Vertovec, S. (1992). *Hindu Trinidad: Religion, ethnicity and socio-economic change*. London: MacMillan Caribbean.

Wiser, W., & Wiser, C. (1971). *Behind mud walls: 1930–1960*. Berkeley: University of California Press.

Yagi, Y. (1999). Rituals, service castes, and women: Rites of passage and the conception of auspiciousness and inauspiciousness in northern India. In M. Tanaka & M. Tachikawa (Eds.), *Living with Sakti: Gender, sexuality and religion in South Asia* (pp. 243–282). Osaka: National Museum of Ethnology.

Younger, P. (2010). *New homelands*. New York: Oxford University Press.

Chapter 102
Liberation Theology in Latin America: Dead or Alive?

Thia Cooper

> What takes my breath away is when people keep saying that liberation theology has gone out of fashion. … Oppression is not a fashion. … If those doing liberation theology are not doing it well, let others do it and do it better, but someone must keep on doing it. (Sobrino[1] in Rowland 1999: 250)

102.1 Introduction

This chapter will explore the emergence of liberation theology in Latin America[2] in the 1960s and assess its successes and failures in the next four decades. Liberation theology emerged from Christian communities (majority Catholic),[3] first in Brazil, and spread throughout Latin America. This theology is a combination of action and reflection that argues that Christianity should free people rather than oppress them. It is a theology that blends political, economic, ecological, and other reflections in order to create just practices. This action and reflection is based on faith, scripture, tradition, reason and experience. Liberation theology emerged from communities trying to survive in economic, political, ecological, racial, and other forms of poverty.

---

[1] Jon Sobrino is a Spanish Jesuit and liberation theologian who has lived and worked in El Salvador much of his life.

[2] The countries of Latin America include: Brazil, Uruguay, Paraguay, Argentina, Chile, Bolivia, Peru, Ecuador, Colombia, Venezuela, Panama, Costa Rica, Honduras, Guatemala, Belize, Nicaragua, El Salvador and Mexico. I cannot address them all in this article. I am also not going to address the Caribbean or Suriname, Guyana or French Guiana, although liberation theology has spread through all of these areas. For details, see Enrique Dussel 1981 and 1992.

[3] The CIA Factbook lists Belize as the country with the lowest percentage of Catholics at 39 % with 32 % Protestant followed by Uruguay at 47 % Catholic, 11 % Protestant, and 23 % "non-denominational." The majority of these countries are between 85 and 97 % Catholic. See https://www.cia.gov/library/publications/the-world-factbook/fields/2122.html#bh. The numbers were even higher in the 1960s. In Brazil, 73 % of Brazilians describe themselves as Roman Catholic and 15 % consider themselves to be Protestant.

T. Cooper (✉)
Department of Religion, Gustavus Adolphus College, St. Peter, MN 56082, USA
e-mail: tcooper@gustavus.edu

S.D. Brunn (ed.), *The Changing World Religion Map*,
DOI 10.1007/978-94-017-9376-6_102

I will outline the political and economic situations in three countries of Latin America as liberation theology emerged and coalesced: Brazil, El Salvador, and Nicaragua. These three countries cover the spectrum of liberation in Latin America. Brazil is both the richest country but the one with the largest social gap. It was the locus from where liberation theology emerged and has remained strong. It emerged more slowly in El Salvador and Nicaragua, but played a large role in the political situations. Next, I will explore the response of the Roman Catholic Church. By 1990 many in the global North assumed that liberation theology was dead. Socialism and communism were deemed to have failed after the fall of the Berlin Wall as liberationists had often supported socialist economic policies. Furthermore, the Catholic Church had condemned liberation theology. However, it is vibrant again in parts of Latin America today. Hence, I will explore its re-emergence in this century.

Liberation theology did not die; it moved from its intense focus on the political and economic repression to addressing the wider scope of oppression including issues of race, gender, and sexuality among others. In so doing, the global North assumed it had died, although the Vatican still investigates and condemns liberation theologians associated with the Church. For example, the Vatican criticized theologian Jon Sobrino in 2006, although it did not silence him. And in 2009, the Pope criticized liberation theology in a meeting of Brazilian Bishops.

Latin America suffers from a large gap between the rich and poor. Industrialization early in the twentieth century had led to some economic successes but WWI, the Great Depression, and WWII combined to reduce other countries' purchasing of Latin American exports. By the 1960s there was a debate in Latin America between those in favor of development and those who wanted to search for alternatives, often termed liberation. Development aimed to raise the living standards of the poor to those of the wealthy through economic growth. Yet, for example, in Brazil, some got richer but others got poorer. In contrast, those in favor of liberation argued that the poor could improve their own situations. What poor communities needed was for the powerful to lessen their control. Those critiquing development defined it as economic growth with increased economic, political, racial, gender, sexual, environmental, and other forms of poverty. Liberation, in contrast, for the marginalized communities was "freedom from" oppression and "freedom to" empower themselves to improve their situations. Liberation practices tended toward socialist economics whereas development practices tended to be capitalist.[4]

Politically, many Latin America countries shifted violently from democracy toward dictatorship (called National Security Regimes), often to enforce a focus on capitalist development. Some of these shifts include: the 1964 military coup in Brazil which lasted through 1985, the Somoza dictatorship in Nicaragua from 1933 to 1979, a 1971 coup in Bolivia lasted until 1978, Peru 1975–1985, Ecuador throughout the 1970s, and Argentina 1976–1983. In 1973, Uruguay dissolved its congress (through 1984) and in Chile, the same year, a coup brought Pinochet to power until 1987.

[4] See Cooper (2007) for the history of this conversation and that of the associated theologies.

Brief hints at a theology of liberation emerged from Latin America throughout the 1960s and formalized later in the decade. In 1968, the Latin American Bishops met in Medellín, Colombia to discuss Vatican II and articulated their support for liberation theology. The same year a Brazilian Protestant Rubem Alves finished his PhD in the United States entitled "Towards a Theology of Liberation." From the very beginning Protestants were involved, although in smaller numbers than Catholics. From 1969 onwards, there were national and international meetings across the continent discussing a theology of liberation, with no overt criticism from the Vatican until the next Bishops' Conference in 1979. For example, in 1969 in Mexico 800 church people held a congress on "Faith and Development," which ended up discussing liberation theology. So when Gustavo Gutiérrez, a Peruvian liberation theologian and parish priest, published *A Theology of Liberation* in English in 1974, it was the culmination of work with many poor communities and of many of these meetings.

People formed what came to be known as base ecclesial communities. These communities of 15–20 people each were poor Brazilians meeting together to discuss the Bible and life. *Base* denoted they were the poorest of society. *Ecclesial* denoted the church-relatedness and *community* signified it was not an individual theology. Although mainly Catholic, BECs also emerged to a lesser extent in Protestant communities.[5] The communities were voluntary, smaller than the parishes themselves, which often contained thousands of worshippers. The churches tended to support these BECs because they encouraged people to remain involved with Catholicism despite the lack of priests. Second, the church hoped that active involvement would also counter the growth of Protestantism. And finally, many of the priests, nuns, and bishops supported the emerging theology that worked to reduce poverty.

As the poor communities read the Bible and discussed life, liberation theology[6] and its new method, the hermeneutical circle, emerged. First, it was a process of *self-awareness* on the part of the community. How and why are we poor? Then they set this reality into the context of *faith*. And they set faith in the context of this reality. The *Bible and life* were read side by side. This analysis led to the fourth step, which is *action*.[7] Three characteristics of liberation theology emerged. Practice is first and theology comes from reflection on the practice. Second, there is no search for right rule, or orthodoxy; instead, the focus is on right practice. And finally, every Christian is a theologian; there is no formal training required.

Reflecting on faith and life, the communities articulated that God is present now, working in the world through human beings. Second, they saw Jesus and God on the side of the poor. Third, salvation is not a purely spiritual concept; it has a material component. It is the end of injustice, oppression. Fourth, then, injustice was sin.

---

[5] See Guillermo Cook's excellent article on "The Genesis and Practice of Protestant Base Communities in Latin America," in Cook, 1994: 150–155. See also Faustino Teixeira's excellent article on "Base Church Communities in Brazil" in Dussel 1992: 403–418.

[6] For an introduction to liberation theology see Boff and Boff 1987.

[7] See Cardenal (1978) for an example of the hermeneutical circle in action.

Fifth, the struggle for justice is the struggle to bring about God's kingdom. There will be a new heaven and a new earth, spiritual and material. And its central value is justice. Let's look at the emergence and expansion of liberation theology in three countries.

102.2 Liberation Theology in Brazil

Liberation theology first emerged in Brazil in the early 1960s and it remains strong there today. In the 1950s, Brazil worked to address its economic poverty,[8] by following policies of development. In 1961, Janio Quadro was elected president and João Goulart was vice-president. Quadro resigned 6 months later and Goulart ascended to the presidency. A progressive government was in place, so progressive that the United States feared Brazil would follow Cuba's move toward socialism. In 1959, a revolution led by Fidel Castro, overthrew the Batista dictatorship in Cuba. Cuba then began to follow a socialist economic model. President João Goulart emphasized that Brazil would remain neutral in the East/West controversy, arguing that the real divide was between developed – developing nations.

In northeast Brazil, Paulo Freire, a philosopher of education worked to create a Basic Education movement which focused on adult literacy. Freire (1970) advocated a problem-solving style of education where the teacher and the students work together sharing their knowledge to solve problems. He disapproved of the traditional "banking-style" of education where knowledge was deposited into the students to be withdrawn later on. He co-ordinated the National Literacy Program of the Ministry of Education and Culture. As this education movement spread throughout Brazil, others worked to reduce economic poverty as well. For example, Francisco Julião, a Brazilian lawyer and politician, founded the Peasant League in Brazil in 1955, working on land reform.

There were many movements toward popular education in the churches in the late 1950s and early 1960s. Archbishop Agnelo Rossi introduced lay people's catechesis in 1956 in northeast Brazil. In 1961, Fr. Antonio Melo led 2,000 peasants in occupying land.[9] In 1962 the Brazilian Catholic Church implemented an Emergency Plan, shifting its focus from the hierarchy to local communities. It presaged many of the changes to emerge in Vatican II, including holding mass in the local languages. And by 1963 the National Conference of Bishops in Brazil had supported the spread of radio programs on adult education including a religious component, following their success in Colombia. Helder Camara, who became archbishop in

---

[8] Even today, as Brazil's economy has grown to become the sixth largest on the planet, Inequality still remains. In 2005, the World Bank stated that the top 20 % of the population had 64 % of the total country income and the bottom 29 % only had 2 %. The situation was even more extreme in the 1960s.

[9] In 1962 in Chile a Bishop named Manuel Larrain gave away 342 ha (845 acres) of church land to the Chilean Institute of Land Reform.

northeast Brazil in 1964, also worked to reduce economic poverty and called for land reform.

These moves toward addressing poverty were abruptly halted when the military in Brazil overthrew the democratically elected government in 1964.[10] This dictatorship lasted until 1985. Under General Humberto Castelo Branco's reign, congress was suspended for 2 years and in his first month in power 50,000 people were questioned and detained by police. The Basic Education Movement was disbanded and Freire and Julião were exiled. The military took over the universities and other organizations. Any opposition to the government was labeled communist.

Yet while the hierarchy could no longer act freely, people continued to meet and the base communities developed. They spread to more than 80,000 in both rural and urban areas. These communities formalized to a certain extent throughout the 1970s and 1980s, even holding national and international meetings. For example, the Latin American Meeting of Base Ecclesial Communities was held in Brazil in 1980.

Camara was one of the first to speak out against the repressive tactics of the military government and its support in the West. He refused to celebrate a mass commemorating the military coup and also formed a Movement for Moral and Liberating Influence. Camara's life was threatened by several militia groups and his assistant Antonio Neto, a 28 year old priest and university chaplain, was tortured and killed in 1969. Despite having previously served as Minister of Education, Camara was labeled a communist.

By the 1970s the military government and its repressive tactics had settled into systematic violence.[11] While the Catholic hierarchy was split between supporting the majority who were poor and the minority who were in charge, many priests and church organizations began to support the poor in their attempts to improve the economic and political situations.[12]

Some of these include CIMI (Missionary Council for Indigenous People), which worked for indigenous rights, the Pastoral Commission on Land, which worked for land reform, and the Workers' Pastoral Commission, which argued for workers' rights. One European-born priest, Pedro Casaldaliga, became a Brazilian bishop in 1971 (–2004) in the Amazon. The same year he became bishop, he wrote a letter on the need for land reform in Brazil. Later in the decade, he attended a protest of the arrest and torture of two women, where one of his colleagues Father João Burnier was shot to death standing beside him. By the 1980s he was deeply involved in

[10]The United States supported this coup. See: www.gwu.edu/~nsarchiv/NSAEBB/NSAEBB118/index.htm for details.

[11]The Church in Argentina suffered similar repression from 1973, particularly violent among the priests who openly supported liberation theology. In Paraguay and Uruguay both, priests speaking out and supporting the poor were also repressed from 1969 onward. Peru, in contrast, had liberationist priests like Gutiérrez supporting the leftist government and vice-versa until the coup in 1975.

[12]Often the communities themselves demanded this action. In Argentina, in 1966 parishioners occupied a church demanding church reform. A similar event occurred in Peru in 1970.

CIMI and the Pastoral Commission on Land. The Vatican censured him for his liberationist stance in 1988. He continued his work however, and now 84, he is still active and creating poetry, although no longer serving as Bishop.

In 1985, the military dictatorship transitioned back to a democratic republic, with a new constitution. And yet the vast economic and political poverty continued. The shift to democratic rule was led largely by business elites and the military leaders themselves; it was not a response to the push from communities of the poor. There have been continued moves towards aspects of liberationist policies on the ground.[13] In Brazil less than 3 % of the population own more than 60 % of the productive land. The 1984 Brazilian Constitution deemed it illegal to own productive land that was not being cultivated. A person or group who does use it productively can apply for the title to it.

In 1985 with the support of the Catholic Church, a group of landless workers started working an unused piece of land. After finally gaining the title to the land, these workers farmed and owned it co-operatively. From this the Rural Landless Workers' Movement (MST)[14] began. Since then the MST has successfully helped over 350,000 families to work unused land, apply for title, and eventually settle. Groups of 250 or so families tend to own the land cooperatively. This has become a holistic alternative to the global capitalist economy, with innovations in politics, education, sustainable farming, etc. The work of the MST continues today despite moves towards conservatism in the hierarchy of the Church.

In 1985, the Vatican replaced Camara with a very conservative successor. Still Brazilians like Leonardo and Clodovis Boff, priests, and Ivone Gebara, a nun and feminist theologian, continued to struggle against the repressive economic and political system and considered other aspects of oppression too in connection with communities on the ground. For example, as the base communities began to deal with issues of gender, so too did Gebara.[15]

> A society which obliges women to choose between keeping their jobs and terminating a pregnancy is an abortive society.... A society which is silent about the responsibility of men and blames only women... is [an] exclusive, sexist, and abortive society (Gebara in Brown 1987: 90)

The abortion issue was not simply an individual woman deciding to keep or terminate a pregnancy, but a part of a wider spectrum of the issue of female survival in society. Gabara silenced by the Vatican for 2 years in 1995, but continues to work and write today in Brazil as does Leonardo Boff.

Leonardo Boff began writing about ecological liberation in the 1980s. His writings came at the same time as some of the work of CIMI around the ecology of the Amazon. He argued,

> "Violence against the environment ... affects them [individuals and social groupings] indirectly, but with immediately harmful consequences, since the decline in the quality of

[13] For further examples of liberationist practices in the 1990s see Ottman's excellent book, 2002.

[14] Movimento dos Trabalhadores Rurais sem Terra. See www.mst.org for further details. Or for an English website: www.mstbrazil.org.

[15] For example, see Ottmann (2002) who shows how some liberationist movements in Brazil have turned to addressing gender.

their surroundings produces social tensions, violence, disease, malnutrition and even death" (Boff and Elizondo 1995: xi).

Environmental devastation has been a huge issue in Brazil, particularly around the deforestation of the Amazon. He was silenced by the Vatican in 1985 for one year and then was investigated again in 1992. He resigned as priest in 1993 and has remained a layperson since. Casaldaliga too became involved with ecological issues as bishop in an Amazon region.

By the end of the 1990s, many of the people most involved at the beginning grew old and died: Camara in 1999, Freire in 1997. However, liberation theologians today still work on economic and political issues among many others. They also represent the ethnic and religious diversity of Brazil. For example, Jung Mo Sung, a Korean Brazilian is Dean of the Methodist University in São Paulo and works on issues of theology, economics and the market. Luis Carlos Susin, a priest and university professor in the South of Brazil, works on issues of land, among many others. And Silvia Regina de Lima Silva is an Afro-Brazilian theologian working on issues of race and gender.

At the same time, like the MST, many community organizations still work with the poorest. In Salvador, located in northeast Brazil, for example, the Center for Studies and Social Action, a Jesuit-based organization, founded in 1967, continues to work with communities from its liberationist perspective. They work on rural and urban issues, dealing with the landless, with slavery, and also with street children, HIV positive homeless, prostitutes, primary school teachers, neighborhood associations in the slums etc.[16] They have two foci: to work for social transformation for justice and to work with the most marginalized people.

Politically, the democratic processes have slowly improved in Brazil. For example, in 2002 Luis Inacio da Silva (Lula) was elected President. He was a former trade union leader. In 2009, the Brazilian government announced it was forming a Truth Commission to investigate the alleged crimes committed by those in power during the military dictatorship. Dilma Roussef, the first woman president, was elected successor to Lula on October 31st 2010. She was imprisoned and tortured during the dictatorship years and has continued plans for the Truth Commission.

Brazil's political situation of dictatorship from 1964 to 1984 and then a slow move toward democracy is different from the experiences of El Salvador and Nicaragua.

102.3 Liberation Theology in El Salvador[17]

In El Salvador, liberation theology was slow to take root, especially in the hierarchy of the Catholic Church. Dictators had ruled the country since 1933. In 1932, a peasant uprising led by Augustin Marti resulted in the massacre of thousands of

[16] See www.ceas.com.br for details. They also produce a quarterly journal Cadernos do CEAS, which can be accessed through their site.

[17] Similar to El Salvador, in Guatemala and Honduras, the poor and those in the church who support them have been violently repressed. In Guatemala this repression lasted through the early 1990s as in El Salvador.

peasants. Yet with the emergence of liberation theology in the 1960s, the rural areas began to reorganize. El Salvador was under military rule from the 1960s through 1979. Many priests and nuns worked to reduce the economic and political poverty throughout these years. The hierarchy of the church, however, tended to worry about the Protestants more than poverty.

While Camara and others spoke out beginning in the 1960s in Brazil, the El Salvadoran response really began in the 1970s. Thc example of Archbishop Oscar Romero is instructive here. He was appointed archbishop by the conservative church in 1977 precisely because he himself was conservative. Although concerned about reducing poverty, he argued that if the military government really knew the pain they were causing, they would change their policies. He slowly realized this was not the case and his shift in perspective to liberation theology was finalized when a good friend of his and liberationist priest, Rutilio Grande, was targeted and killed for his outspokenness in 1977.

From this point on, Romero too spoke out against the government repression, arrests and torture, and in 1980 as he delivered a homily in his home chapel in a local hospital he was gunned down. No one has ever been convicted of this crime, although it was clear who was responsible for the killing. While his death is the most famous in El Salvador, the same year more than 600 people were killed in the River Sumupul massacre. At Romero's funeral, 42 of the mourners were killed by sniper fire. The Jesuits as a group had death threats against them. The army occupied many of the churches.

Different from Brazil, El Salvador did not slowly move toward democracy, instead war erupted in 1979. There was a coup d'etat to prevent a revolution and the military government began to heavily repress the citizens, killing more than 10,000 per year over the next few years. The United States strongly supported the coup and opposed the rebels, just as it had in Brazil in the 1960s. The rebel group, the FMLN, named after Marti, gained supported throughout the 1980s, as this government repression continued. Many priests and nuns have worked on the side of the poor and have suffered repression as a result.[18] As repression lessened in Brazil, it intensified in El Salvador, lasting until 1992.

In 1989, 6 days after the fall of the Berlin Wall, the army shot six Jesuit liberationist priests, including Ignacio Ellacuría, while they slept in their home on the grounds of their university. In addition, two women who had taken refuge there because of the curfew and repression were also killed. After they killed the six, the group of 40 armed people destroyed the university archives.[19] Jon Sobrino, an internationally known liberation theologian, who also lived there, escaped this massacre as he was travelling in Thailand at the time. Throughout the 1970s Sobrino and

[18] This is true of many Protestant groups in El Salvador too, for example, the Lutheran Church. There is an excellent article in Cook (1994: 178–89) by Kim Erno, "El Salvador: Liberation and Resurrection" that details some of the church's actions 1988–1992 in the struggle to end the war.

[19] For further details of the event and information on the perpetrators, see Sobrino (1990).

Ellacuría (Rector at the time of his death) had taught at the university and worked on the side of the poor.

Finally, in 1992, a ceasefire was declared. Throughout the 1990s, the conservative party won elections, although liberationists continued to have support and theologians like Sobrino continued to publish, as you can see from the quote with which I began this chapter. His writings since the massacre have often focused on crucifixion and martyrdom. For a decade, the country slowly recovered from the fractures of war. However, in 2001, there were two severe earthquakes within weeks of each other that set the country back economically. In 2007, the Vatican censured Sobrino for his continued work. Further, in 2009, the FMLN won the elections. In Mauricio Funes' acceptance speech, he quoted Romero. There remains the question of how liberation theology will address the intense poverty in the country.

102.4 Liberation Theology in Nicaragua

There was also an emergence of liberation theology in Nicaragua in response to economic and political poverty; it encountered different circumstances than either Brazil or El Salvador. In 1933 the U.S. Marines were expelled from the country by an army led by Augusto Sandino. Sandino was assassinated and the Somoza government took control. The Somoza dictatorship lasted 45 years, similar to the unresponsive political system in El Salvador. In Nicaragua the church and the base communities began to struggle against the dictatorship of Somoza in the 1960s, although their numbers were initially small.

Ernesto Cardenal, a Nicaraguan who became a priest in 1965, lived on the island of Solentiname and was part of, and wrote about, the base ecclesial communities. He was also active in politics, opposing the Somoza dictatorship throughout his lifetime. He published the *Gospel in Solentiname* in the 1970s, which are excerpts from his community's reading and interpreting the gospels and life. During this decade resistance against the government heightened and so did the repression. In 1977, the Solentiname community was attacked and destroyed by the Nicaraguan National Guard. Cardenal fled to Costa Rica. The Sandinistas struggled against the government from 1978 to 1979. In 1979, they gained power, in stark contrast to El Salvador, and Cardenal returned to become Minister of Culture from 1979 to 1987.

Initially, it seemed the church would support the Sandinistas and vice-versa. Priests became ministers of foreign affairs, housing and the official literacy campaign. A literacy booklet published by the new government contained the following: "There is freedom of worship for all churches that defend the interests of the people. The true church must be committed to the people. The church cannot be indifferent to the needs of the people. Glory to our martyrs!" (Penyak and Petry 2006: 284). And even the conservative bishop Miguel Obando celebrated a Mass of Thanksgiving when the Sandinista government began.

However, there were Christians across the spectrum of the revolution from 1978 to 1980[20] and the split intensified during the Sandinista government. Although this government was to counter the military one before it and lessen political oppression, the U.S. government opposed it for being too socialist. The USA funded the Contras to fight against the government. In 1983, the Pope visited Nicaragua and spoke of the need for the people to unify around the bishops. It was not what the people on the ground expected to hear and the shouting of the crowd drowned him out. He subsequently promoted the conservative Bishop Obando to Cardinal. Many of the liberationist priests and nuns were removed from their posts and replaced with conservatives.

This moment of liberationist struggle ended in 1990 when the Sandinistas lost the elections.[21] During the 1980s, as liberation theology diffused its focus in Brazil, it was active but repressed in El Salvador during the war. Liberation theology appeared to be succeeding in blending religion and politics in Nicaragua. However, this was not long-lived and the conservatives then led the government from 1990 to 2006.

The Sandinistas (FSLN) have recently returned to government with the election of Daniel Ortega in 2006 and his re-election in 2011, however, Cardenal is no longer part of the government. He and other priests are still struggling against oppression, however. For example, in 2005 he was part of a group that fasted every Friday to protest the prisoners held at Guantanamo Bay.

102.5 Liberation Theology's Death

Liberation theology increased in stature throughout the 1970s and 1980s. In the 1970s, it appeared to have the support of the Vatican through Pope Paul VI. Institutions formed to study liberation theology, like the *Departamento Ecuménico de Investigaciones* (DEI) in Costa Rica and many liberationist priests also taught at universities. In a few Latin American countries, liberationists gained political power: Cardenal, for example.

In the 1980s, as the political and economic foments continued, then Cardinal Joseph Ratzinger (Pope from 2005–2013) increasingly condemned liberation theology. The Vatican investigated the Boff brothers and Gutiérrez. The Church silenced

[20] There were even evangelicals on the side of the Sandinistas, although by no means the majority. See Adolfo Miranda Saenz's article in Cook (1994: 190–201) "Nicaragua: Political metamorphosis of evangelicals."

[21] And in 1991, while many were declaring liberation theology dead, a liberationist priest Jean-Bertrand Aristide was elected President in Haiti. Again, while it was a tumultuous presidency fraught with problems, the influence of liberation theology remained clear. It also remained clear that the USA was not in support of the type of politics that liberation theology tended to support.

liberation theologians, including Leonardo Boff (see Cox 1988 for details). The Vatican also published an *Instruction on Certain Aspects of the "Theology of Liberation."*[22] The Church disapproved of two aspects of the theology in particular. First, this theology emerged from the reflection and action of the communities on the ground. It did not emerge from the Pope and move downward. The hierarchy of the church was threatened. Second, in Pope John Paul II's experience, socialist and communist economic policies led to atheism. He feared that the liberationists tendencies toward these policies would turn them toward atheism too, although this was never the case in Latin America. For example, after the Cuban revolution, the relationship between the church and the government was fractious to start. However, as liberation theology emerged throughout the 1960s and 1970s Castro became more conciliatory. In a speech in 1980, he stated, "If we bear in mind that Christianity was, in the beginning, the religion of the poor… there is no doubt that the revolutionary movement… would benefit a great deal from honest leaders of the Catholic Church and other religions return to the Christian spirit of the days of the Roman slaves." (Castro in Penyak and Petry 2006: 322). However, many in the North assumed that with the Vatican condemnation, the fall of the Berlin Wall and subsequent breakup of the Soviet Union, liberation theology would quietly die. However, liberation theology had never been a top-down movement and Latin America has had a history of the Church acting independently of the Pope. For example, during the colonial period, the Pope gave power to the Spanish and Portuguese governments to send bishops, priests, and missionaries to the Americas. The Spanish and Portuguese leaders then interacted with Rome. The Latin American church and the Pope were connected indirectly.

While the world's media and churches in North America and across Europe seemed to read the Instruction as a condemnation, Latin Americans were not so sure. While the Instruction was critical of liberation theology, the next couple of interactions with the Vatican were less so. In 1986, the Vatican issued another "Instruction on Christian Freedom and Liberation" that seemed more supportive.[23] Pope John Paul II also wrote a letter to Brazilian bishops in 1986 stating that liberation theology was "not only opportune but useful and necessary" (Dussel 1992: 399). Politicians still draw on liberation theology today. Liberationists are once again active internationally. Further there is still much poverty to be countered, and so liberation theology continues to grow and thrive.

[22] See the Vatican's website for the full text: www.vatican.va/roman_curia/congregations/cfaith/documents/rc_con_cfaith_doc_19840806_theology-liberation_en.html

[23] See the Vatican's website for the text: www.vatican.va/roman_curia/congregations/cfaith/documents/rc_con_cfaith_doc_19860322_freedom-liberation_en.html. For an excellent analysis of both of these Vatican Instructions, see P. Hebbelthwaite's article "Liberation Theology and the Roman Catholic Church" in Rowland 1999, 179–198.

102.6 Liberation Theology's Resurrection and Spread Beyond Catholicism

While several South American countries limped toward democracy in the 1980s and while there was less repression, there was also still extreme poverty. Liberationists' public voice against the dictatorships waned but they were able to more clearly address economic, gender, ecological, racial and other issues. Liberation theology changed in two ways. First, liberation theology is not solely rooted in the Catholic Church, and second, it is no longer solely focused on political oppression.

From its beginnings, there were Protestant strands, particularly in El Salvador and in Brazil. However, these strands have spread throughout Latin America and beyond. The fastest growing religious group in Latin America today is the Pentecostal tradition, which tends to support capitalism and globalization in direct contrast to liberation theology. Pentecostalism is politically aware and active across Latin America, just like Catholicism. For example, Jorge Serrano Elías, a member of a Pentecostal church, was president of Guatemala from 1991 to 1993. While mainline Protestantism tended to evangelize among the elite or remain focused on middle class European immigrants, Pentecostalism has spread amongst the very poor. However, Pentecostalism is not solely associated with conservative politics. In Brazil, for example, the Brazilian Evangelical Association was established in 1991. The Evangelicals have in part supported Lula and Roussef.

Not only is liberation theology no longer solely focused on political oppression, but also the economic and political terminology has changed. Rather than fight against development policies or capitalism per se, liberationists now often argue against "globalization." Capitalism is one aspect of globalization, but the word can also refer to the hegemony of Western culture, Western politics, and so forth. Capitalism was critiqued because it focused on economics. Still today, human beings are considered to be consumers first and citizens later. But, what happens to people who are outside the economic system, those who cannot consume? As Pablo Richard, a Chilean priest living in Costa Rica, noted in 1994: "No longer are we even the Third World. We are now the Last World, the non-World – the accursed world of those who have been excluded and condemned to death." (Cook 1984: 248) With the shift away from development toward "globalization," which assumes a globalization of capitalism, Richard notes, "capitalism proposes to save the lives of a privileged few, even if this means the death of many" (Cook 1984: 248) Humanity, as a whole, is no longer at the center of the discussion. His analysis remains true nearly 20 years later.

Events in Argentina in the past 10 years are exemplary (see Epstein and Pion-Berlin 2008). From 1999 to 2002 there was an economic crisis in Argentina. After a run on the banks, accounts were frozen. The middle class became extremely poor.

The poor became even poorer, excluded entirely. And yet, there were signs of liberation. People protested in the streets; they came to be known as *piqueteiros*. Groups were formed to share soup and bread. Businesses shut down, especially those reliant on foreign investors. In response, some of the workers reopened as worker owned cooperatives and eventually gained legal access to the businesses. The economy has improved in the past few years, but there remains much poverty and inequality.

New liberation theologians were writing and working throughout this time. For example, Marcella Althaus- Reid, a Methodist struggled to be allowed into seminary, completed her Ph.D. in the UK in 1993 and became a professor at the University of Edinburgh where she taught until her death in 2009. Her work on *Indecent Theology* has changed the field of liberation theology and sexual theologies. She argues that if liberation theology is really interested in practice, then liberationists need to talk about sex and sexuality. Her writings reflect her struggles with issues of gender and sexuality as she grew up in Argentina. There are many liberationists engaging with issues of sex and sexuality today, much of it emerging from Argentina and Brazil.[24]

Another Argentinian, Ivan Petrella, received his PhD in religion and law in the U.S. and was a professor. He returned to Argentina in 2010 to work with a ThinkTank: Fundacion Pensar (see www.fundacionpensar.org). In his writings, he has lamented the loss of the "historical project" and argued for its return, something he works toward in practice. He states that "liberation theology can reread the fall of socialism as an opportunity to try out alternative approaches to capitalism" (Petrella 2006: ix) and indeed this is what liberation theologies are now doing. While his work tells us that liberation theologies continue, his own agnosticism also shows that liberation theologies are expanding further out in Latin America, beyond even the Christian faith.

Politically, liberation theology is also returning to Latin America's agenda. I mentioned above Funes quoting Romero in his acceptance speech and Ortega's return to power. So too in Paraguay in 2008, former Bishop and liberation theologian, Fernando Lugo, was elected President. And in Ecuador, President Rafael Corea is an active Catholic Socialist, who often quotes Leonidas Proano, an Ecuadoran liberation theologian.

[24] See for example Althaus-Reid (2009). The liberationists therein do include professors, but also students and ministers teaching and learning in their home countries. For example, Hugo Cordova Quero, who worked in Japan and has returned recently to his native Argentina works on queer theologies and migrant theologies or Norberto D'Amico, born and living in Buenos Aires, working with the Methodist Church or Roberto Gonzalez, a gay evangelical minister in Buenos Aires. In Brazil, for example, there is Andre Musskopf, a Lutheran liberation theologian, who has worked for local government.

102.7 Conclusion

The geography of this movement has shifted in two ways. First, from the 1960s it spread from Brazil throughout the remainder of South and Central America. And second, it spread from within Catholicism to other religious traditions and has moved beyond religion alone in many areas. The issues it addresses and the language within which it works have also shifted. The dialogue between liberation theology and the North disappeared because the North no longer aimed for development; it aims for globalization whereby the human being is no longer at the center. Hence, there is no need to dialogue with human beings in the South. Liberation theology kept working and speaking in Latin America and other regions around the world but the "world" of the North stopped listening. The question for this century is whether the North (or more generally those with power) will be forced to listen to the South (those without power). If so, liberation theologies will return to the fore. If not, they will continue to bubble under the surface, hidden, yet present, even influencing national policy in certain nations.

Petrella (2006), Althaus-Reid (2000, 2009) and many others like them have worked both in Latin American and international contexts, moving back and forth between countries, also enabling them to address wider issues of globalization and to link liberation theology locally and globally. Further, there is a wealth of ongoing conversation at the international level. The American Academy of Religion has had a variety of places where Latin American liberation theologians could present and be in dialogue throughout the years. In 2007, a new group formed: the Liberation Theologies Group, which has created a space for dialogue between Latin American liberation theologians and practitioners and those around the world, including the UK and Europe, the USA, Asia and so forth. Second, even earlier in 2005 there was a World Forum on Theology and Liberation in Brazil. The book *Another Possible World* is the fruit of that conversation between scholars and practitioners around the world. Third, throughout 2011 there were a series of conferences in Latin America, one in Guatemala, another in Chile, a third in Mexico City and a fourth in Colombia, discussing 'The Future of Liberation Theology'. These led up to a Congress on Theology, October 2012 in Brazil, which will celebrate the fiftieth anniversary of Vatican II as well as the 40th anniversary of Gutiérrez's first book on liberation theology. And fourth, there was a conference in Buenos Aires in July 2013 on the "Legacies of Marcella Althaus-Reid" bringing together post-colonial, queer and liberation theologies. Finally, there are still a variety of publications emerging from Latin America including a journal *Cadernos do CEAS* published by CEAS in Portuguese, which works from a liberationist perspective encompassing several disciplines and several publications available through DEI in Spanish, including their journal PASOS. Both of these journals are multidisciplinary.

Based on the return of liberation theologies to the international agenda and the success of politicians associated with liberation theology on some level, combined with liberation theologians addressing economics, politics, race, gender, sexuality, ecology, and indigeneity among other issues, I expect liberation theology to increasingly spread throughout Latin America again over the next 10 years.

References

Althaus-Reid, M. (2000). *Indecent theology: Theological perversions in sex, gender, and politics*. New York: Routledge.
Althaus-Reid, M. (Ed.). (2009). *Liberation theology and sexuality*. London: SCM Press.
Boff, L., & Boff, C. (1987). *Introducing liberation theology*, 1. Liberation and theology (trans: Burns, P.). Tunbridge Wells: Burns & Oates.
Boff, L., & Elizondo, V. (1995). Editorial. In L. Boff & V. Elizondo (Eds.), *Ecology and poverty: Cry of the earth, cry of the poor* (Concilium 1995/5, pp. ix–xii). Maryknoll: Orbis Books.
Brown, R. M. (1987). *Religion and violence* (2nd ed.). Philadelphia: Westminster Press.
Cardenal, E. (1978). *The gospel in Solentiname* (Vol. 2). Maryknoll: Orbis Books.
Cook, G. (Ed.). (1984). *New face of the church in Latin America*. Maryknoll: Orbis Books. 1994.
Cooper, T. (2007). *Controversies in political theology: Development or liberation?* London: SCM Press.
Cox, H. (1988). *The silencing of Leonardo Boff: The Vatican and the future of world Christianity*. London: Collins.
Dussel, E. (1981). *A history of the church in Latin America: Colonialism to liberation*. Grand Rapids: Eerdmans.
Dussel, E. (Ed.). (1992). *The church in Latin America: 1492–1992*. Maryknoll: Orbis Books.
Epstein, E., & Pion-Berlin, D. (Eds.). (2008). *Broken promises? The Argentine crisis and Argentine democracy*. New York: Lexington Books.
Freire, P. (1970). *Pedagogy of the oppressed* (trans: Ramos, M. B.). New York: Herder & Herder.
Gutiérrez, G. (1974). *A theology of liberation: History, politics and salvation* (trans: Inda, S. C., & Eagleson, J.). London: SCM Press.
Ottman, G. F. (2002). *Lost for words: Brazilian liberationism in the 1990s*. Pittsburgh: University of Pittsburgh Press.
Penyak, L. M., & Petry, W. J. (Eds.). (2006). *Religion in Latin America: A documentary history*. Maryknoll: Orbis Books.
Petrella, I. (2006). *The future of liberation theology: An argument and manifesto*. London: SCM Press.
Rowland, C. (Ed.). (1999). *The Cambridge companion to liberation theology*. Cambridge: Cambridge University Press.
Sobrino, J. (1990). *Companions of Jesus: The Jesuit martyrs of El Salvador*. Maryknoll: Orbis Books.

Chapter 103
Africa's Liberation Theologies: An Historical-Hermeneutical Analysis

Gerald West

103.1 Introduction

In the latter part of the 1980s those working within the framework of liberation theologies began to explore other ways of talking about "liberation." While the term "liberation" was still of immense rhetorical importance, given its historical and hermeneutical heritage (Bonino 1975), the terrain from within which the term had arisen was shifting. There was not the same hope that imagined socialist forms of political and economic liberation would materialize. The collapse of the Soviet Union had much to do with this, as did the failure of socialist-inclined movements and parties to establish themselves politically and economically in nation states in Latin America, Africa, and Asia. So those working within this hermeneutic framework began to re-imagine what 'liberation' might look like, both in terms of what liberation was "from" and what liberation was "to." One way of talking about "liberation" which began to emerge during this time was to speak of "the God of life" and "the idols of death" (Hinkelammert 1986). What liberation theology was about, it was argued, was taking sides with the God of life against the forces of death. The task of Christians was to "read the signs of the times," discerning where God was already at work bringing life in the midst of death, and then to become co-workers with God (Hinkelammert 1986).

Liberation theology was to be done, within this revised terminology, in the context of the struggle for life in the midst of death. In the South African context, for example, the key concept became that of "struggle." The emphasis became less on the end product ("liberation") and more on the ongoing process of God's liberatory project (Mosala 1989; Nolan 1988). The advantage of this shift in emphasis and formulation has been that there is a place for liberation theologies after political

G. West (✉)
School of Religion, Philosophy and Classics, University of KwaZulu-Natal,
Private Bag X01, Scottsville 3209, South Africa
e-mail: west@ukzn.ac.za

S.D. Brunn (ed.), *The Changing World Religion Map*,
DOI 10.1007/978-94-017-9376-6_103

liberation, as is the case in South Africa. As long as the God of life is engaged against the idols of death – whether these be the idols of neo-liberal capitalism, or the idols of patriarchy, or the idols of moral and medical discrimination in the HIV and AIDS pandemic – there is a need for forms of liberation hermeneutics which work with and proclaim the God of life.

While liberation theologies continue to call for and hope for the fullness of liberation, what John calls "abundant life" (John 10: 10), liberation theologians in South Africa and elsewhere have recognised that for millions of people the reality is a daily struggle for survival, for basic life (Haddad 2004). "Struggle" is the reality of life for many, and a distinctive feature of "struggle" in South African and other liberation theologies is its systemic nature. "The struggle" is not primarily an internal struggle within individuals, though it does include this; the primary struggle that is the terrain of liberation theologies is, to rephrase Ephesians, the struggle against systems and structures which bring death (Ephesians 6: 12). The struggle is fundamentally against structural sin (West 2005). Given the systemic nature of the struggle for life, the fundamental commitment of liberation theologies is the taking of sides. There is no neutral position in the contestation between the God of life and the idols of death. There may be more than one choice, as is argued in the landmark theological formulation of the *Kairos Document* (which distinguished between 'State', 'Church', and 'Prophetic' theologies) (Kairos 1985), but there is always a choice and a stand to take.

In what follows we will explore the conceptual framework and fundamental choices of liberation hermeneutics, focusing on the biblical dimensions of liberation theology. Though the conceptual framework represents liberation theologies across many sites of struggle, the particular examples used to give flesh and blood to the frame will come from Africa. And because the Bible is so central to African liberation hermeneutics, the biblical dimensions of liberation hermeneutics are at the core of Africa's liberation theologies.

Liberation theologies are constructed around *five* interrelated concepts, which can be found across a range of sites of struggle. Per Frostin, an astute scholar of liberation theologies, including African forms of liberation theology, argues that the five key conceptual choices of liberation theologies are: "the choice of 'interlocutors', the perception of God, social analysis, the choice of theological tools, and the relationship between theory and practice" (Frostin 1988: 6). Frostin's conceptual characterization of liberation theologies is significant because it draws directly on the primary documentation of a wide range liberation theologies in dialogue with each other. The data Frostin uses is drawn substantially from the self-constituted dialogue of Third World theologians working together in forums such as EATWOT, the Ecumenical Association of Third World Theologians.

103.2 The Choice of Interlocutors

With respect to the first and fundamental conceptual feature, the choice of interlocutors, the emphasis in liberation theologies has been on social relations, not ideas or techniques, as has been the tendency in post-Enlightenment Western theology and biblical studies. "Liberation theologians," argues Frostin, "focus on a new issue seldom discussed in established theology: Who are the interlocutors of theology" (Frostin 1988: 6). Or, to use the language used by Musa Dube and Gerald West within the discourse of biblical liberation hermeneutics (West and Dube 1996), "Who are biblical scholars reading 'with', when they read the Bible?" To these questions liberation hermeneutics gives a decisive answer: "a preferential option for the poor" (Frostin 1988: 6). Echoing the words of Gustavo Gutiérrez we can say that while the primary interlocutor of Western biblical scholarship is the educated middle class, the primary interlocutor of liberation hermeneutics is "the poor, the exploited classes, the marginalized races, all the despised cultures" (Gutiérrez 1973: 241). This choice of interlocutors is more than an ethical commitment, it is also an epistemological commitment, requiring an interpretive starting point within the social experience and social analysis of the poor themselves. As Frostin states it, "solidarity with the poor also has consequences for the perception of social reality, as seen in the phrase 'the epistemological privilege of the poor', reportedly coined by Hugo Assmann" (Frostin 1988: 6). The other four conceptual characteristics of liberation hermeneutics each flow from this first and fundamental conceptual feature, which is why biblical liberation hermeneutics must always be more than an interpretive technique. The actual presence and participation of the poor in any interpretive act is pivotal.

However, exactly how the poor are present in biblical liberation hermeneutics is a matter of some debate in liberation hermeneutics. Clearly articulated by Jan Luis Segundo in the early 1980s (Segundo 1985), the debate has continued (Nadar 2009). Key to these disagreements and the debate is the role of the intellectual (in this case the socially engaged biblical scholar or theologian) and their assessment of the ideological (in classical Marxist terms) capacity of the poor and marginalized. Those who hold to a strong sense of the epistemological privilege of the poor and marginalized with respect to their knowledge of their oppressive contexts operate with "thin" conceptions of ideological hegemony. According to such thin notions of ideological hegemony, dominant ideologies never completely erase the knowledges they subjugate. Hegemony "is always threatened by the vitality that remains in the forms of life it thwarts" (Comaroff and Comaroff 1991: 125). Hegemony "is always intrinsically unstable, always vulnerable" (Comaroff and Comaroff 1991: 27).

Furthermore, those who stress the epistemological privilege of the poor and marginalized recognise that domination has a greater impact on the *socio-political opportunities* for resistance of the poor and marginalised than it does on the *religio-psycho-social capacity* of the poor and marginalized to resist domination. Subordinate classes, argues James Scott, are "*less* constrained at the level of thought and ideology, since they can in secluded settings speak with comparative safety, and *more* constrained at the level of political action and struggle, where the daily exercise of power sharply limits the options available to them" (Scott 1990: 91). Contending with traditional Marxist and early Latin American liberation theology notions of hegemony, James Scott argues that there are minimal constraints at the level of recognition, but more serious constraints at the level of action. So the constraint on the poor and marginalized is not a recognition of their oppression nor a desire to resist, but the social space in which to act. The poor already have a language, though the social space in which they can use their language is constrained by the structures of domination. But the *organized* poor, working class, and marginalized do have their safe sequestered sites in which their dignity is given voice and where a thick and resilient "hidden transcript" is constructed (Scott 1990: 1–16). They have their heterotopias in which "all the other real sites that can be found within the culture, are simultaneously represented, contested, and inverted" (Foucault 1967: 3).

The role of the socially engaged biblical scholar in such *organized* contexts is that of the facilitator or animator, one who comes alongside those who are struggling to assert their dignity and its concomitant ideology/theology. Among *unorganized* individuals, the role of the intellectual may appear to be more interventionist, though it remains fundamentally facilitatory (Freire 2006: 65). Similarly, among those in poor and marginalized communities who have been atomized or kept under constant surveillance, it may appear that their human dignity has been so damaged that a "false-consciousness" has enveloped them. But this apparent silence is not the silence of a consent to hegemony, but the silence of an embodied, but yet to be articulated ideology (West 2009b). Those socially engaged biblical scholars who work with a strong sense of the epistemological privilege of the poor and a weak sense of social hegemony view their contribution as providing those particular sets of critical analysis characteristic of biblical studies alongside the critical resources already being used by organized communities of the poor and marginalized. These additional critical resources, the tools of the biblical studies trade, are not primarily of use because they conscientize the poor, but because they offer other ways of forging lines of connection between the embodied struggle of today's poor and the kindred struggles of the poor within and behind the biblical text.

However, there are those socially engaged biblical scholars and theologians, including Frostin and Segundo, who embrace strong notions of hegemony, arguing that the poor and marginalized have been "colonized" by the dominant ideology and are trapped in "a culture of silence" (alluding to Freire 1985: 72; Frostin 1988: 10). And, in the words of Segundo, because it is the social sciences that "provide the theologian who wants to carry out a de-ideologizing task with valuable cognitive tools" and because these are "tools which … are beyond the grasp of the majority of people" (Segundo 1985: 28), the role of the socially engaged theologian or biblical scholar must be seen as pivotal. In Frostin's terms, it is the socially engaged biblical scholar (whether an organic intellectual from among the poor, or a socially engaged middle class intellectual who works in solidarity with the poor) who enables the poor to break their "culture of silence" and go on to "create their own language" (Frostin 1988: 10). The South African biblical scholar Sarojini Nadar takes up a similar position while standing firmly within liberation hermeneutics, advocating for a critical interventionist contribution from the activist-intellectual (Nadar 2009).

103.3 The Perception of God

Turning to the second conceptual category of biblical liberation hermeneutics, the perception of God, Frostin notes that the first emphasis, the choice of interlocutors, "has important consequences not only for the interpretation of social reality but also for the understanding of God" (Frostin 1988: 7). For while the interlocutors of many forms of progressive (including postcolonial) biblical hermeneutics question or bracket faith, the interlocutors of liberation hermeneutics "'share' the same faith as their oppressors, but they do not share the same economic, social, or political life" (Gutiérrez 1973: 241). The core 'theo-logical' question liberation hermeneutics asks is not whether God exists or not, but whose side God is on. As the Ecumenical Association of Third World Theologians (EATWOT) so aptly expressed it, "The question about God in the world of the oppressed is not knowing whether God exists or not, but knowing on which side God is" (Fabella and Torres 1985: 190). So the central problem in Third World countries is not atheism but "an idolatrous submission to systems of oppression" and the central question posed to any faith tradition is whether and to what extent that tradition deifies the ideologies used to sacralize structures of oppression (Fabella and Torres 1985: 190). So while oppressor and oppressed may be said to share the same faith, the faith tradition itself is a site of struggle, and is profoundly contested.

The South African *Kairos Document*, a prophetic theological response to the State of Emergency in 1985, is perhaps the clearest articulation of a faith tradition as a site of struggle. However, while the document provides an incisive analysis of contestation within the Christian tradition in South Africa, there is no similar analysis of the Bible. The Bible is assumed to be on the side of the liberation struggle. The apartheid state, the *Kairos Document* argues, is "misusing theological concepts and biblical texts for its political purposes" (Kairos 1985: §2). This "hermeneutic of trust" with respect to the Bible is a feature of many forms of liberation hermeneutics, for liberation hermeneutics in its various forms has tended to work with the Bible as sacred scripture. While liberation practitioners may mean different things by the notion of "sacred scripture," there has been general agreement that the Bible is "substantive" rather than merely "instrumental" in its participation in the liberation project (West 1995: 103–130).

That the Bible has a substantive value is related to what the South African liberation theologian Albert Nolan has called "the shape" of scripture (Nolan 1988). Liberation hermeneutics has tended to argue that the Bible has a liberatory 'shape' or "semantic axis," to use the term used by J. Severino Croatto (Croatto 1987). Until recently (Tamez 2003), Latin American voices has been fairly insistent on a hermeneutics of trust, in which the theologians and their interlocutors can trust the basic liberatory trajectory of scripture. Any distortion of God's liberatory project was put down to the church's misuse of scripture, not to the inherent ideology of scripture itself (Richard 1995: 273).

But this consensus has been vigorously contested by African forms of black (Bailey 1998; Mofokeng 1988; Mosala 1989), feminist (Dube 2001), and postcolonial biblical hermeneutics (Dube 1997; Nzimande 2008). From these perspectives the Bible is itself a site of struggle, representing contending discourses, whether at the socio-historical (Mosala 1993), or literary (Nadar 2001), or symbolic-metaphorical (Okure 1995) level. For some African liberation theologians and biblical scholars, albeit a minority, the voices of the marginalized within the text are so compromised by layers of dominating discourse that the Bible can only be an instrumentalist resource, not a substantive resource (Mosala 1989; see also West 1995: 103–130). Socially engaged biblical scholars who adopt an instrumentalist position continue to use the Bible only because the masses consider it a sacred and substantive resource, even though the biblical scholar or theologian does not (Mofokeng 1988).

The increasing attention given to the ideological agenda of the biblical text itself within biblical studies makes it difficult for a biblical hermeneutics of liberation to deny that the Bible has a multiplicity of often contending voices. However, for the socially engaged scholar who works in solidarity with the poor and marginalized it is also difficult to deny that the Bible is a sacred, significant, and substantive resource for the majority of ordinary readers in contexts of struggle. While ordinary readers do not usually have the particular critical capacities of the discipline, they have their own strategies for discerning liberatory strands among the more oppressive voices of scripture; they have forged their own ways of "conjuring" with scripture (Maluleke 2000; Smith 1994). But while ordinary readers can and do either bracket or theologically reconstitute those details of scripture that do not fit within their ideo-theological orientations, scholarly readers are more constrained, being both accountable to the communities of struggle with whom they work and responsible to the ideological detail their discipline discerns in the Bible.

Discerning the God of life from the idols of death in the Bible is difficult to do when confronted with an ambiguous Bible (Mofokeng 1988) or a Bible consisting of ideologically contending redactional layers and textual voices (Mosala 1993). Mario Aguilar poses an even deeper question within liberation theology, that of "dealing with 'the absence of God' or 'the death of God' as markers of hermeneutical significance" (Aguilar 2009: 1). After the Rwandan genocide, he argues, African (as well as Latin American and Asian) liberation theologies in doing their "theologizing from the side of the poor must engage with a phenomenon previously considered European: the absence or silence of God in times of war, ethnic cleansing or genocide" (Aguilar 2009: 1). Though his challenge has not yet found a foothold in Africa's liberation theologies, it must, like the Rwandan genocide that precipitated it, haunt our work.

103.4 The Analysis of Social Struggle

The third conceptual category, that of the necessity for social analysis, also derives from the first, for the option for the poor as the chief interlocutors of biblical liberation hermeneutics is based on a conflictual perception of social reality, affirming that there is a difference between the perspectives of the privileged ("from above") and of the poor ("from below"). EATWOT reports characterize the world as "a divided world," where doing theology and biblical interpretation can only be done "within the framework of an analysis of these conflicts" (Frostin 1988: 8).

The historical emergence of liberation theology was within the context of economic struggle. The Dar es Salaam report of EATWOT (1976) was unequivocal, stating that "the division among rich and poor was perceived as the major phenomenon of contemporary history" (Frostin 1988: 8). Central to the social analysis of liberation hermeneutics is the relationship between theo-logy and economy. The struggle against Mammon is the core struggle for liberation hermeneutics (Frostin 1988: 7). As Jesus put it, "You cannot serve God and Mammon" (Matthew 6: 23). The pursuit of economic profit, the kernel value of capitalism, it was argued, "is incompatible with faith in God" (Frostin 1988: 7). The cult of Mammon distorts Christian faith, substituting faith in the true God with faith in the false gods of an idolatrous economic system. And it is the poor who "have unsubstitutable insights into the difference between God and the idols" (Frostin 1988: 7), both in their own contexts and, as we will examine more fully below, biblical contexts.

However, as Frostin notes, though subsequent EATWOT meetings confirmed the priority of the economic domain, other domains of struggle gradually pushed their way onto the EATWOT agenda. By 1988 when Frostin published his analysis of liberation theology he discerned six "different levels of oppression": economic (rich-poor); classist (capitalists-proletariat); geographic (North-south); sexist (male-female), ethnic (e.g. white-black); and cultural (dominant-dominating cultures) (Frostin 1988: 8). As is evident from these formulations, the social analysis of early forms of liberation hermeneutics tended to adopt a binary format (which is where postcolonial hermeneutics has difficulties with liberation hermeneutics (Sugirtharajah 2006: 77–80)).

While EATWOT consistently stressed the interrelatedness of these struggles, many of the debates within EATWOT centred around the priority given to and the relationships between different levels of oppression. Generally speaking, says Frostin, the discussion "followed continental lines of divisions, where Latin Americans have emphasized the value of socioeconomic analysis while Africans and Asians have tended to stress religio-cultural analysis," and where women from each of these continents consistently emphasised "the male-female contradiction, virtually neglected in Dar es Salaam" (Frostin 1988: 8). By the late 1970s the privileging of Marxist socio-economic analysis was being contested on a broad front.

While Frostin is correct in his analysis of the general tendencies of particular continents, the African context was itself divided on what aspects of the African reality should form the focus of analysis and what analytical concepts should be used. Africa north of the Limpopo River sought to emphasize the cultural (including the religious) and political dimensions of the African struggle against colonialism, while African south of the Limpopo River prioritized the economic (including the racial) and political dimensions of the African struggle against colonialism and apartheid (Balcomb 1998). The use of Marxist social analysis was a source of regular contestation among African liberation theologians, with African Theology (the most prevalent form of liberation theology in West, East, and Central Africa) and Black Theology (the most prominent form of liberation theology in Southern Africa) disagreeing on the appropriateness of Marxist analytical categories and their ideological infrastructure (Tutu 1979).

As indicated, African women had to push hard in order to place gender on the agenda of African liberation theologies (Oduyoye 1983). The recovery of African culture after centuries of colonial denigration made it difficult "to hear" the "internal" critique of African culture posed by African Feminist/Womanist/Women's liberation hermeneutics. But gradually each of the various forms of African liberation theology (Maluleke 1997) came to recognise and engage with patriarchy, masculinity, and sexuality (West 2009a). The advent of HIV and AIDS in the 1980s has been a considerable catalyst in this regard (Dube and Kanyoro 2004). And other forms of marginalization, to do with, for example, disability (Bruce 2010), ethnicity (Nyirimana 2011), and ecology/land (Mugambi and Vähäkangas 2001), are also finding a place in African theologies of liberation. This multiplicity of intersecting marginalizations is forging important intersections between liberation, postcolonial, and queer hermeneutics, particularly in biblical hermeneutics (Punt 2011). However, the question remains of to what extent each of Africa's traditional "liberation" theologies will cohere around their own core axis: African Theology, with its emphasis on culture and religion; Ujamaa Theology, with its emphasis on economics and culture; Black Theology, with its emphasis on race and class; Contextual Theology, with its emphasis on theology and politics; African Women's Theology, with its emphasis on gender and culture; and Reformed Confessing Theology, with its emphasis on reformed doctrine and race (West 2009a). More importantly, the question remains as to what extent the defining 'issue' of liberation theology, economics, will play in emerging forms of African liberation theologies.

Given the importance of Marxist categories and concepts in the historical formation of liberation hermeneutics, including South African Black Theology (Mosala 1993), Tanzanian Ujamaa Theology (Frostin 1988: 41–44), and South African Contextual Theology (Nolan 1988: 219), and the ongoing neo-colonial economic profile of African nation states and our globalised world, an economic orientation remains at the core of liberation hermeneutics, in Africa and elsewhere (Míguez-Bonino 2006), even though there is growing recognition of the many other forms of marginalization that intersect in various ways with class (de Wit et al. 2004). Liberation hermeneutics has been impacted by postmodern, postcolonial, and queer conceptual frameworks that have deconstructed rigid binary forms of social analysis (Althaus-Reid 2000; Nzimande 2008), but has been wary of any form of analysis that downplays the economic (Boer 2005; Croatto 1995: 235; Mosala 1989; Soares-Prabhu 1991). This is the case too with Southern African forms of African Feminist liberation hermeneutics, which explicitly link the cultural and the economic domains (Dube 2000; Nzimande 2008).

103.5 The Tools of Analysis and Interpretation

The fourth conceptual category in Frostin's analysis of the methodology of liberation hermeneutics is closely related to the third and has to do with the choice of interpretive tools with which to engage the theological tradition and the biblical text. "With a different interlocutor and a different perception of God, liberation theologians need different tools for their theological reflection," and so while "in the past theological tools have usually come from philosophy, the social sciences are assuming this role in the new paradigm" (Frostin 1988: 9). As we have seen above, among the first tasks of a theological deployment of the social sciences is economic analysis of contemporary contexts of struggle. A second, related task, for liberation hermeneutics is the use of tools from the social sciences to analyze the theological tradition and the biblical text.

A key relationship in liberation hermeneutics is the relationship between the actual experience/knowledge of struggle of the poor and marginalized and the forms of social analysis provided by the social sciences. The starting point for biblical liberation hermeneutics, in the words of the South African liberation theologian Itumeleng Mosala, are "eyes that are hermeneutically trained in the struggle for liberation today" as they probe "the kin struggles of the oppressed and exploited of the biblical communities" (Mosala 1986: 196). This primary experience is then "resourced" with socio-historical tools, which are used to interrogate past and present power structures. While socio-historical modes of reading have been the preferred choice in liberation hermeneutics, literary and semiotic modes of reading have also found a place within biblical liberation hermeneutics (West 1995: 131–173).

The dominant modes of analysis in the discipline of biblical studies historically and institutionally have been historical and sociological (Lategan 1984), and so the formative years of liberation hermeneutics, whether in Brasil, South Africa, or the Philippines, were dominated by socio-historical analysis of the Bible. The consequence of privileging socio-historical analysis in liberation hermeneutics has been the foregrounding of the socially engaged "scholarly" reader and the "marginalizing," to some extent, of the "ordinary" reader. However, partly in recognition of the non-egalitarian effects of socio-historical modes of reading the Bible and partly because of poststructuralist critique (C. West 1984: 17, 1995), socially engaged biblical scholars from many contexts of struggle, including African, have gradually embraced the full range of other modes of reading, including a diverse array of literary-semiotic and thematic-metaphoric modes (Croatto 1987; Dreher 2004; Míguez 1995; Nadar 2001).

Literary-semiotic-type and thematic-metaphoric-type scholarly modes of reading, though not how most ordinary readers interpret the Bible, do offer more egalitarian entry points for ordinary readers to participate on more equal terms with scholarly readers. However, as already indicated, there is something of a tension here, for central to liberation hermeneutics is the contribution of structured and systematic – critical – modes of analyzing the Bible to communities of ordinary readers of the Bible. This is because biblical liberation hermeneutics is not only about an analogy of struggle (the biblical text in its context//we in our context) but also about an analogy of method (critical modes of interpreting text//critical modes of interpreting context) (Cavalcanti 1995); and the latter has been a particular emphasis of South African Black Theology (Mosala 1989). But notwithstanding the importance of these scholarly critical resources, there seems to be a growing recognition in liberation hermeneutics of the presence of already existing critical forms of discourse among the poor and marginalized, so that socially engaged biblical scholars should understand their collaboration as offering their biblical critical reading resources alongside the local resources already present in communities of ordinary readers of the Bible (West 2008).

103.6 The Relationship Between Theory and Practice

Frostin's fifth and final conceptual feature follows from this, focusing on the dialectics between praxis and biblical interpretation. In biblical liberation hermeneutics, biblical interpretation is "a second act" (Frostin 1988: 10). The first act is the praxis of action and reflection. The "action" envisaged here is actual action in a particular social struggle; integrally related to this action is reflection on the action; and integrally related to this action-induced reflection is further action, refined or reconstituted by the reflection on and reconsideration of theory (and so the cyclical process continues). Out of this first act of praxis a second order liberation biblical interpretation and liberation theology is constructed. In the language of the South African Kairos Document, "peoples' theology" precedes "prophetic [liberation] theology" (Kairos 1986: 34–35, note 15), and the challenge for liberation theologians is to locate and collaborate with the emerging forms of people's theology across the African continent (see for example Gibson 2011).

Liberation hermeneutics is fundamentally a process, and this process can be described as having three movements: See-Judge-Act. Shaped extensively by the "worker priest" movement championed by Fr. Joseph Cardijn in the 1930s in Belgium and embodied in Latin America and South Africa by the Young Christian Workers movement (Cochrane 2001: 76–77; West 1995: 188–193), the threefold See-Judge-Act process provides a structure to liberation hermeneutics, reiterating and integrating each of the distinctive conceptual features discussed. The process of liberation hermeneutics begins with social analysis of a context of struggle (See), moves into a similar systemic analysis of the Bible, bringing text and context into dialogue (Judge), and then moves into community controlled action (Act). Community-based action provides a new site of contestation and struggle, and so further social analysis is required (See), and so the process continues.

This overall shape of liberation hermeneutics is relatively uncontested among practitioners, but, as we have indicated, there are differences in how each moment of the See-Judge-Act process is understood and practiced. Liberation hermeneutics begins with an analysis among the organized poor of the systems that oppress them, giving priority to the economic, but recognizing the multidimensional and interrelated character of structural domination. While local categories and concepts of social analysis are always the starting point for social analysis, the social sciences are brought alongside these local emic understandings by organic intellectuals and activist-intellectuals. Integral to the participatory process of liberation hermeneutics are forms of "liturgical" scaffolding (including prayer, singing, drama, music, and small group work) and collaborative community-based work (including the forms of social engagement that have led to this "new" moment of social analysis).

The reality "uncovered" by social analysis is then interrogated or "judged" by the shape of God's project of liberation that has been discerned within the biblical and theological tradition. Here too there is a place for both local and scholarly interpretive tools as the Bible is brought into dialogue with social reality. In this phase of the process the Bible is read more closely and carefully than is commonly done by people of faith (Riches et al. 2010: 41), which is part of the contribution of the sets of critical resources of biblical studies. More importantly, the resources of biblical scholarship provide additional access to the various ideological layers of and the structural systems in and behind the biblical text. The "Judge" moment in liberation hermeneutics enables a dialogue between a structural analysis of the context and a structural "reading" of the biblical text.

As Norman Gottwald has cautioned, there are limits to the analogies of struggle constructed by the "Judge" moment. "Given the reality that economic systems cannot be 'imported' from the Bible to meet our needs," he says, "the ethical force of the Bible on issues of economics will have to be perspectival and motivational rather than prescriptive and technical" (Gottwald 1993: 345). But as Gottwald goes on to argue, this kind of ethical force is considerable. Notwithstanding the very real differences and the distance between the economic systems of the ancient Near East/ Graeco-Roman world and our world, what connects the two "is a common thread of economic inequity and oppression and a common thread of struggle against needless economic suffering" (Gottwald 1993: 346). Even if, as Mosala would argue, this trajectory is not as prominent as liberation theologians have imagined, recognizing and understanding the economic (and the other) systems that shape biblical texts assists communities of struggle from falling into what the *Kairos Document* refers to as "Church theology," where an "unstructural understanding of the Bible" simply reinforces and confirms an "unstructural understanding of the present" (Mosala 1989: 32).

Having analysed contextual realities, having judged these realities against the shape of God's redemptive project, liberation hermeneutics then moves to work towards the "kin-dom" (Philpott 1993) of God "on earth, as it is in heaven" (Matthew 6: 10). Biblical liberation hermeneutics is not primarily about biblical interpretation, but about social transformation. Indeed, as the See-Judge-Act process demonstrates, biblical interpretation is merely one moment of a larger process within liberation hermeneutics, albeit a pivotal moment.

103.7 A Luta Continua

Liberation hermeneutics "understands itself," Per Frostin reminds us, "as an unfinished process," and one that should be evaluated in relation to 'the first act' of praxis (Frostin 1988: 10). And while the debates of biblical scholars about liberation hermeneutics are potentially useful to the larger project (Sugirtharajah 2006: 77–80), African (and other) liberation hermeneutics should not, in the first instance, be evaluated by its biblical interpretations but by the effects of the praxis that biblical interpretation plays a part in producing.

Central to the praxis of liberation theology (in the singular) is the economic. But while South African Black Theology, South African Contextual Theology, and Tanzanian Ujamaa Theology are "properly" liberation theologies, emphasising the economic dimensions of social sites of struggle, the other forms of African theologies, while clearly "liberating" theologies, have had their emphasis elsewhere, as indicated above. Of Africa's liberation theologies (in the plural), African Women's Theology remains the most vibrant, with the Circle of Concerned African Women Theologians providing a continental infrastructure for collaborative work. The others, though important "ancestors" for any theological work in Africa, have lost some of their momentum. The collapse of African Socialism in Tanzania has marginalized Ujamaa theologians, and the political liberation of South Africa (1994) has drawn many black and contextual theologians into government, creating a vacuum in these theological traditions.

The changing profile of African struggles, post liberation, has led to the recognition of a multiplicity of intersecting marginalizations. The interconnected complexity of African reality calls for hybrid forms of liberation theology, already exemplified in Makhosazana Nzimande's "imbokodo" (grinding stone) hermeneutics (Nzimande 2008), in which liberation (focusing on the economic and political), feminist (focusing on gender and culture), and postcolonial (focusing on empire, power, and identity) hermeneutics are creatively combined. But, the enduring economic exploitation of African nation states, both from within and from without, through various manifestations of neo-liberal global capitalism requires us to retain an emphasis on the economic. For the economic intersects with every other marginalization.

References

Aguilar, M. I. (2009). *Theology, liberation and genocide: A theology of the periphery*. London: SCM Press.

Althaus-Reid, M. (2000). *Indecent theology: Theological perversions in sex, gender and politics*. London/New York: Routledge.

Bailey, R. C. (1998). The danger of ignoring one's own cultural bias in interpreting the text. In R. S. Sugirtharajah (Ed.), *The postcolonial Bible* (pp. 66–90). Sheffield: Sheffield Academic Press.

Balcomb, A. (1998). From liberation to democracy: Theologies of bread and being in the new South Africa. *Missionalia, 26*, 54–73.

Boer, R. (2005). Marx, postcolonialism, and the Bible. In S. D. Moore & F. F. Segovia (Eds.), *Postcolonial biblical criticism: Interdisciplinary intersections* (pp. 166–183). London/New York: T&T Clark.

Bonino, J. M. (1975). *Doing theology in a revolutionary situation*. Philadelphia: Fortress Press.

Bruce, P. (2010). Constructions of disability (ancient and modern): The impact of religious beliefs on the experience of disability. *Neotestamentica, 44*(2), 253–281.

Cavalcanti, T. (1995). Social location and biblical interpretation: A tropical reading. In F. F. Segovia & M. A. Tolbert (Eds.), *Reading from this place: Social location and biblical interpretation in global perspective* (pp. 201–218). Minneapolis: Fortress.

Cochrane, J. R. (2001). Questioning contextual theology. In M. T. Speckman & L. T. Kaufmann (Eds.), *Towards an agenda for contextual theology: Essays in honour of Albert Nolan* (pp. 67–86). Pietermaritzburg: Cluster Publications.

Comaroff, J., & Comaroff, J. L. (1991). *Of revelation and revolution: Christianity, colonialism and consciousness in South Africa* (Vol. 1). Chicago: University of Chicago Press.

Croatto, J. S. (1987). *Biblical hermeneutics: Toward a theory of reading as the production of meaning*. New York: Orbis.

Croatto, J. S. (1995). Exegesis of second Isaiah from the perspective of the oppressed: Paths of reflection. In F. F. Segovia & M. A. Tolbert (Eds.), *Reading from this place: Social location and biblical interpretation in global perspective* (Vol. 2, pp. 219–236). Minneapolis: Fortress.

de Wit, H., Jonker, L., Kool, M., & Schipani, D. (Eds.). (2004). *Through the eyes of another: Intercultural reading of the Bible*. Amsterdam: Institute of Mennonite Studies.

Dreher, C. A. (2004). *The walk to Emmaus*. São Leopoldo: Centro de Estudos Bíblicos.

Dube, M. W. (1997). Toward a postcolonial feminist interpretation of the Bible. *Semeia, 78*, 11–26.

Dube, M. W. (2000). *Postcolonial feminist interpretation of the Bible*. St. Louis: Chalice Press.

Dube, M. W. (Ed.). (2001). *Other ways of reading: African women and the Bible*. Atlanta: Society of Biblical Literature.

Dube, M. W., & Kanyoro, M. (Eds.). (2004). *Grant me justice! HIV/AIDS and gender readings of the Bible*. Pietermaritzburg: Cluster Publications.

Fabella, V., & Torres, S. (Eds.). (1985). *Doing theology in a divided world*. Maryknoll: Orbis Books.

Foucault, M. (1967). *Of other spaces: Heterotopias*. Retrieved March 19, 2008, from http://foucault.info/documents/heteroTopia/foucault.heteroTopia.en.html

Freire, P. (1985). *The politics of education: Culture, power, and liberation*. Westport: Greenwood Publishing Group.

Freire, P. (2006). *Pedagogy of the oppressed* (trans: Ramos, M. B., 30th Anniv. ed.). New York/London: Continuum.

Frostin, P. (1988). *Liberation theology in Tanzania and South Africa: A first world interpretation*. Lund: Lund University Press.

Gibson, N. (2011). *Fanonian practices in South Africa: From Steve Biko to Abahlali baseMjondolo*. New York: Palgrave Macmillan.

Gottwald, N. K. (1993). *How does social scientific criticism shape our understanding of the Bible as a resource for economic ethics? In his The Hebrew Bible in its social world and ours* (pp. 341–347). Atlanta: Scholars Press.

Gutiérrez, G. (1973). *A theology of liberation: History, politics and salvation*. Maryknoll: Orbis.

Haddad, B. G. (2004). The manyano movement in South Africa: Site of struggle, survival, and resistance. *Agenda, 61*, 4–13.

Hinkelammert, F. J. (1986). *The ideological weapons of death: A theological critique of capitalism*. Maryknoll: Orbis.

Lategan, B. C. (1984). Current issues in the hermeneutic debate. *Neotestamentica, 18*, 1–17.

Maluleke, T. S. (1997). Half a century of African Christian theologies: Elements of the emerging agenda for the twenty-first century. *Journal of Theology for Southern Africa, 99*, 4–23.

Maluleke, T. S. (2000). The Bible among African Christians: A missiological perspective. In T. Okure (Ed.), *To cast fire upon the earth: Bible and mission collaborating in today's multicultural global context* (pp. 87–112). Pietermaritzburg: Cluster Publications.

Míguez, N. (1995). Apocalyptic and the economy: A reading of revelation 18 from the experience of economic exclusion. In F. F. Segovia & M. A. Tolbert (Eds.), *Reading from this place: Social location and biblical interpretation in global perspective* (Vol. 2, pp. 250–262). Minneapolis: Fortress.

Míguez-Bonino, J. (2006). Marxist critical tools: Are they helpful in breaking the stranglehold of idealist hermeneutics? In R. S. Sugirtharajah (Ed.), *Voices from the margin: Interpreting the Bible in the third world* (3rd ed., pp. 40–48). Maryknoll: Orbis.

Mofokeng, T. (1988). Black Christians, the Bible and liberation. *Journal of Black Theology, 2*, 34–42.

Mosala, I. J. (1986). The use of the Bible in black theology. In I. J. Mosala & B. Tlhagale (Eds.), *The unquestionable right to be free: Essays in black theology* (pp. 175–199). Johannesburg: Skotaville.

Mosala, I. J. (1989). *Biblical hermeneutics and black theology in South Africa*. Grand Rapids: Eerdmans.

Mosala, I. J. (1993). A materialist reading of Micah. In N. K. Gottwald & R. A. Horsley (Eds.), *The Bible and liberation: Political and social hermeneutics* (pp. 264–295). Maryknoll: Orbis.

Mugambi, J. N. K., & Vähäkangas, M. (2001). *Christian theology and environmental responsibility*. Nairobi: Acton Publishers.

Nadar, S. (2001). A South African Indian womanist reading of the character of Ruth. In M. W. Dube (Ed.), *Other ways of reading: African women and the Bible* (pp. 159–175). Atlanta/Geneva: Society of Biblical Literature/WCC Publications.

Nadar, S. (2009). Beyond the 'ordinary reader' and the 'invisible intellectual': Shifting contextual Bible study from liberation discourse to liberation pedagogy. *Old Testament Essays, 22*(2), 245–252.

Nolan, A. (1988). *God in South Africa: The challenge of the gospel*. Cape Town: David Philip.

Nyirimana, E. (2011). Patrimonialism in the causes of the division of the kingdom in Israel: A reading of the division narrative from the perspective of the Rwandan context of social conflict. *Old Testament Essays, 24*(3), 708–730.

Nzimande, M. K. (2008). Reconfiguring Jezebel: A postcolonial imbokodo reading of the story of Naboth's vineyard (1 Kings 21:1–16). In H. de Wit & G. O. West (Eds.), *African and European readers of the Bible in dialogue: In quest of a shared meaning* (pp. 223–258). Leiden: EJ Brill.

Oduyoye, M. A. (1983). Reflections from a third world woman's perspective: Women's experience and liberation theologies. In V. Fabella & S. Torres (Eds.), *Irruption of the third world: Challenge to theology* (pp. 246–255). Maryknoll: Orbis Books.

Okure, T. (1995). Reading from this place: Some problems and prospects. In F. F. Segovia & M. A. Tolbert (Eds.), *Reading from this place: Social location and biblical interpretation in global perspective* (pp. 52–66). Minneapolis: Fortress.

Philpott, G. (1993). *Jesus is tricky and God is undemocratic: The kin-dom of God in Amawoti*. Pietermaritzburg: Cluster Publications.

Punt, J. (2011). Queer theory, postcolonial theory, and biblical interpretation. In T. J. Hornsby & K. Stone (Eds.), *Bible trouble: Queer reading at the boundaries of biblical scholarship* (pp. 321–341). Atlanta: Society of Biblical Literature.

Richard, P. (1995). The hermeneutics of liberation: A hermeneutics of the Spirit. In F. F. Segovia & M. A. Tolbert (Eds.), *Reading from this place: Social location and biblical interpretation in global perspective* (Vol. 2, pp. 263–280). Minneapolis: Fortress.

Riches, J., Ball, H., Henderson, R., Lancaster, C., Milton, L., & Russell, M. (2010). *What is contextual Bible study? A practical guide with group studies for Advent and Lent*. London: SPCK.

Scott, J. C. (1990). *Domination and the arts of resistance: Hidden transcripts*. New Haven/London: Yale University Press.

Segundo, J. L. (1985). The shift within Latin American theology. *Journal of Theology for Southern Africa, 52*, 17–29.

Smith, T. H. (1994). *Conjuring culture: Biblical formations of black America*. Oxford/New York: Oxford University Press.

Soares-Prabhu, G. M. (1991). Class in the Bible: The biblical poor a social class? In R. S. Sugirtharajah (Ed.), *Voices from the margin: Interpreting the Bible in the third world* (pp. 147–171). Maryknoll: Orbis.

Sugirtharajah, R. S. (2006). Postcolonial biblical interpretation. In R. S. Sugirtharajah (Ed.), *Voices from the margin: Interpreting the Bible in the Third World* (3rd ed., pp. 64–84). Maryknoll: Orbis.

Tamez, E. (2003). 1 Timothy: What a problem! In F. F. Segovia (Ed.), *Toward a new heaven and new earth: Essays in honor of Elisabeth Schussler Fiorenza* (pp. 141–156). Maryknoll: Orbis Books.

Theologians Kairos. (1985). *Challenge to the Church: The Kairos document: A theological comment on the political crisis in South Africa*. Braamfontein: The Kairos Theologians.

Theologians Kairos. (1986). *The Kairos document: Challenge to the church: A theological comment on the political crisis in South Africa* (Rev. 2nd ed.). Braamfontein: Skotaville.

Tutu, D. M. (1979). Black theology/African theology: Soulmates or antagonists? In G. S. Wilmore & J. H. Cone (Eds.), *Black theology: A documentary history, 1966–1979* (pp. 385–392). Maryknoll: Orbis Books.

West, C. (1984). Religion and the left: An introduction. *Monthly Review, 36*, 9–19.

West, G. O. (1995). *Biblical hermeneutics of liberation: Modes of reading the Bible in the South African context* (2nd ed.). Maryknoll/Pietermaritzburg: Orbis Books/Cluster Publications.

West, G. O. (2005). Structural sin: A South African perspective. *Ung Teologie, 1*(5), 15–26.

West, G. O. (2008). The poetry of Job as a resource for the articulation of embodied lament in the context of HIV and AIDS in South Africa. In N. C. Lee & C. Mandolfo (Eds.), *Lamentations in ancient and contemporary cultural contexts* (pp. 195–214). Atlanta: Society of Biblical Literature.

West, G. O. (2009a). Liberation hermeneutics after liberation in South Africa. In A. F. Botta & P. R. Andiñach (Eds.), *The Bible and the hermeneutics of liberation* (Vol. 59, pp. 13–38). Atlanta: Society of Biblical Literature.

West, G. O. (2009b). The not so silent citizen: Hearing embodied theology in the context of HIV and AIDS in South Africa. In T. Wyller (Ed.), *Heterotopic citizen: New research on religious work for the disadvantaged* (pp. 23–42). Göttingen: Vandenhoeck & Ruprecht.

West, G. O., & Dube, M. W. (Eds.). (1996). *"Reading with:" African overtures*. Atlanta: Society of Biblical Literature.

Chapter 104
Asian Liberation Theologies: An Eco-feminist Approach for a More Equitable and Justice Oriented World

Kathleen Nadeau

104.1 Theoretical Background

Late twentieth century Asian liberation theology movements, successfully, overthrew dictatorships such as that of Philippine President Ferdinand Marcos (1965–1986), Indonesian President Suharto (1967–1998) and Korean President Chun Doo-hwan (1981–1988). They were reacting against the kind of top-down modernization and development theory that was associated with top-down globalization processes and authoritarian dictatorships. These movements transcended class-based party politics and hierarchical organizations to organize bottom-up people's power movements. Liberation theology aimed to transform the structures of top-down development models from the perspective of the disenfranchised and poorer classes for whom development was intended to serve.

Currently, liberation theology movements, coming out of the different world religions, continues to struggle against top-down development and globalization processes by counterpoising bottom-up solutions that identify the interdependent relationship between culture and nature, especially at the small agricultural village level. It concerns itself with issues of gender, class, culture, literacy, and human rights. Liberation theology aims to effect changes in ordinary people's lives, such as by helping to provide needed medical services, job security, housing, and a safe and healthy environment. It also embarks from an eco-feminist perspective that calls for men and women to live together harmoniously with each other, as equal partners, and with nature. Its various cultural and religious movements gain momentum by expounding on the idea of growing economic disparity between the rich and poor and the latter's lack of real economic and political power. These movements are indicative of a new millennial trend that finds expression in the eco-feminist

K. Nadeau (✉)
Department of Anthropology, California State University, San Bernadino, CA 92497, USA
e-mail: knadeau@csusb.edu

S.D. Brunn (ed.), *The Changing World Religion Map*,
DOI 10.1007/978-94-017-9376-6_104

partnership between males and females as demonstrated in the international slow food movement or occupy movements that seek to change the current world system into a more egalitarian and justice oriented system. They strive to re-imagine and re-invent a new social and economic world order that supports community based relationships born of mutual respect for the innate dignity of the other, rather than on competition and rapacious profit making (see Reuther 1975, 2012: 1).

Liberation theology movements overlap and intersect with ecological village movements influenced by the teachings of Mahatma Gandhi in South and Southeast Asia (see Batchelor and Brown 1992; Prime 1994; Darlington 1998; Walter 2007). They also find impetus by looking back at past traditional ways of tendering and caring for the environment, animals, and mother earth. These movements largely are inspired by ancient Hindu and Buddhist societies, among others, that strove to live in harmony with the natural world. Like all theology, liberation theology, at root, is God talk. It concerns itself with a God that is embodied and expresses itself through nature and people, the latter who are considered to be part of nature, not as existing somehow independently or apart from it. Liberation theology is an unfolding inductive methodology, or praxis, that combines theory and practice. It discerns God's presence in the world in concrete tangible ways that can be experienced and felt by those involved, by taking into consideration the concerned people's aspirations and, then, looking to see what the Bible has to say about that. It is biased for the poor and oppressed because the God of the Bible is on the side of the poor who comes down to live with them (Boff 1986, 1995). It does not begin through the entry point of some universal, concept of religion that is applicable cross culturally, rather, like engaged Buddhism or engaged anthropological fieldwork, it goes back to the people to think and reflect upon their experiences and realities, to better understand and identify their problems in solidarity with the people concerned (Nadeau 2002).

As Sharon Welch (1985) explains, although biblical texts and scriptures are important in the culturally various and different theologies of liberation, the formative basis of a particular people's faith is hermetically oriented in terms of the present communities of readers who are doing the interpretation. For their faith is not based on some abstract theory divorced from real social life. Rather, it is based on a real and tangible relationship with a God who lives in people's history. Liberation theology is reflected theologically as it is lived in practice. It emerges from organized communities that strategically and politically side with the poor and oppressed. Welch cautions, however, that liberation theologies are not variant strains of thought existing within a traditional theology (for example, progressive theologies versus conservative theologies, within an overriding Catholic theology). Rather, they represent a disruptive break, a new paradigmatic shift, from traditional theology. They are continuous, though, with one tradition, within Christianity, that is critical of social injustices, for example, of hypocrites who feign to follow Christ or to be public servants but, actually, are only interested in their own selfish-aggrandizement (Welch 1985: 24, 34).

According to Pieris (1988), liberation theology practitioners begin by identifying with the poverty and ethics of religious founders such as Christ and Buddha. Adept leaders in such faiths voluntarily take a vow of poverty serving a double purpose as a powerful spiritual and political weapon against religious and secular elites who act in a greedy and selfish manner. According to Pieris (1988: 32), the anti-religious roots of capitalism hinder such leaders from seeing into the religious depths of human nature. He has written extensively about traditional knowledge systems in Christianity and Buddhism that have produced methods for liberating the mind of selfish desires and ideological projections – notably, monks who renounce the world to practice meditation techniques to clarify thinking. Evidence also indicates world religions have been challenged from within the context of their own faith by theologians who call for ideological, socioeconomic, and ecological changes. These liberation theologians encourage people to think clearly and confront their problems collectively as exemplified in Asian people's movements influenced by ethical considerations coming out of Islam, Buddhism, and Christianity.

104.2 Liberation Theology and World Religions

In Asia, before colonialism, world religions were aspects of the different and diverse cultures that had local conceptions of sacred geographical and cosmological space that were perceived to be part of the reproduction of society. Ancient Chinese and Indic cities, for example, were, deliberately and consciously, built to elevate the human spirit to get in touch with the divine (see Wu 1963). An idealized plan of a city is given in the "Code Book of Works" preserved in the *Book of Rites*: "The capital city is a rectangle of nine squares *li. Li* refers to the rational and the official version of what is proper behavior. The essence of *li* refers to social norms or etiquette, which means to humble oneself so as to honor others. The use of *li* is to ensure security, orderliness, or correct conduct. Li-lessness means danger (Wu 1963: 33). On each side of the wall of the capital city are three gates. The Altar of Ancestors is to the left (east), and that of Earth, right (West). The court is held in front, and marketing done in the rear," forming a "Chinese mandala of nine squares with the human being in the center" (quoted in Wu 1963: 37).

Confucius taught that younger generations should respect and obey their elders, women should be subservient to men, and everyone should be obedient to the emperor, who was a parent figure. But, even the emperor who violated the moral order of the universe could be overthrown, legally, and the legitimacy of the coup was determined by whether, or not, it was successful. Confucianism is based on five relationships: ruler-subject; father-son; husband-wife; older brother-younger brother; and friend-friend. Except for the last, all of these relationships are based on differences in status and exemplify different power relationships in Confucian societies. However, Confucius taught that those in superior positions were supposed to be benevolent and caring. This ethical principle provides yet another way for the promotion of greater gender equality by exposing exploitative and oppressive relationships, in the cultural contexts in which they are situated, for violating their own sacred precepts that opens way for culturally sensitive applications of a more universal approach to human rights (Shafer 2000: 108).

All world religions, not only in Asia, carry a similar kind of notion that with rights come responsibilities (Aziz 1999). This notion of rights and duties is replete in the ancient epic stories and oral traditions of India and Indonesia. The ancient Chinese conception of a middle kingdom as consisting of a multiplicity of states bound together by an unwritten code of virtuous conduct is yet another example. In Buddhism there are Bodhisattvas who, like Jesus, upon living virtuous lives reach nirvana but opt, instead, to come down from the mountain to work in the world to guide others along the path to enlightenment. Similarly, we can look at Judaic-Christian parallels in the Old Testament: Psalms, Proverbs, and Ecclesiastics 14: 31–35: "He that opposes the poor blasphemes his maker: but he that honors him is gracious to the poor"; as well as, in the Koran: Surah 12, 168–242: "The Society thus organized must live under laws – based on eternal principles of righteousness and fair dealing." Such ancient cultures made a conscious choice to co-exist in a mutually interdependent relationship.

104.3 Contemporary Contexts

In the Philippines, some liberation theology practitioners are organizing Basic Christian Communities that are resisting the ideological distortions, false consciousness, and fetishisms of world capitalism (see Escobar 1995). It is for this reason that Philippine ecological liberation theology movements, at least ideally, concentrate their training programs not only on increasing peoples' awareness about their basic human rights and on providing support for those whose fundamental rights have been violated, but also on raising peoples' consciousness about their rich cultural heritages in the past and the inequitable roots of their own poverty in the present as a way to move them to action. Their training programs, at best, aim to transform the Philippines, by developing a new social consciousness that is concerned with promoting organic farming, mutuality and respect for one another, in terms of gender equality, and the promotion and care of the natural world (Versola 1993: 12).

For example, in the early 1990s, I conducted fieldwork on the Basic Christian Community movement in an anonymous upland farm community located in the central Philippines for 1 year (Nadeau 2002). My primary methods used were formal and informal interviews and participant observation. From the moment I met the Basic Christian Community farmers, I decided to focus on this community because of its comprehensive program, which included activities ranging from social analysis and creative theater to health care and sustainable agricultural development.

In the late 1980s, when the Basic Christian Community organizers first arrived in the area, they found the farmers impoverished and struggling to survive. Many had forgotten traditional farming techniques such as contour farming and the use of organic fertilizers that were practiced by their predecessors. Instead, they had grown dependent on using expensive artificial inputs to grow their crops. The soil they cultivated was rocky and eroded. Also, they were growing a costly new hybrid variety of yellow corn that attracted a lot of insects and required chemical fertilizers and artificial pesticides. In 1991 the Basic Christian Community organizers challenged the farmers to solve their problems by using resources available in their immediate environment. They introduced a traditional white variety of corn and organic farming techniques that were used by their forbearers. The farmers quickly adopted the church's organic farming program because the traditional white corn could be stored and used longer than the yellow hybrid variety. It also was more pest resistant and did not require costly artificial inputs. The decision of the farmers to adopt the program and maintain their livelihood in terms of "use-value" as opposed to "exchange-value" can be seen as a form resistance based on cultural differences. As elsewhere in Asia, traditional Filipino farmers differ from the dominant agriculture of capitalism in regard to land, food, and the economy.

The farmers also used to think that being religious meant to attend Mass regularly, and keep the sacraments. Those who were perceived to be devout Catholics practiced outward forms of religious behavior. However, the Basic Christian Community organizers introduced the farmers to a new way of practicing their religion by reading and applying lessons learned from the Bible to their own life experiences. They used local metaphors and real life examples to explain what Jesus taught from their own bottom up perspectives. Whereas the farmers used to rely, exclusively, on priests and religious teachers to read and interpret the Bible for them, the organizers now empowered them to discern the meaning of the scriptures for themselves and in conjunction with the clergy. This new way of practicing their religious faith was derived mainly from liberation theology and post-Vatican II social teachings.

Philippine Basic Christian communities also intersect with indigenous struggles for the right of tribal societies to live in their natural habitats (Dove 1998; Vitug 1998). Tribal communities, for example, the Hindagaon tribe of Mindanao, largely seek to protect their environments from being irreparably damaged by the influx of unwanted forms of development such as mining operations and logging concessions that pollute and denude forested areas. Holden and Jacobsen (2012) exhaustively document some of the most detrimental effects of modern mining practices on the surrounding natural environment and communities in the Philippines. These Philippine communities, as well, are actively participating in the ecological village movement being organized by eco-liberation theology practitioners. Tribal people are cultural bearers of indigenous knowledge systems that offer important models for sustainable forestry and agro-forestry practices. The 1993 United Nations Vienna Declaration (see Ishay 1997: 485) recognizes

> the inherent dignity and the unique contribution of indigenous peoples to the development and plurality of society and strongly reaffirms the commitment of the international community to their economic, social, and cultural well-being and their enjoyment of the fruits of sustainable development. States should ensure the full and free participation of indigenous people in all aspects of society, in particular, in matters of concern to them.

Buddhist ecology movements in Thailand, similarly, are focused on teaching sustainability and environmental stewardship to help alleviate the suffering of those who experience some of the ill effects (for example, poverty; displacement; environmental pollution, and water pollution) of industrialization and globalized capitalism. Buddhist monks, who lead this movement, are criticized by those who wish to profit from the destruction of the natural environment, for their political involvement. Despite being subject to threats of violence and, sometimes, even murder, these monks continue their anti-globalization activism with the goal of helping people everywhere to live harmoniously with nature.

Darlington (1998: 5) explains that the reinterpretation of Buddhism for ecology movements is "an effort to put the basic ideas of religion in terms that meet the needs of the modern world." Buddhist ethical values and principles include loving kindness, respect, and compassion for one another. Another refers to the environment and all living things for "every form of sentient life participates in a karmic continuum" (Swearer 2001: 227). Buddhism seeks to end human suffering and some of the most pressing causes of human suffering are directly connected with the destruction of the environment wrought by capitalist development processes. For example, the rate of deforestation in Thailand is higher than anywhere else in Asia, except Nepal, and possibly Borneo, states Darlington (1998: 2). This deforestation and destruction is caused by human vices such as selfishness, greed, and desire, which, ironically, are values apparent in capitalism. The most toxic environmental problems result from the practice of capitalist globalization (for example, multinational corporatization and consumerism), which largely is responsible for increasing inequality, environmental degradation, poverty, and war (Walter 2007: 331). Since, Buddhist monks perceive the causes of human suffering to be derived from capitalist development processes that degrade nature, they feel duty bound to oppose it.

Many villagers around South and Southeast Asia, and beyond, have lost access to much needed land and resources for sustainable livelihoods. Instead, they are being encouraged by capitalist development processes to clear forests and participate in the market economy by increasing cash crop production and building roads to make the forests more accessible for clearing (Darlington 1998: 3). Instead, Buddhist monks have focused on "planting for subsistence, rather than for sale," which poses serious difficulties in practice, since much of the cash earnings villagers stand to gain comes from the forest products. A Buddhist way of living, however, focuses on encouraging people to work together collectively, instead of competing with each other individually, by building communal rice paddies and animal husbandry projects, irrigation projects, and other mutual self help industries, such as the provisioning of alternative herbal medicines, healthcare, and social welfare programs, among the villagers (Walter 2007: 334).

These activist monks teach villagers to view the land as sacred. For example, they symbolically ordain the trees and forests and offer protection for the environment by ritually sanctifying it, while encouraging local people to be spiritually committed to conserving the forests. Darlington (1998: 7) documents examples of prominent monks such as Phra Prajak, Phra Khamkian, and Phrakhru Pitak Nanthakhum who conduct these ceremonies on trees and forests: Phrakhru Manas of Phayao province has been credited for being the first to ordain a tree for ecological purposes (Darlington 1998: 6). While monks are not ordaining trees, in the fullest sense of the term, as ordination ceremonies are reserved for human beings, "the ceremonies are used, symbolically, to remind people that nature should be treated as equal with humans, deserving of respect and vital for human as well as all life" (Darlington 1998: 9). During the ritual ceremony, the tree is wrapped in orange robes, marking its sanctification. In Wang Pa Du village in Northeastern Thailand, a whole mountain was so ordained by "wrapping a three kilometer strip of saffron cloth around its base" to prevent blasting and mining by a quarry company (Walter 2007: 334).

From 1970 to present, some Thai Buddhist monks have been falsely labeled as communist insurgents by government officials and other elites with an interest in local logging concessions and associated industries for their reforestation and agricultural-development work. In 1991, for example, Phra Prajak, a Buddhist monk known for conducting ordination of trees, was jailed for his environmental activism. According to Walter (2007: 339), this was "the first time a robed Buddhist monk had been imprisoned in Thailand." Phra Prajak had led villagers to oppose a eucalyptus plantation that was being developed on their land. He was accused of being a so-called communist monk and underwent intimidation and harassment by local paramilitary and military forces prior to his arrest (Walter 2007: 340). Another example, Phra Sopoj Suwagano was stabbed in 2005. His murderers have yet to be brought to justice but, at the time of his death, he was involved in conserving some 280 acres (113 ha) of forested land that was coveted by a group of local businessmen. Phra Pongsak, the abbot of Wat Palad near Chiang Mai in northern Thailand, is another example of a monk who has continued over the years to work with villagers to reforest and irrigate their rapidly desertifying land in the face of obstacles such as police raids (Brown 1992: Chap. 8).

Many Thai Buddhist monks who engage in social action work originally came from poor rural families. They avail themselves of a monastic education to obtain degrees at large urban Buddhist universities. As part of their degree program, they are expected to participate in development projects in rural communities. Some of these projects were financed in the 1970s and 1980s by the Thai government in an effort to help counter communist insurgency efforts along the Kampuchean and Laos borders and to provide alternatives to opium production in the north. But many monks remained in the villages long after graduation because they became increasingly immersed in the social, cultural, political, economic, environmental, and ecological aspects of sustainable rural development (Gosling 2001: 105).

Similarly, some Buddhist monks in Sri Lanka have participated in a village self-help movement referred to as the Sarvodaya. The Sarvodaya movement emerged in 1958 under the guidance of Professor Ariyaratne when a group of Nalanda College students decided to get in touch with their cultural roots by living with and learning from local farmers. Their experiences were so rewarding that their ideas became popular with other students who followed them. The movement spread rapidly. According to current estimates (Sarvodaya.com), the Sarvodaya movement has a network of about 15,000 ecologically concerned villages operating healthcare programs, educational programs, agricultural projects, and small industries. These programs accomplished much not because organizers employed solutions coming from above, but because they were willing to listen and learn from the local people. They recognized that their greatest resources come from the spirituality and culture of local people. They also practiced Gandhi's teachings of active non-violence as a revolutionary means to social change. This Gandhi-inspired development movement provides an indigenous alternative to the top-down development program of the local government. Since 1958, the Sarvodaya movement has grown from a small group of pioneers working alongside the outcast poor to a people's self-help movement. Its program emphasizes the full range of human well-being; the needs of the whole person must be met-satisfying work, harmonious relationships, a safe and beautiful environment, a life of the mind and spirit, and food, clothing, and shelter.

The Sarvodaya movement (Narayanasamy 2003), like other ecological villages in Asia, finds inspiration in a pre-colonial past. Long before the arrival of the European colonizers, Ceylon, now Sri Lanka. had developed an extensive and elaborate irrigation system of reservoirs and canals. This irrigation system was built around a network of temple communities that were overseen by monks. Beginning with a head monk of a high temple situated on a peak near a major water source, monks based in every community circulating this mountainous island collaborated to determine when and whose agricultural fields would be watered first so that every community was ensured of ample supply of water and bountiful harvest (Ariyaratne and Macy 1992: 78; for a corresponding Indonesian Hindu example, see Lansing 1994).

104.4 Conclusion

Finally, ecological village movements discussed above are indicative of another kind of globalizing trend developing from the ground up that is paving the way for more equitable social relationship between men and women, and humans and nature and new cultural alternatives to the dominant ideology of global capitalism. South and Southeast Asia, for example, are seeing numerous religious movements building networks with each other, locally and internationally, and in the process helping to build alternative forums for peace and justice. As already discussed previously, some religious monks in Thailand are actively engaged in ecology and development activities that coincide with an emphasis on Buddhism's this-world teachings. Such is also the case for some Buddhist monks in Sri Lanka and Christian nuns and priests in the Philippines. These movements seek to develop peripheral communities into self-reliant communities of interpretation and action by synthesizing cultural tolerance, religious pluralism and the practical concern for equity and social justice into a unifying theme. They typically lack substantial financial resources and are hesitant to impose top-down development schemes. To constructively understand them, they need to be studied specifically in relation to the wider societies in which they are situated. The various Asian religious movements, like the Philippine liberation theology movement, have emerged in reaction to the type of development ideology associated with capitalist globalization processes. The justice oriented and ecologically concerned characteristics of these movements, however, each coming out of the different Asian world religions, are ideologically closer to one another in the ecumenical movement than to religious fundamentalism as practiced in their own particular church organizations when viewed from the theological standpoint that creation is an open-ended process for which we share responsibility for the future (Pieris 1988; Gosling 2001).

Humans everywhere are called upon to participate in women's struggle for equal rights, without regard to sex or gender, and this is especially so in Asia. In Confucian based societies (e.g., Korea and Japan), for example, women are addressed less respectfully in relation to men, in the public and domestic spheres, in terms of language usage, as well as the inordinate pressure on young couples to produce male heirs, which results in a greater preponderance of aborted female fetuses (LaFleur 1994). In Buddhist and Hindu societies (e.g., Sri Lanka and India), as well, females, traditionally, have been less valued than males (Risseeuw 1988; Bhatnagar et al. 2005). While males and females are accorded a more equal and interchangeable status in many Southeast Asian contexts (Atkinson and Errington 1990), they still have to struggle for greater equal protection under the law (e.g., there is no divorce in the Philippines, while Indonesian marriage law considers men to be the head of the house. Males also can have more than one wife, according to Islamic law, which allows for divorce, but the Indonesian Family Code requires her divorce be formally recognized in a civil court if she is to avail of the many benefits offered by the government to heads of households). There are other examples. Even so, men and women in Asia, as discussed in the previous section, continue to participate in the international movement for women's empowerment by getting involved in eco-feminist and liberation theology projects at the local village levels.

Finally, the economic aspect of development cannot be separated from socio-cultural, political, and ecological aspects of development. Capitalist models of development that solely emphasize production and exchange of goods for profit are inadequate models when viewed from the ecologically concerned village perspective (Ariyaratne and Macy 1992). Ecological and eco-feminist village movements have changed their overarching strategy from an orientation based on transferring technology and services to the poor toward an orientation based on changing social structures from within. These religious activists offer a new alternative platform for a just and sustainable development paradigm to emerge. In light of the real success stories of such bottom up movements, states that exhibit a strong military arm in collaboration with world powers have subverted and used many popular symbols developed by these grassroots organizations. For example, sustainable development in the early 1990s referred mainly to grassroots initiatives such as organic farming movements, but since has become a popular byword of big agro-industrial complexes. Another example: non-government organizations emerged to channel financial support and practical skills training to grassroots self-help movements, but now the term "non-government organization" is used indiscriminately to apply to any and every organization in the non-profit sector, even to big funding organizations like the Ford Foundation.

Van Ness (1999) illustrates the example of the indigenous concept of Asian values that has been badly misused by Asian dictators in Singapore, Malaysia, and Indonesia under Suharto, to support their authoritarian regimes. In the Philippines, Marcos bandied around Asian values as a pretext to declare Martial Law. He argued that Filipinos needed a strong father figure for them to follow in order to create an ordered society. Some local technocrats and Western scholars wrote about Asian values to prop up his authoritarian regime. Nonetheless, people's movements and grassroots initiatives will not be stopped. Like grass they will grow and grow. People empowerment comes from below. If we really want a more equitable, justice oriented, peaceful, and culturally tolerant world, it will come by working in solidarity with those who find hope in their rich and civilizing ancient traditions.

References

Ariyaratne, A. T., & Macy, J. (1992). The island of temple and tank, Sarvodaya: Self-help in Sri Lanka. In M. Batchelor & K. Brown (Eds.), *Buddhism and ecology* (pp. 78–86). London: Cassell.

Atkinson, J. M., & Errington, S. (1990). *Power and difference in island Southeast Asia*. Palo Alto: Stanford University Press.

Aziz, N. (1999). The human rights debate in an era of globalization: Hegemony of discourse. In P. Van Ness (Ed.), *Debating human rights: Critical essays from the United States and Asia* (pp. 32–55). London/New York: Routledge.

Batchelor, M., & Brown, K. (Eds.). (1992). *Buddhism and ecology*. London: Cassell.

Bhatnagar, R. D., Dube, R., & Dube, R. (2005). *Female infanticide in India: A feminist cultural history*. New York: State University of New York Press.

Boff, L. (1986). *Ecclesiogenesis: The base communities reinvent the church*. New York: Orbis Books.

Boff, L. (1995). *Ecology and liberation: A new paradigm*. New York: Orbis Books.

Brown, K. (1992). In the water there were fish and the fields were full of rice. In M. Batchelor & K. Brown (Eds.), *Buddhism and ecology* (pp. 87–99). London: Cassell.

Darlington, S. (1998). The ordination of a tree: The Buddhist ecology movement in Thailand. *Ethnology, 37*(1), 1–15.

Dove, M. (1998). Local dimensions of 'global' environmental debates. In A. Kalland & G. Persoon (Eds.), *Environmental movements in Asia* (pp. 44–64). Richmond: Curzon Press.

Escobar, A. (1995). *Encountering development: The making and unmaking of the third world*. Princeton: Princeton University Press.

Gosling, D. (2001). *Religion and ecology in India and Southeast Asia*. London: Routledge.

Holden, W. N., & Jacobson, R. D. (2012). *Mining and natural hazard vulnerability in the Philippines*. London: Anthem Press.

Ishay, M. (Ed.). (1997). *The human rights reader: Major political essays, speeches, and documents from the Bible to the present*. London/New York: Routledge.

LaFleur, W. (1994). *Liquid life: Abortions and Buddhism in Japan*. Princeton: Princeton University Press.

Lansing, S. (1994). *The Balinese*. Balinese: Harcourt Brace.

Nadeau, K. (2002). *Liberation theology in the Philippines*. Westport: Praeger Press.

Narayanasamy, S. (2003). *The Sarvodaya movement: Gandhian approach to peace and non-violence*. New Delhi: Mittal Publications.

Pieris, A. S. J. (1988). *An Asian theology of liberation*. New York: Orbis Books.

Prime, R. (1994). *Hinduism and ecology*. London: Cassell.

Reuther, R. (1975). *New women/new earth: Sexist ideologies and human liberation*. Boston: Beacon.

Reuther, R. (2012). Ecofeminism. www.spunk.org/texts/pubs/openeye/sp000943.txt

Risseeuw, C. (1988). *Gender transformation, power, and resistance among women in Sri Lanka: The fish don't talk about the water*. New York: Brill.

Shafer, I. (2000). From Confucius through ecofeminism to partnership ethics. In C. Li (Ed.), *The sage and the second sex: Confucianism, ethics, and gender* (pp. 97–112). Chicago: Open Court.

Swearer, D. (2001). Principles and poetry, places and stories: The resources of Buddhist ecology. *Daedalus, 130*(4), 225–242.

Van Ness, P. (1999). Introduction and conclusion. In P. Van Ness (Ed.), *Debating human rights: Critical essays from the United States and Asia*. London/New York: Routledge.

Versola, R. (1993). The Ramos industrialization program: A contrary view. *Hagit, Quarterly Newsletter of the BCC-CO Visayas, 8*(3), 4–8.

Vitug, M. (1998). The politics of logging in the Philippines. In P. Hirsch & C. Warren (Eds.), *The politics of environment in Southeast Asia: Resources and resistance* (pp. 122–138). New York: Routledge.

Walter, P. (2007). Activist forest monks, adult learning and the Buddhist environmental movement in Thailand. *International Journal of Lifelong Education, 26*(3), 329–345.

Welch, S. (1985). *Communities of resistance and solidarity: A feminist theology of liberation*. New York: Orbis Books.

Wu, N. I. (1963). *Chinese and Indian architecture: The city of man, the mountain of God, and the realm of the immortals*. New York: G. Braziller.

Chapter 105
Cuba's Distinct Religious Traditions: Better Social Changes Come Oh Soooo Slowly

Jualynne Dodson

105.1 Introduction

This book proposes to explore global religious manifestations with the widest representation of human groups on our planet. Of course it will not accomplish the goal because, as a species, humankind is exceptionally complex and heterogeneous but the idea of producing an inclusive "mapping of the world's religion" is commendable. In such an attempt, my chapter focuses on Cuba's distinct religious traditions, some of whose practice components have been on the island since before Europeans arrived and they are traditional practices that have maintained viability as strong or stronger than the imported Catholicism. Indeed, contemporary mapping of religion in Cuba shows that some 80–90 % of the island population adheres to one or more of the seven distinct religious lifestyles. Christian traditions, Catholic and Protestant combined barely equal 10 % of Cuba's religious practitioners even as most statistics from island agencies reflect an increase in religious affiliation for the recent years of economic constraints (Ayorinde 2004). Of equal significance is that more than 60 % of the Cuban population is of African decent and all but one of the distinct religious practices are Africa inspired. However, most literature that purports to engage "religion" in Cuba focuses on the Catholic Church and/or Protestant practitioners, even though the combination of these groups do not yet represent 50 % of religious devotees in the country. This chapter presents a picture of Cuba's religious geography that more reflectively mirrors practice realities of the majority of the island's religious adherents and that would be Cuba's distinct traditions.

J. Dodson (✉)
Department of Sociology, American and African Studies Program, Michigan State University, East Lansing, MI 48824, USA
e-mail: dodsonj2@msu.edu

S.D. Brunn (ed.), *The Changing World Religion Map*,
DOI 10.1007/978-94-017-9376-6_105

If one were to rely on the media for information on religion in Cuba, one also would be impressed that Catholicism is the dominant tradition practiced by island citizens. The extraordinary coverage of the 1998 and 2012 visits from Popes of the Roman Catholic Church, for example, contributes to that logical misimpression. The incorrect impression is equaled to the incorrect assumption that only Christian practitioners attended events of the two Popes. Cuba's more than 400 years of dynamic historical religious interplay has produced an island of "integrated religious multiplicity." This means that any given set of Cuban ritual practices, or any given practitioner, can and does reflect elements from another(s). The distinct religious traditions particularly are not bound exclusively to rituals of a single set of religious practices. This is less true for Christian adherents but then they are not the most popular traditions for the majority of Cuba's 11.5 million citizens. The sum result is that Cubans attending events for a Catholic Pope would easily be practitioners of any of the island's religious traditions, Christian or otherwise.

Despite a religious geography where practitioners share sacred rituals, when leaders of the Roman Catholic Church visited the island, neither chose to meet with leaders of the more popular and/or Africa inspired traditions. The dismissive behavior toward representatives of the majority of religious devotees was a continuation of historical practices toward non-Catholic, especially non-Christian religious activities that began long before the twentieth and twenty-first centuries. The history-laden posture ignored and tried to eliminate Cuba's more widespread and diverse religious geography, even as all the activities shared the same land spaces. Nevertheless, the more popular sacred traditions persisted throughout four centuries of Cuba's central participation in the construction of our modern/post-modern and re-globalized era.[1]

105.2 Beginnings

It was Christopher Columbus' exploratory voyages and landings in the Caribbean that initiated the relocation of new people and materials to the western hemisphere and throughout the globe. By 1493, a small settlement had been established in Haiti and Cuba's coastline geography had been explored. The Haitian site did not succeed and Cuba's strategic location as part the Atlantic-Caribbean Gulf Stream, that huge ocean river that circulates clockwise off the east coast of the Florida Keys through to the Yucatan Channel between Mexico and Cuba, made the island an early and important link between the Spanish metropolis and its Americas' colonial empire. Cuba was the largest island on the shipping highway, had a large number of viable ports and gave Spain and Europe an early ability to ship exploited gold, silver, and other riches of the Americas east to their continental capitals.

[1] I say re-globalized world because the European colonization and management of the trade in enslave Africans established the first globalization phenomenon of modernity.

Cuba's AmerIndian populations were well established in Haiti and had several out-post settlements in Cuba as the Spanish arrived in the fifteenth and sixteenth centuries. The indigenous population also possessed a practiced religious reality that predated the importation of Spanish Catholicism. All members of this Taíno group were not annihilated by European invasions and conquest but in the first generations of contact with the Spanish, the indigenous population definitely lost most of its social organizations and cultural communities. Nevertheless, their religious practices were an alternative to the imported Catholicism and the two sets of religious approaches were expanded yet again in the first decades of the 1500s. Enslaved Africans were imported to replace the thinning numbers of Taíno AmerIndians. Now the Cuban landscape had three religio-cultural traditions and the island's contemporary distinct practices are rooted in this history.

Today, Cuba has seven distinct religious traditions – *Muertéra Bembe de Sao, Palo Monte/Mayombe, Vodú, Regla de Ocha/Lucumí, Ifá, Abakuka, Espiritismo* – and they have been adapted, adjusted, transformed, and sustained from the earliest colonial period to twenty-first century viability. They have survived some of the harshest constraints, overtly oppressive regulations, multiple national campaigns of elimination from several governing authorities, and undeniable systemic racism. The following is an introduction to these traditions and social changes that have affected them. Significantly, the chapter does not give detailed attention to Cuba's Christian practices but allows more knowledgeable others to examine how those too are part of the religious map.

105.3 Distinct Religions Traditions

Six of Cuba's seven sets of distinct practices are relatively well known to researchers and to some from the general public while one tradition requires further study to verify its history and development. The six are *Reglas Congo*, *Vodú*, *Abakuka*, *Espiritismo*, *Regla de Ocha/Lucumí*, and *Ifá*. *Abakuka* is understood as a secret fraternal organization exclusively for men though several of its ritual practices are derived from the other distinct traditions. Customs of the *Reglas Congo* are linked to contact and exchange between sixteenth century Taíno AmerIndians and enslaved Kongolese Africans brought to Cuba. These *Reglas* – rules comprise a family of Africa inspired religious observances that is subdivided into several lines of established rituals. *Palo Monte/Mayombe* is the more well known of the *Reglas Congo* and it is practiced throughout Cuba, though most communities are in urban and suburban areas. The signifying characteristic of Congo practices is the emphasis on working with spirits and material objects of the dead. Concomitantly, most Cubans, whether or not they are religious practitioners, acknowledge that *Palo Monte/Mayombe* is the strongest, more powerful of the religions.

In this sequence of arrivals, what would become Cuba's *Vodú* was the next to be sustained in the colonial environment, brought by enslaved Africans of Haiti who themselves were brought by Haitian French colonials fleeing agricultural changes as well as the tumultuous Haitian Revolution that closed the eighteenth century. This is not meant to suggest that only enslaved Kongo or Dahomey – the African spaces from which Vodú is inspired – Africans were the only ones imported to Cuba and/or the Americas. However, among the larger mix of enslaved Africans, it was Kongo and Dahomey practices that held cultural sway on the island.

Cuba's religious map of distinct religions was expanded yet again during the last decades of the eighteenth century and the first of the nineteenth century as this period marked the arrival of two religious traditions derived from Africa's Yoruba-land, areas we now know as Nigeria. Ifá is a religious system of communicating with the otherworld of divine spirits and the divination practices that accompany Cuban devotees' of the Yoruba inspired tradition of Regla de Ocha/Lucumí (Fig. 105.1). The latter is a ceremonial and sacrificial practice that arrived with the massive number of enslaved Africans of Yoruba ancestry, or of contact with that tradition, who were imported to Cuba after 1750. Ocha/Lucumí is known worldwide as

Fig. 105.1 Priest and fully decorated sacred space (Photo by Jualynne Dodson)

Fig. 105.2 A ceremony of espiritismo de cordon (Photo by Jualynne Dodson)

"santeria" though my research experiences revealed that practitioners use that label when outsiders, particularly foreigners ask about their rituals.

Espiritismo is the least Africa inspired set of Cuba's distinct practices and it originates from U.S. Spiritualism as adapted from work of the Frenchman, Hippolyte Léon Denizard Rivail, known as Allan Kardec. The mid-nineteenth century juncture of socio-political relations between Cuba and its North American neighbor brought Kardecian influenced U.S. Spiritualism to the Caribbean island. The tradition has at least four distinct families or pathways though the largest number and varieties of practicing communities is in the eastern Oriente region, contemporary provinces of Guantánamo, Santiago, Holguín, Las Tunes, and Gramma (Fig. 105.2).

Cuba's seventh set of practices, Muertéra Bembe de Sao, is less known among researchers and has been reported in one research project. Several Cuban pamphlets engage the idea of *muerte* – death as fundamental to all of the distinct traditions and customs of Muertéra appear to be strongly linked to the earliest colonial religious history. However, until we have additional evidence, we can only speculate that Muertéra is a cohesive set Cuba's distinct religious traditions.

105.4 Changes

Sociologically, Cuba's position within Spanish colonial structures and infrastructure of the Roman Catholic Church meant that the colony functioned with Catholic practitioners as predominate administrators and that Christian religious tradition circumscribed the island's sacred and political geography. However, geography is more than outer-rims of a space, but includes internal variations, densities, and dynamics.

The structure and dynamics of colonial subjugation of enslaved African descendants saw another important change at the close of the eighteenth century. Cuban authorities reactivated an organizational tool for the social order of the population. The tool was *cabildos* and they were imported from Spain's continental civil infrastructure. The intent was to draw organized activities of African descendants into Cuba's civil society as a means of thwarting dissatisfaction with their enslaved status and preventing rebellions like the Haitian Revolution. Despite its intent, the change positively altered social circumstances of Africa inspired religious traditions.

Practitioners had already constructed social arrangements – called *nacíones* that followed continental lines of real or perceived ethnic backgrounds: Congo, Dahomey, Mandingo, and others. Authorization of *cabildos* now allowed the informal Africa based collectivities to register as official Cuban organizations, to purchase property, to provide resources for members' burial, and to distribute the collective wealth among their members. More significant, *cabildos* converted the official social arrangements into self-governing arenas where Africa inspired customary practices were actively and purposely continued.

Cabildos and distinct religious observances transferred into the nineteenth century despite on-going enslavement, social exclusion, and horrific maltreatment. Freedom fighters in Cuba's 1895 second War of Independence[2] even performed Africa derived rituals as they waged military resistance against Spain. However in 1898, the United States interrupted the independence struggle and concluded the Cuban war by annexing military, economic, political, social, and religious affairs of the island. The domination changed social circumstances for practitioners of the distinct religions and the North American brand of racism simultaneously was imported to Cuba. The changes lasted through presidential administrations of Tomás Palma Estrada, Miguel Mariano Gómez, Mario G. Menocal, Alfredo Zayas, Gerardo Machados, and Fulgencio Batista, from 1902 until 1959.

Immigration laws that restricted non-whites from moving to the island were coupled with an *anti-brujo*, anti-witch campaign that targeted Africa inspired practices and Black people generally. The nation's penal code reinforced the developments, as police raids on registered *cabildos* were common, ritual artifacts were confiscated, and many leaders, if not members were arrested. Drumming, that is an intricate part of worship in most traditions, was outlawed with the intent of further eliminating African influences. Negative attitudes against descendants and the religions escalated as the Cuban press reported white children abducted and killed ritualistically, as part of the presumably "witchcraft" character of Africa inspired traditions. Few if any such reports were proven to be essential to the religious practices while several deaths were shown to have been committed by whites and made to appear the

[2]The first military campaign for Cuban independence from Spain was the Ten Year War – 1868–1878. Some Cuban leaders capitulated.

work of Africa inspired customs. Clearly Cuba's religious traditions were being criminalized and demonized, circumstances that persisted beyond the first half of the twentieth century.

The circumstantial changes, coupled with overtly racist attitudes and activities altered Cuba's geographic contours of where and how the traditions were practiced. African descendants particularly took ritual performances out-of-sight, deeper underground, to clandestine areas of the island and the pattern became routinized after the 1912 government massacre of more than 2,000 African descendants who organized to protest racialized social inequities. Elder practitioners with whom I have had extensive conversations, and who experienced this period of persecution, never addressed reasons but spoke of rituals performed in wooded forests and other secret locations. For example, I attended a particular re-activated outdoor religious ceremony that devotees said had not been conducted for more than 40 years for fear of persecution. Now, in the twenty-first century, it has been brought back into the liturgical-like calendar of their tradition. It also has taken time to find individuals with enough knowledge to perform the seaside event (Fig. 105.3).

For Africa inspired observers, success of the 1959 Cuban Revolution brought socio-political changes of a different sort though basic social circumstances remained steady for the immediate period. The new government was self-defined Marxist-Leninist and subscribed to scientific atheism but, unlike some socialist countries, Cuba did not outlaw religion. Indeed the new President, Fidel Ruiz Castro repeatedly stated that the government should not appear as an enemy of religion or give practitioners reasons to claim persecution. However, if individuals were known

Fig. 105.3 Revitalized ceremony of regla de ocha at the sea (Photo by Jualynne Dodson)

to be religious believers and/or devotees, they could not join the Communist Party and could not hold positions that led to Party membership. Also, religious associations were required to register and were restricted only to visits devotees whose names and locations were registered. Most practitioners did not comply.

While some in the larger religious sector of Cuba openly resisted the new government, believers of Africa inspired traditions were more inclined to support the new order. Compared to past regimes, it brought free education to all, free health care and medical services; distribution of social resources that made them more accessible; repair of roads; delivery of water and electricity, though slow, was no longer reserved for whites and the rich, and the new government elevated Cuba's international reputation as a sovereign nation committed to justice. As I conducted my field researches, one practitioner related that after the triumph of the Revolution, as poor as he was, he still was able to complete his education, secure the education of his six children– some in foreign languages, and see his oldest daughter become a medical doctor, an occupation few African descendant Cubans could even dream about before the Revolution.

The decade of the 1960s was rife with internal and external sabotage and counter-revolutionary activities, including the infamous attempt by U.S. sponsored guerrillas to over throw the Cuban government with an invasion at *Playa Grion.*[3] In response, government authorities implemented re-education encampments, called Military Units to Aid Production (UMAP), for counter-revolutionary and/or potentially threatening individuals. I have not encountered a leader of Cuba's distinct religious traditions who acknowledges spending time in a UMAP.

As the decade of the 1970s opened, and the Cuban government paid close attention to positive outcomes of Liberation Theology in countries such as Nicaragua, authorities began to differentiate among their island's religious traditions. The assessment was that some religions could make a positive contribution to Cuban society when and if their fundamental values were compatible and practiced in support of national goals. There was no outlawing of any religion but traditions opposed to national allegiance, for example, were forbidden – Seventh Day Adventists, Jehovah Witnesses, and the Gideons.

Foreign travel restrictions for religious practitioners were loosen, particularly for Protestant who were allowed and supported in international travel to religious convocation. I encountered Protestant leaders and laypersons at such gatherings in China, Korea, and Nicaragua. This tradition was even able to negotiate visits to the United States, travel to which contained barriers in both countries. I participated in the installation of a major museum exhibition in Spain where leaders and practitioners of Cuba's distinct religions were in attendance in large numbers. There was even a dance troupe that performed a specially choreographed presentation on Cuba's African heritage's significant place in the island's history and national identity.

The government also authorized studies of core values and practices of several religious traditions. Authorized-study of African inspired religions was a major social change as was an international conference convened in Cuba to explore the

[3] Playa Grion is the island location of the invasion. In the U.S. it is known as the "Bay of Pigs."

presence of the island's Africa-based cultural values and practices. An organization of one tradition also was approved to repair and open to the public, a four-story structure that displays larger than life replicas of many of the religion's sacred icons. Several leaders traveled to Africa to gather soil for the instillation and a renowned international African religious official participated in consecrating the structure and installations. The Yoruba Center has become a major tourist attraction in Havana as well as a national gathering place for essential ritual ceremonies. On each occasion that I visit the Center, there are Cuban practitioners openly engaged in devotional activities. In the geography of the island's Africa inspired religious practices, this is a major social change; a public site of devotion, known to all, certified by government authorities, managed by religious practitioners.

Although the 1980s saw the island economy still dependent on trade relations with nations of the Soviet Union, and Cuba's beginning of the Special Period in times of peace, there was a watershed event that initiated a series of dialogues that once again changed the religious map of Cuba. It was a symposium and worship honoring the Reverend Dr. Martin Luther King Jr. The event was held at a Protestant church in the capital city of Havana, organized by Protestant leaders, attended by Cuba's President Fidel Ruiz Castro and the Reverend Jesse Jackson, then candidate for U. S. President (Fig. 105.4). It was a successful celebratory event of international proportions but, like subsequent visits by Catholic Popes, leaders of Africa inspired traditions were not visibly included.

Those who organized and participated in the seminal worship, and other sanctioned activities of those 10-days, were able to negotiate public conversations

Fig. 105.4 President Fidel Ruiz Castro and the Rev. Jesse Jackson at the seminal worship honoring Rev. Dr. Martin Luther King (Photo by Jualynne Dodson)

with state authorities. Nationally televised conversations, public dialogues, and "roundtables" followed rather immediately and the discussions focused on widespread and official religious discrimination. In 1991, the Cuban Communist Party changed the constitution to allow religious believers to become members of the Party and subsequently participate in other governing activities.

As restrictions against religious believers were lifted, and their visible and acknowledged participation was more welcome within Cuban civil society, the island's distinct religions also experienced significant changes in their social circumstances. Between 1992 and 1994, for example, leaders of at least one religion convened authorized international conferences. Participation was quite strong and a number of Cuban practitioners referred me to the events as I was not able to attend.

Similarly, Cuban police no longer raid local and neighborhood ritual activities as a routine course of action. Many sacred celebrations were no longer conducted "underground," though they could not be described as fully public events. At the same time, I have never observed anyone turned away from a ceremony because they were unknown. There has even been more acceptance of the traditions' more rousing activities, drums of worship and rituals of animal sacrifice. On a regular basis, drums can be heard in many neighborhoods and animal sacrifices only need be registered with local health and sanitation authorities. Many Cuban bookstores now carry a variety of new materials about several of the distinct traditions, even though others than the Cuban publishing house are not represented. Even university curriculum now includes courses on aspects of Africa inspired practices. National and regional newspapers too can be found carrying articles about the traditions and some of the celebratory events are televised. Notwithstanding such coverage, Protestant activities are more regularly included in Cuban newspapers. In all, there have been definite changes in the last 15–20 years and the transformations have brought Cuba's distinct religious traditions much closer into the public vision.

The ability to publish and disseminate instructional statements about how devotees should practice their religion is an additional change for the distinct traditions. Over the years of my visits, I have purchased many materials about the religions, but if local Cuban practitioners wrote the documents, I could not make purchases in public but had to do so quietly and in private. Such purchases were few and far between. Recently however, one leader of the distinct religious has taken to distributing to all new initiated practitioners a small document that outlines their new status and responsibilities. This is a major change and one that coincides with current national commercialization of other Cuban commodities for visiting tourist. It seems to suggest that even the distinct religious practices are being observed and practitioners have begun to seize tourist opportunities to improve their income situation.

For the most, social changes that affect all of Cuba's religious believers continue to be in place, although practitioners of the distinct traditions have felt the changes at a much slower pace than Christian believers. This is especially true for the distribution of resources that are in short supply nationally. Since the dissolution of the Soviet Union, economic shortages throughout the country are real and no one is excluded from scarcities of medicine, water, gasoline, food, items of clothing, pens, paper, light bulbs, and other life necessities. Perhaps because it members have been more aligned with Cuban political leaders, Christians have tended to have a higher quality of social life than devotees of the distinct religions, especially those Africa inspired practices. What is not known is whether the current government, or those to come will be able, or will choose to continue incorporating practitioners of the distinct traditions into the society's social fiber. There is historical precedent for a change in religious sector, particularly those that are not Christian. However, current changes are in place and must be acknowledged, as incomplete to full social participation as they may be.

105.5 Summary Thoughts

Despite the fact that core practices of many of Cuba's distinct religious traditions began as part of the organizational nucleus of the island as a social order, historical understandings and presentation of a religious map have omitted them. The exclusion of these practitioners from participation in the island's civil order also has been an historical reality for more than four centuries. The result has been that most who wish to understand religion in Cuba have been presented a distorted picture of religion, further complicating an already complex social and political reality.

In no way does this brief presentation fully correct the void but it has been put forth as a focused exploration of Cuba's seven sets of distinct religious traditions that continue to be practiced. The intent has been to acknowledge major social changes that have affected devotees and their practices since the traditions began developing in the earliest decades of the 1500s. At the same time, it has taken some 500 years for Africa inspired religions to be incorporated into societal activities of the nation. The pivotal question is whether the long historical experiences of persecution, discrimination, ridicule, criminalization, and racism will reappear if and when the island's socio-political changes neither need to incorporate this component of the religious sector and/or the country returns to a prior ideological disrespect of these practitioners' human expressions?

Reference

Ayorinde, C. (2004). *Afro-Cuban religiosity, revolution and national identity*. Gainesville: University Press of Florida.

Recommended Readings

Brown, D. (2003). *Santeria enthroned: Art, ritual, and innovation in an Afro-Cuban religion*. Chicago: University of Chicago Press.

de La Fuente, A. (2001). *A nation for all: Race, inequality, and politics in twentieth century Cuba*. Chapel Hill: University of Carolina Press.

Dodson, J. E. (2008). *Sacred spaces and religious traditions of Oriente Cuba*. Albuquerque: University of New Mexico Press.

Chapter 106
Global Networks and the Emergent Sites of Contemporary Evangelicalism in Brazil

Jeff Garmany and Hannes Gerhardt

106.1 Introduction

Latin America is a region increasingly in religious flux. Though still dominated by Catholicism, evangelicalism continues its rapid growth in the region both in terms of its adherents and power. This situation is particularly true for Brazil, which has seen some of the most striking shifts to Protestantism in the last two decades. Although making up only 15 % of the population, if current growth rates persist it is estimated by some that Brazil could be a predominately Protestant nation by 2020 (Reel 2005). There are, of course, many factors that could limit this growth trajectory, including the fact that Brazilian evangelicalism has so far been limited to making inroads primarily with poorer sectors of the population. Nevertheless, the evangelical movement in Brazil is formidable and its impact is increasingly being registered in Brazilian society and politics.

In this chapter we ask how it is that evangelicalism has become so pervasive and established in Brazil. We consider how a number of global forces and influences over time have come to create a context in which evangelicalism, and particularly Pentecostalism, has been able to thrive. More specifically, we draw on a case from the city of Fortaleza on the northeastern coast of Brazil to provide a qualitative evaluation of how various global events and processes have actually come to shape the religious environment in the sprawling favela (informal housing settlement) of Pirambu.

J. Garmany (✉)
King's Brazil Institute, King's College London, London WC2R 2LS, UK
e-mail: jeffrey.garmany@kcl.ac.uk

H. Gerhardt
Department of Geosciences, University of West Georgia, Carrolton, GA 30118, USA
e-mail: hgerhard@westga.edu

S.D. Brunn (ed.), *The Changing World Religion Map*,
DOI 10.1007/978-94-017-9376-6_106

The emergence and blossoming of Protestantism in Brazil is a quintessential case of the geographical reach and interconnectivity that characterizes the history of religion more broadly. From the earliest Calvinist missionaries arriving from Europe in the 1500s to the advent of Brazil becoming a mission sending country in the present, the Protestant story in Brazil is one of complex geographical interactions.

106.2 Religious Transformation in Brazil

The Roman Catholic Church in Brazil has long been the dominant religious force; its promotion and protection during Portuguese colonial rule ensured the Catholic Church's powerful position in Brazilian society and government upon independence in 1822. Nevertheless, post-independence Protestant missions, primarily from the U.S., were able to establish a foothold in Brazil as early as the mid 1800s. It is in this early influence that Brazil's evangelical roots became closely linked to the United States and, more particularly, to the Second Great Awakening (see, for instance, Bosch 1991; Chaney 1976). The spiritual and theological outlook offered by these early American missions, primarily fervent believers within Baptist, Methodist, and Presbyterian denominations, came to greatly form the evangelical tone of Brazilian Protestantism. In this sense the evangelical focus on the literality of scripture, "true" conversion, visible piety, proselytizing, and the assigned importance of a "free" decentralized church can all be traced to this earlier contact.

It was, however, not until the advent of Pentecostalism in the early 1900s that the beginnings of a noticeable shift away from Catholicism began to materialize. Pentecostalism embraces basic evangelical values, yet additionally emphasizes a more experiential approach to God, with a considerable focus on the presence and interaction of the Holy Spirit with believers. A critical development in the spread of Pentecostalism was the missionary work of the Indiana-based Assemblies of God Ministries, which today accounts for one-third of all Protestant adherents in Brazil. The Assemblies of God, which was established in Brazil in 1911, focused primarily on the poor residents of the Brazilian countryside, hence providing waves of converts who eventually made their way into the cities (Stoll 1991). From the 1950s onward, Rio de Janeiro became the source of a number of home grown Pentecostal ministries that have been growing at a tremendous rate. The most important of these, the Universal Church of the Kingdom of God, emerged in the late 1970s.

Thus, traced back to its first serious foothold in the mid-nineteenth century, Protestantism in Brazil has had over a century's time to mature. In this sense, despite the Brazilian Catholic Church's view that the emergence and spread of evangelicalism in Brazil is essentially a North American invasion (Jenkins 2002), by the latter half of the twentieth century Brazilian evangelicalism had developed a distinctively national identity that took into account the particular culture, language, economy, and politics of Brazil. In other words, the links connecting Brazilian evangelicalism to North America have weakened considerably in terms of control and direction as the Protestant movement has found its own domestic logic (Freston 2001).

Much of this nationalization process occurred with the encouragement and blessing from U.S. evangelical ministries and mission groups, including, among others, the Assemblies of God. Indeed, although money, supplies, and some mission groups still flow from North America to Brazil in support of evangelization, for the most part the focus has been on empowering local churches and ministers to expand the movement through church planting and proselytizing. A sign of the movement's national strength and independence can be gleaned from the fact that Brazil is increasingly becoming a mission sending country. Driven by the Great Commission to spread the Gospel to all unreached peoples of the world, Brazil now sends more missionaries abroad (about 5,000) than Britain or Canada (Noll 2009; Smither 2010).

And yet, even with the evangelical movement in Brazil having gained considerable autonomy, it would be wrong to say it has become insular, or unaffected, by forces beyond its boundaries. Indeed, there are a number of larger transnational structures within which the evangelical movement functions and ultimately flourishes. These structures can be grouped into religious and political-economic realms. In terms of religion, emphasis must fall on the transformation of the Catholic Church's influence in Brazil over the past 30 years. Historically, the Vatican maintained relatively little contact with Brazil, which, in turn, allowed for two key developments. First, it opened the possibility of a much more folkloric and locally adapted Catholicism throughout the country (Page 1995). Second, it enabled the flourishing, albeit unsanctioned, of what is today identified as liberation theology, that is, a set of discourses and practices rooted in the idea that Christianity can combat social injustice and oppression.

Liberation theology took many forms during the twentieth century. In some cases, the appeals made by Catholic leaders were deeply entrenched in theological debates and biblical texts; in other cases, responses were far more imminent and drawn from daily life and personal experience. Politically, clergy became particularly concerned with the deteriorating state of civil and human rights in Brazil (particularly during the dictatorial regime between 1964 and 1985) and sought to address social and economic inequality through their services and interactions with parishioners (Nagle 1999).

However, beginning in the 1980s, as the world was becoming increasingly interconnected and globalized, the Vatican expanded its control over Catholic churches worldwide, leading to efforts to standardize religious services and practices. The Vatican also worked to rein in leftist and radical clergy members by discouraging debates of liberation theology (Page 1995). By then relocating clerical leaders with strong and popular followings, the Church ceased to be the obvious guardian and advocate for the poor. At the same time, however, the traditional ties between the Catholic Church and the Brazilian state were faltering as the Vatican began emphasizing individual rights and resistance to authoritarian rule associated with The Second Vatican Council (Burdick 1993).

As a result, when Brazil did achieve democracy in 1985, riding what has been termed the third wave of global democratization (Huntington 1993), the Catholic Church actually emerged in many ways as a much more disconnected and weakened entity.

More specifically, the Catholic Church in Brazil had fallen out of touch with a Brazilian culture that seemed to demand a more animated and personalized religion, while at the same time losing its standing both with the poor (through the undercutting of liberation theology) and the ruling elite, as the Church's cozy relationship with Brazil's undemocratic rulers came to an end.

It is within this context that Brazilian evangelicalism began to flourish. Rather than becoming a more obviously secular society, people began to explore the "new" religious options that had become available to them. Being dedicated to the "free" church and open proselytizing, the numbers of Pentecostal and other evangelical churches skyrocketed during this period, and with their already established domestic infrastructural networks, in addition to foreign (mostly American) missionary efforts, non-Catholic Christian churches[1] sprouted in cities all over Brazil (Stoll 1991). Furthermore, the embrace of a competitive religious landscape by these new Protestant churches allowed them to heed the intricacies of Brazilian culture, which has largely resulted in the blossoming of much more passionate and personalized church services associated with Pentecostalism (MacHarg 2011).

Related to this is the fact that evangelical ministries have seized on their new found freedoms by taking a page from U.S. televangelism, viz., accessing technology to reach an increasingly plugged-in Brazilian society. The Universal Church of the Kingdom of God, for example, owns a major television network as well as maintaining a host of radio stations that span the entire country. As a result of this rapid growth and outreach, the evangelical movement in Brazil has also gained significant political clout, both in terms of swaying the agendas of non-religions politicians as well as placing outspoken religious representatives in national and state offices (Freston 2001). This political power, it should be noted, has been primarily used to solidify religious freedoms and to push for a greater Christian presence in society, while also generally avoiding class-based politics (Freston 2001; Stoll 1991). Indeed, it is the avoidance of class-based politics in favor of an emphasis on personal responsibility that points to the synergies between the evangelical movement and Brazil's broader political economic context over the past 20 years.

When we turn to the role of the political economic structures that have influenced the emergence of evangelicalism in Brazil, it must be understood that the return to democracy in Brazil coincided with a move to replace the country's inward looking import substitution strategy for growth with a neoliberal model, grounded in the idea that free market forces are best at allocating capital while the state should take, at most, a passive role in terms of development. It is in this context that the prevalent evangelical discourses and practices pertaining to piety and self-transformation are particularly interesting. In Brazil, the emphasis on this personalized message of self-improvement to the poor, who make up the vast majority of evangelical adherents, has resonated strongly in impoverished communities. This message, not surprisingly,

[1] Today, in Brazil, there are a multitude of Christian churches and faiths. Some are easily identified as Pentecostal, or Protestant, or Presbyterian, and so on, but many are more ambiguous. In this chapter, we refer to this entire amalgamation of Christian-based churches and faiths with the term "non-Catholic Christian."

diverges significantly from the liberation theological focus on unjust socioeconomic structures. Pentecostals, for example, place their entire message on the power of personal conversion and piety to bring about the reward of being blessed by the Holy Spirit, which, in turn, can improve one's real world chances of achieving success and wealth.

This dimension of the evangelical movement has also been remarked on and promoted within broader international development discourses which increasingly emphasize the need for societal transformation to ensure the ability to prosper in a globally competitive climate (Rojas 2004). For example, in academic work emanating from the Institute on Culture, Religion and World Affairs at Boston University, the transformative potential of evangelic Christianity in the developing world is highlighted, especially in relation to economically relevant topics such as entrepreneurialism and general work ethics. Emphasis here again falls on the positive practices promoted by evangelical piousness, such as the avoidance of drinking and gambling as well as toning down the violence associated with machismo identity. These practices are then seen as coalescing to create more successful and well-adjusted subjectivities in an increasingly homogenized world characterized by a dominant, global market rationality (Council on Foreign Relations 2006; Grier 1997; Martin 1990; Sherman 1997).

Yet evangelical churches are not limited only to the message of personal responsibility. The devotion and piety expected of the evangelical congregation also creates cohesion within the group as well as the desire to reach out to the community at large. These community networks of support and service are also natural responses to the realities of a streamlined state that has become the global norm. It should be noted here that even with the rise to power of the Workers Party in 2003, which has invested in significant government programs intended to assist the poor, the current regime in Brazil continues its commitment to a globally competitive and market-driven economy. This is a commitment in which poverty, and hence needs-based church outreach programs, will continue to exist on a large scale for the foreseeable future.

106.3 Fortaleza, Pirambu and the Protestant Boom

Taking into consideration the broader religious and political economic contexts in which Brazil exists, we can glean a better understanding of the contours and dynamics of the current rise of evangelicalism there. To enhance this broad analysis with more place-specific detail we now turn to the case of Forteleza. A close examination of this site shows that the processes of evangelical and Pentecostal growth are indeed linked to and bound within patterns of globalization and international networks. At the same time, however, they are also deeply grounded in local geographic, historical, socio-political, and economic milieus.

Fortaleza, capital of the state of Ceará, sits along the Atlantic coast in the Northeast region of Brazil (Fig. 106.1). As one of the largest and fastest growing cities in Latin America today, the Fortaleza metropolitan area is now home to roughly

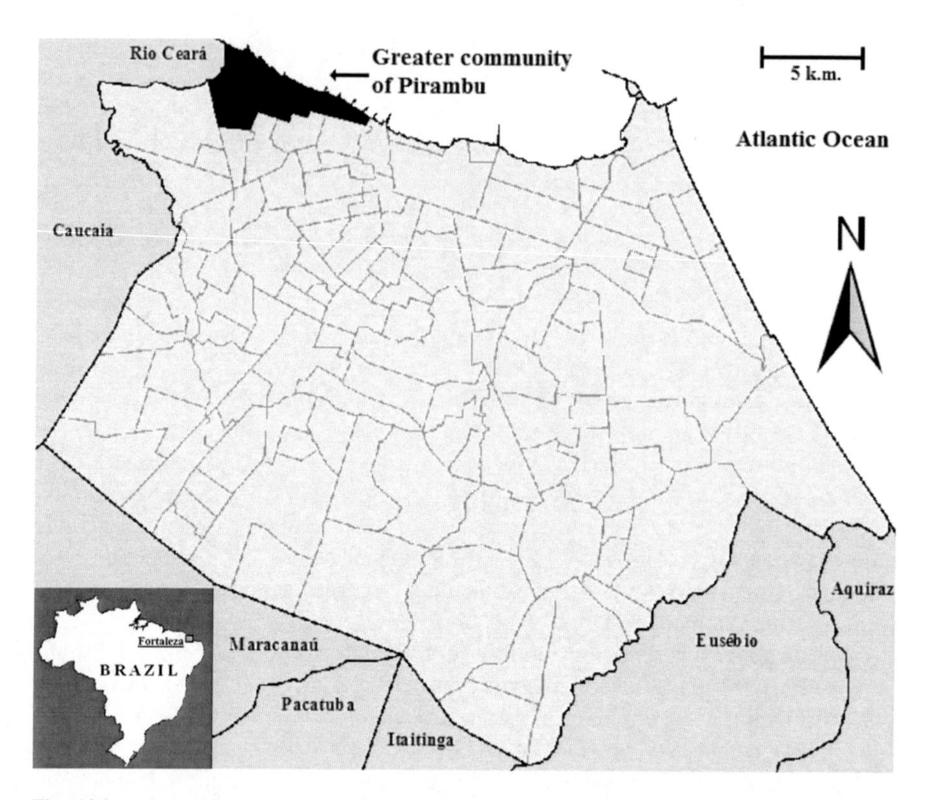

Fig. 106.1 Map of Fortaleza showing the community of Pirambu (Map by Jeff Garmany and Hannes Gerhardt, adapted from Garmany 2009)

four million people (Garmany 2009). Known for a long history of impoverishment and underdevelopment, northeastern Brazil is also often recognized as a region of intense spiritual and religious faith (Page 1995). Fortaleza characterizes this reputation in many ways, especially with an active and very broad religious population that includes a variety of faiths and syncretistic practices (e.g., Catholicism, Christianity, Candomblé, Umbanda, Spiritism, Islam and Judaism). While the Catholic Church still holds sway as the single most powerful religious institution, the growth of evangelical/Pentecostal churches has undeniably altered the social, political, and religious landscapes of Fortaleza and other cities throughout Brazil.[2]

Like other urban contexts in Brazil and Latin America, conversionary trends from Catholic to non-Catholic Christian churches in Fortaleza have been most pronounced

[2] Unless otherwise noted, data in this chapter come from a collection of field notes gathered over a period of several years. Fieldwork was begun in the city of Fortaleza in 2000 and then formally within the community of Pirambu in 2005. Since then, extended return visits were made in 2006, 2007, 2009, and 2010. The dataset contains more than 150 semi- and unstructured interviews, significant archival research, and months of participatory observation notes gathered while conducting fieldwork in Pirambu (Garmany 2010).

in lower income, favela neighborhoods (Burdick 1993; Oosterbaan 2009). The largest and one of the oldest of these communities in Fortaleza is Pirambu,[3] situated along the coastline in the northwest sector of the city (see Fig. 106.1). First inhabited by squatter settlers in the 1930s and 1940s, Pirambu grew rapidly throughout the twentieth century, expanding northwestwardly along the cearense coastline (Costa 1999). Today well over 100,000 people currently reside there; it has a mixture of working class and lower income families, small businesses, schools, community centers, non-profit organizations, and religious centers.

Soon after its establishment, Pirambu became organized around a number of demands being made by the low-income residents there. Hence, by the late 1950s, at which point Pirambu was already home to roughly 40,000 people, residents began to collectivize around campaigns for basic public services (e.g., water, sewage, garbage, healthcare, schools). Working with labor activists and members of the Brazilian communist party, Catholic priest Hélio Campos emerged as the neighborhood's most visible public leader, organizing residents in a multiyear push for citizen's rights. Padre Campos and the community were successful in their efforts, and finally, in 1962, Pirambu was officially incorporated into the city of Fortaleza, providing residents the same basic services privy to all other city inhabitants. Under Padre Campos the community had found its voice and a sense of identity and Pirambu became a bastion for labor, leftist politics, and liberation theology (Costa 1999).

Pirambu was not alone in this social transformation as throughout Brazil calls for worker's rights, political economic transformation, and agrarian reform grew increasingly louder. The modest reforms of President João Goulart, however, ultimately proved too radical for ruling elites, and in 1964 the military seized power, establishing a dictatorship that would last for more than 20 years. Under the military dictatorship, opposing political and social groups were silenced, and dissenters were frequently jailed, tortured, and even murdered (Skidmore 2010). The Catholic Church was often the only "safe" space from military police. In Pirambu, the left-leaning Catholic Church became the center for community action and organization, and while evangelical/Pentecostal churches began to take root and spread elsewhere in Brazil, in Pirambu, evangelical (or crente[4]) churches remained few and far between during much of the 1960s and 1970s.

It was not until the early 1980s, with the Vatican's concerted efforts to homogenize church services and undermine liberation theology, when an opening developed for religious alternatives. For one, new social movements (e.g., the MST, O Movimento dos Trabalhadores Rurais Sem Terra – The Landless Rural Workers Movement)

[3] Throughout this chapter we refer to the community (or neighborhood) of "Pirambu." The greater Pirambu area is actually a very large collection of individual favela neighborhoods, and for purposes of confidentiality, we refrain from using the name of the specific neighborhood within Pirambu where this research was conducted.

[4] The word crente literally means "believer." In Brazil, the term crente (or crentes) is generally considered synonymous with evangelical (or evangelicals) and is regularly used as an umbrella term for all those of evangelical/Protestant/Pentecostal faiths.

and political parties (e.g., the PT, the Worker's Party) emerged to fill the leftist void in Pirambu. Yet it was also at this point that crente churches began to vie for the spiritual and practical needs of favela communities. The success of these churches is evidenced in that they exist on nearly every street in Pirambu (Garmany 2010). Still, despite their current strong presence, Fortaleza was by no means a major birthplace of non-Catholic Christianity in Brazil. These churches existed in small numbers for years in Fortaleza, but unlike São Paulo, Rio de Janeiro, and even Belém, it was not until the early 1990s that non-Catholic Christian churches became ubiquitous in low-income neighborhoods throughout the city. Reasons for this delayed emergence are myriad, but in general, Fortaleza was not a significant nodal point for missionary groups, either domestic or foreign, until the 1980s. But by the end of that decade, Fortaleza's population had doubled in only 20 years time (reaching nearly two million in the city proper), and with soaring levels of poverty and unemployment (Gondim 2006), and Catholic dioceses withdrawing from issues of social injustice, Pentecostal and evangelical congregations began to multiply at an astonishing rate.

Much work exists explaining the appeal of non-Catholic Christian churches to Brazil's poor (Goldstein 2003; Lavalle and Castello 2004; Vásquez 1998), and within this body of research there exist many contentious debates. What these authors tend to agree upon, however, is that the non-Catholic Christian churches in Brazil's favelas tend to address the social problems confronted by community residents at a very direct level. For those who live in environments where the effects of poverty, drug abuse, violence, and prostitution are glaringly obvious, these churches offer an unambiguously clear message: join us, be saved, and you will not succumb to the vices that surround you. As Pirambu and other favelas in Fortaleza continued to grow rapidly through the late 1980s, evangelical/Pentecostal churches saw hundreds of new attendees on a weekly basis.

Accompanying the above changes were state-level development initiatives, national migration flows and international markets of investment and consumption. For example, as southeastern Brazil began to grow and industrialize in the first half of the twentieth century, migrants from northeastern Brazil flowed southward to fill the labor demands of booming cities like Rio de Janeiro and São Paulo (Thery 2009). Patterns of chain migration concentrated families from Fortaleza and other parts of Ceará in the São Paulo metropolitan region where, in many cases, they lived in impoverished communities on the periphery of the city. Here they frequently came into contact with evangelical/Pentecostal proselytizers. Moreover, in addition to confronting problems of poverty and destitution, the community and familial networks that had formerly anchored these new migrants had been disconnected. The establishment of crente churches, which were much smaller and more "personal" than their Catholic counterparts, offered an immediate sense of community and belonging. Non-Catholic Christian churches won thousands of converts among "northeasterners" during this period. By the 1980s, when democracy returned to Brazil and liberation theology waned, São Paulo and Rio were already home to hundreds of thousands of crente followers (Stoll 1991).

Through the efforts of Brazilian and international missionaries, and the processes of return migration (where, for example, small but consistent numbers of non-Catholic Christian converts returned to Fortaleza from cities like São Paulo), Pentecostal and evangelical congregations slowly expanded in Fortaleza throughout the 1980s. Also just as migrants from the Northeast had been attracted to smaller, more intimate non-Catholic Christian churches in São Paulo, so too did these same churches in Fortaleza fill with new migrants from the countryside who gravitated towards the city by the thousands during this same period. With a deteriorating agricultural sector prone to drought and low levels of capital investment in the state of Ceará, rural-to-urban migration fed much of Fortaleza's population surge between 1970 and 1990 (Souza and Neves 2002). As the 1990s began, local officials, faced with high population growth, low standards of urban infrastructure, and soaring unemployment numbers, unveiled a new roadmap for development in the cearense capital: tourism was to be the new pathway for economic development, and Fortaleza promoted itself both in Brazil and abroad as an accessible and affordable tropical getaway (Gondim 2004).

The tourism industry was, indeed, successful in bringing economic impacts to the region along with a host of other consequences (Garmany 2011a, b). Not only did Fortaleza capture the attention of vacation goers, but of missionary organizers as well. In a sense, the city had become a logical setting for major religious transformations. First, favela neighborhoods were expanding rapidly with poor migrant families (as neoliberal economic policies and the tourism industry did little to alleviate socioeconomic inequality). Second, Fortaleza had grown quite haphazardly to become one of the largest cities in Brazil, attracting with it a variety of national and international interests. And, third, unlike the Catholic Church, evangelical/Pentecostal groups challenged head on the social ills that plagued low-income communities throughout the city; not at a structural or political economic level as the Church had previously done, but, as Manuel Vásquez (1998) notes, perhaps at a level much more resonant with impoverished families. The effects were unmistakable, and throughout the 1990s, non-Catholic Christian churches, primarily evangelical/Pentecostal in denomination, proliferated across the urban landscape (Fig. 106.2). By 2000, these churches had already become a standard fixture in poor communities across the metropolitan region, and, like nearly every city elsewhere in Brazil, Fortaleza's population of non-Catholic Christians hovered somewhere around 20 %.

Not to be overlooked in this process, though, have been the adaptations and changes in protocol adopted by many Pentecostal and evangelical groups. As recently as the 1990s, crente followers were often easily identified by their dress and unique codes of behavior. They frequently wore white clothing, avoided jewelry, make up, and hair styling, and were forbidden to drink alcohol, dance, or even listen to secular music. These groups were easily identifiable in favela neighborhoods. As anthropologist Donna Goldstein hints (2003), these qualities were likely one of the major appeals of evangelical/Pentecostal churches. Church members were clearly set apart from other community residents, providing them, *perhaps*, with some level of extra security. Crentes, even today, are known for strict codes of behavioral protocol, and in settings where drug trafficking and violence are rife, this reputation helps to

Fig. 106.2 A Deus é Amor (God is Love) Pentecostal church in Pirambu (Photo by Jeff Garmany and Hannes Gerhardt)

disassociate them from sordid and criminal activities. More bluntly, both drug traffickers and the police (a fuzzy line in many circumstances) are constantly vigilant, and favela residents unfortunate enough to be (mis)identified with an oppositional group or faction face potentially lethal retribution. Dressed in white and brandishing bibles, crentes made it unequivocally clear they had no involvement with drugs, criminality, or violent activity, offering them a social shield against their immediate surroundings.

Today, however, the lines between non-Catholic Christians and "everyone else" are much less obvious. In Pirambu, for example, very few evangelical/Pentecostal residents dress in white or outwardly distinguish themselves from others. Churches seldom prohibit make up, jewelry, and beautification practices, and while secular music and dancing are seriously cautioned, churchgoers are rarely prohibited from indulging in these activities altogether. Alcohol remains taboo for nearly every evangelical/Pentecostal group, but wine, in moderation, is now permitted by some churches. Unless crentes are on their way to or from church services in Pirambu, where they will dress neatly (though very rarely in white) and sometimes carry a bible, there is little to obviously identify them as non-Catholic Christians. And in most favela neighborhoods, where levels of violence and criminal activity are nowhere near as pronounced as they are in the oft-publicized favelas of Rio or São Paulo, these attributes may help to explain the rising popularity of evangelical/Pentecostal churches in recent years: personal codes of conduct have been relaxed, offering broader appeal to community residents. In truth, the majority of favelas in

Brazil bare little resemblance to the violent, urban warfare zones that capture so much media attention in Rio de Janeiro. Consequently, the stigma effects of a strict evangelical lifestyle may outweigh the possible safety benefits that accompany it. Becoming crente, in many respects, is not the radical practice today it once was, a change that has much to do with increasing conversionary trends and mainstream appeal.

With the above changes have come many more heterogeneous landscapes of non-Catholic Christian churches. Whereas a small collection of nationally centralized denominations, often Pentecostal (for example, the Assembly of God and The Universal Church of the Kingdom of God), once characterized what it meant to be crente, today there are dozens of individual Christian churches and denominations in favela communities throughout Brazil. Some are connected to international missionary efforts, but many have emerged from splintered, domestically founded religious groups. To say that Evangelical/Pentecostal growth in contemporary Brazil today is linked primarily to international evangelizers, as it is, for example, in certain African contexts (Brouwer et al. 1996), would be wrong: Brazil has a very robust non-Catholic Christian following and increasingly the country is a major "exporter" of international proselytizing efforts (Freston 2008). In communities such as Pirambu, these evangelical and Pentecostal churches quite often spring from humble origins (for example, on a street corner or in someone's home), and church leaders are frequently more concerned with their adherence to spiritual practice than their allegiance to pre-existing institutions and formulas. Some of these churches grow rapidly, attracting large congregations and financial support, while many others flame out as suddenly as they emerge. A constant ebb and flow of non-Catholic Christian churches typifies Pirambu and many favelas like it. The "boom" of evangelical and Pentecostal growth in Brazil has in many ways been replaced by a steady and dynamic presence of multiple and diverse non-Catholic Christian churches.

106.4 Evangelicalism in Twenty-First Century Brazil

Brazil is truly a fascinating melting pot of religious practices. Syncretistic traditions have blended multiple spiritualities for centuries in Brazil, making it, without doubt, a key site in processes of "religious globalization" (Jenkins 2008). And when considering recent spiritual change and the growth of evangelical/Pentecostal churches, it is important to remember and reflect upon this history. Syncretism and religious flexibility (that is, the frequenting of multiple churches) is remarkably common in Brazil (e.g., Ferretti 2001), and religious conversion, therefore, does not signify a singular or fixed religious future. With a globalized religious smorgasbord, there exist more religious options today than ever before. Brazilian spiritual identities remain dynamic in the face of these processes, and to be sure, Pentecostal and other evangelical churches are not the only rapidly growing religious sects in Brazil. Spiritism, for example, once a more middle to upper-class spiritual movement, is growing rapidly in lower income communities (Clarke 2006). In Pirambu, residents increasingly discuss their interest in this religious philosophy with others, and many now openly identify as Spiritists. Moreover, while growing numbers of Brazilians may identify as evangelical, what this means for their daily practice, political affiliations, and socio-cultural principles (for example, their position on such issues as the environment, education, healthcare, gender, social injustice, premarital sex, and so on) becomes increasingly ambiguous. In short, Brazil's evangelical future, presuming it continues to grow, may differ greatly from its evangelical past.

Today, Brazilian evangelical/Pentecostal missionaries can be found in countries all over the world. In this sense, Brazilians have long been on the front lines of global proselytizing work: as Luis Bush declared in 1987 at the first Ibero American Missionary Congresses (COMIBAM), "From a mission field, Latin America has become a mission force" (cited in Smither 2010: 89). In Brazil, the Associação de Missões Transculturais Brasileiras (Association of Transcultural Brazilian Missions), which houses dozens of missionary organizations, has been a significant actor in fostering this vision. Interestingly, it has been noted that Brazilian evangelical missionaries distinguish themselves from many others with their open ecumenicalism and embrace of the "whole Gospel," which calls for a greater concern for the poor and social issues (See Freston 1994; Ekström 2009). This latter focus, largely influenced by the Latin American Theological Fraternity, can in many ways be seen as an evangelical response to the legacy of liberation theology (Smither 2010).

Yet the extent of Brazil's missionary activity is difficult to gauge since much of it is occurring unofficially, led by Brazilians who emigrate for reasons over and beyond mere missionary work (Salinas 2008). This is particularly true for Brazilians evangelizing in the US and Europe. With regard to official missionary work, most activity occurs within Latin America. However, Brazil has also maintained a Protestant missionary presence in the Portuguese speaking parts of southern Africa dating back to the 1970s. More recent African missionary activity has seen a move northward, shifting towards societies with significant Muslim populations. In the limited proselytizing that Brazilians conduct in Asia, work has focused primarily upon Japan, which is due mostly to the presence of evangelized Brazilians of Japanese descent.

In Brazil itself, the frontier of religious conquest has shifted slightly, moving from urban shantytowns in major coastal cities to the Amazon region in the north of the country. Eduardo Gomes International Airport in Manaus, located in the heart of the Amazon rainforest, receives thousands of missionary workers every month from both the United States and Brazil. The projects these groups undertake range broadly in terms of focus and scope, but in general, the primary foci are human development and religious instruction. As such, their influence on societal and cultural change is often quite profound. Catholic missionaries have also been active within this region for centuries, and today it is a major site for global missionary convergence, representing one of the most significant nodal points for Christian social work anywhere in the world (Silva 2009).

Like many countries, networks of global exchange and international missionary work have played a significant role in Brazil's religious development. In addition, many of the same factors that allowed for the growth and development of

Afro-Brazilian religions and folkloric, mystic forms of Catholicism (for example, religious tolerance and syncretistic adaptations) have also given space to evangelical/ Pentecostal groups. Though international missionary efforts should not be overlooked, Brazilian evangelical sects have existed for decades, and a significant portion of the evangelizing work that has taken place over the past 50 years has been financed and carried out by Brazilian groups. Recent developments in information technology and global communication have helped to fund and manage these activities (e.g., Oosterbaan 2011), and political and economic changes in Brazil have factored importantly into the growth of these churches. Brazil's contemporary religious landscape reflects more than five centuries of globalization and spiritual flux, and with increasing rates of in-migration and socio-economic development, continued trends of religious change and spiritual dynamism are likely to define the future just as much as they have the past.

References

Bosch, D. (1991). *Transforming mission: Paradigm shifts in theology of mission*. Maryknoll: Orbis Books.

Brouwer, S., Gifford, P., & Rose, S. D. (1996). *Exporting the American gospel: Global Christian fundamentalism*. New York: Routledge.

Burdick, J. (1993). *Looking for God in Brazil: The progressive Catholic Church in urban Brazil's religious arena*. Berkeley: University of California Press.

Chaney, C. (1976). *The birth of missions in America*. Pasadena: William Carey Library.

Clarke, P. B. (2006). *New religions in global perspective: A study of religious change in a modern world*. New York: Routledge.

Costa, M. G. (1999). *Historiando o Pirambu*. Fortaleza: Seriates Edições.

Council on Foreign Relations. (2006). *The nexus of religion and foreign policy: The global rise of Pentecostalism*. Retrieved July 30, 2008, from www.cfr.org/publication/11758/nexus_of_religion_and_foreign_policy.html

Ekström, B. (2009). Brazilian sending. In R. D. Winter & S. C. Hawthorne (Eds.), *Perspectives on the world Christian movement: A reader* (pp. 371–372). Pasadena: William Carey Library.

Ferretti, S. F. (2001). Religious syncretism in an Afro-Brazilian cult house. In S. M. Greenfield & A. Droogers (Eds.), *Reinventing religions: Syncretism and transformation in Africa and the Americas* (pp. 87–98). Lanham: Roman and Littlefield.

Freston, P. (1994). Brazil: Church growth, parachurch agencies, and politics. In G. Cook (Ed.), *New face of the Church in Latin America: Between tradition and change* (pp. 226–242). Maryknoll: Orbis.

Freston, P. (2001). *Evangelicals and politics in Asia, Africa and Latin America*. Cambridge: Cambridge University Press.

Freston, P. (Ed.). (2008). *Evangelical Christianity and democracy in Latin America*. Oxford: Oxford University Press.

Garmany, J. (2009). The embodied state: Governmentality in a Brazilian favela. *Social and Cultural Geography, 10*(7), 721–739.

Garmany, J. (2010). Religion and governmentality: Understanding governance in urban Brazil. *Geoforum, 41*(6), 908–918.

Garmany, J. (2011a). Situating Fortaleza: Urban space and uneven development in northeastern Brazil. *Cities, 28*(1), 45–52.

Garmany, J. (2011b). Drugs, violence, fear, and death: The necro and narco-geographies of contemporary urban space. *Urban Geography, 32*(8), 1148–1166.

Goldstein, D. M. (2003). *Laughter out of place. Race, class, violence, and sexuality in a Rio shantytown*. Berkeley: University of California Press.
Gondim, L. M. P. (2004). Creating the image of a modern Fortaleza: Social inequalities, political change, and the impact of urban design. *Latin American Perspectives, 135*(2), 62–79.
Gondim, L. M. P. (2006). *O Dragão do Mar e a Fortaleza Pós-Moderna: Cultura, Patrimônio e Imagem da Cidade*. São Paulo: Annablume.
Grier, R. (1997). The effect of religion on economic development: A cross national study of 63 former colonies. *Kyklos, 50*(1), 47–62.
Huntington, S. P. (1993). *The third wave: Democratisation in the late twentieth century*. Norman: University of Oklahoma Press.
Jenkins, P. (2002). *The next Christendom: The coming of global Christianity*. Oxford: Oxford University Press.
Jenkins, P. (2008). The Christian revolution. In F. Lechener & J. Boli (Eds.), *The globalization reader* (pp. 379–386). Oxford: Blackwell.
Lavalle, A. G., & Castello, G. (2004, March). As benesses desse mundo: Associativismo religioso e inclusão socioeconômica. *Novos Estudos, 68*, 73–93.
MacHarg, K. D. (2011). *Evangelical Christianity thriving in Brazil, Latin America Mission*. www.jesusforlife.net/Documents/Evangelical%20Christianity%20thriving%20in%20Brazil.pdf
Martin, D. (1990). *Tongues of fire: The explosion of Protestantism in Latin America*. Oxford: Blackwell.
Nagle, R. (1999). Liberation theology's rise and fall. In R. M. Levine & J. J. Crocitti (Eds.), *The Brazil reader: History, culture, politics* (pp. 462–467). Durham: Duke University Press.
Noll, M. (2009). *The new shape of world Christianity: How American experience reflects global faith*. Downers Grove: InterVarsity Press.
Oosterbaan, M. (2009). Sonic supremacy: Sound, space and charisma in a favela in Rio de Janeiro. *Critique of Anthropology, 29*(1), 81–104.
Oosterbaan, M. (2011). Virtually global: Online evangelical cartography. *Social Anthropology, 19*(1), 56–73.
Page, J. A. (1995). *The Brazilians*. Reading: Perseus Books.
Reel, M. (2005). Brazil's priests use song and dance to stem Catholic Church's decline. *Washington Post Foreign Service*. Retrieved April 14, 2005, from www.washingtonpost.com/ac2/wp-dyn/A51511-2005Apr13?language=printer
Rojas, C. (2004). Governing through the social: Representations of poverty and global governmentality. In W. Larner & W. Walters (Eds.), *Global governmentality: Governing international spaces* (pp. 97–115). New York: Routledge.
Salinas, J. D. (2008). The great commission in Latin America. In M. I. Klauber & S. M. Manetsch (Eds.), *The great commission: Evangelicals and the history of world missions* (pp. 134–148). Nashville: B & H Academic.
Sherman, A. L. (1997). *The soul of development: Biblical Christianity and economic transformation in Guatemala*. Oxford: Oxford University Press.
Silva, A. V. (2009). Migração e segmentação evangélica: as dinâmicas de um processo. *Religião e Sociedade, 29*(2), 225–228.
Skidmore, T. E. (2010). *Brazil: Five centuries of change*. Oxford: Oxford University Press.
Smither, E. (2010). *Brazilian evangelical missions among Arabs: History, culture, practice, and theology*. Ph.D. dissertation, Faculty of Theology, University of Pretoria.
Souza, S., & Neves, F. C. (Eds.). (2002). *Seca*. Fortaleza: Edições Demócrito Rocha.
Stoll, D. (1991). *Is Latin America turning Protestant? The politics of evangelical growth*. Berkeley: University of California Press.
Thery, H. (2009). A cartographic and statistical portrait of twentieth-century Brazil. In I. Sachs, J. Wilheim, & P. S. Pinheiro (Eds.), *Brazil: A century of change* (pp. 1–19). Chapel Hill: University of North Carolina Press.
Vásquez, M. (1998). *The Brazilian popular church and the crisis of modernity*. Cambridge: Cambridge University Press.

Chapter 107
Legacy of a Minority Religion: Christians and Christianity in Contemporary Japan

Christina Ghanbarpour

107.1 Introduction

Though Christianity has been practiced in Japan for over 400 years, its status as an oft-persecuted religion practiced by a minority of Japanese has long presented challenges for Japanese Christians. Since the early years of its establishment in the mid-sixteenth century, government officials have tended to view Christianity with suspicion, as they question the compatibility of this "foreign" religion with loyalty to the state. Restrictions on Christianity were lifted in the late nineteenth century, but it once again became a target of political suppression when Japan began a disastrous 15-year war that brought it into conflict with Europe, the United States, and much of Asia-Pacific region. Though the past 60 years of peaceful government have allowed Christians to assume a more or less untroubled place within mainstream Japanese culture, they continue to face challenges ranging from oppressive policies to new social and political currents that compel them to examine their beliefs.

This chapter traces the historical trajectory of Japanese Christianity by examining its emergence, evolution, and contemporary role in Japanese society and culture. From the origins of Christianity in Japan to the present, Japanese Christians' innovative responses to changing circumstances have helped Christianity remain relevant in Japanese society and established its place as a source of social change and cultural influence. Christians today continue to respond to evolving national and international trends even as Christian symbols and rituals have come to permeate mainstream society.

C. Ghanbarpour (✉)
History Department, Saddleback College, Mission Viejo, CA 92692, USA
e-mail: cghanbarpour@saddleback.edu

S.D. Brunn (ed.), *The Changing World Religion Map*,
DOI 10.1007/978-94-017-9376-6_107

107.2 The Spread of Christianity in Japan: Origins to 1868

Francis Xavier (1506–1552), a Jesuit missionary from the Basque region of northern Spain, was the first person to establish a Christian mission in Japan. Xavier spent many years proselytizing in the port cities of Asia and Africa before arriving at Kagoshima in southwestern Japan in 1549. Though he died of illness only a few years after his arrival, his initiative provided the point of departure from which the Christian population of Japan would eventually grow to its current 2.1 million members.

Having little or no knowledge of the Japanese language, early missionaries relied largely on art, plays and gestures to convey Christian beliefs. Because of this and Japanese Christians' requests for religious art, Christian missionaries such as Alessandro Valignano (1539–1606) and Giovanni Niccolo (1563–1626) established schools that taught Japanese about Christianity while training them to reproduce religious images (Fig. 107.1). Niccolo in particular is thought to have been a talented artist,

Fig. 107.1 Portrait of St. Francis Xavier, sixteenth century (Photo by anonymous Japanese artist affiliated with the Kanô School of Art. https://en.wikipedia.org/wiki/Francis_Xavier)

and likely produced several large pieces for churches in southwestern Japan. However, few of the works attributed to him have survived to the present. This early production of Christian art created a material culture that helped spread Christianity and embed it in Japanese culture.

Jesuit missionaries became able to communicate more directly with Japanese through the work of other missionaries, such as Joao Rodrigues (ca. 1561–1633). Though little is known about Rodrigues' early life, he first came to Japan at around the age of 15 on board a ship carrying 14 Jesuit missionaries. At age 18, he entered the Jesuit order as a novitiate and, after several years of study, was ordained in 1596. During the more than three decades that Rodrigues lived in Japan, he became a renowned interpreter, taught Latin to Japanese students, and published at least two books on Japanese life and culture as well as a book on Japanese grammar that quickly became essential. Rodrigues' growing importance to the missionary community coincided with its early expansion; by 1581, it is thought that over 80 Jesuit missionaries were active in Japan, and other orders, such as the Franciscans and Dominicans, soon followed.

Once established, Christianity spread quickly. This was partly due to the support of major political figures such as Oda Nobunaga (1534–1582),[1] a warlord who rose from humble origins to become one of the most powerful rulers in Japan's history. Oda's ruthless efforts to eliminate his opponents ranged from employing new weapons in battle, such as Portuguese arquebuses (an early type of firearm), to encouraging Jesuit missionaries to spread the Catholic faith. This latter effort was intended to undermine the major Buddhist temple complexes, such as Enryakuji and Honganji, whose warrior monks were among Oda's most tenacious adversaries. His support contributed to the construction of an estimated 200 churches throughout western Japan and led to the growth of the Christian population to approximately 150,000 members by the time of his death.

Oda's successor, Toyotomi Hideyoshi (circa 1536–1598), initially allowed Christianity to spread before waging a campaign to suppress it. Hideyoshi may have been reluctant to ban Christianity at first because some of the women of his household had adopted Christian nicknames, jewelry and icons. In addition, at least one of them—his adopted daughter—converted to Christianity. As Christianity spread among the upper classes, however, he appears to have become concerned that his rivals would use it as an ideological means of increasing control in their territories, thus threatening his rule. In 1587, Hideyoshi banned Christianity, and 10 years later began persecuting Christians. These campaigns focused less on eliminating Christianity among peasants in rural areas than on eliminating missionaries and converts of high rank—precisely those who were most likely to threaten his power base.

An incident in 1597 marked this gradual worsening of tensions. As a warning to secret practitioners of the faith, Hideyoshi had 26 Christians mutilated and paraded through the streets of Kyoto before sending them off to Nagasaki to be stabbed,

[1] All Japanese names are given with the last name before the first name, as is Japanese custom, unless generally accepted alternatives are available.

crucified, and left to hang for 80 days. Rather than discourage Christianity, however, stories of the Christians' bravery spread, as did reports of miracles associated with the martyrs' bodies. This not only inspired others to convert, but helped establish Nagasaki as a major center for Christian activities; the 26 were later canonized. Nevertheless, a year after the crucifixions, authorities fearful of further state suppression ordered the destruction of 137 Christian churches throughout the island of Kyushu—an area which includes the Christian communities of Nagasaki and its environs.

Hideyoshi's successor, Tokugawa Ieyasu (1542–1616), would take these prohibitions even further. After promising the dying Hideyoshi that he would administer the realm with a coalition of regents until Hideyoshi's son, Hideyori, came of age, Ieyasu instead began consolidating his own base of power. A year after Hideyoshi's death, Ieyasu had defeated his rivals in a major battle for control of Japan, and in 1603, the emperor officially recognized Ieyasu's power by granting him the title of *se-i-tai shogun*, or "barbarian-subduing generalissimo." This title would be used by all of the rulers of the Tokugawa dynasty until 1867. To ensure that the power he had won would remain in his family, Ieyasu officially withdrew from power and ceded the title of shogun to his son, Hidetada, in 1605, but continued to rule from behind the throne until his death in 1616.

The Tokugawa rulers put in place a variety of measures designed to suppress Christianity. In 1606, they issued a series of decrees that limited Christian activities and forbid new converts; 4 years later, they banned all missionaries from Japan; and in 1614, all Christian activity in Japan was prohibited. A major uprising that occurred near Nagasaki in 1637 brought the conflict to a head. Known as the Shimabara Uprising, the revolt began as a protest of local farmers against high taxes and a lack of famine relief. Because many of the farmers were Christian, the revolt soon adopted Christian overtones. Approximately 40,000 farmers held off 100,000 government troops for 4 months before being forced to surrender. The troops spared no one, and to assure that a Christian revolt would never again occur, the Tokugawa strengthened its state-wide system of surveillance and banned virtually all foreign trade until the mid-nineteenth century.

Despite the Tokugawa's efforts to root out any remaining Christians, a group of Christians called *kakure kirishitan* (Hidden Christians) continued to practice their faith. In the early years of the ban, their tactics included hiding foreign missionaries and worshipping at home rather than in a church. As the ban became more stringently enforced, they added other strategies, such as performing Christian burial rites before or after the requisite Buddhist ceremonies and sculpting Christian figures and cross-like shapes into the bases of "Buddhist" stone pagodas and lanterns (Fig. 107.2). To further hide Christian activities, *kakure kirishitan* formed faith-based villages and manipulated state-mandated organizations such as the *gonin gumi*, a five-family system of mutual surveillance, so that they included only Christian families. Though the constraints that secrecy imposed on them necessitated divergences from Church doctrine, Christian communities were able to preserve fairly accurate versions of the religious calendar, liturgy and prayer without outside contact for over 200 years.

Fig. 107.2 Hidden Christian iconography. This stone lantern at Kanchiin, Toji, a Buddhist temple in Kyoto, illustrates how Christians hid their beliefs after Christianity was banned in Japan (Source: http://en.wikipedia.org/wiki/File:TojiKanchiinKirishitan.jpg)

Kakure kirishitan who survived the purges passed down their faith to their descendants. These were able to reunite with the larger Christian community when missionaries were allowed to return to Japan in the mid-nineteenth century. Once reunited, many *kakure kirishitan* joined the Roman Catholic Church. However, approximately 35,000 of them refused to join, believing that they had to continue to practice the original faith as communicated to their ancestors by the sixteenth century Jesuits. About 20,000 Japanese Christians maintain the *kakure kirishitan* faith today, and many continue to live in the areas near Nagasaki that have traditionally been Christian centers (Fig. 107.3). Yet, due to factors such as a lack of leadership,

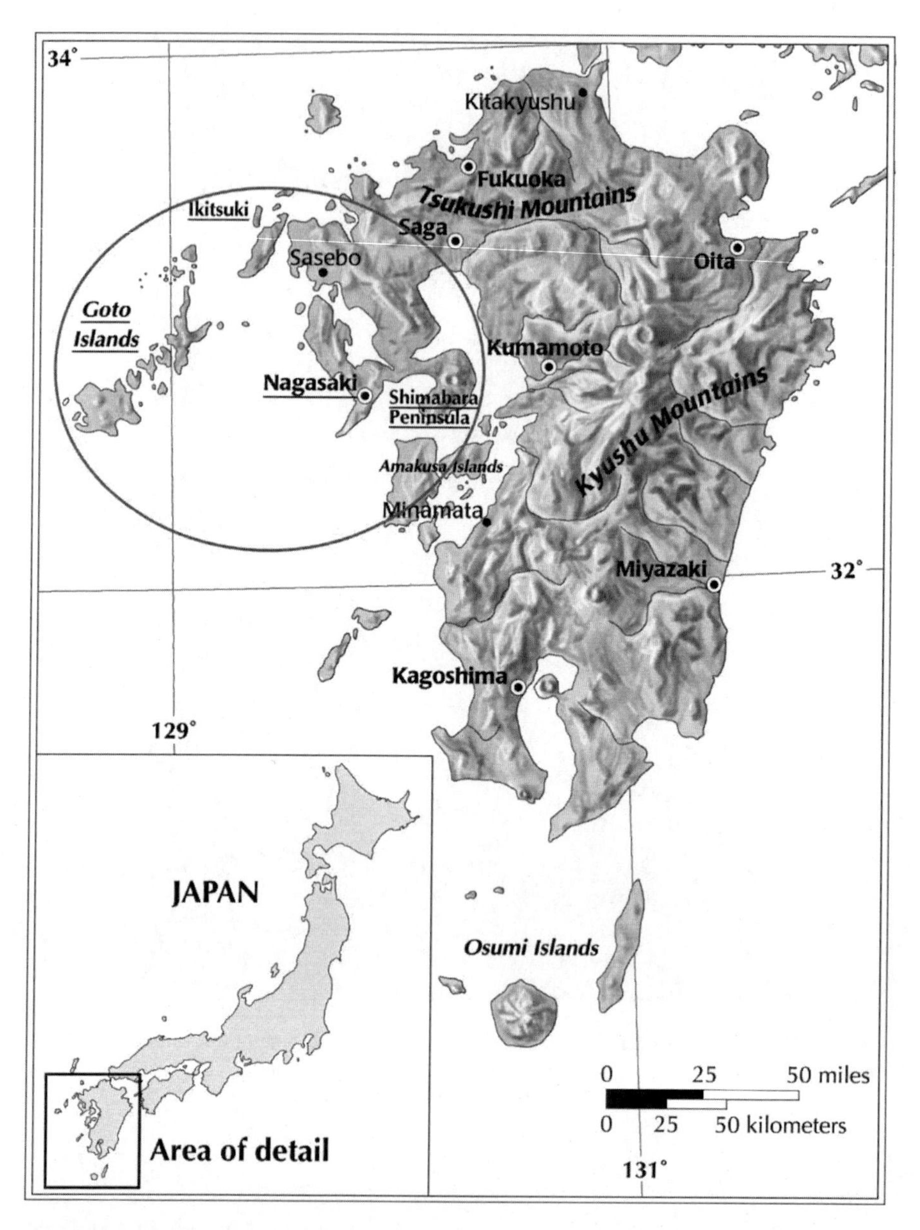

Fig. 107.3 Sites of Christian activity when Christianity was banned in Japan (ca. 1614–1872). Areas such as the Goto Islands, Ikitsuki, and Shimabara (*circled*) are home to the majority of kakure kirishitan (Hidden Christians) today (Map by Dick Gilbreath, University of Kentucky Gyula Pauer Center for Cartography and GIS, commissioned by the editor)

competition with other religious groups, and the urban migration of younger family members away from the rural areas where the faith had thrived, the number of *kakure kirishitan* has declined considerably in the past 30 years.

107.3 Christianity from the Meiji Period to 1945

After two centuries of suppression, the first major changes in the state's response to Christianity began to occur after the fall of the Tokugawa dynasty and the establishment of a new government. The Meiji government (1868–1911) initially upheld the ban on Christianity that it had inherited from its Tokugawa predecessors, and in the first few years of its establishment posted signs that warned Japanese that Christians would still be prosecuted. However, in the face of Western pressure, in 1871 the government rescinded the system that required every family to register as members of a local Buddhist temple (*danka seido*), and 2 years later it allowed a limited religious freedom for the first time in over 200 years. An enduring commitment to religious freedom seemed possible when, in 1889, several articles that granted a degree of religious freedom were written into the Constitution.

Large numbers of Christian missionaries flocked to Japan, re-introducing the Roman Catholic faith and bringing in Christian denominations that had arisen after the ban. As a result, the number of Christian converts increased dramatically, fueled in part by the desire of both Christian and non-Christian Japanese to study Western languages, science and customs at the missionaries' schools, churches, hospitals, and theological seminaries. By 1890, the Roman Catholic Church claimed 44,505 members, including *kakure kirishitan* converts. Protestant denominations fared even better, claiming only 59 converts in 1873 as compared to 31,361 by 1891. Both groups maintained over 50,000 members by 1901. Though the overall number of converts never grew to more than a small percentage of the population—at best, perhaps 200,000 out of a turn-of-the-century population of roughly 44 million—the rapid growth of Christianity illustrates how missionaries' activities and institutions provided a vital means for many Japanese to access knowledge of the Western countries that both threatened them and offered them the tools to build a modern nation.

The 60 years between 1870 and 1930 was thus a kind of renaissance for Japanese Christianity as the number of converts increased and Christian institutions spread. These include many of the Christian-based universities that still exist today. For example, Meiji Gakuin University, the oldest Christian university in Japan, grew from a small mission school founded by Presbyterian missionaries in 1863 to a prestigious private institution by 1886, the same year it adopted its current name. A Japanese Protestant named Joseph Hardy Neesima established Doshisha Academy, which is now Doshisha University, in 1875. Doshisha added several branch schools, including a girls' technical school, during the late 1800s and early 1900s before gaining university status in 1912. Sophia University, a Jesuit institution in Tokyo, was founded by three missionaries and opened in 1913. These illustrate how the close relationship between Christianity and the pursuit of knowledge—particularly

Western knowledge—continued to grow and develop as Christianity spread throughout Japan.

As Christian institutions spread, so did schools for girls and women. The Carolyn Wright Memorial School, now Iai Gakuin, is the oldest girls' school in Hokkaido and was founded by the American Methodist Episcopal missionaries Dr. and Mrs. M. C. Harris. Women's universities founded by missionaries include Seiwa College, which was founded as a Bible school for girls in 1880 by American missionaries Julia E. Dudley and Martha J. Barrows; Tokyo Christian University, which was founded in 1881 as a women's theological seminary by the Women's Union Missionary Society; and Tokyo Women's Christian University, which opened in 1918 under the auspices of a coalition of Protestant Christian participants in the 1910 Edinburgh Missionary Conference. One of the most famous of the Christian-based girls' schools, the prestigious women's higher school Joshi Eigaku Juku (now Tsuda College), was founded by the prominent women's educator and Christian convert Tsuda Ume in 1900. These schools and colleges helped spread the idea that men and women should be educated equally, an idea they continue to promote today.

The trend towards institutionalization likely contributed to Christianity achieving the status of being recognized on a national level. In 1912, the same year that Doshisha Academy became a university, the Meiji government sponsored a major religious conference called the Three Religions Conference (*Sankyô kaidô*). This event, which was designed to address the social problems that had arisen in the wake of an economic downturn, marked the first time that Christian churches were invited to send their representatives on an equal footing with those of Japan's two major religions, Buddhism and Shinto. The event thus confirmed official acceptance of the religion and acknowledged that Christianity had achieved a place in Japanese society despite its still relatively small number of adherents. The public acceptance of Christianity only a few decades after Christians had once faced certain death for openly practicing their beliefs must have been a satisfying, though unprecedented, turn of events for the Japanese Christian community.

Christian reformers rose to become prominent social activists whose work with the Japanese government shows both the kinds of issues Christians supported and how close the Christian-state relationship had become. Reformers such as Yamamuro Gunpei (1872–1940), the founder of the Japanese Salvation Army, focused on issues such as ending prostitution and improving the lives of the poor (Fig. 107.4). He and Tomeoka Kôsuke (1864–1934), a Christian who focused on efforts to improve children's welfare, were temporarily employed by the government to lead reform efforts. The Japanese government also sought out Christian support to achieve goals such as encouraging austerity and eliminating communism.

One of the most prominent Christian reformers to emerge in the early twentieth century was Kagawa Toyohiko (1888–1960), a writer and activist who earned an international reputation for his work to end poverty and promote world peace. Believing that it was necessary to live among the poor to understand how best to help them, Kagawa lived in a 6 ft^2 shed in the slums of Kobe for 14 years as he worked to promote industrial and agrarian reforms. He helped organize unions among shipyard workers, farmers and factory workers, and went on to help found

Fig. 107.4 Yamamuro Gunpei (1872–1940), Christian writer, preacher and founder of the Salvation Army in Japan (Photo from Christina Ghanbarpour's collection)

the Japanese Federation of Labor in 1921. His efforts to promote peace included participating in peace missions to the United States and trying to organize a peace agreement between the Japanese government and Franklin D. Roosevelt before the outbreak of war in 1941. In 1955, he was nominated for the Nobel Peace Prize. His legacy continues to be honored through events such as the 100th anniversary celebration of his decision at age 21 to dedicate his life to the poor by living in the Kobe slums, which was held in 2009. A memorial hall and museum in Naruto City, the ancestral seat of his father's family and the area where Kagawa spent his childhood, houses artifacts from his life and work.

Other organizations, such as the Women's Christian Temperance Society, highlight Christian women's roles in social activism. The organization attracted women who were dissatisfied with traditional notions that denigrated women, allowed men to keep concubines and mistresses, and focused more on continuing family lineages than on women's and children's welfare. For these women, Christianity's messages of monogamous, love-based marriages and equality before God became rallying points that led them to promote an array of causes in addition to the group's focus on banning alcohol. These included efforts to remove the ban on women's political activism and to end legal prostitution.

The new era of Christian-state accommodation was not without its ideological conflicts, however. Uchimura Kanzô (1861–1930), a prominent Christian writer and educator, was a reformer who tried to combine Christian practice with national identity. The movement that he founded, which is called the Non-church Movement (*mukyôkai*), focused on the study of the Bible in small groups in the home of a priest or an adherent rather than in a church. In 1891, he was dismissed from his position as an instructor at the First Higher School in Tokyo when he refused to bow to a copy of the Imperial Rescript on Education, a practice which seemed to affirm the godlike status of the emperor. His belief in pacifism also led to government pressure and censorship of his work on the eve of the Russo-Japanese (1904–5) War. Though he died in 1930, his movement inspired several indigenous branches of Christianity, such as the Christ Heart Church, and continued to attract several thousand followers as late as the mid-1970s.

Other Christians ran into similar problems. In 1900, Yamakawa Hitoshi (1880–1958) was charged with the crime of lèse-majesté, that is, a crime against the dignity of the emperor, for describing the marriage of the Crown Prince as "a great tragedy" in the Christian magazine *Gospel for Youth*. He spent over 3 years serving a prison sentence with hard labor. Both Christian and Buddhist groups also had to fight the government's continuous efforts to disseminate the state-sponsored version of Japan's indigenous religion, known as State Shinto. These included the efforts of local and regional governments to compel all households to maintain a Shinto altar and worship at Shinto shrines in the 1910s.

Christianity also became a starting point for anti-government activists who would later seek out more radical means to political and economic change. The anarchist Osugi Sakae (1885–1923), the nihilist Kaneko Fumiko (1903–1926), and Yamakawa himself, who later adopted socialism, all initially became involved with Christian groups before turning to the more radical forms of social change with which they are closely associated today. Thus, while Christians were able to work with the state on a number of issues, Christianity maintained a radical reputation that continued to attract activists well into the 1920s.

Japan's 15 Year War (1931–1945), a war that included Japan's involvement in World War II, marked yet another turning point in Christian-state relations. In the 1930s, a series of legal decisions restricted religious freedom for all religions outside of State Shinto. The government's gradually worsening attitude towards so-called "foreign" religions culminated in the revised Peace Preservation Law of 1941, which allowed the state to punish members of any religion that appeared to contradict the principles of emperor worship and State Shinto. This law led to the imprisonment of some Christian teachers and students and forced the merger of Protestant churches into the United Church of Christ in Japan (UCCJ).

Christians faced increasing pressure to reconcile their Christian faith with the jingoist nationalist rhetoric that dominated public life. Throughout the late 1930s and early 1940s, the pacifist, Kagawa Toyohiko was pressured to resign from international peace organizations and to denounce the United States in radio broadcasts, even to the extent of justifying the bombing of Pearl Harbor. He was nevertheless arrested several times and kept under surveillance by the Special Police. Others tried to use their common Christian faith as a way of convincing Christians in the colonized territories of Taiwan and Korea to accept the Japanese empire. When the state singled out a small Christian sect called the Holiness Band as a seditious group in the 1940s, the UCCJ refused to support them out of fear that the state would ban all Christian groups. Both Kagawa and the UCCJ later apologized for their actions during the war.

107.4 Christian Growth and Identity After 1945

When the war ended in August 1945, a foreign government was put in charge of Japanese affairs for the first time in Japan's history. The emperor was compelled to renounce his divinity, and wartime values were generally discredited. These factors helped Christianity grow exponentially in the early postwar years (1945–1952) as the Christian population more than doubled from less than 250,000 to over 600,000. Some scholars have argued that this rapid growth was partly the result of the occupying army's policies, which sought to fill what it saw as a growing and potentially dangerous "spiritual vacuum." This is particularly true of General Douglas MacArthur, the American general who headed the Occupation and encouraged Christian missionaries to return to Japan. At the same time, legal protections against discrimination and the guarantee of religious freedom helped create a climate of religious tolerance that the Japanese government had sought to eliminate during the war.

Yet, the Christianity that has grown and spread since 1945 has come to differ somewhat from that which spread before the war. Uchimura's Non-church Movement illustrates this process. Before 1945, Christians such as Uchimura had tried to reconcile Christianity with the explosion of nationalistic sentiment in Japan by claiming that Japanese Christianity was a unique form of Christianity. He argued that Japanese Christianity should differ from that practiced in other countries, particularly in Europe and America, to better fit the needs of Japanese followers. In order to create this new type of Christianity, he emphasized the personal relationship between preacher and adherent—a form of self-cultivation more typical of traditional Japanese teaching than international Christianity. Uchimura's work thus exemplifies one way that Japanese Christians before 1945 had tried to temper the perception of Christianity as a foreign religion by emphasizing their loyalty to the state.

After 1945, the mainstream Christian community has sought to re-orient itself towards embracing the international community. Japanese Christians today are active in international efforts that range from generating support for a cause within the same denomination to working with many different churches to organize charitable programs, such as relief aid for the 2011 Tohoku earthquake and nuclear disaster. In July 2012, the ministerial board of the Japan Alliance Christ Church (*Nihon dômei kirisuto kyôdan*) released a statement that nuclear power went against the Bible, citing, among other points, that nuclear power plants go against Christ's injunction to love others as one would oneself by putting plant workers' lives in danger. This illustrates how Japanese churches have joined together to protest nuclear power plants.

Christians also appear to have become more involved in international efforts. Christian universities sponsor lectures and research groups on world peace, worldwide nuclear disarmament and international movements, such as the Peace Now movement calling for a peaceful resolution to the Israel-Palestine conflict. Almost all of the Japanese academic institutions that are listed as participants in the United Nations Global Compact, a global platform that has worked to encourage private industries to support human rights and environmentally sustainable practices since 2000, are Christian universities, and include: Doshisha University, Doshisha Women's College of Liberal Arts, International Christian University, Kwansei Gakuin University, School of International Studies—Keiai University, and St. Viator Rakusei Junior and Senior High School (United Nations 2011). These illustrate how Christians have sought to define themselves as part of an international community.

In addition, though the number of Japanese Christians remains low compared to the overall population, Christian symbols and rituals have spread widely since 1945. This is nowhere more visible than in the Christian marriage ceremony, which has been adopted by both Christian and non-Christian Japanese and has largely replaced the indigenous Shinto ceremony. Christian symbols and concepts have spread as widely as to be incorporated into the religious ideology of some of Japan's "new religions," that is, the primarily Buddhist- or Shinto-based religions that were founded in Japan after 1868 and that have attracted many adherents since 1945.

Nevertheless, Christians in contemporary Japan still face many challenges. These include the difficulty that minority groups continue to face in being accepted in Japanese society, the fractured nature of the Japanese Christian community, and declining membership in some Christian denominations. Whereas Christians had once been persecuted for their beliefs, it has become more common for Christian churches to lose membership due to factors such as competition with Japan's new religions. The challenges that many Christian communities now face on a global level, such as the spread of new types of lifestyles and the ongoing question of women's place in religious organizations, have also come to shape Japanese Christianity. These issues, which are discussed in the last section of this chapter, represent some of the concerns that Japanese Christians must face if their communities are to survive into the future.

107.4.1 Legal Rights and the Constitution

The 1947 Constitution differed from its predecessor in that, under the 1889 Meiji Constitution, the right to freedom of religion was limited and made subordinate to state goals such as preserving peace and order. This had allowed the state enormous freedom to promote the emperor's divinity and its own religion while suppressing any religious activities that it considered detrimental. In contrast, Article 20 of the 1947 Constitution not only granted freedom of religion, but prevented the state from exerting its authority to promote one religion above any others. The state could no longer act as a religious authority, nor could it compel anyone to participate in religious activities against their will. Lastly, Article 20 stipulated that religious beliefs could no longer be taught in schools, nor could the state become involved in any religious activities. Thus, the new constitution was designed to prevent many of the legal problems that had contributed to the rise of religious persecution during the war, and established, at least in principle, legal protections that benefited Japanese citizens who followed minority religions such as Christianity.

Though the Constitution helped Christians practice their faith more openly, a 1988 Supreme Court case revealed the limits of the constitutional guarantee. The court case concerned a Japanese Christian widow's refusal to have her husband enshrined as a deity in a local national defense shrine. Her husband was a member of the National Police Reserve, a police force that replaced Japan's military after 1945.[2] When a member of the National Police Reserve (now the National Self-Defense Force) died, the member was usually enshrined as a type of standard procedure. Mrs. Nakaya refused to participate in the enshrinement and informed the Self-Defense Forces that she did not want her husband enshrined because of her religious beliefs (her husband had no clear religious beliefs). However, the group, acting in tandem with the Veterans' Association, went ahead and enshrined him anyway. The court case revolved around the legality of a governmental organization's participation in a Shinto enshrinement, which appears to contradict Article 20's stipulation that the state is not to favor any religion. The 14–1 decision upheld the enshrinement on the grounds of what might be construed as a technicality: that the Self-Defense Force was only indirectly involved in the enshrinement and that its goals were to improve morale by honoring a fallen comrade rather than to promote religious beliefs. More damaging, however, was the Supreme Court's interpretation of the separation of church and state clause in the Constitution as specifying activities prohibited to the state rather than guaranteeing religious freedom as such.

Japanese prime ministers and other government officials have also chosen to make semi-official visits to Yasukuni Shrine, the national Shinto shrine that houses the war dead. The problem for Christians is twofold: first, state officials' informal yet extremely public demonstrations of worship undermine the separation of church and state and seem to maintain Shinto above other religions. Second, visits to the shrine show respect not only to the soldiers killed in combat, but to the war criminals who are also enshrined at Yasukuni, thus undermining the Constitution's renunciation of war and conflicting with the pacifist and internationalist trends that permeate contemporary Christian culture. Though the past few prime ministers have chosen not to visit the shrine, the 2010 decision by Naoto Kan, Japan's prime minister from 2010 to 2011, to prevent anyone in his cabinet from visiting the shrine marked the first time since 1980 that the entire cabinet stayed away.

[2] Article 9 of the Constitution prohibits Japan from maintaining a military force.

A survey conducted from April to June 2012 for the purposes of inclusion in this chapter indicates how some Christians feel they are treated under current laws and institutions.

The survey, which gathered information from 83 Christians of various denominations in the Tokyo area, asked Christians whether they thought that laws and policies in schools, government and other institutions treated Christians fairly. In both cases, the majority of respondents indicated that they felt that laws and institutional policies were fair, though a high number of respondents felt that they did not have enough information about institutional policies to form an opinion.

In addition, some respondents found it difficult to give an unqualified response. For example, with regards to whether laws were fair, a few Christians seemed reluctant to single out Christianity as a target of legal action, stating that religious freedom is protected, that the law is fair to all religions, and that there are no particular laws that target Christians. Others qualified their responses that laws were fair, stating, "There are no particular problems." One respondent, Mrs. K, stated that although religious belief is protected under the Constitution, "based on differences in how the national anthem is interpreted, the problem has arisen recently that some priests won't sing the anthem or participate in the ceremony because they think [their religious beliefs] are being treated unfairly." Another respondent, Mr. O, stated, "From the perspective of constitutional protections, there is no problem. However, when an issue has arisen in the past regarding a person in a particular religious group, it has been very frightening to see, in the media, public opinion and elsewhere, that the national tides seem to shift towards changes in the whole legislative structure." These responses suggest that, though most respondents were satisfied with current laws and policies, a few remained concerned about whether their freedoms would continue to be protected.

Survey respondents were also asked whether they felt there was any discrimination against Christians and whether attitudes towards Christians—whether positive or negative—had changed. With regards to changes in attitudes, a slight majority indicated that they thought that attitudes had not changed noticeably, while an almost equal number thought that attitudes and policies had changed for the better. The majority of participants also did not express a feeling of being discriminated against. Referencing the negative attitudes that Japanese tend to have towards the "new religions," one of which was implicated in the 1995 sarin gas attack on a Tokyo subway, one respondent seemed to think that Christianity was not viewed as negatively. She wrote, "Although there is an innumerable number of 'new religions' that have entered Japan, I think people tend to view Christianity a little bit differently."

Representing the minority opinion, that is, that Christians are discriminated against, one respondent wrote, "I think there is discrimination. Christians are not accepted because they do not participate in the seasonal festivals, weddings and funerals that are Japanese, Shinto and Buddhist ceremonies." Another wrote that there is simply no overt "appearance" of discrimination. A third thought that Japan's history of oppression has had a tendency to stick in people's minds when they think about Christianity, thus creating a negative impression of the religion; while yet another respondent believed that, though there was no discrimination against particular Christians, some people feared that worldwide conflicts between Christians and Muslims would draw Japan into war. None of the respondents mentioned discrimination against a particular sect, rather viewing the Christian community as a whole as a target for Japanese prejudices; this undoubtedly reflects Christians' awareness of the mainstream culture's tendency to gloss over differences between different branches of Christianity. The responses nevertheless reflect persistent fears that Christians are not truly accepted in Japanese society, and suggest how world events, particularly U.S. intervention in the Middle East, have influenced how Christians feel they are viewed by Japanese society in general.

Thus, though postwar changes in laws and policies have helped Japanese Christians gain more rights and have increased Christians' sense that they are no longer being discriminated against, a few issues have arisen that continue to reflect Christians' status as members of a minority religion. These include persistent ties between the government and the national Shinto religion and Christian fears that negative attitudes still exist, but have become hidden, latent, or more abstract.

107.5 Spread of Christian Ceremonies and Symbols in Japanese Culture

On a cultural level, Christianity appears to have done quite well, having permeated mainstream culture through vehicles such as the Christian wedding ceremony. Christian-style weddings tend to include all the trappings of Western ceremonies, such as wedding cake and the white bridal gown. More religious types of symbolism may include the fact that the wedding is held in a chapel or church and is presided over by a "priest." This priest, however, is not usually a member of a Christian religious order, nor are Japanese who participate in such ceremonies doing so out of Christian faith; rather, these "priests" are often white European, Australian, New Zealander or American men who simply fit people's expectations of what a Christian priest should look like. As demand for the ceremony has increased, so too has the number of non-ordained Western men who have been hired for this purpose. As late as 1982, over 90 % of weddings were Shinto ceremonies, while only slightly over 5 % were Christian. By 1998, however, Christian ceremonies increased to over 50 % of all ceremonies, while Shinto wedding ceremonies declined by almost two-thirds to approximately 32 % of all weddings.

Reasons for the increase in Christian-style weddings have as much to do with American cultural influence and changing perceptions of what constitutes an attractive wedding ceremony as with the ideas that Japanese people associate with the two religions. In the case of the modern Shinto wedding, which is a formalized ceremony that was not practiced widely in Japan until the twentieth century, younger Japanese seem to feel that it represents traditional values that have become less appealing, such as the idea of a wedding as the union of two families. In contrast, the Christian ceremony is associated with Western ideas of a love-based marriage that is the personal choice of two individuals. The latter interpretation has gained increasingly among Japanese who now have more control over their choice of spouse than had previous generations and who are less likely to view the goal of marriage as having children who will carry on the family name.

In a larger context, the Christian ceremony reflects longstanding Japanese attitudes towards religion, which tend to divide spiritual work according to the religion that best seems to fit it. Thus, whereas Buddhism is used for rituals involving death, such as funerals and memorial services, Shinto is used for ceremonies celebrating life, such as births and coming-of-age ceremonies. Christianity seems to have become incorporated into this tradition as the appropriate religion for wedding ceremonies. The growth of Christian weddings therefore shows how ideas about love and marriage have changed rather than indicating a decline in traditional Japanese religion.

Christian symbols and ideas have not only influenced mainstream culture, but have been incorporated into some of the religious subcultures of Japan's so-called "new religions." The millenarian polytheistic religion Mahikari, for example, uses symbols such as the cross and the Star of David to assert an ancient relationship between Japan and Middle Eastern cultures (Fig. 107.5). Crosses are included in the religion's sacred religious symbols, the construction of the main sanctuary, and the shape of the main stage, which is itself in the form of a cross. Mahikari also refers to their spiritual leader as a "messiah," adopting both the English word and the idea of a savior.

Indeed, some of the newer branches of Christianity—that is, local branches founded by Japanese individuals in the mid-twentieth century—can themselves be

Fig. 107.5 A symbol of the Mahikari religion. Mahikari combines symbols from other religions, such as the Jewish star and the Christian cross (center of symbol), to illustrate the ancient relationship between Japan and Middle Eastern cultures (Image from Christina Ghanbarpour's collection)

considered "new religions." Like their Buddhist and Shinto counterparts, the newer Christian sects have attracted a wide following among Japanese even as older Christian denominations, such as the *kakure kirishitan*, have declined. In addition, like other "new religions," they tend to be based around the personal charisma of the original spiritual leader, with membership declining after the leader's death. It is estimated that 10 % of all Christians in Japan are members of one of these newer denominations. Though these newer branches of Christianity tend to assert the same basic beliefs and texts as mainstream Christianity, they favor Japanese, local and regional values and traditions over universalistic principles. Moreover, they often place special emphasis on a founder whose prophecies or visions are believed to have been directly received from God.

One of the largest of these churches, the Holy Ecclesia of Jesus, offers an example of this phenomenon. The Holy Ecclesia of Jesus, which was founded in Hiroshima in 1946 by Ôtsuki Takeji, is based on the belief that God sent the founder special revelations regarding Israel and the Second Coming of Christ. Baptism and apostolic faith, which adherents believe is not sufficiently developed in mainstream Christianity, are of central importance in the religion. As of the late 1990s, the church consisted of over 100 churches that had a following of approximately 10,000 people; about 7,000 people attended with weekly services, and the church was thought to be growing thanks to the charismatic leadership of the current head.

However, even here membership has changed as what was once the largest group of attendees—children in elementary and junior high school—has declined, largely due to the low birth rate and pressure to do well on school exams. The decline in the number of children at Sunday services is thought to have had a negative impact on all Christian churches in Japan, and is one of the problems that churches must resolve if they are to stay relevant.

107.6 The "Spiritual Boom"

Over the past 20 years, the growth of nontraditional religions and an increased interest in spirituality in general has led to a phenomenon that some have described as a "spiritual boom," or a sudden increase in interest in and the practice of various religions. Evidence for the boom comes from several sources. For example, a 2011 article in the *Asahi Shimbun* reported the results of a survey in which the number of college students who indicated that they had some form of religious belief had increased from slightly over 30 % in 1996 to over half of all college students in 2010. A 2007 NHK investigation into the causes for the boom showed that it affected a wide range of individuals, and reflected both widespread discontent caused by the continuing recession and what some perceive to be a spiritual vacuum in contemporary Japanese life. Indeed, the most stressful parts of contemporary life in Japan—school and work—demand loyalty, dedication, and complete identification with the goals of the institution, yet these no longer guarantee the job security and other benefits for which many Japanese had been willing to sacrifice in the booming 1980s.

As a result, many Japanese have been left with a sense of a loss of purpose as they no longer identify with their education or careers, find it difficult to find fulfilling work, or cannot find any work at all.

Most of the religious institutions that are thought to have benefited from the boom are the newer sects of Shinto and Buddhism, such as Tenrikyô. Other groups that are thought to have benefited are not affiliated with a religion *per se*, but revolve around longstanding folk practices such as fortune-telling. The April-June 2012 survey conducted in the Tokyo area, which was cited above in relation to laws and policies regarding Japanese Christians, asked Christians whether they had heard of the "spiritual boom," and if so what effect, if any, it has had on either Christians or Christianity in Japan. The majority of respondents had heard of the "spiritual boom," and about a quarter thought that it had affected Christianity though none identified a specific church that they thought had benefited. Rather, of those that thought it had affected Christianity, the majority believed that this effect was largely negative. Several respondents believed that people had become suspicious towards religion in general because the spiritual boom had led people to join religious groups such as Aum Shinrikyô, the group responsible for the Tokyo subway incident. One woman thought it had led people to lose their faith, while a man thought that there was a danger that people would stop following the Bible.

The minority opinion was that it had had a positive effect on Christianity. One woman wrote that she thought it had led people who were not Christian to become more familiar with Christianity. Mr. F, a 31-year-old Protestant, wrote the following:

> I have no idea what effect [the spiritual boom] will have. Differences within society have increased. Unlike in the past, we have reached a point where one can't rely on anything. Because of this, I think that there are people who are crying out for help [sukui wo motomeru]. If we could make use of this condition which has become so noticeable, there is a possibility that the image of religion would improve.

While Mr. F's statement is less optimistic than the former, it highlights the difficulties of predicting where people's need for guidance will lead them. Moreover, it points up what might be considered one of the underlying causes of the spiritual boom: the birth of a complex, uncertain world in which people must struggle to find their paths in life. Whether Christianity will become the solution for these people remains to be seen.

107.7 Continuing Challenges for Christians in Contemporary Japan

Though statistics on the number of Christians and their denominations in the general population vary, surveys indicate that their numbers today are still small. Japanese Christians make up only about 1 % of the Japanese population, and there are many divisions within that group. For example, less than a quarter of Japanese Christians

(467,865 out of 2,121,956) identify as Catholic, a quarter identify as Protestant (539,832), and about half identify with a variety of other Christian groups, such as the Jehovah's Witnesses and certain branches of the Methodist, Baptist and Evangelical churches that do not consider themselves Protestant. Of those that identify as Protestant, 54,258 are Anglicans, 41,716 are Baptists, 2,958 are Methodists, and 5,270 belong to an Evangelical church—an indication of how diverse the Christian population is, and how small the number of members in each denomination may be. Neither Christianity nor either of the two major religions, Shinto and Buddhism, have gained or lost a significant amount of members over the past 40 years, but some Christian denominations have all but died out over the past 30 years as a result of declining membership. As for other world religions, such as Islam, numbers of adherents are thought to be less than 100,000.

Christians in Japan, like Christians in other parts of the world, also face challenges regarding Christian doctrine. One of these issues concerns the compatibility of homosexuality and church doctrine. An incident involving the United Church of Christ, which is the largest Protestant denomination in Japan, illustrates this conflict. In 1998, a theological seminarian revealed his homosexuality during an exam to become a minister. In response, a member of the Church's Executive Committee stated that homosexuals should not be allowed to become ministers, a sentiment repeated later the same year by a minister and professor of the prestigious Tokyo Union Theological Seminary. Though the young man was eventually ordained, neither of the statements was retracted despite pressure from gay and non-gay members of the Church. The incident remains unresolved though activists, led by a female minister and lesbian member of the Church, continue to protest and raise awareness through a support group for gay members.

Women's roles in the church are another concern. As in many aspects of Japanese life, women are rarely considered for leadership positions and have few means to promote their interests. The Non-church Movement, for example, which has a reputation for being a "manly" version of Christianity and has long been dominated by male preachers and adherents, has in recent years opened a Women's Planning Group and invited female speakers to give talks. Other groups, such as the Japan Evangelical Association, sponsor organizations for female members. One of the most progressive churches, the United Church of Christ, allows female and transgender ministers to preach. Overall, however, men continue to dominate leadership positions.

The Internet has also offered new opportunities for Japanese Christians. Online newsletters, listservs, Bible discussion groups, blogs, online newspapers, and eLearning pages have helped Japanese Christians promote causes, find followers, and communicate information. The Holy Hope Project, an online blog and activist organization run by Takeshita Chikara, posts explanations of passages from the Bible, informs viewers of other Internet resources such as other Christian blogs, and helps promote causes such as nuclear disarmament. Other websites, such as Nikomaru Tours (also run by Takeshita), help arrange visits to the Holy Lands. Because streets are rarely labeled in Japan, many of these websites provide maps and lists of locations where services are available, thus making it easier to locate churches. In these ways, websites have allowed this relatively small community to become more integrated, connected, and international.

These examples highlight the challenges that Japanese Christians must face as they advance into their fifth century as members of one of Japan's minority religions. Through perseverance and flexibility, Japanese Christians have been able not only to maintain their place in Japanese society, but have managed to spread their beliefs and symbols among Japan's many non-Christians. These qualities indicate the degree to which Japanese Christians have been successful in maintaining their religious beliefs despite centuries of oppression and consequently low numbers of adherents. Christianity has thus contributed an important legacy to Japanese history and continues to influence contemporary Japanese culture and society.

Reference

United Nations. (2011). *United Nations global compact website.* Retrieved July 20, 2011, http://www.unglobalcompact.org/ParticipantsAndStakeholders/academic_participation.html

Chapter 108
The Chinese Church: A Post-denominational Reality?

Chloë Starr

108.1 Introduction

The Chinese church is officially "post-denominational." This is both a statement of theological aspiration and a description of the official state of church affairs. In the eyes of many Chinese Christians, it is a wholly good thing, and a positive ecumenical move from which the rest of the world church could learn much. China's post-denominationalism has its roots in early twentieth century anti-Imperialist movements and the drive by Chinese Protestants towards an autonomous church, but its institutionalization was engineered by the demands placed on the church in the late 1950s. Post-denominationalism, with its positive rejection of the western divisions of the imported church, as well as its tinges of spiritual and national pride, is, however, not the whole story of the contemporary Chinese church. Unregistered churches of indeterminate affiliation probably form the majority of Christian worship places in contemporary China, and "heretical" sects, denounced by both the government and the official church, have sprung up with surprising regularity since religious regulation was relaxed in the early 1980s. As different theological and ideological affiliations have re-surfaced in the reform era, new networks of churches, both local and trans-regional, have grown up alongside the state churches. This brief article traces some of the competing descriptions of "church" in China and places them in the context of historical developments in the twentieth century. It surveys the ideology behind the post-denominational stance of the Chinese church, and considers how recent developments are complicating the official discourse. The aim here is to trace intellectual lineages and to assess the rhetoric of church structures against their reality, not to map the church numerically or spatially.

C. Starr (✉)
Asian Christianity and Theology, Yale Divinity School, New Haven, CT 06511, USA
e-mail: chloe.starr@yale.edu

S.D. Brunn (ed.), *The Changing World Religion Map*,
DOI 10.1007/978-94-017-9376-6_108

108.2 Ecclesial Identities

As is widely known, the Roman Catholic and Protestant branches of the Christian church were effectively designated two separate religions by the Chinese government in its recognition of five official religious bodies. Within each Christian body, there exists a state-approved church, with separate organs for liaison with the government and with church communities, and an unregulated sector beyond the state-church reach. This sector, variously called the "underground," "house," "family," or "unregistered" church (depending on point of view and denomination), is believed by many to be home to the majority of Chinese Christians. The Chinese government used to designate these "illegal" churches, but has more recently used the term 'extra-legal' for this gray sector. Estimates for numbers of Chinese Christians vary greatly, from the c25 million of government statistics to western estimates of c70 million to 115 million (see, for example, Stark et al. 2011; Johnson and Ross 2010: 140), with most scholars assuming that around two-thirds of the total worship in unregistered churches. Both the churches and the government recognize that the anomaly where a great number of worshippers meet below the official radar cannot exist indefinitely, and are actively seeking ways forward, with academics and Hong Kong church leaders involved in mediating on the Protestant side, and Vatican representatives on the Catholic side ("underground" Catholic church members foreground loyalty to Rome). In the meantime, this gives us four macro categories of church—official Catholic, unofficial Catholic, official Protestant and unofficial Protestant, leaving aside smaller Orthodox, Adventist and other communities—with numerous streams within the unofficial sectors, and a fifth category of unaffiliated, often academic, Christians.

A historical re-cap is necessary to understand the present structures of the church in China. The Three-Self Patriotic Movement, the official Protestant body, is seen as a monolith of the PRC era, but was not at the outset a movement to unify the church. Modeled on missionary formulations of a self-funding, self-governing and self-propagating indigenous church, and actively promoted by Chinese theologians in the 1920s and 1930s, the movement began as a call for autonomy for Chinese churches. By the late 1940s it had allied with anti-imperialist ideologies, acknowledging the churches' involvement in colonial oppression, and in the early 1950s adopted the language of "socialist reconstruction" as a guiding principle for action. During this period the new government took steps to enforce self-support and self-government through a series of regulations on foreign income and foreign personnel. The development of the Three-Self Patriotic Movement in effect occurred as a series of steps between state and church; as an advocacy movement, the TSPM was quite different to its later function as the sole recognized Protestant church (Fig. 108.1). Few expected the instant unification of the church. As late as 1956 the Anglican bishops meeting in Shanghai dedicated Holy Trinity church a new "National Cathedral" of Chinese Anglicanism (Merwin and Jones 1963: 145). It was only in 1958, during Mao's Great Leap Forward, that the final steps towards unity of worship were taken in the major cities, with church property turned over to

Fig. 108.1 Entrance to Haidian church, which has four Chinese services and one English language service each Sunday (Photo by Chloë Starr)

common administration by the TSPM. The "socialist study" which Christians had been engaged in through the 1950s had enabled (some of) them to see how combining churches might free up pastors' labor and church estate for the good of the nation.

On the Catholic side, the issue was never one of multiple denominations, but similar questions of Chinese leadership and authority did arise, and of unity in a bifurcated church. The Chinese Catholic Patriotic Association was brought into being in 1957, against greater opposition, and following a war of words with the Vatican over foreign mission, the arrest of priests and nuns, and campaigns to discredit members of organizations such as the Legion of Mary. Just as Protestants who rejected the new regime continued to worship illegally and outside official bounds, Roman Catholic dissenters came to form an underground Catholic church. No formal schism has ever been declared with the Chinese Catholic church, and the church remains one church. But the ordination of bishops in particular has caused deep rifts between the "Open" Catholic church and Rome. By the late 1950s, following expulsions of foreigners and the imprisonment of bishops unwilling to cooperate with the CCPA, fewer than one-fifth of Chinese dioceses had bishops, and clergy began the process of selecting replacements. When Rome refused to sanction these, a series of unsanctioned ordinations took place within the CCPA, producing bishops whose orders were canonically "valid" but "illegitimate." Many unsanctioned bishops subsequently sought (and later received) the blessing of Rome, leading to significant confusion, if not outright mistrust (Fig. 108.2).

Both national organizations, the TSPM and the CCPA, were dormant for much of the 1960s and 1970s, and were reconstituted in the late 1970s after the end of the Cultural Revolution. Since then, both have run national and provincial seminaries, published church journals, and provided the infrastructure and representation to be expected of national churches. These are the churches which have represented Chinese Christianity to the Chinese people, to the outside world, and to the government, for most of the history of the Peoples' Republic. What is sometimes surprising and shocking to outsiders is the vehemence of the rejection of co-religionists within the Catholic/Protestant divisions, which in some cases has persisted to the

Fig. 108.2 Underground Catholics kneel at a shrine outside Sheshan Basilica to pray while Catholics who have been to Mass inside the Basilica stream past (Photo by Chloë Starr)

present. The depths of divisions among Protestants can be seen, for example, in a document from the Word of Life church, one of the largest house church networks in China, whose formulation in the 1980s of seven principles of theological education included an article requiring members to "clearly discern that the TSPM does not represent the true church" but exists as an "adulterous political organism" serving atheism (Xin Yalin 2009: 89). This rejection of believers by believers confirms the deep wounds caused by a sense of betrayal among those who, in conscience, could not compromise and work within the religious framework established by the atheist, communist state in the 1950s. Those Christians who remained outside the official churches in general suffered more, and for longer, than other Christians during the years of persecution. We can trace the detail of theological debates among proponents of the differing positions, such as between Bp. Ding Guangxun on the official Protestant side and Wang Mingdao on the dissenting side, or between Roman Catholic Archbishop Bi Shushe of the CCPA and Bp. Fan Xueyan of the underground church, but one group branding another anathema still shocks.

The painful, and confusing, split that the creation of the Chinese Catholic Patriotic Association caused among believers lasted half a century, and was particularly visible in contested episcopal ordinations, which left the faithful unsure whose authority to follow. The decrees issued by the Vatican in 1957 and 1958 fostered a fierce loyalty: denouncing the new CCPA, they encouraged believers to remain loyal to Rome even to the point of martyrdom, and implied the risk of excommunication for those who joined the new association. As the church reopened through the 1980s, underground leaders received support from John Paul II, even in the consecration of bishops without Vatican approval (see, for example, Bays 2012: 192), leaving many dioceses with two functioning bishops. Pope Benedict XVI's carefully worded 20 page letter to Chinese Catholics in 2007 marked a new point in relations, encouraging moves to unity by removing certain political blocks to dialogue with Rome, and, more importantly for believers, rescinding the outright rejection of the CCPA (Benedict XVI 2007). To the difficult question of whether recognition from the civil authorities compromised "communion with the universal Church," Pope Benedict deferred to the judgment of local bishops. While restating that an autonomous Chinese Catholic church was "incompatible" with Catholic doctrine, the letter took cognizance of the pastoral difficulties the split had caused, and reiterated the validity of the sacraments of illegitimately ordained bishops or priests, making clear that believers could receive from them with a clear conscience where necessary. In ceasing further "underground" ordinations, Pope Benedict asked much of the Catholics who had remained loyal to Rome: they were the ones to make the moves towards unity with the government-recognized hierarchies (Fig. 108.3).

If the official churches remain divided from their unregistered brothers and sisters, there have been signs of rapprochement recently. The younger generation of church-goers knows little of the personal animosity their elders may feel, and has not experienced the same level of fear or suffering. For Roman Catholics, the strengthening of the Bishops' Conference vis à vis the CCPA, and measures which allowed for the recognition of the Pope as the spiritual leader of the church, meant

Fig. 108.3 Hou Sangyu village has had a Roman Catholic presence since the fourteenth century. The present church was destroyed in the Cultural Revolution and rebuilt in 1987 (Photo by Chloë Starr)

that from the 1990s on the grounds for division between the open and underground churches were much reduced. The rapid growth of Protestant Christianity has primarily been through conversion, and does not come with a family solidarity to one group of the other. In cities, some attend a TSPM church on a Sunday and a house church group in the week. Gao Shining's study of Beijing church-goers (Gao Shining 2006) shows that for many urban Protestants, church choice is a practical as much as theological decision. With only 57 of 827 church "meeting points" in Beijing registered, proximity to home is one of the single strongest factors determining place of worship. (While the constitution guarantees freedom of religious belief, subsequent directives, such as Document 19 issued in 1982 have regulated where religious gatherings may take place, and under whose authority. Churches and meeting-points which operate outside of the sphere of the national patriotic religious bodies are referred to as "unregistered.") Other factors in opting for an unregistered meeting-point included the desire for a smaller or more close-knit fellowship and the availability of weekday worship, as well as sermon quality. One important fact that Gao's data highlights is that it is becoming increasingly acceptable to be both Chinese and Christian. If the original reason for the Protestant split with the mission-run churches was to get away from the "foreign" label and its hindrance to evangelization among Chinese, Christianity is at last shedding, among the populace if not politicians, its reputation as an alien religion.

If the most pressing problem for the church in the early twentieth century was its "western" nature, the equivocal position of foreign missionaries is still a live issue in

China. Whether expelled in the early eighteenth or mid-twentieth century, missionaries have always returned to China as soon as possible. There has been a huge influx of mission partners of all denominations since the 1990s, firstly as language teachers and more recently in a range of professions. Mission groups in China include overseas and diasporic Chinese Christian groups, international organizations like OMF and specialist groups like International Medical Relief, as well as individual church outreach team. Just as the western Church in its missionary reports, from the Jesuits of the Ming through to the Victorians of the late Qing, tended to overestimate its part in the conversion of China and greatly underestimate the role of local Chinese Christians in the process (see Tiedemann 2008), it is easy to gain the impression from mission organizations working in China that much of the current growth in the church has been through foreign witness. Data which separates out Chinese or foreign influence on coming to faith is difficult to access; one data set (Gao Shining 2010) shows that 28 % of respondents converted through the influence of another Christian, excluding family members, without giving further specifics. There has certainly been much material and personnel support from outside China, particularly to the unregistered churches, but, aside from some urban student groups, it might be difficult to substantiate a claim that the spectacular growth in China has come through anything other than an overwhelmingly Chinese witness (Fig. 108.4).

Evangelical mission groups have brought another complication back to China, according to Yalin Xin's study of house church networks (Xin Yalin 2009): their denominationalism. In his analysis of the disunity among house church members in the early 1990s, Xin quotes one leader laying the blame firmly with the influx of

Fig. 108.4 Transport to rural Nanjing church (Photo by Chloë Starr)

ideas from foreign and overseas missionaries. As individual house church leaders began to travel and seek outside teaching materials, some were deeply affected by the particular brand of Christianity with which they came into contact, espousing the new ideas they encountered, and hastening splinter movements among house church networks. On a more positive note, one thing that the outside church, including missionaries on the ground in China, has done, is to document abuse: abuse of government regulations where local officials have acted beyond their powers in persecuting Christians, and more widespread persecution of Christians and others, such as after 1989 (see, for example, Dunch 2010).

108.3 Sects and Syncretists

There have been some excellent recent studies in English of the development of unofficial church groups, or 'popular' Christianity in China, and an acknowledgement of the range of different Christianities present today, particularly in rural areas. Two lines of research have emerged: of individual church groups on the spectrum between independent indigenous movement and sect, and of Christianity itself as a folk religion (both Catholic and Protestant versions). As Daniel Bays writes, 'for Protestants, it was in the 1980s that some of the inherent similarities between Protestant radical millenarianism and the eschatological features of traditional Chinese popular religion became visible and interactive' (Bays 2012: 193). Like Christianity, Chinese folk religions were strongly attacked as feudal superstition in the 1950s, and began their recovery in the 1980s. Some indication of the rapidity of the resurgence of folk religion can be seen, for example, in a report documenting the destruction of nearly 18,000 temples built without legal permission, alongside hundreds of thousands of luxury tombs for the dead (Gao Shining 2010: 171). The paradox emerges that Christianity grew in rural areas in part by filling the void caused by the suppression of folk religions, but Christianity itself is increasingly imbued with folk aspects. These include, naturally, an interest in ancestor worship, or Christian versions of that ritual, but also a strong emphasis on healings, with conversion because of illness the highest reported factor in conversions (Gao Shining 2010: 178), as well as superstitious practices such as investing Christian objects with talismanic powers, color taboos, or restrictions on church participation of menstruating women.

As Lian Xi has shown in his pioneering study, the unofficial church groups of the late twentieth century by and large grew out of the independent churches of the pre-1949 era (Lian Xi 2010: 206)—and the sects grew as offshoots of these too. A combination of government pragmatism—legitimizing groups is a prime way to keep an eye on them—and the theological ideal of unity led to the two largest indigenous church groups which had re-emerged in the post-Mao era being brought back into the Three Self fold in the 1980s. These were the True Jesus church (founded by Wei Baoluo, Paul Wei c.1919) and the Little Flock church (which came to prominence in the 1930s under its charismatic leader Ni Tuosheng, or Watchman

Nee). Other pre-49 church groups have fared less well at gaining recognition, and remain "underground."

Independent Chinese churches had co-existed reasonably peaceably with denominational churches and with the newly forming national churches in the early decades of the twentieth century. The independent churches (which were *more* self-supporting, self-governing etc. than any other group) had hoped for tolerance of their activities under the new government, but beginning in 1952, the major independent players were targeted, and by 1958 when denominations became defunct, the leaders of all of the independent, evangelical, Pentecostal or communal Christian groups were imprisoned or removed from their posts. From 1979 onwards, just as the TSPM and the CCPA were being re-established, the re-emerging and newly forming independent churches were taking stock of their situation, and beginning to consider leadership training, theological education, and the needs of a growing church body. Xin's study (Xin Yalin 2009) of one of the largest house church networks, the Word of Life group, provides a helpful documentation of one such movement, showing how it developed from an individual itinerant leader joining together re-emerging congregations in the early 1980s, through a period of rapid growth, with over 3,000 churches by the late 1980s, to an organization concerned in the late 1990s with structural hierarchies, pastoral regions, and a four-stage program of theological education. Xin points to a subsequent trend, as several of the major new house church networks, including the some of the largest like the Word of Life, the Fangcheng Fellowship, and the China Gospel Fellowship, began to take steps towards unity among themselves in the late 1990s. One of the best known examples of this was the "Sinim Fellowship" of like-minded groups (although website reports indicate that not all member groups have remained within the fellowship since).

The apocalypticism of Ni Tuosheng, the miraculous healings of John Sung and revival speeches of Dora Yu, the black-and-white certainty of Wang Mingdao—all of these elements of the vibrant independent Christianity of the 1920s–1940s began to re-surface in the post-Mao Christian resurgence. If the great orators and martyrs of the Chinese churches inspired both the re-forming independent churches and the newer house church movements, they also provided a vision for more extreme groups. The unchartered space beyond the official church and the emerging unregistered mainstream has played host to a number of fervent congregations, some led by former prisoners/martyrs, some by former Pentecostal leaders, many of whom were opponents of both the TSPM and the government. Most were lay, few were trained. The more extreme congregations fostered cultic or messianic leaders who became the first casualties in the new crackdowns of the Deng Xiaoping era, beginning with the "Spiritual Pollution" campaign of 1983. In various cases, the new movements can be traced directly back to earlier indigenous groups, such as the Spirit Sect founded by a true Jesus Church member in Jiangsu in the early 1980s, where practices such as faith healings and exorcisms were combined with prophecies of the imminent end of the world, or the Shouters, whose bellowing expressions of life in the Spirit came back into China via evangelists from Taiwan and California, offshoots of the Little Flock under Ni's successor Li Changshou (and in whose name people were baptized; see Lian Xi 2010: 214–16). Other sects, such as Eastern

Lightning with its female Christ, Ms. Deng, or the reincarnated Jesus, Wu Yangming of the Established King sect, were more evidently cultic—the penalties for which, in some cases, were state executions. As Bays, Lian and others have noted, there was a serious dearth of both education and theological training for church leaders at this period; this, combined with a sense of freedom and a spiritual battle against government repression, meant that conditions for cult development, and for the continual splintering of new groups, were ripe. One headache for local government officials has been trying to separate out acceptable unregistered churches from the more extreme, and potentially dangerous, types.

108.4 Theological Identities

Alongside the resurgence of the churches in the reform era, a remarkable growth has occurred in academic studies of Christianity. Scholars within the official church systems, working out of Roman Catholic or Protestant seminaries, have developed new frameworks of theological thought for the era, such as the "theological construction" program initiated by Ding Guangxun. But beyond these, the greater change has been in the development of academic theology not affiliated to any church stream or theological position. There are thriving MA (and also PhD) programs in Christian Studies at dozens of Chinese universities, specializing in different aspects of Christian philosophy, mission history, biblical studies, literary criticism and cultural studies. A range of Chinese academic journals supports this enquiry, and e-networks circulate announcements daily of new articles on Chinese Christianity in both Chinese and English, on topics ranging from contemporary Chinese Christian art to Ming dynasty Confucian-Christian dialogue. The Institute of World Religions at the Chinese Academy of Social Sciences hosts an annual large scale conference and gathering of scholars working on Christianity, and proceedings from this and other conferences are published by leading university and research presses. Secular academics in state universities have been pivotal to the greater acceptance of Christianity in China by politicians and officials, and their teaching of a new generation of academic theologians and scholars of Christianity has provided the foundation for the broader development of Chinese Christianity.

In a survey article on the status of Christian theology in China today, Zhuo Xinping, the Director of the Institute of World Religions, argues that theology in China forms three separate streams, "Chinese theology," "Sino-Christian theology," and "academic theology" (Zhuo Xinping 2013). "Chinese theology" here indicates the theology of the church. Zhuo hold that Christian Studies in China meanwhile form part of the discipline of Religious Studies, and as such are 'faith-neutral.' Zhuo's second category of "Sino-Christian theology," is arguably the most important and interesting movement of the last two decades. Sino-Christian theology (*Hanyu shenxue*, lit. 'Chinese-language theology') refers to a movement which came into being in the mid-1990s in China, initiated by the philosophers Liu Xiaofeng and He Guanghu, and drawing support from Hong Kong. The Sino-Christian theology movement as it developed came to include scholars working on disparate aspects of Chinese Christianity, who sought to carve out institutional space for Christian theology to become an academic discipline in China. The movement aimed to bring theology into public discourse, especially on values and ethics (see Yang and Yeung 2006). To give some scope of topics covered, a recent English language compendium (Lai Pan-Chiu and Lam 2010) of Sino-Christian theology covers articles from "Cultural Christians" in China to the value of theology in the humanities, and from the unfolding of the Dao as both word and Word, to messianic predestination and classical Confucianism. Zhuo Xinping argues that Sino-Christian theology exists "between" church and academia, and the movement should be distinguished from that neutral enquiry which properly comprises "academic" theology. While scholars may disagree on these distinctions and on the place of confessional theology in the academy, Zhuo and many of the scholars who have written on aspects of *Hanyu Shenxue* would agree that denominational and church theologies remain peripheral in China, both because of their relative lack of academic rigor and the institutional limitations on the interaction of the church in society.

The existence of a "fifth" sector of Christians unaffiliated to any particular church was alluded to above. Many in this sector, whom some have termed "Cultural Christians" because of their humanistic values and lack of denominational belonging (see Fällman 2004), are also academics. The cultural Christian phenomenon has been confused by its use as a category for scholars studying and writing on Christianity who may themselves not be Christians—but for those who are theists, the reasons for avoiding institutional churches have included the problems of church divisions and state oversight, and the poor quality of much teaching available. Some, such as Liu Xiaofeng himself, have taken a theological stand to argue that the church has always been conditional, and may be dispensed with (Liu Xiaofeng 2006). A new group has joined this fifth sector in the twenty-first century: China's virtual church communities. As Gerda Wielander documents (Wielander 2013), numerous religious and quasi-religious groups have taken to the internet to distribute religious texts, provide training for pastors and create virtual communities. Some are established churches using the web to post recordings of entire services; others are new groupings formed through online media. Wielander's study shows the range of different Christian publications available online, from the more literary to those addressing questions of church-building, rights' awareness or Christian news services, and points to the different denominational audiences for the various sites (some of which more critical of the TSPM; others of the sects etc.). Wielander notes how these virtual communities in China have allowed believers to feel part of a worldwide community of Christians, and how many Chinese Christian web platforms were linked to North America, either by funding or through diasporic writers. Given that China has well over 500 million web users, the intersection and overlapping of communities of belief and belonging, virtual and corporal, can only grow.

108.5 Conclusions

Chinese post-denominationalism is truly alive and well: communities of Baptists, Methodists, Anglicans, Brethren and others who streamed to China in the late nineteenth century are no longer in evidence. Their colleges and seminaries have been incorporated into university campuses, and their churches, for the most part, handed back to the TSPM. That is not to say that China does not have denominations, as the official church and state rhetoric would have us believe, but that the denominations which exist—including the official Protestant and Catholic churches, as well as the underground Catholic and unaffiliated Protestant churches—are now thoroughly Chinese in identity (Fig. 108.5). China's brand of postdenominationalism does not mean that some house churches meetings do not closely resemble evangelical gatherings elsewhere in the world, or that local Catholic practices do not look increasingly like Catholicism in the rest of the world (see Harrison 2013), but that these are post-western, Chinese-chosen forms of being church. The official "unity" may turn out to be a kaleidoscope of different allegiances, groupings, networks and even sects, but unity was ever, on the Protestant side at least, a secondary goal to Chinese autonomy. For Roman Catholics, the tension between the different variants of

Fig. 108.5 Catholic village church (Photo by Chloë Starr)

Catholic identity and a shared desire for oneness remains acute. Whether differences in the vision of unity, within the Chinese church and in a trilateral conversation between the Chinese government, the Chinese church and Rome, can be reconciled in the short or medium term is unclear. Roman Catholic unity cannot be unilaterally imposed; after years of improvement, relations between Beijing and the Vatican dipped to a nadir in 2011. Pope Francis has much in his Chinese in-tray.

At another level, of course, finessing the definition of postdenominational only pinpoints the problem. The challenge for the government is to find new ways to legitimize and manage the array of evolving church groups. The numerical strength and health of the unaffiliated Protestant churches is now such that all sides must reassess the rhetoric and self-understanding of the Chinese church in its entirety. The most ardent supporters of postdenominationalism have been those within the mainstream church, from where it is easy to surmise that it is others who are the schismatics and the separatists. Given the interconnection of the rejection of western denominations with the rejection of western authority, the cause of Chinese church unity has never been dissociated from national politics. Confidence to believe in the Chinese-ness of the contemporary Chinese church still seems to be lacking in government authorities. It is not surprising that ecclesial unity, and even homogeneity, was seen as a positive value in a church which had lived through the disindividualized socialist and communist politics—or that new structures are evolving in an era of greater individualism. There are still voices holding out for the older vision of postdenominationalism, for a single Chinese church which enfolds all other divisions into itself, which continues to argue that "our theological thinking should transcend denominationalism, promote church unity, mutual respect, mutual learning and mutual enrichment," (Chen Zemin 1986) but events have surely overtaken such ideology. Given how strongly formative rhetoric is in authoritarian systems, the sooner denominations are detached from the negative connotations of "western" and come to mean something akin to "churches with Chinese characteristics," the better for the plethora of unregistered groups, but also for the accurate representation of the church corporate to the nation.

108.6 Sample Web Sites

- www.ccctspm.org/english (official website of the TSPM and CCC, Protestant organizations in China); see also
- www.bibleinchina.org
- www.chinacatholic.org (official website of the CCPA and Bishops' Conference, Catholic organizations in China) [no English]
- www.catholicnews.com/data/china.htm (international Roman Catholic site on Chinese Catholicism)
- www.cardinalkungfoundation.org (US-based website supporting sector of underground Catholic church)

- Individual churches, e.g. http://www.english.hdchurch.org (TSPM church, Beijing); https://shwchurch.org (unregistered/house church, Beijing) [no English]

References

Bays, D. H. (2012). *A new history of Christianity in China*. Oxford: Wiley-Blackwell.

Benedict XVI, Pope. (2007). *Letter of the Holy Father Pope Benedict XVI to the bishops, priests, consecrated persons and lay faithful of the Catholic Church in the People's Republic of China*. Retrieved March 22, 2013, from http://www.vatican.va/holy_father/benedict_xvi/letters/2007/documents/hf_ben-xvi_let_20070527_china_en.html

Chen Zemin. (1986). Self-propagation in the light of the history of Christian thought (trans: Wickeri, J. K.). *Chinese Theological Review, 1986*, 18–27.

Dunch, R. F. (2010). Autonomous churches and the question of religious freedom. In R. G. Tiedemann (Ed.), *Handbook of Christianity in China: 1800 to the present* (Vol. 2, pp. 882–889). Leiden: Brill.

Fällman, F. (2004). *Salvation and modernity: Intellectuals and faith in contemporary China*. Stockholm: Stockholm University.

Gao Shining. (2006). Cong shizheng yanjiu kan Jidujiao yu dangdai Zhongguo shehui. *Zhejiang xuekan* (Vol. 5., trans: Chi Zhen & Caroline Mason). Christianity and Contemporary Chinese Society: An empirical perspective.

Gao Shining. (2010). The impact of contemporary Chinese folk religions on Christianity. In M. Ruokanen & P. Huang (Eds.), *Christianity and Chinese culture* (pp. 170–181). Grand Rapids: W. Eerdmans.

Harrison, H. (2013). *The missionary's curse and other tales from a Chinese Catholic village*. Berkeley: University of California Press.

Johnson, T., & Ross, K. (2010). *Atlas of global Christianity*. Edinburgh: University of Edinburgh Press.

Lai Pan-Chiu, & Lam, J. T. S. (2010). Retrospect and prospect of Sino-Christian theology: An introduction. In Lai Pan-Chiu & J. Lam (Eds.), *Sino-Christian theology: A theological qua cultural movement in contemporary China* (pp. 8–9). Frankfurt am Main: Peter Lang.

Lian Xi. (2010). *Redeemed by fire: The rise of popular Christianity in modern China*. New Haven: Yale University Press.

Liu Xiaofeng. (2006). Sino-Christian theology in the modern context. In H. Yang & D. Yeung (Eds.), *Sino-Christian studies in China* (pp. 52–89). Newcastle-upon-Tyne: Cambridge Scholars.

Merwin, W. C., & Jones, F. P. (1963). *Documents of the three-self movement*. New York: National Council of the Churches of Christ in the U.S.A.

Stark, R., Johnson, B., & Mencken, C. (2011, May). Counting China's Christians. *First Things*. Retrieved from http://www.firstthings.com/article/2011/05/counting-chinarsquos-christians

Tiedemann, R. G. (2008). Indigenous agency, religious protectorates and Chinese agency: The expansion of Christianity in nineteenth century China. In D. L. Roberts (Ed.), *Converting colonialism* (pp. 206–241). Grand Rapids: W. B. Eerdmans.

Wielander, G. (2013). *Christian values in communist China*. London: Routledge.

Xin Yalin. (2009). *Inside China's house church network: The word of life movement and its renewing dynamic*. Lexington: Emeth.

Yang, H., & Yeung, D. (Eds.). (2006). *Sino-Christian studies in China*. Newcastle-upon-Tyne: Cambridge Scholars.

Zhuo Xinping (Ed.). (2013). *Religious studies in contemporary China collection*. Boston: Brill.

Chapter 109
Protestant Christianity in China, Urban and Rural: Negotiating the State and Propagating the Faith

Teresa Zimmerman-Liu and Teresa Wright

109.1 Introduction

Protestant Christianity is one of the fastest growing religions in mainland China today. In 1949, at the time of the Communist Revolution, there were only 623,000 Protestant Christians in the mission churches in China, and they made up less than one tenth of 1 % of the entire population (Xi 2010: 10). Today, there are at least 50 million Protestants in China, and some scholars estimate that there are as many as 100 million (Madsen 2010: 62). Roughly 20 % of China's Protestants worship in churches affiliated with the government's Three Self Patriotic Movement (*Zhongguo jidujiao sanzi aiguo yundong weiyuan hui* or *sanzi jiaohui*, hereafter TSPM). The rest worship in unregistered "house" churches that are not recognized by the government authorities, and thus are not officially tracked (Xin 2009: 31) (Fig. 109.1).

These figures illustrate the unique context of Protestant churches in China: the country's ruling Chinese Communist Party (CCP) allows certain religious groups to legally exist, but only under the Party's oversight and regulation. Although somewhat unusual from a global perspective, such political controls are typical of totalitarian and quasi-totalitarian regimes that seek to dominate all forms of social organization. Generally speaking, as in the case of China, regimes of this type have been led by communist parties that have viewed religion not only as the "opiate of the masses" (the Marxist notion that religion is designed to distract the oppressed

T. Zimmerman-Liu (✉)
Departments of Asian/Asian-American Studies and Sociology, California State University, Long Beach, CA 90840, USA
e-mail: teresacalalpha@gmail.com

T. Wright
Department of Political Science, California State University, Long Beach, CA 90840, USA
e-mail: teresa.wright@csulb.edu

S.D. Brunn (ed.), *The Changing World Religion Map*,
DOI 10.1007/978-94-017-9376-6_109

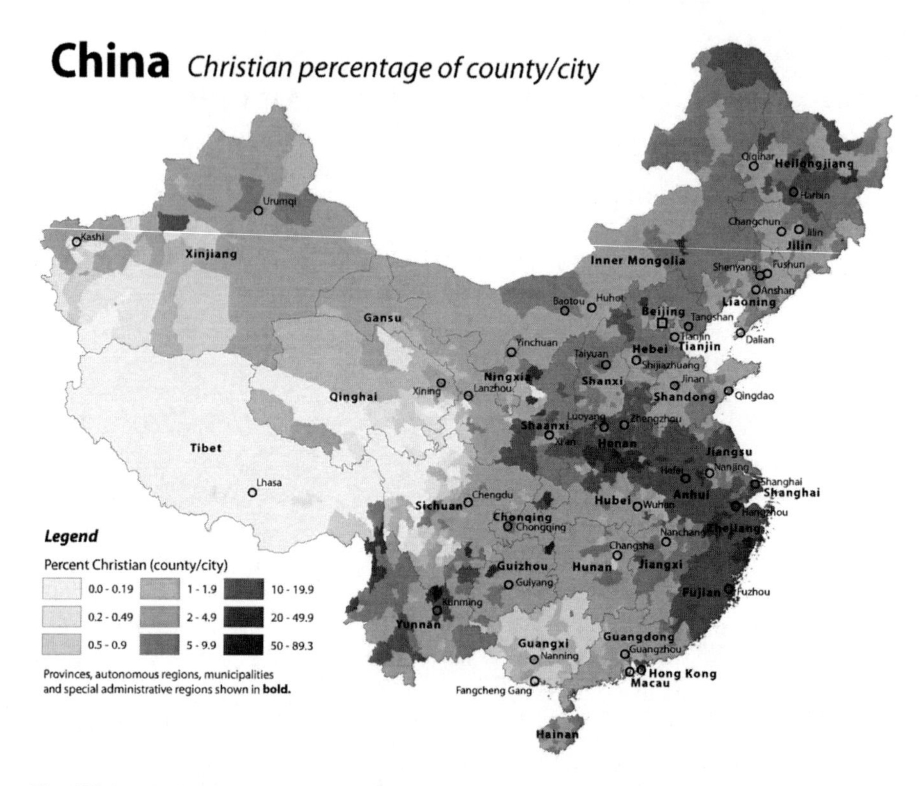

Fig. 109.1 Distribution of Christians in mainland China (Map by Global Mapping International. www.asiaharvest.org/pages/Christians%20in%20China/Map%20of%20China%27s%20Christians.pdf, reproduced with permission)

from their earthly miseries and self-identification as workers), but also as a threat to the populace's primary loyalty to the Party.

For both officially-sanctioned and unregistered churches, a critical challenge that results from the restrictive political context within which religious groups exist is leadership training. As will be discussed, this challenge is particularly daunting in rural areas. Some churches train their leaders in "guerilla seminaries" (Xin 2009: 91–96; Lambert 2006: 191–192; Aikman 2003: 120–125). Another solution used by both TSPM and house churches has been a radio-broadcast ministry complete with a radio seminary based in Hong Kong—the Far East Broadcasting Company's (hereafter FEBC) ministry to China and its Voice of Friendship Seminary (Fig. 109.2).

The FEBC is a "non-denominational, international Christian radio network that broadcasts the Good News in more than 130 languages from 128 transmitters located throughout the world" (www.febc.org/about/index.html 2011). The FEBC was founded by John Broger and Bob Bowman in 1945, and its first broadcast aired from Shanghai that same year. As mission work in China came to an end in 1948, FEBC moved to the Philippines, and its broadcasts to China resumed in 1949 (www.febc.org/about/history.html 2011). FEBC currently has three broadcast towers directed towards China—in the Philippines, in Korea, and in Saipan—enabling their

Fig. 109.2 Far East Broadcasting Company leaflet (Source: www.febc.org, reproduced with permission)

programming to be heard in every Chinese province (FEBC presentation 2011). In 1983, FEBC's Hong Kong office started the radio-broadcast Voice of Friendship Seminary to help meet the growing need for leadership training among rural churches in China (FEBC presentation 2011; Interviews with FEBC leaders 2011).

The efforts of the FEBC have been facilitated by the special political status of its Hong Kong office. As will be further discussed below, because Hong Kong is a

Special Autonomous Region of the People's Republic of China, Hong Kong-based organizations have enjoyed a rather unique ability to support mainland Chinese Protestant religious groups. The FEBC's activities also demonstrate the usefulness of online and distance learning in overcoming the challenges of China's restrictions on religious practice. Yet even though the FEBC has had some success in its work with mainland Chinese Protestants, the restraints on Protestant churches that derive from China's authoritarian, Chinese Communist Party-led political environment in China have continued to threaten the growth and health of Protestant religious groups.

109.1.1 Data Sources

In addition to relying on the increasingly rich literature about Chinese Protestantism, this paper is based on the contents of a random sample of 800 letters written by listeners in China to FEBC during 2002–2005. These letters represent a small portion of FEBC's 300,000 archived letters from listeners received between 1949 and 2005. During a July 2011 research visit to the Chinese University of Hong Kong, the co-author was given access to several thousand of the letters that have been scanned into pdf format.[1] She read through 800 of them, 200 each from the years 2002, 2003, 2004, and 2005 and sent from Guizhou, Jiangxi, Shandong, Liaoning, Anhui, Inner Mongolia, Shenyang, Shaanxi, Hunan, Jilin, Fujian, Jiangsu, Tianjin, Guangxi, Gansu, Hubei, Hainan, Hebei, Xinjiang, Heilongjiang, Sichuan, Guangdong, Henan, Yunnan, Shanghai, Shanxi, and Zhejiang. The authors also conducted interviews with students and leaders in FEBC's Hong Kong office in January and July 2011, took notes on statements made by Voice of Friendship Seminary students visiting Hong Kong at an FEBC event in July 2011, and conducted other interviews with former Chinese house church leaders, who currently reside in the US, during 2009–2010. Overall, these data demonstrate the ways in which religious organizations outside of mainland China have helped both TSPM and house churches in China address the need for leadership training, especially in rural areas.

---

[1] The letters are currently in the possession of the Chung-Chi Divinity School of the Chinese University of Hong Kong. The Divinity School is applying for grants to process the letters and to make them publically available to scholars. Because the letters have not yet been redacted, no copies could be made or taken away from the Divinity School. There are two Chinese language studies of these letters, copies of which were provided by Chung-Chi Divinity School for reference: a paper by Ying Fuk-Tsang entitled 1950–60 niandai neidi fuin guangbo tingzhong xinhan fenxi (Analysis of Letters from Listeners to Gospel Radio Broadcasts from the 1950s and 1960s) and presented December 2010 at a conference at Huadong Normal University in Shanghai, PRC; and a June 2010 Master's thesis by Wu Jianli of Southeast Asian Divinity School entitled Zhumu juanqi—cong liangyou diantai neidi tingzhong laixin (1959–1983) kan damen chongkai qianhou de zhongguo jiaohui (Lifting the Bamboo Curtain—Looking at China's churches before and after the door opened through the letters from listeners (1959–1983) to Voice of Friendship Broadcasting).

109.2 The History of Protestantism in China

An overview of the history of Protestantism in China is important in understanding today's situation. Protestantism is not new to China, but it was not a significant social phenomenon before the post-Mao era (1978–present). The earliest Christians to arrive in China were the Nestorians, who came in 635 CE during the Tang Dynasty (618–907). Nestorianism, or *Jingjiao*, was briefly popular among elite nobles, but it waned after an imperial purge of foreign religions in 845 CE. Nevertheless, a small group of Nestorian believers persisted until the beginning of the Ming Dynasty (1368–1644), when the new emperor nearly eradicated all foreign religions from China. The second group of Christians to enter China was comprised of Catholic Franciscans, who came as emissaries from the Pope and later from King Louis IX, during the Mongolian Yuan Dynasty (1271–1368). The Franciscans were followed by the Jesuits under Mateo Ricci, who in 1601 was invited to be an advisor to the Imperial Court of the Ming Dynasty's (1368–1644) Wanli emperor (Aikman 2003). The first Protestant missionary to China was Robert Morrison (1782–1834); he based himself in the Portuguese colony of Macau and did missionary work in the Guangzhou area beginning in 1807. Most Protestant missionaries did not come to China until the mid-nineteenth century, after the Qing Dynasty (1644–1911) granted them access as part of concessions to European nations in the aftermath of China's losses in the Opium Wars (1839–1842 and 1856–1860). Protestant missionaries from Europe and the United States labored for 150 years with seemingly negligible results. As noted above, at the time of the 1949 Communist Revolution, only a tiny portion of China's population espoused Protestantism (Xi 2010).

109.2.1 Chinese Christianity During the Maoist Era (1949–1976)

After Mao Zedong (1893–1976) and the Chinese Communist Party (CCP) gained control of China in 1949, Mao expelled all foreign missionaries. In 1954 the China Christian Three Self Patriotic Movement National Committee (TSPM) was founded to help the government oversee Protestant Christian churches (Xin 2009: 28). The term "Three Self" came from Mao's insistence that all religions in China be "self-governing, self-propagating, and self-supporting" to keep China free of foreign hegemonic influences, masquerading under the cloak of religion. Protestant churches that registered with the TSPM and accepted government control continued to operate legally until the Cultural Revolution (1966–1976), when all religions were banned. TSPM leaders and official TSPM seminaries taught liberal or modernist theology that emphasized Christianity as a religion of social good works.

Many Chinese Protestants refused to join the TSPM, for a number of reasons; they disagreed with its liberal theology, they believed in the separation of church and state, and they refused to affiliate with the atheistic CCP (Interviews with former house church leaders 2009). Prior to 1949, there were a small number of indigenous Chinese churches with well-established, small-group Bible studies led by lay believers (Bays 1996); these groups easily made the transition to worshipping illegally as "underground" churches (*dixia jiaohui*). Protestants from evangelical mission churches who did not agree with the liberal theology of the TSPM churches often joined the underground churches. These Protestants met secretly in groups as small as two or three individuals throughout the Maoist era; in rural areas, some churches even continued to practice their faith during the Cultural Revolution (Zhong and Chan 1993). The influence on contemporary Chinese Christianity of the Chinese theologians, who founded the pre-1949 indigenous churches, has continued to be significant (Xi 2010; Bays 1996). The various denominations founded by Western missionaries were eradicated during the Maoist era, and the Christianity that has survived has been principally that of the indigenous theologians (Lambert 2006: 59). Those theologians worked in an era of strong anti-imperialistic sentiment; hence, to contextualize Christianity to their region and era, they advocated leaving mission denominations and returning to the practices of the primitive church based on the fundamental teachings[2] of the Bible (Xi 2010; Bays 1996; Lambert 2006). As a result, most house churches have not affiliated with any one denomination. However, they generally have been fundamental and evangelical. Frequently, they also have promoted spiritual healing and other charismatic gifts.

109.3 Contemporary TSPM Churches

After Mao's death and the end of the Cultural Revolution in 1976, the CCP relaxed its restrictions on religion. Since the early 1980s, TSPM churches have been allowed to operate legally. However, although China's 1982 Constitution affirms the citizenry's freedom of religious belief, it protects only "normal" religious activities. The CCP Central Committee's "Document 19" (1982) grants legal existence to five religions (Buddhism, Daoism, Islam, Protestantism, and Catholicism), but only under government-affiliated "patriotic" associations. For Protestants, this association is still the Three-Self Patriotic Movement.[3] In order to register with the government, a religious group must submit a preliminary application, a document of

[2] Stump notes that two common strategies for minority religious groups attempting to escape hegemonic pressures are (1) a return to fundamentalism and (2) a reduction of scale (2008: 112, 279–280). Indigenous Chinese Christianity in the early twentieth century exhibited both these tendencies with its return to scriptural fundamentalism and its rejection of international denominations with their clerical hierarchies in favor of small, local congregations with lay ministers and universal service. These two factors have continued to characterize China's house churches today, and have enabled them to more easily negotiate their nebulous legal status under Communist rule.

[3] While the TSPM oversees relations between the government and Protestant religious groups, the official Chinese Christian Association (Zhongguo jidujiao xiehui) is charged with managing relations among Protestant religious groups.

approval from the TSPM, records of assets and proof of the right to its meeting place, a membership list, and a constitution (Homer 2010: 57). In addition, the group must agree to the "three fixes" (*san ding*): a fixed meeting place, leader, and area of coverage. The group also must pledge to eschew the inclusion of individuals below the age of 18.

TSPM churches receive government funding, and their clergy are trained in 23 official seminaries nationwide. There is the main national seminary: Nanjing Union Theological Seminary; five regional seminaries, including the East China Seminary in Shanghai, and the Guangdong Union Theological Seminary; and other smaller seminaries scattered around the country (www.bibleinchina.org/index.html 2012). The theology taught in most government-approved seminaries has continued to be liberal and progressive (Lambert 2006: 45–46; Aikman 2003: 152). Promoting Protestantism has not been a priority for the government, and the seminaries have not produced sufficient clergy to adequately serve all of the TSPM church members. One radio-broadcast seminary student from Shenzhen described the need for leadership in her TSPM church: although the church has more than 30,000 members meeting in five congregations, it has only five official pastors. The church holds prayer meetings, Bible studies, and other activities every night of the week (except Monday), with hundreds of attendees in each meeting. Because the church is dynamic and growing, the lead pastor encourages members to obtain certification as deacons from the FEBC so that they can help care for the needs of the congregation (FEBC student statements 2011).

109.4 Contemporary House Churches

The underground churches that had continued even during the Cultural Revolution experienced a great expansion with the relaxing of social controls in the post-Mao era. Former church members who had ceased their practice during the Maoist era became active again, and new members joined. They found solace in the Protestant message after becoming disillusioned with Communism during the excesses of the Cultural Revolution (Zhong and Chan 1993: 252; Lambert 2006: 90; Aikman 2003: 77–78). The groups continued meeting in homes, and most refused to register with the TSPM. Many of these groups grew large enough to move out of homes. Some churches acquired large halls that seat hundreds of people. Others would hold worship services in businesses after hours (Interview with Chinese house church leader 2009). Because they do not register with TSPM, these groups exist in a gray area of the law, as Document 19 does not provide clear instructions regarding how these groups should be treated. As long as they avoid affiliation with foreign churches and worldwide denominations, house churches have been able to practice their faith reasonably freely. Most Chinese now refer to these unregistered churches with the value-neutral term of "house churches" (*jiating jiaohui*), even though many such churches no longer worship in private homes. The earlier pejorative term of "underground churches" is no longer widely used (Fig. 109.3).

Fig. 109.3 House church gathering, mainland China (Source: China Aid, used with permission)

Although Protestant house churches generally do not affiliate with any one denomination, they are almost universally fundamental in their theology. They emphasize spiritual gifts such as faith healing, even if they do not practice glossolalia. They emphasize spiritual experience in the believers' daily life as evidenced by holiness in their living. They also encourage participation and active service from all the believers (Zhong and Chan 1993: 253–257; Bays 2003: 493–494; Xin 2009: 84–85). Because these churches do not receive any resources from the state, they have a greater need for leadership training than is the case even for the under-staffed TSPM churches.

A November 2002 letter to FEBC describes a house church that started in 1999 with "no building, no preacher, and no books." The 12 founding believers went out regularly to the 500 villages in their county, and within 3 years they had 114 members. They built their chapel by themselves on some land near the railroad. The letter notes that they have a church but no preachers, as only ten of the members are high school graduates and only half are junior high graduates. Despite the lack of leadership, their numbers continue to grow, and their new church building is already too small (FEBC letter, 14 November 2002).

109.4.1 Differences Between Urban and Rural Churches

As is true for Chinese society as a whole, there is a marked difference between urban and rural Protestant churches. People in rural areas are poorer and less literate than Chinese urbanites. Rural areas of China tend to have fewer government services and fewer officials. Although standards of living are rising all over China, there are still significant rural areas in which the people live at the level of bare subsistence, especially in China's central and western regions.

109.4.2 Urban and Rural TSPM Churches

Among TSPM churches, urban churches tend to be better-funded and better-staffed than rural churches.[4] Unless there is foreign investment in a rural county, the TSPM usually establishes only one church per county. Thus, it frequently is difficult for rural residents to travel to services at their designated TSPM church (Interviews with FEBC leaders 2011). One letter to the FEBC's Hong Kong office described a rural TSPM church of 120 members. Most were old and infirm, and only half could make it to the Sunday services. Three women decided to serve the house-bound members with a series of home meetings, which consisted of an hour of hymn-singing and an hour of Bible reading and exposition. One woman led the singing, another read the Bible, and the third—the letter writer—was responsible for expounding the Bible. The woman was not trained in scriptural exegesis, and in the end, she resorted to repeating what she had heard on the latest FEBC radio broadcast (FEBC letter, 18 November 2002). The women's TSPM pastor did not attend these home services because the TSPM's "three fixes" principle prohibits pastors from engaging in church work outside their church buildings.

The limitations on TSPM pastors tie them to one building in the county seat and make it difficult for them to minister to the practical needs of their congregations. This situation seems to be engendering a new kind of hybrid TSPM/house church in rural areas. Many letters to FEBC from listeners in rural areas mention the "big church pastor" far away and the "house church pastor" near the writer's home, both of whom are in communication (FEBC letters, various dates).[5] House church leaders in villages outside the county seat can be deputized by the county's TSPM pastor as lay ministers, and they can continue their church activities under the umbrella of the TSPM. Some provinces, such as Gansu and Inner Mongolia, require all such lay ministers to receive FEBC certification before they can legally operate under the auspices of the TSPM churches (Interviews with FEBC leaders 2011; FEBC student statements 2011). Thus, the lines between TSPM and house churches may be blurring in some rural areas, and FEBC's Voice of Friendship Seminary seems to be a factor in this change (Fig. 109.4).

---

[4] Several letters in the random sampling perused by the co-author were written by rural TSPM pastors asking for donations from Hong Kong to build church facilities or to purchase books and other church supplies.

[5] There were more than five letters using terminology similar to these. The letters came from different places and were written throughout the various years reviewed by the co-author.

Fig. 109.4 Many urban house churches are simply rooms in people's homes, as in the apartment complex pictured here (Photo by Teresa Zimmerman-Liu and Teresa Wright)

109.4.3 Urban House Churches

House churches in urban areas tend to be smaller and more discreet than those in rural areas. Because China's urban areas are densely populated, house churches in the cities risk getting caught if they grow too large or too noisy, as their neighbors may report them to the police (Interviews with Chinese house church leaders 2009). But since many urban Protestants are wealthy, they typically sound-proof their homes or businesses before holding church functions. Some of them also cultivate good relationships (*guanxi*) with police officials or neighborhood administrators. Wealthy Protestant businessmen in Wenzhou, an eastern coastal city, have so much influence with local officials that unregistered Protestant churches can operate freely and openly there (earning Wenzhou the nickname of "China's Jerusalem") (Aikman 2003: 179–190).

Despite their urban advantages, house churches in China's cities also face a great need for trained leaders. One Voice of Friendship seminary student from the city of Nanjing reported that because she is well-educated, her house church requested that she begin seminary training just 3 months after her conversion to Protestantism. She began preaching on Sundays long before she finished her course of study, and she received her graduation certificate from the seminary well before the third anniversary of her baptism as a Christian. FEBC leaders indicated that this student's story is a common one among their students (FEBC presentation 2011).

109.4.4 Rural House Churches

In contrast to the cities, some rural Protestant house churches operate freely and openly because there are not enough officials in the region to regulate them. In other rural areas, Christians in unregistered Protestant church groups are targeted for harassment by officials, and when they are repressed, typically are treated more violently than is the case in urban areas. For example, rural house church members, and even their lawyers, are routinely beaten by the police when they are taken in for questioning. Rural police often levy arbitrary fines against house church members and are more apt than urban officers to send house church members directly to labor reeducation without a trial or hearing (Interviews with Chinese house church leaders 2009). As in the rural TSPM churches, members of rural house churches tend to be poorer and less well-educated.[6] Since the beginning of the reform era, church numbers have swelled with new members who have no background in Protestantism and who are not sufficiently educated to study for themselves (Bays 2003: 495). This has created a crisis of leadership in rural churches.

Letters from FEBC listeners attest to the great need for leadership in rural Chinese house churches. One letter laments, "if our pastors were half as good as you Hong Kong preachers, our church would be so much better" (FEBC letter, 26 November 2002). Another listener asks in a letter for outlines from the radio programs because his church is declining and he wants to help; his only source of information about the Bible is the radio broadcasts (FEBC letter, 26 November 2002). A third listener says that the only two preachers in her church have left the countryside to make money in the city, reducing the group's church services to merely singing hymns and reading the Bible, upon which none of the remaining members can expound (FEBC letter, 4 February 2003). In one letter, a church member asked the FEBC to help the leaders of her church because none of them have much education (FEBC letter, 13 February 2004). These are but a few of the many letters to FEBC showing the desperate need for trained leaders in rural house churches. These letters also show how valuable outside resources and distance learning possibilities are perceived to be by mainland Chinese Protestants.

109.5 Leadership Training

109.5.1 "Guerilla Seminaries"

Several of the larger house church networks hold their own training programs in secret. In urban areas, the trainees are frequently cloistered in one building for weeks or months while they attend intensive classes in Bible study and church service. In rural areas, church networks hold seminary trainings in remote,

[6] An FEBC letter received 13 February 2004 described a rural church in which most people had a sixth grade education or less.

mountainous areas. The curriculum is similarly intensive, and the students are restricted to a small area in order to escape detection (Xin 2009: 92–23; Aikman 2003: 120–121). Students in these programs call them "guerilla seminaries" because in the event of police raids, the staff and students scatter into the mountains and meet up again at a previously agreed upon spot to resume their studies in a different place. One former student of a "guerilla seminary" in Hubei Province reported disruptions every other week during her 18-week course of study, one of which forced students to hide without food or water in the hills for almost a week (Interview with Chinese house church leader 2010).

109.5.2 Far East Broadcasting Company's Voice of Friendship Seminary

Given the dangers inherent in the "guerilla seminaries," FEBC's radio ministry provides a viable alternative for both TSPM and house church Christians who want more in-depth training in Bible study and church service. FEBC's Hong Kong office produces all of its Chinese programming. Hong Kong's location and its unique relationship with China give Christians in Hong Kong a way to discover and understand the needs of Chinese Protestants and Chinese society as a whole. From the mid-1800s through 1997, Hong Kong island and the mainland territory roughly 17 mi (28 km) to the island's north were a British colony. Due to its extremely close proximity to the People's Republic of China (PRC), Hong Kong received news from China's interior even during the Maoist period (1949–1976). In 1997, Hong Kong reverted to control by China, under the leadership of the CCP. Since then, Hong Kong has been a Special Administrative Region of the PRC. Because they are located in a special region, Hong Kong's churches and ministries can accept money and aid from foreign church groups, unlike churches within China proper. At the same time, because Hong Kong is part of the PRC, Hong Kong's Protestant organizations can make donations to PRC church groups without violating the TSPM principle that religious bodies must be "self-governing, self-propagating, and self-supporting." Thus, FEBC's Hong Kong office can legally use donations from abroad to administer and produce programs that help Christians in mainland China.

FEBC's online and distance learning resources further work to benefit mainland Chinese Protestant groups and members. As noted above, the FEBC began its radio-broadcast Voice of Friendship Seminary in 1983 for the purpose of training church leaders in China's rural areas. The seminary's programming falls into three categories: training for church volunteers, basic training for church deacons (4 h per day/6 days a week), and advanced training for pastors (1.5 h per day/6 days per week) The program for church volunteers does not require registration or homework; FEBC does not know how many people listen to and receive training from this line of programming. The basic course requires registration, and students must turn in weekly homework. The Voice of Friendship Seminary mails a certificate and a Bible to each student who finishes the 2-year program (FEBC presentation 2011).[7] Students in the advanced course must have completed the basic course and must also prove that they are actively serving in a mainland Chinese church (FEBC presentation 2011). FEBC Hong Kong's executive director, Pastor Raymond Lo, states that the FEBC currently emphasizes the Voice of Friendship seminary as its best way to put qualified church workers on the ground in China to help with ministerial problems locally (Interview with FEBC leaders 2011).

The content of Voice of Friendship seminary courses and radio segments is written by pastors and seminary students from Chinese churches around the world and is coordinated by FEBC's Hong Kong staff. Because the FEBC is non-denominational, the theology of its programming is also non-denominational. It is generally fundamental and evangelical, adhering closely to the Bible (Interview with FEBC leaders 2011). FEBC staff does its best to communicate in terms simple enough for people with all kinds of educational and cultural backgrounds to understand (FEBC presentation 2011).

Since 2005, Voice of Friendship Seminary classes have been available as streaming audio online. FEBC leaders note that even the most remote Chinese villages have Internet cafes and these leaders feel that putting their courses online has increased accessibility (FEBC presentation 2011). One student stated that when he started the seminary courses, it was difficult for him to arrange his schedule around the radio broadcasts, especially because his area did not always have clear reception. His difficulties were so severe that he almost gave up on the program. After the courses went online, this student was able to finish his course of study more successfully because he could hear the courses and replay them as often as necessary until he understood the concepts (FEBC student statements 2011). The Voice of Friendship Seminary website has a secure section available to only students and faculty. Students can download their course modules, upload their homework, and participate in the course discussion boards with faculty and other students (FEBC presentation 2011). Adding the online option to its curriculum seems to have enhanced Voice of Friendship Seminary's ability to serve its students (Fig. 109.5).

The Voice of Friendship Seminary currently has 2,800 students enrolled in both the basic and advanced classes. The students come from 15 different provinces all across China, including Guangdong, Inner Mongolia, Gansu, Zhejiang, Anhui, and Shandong. In addition to the broadcast lessons and homework, FEBC arranges for groups of students to come to Hong Kong for visits to area churches so that they can see models of established church service. Some students travel as long as 48 h by train to attend such field trips. If students are unable to come to Hong Kong, FEBC sends groups of teachers to rural areas in China for training (FEBC presentation 2011).

[7] The certificates and Bibles often get lost in the mail and have to be sent out several times before they are delivered.

Fig. 109.5 Voice of Friendship Seminary students (Photo from www.febc.org, used with permission)

109.5.3 FEBC's Correspondence Counseling Ministry

In addition to its Voice of Friendship Seminary, FEBC's ministry includes daily gospel outreach and pastoral programming broadcasts to all areas of China. The gospel programs are designed to "introduce the Christian faith to non-believers through life testimonies and sermons related to daily life," while the pastoral programming encourages believers to be well-founded in the scriptures and to "build a good relationship with God" (www.febchk.org/News/806/leaflet1109.pdf 2012). A number of the pastoral programs are produced in a discussion format. Some of the general programs are available online through streaming audio. FEBC's Hong Kong office counts 13,500 hits to its website every day from people accessing the website to listen to its Christian programming (www.febchk.org/News/806/leaflet1109.pdf 2012). In addition to the programming in Mandarin Chinese, FEBC also broadcasts its general Christian programming in the seven minority languages of China (www.febc.org/content/Hong%20Kong 2012).

In response to the programming, listeners write in with questions about the Christian faith, prayer requests, or requests for pastoral care. FEBC responds to these letters from listeners, and sends out booklets with further information on Biblical principles about the areas of concern expressed in each letter. Pastor Lo notes that prior to 2006 every letter was answered personally. Since FEBC went online in 2003, the number of letters from listeners has increased exponentially to 6,000 or 7,000 letters every month; hence, according to Lo, FEBC has been forced to adopt a triage system, answering only letters with serious questions. Nevertheless, the counseling section at FEBC Hong Kong has increased from 2 to 16 or 17 employees, and those employees continue to work full-time sending out emails or letters of encouragement and advice with enclosed counseling booklets. FEBC

Hong Kong ministers have even entered China proper, where they have held 35 seminars counseling more than 5,000 attendees on matters related to marriage and family (www.febc.org/content/Hong%20Kong 2012). This increasingly large demand for the services of its counseling section contributes to FEBC's conviction that the Voice of Friendship Seminary is a necessary and viable method of training more lay ministers and clergy to meet the growing needs of China's Protestant churches (Interview with FEBC leaders 2011).

109.6 Conclusion

Protestant Christianity has seen rapid growth in China since the beginning of the post-Mao era. Most Christians in China attend unregistered house churches instead of government-affiliated TSPM churches. TSPM and house churches in urban areas tend to have more resources than those in rural areas, where the populace as a whole is poorer and less educated; nevertheless, urban and rural TSPM and house churches all report a shortage of qualified leaders.

The FEBC's Voice of Friendship Seminary works to meet the need for trained church leadership with its radio-broadcast and streaming online audio classes, which are accessible in all regions of China. The FEBC enjoys an advantage due to its status as a Hong-Kong-based ministry. Located in a Special Administrative Region of the PRC, Hong Kong groups such as the FEBC can access resources from Protestants around the world. These resources can then be utilized when ministering to Protestant groups in China's interior without violating the TSPM prohibition on foreign support for domestic religious groups. In addition, official TSPM churches recognize the Voice of Friendship Seminary's certification for deacons, preachers, and lay-ministers, all of whom can assist in understaffed TSPM congregations. House church leaders also benefit from the Voice of Friendship Seminary because, in addition to receiving needed training in scriptures and church service, once they are FEBC-certified, some provinces allow house church leaders to come under the umbrella of their county's TSPM pastor and to operate legally in a kind of TSPM/house church hybrid. Such a hybrid seems to be beneficial to all involved: the house churches become legitimate; the TSPM leaders have a way to harmonize with the house church Christians in their jurisdiction; and TSPM Christians living outside the county seat gain closer options for worshipping and obtaining church care.

In this way, the FEBC has been able to capitalize on Hong Kong's unique status, harnessing global resources and utilizing radio and online methods of communication to help meet the critical need for leadership in China's Protestant churches. Even so, trained leadership remains in shortage, especially in rural areas. And as long as China's ruling CCP continues both to control all religious groups and to condemn unregistered worshippers, this shortage is unlikely to be resolved—particularly if the number of Protestants in China continues to grow.

References

Aikman, D. (2003). *Jesus in Beijing: How Christianity is transforming China and changing the balance of power*. Washington, DC: Regnery Publishing.

Bays, D. (1996). The growth of independent Christianity in China, 1900–1937. In D. Bays (Ed.), *Christianity in China from the eighteenth century to the present* (pp. 307–316). Palo Alto: Stanford University Press.

Bays, D. (2003). Chinese Protestant Christianity today. *The China Quarterly, 174*, 488–504.

FEBC letters received. (Various years). *Archives*. Chung-Chi Divinity School, Chinese University of Hong Kong.

FEBC presentation. (2011, July 30). Hong Kong.

FEBC statement to co-author. (2011, July 30). Hong Kong.

FEBC student statements. (2011, July). Hong Kong.

Homer, L. B. (2010). Registration of Chinese Protestant house churches under China's 2005 regulation on religious affairs: Resolving the implementation impasse. *Journal of Church and State, 52*(1), 50–73.

Interview with Chinese house church leader. (2010, May). Southern California.

Interviews with Chinese house church leaders. (2009, October). Southern California.

Interviews with FEBC leaders. (2011). Hong Kong, January, July.

Lambert, T. (2006). *China's Christian millions*. Toronto: Monarch Books.

Madsen, R. (2010). The upsurge of religion in China. *Journal of Democracy, 21*(4), 58–71.

Stump, R. W. (2008). *The geography of religion: Faith, place, and space*. Lanham/Boulder/New York/Toronto/Plymouth: Rowman & Littlefield Publishers.

www.bibleinchina.org/index.html. (2012). Accessed 15 January.

www.febc.org/about/history.html. (2011). Accessed 15 August.

www.febc.org/about/index.html. (2011). Accessed 15 August.

www.febc.org/content/Hong%20Kong. (2012). Accessed 16 January.

www.febchk.org/News/806/leaflet1109.pdf. (2012). Accessed 16 January.

Xi, L. (2010). *Redeemed by fire: The rise of popular Christianity in modern China*. New Haven: Yale University Press.

Xin, Y. (2009). *Inside China's house church network: The word of life movement and its renewing dynamic*. Lexington: Emeth Press.

Zhong, M., & Chan, K. K. (1993). The apostolic church: A case study of a house church in China. In B. Leung & J. D. Young (Eds.), *Christianity in China: Foundations for dialogue* (pp. 250–265). Hong Kong: University of Hong Kong Centre of Asian Studies.

Chapter 110
Analysis of the Emergence of Missionary Territorial Strategies in a Mexican Urban Context

Renée de la Torre Castellanos and Cristina Gutiérrez Zúñiga

110.1 Introduction

Until a few years ago, the predominant approach used in studies of religious diversity in Mexico was ethnographic with case studies of particular congregations. Only a few collective projects conducted on the northern and southern borders of the country had proposed studying the transformation of the regional dimension, which is discussed by Casillas (1996). However, an interest in exploring this question in a more integrated way, making use of census data (which are available for a hundred years in Mexico, as a source of information on religious affiliation) combined with the technical possibility of spatial analysis provided by Geographic Information Systems (GIS), has opened up new ways of exploring the territorial dimensions of the religious change that we, along with other societies in Latin America, see taking place (INEGI 2005; De la Torre and Gutiérrez 2007; Hernández and Rivera 2009).

In this context we undertook to produce a religious cartography for Guadalajara, the second largest city of Mexico. From the data obtained on non-Catholic places of worship, we were able to establish that in 2007 there were within the Metropolitan Area of Guadalajara (MAG) 449 centers of worship belonging to religious denominations other than the Roman Catholic, that were Protestant, Evangelical and Pentecostal (Fig. 110.1).

R. de la Torre Castellanos (✉)
Centro de Investigaciones y Estudios Superiores en Antropologia Social-Occidente,
Av. España 1359, Col. Moderna, C.P. 44190 Guadalajara, Jalisco, Mexico
e-mail: renee@ciesas.edu.mx

C. Gutiérrez Zúñiga
Centro Universitario de Ciencias Sociales y Humanidades, El Colegio de Jalisco,
Calle 5 de Mayo 321, Zapopan Centro, C.P. 45150 Zapopan, Jalisco, Mexico
e-mail: mcgz@coljal.edu.mx

S.D. Brunn (ed.), *The Changing World Religion Map*,
DOI 10.1007/978-94-017-9376-6_110

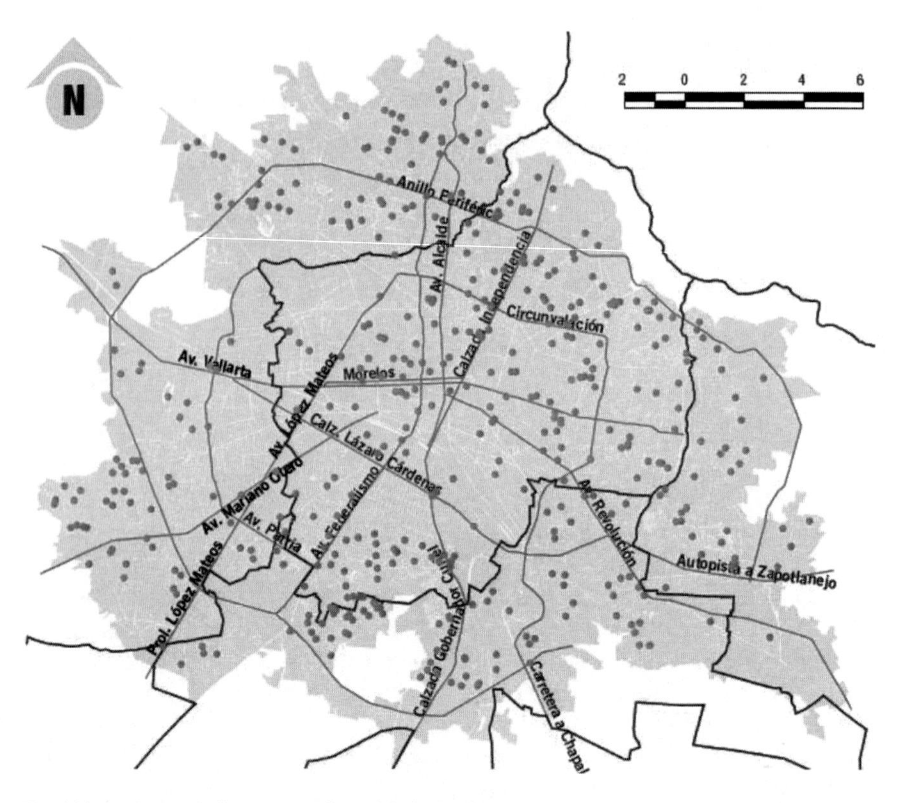

Fig. 110.1 Non-catholic centers of worship in the Conurbated Area of Guadalajara, 2007 (Map by C. Gutiérrez Zúñiga, R. De la Torre and C. Castro, 2011)

This map is important because the city of Guadalajara, which has 472 Catholic churches (of various size and capacities) is considered to be the center of Catholic operations in the center-west region of the country, a region characterized as the most resistant to religious change in Mexico (De la Torre and Gutiérrez Zúñiga 2007). In effect, according to data from INEGI, Catholic believers account for more than 90 % of the population. The continuity of Catholicism and its strong presence is explained by the fact that a large amount of Catholic infrastructure is concentrated in Guadalajara (seminaries, convents, private colleges, religious hospitals, church buildings and one of the most imposing sanctuaries to Catholic martyrs of the Cristera War (1926 and 1929) is being built here) (De la Torre 2006).

So in spite of the hegemony of Catholicism in Guadalajara, these figures show us that the religious landscape of the city is undergoing a transformation and that the city has become a metropolis with an increasing diversity of religious options. The same thing has happened in distinct neighborhoods of the city where we were able to detect some *colonias* that had up to ten different denominations apart from the Roman Catholics living in them. This not only has repercussions on transforming the religious landscape of the city, but also contributes to an internal reorganization of urban territory, which as recently as the early 1980s neighborhood life was still under the social/territorial control of the Catholic parishes (De la Peña and De la Torre 1994).

The parish is the basic cell of Catholic territorial organization. More than just a church building, it is the operational center from which an ecclesiastical territory is managed. The system of parishes was designed to organize the daily life of a territory inhabited by a religiously homogeneous population (everyone being Catholic) around the religious activities of the Catholic calendar, which accounted for nearly all of the social and cultural activities of the neighborhood.[1] Without complete religious homogeneity in the community, the parish system does not work (De la Torre 2002).

However, in the course of the twentieth century, the city of Guadalajara experienced an accelerated growth of its population and an expansion of the urban zone.[2] By the start of the twenty-first century, these elements had had an impact on the territorial dispersion of the city; affecting where the actors (churches and believers) reside, how they travel and identify with their urban territory.

With these changes in mind we conducted a study of residential patterns, attendance at religious services, and strategies of territorialization undertaken by non-Catholic denominations in the city of Guadalajara. Our main objective is to demonstrate that religious diversification is transforming the way in which the territory of societies with a Catholic monopoly is organized. First, we note that there has been a transformation of the meaning, materiality and temporality of the center of worship, that is, from buildings charged with identity and memory, whose architecture was fixed and permanent in time, to ever less permanent and more provisional constructions. We also find that new logics are being introduced to the sense of belonging to the organization of territory and to the identification of believers with those spaces.

From this perspective we were able to formulate the following questions

1. How do religious communities behave in relation to urban dispersion?
2. What are the patterns of residence and mobility for believers attending services?
3. Can we think of missionary strategies as territorial strategies and if so, what types of territorial strategy?
4. Do the various denominations apply in relation to these transformations of the urban space?
5. And impact or influence do they have on the sense of identity of the faithful or followers?

---

[1] For accounts of the historical development of the parochial order, see Palard 1999 and Courcy 1999.

[2] Guadalajara is the second most highly populated city in the country, exceeded only by Mexico City. According to census data, in 2005 the municipalities of the Guadalajara conurbation had a population of 3,728,465. These inhabitants accounted for 55.21 % of the population of the state of Jalisco (Cruz and Jiménez 2011: 164). The area covered was of 39,257 ha (approximately 152 sq mi).

110.2 Methodology

To answer these questions, we designed the survey "*Patrón de residencia y comportamiento territorial de centros de culto no católicos en el AMG.*"[3] The purpose was to interview those attending Saturday and Sunday services, in a random sample that would be representative of the principal Christian denominations in the city, that is, the places of worship listed in the directory "*Directorio de centros de culto no católicos en la ZMG*" or Directory of Non-Catholic Centers of Worship in the Metropolitan Zone of Guadalajara. The 1912 questionnaires were distributed anonymously after religious services, outside the centers of worship, at a sample of 29 centers of worship selected from the 449 places of worship previously identified in directory. The centers of worship chosen represented both the most common denominations in the city and that had contrasting architectural features noted at first sight, for example, those located in the city center, neighborhood churches, mega temples on the periphery, large sanctuaries, and low income neighborhood chapels.

110.3 General Tendencies of Patterns of Residence, Attendance at Services, Identification and Affiliation

Initially we were interested in knowing if the religious community, conceived of as one celebrating collectively the same rite in the same place at the same time, could or could not be considered: (1) a community identified with a territory and (2) a religious community whose congregation shared the same religious affiliation.

In regard to the territorial community, we decided our first priority was to describe the patterns of residence and mobility that applied to those attending the places of worship. A number of specific questions were asked: time taken to arrive, means of transport, place of residence, and some sense of the distance to the center of worship; information on the place of residence allowed us to obtain the physical distance in meters to the center of worship.[4] We were also interested in testing whether the centers of worship operated like centers of urban territory (as in neighborhoods) or whether they worked more like points of attraction to inhabitants whose residences were dispersed in the urban space and whose only community link, hence point of common identity would be attendance at the service (where distinct

[3]The questionnaires were applied individually, without compromising the anonymity of those interviewed, after religious services on the 19th, 25th and 26th of July, and the 2nd of August 2009, at out of the way places of worship. The sample consisted of a scheme of random subsamples with a margin of error of around 2.5 % and a degree of reliability of 95 %, in respect of the 449 places of worship detected previously (Gutiérrez Zúñiga et al. 2011) as a sample framework.

[4]In order to control the variable "distance," a record was made of whether those attending the religious service came from their homes. The distance could also be checked from the time taken to arrive, but this might depend on the means of transport employed.

and contrasting social or professional classes and neighborhood identities might or might not coincide).

The findings revealed that most of those interviewed (94.3 %) travel from their house to the center of worship and that 77 % use motorized transport. In regard to the time devoted by the faithful traveled from their homes to the place of worship, the average was 21 min using different means of transport. This fact shows that those interviewed usually travel outside their neighborhood to attend religious services.

In effect, calculating the approximate distance in meters from home to place of worship[5] confirmed that half of those interviewed (52.3 %) lived at a distance considered fairly great: between 4 and 6 km (2.5–6.7 mi) away. Nevertheless, it should be pointed out that there are qualitative differences between the various denominations that are so significant that we were led to produce a more detailed description, as we discuss below.

Finally, the subjective sense of distance was also taken into account, an element that proved to be fundamental in the community notion of territory. Though not really close by, half (50.1 %) of those interviewed said they "live near" the place of worship they attend, and 21.5 % said they "live very near." This subjective sense of territorial distance is what lets them maintain the idea of a community. We were also able to detect that the various religious communities that comprise the diversity do not necessarily share the same ideas or ways of constructing territoriality. This diversity is projected in the way in which the religious community is conceived (Gutiérrez Zúñiga et al. 2011). These tendencies lead us to agree with the proposal developed by Segato (2007) on the role of churches in the construction of territoriality:

> That persons take up territorial markers by themselves and their territories can stretch, growing in size as their respective loyalties expand. Gradually, a people (pueblo) would seem to come to be defined not by all the inhabitants of a geographically limited territory, but as a group carrying the symbol of a common loyalty which then marks out a territory for itself in the space that it occupies. Thought of in this way, a church is formed in a territory that consists of its congregation. (Segato 2007: 312)

One of the characteristics of contemporary cities is the ambivalence they show with respect to the territorial segmentation of urban identities. While cities originally provided a pattern of concentric relations, today they not only host diversity and inequality, but also generate a greater spatial mobility, which has an impact on the tendencies towards fragmentation and dilution of territorial community. This means that:

> On the one hand, the exterior is strenghtened in relation to the rest of the city, and on the other, it is diluted in relation to the global flows, which interconnect with other realities and from within each household. The proportions of interaction with otherness, which is what defines the utopian sociability of the city, are made smaller. (Aceves et al. 2004: 292)

---

[5] Distances were classified into eight ranges: (1) intra zone which includes those living in the same Basic Geostatistical Area (AGEB in Spanish, a territorial unit equivalent to the area covered on foot by pollsters in a day) as the center of worship; (2) local (–<1 km); (3) short distance (1–2 km); (4) fairly short distance (2–4 km); (5) fairly long distance (from 4 to 6 km); (6) long (6–10 km); (7) very long (>10 km) and (8) far away.

In this new way of inhabiting, travelling across and belonging to the city, the organic nature of the religious community is also transformed, though it continues to be an important point of reference in the urban group identity. As Giménez Beliveau (2004) warned, the ways in which the relationship between religious communities and their cities are presented have also been dispersed. She recognizes at least:

> Those who claim to belong to a community and those who identify with processes of the individual; those who recognize themselves in cultural identities with a strong historical density and those whose priority is emotional spaces where the here and now are central. (Giménez 2004 cited in Forni et al. 2008: 23)

This phenomenon of dispersion invalidates the premise of "neighborhood" implicit in the use of geographical instruments, such as buffers, which are used repeatedly to evaluate territories with different types of social behavior. The data become even more interesting when the distinctive socioeconomic composition of the community of those who attend the same religious services is observed and measured, especially in terms of the degrees of marginality of their places of abode and the location of their centers of worship. To determine these locations we qualified the location of places of worship and that of the places of residence of members of the congregations according to the five degrees of marginality established by CONAPO.[6] In general terms we may state that the non-Catholic places of worship are mostly located in areas of low and medium marginality, while those attending come from all areas, with notable differences according to their denomination. For example: in the center of worship of the Church of Jesus Christ of Latter-day Saints located in an area of very low marginality, 68.40 % of the congregation lives in area with the same degree of marginality, while of those attending a chapel of ICIRMAR, *la Iglesia Cristiana Interdenominacional de la República Mexicana* (Interdenominational Christian Church of the Mexican Republic) located in an area of medium marginality, only 38 % of the faithful reside in an area of the same quality. In general, we may say that in over half the cases there is no correspondence between the degree of marginality of the center of worship and the places of residence of those attending. The findings of this particular analysis indicated to us that it is not valid to infer the socioeconomic status of the congregation from the location of the centers of worship they attend.

---

[6] Margination is understood as "the lack of opportunities and of a minimum of conditions of well-being for the population inhabiting a particular territory, [which] is evidence for the material and human shortages that exclude wide sectors of the population from the considerable advances recorded in other areas" (Rubalcaba and Chavaría 1999: 73). El Consejo Nacional de Población, CONAPO, has established an Index of Margination using the principal relevant sociodemographic indicators which are obtained from the General Census of Population and Housing every 10 years. These indicators cover: illiteracy, level of schooling, quality of home (sewerage, electricity, piped water, type of floor, crowding) and the level of income. The results per territorial unit (in Spanish, Basic Geostatistical Area, or AGEB, see note 6, supra) have been layered into 5° of marginality: very high, high, medium, low and very low. Taken from www.conapo.gob.mx, "Índices de marginación 2000" Anexo C, Metodología del índice de marginación: p. 196.

In order to answer the question whether there is homogeneity in religious attendance and affiliation, we began with the application of a filter question to identify the regular and the occasional attendants of religious services at each of the places of worship included. 90.2 % said they attended the religious service at the center of worship in question on a regular basis. Only 6.6 % attended occasionally and only 2.3 % said they were there for the first or the second time. Also 0.8 % said they came to keep someone else company.

Although in general terms the data show us that non-Catholic Christians are consistent in fulfilling their commitment to attend on the Sabbath (Saturday or Sunday), there are marked differences between the denominational groups in the sample belong. Some congregations had a high proportion of regular attendants: at 23 out of 29 centers of worship those interviewed reported regular attendance in 90 % of the cases. In general terms most of those attending places of worship said they attended exclusively services held by their own churches (87.5 % of those interviewed do not attend religious services other than those of their own affiliation), but 12.2 % do attend services held by other denominations. Through another question we were able to confirm new tendencies with regards to attendance, viz., exclusivity and religious identity. In order to check the correspondence between those attending a particular center of worship with their affiliation to the religion running it, the following question was asked: are you a member of the religion that is practiced here? The answer was affirmative in 95 % of the cases, but with significant variations in some denominations. Once again the findings by denomination indicated contrasting patterns of exclusivity. And we distinguished emerging non-exclusive types of behavior, either because of the existence of interdenominational collaborations or because the congregations are developing a different kind of link to others attending the service, which is a mark neither of membership nor of exclusiveness and brings attendance at religious services into individual patterns of consumption from the multiplicity of religions offered.

110.3.1 Homogeneous Membership of a Religion with Nodal Territoriality

We define nodal territoriality as that which is devoted exclusively to a single religious institution, capable of organizing and giving meaning to the set of practices and relations that are conducted in that territory. This model works in territories inhabited by populations with a high degree of internal homogeneity, that is, sharing the same creed. It intensifies the concentration of resources (material and symbolic) and exercises power in nodal spaces that produce a functional interdependence in the interaction between central nuclei and peripheral areas (Palacios 1983).

This model has been exercised by various churches. A classic example is the parochial territories managed by the Catholic Church, but other denominations also have these parishes. Other territorial communities ruled by a single religious principle that would fit here include ashrams (for example, Hindu) or kibbutz (Israeli-Jewish).

In the case of Mexico, this type of territorial community has been established by *La Iglesia La Luz del Mundo* (The Church of the Light of the World, TLW),[7] with their "Hermosa Provincia" (Beautiful Province) and by *la Nueva Jerusalem* (The New Jerusalem), a syncretic Mexican religion that grew up in the state of Michoacán, to give just two examples.

We illustrate this ideal type of territorial-identity strategy with the example of "*La Luz del Mundo. Columna y Apoyo de la Verdad*" (The Light of the World, Pillar and Support of the Truth) (Fig. 110.2). The number of TLW centers of worship is 33. These include four "mega temples," one "large temple," 12 "medium temples," 11 "small temples" and 5 "missions." It may be seen from Fig. 110.2 that this church has built a territorial concentrations on the eastern side of the city where they have 18 centers of worship. These appear to be grouped around their international headquarters, in the *colonia* Hermosa Provincia. It is important to note that in recent years the church has started to venture into other urban zones, while keeping its preference for the periphery. In fact, Fig. 110.2 shows us a kind of "missionary route" starting in the east of the Conurbated Area of Guadalajara (CAG) and moving northwards. In the course of its history this church suffered from expressions of intransigence on the part of Catholics, so in 1958 it adopted the strategy of constructing a distant *colonia* exclusively for its own members, known as the Hermosa Provincia community. TLW has maintained a pattern of growth since the 1970s based on the construction of this type of urban community. Outside these territorial communities, the new evangelists would meet in private homes, almost clandestinely. Today, however, it is more common to find temples and houses of prayer belonging to different denominations in the same neighborhood, sharing the territory and sharing the same beliefs with other residents.

Since their central temple was built in the 1980s, the church has erected new temples whose architectural designs vary but basically follow the same distinctive style. These have also developed their own symbology; at first it was emblems of the Israelites like the Star of David, the candelabrum and the Lion and now Aaron's Rod. The outline of its own temple has become a new symbol, that is, like a kind of logotype. Further, all their places of worship can be identified, as they have their names and the denomination they belong to inscribed outside. Also, starting in the

---

[7]The church of TLW is the most important Mexican Pentecostal denomination. It was started by a teacher (ex-soldier) called Joaquín González in 1926, in the middle of the Cristera War, in the city of Guadalajara, where its international headquarters is currently based. The colonia Hermosa Provincia was established in the East of Guadalajara as an urban space concentrating the social, working and religious lives of the church's members, and it has served as a model for founding Hermosa Provincia communities elsewhere. In the middle of the colonia they built a giant temple (with seating for 12,000 people), which is said to be the largest temple in Latin America. Currently, the director of the church is Samuel Joaquín González, son of the founder, who has a strong charismatic hold over his followers. TLW is characterized by having built up a church with a strong nationalist feeling, which has led it to expand significantly in Mexico (with 188,326 members in this country, according to census data from INEGI 2010). Since the 1980s it has also expanded internationally (according to the church's own figures, it has a presence in 22 countries). See De la Torre 2000.

Fig. 110.2 Centers of worship of "La Luz del Mundo" in the Conurbated Area of Guadalajara, 2007 (Map by C. Gutiérrez Zúñiga, R. De la Torre and C. Castro, 2011)

1990s, temples have been built on the main arteries or roads with the heaviest traffic in order to guarantee maximum visibility.

It may be noted that the closer they are to the church's international headquarters, the better the material conditions of the temples as opposed to the situation in the more distant municipality of Tlaquepaque, a municipality where TLW has only three centers of worship and its missionary effort is only just starting. Here it can be seen that the buildings are houses adapted for the use of missions.[8] It is interesting to compare the missions with the central temple of the church and see how different they are (Fig. 110.3).

It is worth stressing that TLW is the only church with a temple in Guadaljara that presumes to be the largest in Latin America with a seating capacity of 12,000 persons and enough room for another 20,000 standing. The building of the temple was completed at the end of the 1990s and is located at the central roundabout of

[8] These missions are not registered by the institution. We understand the fact that church headquarters does not exercise specific control of all its meeting places as they are part of the church's strategy for growth in the city, that is, seeking to establish a presence in those places where there are believers already, but not as yet a meeting place nearby.

Fig. 110.3 House of prayer of La Luz del Mundo in Pedro Moreno Street (Photo by C. Gutiérrez Zúñiga, R. De la Torre and C. Castro, 2011)

the Hermosa Provincia *colonia*. The community of "brethren" who live there make their way daily and uninterruptedly to the religious services held in the temple. The height of the temple is 58 m (190 ft) and at the pinnacle it is crowned by an enormous golden statue representing the Rod of Aaron. At night the temple is illuminated with reflections of colored lights that highlight the contrasts between its platforms and the rising curve of the building. It has also introduced the use of laser beams to lighten the way of pilgrims to the grand temple, which has gained a high degree of visibility from different points in the city (Fig. 110.4).[9]

As this temple is the international headquarters of the church, it serves also as an international sanctuary (equivalent to Mecca or the Vatican for the faithful) which is visited on the 14th of August every year by more than 300,000 followers from different parts of Mexico and the world. They congregate for the celebration of the

[9]The construction of the temple started in 1983 and was finished in 1990. The height of the temple is 58 m (190 ft) and the central nave is 56 m (183 ft) long. The building also includes two subterranean mezzanines, one for holding religious services and the other is used as an office. The temple has become an emblem of the greatness of the church. On the pinnacle a bronze sculpture representing Aaron's Rod has been erected, and according to the architect who designed it, "it is revivified by God as a symbol of being born into a better life, and crowns the structure like a luminous way to heaven, an effect that is amplified by a laser beam that communicates with the city of Guadalajara" (Silva Orozco 1988: 89).

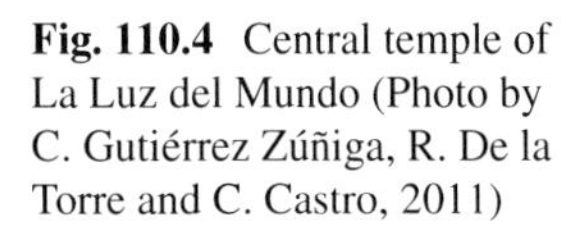

Fig. 110.4 Central temple of La Luz del Mundo (Photo by C. Gutiérrez Zúñiga, R. De la Torre and C. Castro, 2011)

Holy Supper. This "mega temple" is the only one of this size and with these features in Guadalajara. The other "mega temples" of this church located in Guadalajara have room for approximately 4,500; 3,000; and 2,000 persons respectively.

According to the Survey "*Patrón de residencia y comportamiento territorial de centros de culto no católicos en el AMG,*" those attending temples of TLW show two clear signs of a territorial community: they live very close to their center of worship, and identify themselves plainly as belonging to the religion of that temple: "*La Luz del Mundo*" (Fig. 110.5).

In effect, while the average distance for non-Catholics interviewed from their home to their place of worship, is 4.1 km (2.5 mi); those attending the two temples of TLW included in the sample reported a distance of 0.7 km (0.434 mi) (in the case of the central temple) and 1.2 km (0.74 mi) (to the House of Prayer). Between them nearly 80 % (78 % to be exact) live within one kilometer (0.6 mi) of the center of worship. Questioned as to which religion they belong to, 95 % and 100 % respectively they said they belonged "to the religion that is practiced in this place."

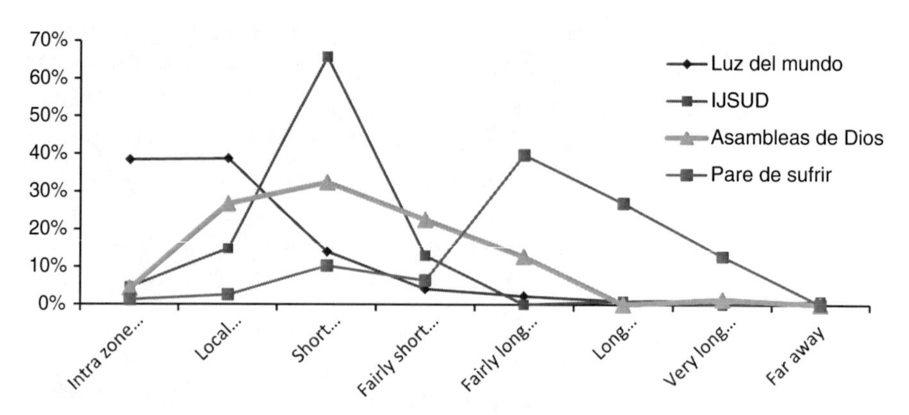

Fig. 110.5 Distance between places of residence of the congregation and their center of worship by denominational group (Source: C. Gutiérrez Zúñiga, R. De la Torre and C. Castro, 2011)

110.3.2 Heterogeneous Membership of a Religion with Territorial Hegemony

The modern age had an impact, from the nineteenth century onwards on the division into specialized spheres (or fields, using the terminology of Pierre Bourdieu 1983) that marked a competition between religion and the state for the management of territory. This was called secularization and it had a strong impact, unraveling the nodal and almost total control that the churches had exercised over urban territory, that is, homogeneous membership of a religion with nodal territoriality. On the other hand, the urbanization implied internal diversification, both in cultural and religious terms.

Today we face not so much processes of religious homogenization in a particular territory as processes of religious diversification which invite us to pay attention to the plurality of sources for producing and distributing knowledge, values, and meanings that come from different religious institutions which are characteristic of contemporary society.

Today, in most of the neighborhoods or *colonias* of the city, we have different religious options available and they compete with each other to gain adherents for their creed.[10] However, as noted by Lomnitz (1995: 41), the "urban-neighborhood" culture should indeed be seen as an internally differentiated cultural space, implying the existence of a common culture with specific categories of understanding. We consider that religious affiliation is currently one of the most important institutional ways of producing categories of understanding and differentiation within an urban space.

[10] Most of the surface of the Conurbated Area of Guadalajara might be characterized as "areas of mixed religious influence." See González et al. (2011).

The results from our survey allow us to appreciate that the believers move across the territory of the city to attend the religious community they belong. We, therefore, consider that some religious communities exercise an urban hegemony cuts across those belonging to urban neighbourhoods or street blocks. Using this model, we feel it is of the utmost importance to take up the proposal made by Lomnitz (1995) in his study of hegemony as a fundamental tool of analysis the study of religious identity as part of a regional culture in urban space. In the hegemony model referred to, one may observe "a certain unity of meanings within the cultural diversity" (Lomnitz 1995: 43). According to our own figures, some churches have characteristics that reflect a possible model for these new urban and religious situations. These are two of the most important para-Protestant churches with an international range that have grown the most in Latin America in recent times: the Jehovah's Witneses and the Church of Jesus Christ of Latter-day Saints, or Mormons. In effect, although the faithful do not necessarily share the same place of residence or neighborhood identity, their communities of believers do constitute spaces where the sense of belonging is identified and associated with boundaries of identity that are created. These denominations produce both the dynamics of belonging and of territorial hegemony in vast urban areas due to their competence in exercising a cultural system made up of cultural processes that are internally segmented but also hierarchically interrelated.

To exemplify an ideal type of this, we only look in detail at the Church of Jesus Christ of Latter-day Saints (CJCLS), which has 19 centers of worship in the Conurbated Area of Guadalajara (Fig. 110.6).[11]

This denomination has managed to gain a place in the city and might be considered one of the most solid in terms of physical infrastructure, which is easy to identify because of its architectural style. According to Fig. 110.6 it is evident that their strategy is not to build a large number of churches, but churches with a large capacity, located especially in a ring outside the old center of the city. Although the information provided to us by the church is incomplete, with the help of a photographic record of their centers of worship it was possible to calculate that four of these would be classified as "mega temples" (room for over 1,000 people), seven would be "large temples" (between 500 and 1,000 people) and eight would be "medium temples" (for 100–500 people). We do not find the church all over the city, but for the most part are in the northeast.

The CJCLS has its own idea of territorial-ecclesiastical organization. It is organized into *stakes,* with smaller territorial divisions called *barrios* (translated here

[11]The CJCLS is a para-Christian church that arose in the United States of America with the Great Awakening of that country in the first years of the nineteenth century, based on the appearance of numerous churches seeking to restore the primitive church of the Christian apostles. It was founded in 1830 by Joseph Smith, who published the Book of Mormon, on the pages of which he had written the revelations received by the leader himself from the angel Moroni. The first Mormon missionaries arrived in Guadalajara in 1935 but did not obtain positive results. They came back in 1950 and in 1960 began to have converts. Their principal missionary activity was preaching on the doorstep to householders who opened the door to them. See Fortuny and Ortíz 2005: 25–126.

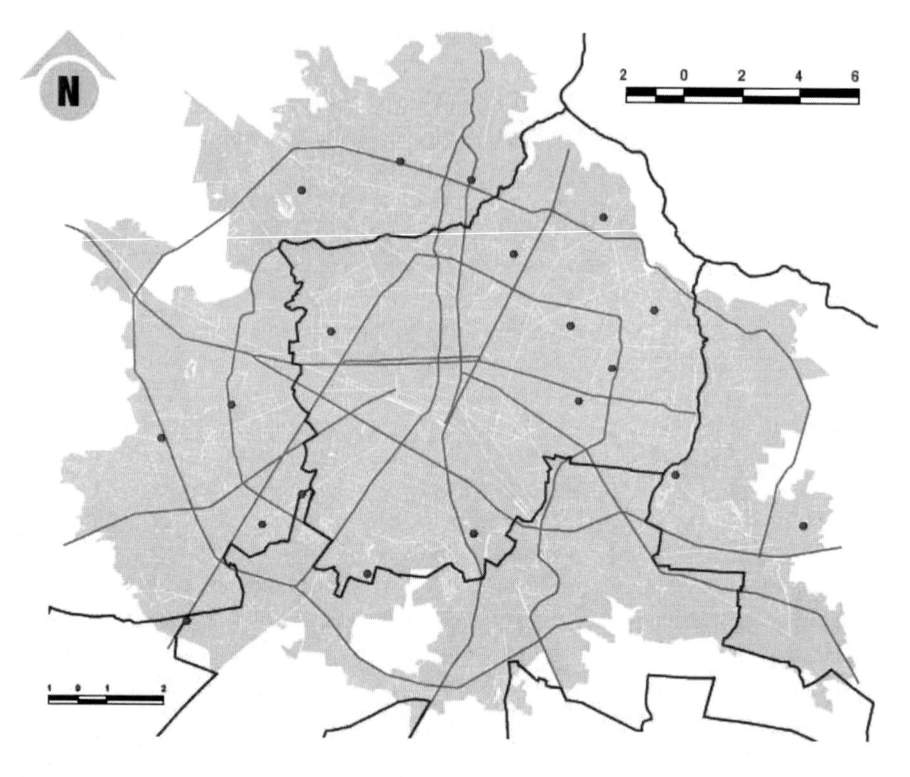

Fig. 110.6 Centers of worship of the "Church of Jesus Christ of Latter-day Saints" in the Conurbated Area of Guadalajara, 2007 (Map by C. Gutiérrez Zúñiga, R. De la Torre and C. Castro, 2011)

as neighborhoods).[12] In the case of the city of Guadalajara there are four stakes, directed by a president or a group of 12 priests, who run the neighborhoods and 70 supervisors who oversee the internal workings of these. Each stake has eight *barrios* under its jurisdiction. These work a bit like parishes and are presided over by a bishop (with approximately 330 members in each). These in turn are divided into branches, which link the small congregations together; they are called small because they have few members and do not have chapels to hold services in. Fortuny reports that according to the figures estimated by the church itself, the four stakes have over 10,000 members (2005: 126). The CJCLS has erected large buildings that place it, at least on the level of physical infrastructure, as significant competition for the regional hegemony of the Roman Catholic church. And, without doubt, it is the only church whose architectural style makes it possible to identify their temples wherever

[12] "A territory becomes a stake when it has a number of priests considered sufficient to sustain the church in all its spiritual, administrative and material aspects. It is the number of priests rather than the number of members that is considered, because it is they can occupy the posts of direction and conduct the ordinances" (Zalpa 2007: 281).

Fig. 110.7 CJCLS located on Castellanos y Tapia Street, number 9154, in the colonia Talpita, in Guadalajara (Photo by C. Gutiérrez Zúñiga, R. De la Torre and C. Castro, 2011)

they are in the world (a central tower and a double sloping roof). Keeping to a recognizable style, their temples are similar to worldwide brands like Sheraton or Marriot hotels. Their temples tend to be built on large plots, where various modules are built for different activities, and they are always surrounded by low railings that make it possible to see what is there (Fig. 110.7). Without a doubt, the monumental work of this church in Guadalajara is the marble temple they finished constructing in 2001. It has room for a congregation of 6,000 and is built on a platform of 10,700 m^2 (roughly 2.64 acres) (Fig. 110.8).[13] We can see from a map that the temples of this church are more widely dispersed over the urban zone than those in the previous case studied, that of *La Luz del Mundo*. The faithful also live farther away from their center of worship. The average distance to the two Mormon temples included in our sample for the Survey are 1.5 and 1.7 km (0.9 and 1.0 mi) (see Fig. 110.5). The self-identification of those attending services who were interviewed is clear: 98 % say they belong to the religion practiced in this place.

[13] According to the church's own information, 17 stakes and 9 districts form the new district temple and there are plans to build a new stake next to the temple. Elder Eran (1999), president of the northern area of Mexico, explains: "What a blessing it is that I was a mission president here just 29 years ago," he said. "We had four little branches here then. Now there are eight stakes. This is a great time to have a temple in Guadalajara," he told those assembled. "We have many faithful saints in Guadalajara."

Fig. 110.8 "Mega temple" located on Avenida Patria number 879, in the colonia Chapalita Occidente, in Zapopan (Photo by C. Gutiérrez Zúñiga, R. De la Torre and C. Castro, 2011)

110.3.3 Belonging Emotionally to a Community of Believers That Is Territorially Dispersed and Institutionally Diffuse

Our figures show that many of the churches we recognize as Evangelical or Pentecostal do not have a denominational identification with their own institution or territory. Rather, they identify themselves as an imagined community of Christians in general. However, most of these churches conduct highly emotional rituals and ceremonies, in which manifestations of receiving the gifts of the Holy Spirit are intensely lived, for example, glossolalia (commonly referred to as speaking in tongues), miracles of healing, expressions of exorcism, musical praises etc. Both ministers and members of the congregation relate themselves to religion through the dynamics of belonging that are unstable, multiple, and inclusive. They may move from one denomination to another without any trouble or attend ceremonies convened by other congregations. In many cases they even reject the idea of belonging to a religion and obeying institutional rules, emphasizing that they are Christians or Evangelicals, or Pentecostals, indistinctly from one another and also interchangeable. As was noted by Giménez Beliveau (2004: 23), one should not

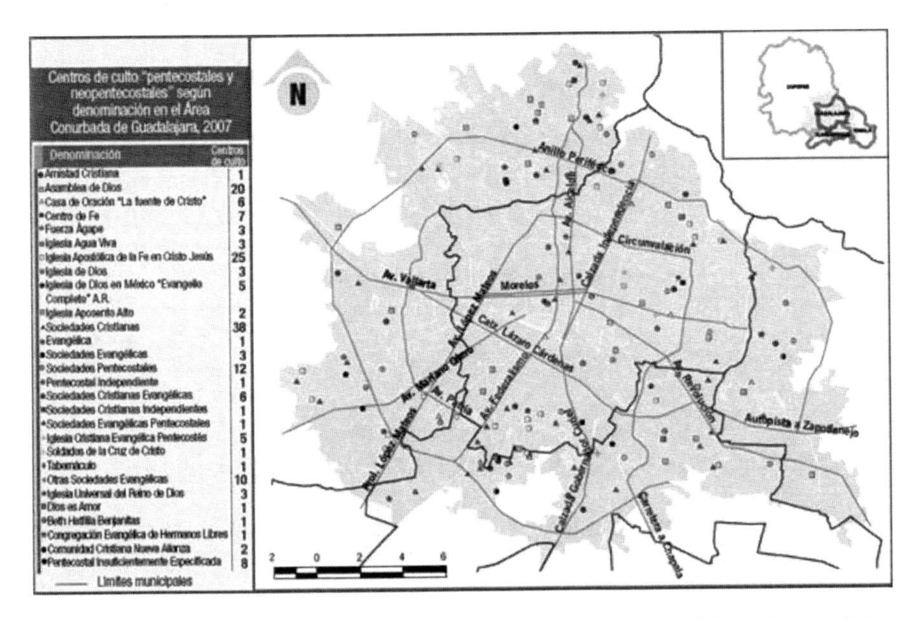

Fig. 110.9 Centers of worship of various "Pentecostal and NeoPentecostal" denominations in the Conurbated Area of Guadalajara, 2007 (Map by C. Gutiérrez Zúñiga, R. De la Torre and C. Castro, 2011)

forget that their identity is anchored in an emotional community and that "they prioritize emotional spaces where the here and the now are central."

These denominations represent another ideal type, one which provides a solution to the dilemma of religious pluralization and urban expansion and segmentation. As an example, we analyze the Pentecostal and NeoPentecostal churches as a whole,[14] and then approach the specific case of one of their most prominent and best known international exponents, the Assemblies of God (Fig. 110.9).

If we look at the distribution of Pentecostal places of worship shown in Fig. 110.9, we can say that in general terms they cover virtually the whole territory of the CAG, although they are not so concentrated in the semicircle of zones inhabited by the middle class and upper middle class sectors. We can also assert that they tend to be located towards the edges of the metropolitan area rather than concentrating in the historical city center. They even choose to establish themselves in the least well-off areas beyond the outer periphery, where 50 of them are to be found – corresponding to almost one-third of all Pentecostal meeting buildings. Within this denominational group, the Assemblies of God are a good example. Originally from Arkansas,

[14] The Pentecostal type of religion is based on the experience of the gifts of the Spirit. It has been identified as a clear tendency in Latin American Protestantism, which has achieved massive expansion both in urban and rural areas in the last decades of the twentieth century. See Martin 1990; Garma 2004 and Fortuny and Ortíz 2005.

in the United States, they first came to Mexico in the early decades of the twentieth century and formed themselves into a seedbed for local Pentecostal congregations all over the country.[15] In Guadalajara they have 20 places of worship, which range from two "large temples" and 14 "medium temples" to three "small temples" and one "mission" installed in people's houses. From an examination of the distribution of its temples, it is possible to state that this church is directed at working with the most diverse socioeconomic sectors. Considering that the Church is financed by its members, some of its missions are better positioned in socioeconomic terms than others.

Five meeting centers identified as "medium temples" in process of construction are half way to completion. As happens with other denominations, the buildings used for prayer tend on average to operate at 50 % of their capacity. There is not an architectural style that makes it possible to identify the denomination that the Assemblies' churches belong and although there is an emblem using the letters "AD" (*Asambleas de Dios*), it does not always appear on their sign boards. The churches located in the most low income *colonias* are the ones most generally found half built or just with foundations waiting for construction (Fig. 110.10). Those attending services in temples of the Assemblies of God show a residential pattern of living farther away than is the case of the two churches discussed above: the average distance between home and temple for the two places of worship chosen for our study of this denomination is 1.7 km (1.0 mi) in one case and 2.2 km (1.37 mi) in the other, while very few of those attending live locally (less than 1 km away); the average distribution is between 1 and 6 km (0.6–3.7 mi) (see Fig. 110.5). The congregation of the Assemblies of God provides a representative example of the particular features of identity proper to this ideal type. According to our Survey, those attending temples of the Assemblies of God do not identify themselves as belonging to this denomination in particular: only 7.9 % say they belong "to the religion that is practiced in this place," while 65 % identify themselves as Evangelical Christians, and 23.7 % just as Christians. A small percentage of 2.6 % say they are Evangelical Pentecostal Christians.

110.3.4 Multiple Religious Belonging with Territorial Dispersion

The religious plurality of contemporary cities is multiplying not only the distribution of the religious membership of the population, but also the individual believers' membership over the course of their lives and even the meaning of this membership.

[15] The Assemblies of God is a Christian denomination of a Pentecostal type that originated in the U.S. where it has its headquarters. It has expanded into over 186 countries and has about 35 million followers (Corten et al. 2003: 20). It is very important to the spread of Pentecostalism in Mexico because it has been the "stem cell" for subsequent versions of Pentecostalism in Mexico such as la Iglesia Apostólica de la Fe en Cristo Jesús y la Iglesia de Dios (the Apostolic Church of Faith in Jesus Christ and the Church of God). It was through the Assemblies that Mexicans living in the U.S.A. first heard the message of the Bible and experienced the manifestations of the Holy Spirit. Pentecostal missionary activity in Mexico started between 1905 and 1914, mainly choosing states on the northern border (De la Torre and Castro 2011: 135).

Fig. 110.10 "Eben Ezer" Church of the Assemblies of God in the colonia Mezquitera (Photo by C. Gutiérrez Zúñiga, R. De la Torre and C. Castro, 2011)

There is not necessarily a correspondence between belonging to a church and the practice and symbols that they identify with; frequently their membership is not exclusive. This new behavior was conceptualized by Lísias Nogueira (2008), who proposes that in the city of São Paulo, Brasil there is a new tendency towards pluralism and belonging to a multiplicity of religions.

At present, religious identities are formed from networks and affinities with creeds and religious practices that are defined in many cases more by social relations that are dispersed and are not contained within a clearly defined physical space.[16] To speak of religious diversity in Brazil, it is not enough to take note of the denominations and movements with a growing number of followers. One will also have to speak of the mixtures producing double identities, such as: Catholic-spiritualist, Catholic-afrobrazilian; Catholic-Protestants; orientalists-indigenist; esoteric-Afrobrazilian, and, as the author points out, it is even common to find combinations of more than two religions. This pattern of multiple belonging and promoting religious services does not imply an exclusive commitment on the part of those attending. This pattern is already present in the case of Guadalajara and

[16] For more on this insight, see the work of Renato Rosaldo "Ciudadanía cultural en San José, California" in Nestor García Canclini, et al., De lo local a lo global. México: Universidad Autónoma Metropolitana, 1995, 71.

la Iglesia Universal del Reino de Dios, the Universal Church of the Kingdom of God (UCKG). This church was born in Brazil in the 1970s,[17] and is promoted under the slogan "*Pare de sufrir*" (Stop Your Suffering). As in other parts of the world, the strategy of the church has been to adapt old cinemas and refurbish them for their religious services.

The UCKG has been very controversial all over the world. It is known to have been engaged in pastoral work since the end of the 1990s. The Brazilian pastors in Mexico were expelled in 1996 as their papers were not properly brought up to date, but in spite of the fact that they had no permission and were not legally registered, they have continued their evangelizing work in the country. In the specific case of Guadalajara, their presence was first registered in 1997 (Garma 2004: 142).

This church presents an outstanding example of two of the most important tendencies of what is called "NeoPentecostalism," viz., the intensive use of marketing methods applied to expand membership of the church, known as "Church Growth" techniques ("*iglecrecimiento*" in Spanish) and the appearance of an ethical orientation denominated "theology of prosperity" which legitimizes the pursuit of economic well being as a good favored by God and generosity in donations as a means of obtaining greater goods and blessings.[18] This tendency has not only traversed several churches, but has also been characterized as having stimulated interdenominationalism.

In line with these elements, our Survey shows that, of those attending services of "*Pare de sufrir, Santuario de la Fe*" (Stop Your Suffering, Sanctuary of Faith), only a one-third identify themselves with this religion (33.3 %), while the majority presented themselves as Catholics (39 %) or Without Religion (12 %), and only 14.1 % saw themselves as "Christians." We consider that this church is developing an alternative modality in which those who attend feel invited by the media promotions to take part in the services, without, however, having to attach their commitment or their identity exclusively to the host institution. We can observe, as the researchers in Brazil have sketched out, that this denomination works under a double model (a) that of a religion with a certain degree of rigidity in its hierarchy and internal organization, recruiting a relatively permanent congregation which undergoes a

[17] The Universal Church of the Kingdom of God (UCKG) started in Brazil and developed thanks to the use of the mass media for its missionary work (especially programs on television). It has managed to establish itself as a multinational church with a presence in 80 countries and has recruited two million adherents (Corten et al. 2003: 13). It is commonly known as "Pare de sufrir," as its temples are announced with this successful phrase. It is also recognized as the principal promoter of the theology of prosperity. It came to Mexico in 1993, and 2 years later already had 11 temples, and 10 years later, 24 (Doran 2003: 87–88). This church has worked successfully at restoring big cinemas, where activities are held every day of the week, specializing in daily prayers to solve current problems, such as financial problems, incurable diseases, family trouble, exorcism and liberation, love therapy, and prosperity, etc.

[18] Whether the term "Neopentecostalism" refers to a concept clearly distinguishable from Pentecostalism is debated by researchers into the area. Where such movements have probably been studied the most to date is in Brazil, especially because of the importance acquired by the Universal Church of the Kingdom of God which started in that country. See Pereira (2006) and for the Mexican case, Jaimes (2007).

process of conversion that detaches itself from its previous religion, but at the same time, (b) having a model of occasional participation, open to the public in general, where the meetings are experienced as a kind of spiritual consumption. The large number of participants who are neither undergoing conversion, nor converted, and do not identify themselves as Christians, shows us a kind of identification similar to that of consumers with the products they acquire, viz., one that does not imply that they must become exclusively or totally identified with the product (cf. Nogueira 2008).

We can find three places of worship belonging to this denomination in Fig. 110.9 each aligned equal distances apart on an east-west axis. But the whole of the congregation is not limited to local residents. In line with their own model of a non-exclusive type of "spiritual consumption" identity, we were not surprised to find the average distance between home and temple in this case was the longest of all churches discussed above: 6.3 km (3.9 mi) (see Fig. 110.5).

110.4 Final Reflections

We can agree with Chris Park (2004) that religion is a cultural phenomenon with an impact on the organization of space, because religions are forms of social and institutional organization generating different patterns for gaining territory. But it is also true that the contemporary characteristics of urban space have an impact on the new relation between believers and churches. The findings of our study show us an increasing complexity in the relations between denominations, believers, and territory.

Summing up what we have learned about this complexity in Guadalajara would suggest that:

- The socioeconomic level of a congregation cannot be deduced from the socio-economic characterization of their place of worship.
- The socioeconomic level of the congregation affects their perception of proximity to the place of worship insofar as it determines the means of transport used to get there.
- The churches give different weights to residential proximity in their various models of building a community and forming a religious identity.
- The churches create different models and strategies of territoriality.
- They combine their models of territoriality with different modalities of belonging: the closer to the temple, the greater the identification and exclusive sense of belonging; the farther away, the more multiplicity.

We have come to appreciate that there is a tendency to recreate territorial religious communities which in turn strengthen exclusive membership of a single church. But we have also seen new organizational modalities that tend to create inclusive and generic indentities, no longer of just belonging, but making reference to an imagined community, one that proceeds in a dynamic way in segmented spaces, making the religious services a reference point for the identity of those attending them.

To a large extent, this complexity is due as much to the characteristics of accelerated mobility and the dispersion of a population across urban space as it is to the diversification of religious options that provide differentiated patterns of establishing ways of belonging or of participating, for believers. It is important to stress that one of the principal discoveries of this study was to realize that there is no one way for this relationship to be established. What we find is, rather, a set of modalities, that maintain a correspondence with the strategies of the various religious organizations.

One will need to highlight the fact that our study shows there are signs of a growing tendency towards urban territorial segmentation and the deterritorialization of identities. However, the religious institutions are seen as being capable of facing these tendencies with differentiated strategies, for example, *La Luz del Mundo* promotes religious communities of a medieval type while the new Neopentecostal churches like the UCKG promote activities that allow a new type of identity, not circumscribed in space, nor to exclusive membership, to be adopted by individuals as they come and go.

Acknowledgements Our thanks to el Instituto de Mercadotecnia y Opinión for the design of the sample and the application of the questionnaires. In particular to Yasodhara Silva for technical advice in designing the card, and her professionalism in conducting the field work and processing the information. We would also thank Nicholas Barrett for the English translation of this paper, originally written in Spanish.

References

Aceves, J., De la Torre, R., & Safa, P. (2004). Fragmentos urbanos de una misma ciudad: Guadalajara. *Espiral. Estudios sobre estado y Sociedad (X Aniversario)*. Guadalajara: Universidad de Guadalajara, *31*(11), 277–322.

Bourdieu, P. (1983). *Campo del poder y campo intelectual*. Buenos Aires: Folios Ediciones.

Casillas, R. (1996). La pluralidad religiosa en México. In G. Giménez (Ed.), *Identidades religiosas y sociales en México* (pp. 67–102). Mérida: IFAL-IIS-UNAM.

Consejo Nacional de Población. (2000). Índices de marginación, 2000. Anexo C, Metodología del índice de marginación. Retrieved April 18, 2012, from www.conapo.gob.mx

Corten, A., Dozon, J., & Oro, A. (2003). *Les nouveaux conquérants de la foi. L'Église universelle du royaume de Dieu (Brésil)*. Paris: Éditions Karthala.

Courcy, R. (1999). La paroisse et la modernité. Lieu fondatour et arguments actualisés. *Archives de Sciences Sociales de Religions*. Paris: Centre National de la Recherche Scientifique. *44*(107), 21–40.

Cruz, H., & Jiménez, E. (2011). La distribución de los lugares de culto no católico en el Área Conurbada de Guadalajara desde una perspectiva territorial. In C. Gutiérrez Zúñiga, R. De la Torre, & C. Castro (Eds.), *Una ciudad donde habitan muchos dioses. Cartografía religiosa de Guadalajara* (pp. 159–184). Mexico: El Colegio de Jalisco/CIESAS.

De la Peña, G., & De la Torre, R. (1994). Identidades urbanas: Guadalajara al final del milenio. In *Ciudades* (pp. 24–32). Mexico: RNIU.

De la Torre, R. (2000). *Los hijos de la Luz. Discurso, identidad y poder en La Luz del Mundo*. Guadalajara: Universidad de Guadalajara/CIESAS/ITESO.

De la Torre, R. (2006). *La Ecclesia Nostra, el catolicismo desde la perspectiva de los laicos: el caso de Guadalajara*. Mexico: FCE/CIESAS.

De la Torre, R., & Castro, C. (2011). El mapa de la diversidad religiosa no católica en el Área Conurbada de Guadalajara en la década de 2000. In C. Gutiérrez Zúñiga, R. De la Torre, & C. Castro (Eds.), *Una ciudad donde habitan muchos dioses. Cartografía religiosa de Guadalajara* (pp. 159–184). Mexico: El Colegio de Jalisco/CIESAS.

De la Torre, R., & Gutiérrez Zúñiga, C. (Eds.). (2007). *Atlas de la diversidad religiosa en México*. Mexico: CIESAS/UQRO/COLEF/ELCOLJAL/COLMICH/SEGOB.

Doran, M.-C. (2003). A Igreja Universal no México. Em Oro, Ari Pedro, Corten, André e Dozon, Jean-pierre (orgs.) Igreja Universal do Reino de Deus: os novos conquistadores da Fé. Sao Paulo, Paulinas, pp. 93–100.

Eran, E. (1999). *Church News*. Retrieved June 26, 2011, from http://es.mormonwiki.com/Templo_Guadalajara_M%C3%A9xic

Fortuny, P., & Ortiz, R. (2005). *Los 'otros hermanos'. Minorías religiosas protestantes en Jalisco*. Guadalajara: Secretaría de Cultura.

Garma, C. (2004). *Buscando el espíritu. Pentecostalismo en Ixtapalapa y la ciudad de México*. Mexico: Universidad Autónoma Metropolita/Plaza y Valdés Editores.

Giménez, B. V. (2004) cited In F. Forni, M. Fortunato, & L. Cárdenas (Eds.). (2008). *Guía de la diversidad religiosa de Buenos Aires* (p. 23). Buenos Aires: Editorial Biblos.

González, E., Gutiérrez Zúñiga, C., & De la Torre, R. (2011). Análisis comparativo de la distribución de los templos católicos y los lugares de culto no católicos en el Área Conurbada de Guadalajara. In C. Gutiérrez Zúñiga, R. De la Torre, & C. Castro (Eds.), *Una ciudad donde habitan muchos dioses. Cartografía religiosa de Guadalajara* (pp. 205–237). Mexico: El Colegio de Jalisco/CIESAS.

Gutiérrez Zúñiga, C., de la Torre, R., Castro, C., Cruz, H., Gonzalez, E., & Jimenez, E. (2011). *Una ciudad donde habitan muchos dioses. Cartografia religiosa de Guadalajara*. Mexico City: Centro de Investigaciones y Estudios Superiores en Antropologia Social.

Hernández, A., & Rivera, C. (Eds.). (2009). *Regiones y religiones en México. Estudios de la transformación sociorreligiosa*. Mexico: El Colegio de la Frontera Norte/CIESAS/El Colegio de Michoacán A.C.

Instituto Nacional de Estadística, Geografía e Informática. (2005). *Diversidad religiosa en México*. Mexico: INEGI.

Jaimes, R. (2007). Neopentecostales en Tijuana. In R. De la Torre & C. Gutiérrez Zúñiga (coords.). *Atlas de la diversidad religiosa en México* (pp. 305–311). Mexico: CIESAS/UQRO/COLEF/ELCOLJAL/COLMICH/SEGOB.

Lomnitz, C. (1995). *Las salidas del laberinto: cultura e ideología en el espacio nacional mexicano*. Mexico: Joaquín Mortiz Editorial.

Martin, D. (1990). *Tongues of fire. The explosion of Protestantism in Latin America*. London: Basil Blackwell.

Nogueira, L. (2008). Pluralismo e multiplicidades religiosas no Brasil contemporáneo. Dossiê: Pluralidade religiosa na América Latina, Revista *Sociedade e Estado*. Brasilia: Universidad de Brasilia. *23*(2), 261–280.

Palacios, J. (1983). El concepto de región: La dimensión espacial de los procesos sociales. *Revista Interamericana de Planificación*. Mexico: SIAP. *66*(XXII).

Palard, J. (1999). Les recompositions territoriales de l'Église catholique entre singularité et universalité. *Archives de Sciences Sociales de Religions*. Paris: Centre National de la Recherche Scientifique. *44*(107), 55–76.

Park, C. (2004). Religion and geography. In J. Hinnels (Ed.), *Routledge companion to the study of religion* (pp. 439–456). London: Routledge.

Pereira, M. (2006). *Aspectos psicossociais da conversão religiosa* (Um estudo de aso na Igreja Universal Do Reino De Deus). Goiânia: Universidade Católica de Goiás. Retrieved May 16, 2012, from http://tede.biblioteca.ucg.br/tde_busca/arquivo.php?codArquivo=377

Rubalcava, R., & Chavarría, J. (1999). La marginación en la zona metropolitana de la Ciudad de México. In G. Garza (Ed.), *Atlas demográfico de México* (pp. 59–83). Mexico: Progresa/Conapo.

Segato, R. (2007). La faccionalización de la república y el paisaje religioso como índice de una nueva territorialidad. In A. Alonso (comp.). *América Latina y el Caribe. Territorios religiosos y desafíos para el diálogo* (p. 312). Buenos Aires: Prometeo Libros. Retrieved May 16, 2012, from http://bibliotecavirtual.clacso.org.ar/ar/libros/grupos/alonso/Segato.pdf

Silva Orozco, J. (1988). Símbolo y realidad. *Revista de La Luz del Mundo.* Guadalajara, agosto, núm 10.

Zalpa, G. (2007). La Iglesia de Jesucristo de los Santos de los Últimos Días en Aguascalientes. In R. De la Torre & C. Gutiérrez Zúñiga (coords.). *Atlas de la diversidad religiosa en México.* Mexico: CIESAS/UQRO/COLEF/ELCOLJAL/COLMICH/SEGOB.

Printed by Books on Demand, Germany